Geography
An Integrated Approach

David Waugh

Nelson

Thomas Nelson & Sons Ltd
Nelson House Mayfield Road
Walton-on-Thames Surrey
KT12 5PL UK

Nelson Blackie
Wester Cleddens Road
Bishopbriggs
Glasgow
G64 2NZ UK

Thomas Nelson Australia
102 Dodds Street
South Melbourne
Victoria 3205 Australia

Nelson Canada
1120 Birchmont Road
Scarborough Ontario
M1K 5G4 Canada
© David Waugh 1990, 1995

First published by
Thomas Nelson and Sons Ltd 1990
second edition 1995

I(T)P Thomas Nelson is an
International Thomson Publishing
Company

I(T)P is used under licence

ISBN 0-17-444072–3

NPN 9 8 7 6 5 4 3 2

Acknowledgements

The author and publishers are grateful to the following
for permission to reproduce copyright material. If any
acknowledgements have been inadvertently omitted,
this will be rectified at the earliest opportunity.

Chapter 1
1.21 David Waugh, 1.22 The Image Bank, 1.23 Topham
Picturepoint, 1.24 Frank Lane Photo Agency, 1.25
Robert Harding Picture Library, 1.31 David Waugh, 1.33
David Waugh, 1.39 Alberto Garcia/Saba Katz Pictures,
1.40 Sheila Morris, 1.43 Nigel Dickenson/Leader
Photos/Eye Ubiquitous.

p8, BBC Enterprises, extract from *The Making of a
Continent* by Ron Redfern, 1983; p9, Nigel Calder, extract
from *The Restless Earth*, 1972; pp30,31 Newspaper
Publishing plc, articles in *The Independent*, 13 June
1991 and *The Independent on Sunday*, 10 May 1992

Chapter 2
2.1 David Waugh, 2.2 David Waugh, 2.3 The Image
Bank, 2.5 David Waugh, 2.6 David Waugh, 2.7 David
Waugh, 2.8 Eye Ubiquitous, 2.14 Tony
Waltham/Geophotos, 2.18 J Allan Cash, 2.19 the
Hutchison Library, 2.25 The Hulton Deutsch Collection
2.27 J Allan Cash, 2.28 Courtesy of Geotechnical Control
Office, Hong Kong, 2.29 Topham Picturepoint.

Chapter 3
3.8 Tony Waltham/Geophotos, 3.17 Topham
Picturepoint, 3.22 David Waugh, 3.26 Eric R Delderfield,
3.29 David Waugh, 3.31 Penni Bickle, 3.33 Bruce
Coleman Ltd. 3.34 Zefa Pictures, 3.36 David Waugh, 3.40
David Waugh, 3.42 J Allan Cash, 3.44 Oxford Scientific
Films, 3.47 David Waugh, 3.51 David Waugh, 3.62 The
Science Photo Library, 3.64 The Science Photo Library,
3.68 Sheila Morris, 3.71 Sheila Morris

Chapter 4
4.4 David Waugh, 4.7 Tony Waltham/Geophotos, 4.12
Tony Waltham/Geophotos, 4.14 Geoscience Features,
4.15 Tony Waltham/Geophotos, 4.16 Tony
Waltham/Geophotos, 4.18 Topham Picturepoint, 4.19
David Waugh, 4.20 Tony Waltham/Geophotos, 4.25 Tony
Waltham/Geophotos, 4.27 Tony Waltham/Geophotos,
4.28 Tony Waltham/Geophotos, 4.38 Topham
Picturepoint, 4.43a Topham Picturepoint, 4.43b Topham
Picturepoint, 4.43c Robert Harding Picture Library,
4.43d J Allan Cash.

Chapter 5
5.7 Tony Waltham/Geophotos, 5.9 Professor Peter
Worsley, University of Reading, 5.10 Tony
Waltham/Geophotos, 5.12 David Waugh.

p116, David and Charles, extract from *The Unquiet
Landscape* by Denys Brunsden and John Doornkamp,
1977

Chapter 6
6.9 David Waugh, 6.13 David Waugh, 6.15 Tony
Waltham/Geophotos, 6.16 David Waugh, 6.17 David
Waugh, 6.19 David Waugh, 6.21 Landform Slides, 6.22
Aerofilms, 6.24 Landform Slides, 6.25 Tony
Waltham/Geophotos, 6.27 Nasa, 6.28 Paul Jeffrey
Godfrey, University of Massachusetts, 6.30 David
Waugh, 6.31 David Waugh, 6.32 David Waugh, 6.33
Landform Slides, 6.46 Tony Waltham/Geophotos, 6.48 J
Allan Cash, 6.49 Landform Slides. 6.56, 6.57, 6.59, 6.60,
6.62, 6.63, 6.64, 6.65, 6.67, 6.68 Gary White.

p124, Cambridge University Press, extract from *The
Coastline of England and Wales* by J A Steers, 1960; p161,
Centurion Publications, article in *Geographical
Magazine*, March 1994

Chapter 7
7.3 Penni Bickle, 7.4 David Waugh, 7.5 David Waugh, 7.6
Landform Slides, 7.7 Tony Waltham/Geophotos, 7.9
David Waugh, 7.12 Landform Slides, 7.15 Landform
Slides 7.16 Landform Slides, 7.17 Landform Slides, 7.18
Landform Slides, 7.19 Landform Slides 7.20 David
Waugh, 7.23 Tony Waltham/Geophotos, 7.24 Tony
Waltham/Geophotos, 7.28 Landform Slides, 7.29 a and
b Still Pictures.

Chapter 8
8.2 Landform Slides, 8.3 Landform Slides, 8.4 David
Waugh, 8.5 Tony Waltham/Geophotos, 8.6 Landform
Slides, 8.7 David Waugh, 8.10 Landform Slides, 8.11
Landform Slides, 8.12 Tony Waltham/Geophotos, 8.13
David Waugh, 8.15 Landform Slides, 8.16 Tony
Waltham.

p178, Unwin Hyman, extract from *The Physical
Geography of Landscape* by Roy Collard, 1988

Chapter 9
9.21 (1) Frank Lane Picture Agency, 9.21 (2) Science
Photo Library, 9.21 (3) C S Broomfield/Meteorological
Office, 9.21 (4) Frank Lane Picture Agency, 9.21 (5) C S
Broomfield/Meteorological Office, 9.21 (6) B J
Burton/Meteorological Office, 9.21 (7) Frank Lane
Picture Agency, 9.21 (8) J FP Galvin/Meteorological
Office, 9.21 (9) Frank Lane Picture Agency, 9.21 (1)
Science Photo Library, 9.24 David Waugh, 9.27 W S
Pike/Meteorological Office, 9.37 Nigel Press
Associates/Meteosat, 9.48 Science Photo Library, 9.49
Dundee Satellite Station/Meteorological Station, 9.52
State University of New York/Meteorological Office,
9.634 Dundee Satellite Station.

Chapter 10
10.25 Landform Slides, 10.26 Greg O'Hare, 10.27 David
Waugh, 10.33 Oxford Scientific Films, 10.34 Tony
Waltham/Geophotos, 10.35 Still Pictures, 10.36 Frank
Lane Picture Agency, 10.41 Still Pictures.

p238, Unwin Hyman, extract from *Soil Processes* by
Brian Knapp; p258, Guardian News Service Ltd, article in
The Observer, 29 January 1989

Chapter 11
11.6 a and b David Waugh, 11.9 Landform Slides, 11.10
Heather Angel/Biofotos, 11.1 Heather Angel/Biofotos,
11.13 Heather Angle/Biofotos, 11.5 David Waugh, 11.16
Heather Angel/Biofotos, 11.30 David Waugh, 11.31
David Waugh, 11.32 David Waugh, 11.35 Telegraph
Colour Library, 11.43 Mary Frith, 11.45 Mary Frith.

p264, Pan Macmillan, extract from *The GAIA Atlas of
Planet Management* by James Lovelock, 1985

Chapter 12
12.4 Tony Waltham/Geophotos, 12.5 Frank Lane Picture
Agency, 12.6 Heather Angel/Biofotos, 12.9 David
Waugh, 12.14 David Waugh, 12.15 Philip Craven, 12.18
Tony Waltham/Geophotos, 12.19 J Allan Cash, 12.23
Spectrum Colour Library, 12.24 David Waugh, 12.32
Frank Lane Picture Agency, 12.36 Eye Ubiquitous, 12.27
The Hutchison Library, 12.37 David Waugh, 12.42 Tony
Waltham/Geophotos, 12.43 Landform Slides, 12.45
David Waugh, 12.46 David Waugh, 12.49 Frank Lane
Picture Agency, 12.50 David Waugh , 12.51 David
Waugh, 12.52 David Waugh, 12.53 David Waugh, 12.56
Oxford Scientific Films/Earthscenes, 12.57 Oxford
Scientific Films, 12.58 The Hutchison Library.

p290, Oxford University Press, extract from *The Ages of
GAIA* by James Lovelock, 1989

Chapter 13
13.35 Still Pictures, 13.37 The Hutchison Library, 13.38
Colorific/Telegraph Colour Library, 13.43 The Hutchison
Library, 13.44 Starfoto/Zefa Picture Library, 13.45 The
Hutchison Library, 13.48 David Waugh, 13.50 Still
Pictures, 13.51 Impact Photos, 13.57 The Hutchison
Library.

p320, Pan Macmillan, extract from *North-South: A
Programme for Survival* by Willy Brandt, 1980;
Newspaper Publishing plc, map in *Independent on
Sunday*, 28 August 1994

Chapter 14
14.7a Oxford Scientific Films, 14.7b Aerofilms, 14.7c
James Davis Travel Photography, 14.10 The Hutchison
Library, 14.11 Martin Chillmaid/Oxford Scientific Films,
14.12 James Davis Travel Photography, 14.13 Aerofilms,
14.14 The Hutchison Library, 14.15 Photoair Ltd. 14.35 J
Allan Cash, 14.36 Eye Ubiquitous.

p381, J M Dent & Sons, extract from *Villages in the
English Landscape* by Trevor Rowley, 1978

Chapter 15
15.15a James Davis Travel Photography, 15.15b James
Davis Travel Photography, 15.21a1 The Telegraph Colour
Library, 15.21a2 James Davis Travel Photography,
15.21b1 The Telegraph Colour Library, 15.21b2 David
Waugh, 15.21b3 Stock photos/The Image Bank,
15.21b4 The Hulton Deutsch Collection, 15.21c1 Penni
Bickle, 15.21c4 Aerofilms, 15.21d1 The Image Bank,
15.21d2 P Thompson/Eye Ubiquitous, 15.21d3 The
Hutchison Library, 15.21d4 Stockphotos/The Image
Bank, 15.25 Mark Edwards/Still Pictures, 15.34 David
Waugh, 15.35 David Waugh, 15.36 David Waugh, 15.37
Mark Edwards/Still Pictures, 15.38 Mark Edwards/Still
Pictures, 15.39 Mark Edwards/Still Pictures, 15.41 Mark
Edwards/Still Pictures, 15.42 Intermediate Technology,
15.43 James Davis Travel Photography, 15.44 David
Waugh, 15.48 The Hulton Deutsch Collection. 15.49
Peter Menzel/Impact Photos, 15.50 Colorific/Telegraph
Colour Library, 15.54 Rex Features, 15.55 Rex Features.
p384, BBC Enterprises Ltd, extract from *Only One Earth*
by L Timberlake, 1987

Chapter 16
16.8 Levon Babayan/SCR Photo Library, 16.9 SCR Photo
Library, 16.12 Landform Slides, 16.13 The Image Bank,
16.26 The Telegraph Colour Library, 16.29 Still Pictures,
16.30 The Hutchison Library, 16.31 James Davis Travel
16.32 Zefa Pictures, 16.33 David Waugh, 16.36 David
Waugh, 16.39 The Image Bank, 16.42 David Waugh,
16.44 David Waugh, 16.45 David Waugh, 16.46 Science
Photo Library, 16.47 Oxford Scientific Films, 16.52 Mark
Edwards/Still Pictures, 16.55 Zefa Pictures, 16.60 Mark
Edwards/Still Pictures, 16.65 Mark Edwards/Still
Pictures, 16.67 16.71 Sheila Morris.

p423, BBC Enterprises Ltd, extract from *Only One Earth*
by L Timberlake, 1987; p 455, Stanley Thornes
(Publishers) Ltd, material from *Geofile*, September 1994

Chapter 17
17.3 Mark Edwards/Still Pictures, 17.4 The Image Bank,
17.5 Zefa Pictures, 17.6 David Waugh, 17.7 Celtic Picture
Library, 17.8 David Waugh, 17.9 David Waugh, 17.13a
Mark Edwards/Still Pictures, 17.13b Mark Edwards/Still
Pictures, 17.13c Zefa Pictures, 17.15 Ed Childs, 17.16
Stephen Cotterell, 17.17 The Hutchison Library, 17.18
Stephen Cotterell, 17.19 Guido Alberto Rossl/The Image
Bank, 17.20 Stephen Cotterell.

Chapter 18
18.10 Aerofilms, 18.13 Caroline Penn/Impact Photos,
18.15 Colorific/Telegraph Colour Library, 18.23 A I
Intermediate Technology.

p488, Penguin Books Ltd, extract from *No Full Stops in
India* by Mark Tully, 1991

Chapter 19
19.15 The Image Bank, 19.19 David Waugh, 19.22
Cambridge Newspapers Ltd. 19.28 Telegraph Colour
Library, 19.30 Sony UK Ltd. 19.34 Intermediate
Technology, 19.35 Intermediate Technology, 19.36
Intermediate Technology, 19.37 Intermediate
Technology, 19.45 J Allan Cash, 19.46 Piers
Cavendish/Impact Photos.

pp502, 526, Abacus Books, extracts from *Small is
Beautiful* by E F Schumacher, 1974; p529, Centurion
Publications, article in *Geographical Magazine*,
September 1994

Chapter 20
20.3 J Allan Cash, 20.4 AirFotos Ltd. 20.5 J Allan Cash,
20.14 Sheila Morris, 20.16 L Fordyce/Eye Ubiquitous,
20.17 20.22 Sheila Morris.

Chapter 21
21.5 Paul Simmons/Impact Photos, 21.6 J Allan Cash,
21.7 J Allan Cash, 21.8 David Waugh, 21.10 Telegraph
Colour Library, 21.9a J Allan Cash, 21.29b J Allan Cash,
21.29c J Allan Cash, 21.32 Mike McQueen/Impact
Photos, 21.33 Hong Kong Airport Core Programme.

p558, Penguin Books Ltd, extract from *No Full Stops in
India* by Mark Tulley, 1991

Chapter 22
22.4 David Waugh, 22.19a Gavin Wickham/Eye
Ubiquitous, 22.19b J Allan Cash, 22.5a
Colorific/Telegraph Colour Library, 22.25b Julia
Waterlow/Eye Ubiquitous, 22.27a Alain Evrard/Impact
Photos, 22.27b Christophe Bluntzer/Impact Photos,
22.27c Julian Calder/Impact Photos.

p568, Penguin Books Ltd, extract from *No Full Stops in
India* by Mark Tully, 1991; p580, Routledge, extract from
Health and Development by David Phillips and Yola
Verhasselt, 1994

Examining bodies
NEAB, pp 61, 122, 175, 200, 256 (Q4), 434, 510
ULEAC, pp 63, 143, 216, 393

Contents

Introduction for the reader

Geography: An Integrated Approach has been written as much for those students who have an interest in Geography, an enquiring mind and a concern for the future of the planet upon which they live, as for those specialising in the subject. The text has been written as concisely as seemed practical in order to minimise the time needed for reading and note-taking and to maximise the time available for discussion, individual enquiry and wider reading. Full colour has been used throughout for photographs, sketches and maps, which aim to illustrate the wide range of natural and human-created environments. Annotated diagrams are included to show interrelationships and to help explain the more difficult concepts and theories. A wide range of graphical skills have been used to portray geographical data, data which are as recent as possible at the time of writing.

It is because Geography is concerned with interrelationships that this book has included, and aims to integrate, several fields of study. These involve physical environments (atmosphere, lithosphere and hydrosphere) and the living world (biosphere); economic development (or lack of it); the frequent misuse of the environment, the long-overdue concern over the resultant consequences and the need for careful management; together with the application, where appropriate, of a modern scientific approach using statistical methods in investigations. It is intended that this single book will (a) satisfy the requirements of existing A-level examination syllabuses and (b) address the statement made in the Geographical Association's booklet, *Geography in the National Curriculum* (1989), that:

'Geography is concerned with people as well as landscapes, with economic as well as ecological systems. The character of places - the subject's central focus - derives from the interaction of people and environments.'

By coincidence, the initial letters of the title of this book form the word GAIA. In ancient Greece, Gaia was the goddess of the earth. Today the term has been reintroduced to mean 'a new look at life on earth', an approach which looks at the earth in its entirety as a living organism. Hopefully the following text reflects aspects of this approach.

You, as the reader, should be aware that there is no rigid prescribed sequence in the order either of the chapters themselves or in their content. Each is open to several routes of enquiry. Terminology can be a major problem because geographers may use several terms, some borrowed from other disciplines, to describe the same phenomenon. When a term is introduced for the first time it is shown by emboldened print, has a list of alternatives (one of which is subsequently retained for consistency) and is defined. Definitions occasionally appear as marginal notes. Alternative terms and specific examples often appear in brackets in order to save space.

The book sets out to provide a background store of information which will help you to understand basic concepts, to be able to enter discussions and to develop your own informed, rather than prejudiced, values and attitudes. Due to the limited space, descriptions of specific examples or, as they are called in the text, Places, have been kept brief. However they should either enable you to build upon your earlier knowledge or stimulate you into reading more widely. New to this edition is the inclusion, at the end of each chapter, of a more detailed Case Study. The Case Studies include natural hazards, problems created by population growth and the mis-use of the natural environment, and the attempts, or lack of, to manage the environment and the earth's resources. References given at the conclusion of each chapter are those to which the author has himself referred and are not intended as a comprehensive list. As the reader, it is essential that you appreciate that Geography is a dynamic subject with data, views, policies and terms which change constantly. Consequently, your own research must not be limited to textbooks, which in any case are out-of-date even before their publication, but should be widened to include newspapers, journals, television, radio and many 'non-academic' media.

The book also includes fourteen Frameworks whose function is to stimulate discussion on methodological and theoretical issues. They illustrate some of the skills required, and the problems involved, in geographical enquiry, e.g. the uses, limitations and reliability of models; quantitative techniques; the collection of data; making classifications; the dangers of stereotyping and of making broad generalisations. Geography is also concerned with the development of graphical skills. The media show an increasing amount of data in a graphical form, and this is likely to grow further as geographical information systems develop. It is assumed that the reader already understands those skills covered by present 16+ examination syllabuses and therefore only new skills are explained in this book. Quantitative and statistical techniques are incorporated at appropriate points although each may be relevant elsewhere in either physical or human-economic chapters. After an explanation of each technique, there is a worked example and follow up exercises. Opportunities for self-assessment are provided throughout the book with the inclusion of examination-style questions.

Finally, I hope that the book (as well as helping you earn a good 'A' level grade) will give you some of the enjoyment which Geography has given me, and that it will encourage you to take every opportunity to travel. The best way to learn about the world and to try to understand the way of life of its different people, is through firsthand experience.

Author's acknowledgements

Before beginning the task of revising *Geography: An Integrated Approach* (written in 1988–89 and published in 1990), several leading geographers were asked to comment upon the present day accuracy and relevance of the first edition, and to indicate recent academic changes in terminology, concepts and approach. I am, therefore most grateful to the following:

David Chester (University of Liverpool) for 'Plate tectonics, earthquakes and volcanoes'

Dr Derek Mottershead (Edge Hill University College, Liverpool) for 'Weathering and slopes' and 'Rock types and landforms';

Professor Malcolm Newson (University of Newcastle upon Tyne) for 'Drainage basins and rivers'

Dr David Evans (University of Glasgow) for 'Glaciation';

Professor John Boardman (University of Durham) for 'Periglaciation';

Professor John Pethick (University of Hull) for 'Coasts';

Professor Andrew Goudie (University of Oxford) for 'Deserts';

Dr Grant Bigg (University of East Anglia) for 'Weather and climate';

Dr Steven Trudgill (University of Sheffield) for 'Soils';

Professor Ian Simmons (University of Durham) for 'Biogeography' and 'World climate, soils and vegetation';

Professor David Phillips (University of Nottingham) and Dr Nicholas Ford (University of Exeter) for 'Population';

Professor W.F. Lever (University of Glasgow) for 'Settlement' and 'Urbanisation';

Dr I Bowler (University of Leicester) for 'Farming and food supply' and 'Rural land use';

Dr Louise Crew (University of Nottingham) for 'Manufacturing Industry'; and

Dr Richard Knowles (University of Salford) for 'Transport and trade'.

I am also indebted to Professor Phillips and Dr Ford for sending and allowing me to use their own material, some of which had as then not been published. My thanks also to the following contributors to the Case Studies:-

Sheila Morris (Case Studies 3b, 11, 12b, 15b, 16 and 20);

Gary White (Case Study 6);

Stephen Cotterell (Case Study 17) and Intermediate Technology (Case Study 18) and to all the important back up team which included my wife Judith (whom I had to rely upon to perform even the most elementary of word processing operations), and Dr Elizabeth Clutton.

Chapter 1

Plate tectonics, earthquakes and volcanoes

". . . how does a supercontinent begin to rift and how do the pieces move apart? What effects do such movements have on the shaping of the continental landscapes, on hot climates and ice ages, on the evolution of life in general and on humanity's relationship with the upper crust of the Earth in particular?"

The Making of a Continent, Ron Redfern, 1983

The history of the earth

It is estimated that the earth was formed about 4 600 000 000 years ago. Even if this figure is simplified to 4600 million years, it still presents a timescale far beyond our understanding. Nigel Calder, in his book *The Restless Earth*, made a more comprehensible analogy by reducing the timespan to 46 years. He ignored the eight noughts and

Figure 1.1

The geological timescale

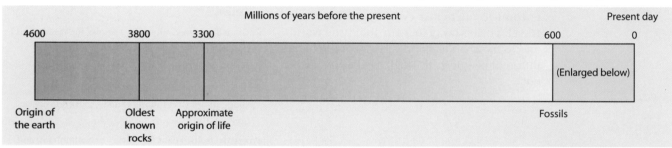

Millions of years before the present			Present day
4600 3800 3300		600	0

Origin of the earth   Oldest known rocks   Approximate origin of life   (Enlarged below)   Fossils

Era	Geological period		Millions of years before present	Conditions and rocks in Britain	Major world events
CENOZOIC	Quaternary	Holocene	0.01	Post ice age. Alluvium deposited, peat formed	Early civilisations
		Pleistocene	2	Ice age, with warm periods	Emergence of the human
	Tertiary	Pliocene	7	Warm climate: Crag rocks in East Anglia	
		Miocene	26	No deposits in Britain	Formation of the Alps
		Oligocene	38	Warm shallow seas in south of England	Rockies and Himalayas begin to form
		Eocene	54	Nearly tropical: London clay	Volcanic activity in Scotland
MESOZOIC	Cretaceous		136	Chalk deposited: Atlantic ridge opens	End of the dinosaurs/Age of the dinosaurs
	Jurassic		195	Oxford clays and limestones: warm	Pangaea breaks up
	Triassic		225	Desert: sandstones	First mammals
PALAEOZOIC	Permian		280	Desert: New Red sandstones, limestones	Formation of Pangaea
	Carboniferous		345	Tropical coast with swamps: coal	First amphibians and insects
	Devonian		395	Warm desert coastline: sandstones	First land animals
	Silurian		440	Warm seas with coral: limestones	First land plants
	Ordovician		500	Warm seas: volcanoes (Snowdonia) sandstones, shales	First vertebrates
	Cambrian		570	Cold at times: sea conditions	Abundant fossils begin
	Pre-Cambrian			Igneous and sedimentary rocks	

"…Or we can depict Mother Earth as a lady of 46, if her 'years' are megacenturies. The first seven of those years are wholly lost to the biographer, but the deeds of her later childhood are to be seen in old rocks in Greenland and South Africa. Like the human memory, the surface of our planet distorts the record, emphasising more recent events and letting the rest pass into vagueness — or at least into unimpressive joints in worn down mountain chains.

Most of what we recognise on Earth, including all substantial animal life, is the product of the past six years of the lady's life. She flowered, literally, in her middle age. Her continents were quite bare of life until she was getting on for 42 and flowering plants did not appear until she was 45 — just one year ago. At that time, the great reptiles, including the dinosaurs, were her pets and the break-up of the last supercontinent was in progress.

The dinosaurs passed away eight months ago and the upstart mammals replaced them. In the middle of last week, in Africa, some man-like apes turned into ape-like men and, at the weekend, Mother Earth began shivering with the latest series of ice ages. Just over four hours have elapsed since a new species calling itself *Homo sapiens* started chasing the other animals and in the last hour it has invented agriculture and settled down. A quarter of an hour ago, Moses led his people to safety across a crack in the Earth's shell, and about five minutes later Jesus was preaching on a hill farther along the fault line. Just one minute has passed, out of Mother Earth's 46 'years', since man began his industrial revolution, three human lifetimes ago. During that time he has multiplied his numbers and skills prodigiously and ransacked the planet for metal and fuel."

N. Calder, *The Restless Earth* (1972)

The 'Moho' discontinuity is the junction between the earth's crust and the mantle where seismic waves are modified. The Moho is at about 35–40 km beneath continents (reaching 70 km under mountain chains) and at 6–10 km below the oceans.

compared the 46 years with a human lifetime (Places 1).

Geologists have been able to study rocks and fossils formed during the last 600 million years, equivalent to the last "six years of the lady's life", and have produced a time chart, or **geological timescale**. Not only have they been able to add dates with increasing confidence, but they have made progress in describing and accounting for the major changes in the earth's surface, e.g. sea level fluctuations and landform development, and in its climate. The timescale, shown in Figure 1.1, should be a useful reference for later parts of this book.

Earthquakes

Even the earliest civilisations were aware that the crust of the earth was not rigid and immobile. The first major European civilisation, the Minoan, based in Crete, constructed buildings such as the Royal Palace at Knossos which withstood a succession of earthquakes. However, this civilisation may have been destroyed by the effects of a huge volcanic eruption on the nearby island of Thera (Santorini). Later, inhabitants of places as far apart as Lisbon (1755), Tokyo (1923), San Francisco (1906), Mexico City (1985) and Kobe (1995) were to suffer from the effects of major earth movements.

It was by studying earthquakes that geologists were first able to determine the structure of the earth (Figure 1.2). At the **Mohorovičič** or **'Moho' discontinuity**, it was found that shock waves begin to travel faster indicating a change of structure — in this case, the junction of the earth's **crust** and **mantle** (Figure 1.2). Earthquakes result from a slow build up of pressure within crustal rocks. If this pressure is suddenly released then parts of the surface may experience a jerking movement. Within the crust, the point at which the release in pressure occurs is known as the **focus**. Above this, on the surface and usually receiving the worst of the shock or **seismic waves**, is the **epicentre**. Unfortunately, it is not only the immediate or primary effects of the earthquake which may cause loss of life and property; often the secondary or after effects are even more serious. These (Places 2) may include fires from broken gas pipes, disruption of transport and other services, exposure caused by a lack of shelter, a shortage of food, clean water and medical equipment, and disease caused by polluted water supplies.

Plate tectonics, earthquakes and volcanoes

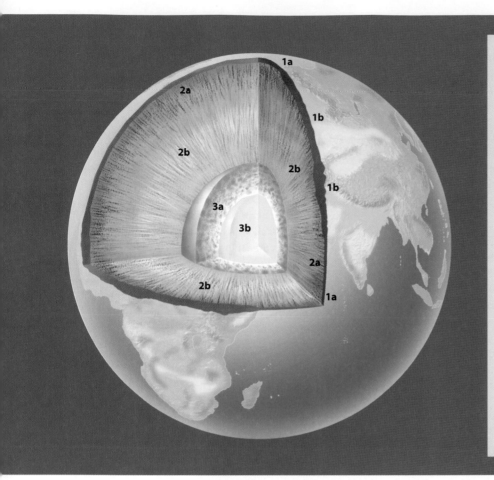

1. **Crust** Relatively speaking, this is as thin as the skin of an apple is to its flesh.
(a) *Oceanic crust* (sima) is a layer consisting mainly of basalt, averaging 6–10 km in thickness. At its deepest it has a temperature of 1200°C.
(b) *Continental crust* (sial) can be up to 70 km thick. The crust is separated from the mantle by the Moho discontinuity. The crust and the rigid top layer of the mantle are collectively known as the lithosphere (Figure 1.6).

2. **Mantle** This is composed mainly of silicate rocks, rich in iron and magnesium. Apart from the rigid top layer (**2a**), the rocks in the remainder of the mantle, the asthenosphere, are kept in a semi-molten state (**2b**). The mantle extends to a depth of 2900 km where temperatures may reach 5000°C. These high temperatures generate convection currents.

3. **Core** This consists of iron and nickel. The outer core (**3a**) is kept in a semi-molten state, but the inner core (**3b**) is solid. The temperature at the centre of the earth (6371 km below the surface) is about 5500°C.

Figure 1.2

The internal structure of the earth

Places 2   The San Francisco earthquakes

1906

At 0512 hours on the morning of 18 April, the ground began to shake. There were three tremors, each one increasingly more severe. The ground moved by over 6 m in an earthquake which measured 8.2 on the Richter scale. Many apartment buildings collapsed, bridges were destroyed — the Golden Gate had not then been built — and water pipes fractured. The worst damage was 'downtown' where the housing density was greatest. Although many people were trapped within collapsed buildings there were relatively few deaths.

Then came the fire! It started in numerous places resulting from overturned stoves or sparked by electricity or the ignition of gas escaping from the broken mains. As the water pipes had been fractured, it hardly mattered that there were only 38 horse-drawn fire engines to cope with 52 fires. As the fire spread, houses were blown up with dynamite to try to create gaps to thwart the flames, but the explosions only caused further fires. It took over three days to put out the fires by which time over 450 people (mainly those previously trapped) had died, 28 000 buildings in 500 blocks had been destroyed, and an area six times greater than that destroyed by the Great Fire of London had been ravaged. Fortunately, the threat of disease and starvation was averted.

1989

During rush-hour on 18 October an earth-quake measuring 6.9 on the Richter scale shook the city for 15 seconds. The early warning system had given no clues. Skyscrapers swayed 3 m, fractured gas pipes caused fires in one residential area and a 1.5 km stretch of the two-tier highway in nearby Oakland collapsed, killing people in their vehicles. The figures of 67 dead and 2000 homeless were low, however, compared to the quake of similar magnitude in Armenia 11 months earlier and which killed over 55 000 — in California, precautions had been taken and fully-equipped and trained emergency services were on hand.

Q Read the eye-witness account of the San Francisco earthquake of 1906 on page 10 and then answer the questions below.

1 What is an earthquake?

2 Refer to the diagram of the Richter scale (Figure 1.3).
   a Why is an earthquake of 7.0 on the Richter scale 100 times more severe than one which measures 5.0?
   b How many times greater was the Mexican earthquake of 1985 than the one which affected Carlisle in 1981?

3 Describe
   a the primary and
   b the secondary effects of the San Francisco earthquake of 1906.

4 What types of damage other than those experienced in San Francisco may result from earthquakes?

5 On an outline map of the world, mark by means of a dot — no need to name the places — the location of each earthquake in the list below.

6 a On a tracing overlay, mark and name all the volcanoes listed.
   b On a separate overlay, mark and name the young fold mountains.

7 Describe the distribution of earthquakes, volcanoes and fold mountains on your three maps.

Figure 1.3

The Richter Scale

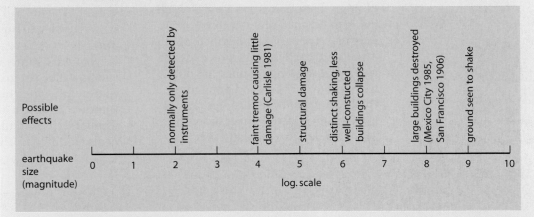

Distribution of recent major earthquakes, volcanoes (active, dormant or extinct) and young fold mountains

■ **Some recent major earthquakes**

1903 Mexico, Java, Aleutians, Greece
1904 Tokyo, Kamchatka, Costa Rica
1906 Japan, San Francisco, Aleutians, Chile
1907 Mexico, Afghanistan
1908 Mexico, Sicily
1909 Fiji, Japan
1916 New Guinea
1917 Java
1918 Tonga
1920 Taiwan, Fiji
1921 Peru
1922 Chile
1923 Japan
1924 Philippines
1925 California
1926 Rhodes
1927 Japan
1928 Chile

1929 Aleutians, Japan
1931 New Zealand
1932 Mexico
1933 California
1935 Sumatra
1938 Java
1939 Chile, Turkey
1940 Burma, Peru
1941 Ecuador, Guatemala
1943 Philippines, Java
1944 Japan
1946 West Indies, Japan
1949 Alaska
1950 Japan, Assam
1953 Turkey, Japan
1956 California
1957 Mexico
1958 Alaska
1960 Chile, Morocco
1962 Iran
1963 Yugoslavia
1964 Alaska, Turkey, Mexico, Japan, Taiwan
1965 El Salvador, Greece
1966 Chile, Peru, Turkey
1967 Colombia, Yugoslavia, Java, Japan

1968 Iran
1970 Peru
1971 New Guinea, California
1972 Nicaragua
1976 Guatemala, Italy, China, Philippines, Turkey
1978 Japan
1980 Italy
1985 Mexico, Colombia
1988 Armenia
1989 San Francisco, Iran
1993 Java, Japan, India, Egypt
1994 Los Angeles
1995 Japan

■ **Volcanoes**

Aconcagua, Chimborazo, Cotopaxi, Nevado del Ruiz, Paricutín, Popocatepetl, Mt St Helens, Fuji, Pinatubo, Mayon, Krakatoa, Tarawera, Erebus, Helgafell, Surtsey, Azores, Ascension, St Helena, Tristan da Cunha; Vesuvius, Etna, Pelée, Mauna Loa, Kilauea

■ **Young fold mountains**

Andes, Rockies, Atlas, Pyrenees, Alps, Caucasus, Hindu Kush, Himalayas, Southern Alps

Plate tectonics

As early as 1620 Francis Bacon noted the jig-saw-like fit between the east coast of South America and the west coast of Africa. Others were later to point out similarities between the shapes of coastlines of several adjacent continents.

In 1912, a German meteorologist, Alfred Wegener, published his theory that all the continents were once joined together in one large supercontinent which he named **Pangaea**. Later, this landmass somehow split up and the various continents, as we know them, drifted apart. Wegener collated evidence from several sciences:

- **Biology** Mesosaurus was a small reptile living in Permian times (Figure 1.1); its remains have been found only in South Africa and Brazil. A plant which existed when coal was being formed has only been located in India and Antarctica.
- **Geology** Rocks of similar type, age, formation and structure occur in south-east Brazil and South Africa, and the Appalachian Mountains of the eastern USA correspond geologically with mountains in north-west Europe.
- **Climatology** Coal, formed under warm, wet conditions, is found beneath the Antarctic ice cap, and evidence of glaciation had been noted in tropical Brazil and central India. Coal, sandstone and limestone could not have formed in Britain with its present climate.

Wegener's theory of **continental drift** combined information from several subject areas, but his ideas were rejected by specialists in those disciplines partly because he was not regarded as an expert himself but perhaps mainly because he could not explain how solid continents had changed their positions. He was unable to suggest a mechanism for drift.

Figure 1.4 shows Wegener's Pangaea and how it began to divide up into two large continents, which he named **Laurasia** and **Gondwanaland**; it also suggests how the world may look in the future if the continents continue to drift.

Figure 1.4

The wandering continents

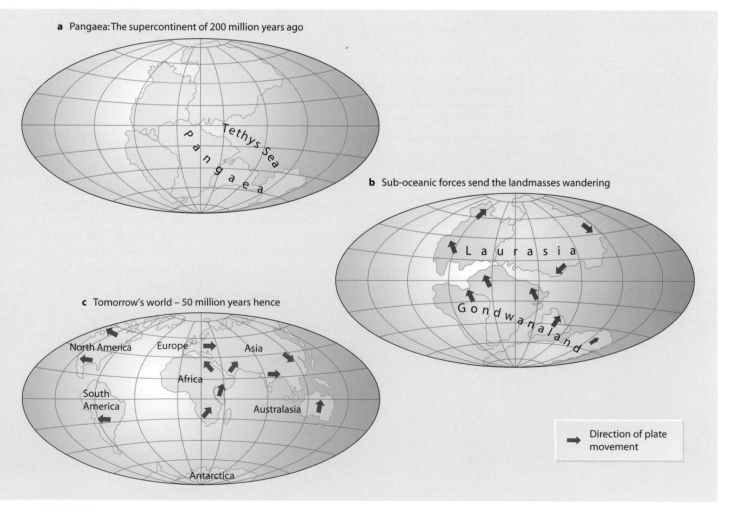

a Pangaea: The supercontinent of 200 million years ago

b Sub-oceanic forces send the landmasses wandering

c Tomorrow's world – 50 million years hence

Direction of plate movement

Plate tectonics, earthquakes and volcanoes

1 Which two present day continents formed Laurasia?

2 Which present day continents formed Gondwanaland?

3 How may the world of 50 million years time be different from that of today?

Since Wegener first put forward this theory, three groups of new evidence have become available to support his ideas.

1 **The discovery and study of the Mid-Atlantic Ridge** While investigating islands in the Atlantic in 1948, Ewing noted the presence of a continuous mountain range extending the whole length of the ocean bed; he also noted that the rocks of this range are volcanic and recent in origin — not ancient rocks as previously assumed.

2 **Studies of palaeomagnetism in the 1950s** During underwater volcanic eruptions, basaltic lava is intruded into the crust and cools (Figure 1.27). During the cooling process, individual minerals, especially iron, align themselves along the earth's magnetic field — i.e. in the direction of the magnetic pole. Recent refinements in dating techniques enable the time at which rocks were formed to be accurately calculated. It was known before the 1950s that the earth's magnetic pole varied a little from year to year, but only then was it discovered that the magnetic field reverses periodically — i.e. the magnetic pole is in the south for a period of time and then in the north for a further period, and so on. It is claimed that there have been 171 reversals over 76 million years. If formed when the magnetic pole was in the north, new basalt would be aligned to the north. After a reversal in the magnetic poles, newer lava would be orientated to the south. After a further reversal, the alignment would again be to the north. Subsequent investigations have shown that these alternations in alignment are almost symmetrical in rocks either side of the Mid-Atlantic Ridge — the name given to the mountain range identified by Ewing (Figure 1.5).

3 **Sea floor spreading** In 1962, Hess studied the age of rocks from the middle of the Atlantic outwards to the coast of North America. He confirmed that the newest rocks were in the centre of the ocean, and were still being formed in Iceland, and that the oldest rocks were those nearest to the USA and the Caribbean. He also suggested that the Atlantic could be widening by up to 5 cm a year.

One major difficulty resulting from this concept of sea floor spreading was the implication that the earth must be increasing in size. Since this is not so, evidence was needed to show that elsewhere parts of the crust were being destroyed. Such areas were found to correspond to the fringes of the Pacific Ocean — the region where you plotted some major earthquakes and volcanic eruptions (page 11). These discoveries led to the development of the theory of **plate tectonics** which is now virtually universally accepted, but which may still be modified following further investigation and study.

Figure 1.5

The repeated reversal of the earth's magnetic field

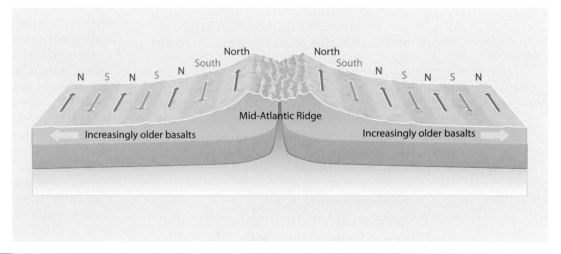

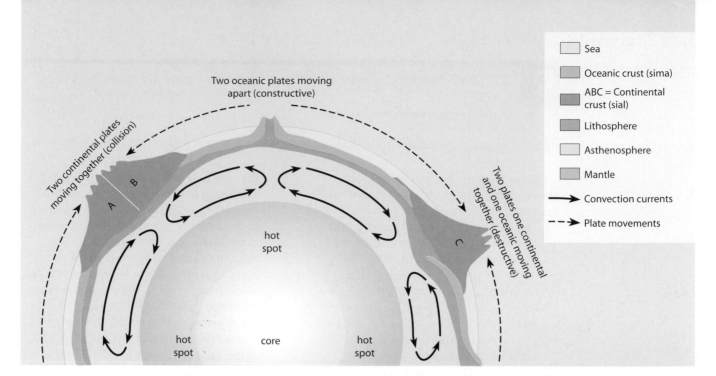

Figure 1.6

How plates move

Figure 1.7

Differences between continental and oceanic crust

	Continental crust (sial)	Oceanic crust (sima)
Thickness	35-40 km on average, reaching 60-70 km under mountain chains	6-10 km on average
Age of rocks	very old, mainly over 1500 million years	very young, mainly under 200 million years
Weight of rocks	lighter, with an average density of 2.6	heavier, with an average density of 3.0
Nature of rocks	light in colour; many contain silica and aluminium numerous types, granite is the most common	dark in colour; many contain silica and magnesium few types, mainly basalt

The theory of plate tectonics

The **lithosphere** (the earth's crust and the rigid upper part of the mantle) is divided into seven large and several smaller **plates**. The plates, which are rigid, float like rafts on the underlying semi-molten mantle (the **asthenosphere**) and are moved by **convection currents** (Figure 1.6). Plate tectonics is the study of the movement of these plates and their resultant landforms. There are two types of plate: **continental** and **oceanic**. However, these terms do not refer to actual continents and oceans but to different types of crust or rock.

Continental crust, or **sial**, is composed of older, lighter rock of granitic type. It is dominated by minerals rich in silica (Si) and aluminium (Al), from which the term is derived. Oceanic crust, or **sima**, consists of much younger and denser rock of basaltic composition. Its dominant minerals are silica (Si) and magnesium (Ma). The major differences between the two types of crust are summarised in Figure 1.7.

Plate movement

As a result of convection currents generated by heat from the centre of the earth, plates may move towards, away from or sideways along adjacent plates. It is at plate boundaries that most of the world's major landforms occur, and where earthquake, volcanic and mountain-building zones are located (Figure 1.8). However, before trying to account for the formation of these landforms several points should be noted.

1 Due to its relatively low density, continental crust does not sink and so is permanent; being denser, oceanic crust can sink. Oceanic crust is being formed and destroyed continuously.

2 Continental plates, such as the Eurasian Plate, may consist of both continental and oceanic crust.

3 Continental crust may extend far beyond the margins of the landmass.

4 Plates cannot overlap. This means that either they must be pushed upwards on impact to form mountains (AB on Figure 1.6) or one plate must be forced downwards into the mantle and destroyed (C on Figure 1.6).

5 No 'gaps' may occur on the earth's surface so, if two plates are moving apart, new oceanic crust originating from the mantle must be being formed.

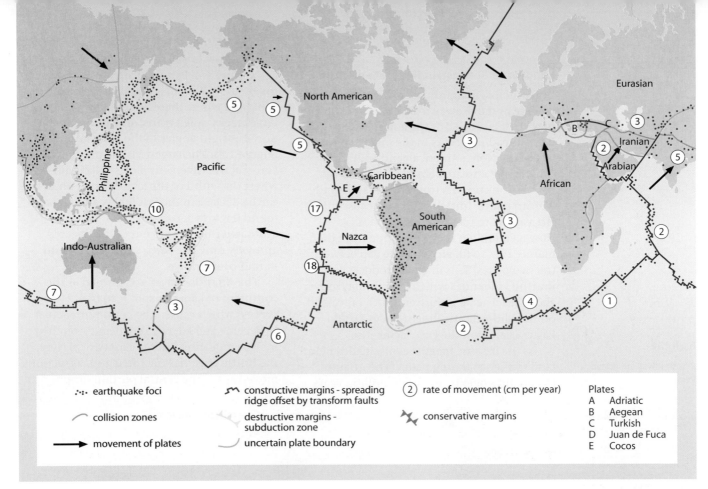

Figure 1.8

Plate boundaries and active zones of the earth's crust

6 The earth is neither expanding nor shrinking in size. Thus when new oceanic crust is being formed in one place, older oceanic crust must be being destroyed in another.

7 Plate movement is slow (though not in geological terms) and is usually continuous. Sudden movements are detected as earthquakes.

8 Most significant landforms (fold mountains, volcanoes, island arcs, deep sea trenches, and batholith intrusions) are found at plate boundaries. Very little change occurs in plate centres (shield lands). Figure 1.9 summarises the major landforms resulting from different types of plate movement.

Figure 1.9

The major landforms resulting from plate movements

Type of plate boundary	Description of changes	Examples
A Constructive margins (spreading or divergent plates)	two plates move away from each other; new oceanic crust appears forming mid-ocean ridges with volcanoes	Mid-Atlantic Ridge (Americas moving away from Eurasian and African Plates) East Pacific Rise (Nazca and Pacific Plates moving apart)
B Destructive margins (subduction zones)	oceanic crust moves towards continental crust but, being heavier, sinks and is destroyed forming deep sea trenches and island arcs with volcanoes	Nazca sinks under South American Plate (Andes) Juan de Fuca sinks under North American Plate (Rockies) Island arcs of the West Indies and Aleutians
Collision zones	two continental crusts collide and, as neither can sink, are forced up into fold mountains	Indian Plate collided with Eurasian Plate, forming Himalayas African Plate collided with Eurasian, forming Alps
C Conservative or passive margins (transform faults)	two plates move sideways past each other – land is neither formed nor destroyed	San Andreas fault in California
Note: centres of plates are rigid...	rigid plate centres form a shields lands (cratons) of ancient worn-down rocks b depressions on edges of the shield which develop into large river basins	 Canadian (Laurentian) Shield, Brazilian Shield Mississippi-Missouri, Amazon
...with one main exception	(Africa dividing to form a rift valley and possibly a new sea)	(African Rift Valley and the Red Sea)

Landforms at constructive plate margins

Constructive margins occur where two plates diverge or move away from each other and new crust is created at the boundary. Sea floor spreading occurs in the mid Atlantic where the North American and South American Plates are being pulled apart by convection currents from the Eurasian and African plates. Initially this may cause huge **rift valleys** to form on the sea floor. However, molten rock or **magma** from the mantle rises to fill any possible gap between the two plates. This magma produces **submarine volcanoes** which in time may grow above sea level, e.g. Surtsey, south of Iceland on the Mid-Atlantic Ridge (Places 3), and Easter Island, on the East Pacific Rise.

As the basaltic magma cools, it adds new land to the separating plates. The Atlantic Ocean did not exist some 150 million years ago (Figure 1.4, Pangaea) and it is still widening by some 2–5 cm a year. When the magma cools and where there is lateral movement, large cracks called **transform faults** are produced at right angles to the plate boundary (Figure 1.8). The largest visible product of constructive divergent plates is Iceland, where one-third of the lava emitted on to the earth's surface in the last 500 years can be found (Figures 1.10a and 1.24). It is estimated that 73 per cent of lava ejected annually onto the earth's surface is found in mid-ocean ridges.

The Atlantic Ocean was formed as the continent of Laurasia split into two. The process which caused the continental crust to become arched, weakened and stretched may be repeating itself in East Africa. Here the brittle crust has fractured and, as sections moved apart, the central portion dropped to

Figure 1.10

A constructive plate margin: Iceland

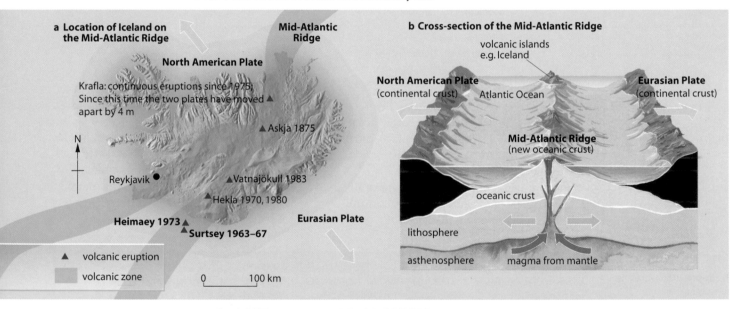

Places 3   Iceland

On 14 November 1963, the crew of an Icelandic fishing boat reported an explosion under the sea south-west of the Westman Islands. This was followed by smoke, steam and emissions of pumice stone. Having built up an ash cone of 130 m from the sea bed, the island of Surtsey emerged above the waves. On 4 April 1964, a lava flow covered the unconsolidated ash and guaranteed the island's survival.

Just before 0200 hours on 23 January 1973, an earth tremor stopped the clock in the main street of Heimaey, Iceland's main fishing port (Figure 1.24). Once again the North American and Eurasian Plates were moving apart (Figure 1.10b). Fishermen at sea witnessed the crust of the earth break open and lava and ash pour out of a fissure 2 km in length. Eventually the activity became concentrated on the volcanic cone of Helgafell and the inhabitants of Heimaey were evacuated to safety. By the time volcanic activity ceased six months later, many homes nearby had been burned; others farther afield had been buried under 5 m of ash; and the entrance to the harbour had been all but blocked.

Plate tectonics, earthquakes and volcanoes

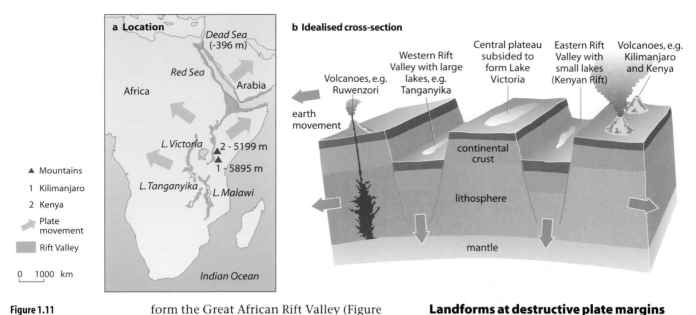

a Location

Dead Sea (-396 m)

Red Sea

Africa

Arabia

L. Victoria

2 - 5199 m

1 - 5895 m

L. Tanganyika

L. Malawi

Indian Ocean

▲ Mountains
1 Kilimanjaro
2 Kenya

Plate movement

Rift Valley

0   1000   km

b Idealised cross-section

Western Rift Valley with large lakes, e.g. Tanganyika

Central plateau subsided to form Lake Victoria

Eastern Rift Valley with small lakes (Kenyan Rift)

Volcanoes, e.g. Kilimanjaro and Kenya

Volcanoes, e.g. Ruwenzori

earth movement

continental crust

lithosphere

mantle

Figure 1.11

The African Rift Valley

Figure 1.12

A destructive plate margin – the Nazca and South American Plate boundary

form the Great African Rift Valley (Figure 1.11) with its associated volcanic activity. In Africa the rift valley extends for 4000 km from Mozambique to the Red Sea. In places its sides are over 600 m in height while its width varies between 10 and 50 km. Where the land has been pulled apart and dropped sufficiently, it has been invaded by the sea. It has been suggested that the Red Sea is a newly forming ocean. Looking 50 million years into the future (Figure 1.4c), it is possible that the east of Africa may detach itself from the remainder of the continent.

saline lakes (e.g. Titicaca), remnants of disappearing former oceans

(ii) some lava reaches surface to form volcanoes 6000 m high e.g. Chimborazo and Cotopaxi

young fold mountains of the Andes, separated by the Altiplano (High Plateau)

Western Cordillera

Eastern Cordillera

Peru – Chile deep sea trench

Amazon and Parana lowlands (sedimentary rocks)

Atlantic Ocean

Coastal Range

Brazilian Plateau, an ancient shield having always been part of a stable continental plate

Pacific Ocean

(i) lava rises, but cools within the crust forming the granite Andean batholith

sea level

Nazca Plate (oceanic crust)

American Plate (continental crust)

earthquake foci

lithosphere

friction from the subduction zone gives extra heat producing either (i) or (ii) above

asthenosphere

Subduction zone, oceanic plate breaks up producing earthquake foci

Landforms at destructive plate margins

Destructive margins occur where continental and oceanic plates converge. The Pacific Ocean, which extends over five oceanic plates, is surrounded by several continental plates (Figure 1.8). The Pacific Plate, the largest of these oceanic plates, and the Philippine Plate move north-west to collide with eastern Asia. In contrast, the smaller, but no less important, Nazca, Cocos and Juan de Fuca Plates travel eastwards towards South America, Central America and North America respectively. Figure 1.12 shows how the Nazca Plate, made of oceanic crust which cannot override continental crust, is forced to dip downwards at an angle to form a **subduction zone** with its associated **deep sea trench**. As oceanic lithosphere descends, the increase in pressure can trigger off major earthquakes while the heat produced by friction helps to convert the disappearing crust back into magma. Being less dense than the mantle, the newly formed magma will try to rise to the earth's surface. Where the magma does reach the surface (Figure 1.13) volcanoes will occur. These volcanoes are likely to form either a long chain of **fold mountains** (e.g. the Andes) or, if the eruption takes place offshore, an **island arc** (e.g. Japan, West Indies). Estimates claim that 80 per cent of the world's present active volcanoes are located above subduction zones. As the rising magma at destructive margins is more acidic than the lava of constructive margins (page 22), it is more viscous and flows less easily. It may solidify within the mountain mass to form large **intrusive** features called **batholiths** (page 26).

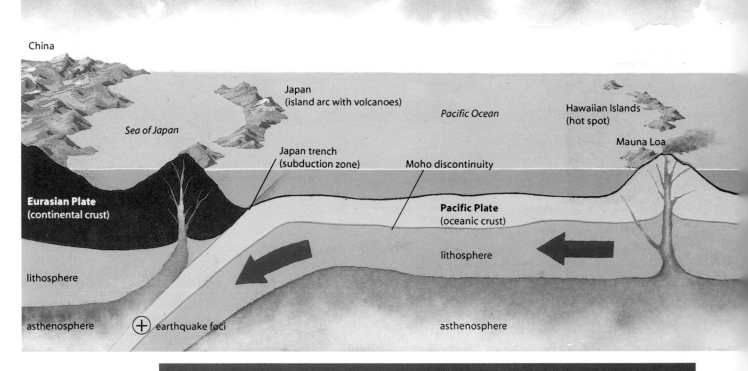

China

Japan
(island arc with volcanoes)

Pacific Ocean

Hawaiian Islands
(hot spot)

Sea of Japan

Mauna Loa

Japan trench
(subduction zone)

Moho discontinuity

Eurasian Plate
(continental crust)

Pacific Plate
(oceanic crust)

lithosphere

lithosphere

lithosphere

asthenosphere

⊕ earthquake foci

asthenosphere

Figure 1.13

Landforms resulting from plate tectonics in the Pacific Ocean

Hot spots Sometimes heat under the earth's crust is localised and creates 'hot spots' (Figure 1.13). The resultant rising magma can produce volcanoes such as those whose summits form the Hawaiian Islands.

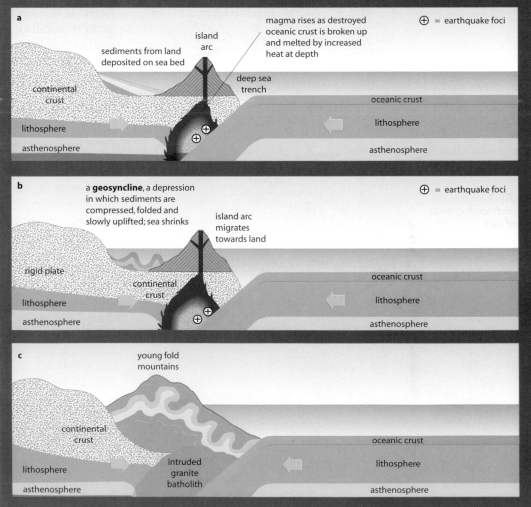

a

sediments from land deposited on sea bed

island arc

magma rises as destroyed oceanic crust is broken up and melted by increased heat at depth

⊕ = earthquake foci

deep sea trench

continental crust

oceanic crust

lithosphere

lithosphere

asthenosphere

asthenosphere

b

a **geosyncline**, a depression in which sediments are compressed, folded and slowly uplifted; sea shrinks

island arc migrates towards land

⊕ = earthquake foci

rigid plate

oceanic crust

continental crust

lithosphere

lithosphere

asthenosphere

asthenosphere

c

young fold mountains

continental crust

oceanic crust

lithosphere

lithosphere

intruded granite batholith

asthenosphere

asthenosphere

Figure 1.14

A collision zone – the formation of fold mountains (orogenesis)

Plate tectonics, earthquakes and volcanoes

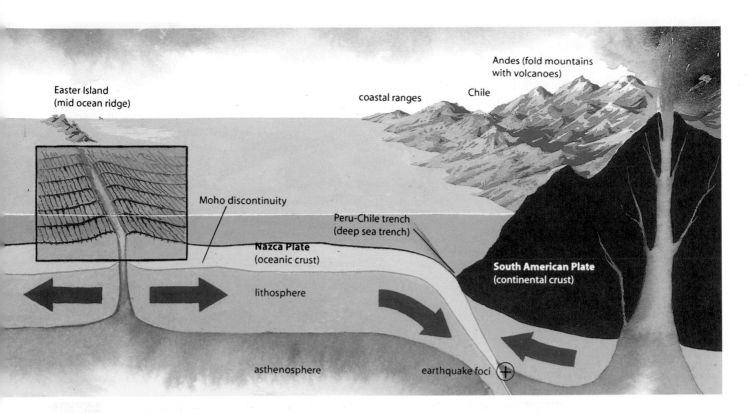

Easter Island
(mid ocean ridge)

coastal ranges

Chile

Andes (fold mountains
with volcanoes)

Moho discontinuity

Peru-Chile trench
(deep sea trench)

Nazca Plate
(oceanic crust)

lithosphere

South American Plate
(continental crust)

asthenosphere

earthquake foci ⊕

Landforms at collision plate margins

The formation of fold mountains is often extremely complex. As has already been explained in the context of the Pacific, fold mountains often occur where oceanic crust is subducted by continental crust (Figure 1.14).

former sediments of the Tethys Sea
(Figure 1.4a) folded upwards
to form the Himalayas

Indian Plate (ancient
shield) moving north
and east

Eurasian Plate (rigid)

collision
zone

continental
crust (sial)

mountain
roots

lithosphere
(sima)

Figure 1.15
....
Mountain building
(orogenesis)

A second, though less frequent occurrence, is when two plates composed of continental crust move together. In Figure 1.15 the Indian sub-continent, forming part of the Indo-Australian Plate, is shown to have moved north-eastwards and to have collided with the Eurasian Plate. Because continental crust cannot sink, the subsequent collision caused the intervening sediments, which contained sea shells, to be pushed upwards to form the Himalayas — an uplift which is still continuing. It is where these continental collisions occur that fold mountains form and the earth's crust is at its thickest (Figure 1.6 and 1.7).

Landforms at conservative plate margins

Conservative margins occur where two plates move parallel or nearly parallel to each another. Although frequent small earth tremors and occasional severe earthquakes may occur as a consequence of the plates trying to slide past each other, the margin between the plates is said to be conservative because crustal rocks are being neither created nor destroyed here. The boundary between the two plates is characterised by pronounced transform faults (Figure 1.16a). The San Andreas fault is the most notorious of several hundred known transform faults in California (Places 4).

The San Andreas Fault forms a junction between the North American and Pacific Plates. Although both plates are moving north-west, the Pacific Plate moves faster giving the illusion that they are moving in opposite directions. The Pacific Plate moves about 6 cm a year, but sometimes it sticks (like a machine without oil) until pressure builds up enabling it to jerk forwards. The last major movement resulted in the San Francisco earthquake of 1989 (page 10). Should these plates continue to slide past each other, it is likely that Los Angeles will eventually be on an island off the Canadian coast.

Figure 1.16

A conservative plate margin

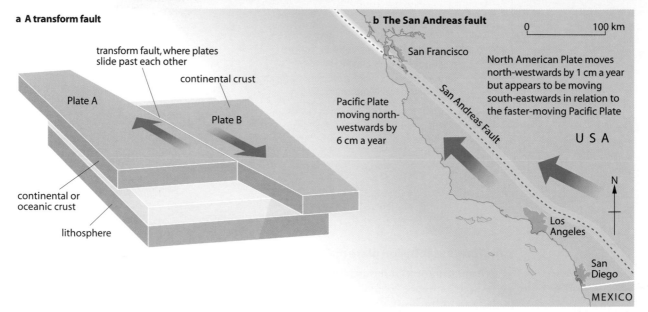

a A transform fault

transform fault, where plates slide past each other

continental crust

Plate A

Plate B

continental or oceanic crust

lithosphere

b The San Andreas fault

0   100 km

San Francisco

North American Plate moves north-westwards by 1 cm a year but appears to be moving south-eastwards in relation to the faster-moving Pacific Plate

Pacific Plate moving north-westwards by 6 cm a year

San Andreas Fault

USA

N

Los Angeles

San Diego

MEXICO

Plate tectonics and the British Isles

During the Cambrian period (Figure 1.1), northern Scotland lay on the American Plate while the rest of Britain was on the Eurasian Plate, as it is today. Both plates are thought then to have been in the latitude of present day South Africa. In the Ordovician and Silurian periods, the two plates began to converge causing volcanic activity and the formation of mountains in Snowdonia and the Lake District (a collision zone). Being continental crust, sediment between the plates was pushed up to form the Caledonian Mountains which linked Scotland to the rest of Britain. During the Devonian period, the locked plates drifted northwards through a desert environment (the present Kalahari Desert) when the Old Red Sandstones were deposited. This northward movement continued in Carboniferous times, accompanied by a sinking of the land which allowed the limestones of that period to form in warm, clear seas (page 180).

As the land began to emerge from these seas, millstone grit was formed from sediments in a shallow sea, and then coal measures were laid down under the hot, wet, swampy conditions usually associated with equatorial areas. It was during the Permian and Triassic periods that the continents collided to form Pangaea. Africa moved towards Europe, and Britain's New Red Sandstones were laid down under dry, hot desert conditions (in the position of the present day Sahara Desert). A further submergence during Jurassic/Cretaceous times enabled the Cotswold limestones and then the chalk of the Downs to form, again in warm, clear seas.

During the Tertiary era, the North American and Eurasian Plates split apart forming a constructive boundary and the volcanoes of north-western Scotland. At the same time, the African Plate moved further north pushing up the Alps and the hills of southern England. Subsequently, although Britain has been located away from the volcanoes and severe earthquakes associated with various plate margins, its landscape has been modified both during and since the ice ages. These, however, have been due to climatic change rather than plate movement.

Q

1 Refer to the geological timescale (Figure 1.1). Copy out the table and add notes to it describing mountain building and volcanic activity which have taken place in the British Isles.

2 Outline the evidence that supports the theory of plate tectonics.

3 How does an understanding of plate tectonics help to explain the distribution of
   a earthquake zones
   b volcanoes and
   c fold mountains?

4 With reference to any continent which you have studied, explain how an understanding of plate tectonics accounts for its major relief features.

5 With reference to the map of North and South America (Figure 1.17), explain how an understanding of plate tectonics helps to account for:
   a the presence of the two landmasses and the Pacific Ocean;
   b the presence of the Atlantic Ocean;
   c the Canadian and Brazilian Shields;
   d Ascension Island, Tristan da Cunha and Easter Island;
   e the four deep sea trenches;
   f the Andes and Rockies;
   g the West Indies and Aleutian Islands;
   h Mt St Helens, Paricutín and Cotopaxi;
   i the two different areas where transform faults occur.

6 Why do earthquakes on the Mid-Atlantic Ridge usually have a shallow focus, whereas those in the Andes and Central America normally have a deeper focus?

7 The Hawaiian Islands are an anomaly as they are volcanoes which do not occur on a plate boundary (Figure 1.13).
   a Why do volcanoes occur in the middle of the Pacific Ocean Plate?
   b Why do the Hawaiian Islands extend north-west?

Figure 1.17

Landforms and major relief features resulting from plate tectonics in the Americas

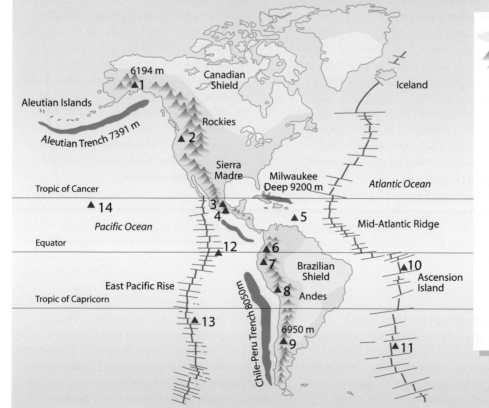

Shield lands (cratons)

Fold mountains

Deep sea trenches

Transform faults

▲ Volcanoes

1 Mt McKinley
2 Mt St Helens
3 Paricutín
4 Popocatepetl
5 Mt Pelée
6 Cotopaxi
7 Chimborazo
8 El Misti
9 Aconcagua
10 Ascension Island
11 Tristan da Cunha
12 Galapagos Islands
13 Easter Island
14 Hawaiian Islands

6194 m ▲1
Canadian Shield
Iceland
Aleutian Islands
Aleutian Trench 7391 m
Rockies
Tropic of Cancer
Sierra Madre
Milwaukee Deep 9200 m
Atlantic Ocean
▲14
▲2
Pacific Ocean
▲3
▲4
▲5
Mid-Atlantic Ridge
Equator
▲12
▲6
Brazilian Shield
▲10
Ascension Island
▲7
East Pacific Rise
▲8
Andes
Tropic of Capricorn
Chile-Peru Trench 8050m
▲13
6950 m
▲9
▲11

Figure 1.18

Basic and acid lava

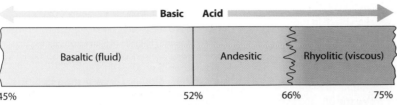

Silica content

Basaltic or basic lava	Andesitic or acid lava
Has low viscosity, is hot (1200°C) and runny, like warm treacle	Viscous, less hot (800°C), flows more slowly and shorter distances
Has a lower silica content	Has a higher silica content
Tales a longer time to cool and solidify, so flows considerable distances as rivers of molten rock	Soon cools and solidifies, flowing very short distances
Retains its gas content which makes it mobile	Loses gases quickly and becomes viscous
Produces extensive but gently sloping landforms	Produces steep-sided, more localised features
Eruptions are frequent but relatively gentle	Eruptions are less frequent but are violent due to the build up of gases
Lava and steam ejected	Ash, rocks, gases, steam and lava ejected
Found at conxtructive plate margins where magma rises from the mantle e.g. fissures along the Mid-Atlantic Ridge (Heimaey) e.g. over hot spots (Mauna Loa, Hawaii)	Found at destructive margins where oceanic crust is destroyed (subducted), melts and rises e.g. subduction zones (Mt St Helens) e.g. as island arcs (Mt Pelée, Martinique)

Volcanology

The term **volcanology** includes all the processes by which solid, liquid or gaseous materials are forced into the earth's crust or are ejected onto the surface. Although material in the mantle has a high temperature, it is kept in a semi-solid state because of the great pressure exerted upon it. However, if this pressure is released locally by folding, faulting or other movements at plate boundaries, some of the semi-solid material becomes molten and rises, forcing its way into weaknesses in the crust, or onto the surface, where it cools, crystallises and solidifies.

The molten rock is called **magma** when it is below the surface and **lava** when on the surface. When lava and other materials reach the surface they are called **extrusive**. The resulting landforms vary in size from tiny cones to widespread lava flows. Materials injected into the crust are referred to as **intrusive**. These may later be exposed at the surface by erosion of the overlying rocks. Both extrusive and intrusive materials cooled from magma are known as **igneous rocks**.

Extrusive landforms

There are several types of extrusive landform whose nature depends on how gaseous and/or viscous the lava is when it reaches the earth's surface (Figure 1.18).

- Lava produced by the upward movement of material from the mantle is **basaltic** and tends to be located along mid-ocean ridges, over hot spots and alongside rift valleys.
- Lava that results from the process of subduction is described as **andesitic** (after the Andes) and occurs as island arcs or at destructive plate boundaries where oceanic crust is being destroyed.
- **Pyroclastic material** (meaning 'fire broken') is material ejected by volcanoes in a fragmented form. Tephra, fragments of different sizes, include ash, lapilli (small stones) and bombs (larger material) which are thrown into the air before falling back to earth. Pyroclastic flows are materials which move down the side of a volcano. The fragments may occur as mudflows (or lahars), e.g. Mt Pinatubo (Case Study 1) and Nevado del Ruiz, or as a fast-moving cloud known as a nuée ardente, e.g. Mt Pelée. The deposited material is **ignimbrite**.

a Draw a diagram to show the differences in shape between a basic volcano and an acid volcano.

b Add labels to show six differences between the two types of volcano.

How can volcanoes be classified?

Because of the large number of volcanoes and wide variety of eruptions, it is convenient to group together those with similar characteristics (Framework 5, page 151). Unfortunately, there is no universally accepted method of classification. One of the two most quoted groupings is according to the **shape** of the volcano and its vent which, because it describes landforms, is arguably of more value to the geographer. The other is the nature of the **eruption**, which has traditionally been the method used by volcanologists.

The shape of the volcano and its vent

1 Fissure eruptions When two plates move apart, lava may be ejected through fissures rather than via a central vent (Figure 1.19a). The Heimaey eruption of 1973 (Figures 1.10, 1.24) began with a fissure 2 km in length. This was small in comparison with that at Laki, also in Iceland, where in 1783 a fissure exceeding 30 km opened up. The basalt may form large plateaux, filling in hollows rather than building up into the more typical cone-shaped volcanic peak. The remains of one such lava flow, formed when the Eurasian and North American Plates began to move apart, can be seen in Northern Ireland, north-west Scotland, Iceland and Greenland. The columnar jointing produced by the slow cooling of the lava provides tourist attractions at the Giant's Causeway in Northern Ireland (Figure 1.25) and Fingal's Cave on the Isle of Staffa.

2 Basic or shield volcanoes In volcanoes such as Mauna Loa on Hawaii, lava flows out of a central vent and can spread over wide areas before solidifying. The result is a 'cone' with long, gentle sides made up of many layers of lava from repeated flows (Figure 1.19b).

Figure 1.19

Classification of volcanoes based on their shape (not to scale)

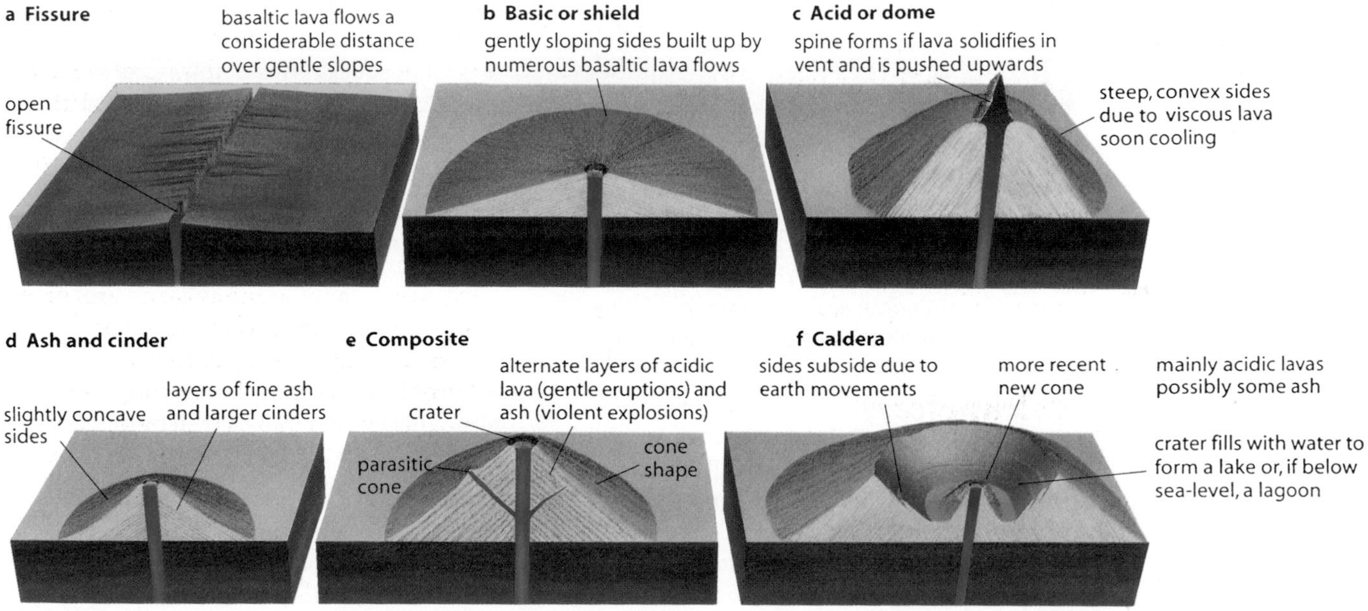

a Fissure — basaltic lava flows a considerable distance over gentle slopes — open fissure

b Basic or shield — gently sloping sides built up by numerous basaltic lava flows

c Acid or dome — spine forms if lava solidifies in vent and is pushed upwards — steep, convex sides due to viscous lava soon cooling

d Ash and cinder — slightly concave sides — layers of fine ash and larger cinders

e Composite — crater — parasitic cone — alternate layers of acidic lava (gentle eruptions) and ash (violent explosions) — cone shape

f Caldera — sides subside due to earth movements — more recent new cone — mainly acidic lavas possibly some ash — crater fills with water to form a lake or, if below sea-level, a lagoon

3 Acid or dome volcanoes Acid lava quickly solidifies on exposure to the air. This produces a steep-sided, convex cone as in most cases the lava solidifies near to the crater (Figure 1.19c). In one extreme instance, that of Mt Pelée, the lava actually solidified as it came up the vent and produced a spine rather than flowing down the sides.

4 Ash and cinder cones (Figure 1.19d) Paricutín, for example, was formed in the 1940s by ash and cinders building up into a symmetrical cone.

5 Composite cones Many of the larger, classically shaped volcanoes result from alternating types of eruption in which first ash and then lava (usually acidic) are ejected (Figure 1.19e). Mt Etna is a result of a series of both violent and more gentle eruptions.

6 Calderas When the build up of gases becomes extreme, huge explosions may clear the magma chamber beneath the volcano and remove the summit of the cone. This causes the sides of the crater to subside, thus widening the opening to several kilometres in diameter. In the cases of both Thera (Santorini) and Krakatoa, the enlarged craters or calderas have been flooded by the sea and later eruptions have formed smaller cones within the resultant lagoons (Figures 1.19f, 1.22).

Minor extrusive features

These are often associated with, but are not exclusive to, areas of declining volcanic activity. They include solfatara, fumaroles, geysers and mud volcanoes (Figure 1.20 and Places 5).

a Mud volcano: hot water mixes with mud and surface deposits

b Solfatara: created when gases, mainly sulphurous, escape onto the surface

c Geyser: water in the lower crust is heated by rocks and turns to steam; pressure increases and the steam and water explode onto the surface

d Fumaroles: superheated water turns to steam as its pressure drops when it emerges from the ground

Magma chamber
(probably solid by this stage)

Figure 1.20

Minor extrusive landforms

Places 5 Solfatara, Italy

Solfatara is a small volcano on the outskirts of Naples. Its crater is 2 km in diameter, making it larger than that of nearby Vesuvius, but there is no volcanic cone. Solfatara takes its name from the gases which escape to the surface; they are mainly sulphurous and can be smelled from a considerable distance. Many rocks are coated with sulphur. **Solfatara** has given its name to all similar features of this type. **Fumaroles**, resulting from superheated water being turned to steam as it cools on its ejection through the thin crust, are numerous in the area (Figure 1.21). Evidence of the thinness of the crust (magma is only 3 m below the surface) is provided by a guide who throws a boulder on to the surface and making groups of tourists jump in harmony to hear the hollowness of the ground. The guide, who is needed to keep visitors safely away from bubbling mud volcanoes and areas too hot to walk upon, also shows volcanic activity by lighting twigs and stirring loose material to cause a miniature eruption.

The only minor feature missing is the **geyser**, an intermittent fountain of hot water (e.g. Old Faithful, Yellowstone National Park, USA).

During the mid 1980s the temperature (160°C), pressure and surface of Solfatara all rose, giving rise to fears of a new eruption — the last was in 1198. Despite the appearance of a small fissure near to the observatory, which led to its abandonment, activity appears to have stabilised.

Figure 1.21

Inside the Solfatara crater, near Naples, Italy

Plate tectonics, earthquakes and volcanoes

Figure 1.22

Krakatoa with Anak Krakatoa

Figure 1.24

The Heimaey eruption, Iceland, 1973

Figure 1.23

Vesuvius: notice the new cone within the old crater of Monte Somma

Figure 1.25

The Giant's Causeway, Northern Ireland

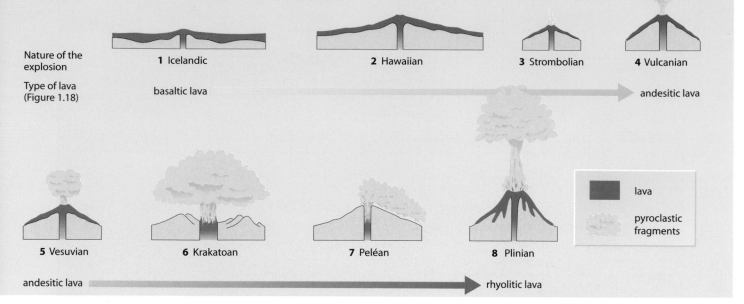

Figure 1.26

Classification of volcanoes according to the nature of the explosion

The nature of the eruption

This classification of volcanoes is based on the degree of violence of the explosion which is a consequence of the pressure and amount of gas in the magma (Figure 1.26). Its categories may be summarised as follows:

- Icelandic, where lava flows gently from a fissure;
- Hawaiian, where lava is emitted gently but from a vent;
- Strombolian, where small but very frequent eruptions occur;
- Vulcanian, which is more violent and less frequent;
- Vesuvian, which has a violent explosion after a long period of inactivity (Figure 1.23);
- Krakatoan, which has an exceptionally violent explosion that may remove much of the original cone (Figure 1.22);
- Peléan, where a violent eruption is accompanied by pyroclastic flows that may include a nuée ardente ('glowing cloud');
- Plinian, where large amounts of lava and pyroclastic material are ejected.

Geomorphology is the study of the structure, origin and development of the topographical features of the earth's crust.

Intrusive landforms

Usually, only a relatively small amount of magma actually reaches the surface as most is intruded into the crust, where it solidifies. Such intrusions may initially have little impact upon the surface **geomorphology**, but if the overlying rocks are later worn away distinctive landforms may then develop (Figure 1.27).

During the Tertiary era, an upthrust of magma was **intruded** into the sedimentary rocks of Arran to form the Northern Granite. As the magma slowly cooled, it formed large crystals (unlike on the surface where rapid cooling forms fine crystals), contracted and cracked resulting in a series of joints. The magma also produced a large, deep-seated, dome-shaped **batholith** as it solidified.

Surrounding the batholith is a **metamorphic aureole** where the original sedimentary rocks have been changed (metamorphosed) by the heat and pressure of the intrusion from sandstones into **schists**. Since then, the overlying rocks have been removed by water, ice and even the sea to leave the granite batholith with its jointing exposed (Figure 1.28). These joints have been widened by chemical weathering (pages 36–37) to form the large granite slabs and tors surrounding Goatfell (Figure 8.12).

Figure 1.27

Diagrammatic model showing intrusive landforms: batholith, dyke and sills

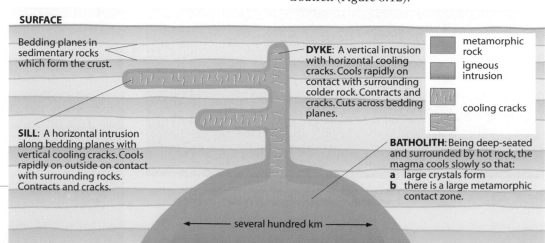

SURFACE

Bedding planes in sedimentary rocks which form the crust.

DYKE: A vertical intrusion with horizontal cooling cracks. Cools rapidly on contact with surrounding colder rock. Contracts and cracks. Cuts across bedding planes.

SILL: A horizontal intrusion along bedding planes with vertical cooling cracks. Cools rapidly on outside on contact with surrounding rocks. Contracts and cracks.

BATHOLITH: Being deep-seated and surrounded by hot rock, the magma cools slowly so that:
a large crystals form
b there is a large metamorphic contact zone.

metamorphic rock

igneous intrusion

cooling cracks

← several hundred km →

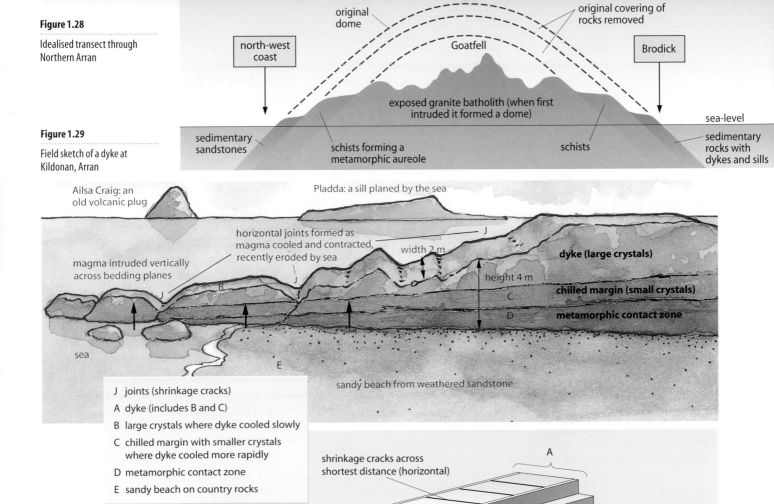

Figure 1.28

Idealised transect through Northern Arran

original dome
original covering of rocks removed
north-west coast
Goatfell
Brodick
exposed granite batholith (when first intruded it formed a dome)
sea-level
sedimentary sandstones
schists forming a metamorphic aureole
schists
sedimentary rocks with dykes and sills

Figure 1.29

Field sketch of a dyke at Kildonan, Arran

Ailsa Craig: an old volcanic plug
Pladda: a sill planed by the sea
horizontal joints formed as magma cooled and contracted, recently eroded by sea
width 2 m
J
dyke (large crystals)
magma intruded vertically across bedding planes
height 4 m
chilled margin (small crystals)
metamorphic contact zone
sea
sandy beach from weathered sandstone
E

J joints (shrinkage cracks)
A dyke (includes B and C)
B large crystals where dyke cooled slowly
C chilled margin with smaller crystals where dyke cooled more rapidly
D metamorphic contact zone
E sandy beach on country rocks

shrinkage cracks across shortest distance (horizontal)
A
Cross-section (viewed from above)
E
D
B
D
E
C
country rock

Figure 1.30

Diagrammatic cross-section of a dyke, Arran

Figure 1.31

Dyke at Kildonan, Arran

If, in trying to rise to the surface, magma cuts across the bedding planes of the sedimentary rock it is called a **dyke** (Figures 1.27 and 1.29). The material which forms the dyke cools slowly although those parts which come into contact with the surrounding rock will cool more rapidly to produce a chilled margin (Figure 1.30). Most dykes on Arran were formed after, and radiate from, the batholith intrusion; they are so numerous that they have been termed a 'dyke swarm'. Most of the dykes are more resistant to erosion than the surrounding sandstones and so where they cross the island's beaches they stand up like groynes (Figure 1.31). Although averaging 3 m, these dykes vary from 1 to 15 m in width.

Plate tectonics, earthquakes and volcanoes

A **sill** is formed when the igneous rock is intruded along the bedding planes between the existing sedimentary rocks. The magma cools and contracts but this time the resultant joints will be vertical and their hexagonal shapes can be seen when the landform is later exposed as on headlands such as that at Drumadoon on the west coast of Arran (Figures 1.32, 1.33) and The Giant's Causeway in Northern Ireland (Figure 1.25). The sill at Drumadoon is 50 m thick.

Figure 1.32

Field sketch of a sill exposed at Drumadoon, Arran

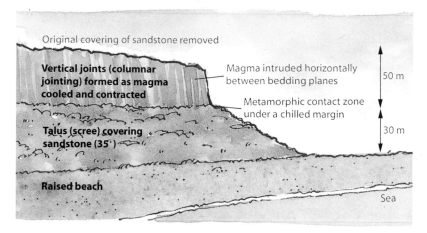

Original covering of sandstone removed

Vertical joints (columnar jointing) formed as magma cooled and contracted

Magma intruded horizontally between bedding planes

50 m

Metamorphic contact zone under a chilled margin

30 m

Talus (scree) covering sandstone (35°)

Raised beach

Sea

Figure 1.33

Sill at Drumadoon, Arran

Q
1 In Figure 1.34 which of the diagrams show sills and which dykes?

2 Using specific examples, discuss the benefits and dangers (table below) faced by people living near the plate boundaries.

Figure 1.34

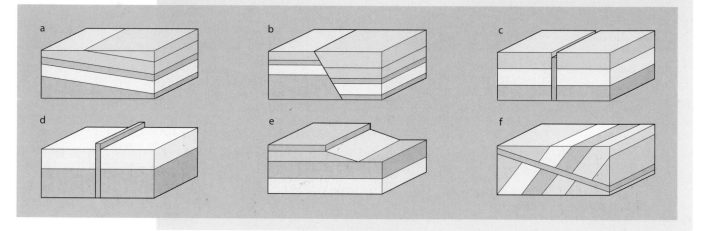

Benefits	Hazards
Lava and ash weather rapidly. Basic lava produces fertile soils ideal for farming, (the region surrounding Mt Etna). The fertility of acid lava is much lower	Earthquakes destroy buildings and result in loss of life
Igneous rock contains minerals such as gold, copper, lead and silver	Violent eruptions with blast waves and gas may destroy life and property, (Mt Pelée, Mt St Helens)
Igneous rock is used for building purposes, (Naples, Aberdeen)	Mudflows/lahars may be caused by heavy rain and melting snow, (Armero, Colombia, and Pinatubo, the Philippines)
Extinct volcanoes may provide defensive settlement sites, (Edinburgh)	Tidal waves/tsunamis, (following the eruption of Krakatoa)
Geothermal power is being developed, (Iceland, New Zealand)	Ejection of ash and lava ruins crops and kills animals
Geysers and volcanoes are tourist attractions, (Yellowstone National Park), generating revenue for local communities	Interrupts communications
Volcanic eruptions may produce spectacular sunsets, (Krakatoa)	Short-term climatic changes occur as volcanic dust absorbs solar energy, lowering temperatures and increasing rainfall

Plate tectonics, earthquakes and volcanoes

Case Study 1

Natural hazards and volcanic eruptions

Natural hazards

What are natural hazards?

Natural hazards, which include earthquakes, volcanic eruptions, floods, drought and storms, result from natural processes within the environment (Figure 1.35). They are, therefore, different from environmental disasters, such as desertification, ozone depletion and acid rain, which are caused by human activity and the mis-management of the environment. It is important, however, to stress the difference between a natural hazard and a hazard event. Natural hazards have the potential to affect people and the environment; it is the hazard event which causes the damage. An event only becomes a hazard if it affects, or threatens, people and property. For example, the submarine volcanic eruption which created the new island of Surtsey (Places 3, page 16) was hardly a hazard event as it posed no threat to people or property. In contrast, the eruption of Mt St Helens was a hazard event as it killed 61 people, destroyed roads and property and interrupted human activities. The impact of a hazard event may be felt over a wide area; the effects may be long-term as well as immediate; and the event can be costly to property and dangerous to people.

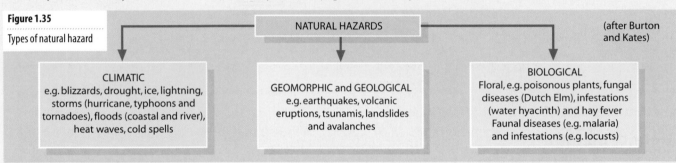

Figure 1.35

Types of natural hazard

NATURAL HAZARDS

(after Burton and Kates)

CLIMATIC
e.g. blizzards, drought, ice, lightning, storms (hurricane, typhoons and tornadoes), floods (coastal and river), heat waves, cold spells

GEOMORPHIC and GEOLOGICAL
e.g. earthquakes, volcanic eruptions, tsunamis, landslides and avalanches

BIOLOGICAL
Floral, e.g. poisonous plants, fungal diseases (Dutch Elm), infestations (water hyacinth) and hay fever Faunal diseases (e.g. malaria) and infestations (e.g. locusts)

How may people react to natural hazards?

The study of natural hazards, involving the relationship and interaction between physical systems and human systems (i.e. people and the environment), falls into the domain of geographers — although not exclusively so.

The following questions need to be asked when studying either the risk of a potential natural hazard or a specific hazard event:

1 What are people's perception of the natural hazard?

Perception is how individuals or groups of people view the hazard risk and often depends upon their knowledge and experience of the event. The inhabitants of Pompeii, many of whom were killed during the eruption of Vesuvius in AD 79, did not realise until that event that the mountain was indeed a volcano. Since then, Vesuvius has erupted on numerous occasions. So why do people continue to rebuild and live in this, and other, hazardous areas? It may be because they

- perceive the area as providing the

best of opportunities to earn a living;
- are too concerned with day-to-day problems to consider the hazard risk;
- have the capital and technology to cope with the hazard event.

2 What are the immediate and long-term effects of the event?

3 How do people respond to the event (Figure 1.36)?

4 How might people adjust to and plan for a future event?

It has been suggested that people have six options. They may try to: prevent the event; modify the hazard; lessen the possible amount of damage; spread the losses caused by the event; claim for losses; or do nothing but pray that the event will not occur again (at least not in their own lifetimes).

5 Can a future event be predicted?

This involves predicting where the next event will take place, when it is likely to occur and how big it is likely to be.

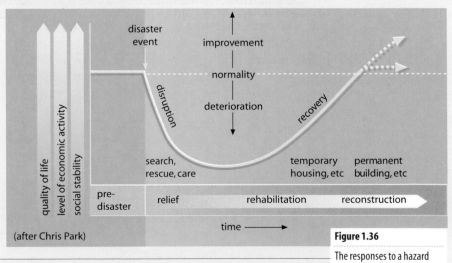

Figure 1.36

The responses to a hazard event

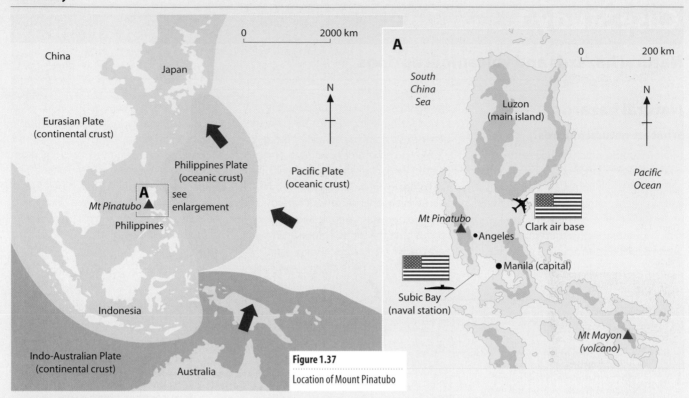

Figure 1.37

Location of Mount Pinatubo

Mount Pinatubo

Why is Mount Pinatubo in a hazard risk area?

Mount Pinatubo is located in the Philippines (Figure 1.37). The Philippines lie on a destructive plate margin where the Philippines Plate, composed of oceanic crust, moves towards and is subducted by the Eurasian Plate, which consists of continental crust. As the oceanic plate is subducted, it is converted into magma which rises to the surface and forms volcanoes. The Philippines owe their existence to the frequent ejection of lava over a period of several million years. Even before Pinatubo erupted in 1991, there were over 30 active volcanoes in the Philippines.

Why did people live in this hazard risk area?

As Mount Pinatubo had not erupted since 1380, people living in the area no longer considered it to be a hazard. During that time, ash and lava from earlier eruptions had weathered into a fertile soil, ideal for rice growing. By 1991, people no longer perceived Pinatubo to be a danger. On the lower slopes of the mountain, the Aeta, recognised as the aboriginal inhabitants of the islands, practised subsistence farming (slash and burn agriculture, Places 52, page 441). Near the foothills was the rapidly growing city of Angeles, together with an American air base and a naval station (Figure 1.37).

What were the nature, effects and consequences of the eruption?

1 Immediate effects

The volcano began to show signs of erupting in early June, 1991. Fortunately, there were several advance-warning signs which allowed time for the evacuation of thousands of people from Angeles and the 15 000 personnel from the American air base. The number and size of eruptions increased after 9 June. On 12 June, an explosion sent a cloud of steam and ash 30 km into the atmosphere — the third-largest eruption experienced anywhere in the world this century (Figure 1.39). Up to 50 cm of ash fell nearby, and over 10 cm within a 600 km radius (Figure 1.38). The eruptions were, characteristically, accompanied by earthquakes and torrential rain — except that the rain, combining with the ash, fell as thick mud. The ash destroyed all crops on adjacent farmland and its weight caused buildings to collapse, including 200 000 homes, a local hospital and many factories. Power supplies were cut off for three weeks and water supplies became contaminated. Relief operations were hindered as many roads became impassable and bridges were destroyed.

Figure 1.38

Eye-witness account of the eruption

Seismologists said a mixture of searing gas, ash and molten rock quickly raced down the mountain's west and northern flanks and into the Marella, Maraunot and O'Donnel rivers. Ash also rained down on seven towns in the region and traces of ash were detected near the Subic Bay naval base, 50 miles to the southwest. Pumice fragments measuring up to 1.2 inches long fell on villages south-west of the volcano. At a refugee centre in Olongapo, about 35 miles south-west of the volcano, survivors said they saw the sky grow dark and then heard a tremendous explosion followed by a rain of ash and stones as big as a man's head.

"The smoke is very thick, like a dark mushroom in the sky,' Gus Abelgas, a reporter for ABS-CBN television news said in a broadcast from Botolan, near the volcano's western slopes. 'It's just like Hiroshima.'

Other reporters described panic as people fled with their belongings and livestock over roads made slippery by the falling ash. Refugees wore cardboard boxes with air and peepholes to protect themselves from the ash. The ash was so thick in the air that at noon motorists were driving with their headlights on and wipers operating to clear the débris.

Figure 1.39

A pyroclastic cloud produced by the eruption of Mt Pinatubo, June 1991

Figure 1.40

A lahar at Angeles, near Mt Pinatubo

2 Longer-term effects

The thick fall of ash not only ruined the harvest of 1991, but made planting impossible for 1992. Over one million farm animals died, many through starvation due to the lack of grass. Several thousand farmers and their families had to take refuge in large cities. The majority were forced to seek food and shelter in shanty-type refugee camps. Disease, especially malaria, chicken-pox and diarrhoea, spread rapidly and doctors had to treat hundreds of people for respiratory and stomach disorders. Soon after the event, and again in 1993, typhoons brought heavy rainfall which caused flooding and lahars (mud flows). Lahars form when surface water picks up large amounts of volcanic ash in mountainous areas and deposits it as mud over lower-lying areas (Figure 1.40). The ash which was ejected into the atmosphere is believed to have caused changes in the earth's climate, including the lowering of world temperatures and ozone depletion (Figure 1.41).

The eruption and its after effects were blamed for about 700 deaths. Of these, only six were believed to have been a direct result of the eruption itself. Over 600 people were to die from disease and a further 70 from suffocation by lahars.

Figure 1.41

The climatic effects of the eruption

IT HAS been described as the world's greatest climatic experiment, but unlike most scientific endeavours it was unplanned. When the tropical tranquillity of the Philippines was shattered last June by a volcanic explosion, Mount Pinatubo was a relatively obscure volcano, known in the scientific community only to a handful of geologists. Having sent more than 20 million tonnes of dust and ash into the atmosphere, altering its heat balance and accelerating ozone depletion over a large part of the globe, Pinatubo has become the focus of several far-reaching studies.

Climatologists now use the term "Pinatubo effect" to describe how volcanic ash and débris, if sent high enough into the atmosphere, can influence temperature and weather for several years afterwards. The dust from Pinatubo was ejected as high as 20 miles above the Earth. From the haven of Earth orbit, satellites observed the plume of volcanic ash as it girdled the globe at speeds approaching 75 miles per hour. A month after the eruption which killed 350 people, a 3,000 mile cloud of ash and sulphur compounds circled the Earth. Satellite temperature measurements confirmed that the dust had effectively shaded the surface of the Earth from the sun's rays, resulting in a lowering of the average global temperature. A Nasa team at the Goddard Institute for Space Studies in New York, led by James Hansen, tried to assess what effect the cooling caused by the dust of Mount Pinatubo would have on global warming caused by man-made emissions of carbon dioxide. They concluded that Pinatubo would in effect delay global warming by several years. While global warming experts argue about the effect of Pinatubo's eruption on average temperatures, ozone specialists are interested in the effect the volcano has had and will have on the ozone layer. The volcano has spewed out huge quantities of sulphate aerosols, particles containing sulphur that remain suspended in the atmosphere for several years. These sulphate particles are important in the chemistry of ozone destruction for two reasons: first, they act as sites where ozone-destroying reactions take place; and secondly, they mop up nitrogen -containing compounds that help to prevent ozone destruction. This winter American and European scientists undertook the most intensive investigation of ozone depletion over the northern hemisphere, including Europe and North America. More than 300 scientists from 17 countries were involved and their work has shown that ozone levels fell by 10 to 20 per cent more than expected. "The eruption of Mount Pinatubo has increased the abundance of natural sulphate particles, potentially enhancing ozone losses due to chemical reactions that occur on particle surfaces," the NASA ozone monitoring team said earlier this month.

The Independent on Sunday, 10 May 1992

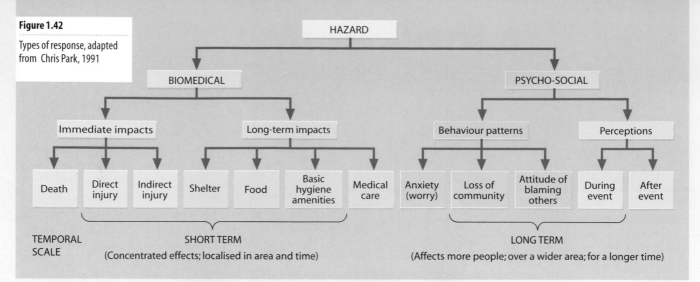

Figure 1.42

Types of response, adapted from Chris Park, 1991

HAZARD
- BIOMEDICAL
  - Immediate impacts
    - Death
    - Direct injury
    - Indirect injury
  - Long-term impacts
    - Shelter
    - Food
    - Basic hygiene amenities
    - Medical care
- PSYCHO-SOCIAL
  - Behaviour patterns
    - Anxiety (worry)
    - Loss of community
    - Attitude of blaming others
  - Perceptions
    - During event
    - After event

TEMPORAL SCALE

SHORT TERM (Concentrated effects; localised in area and time)

LONG TERM (Affects more people; over a wider area; for a longer time)

Figure 1.43a

Members of the Aeta tribe

How did people respond to the hazard event?

Chris Park has divided human responses during and after any hazard event into two categories (Figure 1.42).

Within a few weeks of the major Pinatubo eruption, groups of evacuees from the affected area began to consider their future options and their next move.

Their range of responses included the following:

1 Some members of the Aeta tribe (Figure 1.43) decided not to return to their former homes. As a spokesperson explained, "Everything we have planted has been destroyed. There is no point in going back. The government will have to put us somewhere else".

Figure 1.43b

Members of the Aeta tribe

2 In contrast, the majority of the Aeta tribe decided to return. To them, the mountain slopes, although vastly changed, were still their home and the hard way of life in the hills was preferable to the foreign habits of the lowlanders and to living in urban areas.

3 Most of the people who fled from the city of Angeles have, so far, opted against returning home. To them, life in the shanty refugee camps is safer than returning to an area where eruptions and earthquakes are still occurring and where the heavy rain is likely to cause lahars for several years until the regrowth of vegetation stabilises the slopes.

Can future eruptions be predicted?

It is now possible to modify some natural hazards (e.g. cloud seeding to reduce drought) and to forecast and control other actual hazard events (e.g. implementing flood control schemes, installing early-warning hurricane systems and constructing buildings designed to withstand earthquakes). However, at present, while it may be possible to predict fairly accurately where volcanic eruptions are likely to occur (i.e. at constructive and destructive plate margins, Figure 1.8), there seems little prospect of people being able to predict either the time or the scale of a specific event. Also, in places where a volcanic eruption has not occurred for several centuries, as in the case of Pinatubo, the human perception of the hazard risk decreases and the event, should it take place, catches more people and organisations unprepared.

UN recommendations for the surveillance of volcanoes

1 Regional networks of seismograph stations should be set up in risk areas which are populated.

2 Portable seismographs and tilt-meters should be available for instant installation on any volcano showing signs of activity.

3 Trained volcanologists should be made available to interpret events.

4 Detailed histories of volcanoes should be produced. Some, but by no means all, volcanoes have a cycle of eruptions; if this cycle is known, it can aid prediction of future events.

5 Detailed maps should be produced to show deposits of lava and pyroclastic materials from previous eruptions. These maps can show areas most likely to be at risk from the hazard (such a map would not have predicted the effects on Mt St Helens).

6 Periodic measurements should be taken to show possible changes in temperature and pressure within the volcano. Chemical analyses of fumaroles and hot springs might indicate variations from the norm.

7 Emergency procedures, including evacuation procedures, should be established for all communities.

8 Aerial magnetic and infra-red photographic surveys should be made at regular intervals. It has been shown that, before an eruption, there are local variations in the earth's magnetic field while infra-red photos could show a build up in temperature and any possible swellings (bulging) of the cone.

These procedures are easier to adopt in an economically more developed country with its greater wealth and technology. UN procedures are also more likely to be followed where the expected frequency of the event is high and in risk areas having high population densities.

Choose **either**

a an earthquake in an economically more developed country (e.g. San Francisco or Los Angeles) and in an economically less developed country (e.g. Central India or Armenia)

or

b a volcanic eruption in an economically more developed country (e.g. Mt St Helen's) and in an economically more developed country (e.g. Pinatubo or Nevado del Ruiz).

1 What was the cause of the hazard event?

2 What was the local perception of the hazard risk?

3 What were the
  i immediate effects
  ii long term effects of the event?

4 How did local people respond to the event?

5 Why was the death role and the disruption to every day life less in the economically more developed country than in the economically less developed country?

6 i What can be done to predict a future event in the area?
  ii Why is it so difficult to predict the exact location, time and scale (size) of any possible future event?

References

Calder, N. (1972) *The Restless Earth*. BBC Publications.

Chester, D. (1993) *Volcanoes and Society.* Edward Arnold.

Decker, D. W. and Decker, B. (1991) *Mountains of Fire*. Cambridge University Press.

Earthquakes (1983) Geological Museum Publications.

Eiby, G. A. (1980) *Earthquakes*. Heinemann.

Francis, P. (1984) *Volcanoes*. Penguin Books.

Goudie, A. (1993) *The Nature of the Environment*. Basil Blackwell.

Park, C. (1991) *Environmental Hazards*.Thomas Nelson.

Planet Earth: Continents in Collision (1983) Time–Life Books.

Planet Earth: Earthquakes (1982) Time–Life Books.

Planet Earth: Volcanoes (1982) Time–Life Books.

Prosser, R. (1992) *Natural Systems and Human Responses*. Thomas Nelson.

Volcanoes (1974) Geological Museum Publications.

Chapter 2

Weathering and slopes

'Every valley shall be exalted, and every mountain and hill shall be made low: and the crooked shall be made straight, and the rough places plain.'

The Bible, Isaiah, 40:4

Weathering

The majority of rocks have been formed at high temperatures (igneous and many metamorphic rocks) and/or under great pressure (igneous, metamorphic and sedimentary rocks), but in the absence of oxygen and water. If, later, these rocks become exposed on the earth's surface, they will experience a release of pressure, be subjected to fluctuating temperatures, and be exposed to oxygen in the air and to water. They are therefore vulnerable to **weathering** which is the disintegration and decomposition of rock *in situ* — i.e. in its original position. Weathering is, therefore, the natural breakdown of rock and can be distinguished from erosion because it need not involve any movement of material. Weathering is the first stage in the **denudation** or wearing down of the landscape; it loosens material which can subsequently be transported by such agents of erosion as running water (Chapter 3), ice (Chapter 4), the sea (Chapter 6) and the wind (Chapter 7). The degree of weathering depends upon the structure and mineral composition of the rocks, local climate and vegetation, and the length of time during which the weathering processes operate.

There are two main types of weathering:

1 **Mechanical** (or **physical**) **weathering** is the disintegration of rock into smaller particles by mechanical processes but without any change in the chemical composition of that rock. It is more likely to occur in areas devoid of vegetation such as deserts, high mountains and arctic regions. Physical weathering usually produces sands.

2 **Chemical weathering** is the decomposition of rock resulting from a chemical change. It produces changed substances and solubles, and usually forms clays. Chemical weathering is more likely to take place in warmer, more moist climates where there is an associated vegetation cover.

It should be appreciated that, although in any given area either mechanical or chemical weathering may be locally dominant, both processes usually operate together rather than in isolation.

Mechanical weathering

Frost shattering

This is the most widespread form of mechanical weathering. It occurs in rocks which contain crevices and joints (e.g. joints formed in granite as it cooled, bedding planes found in sedimentary rocks and pore spaces in porous rocks), where there is limited vegetation cover and where temperatures fluctuate around 0°C. In the daytime, when it is warmer, water enters the joints, but during cold nights it freezes. As ice occupies 9 per cent more volume than water, it exerts pressure within the joints. This alternating **freeze–thaw process**, or **frost shattering**, slowly widens the joints and, in time, causes pieces of rock to shatter from the main body. Where this **block disintegration** occurs on steep slopes, large angular rocks collect at the foot of the slope as **scree** or **talus** (Figure 2.1); if the slopes are gentle, however, large **blockfields** (felsenmeer) tend to develop. Frost shattering is more common in upland regions of Britain where temperatures fluctuate around freezing point for several months in winter, than in polar areas where temperatures rarely rise above 0°C.

Salt crystallisation

If water entering the pore spaces in rocks is slightly saline then, as it evaporates, salt

Figure 2.2

Weathering pits caused by salt crystallisation, Arran

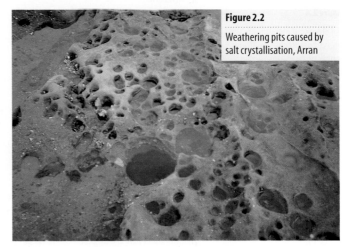

Figure 2.1

The formation of screes resulting from frost shattering, the Honister Pass in the Lake District National Park

crystals are likely to form. As the crystals become larger, they exert stresses upon the rock causing it to disintegrate. This process occurs in hot deserts where capillary action draws water to the surface and where the rock is sandstone (page 166). Individual grains of sand are broken off by **granular disintegration**. Salt crystallisation also occurs on coasts where the constant supply of salt can lead to the development of weathering pits (Figure 2.2).

Pressure release

As stated earlier, many rocks, especially intrusive jointed granites, have developed under considerable pressure. The confining pressure increases the strength of the rocks. If these rocks, at a later date, are exposed to the atmosphere, then there will be a substantial release of pressure. (If you had 10 m of bedrock sitting on top of you, you would be considerably relieved were it to be removed!) The release of pressure weakens the rock

allowing other agents to enter it and other processes to develop. Where cracks develop parallel to the surface, a process called **sheeting** causes the outer layers of rock to peel away. This process is now believed to be responsible for the formation of large, rounded rocks called **exfoliation domes** (Figures 2.3 and 7.6) and, in part, for the granite tors of Dartmoor and the Isle of Arran (Figures 8.13 and 8.14). Jointing, caused by pressure release, has also accentuated the characteristic shapes of glacial cirques and troughs (Figures 2.4, 4.13, and 4.18).

Thermal expansion or insolation weathering

Like all solids, rocks expand when heated and contract when cooled. In deserts, where cloud and vegetation cover are minimal, the diurnal range of temperature can exceed 50°C. It was believed that, because the outer layers of rock warm up faster and cool more rapidly than the inner ones, stresses were set up which would cause the outer thickness to peel off like the layers of an onion — the process of **exfoliation**. Initially, it was thought that it was this expansion–contraction process which produced exfoliation domes. Changes in temperature will also cause different minerals within a rock to expand and contract at different rates. It has been suggested that this causes granular disintegration in rocks composed of several minerals (e.g. granite which consists of quartz, feldspar and mica), whereas in homogeneous rocks it is more likely to cause block disintegration.

Laboratory experiments (e.g. by Griggs 1936 and Goudie in 1974) have, however, cast doubt upon the effectiveness of insolation weathering (page 165).

Figure 2.3

An exfoliation dome: Sugarloaf Mountain in Rio de Janeiro, Brazil

Figure 2.4

The process of pressure release tends to perpetuate landforms: as new surfaces are exposed, the reduction in pressure causes further jointing parallel to the surface

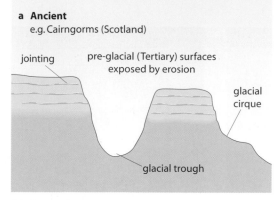

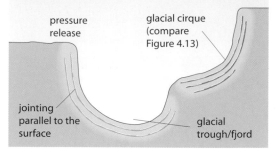

Figure 2.5

Mechanical (biological) weathering caused by expanding tree roots in Geltsdale, Cumbria

Biological

Tree roots may grow along bedding planes or extend into joints, widening them until blocks of rock become detached (Figure 2.5). It is also claimed that burrowing creatures, such as worms and rabbits, may play a minor role in the excavation of partially weathered rocks.

Figure 2.6

Oxidation in Geltsdale, Cumbria

Chemical weathering

Chemical weathering tends to:
- attack certain minerals selectively;
- occur in zones of alternate wetting and drying, e.g. where the level of the water table fluctuates;
- occur mostly at the base of slopes where it is likely to be wetter and warmer.

This type of weathering involves a number of specific processes which may operate in isolation but which are more likely to be found in conjunction with one another. Formulae for the various chemical reactions are listed at the end of the chapter.

Oxidation

This occurs when rocks are exposed to oxygen in the air or water. The simplest and most easily recognised example is when iron in a **ferrous** state is changed by the addition of oxygen into a **ferric** state. The rock or soil, which may have been blue or grey in colour (characteristic of a lack of oxygen), is discoloured into a reddish brown — a process better known as rusting (Figure 2.6). Oxidation causes rocks to crumble more easily.

In waterlogged areas, oxidation may operate in reverse and is known as **reduction**. Here, the amount of oxygen is reduced and the soils take on a blue/green/grey tinge (see **gleying**, page 254).

Hydration

Certain rocks are capable of absorbing water into their structure, causing them to swell and to become vulnerable to future breakdown. For example, gypsum is the result of water having been added to anhydrite ($CaSO_4$). This process appears to be most active following successive periods of wet and dry weather and is important in forming clay particles. Hydration is in fact a physiochemical process as the rocks may swell and exert pressure as well as changing their chemical structure.

Quartz	Mica	Feldspar
not affected by water, remains unchanged as sand (Figure 2.7)	may be affected by water under more acid conditions releasing aluminium and iron	readily attracts water producing a chemical change which turns the feldspar into clay (kaolin or china clay)

Hydrolysis

This is possibly the most significant chemical process in the decomposition of rocks and formation of clays. Hydrogen in water reacts with minerals in the rock or, more specifically, there is a combination of the H+ and OH- ions in the water and the ions of the mineral (i.e. the water combines with the mineral rather than dissolving it). The rate of hydrolysis depends upon the amount of H+ ions, which in turn depends upon the composition of air and water in the soil (page 240), the activity of organisms (page 246), the presence of organic acids (page 250) and the cation exchange (page 247). An example of hydrolysis is the breakdown of feldspar, a mineral found in igneous rocks such as granite, into a residual clay deposit known as kaolinite (china clay). Granite consists of three minerals — quartz, mica and feldspar (Figure 8.2c) — and, as the table above shows, each reacts at a different rate with water.

Carbonation

Rainwater contains carbon dioxide in solution which produces carbonic acid (H_2CO_3). This weak acid reacts with rocks which are composed of calcium carbonate, such as limestone. The limestone dissolves and is removed in solution (calcium bicarbonate) by running water. Carboniferous limestone is well-jointed and bedded (Chapter 8), which results in the development of a distinctive group of landforms (Figure 2.8).

Solution

Some minerals, e.g. rock salt, are soluble in water and simply dissolve *in situ*. The rate of solution can be affected by acidity since many minerals become more soluble as the pH of the solvent increases (page 248).

Chelation

Humic (organic) acid is derived from the decomposition of vegetation (humus) and contains important elements such as calcium, magnesium and iron. These are released by a process called **chelation** (page 250). The presence of organic life (plants, animals, bacteria) increases the concentration of carbon dioxide in soil and thus the level of carbonation. Lichens can extract iron from certain rocks and concentrate it at the surface. However, it should be remembered that the presence of vegetation cover dramatically reduces the extent of mechanical weathering.

Acid rain

Human economic activities (such as power generation and transport) release increasingly more carbon dioxide, sulphur dioxide and nitrogen oxide into the atmosphere. These gases then form acids in solution in rain-water (page 205). Acid rain readily attacks limestones and, to a lesser extent, sandstones, as shown by crumbling buildings and statues. The increased level of acidity in water passing through the soil tends to release more hydrogen and so speeds up the process of hydrolysis.

Figure 2.7

Decomposition of granite by hydrolysis on Goatfell, Arran

Figure 2.8

Carbonation of limestone near Ingleton, North Yorkshire

Climatic controls on weathering

Mechanical weathering

Frost shattering is important if temperatures fluctuate around 0°C, but will not operate if the climate is too cold (permanently frozen), too warm (no freezing), too dry (no moisture to freeze), or too wet (covered by vegetation). Mechanical weathering will not take place at **X** on Figure 2.9a where it is too warm and there is insufficient moisture, while at **Y**, the high temperature and heavy rainfall will give a thick protective vegetation cover against insolation.

Chemical weathering

This increases as temperatures and rainfall totals increase. It has been claimed that the rate of chemical weathering doubles with every 10°C temperature increase. Recent theories suggest that, in humid tropical areas, direct removal by solution may be the major factor in the lowering of the landscape, due to the continuous flow of water through the soil. Chemical weathering will be rapid at **S** (Figure 2.9b) due to humic acid from the vegetation. It will be limited at **P**, because temperatures are low, and at **R**, where there is insufficient moisture for the chemical decomposition of rocks.

Peltier, an American physicist and climatologist, attempted to predict the type and rate of weathering at any given place in the world from its mean annual temperature and mean annual rainfall (Figure 2.9c).

Figure 2.9

Climatic controls on weathering (*after* Peltier)

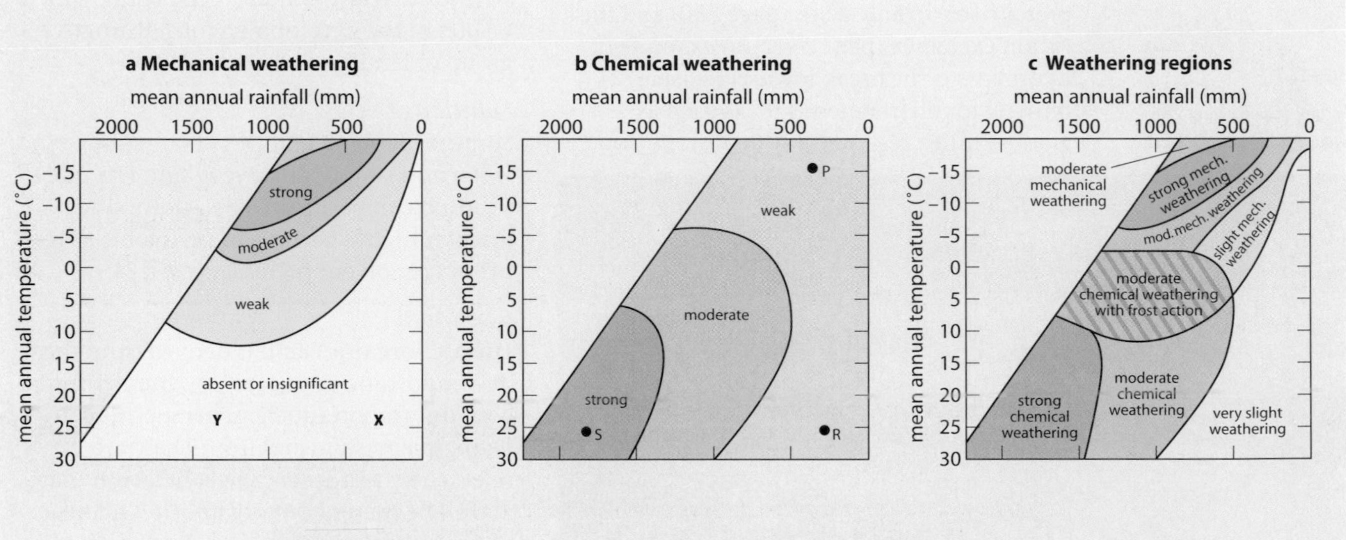

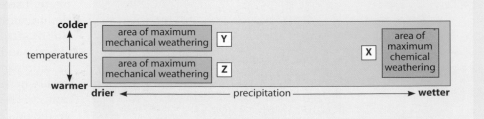

a Mechanical weathering

mean annual rainfall (mm)

strong / moderate / weak / absent or insignificant

Y   X

b Chemical weathering

mean annual rainfall (mm)

P / weak / moderate / strong

S   R

c Weathering regions

mean annual rainfall (mm)

moderate mechanical weathering / strong mech. weathering / mod. mech. weathering / slight mech. weathering / moderate chemical weathering with frost action / strong chemical weathering / moderate chemical weathering / very slight weathering

---

Q

Using Figure 2.9c:

1 What is the minimum mean annual rainfall which allows
   a strong chemical weathering and
   b strong mechanical (physical) weathering to occur?

2 What is the minimum mean annual temperature which enables strong chemical weathering to take place?

3 What are the temperature extremes for significant mechanical weathering?

4 Figure 2.10 is a model different from that of Peltier.
   a Why is chemical weathering more important than mechanical weathering at location **X**?
   b Describe how the processes of mechanical weathering will differ between places **Y** and **Z**.

5 Without referring to precipitation, describe three other factors which affect the rate and type of weathering in the British Isles.

Figure 2.10

A model showing the relationship between climate and weathering

colder

temperatures

warmer

area of maximum mechanical weathering **Y**

area of maximum mechanical weathering **Z**

X area of maximum chemical weathering

drier ◄——— precipitation ———► wetter

A systems approach

One type of model (Framework 9, page 326) widely adopted by geographers to help explain phenomena is the system. The system is a method of analysing relationships within a unit and consists of a number of components between which there are linkages. The model is usually illustrated schematically as a flow diagram.

Systems may be described in three ways.

- **Isolated:** In which there is no input or output of energy or matter. Some suggest the universe is the sole example of this type; others claim the idea is not applicable in geography.

- **Closed:** there is input, transfer and output of energy but not of matter (or mass).

- **Open:** most environmental systems are open and there are inputs and outputs of both energy and matter.

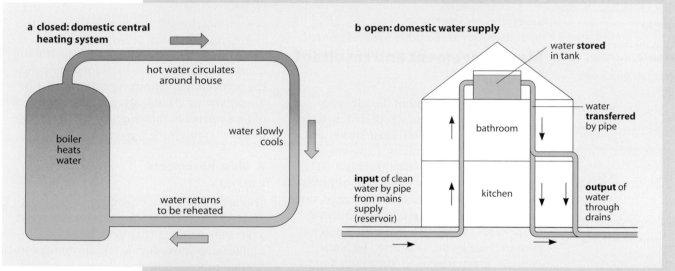

a **closed: domestic central heating system**

hot water circulates around house

boiler heats water

water slowly cools

water returns to be reheated

b **open: domestic water supply**

water **stored** in tank

water **transferred** by pipe

bathroom

input of clean water by pipe from mains supply (reservoir)

kitchen

output of water through drains

Figure 2.11

Closed and open systems in the house

Examples of the systems approach used and referred to in this book (chapter number is given in brackets):

Geomorphological	Climate, soils and vegetation	Human and economic
Slopes (2)	Atmosphere energy budget (9)	Farming (16)
Drainage basins (3)	Hydrological cycle (9)	Industry (19)
Glaciers (4)	Soils (10)	Transport (21)
	Ecosystems (11)	
	Nutrient cycle (12)	

When opposing forces, or inputs and outputs, are balanced, the system is said to be in a state of **dynamic equilibrium**. If one element in the system changes because of some outside influence, then it upsets this equilibrium and affects the other components. For example, equilibrium is upset when

- prolonged heavy rainfall causes an increase in the discharge and velocity of a river which leads to an increase in the rate of erosion.

- an increase in carbon dioxide into the atmosphere causes global temperatures to rise (global warming, Case Study 9).

- drought affects the carrying capacity of animals (or people) grazing (living) in an area as the water shortage reduces the availability of grass (food supplies)(page 354).

- an increase in the number of tourists to places of scenic attraction harms the environment, especially where it is fragile, which was the source of the attraction (page 539).

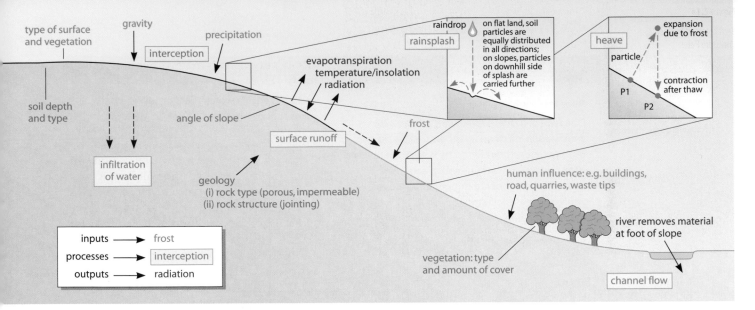

inputs ────▶ frost
processes ────▶ interception
outputs ────▶ radiation

Figure 2.12

The slope as a dynamic open system

Mass movement and resultant landforms

The term **mass movement** describes all downhill movements of weathered material (**regolith**), including soil, loose stones and rocks, in response to gravity. However, it excludes movements where the material is carried by ice, water or wind. When gravitational forces exceed forces of resistance, slope failure occurs and material starts to move downwards. A slope is a **dynamic open system** (Framework 1) affected by biotic, climatic, gravitational, groundwater and tectonic inputs which vary in scale and time. The amount, rate and type of movement depend upon the degree of slope failure (Figure 2.12).

Although by definition mass movement refers only to the movement downhill of material under the force of gravity, in reality water is usually present and assists the process. When Carson and Kirkby (1972) attempted to group mass movements, they

used the speed of movement and the amount of moisture present as a basis for distinguishing between the various types (Figure 2.13). The following classification is based on speed of flow related to moisture content and angle of slope (Framework 5, page 151).

A Slow movements

Soil creep

This is the slowest of downhill movements and is difficult to measure as it takes place at a rate of less than 1 cm a year. However, unlike faster movements, it is an almost continuous process. Soil creep occurs mainly in humid climates where there is a vegetation cover. There are two major causes of creep, both resulting from repeated expansion and contraction.

- **Wet–dry periods** During times of heavy rainfall, moisture increases the volume and weight of the soil, causing expansion and allowing the regolith to move downhill under gravity. In a subsequent dry period, the soil will dry out and then contract, especially if it is clay. An extreme case of contraction in clays occurred in south-east England during the 1976 drought when buildings sited on almost imperceptible slopes suffered major structural damage.
- **Freeze–thaw** When the regolith freezes, the presence of ice crystals increases the volume of the soil by 9 per cent. As the soil expands, particles are lifted at right angles to the slope in a process called **heave** (Figure 2.12). When the ground later thaws and the regolith contracts, these particles fall back vertically under the influence of gravity and so move downslope.

Figure 2.13

A classification of mass movement processes (*after* Carson and Kirkby, 1972)

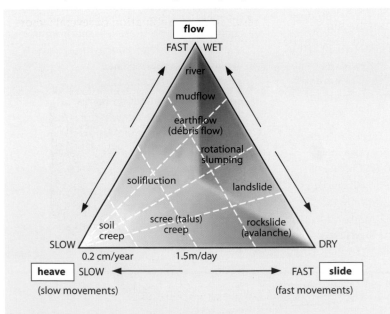

Figure 2.14

Terracettes, Wharfedale, Yorkshire Dales

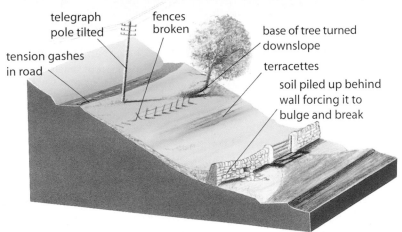

telegraph pole tilted

fences broken

tension gashes in road

base of tree turned downslope

terracettes

soil piled up behind wall forcing it to bulge and break

Figure 2.15

The effects of soil creep

Figure 2.16

Field sketch showing the causes of a mudflow, Glen Rosa, Arran

winter season, both the bedrock and regolith are frozen. In summer, the surface layer thaws but the underlying layer remains frozen and acts like impermeable rock. Because surface meltwater cannot infiltrate downwards and temperatures are too low for much effective evaporation, any topsoil will soon become saturated and will flow as an **active layer** over the frozen subsoil and rock. This process produces **solifluction sheet** or **lobes** (Figure 5.12), rounded, tongue-like features reaching up to 50 m in width, and **head**, a mixture of sand and clay formed in valleys and at the foot of sea cliffs (Figure 5.13). Solifluction was widespread in southern Britain during the Pleistocene ice age; covered most of Britain following the Pleistocene; and continues to take place in the Scottish Highlands today.

B Flow movements

Earthflows

When the regolith on slopes of 5–15° becomes saturated with water, it begins to flow downhill at a rate varying between 1 and 15 km per year. The movement of material may produce short **flow tracks** and small bulging lobes or tongues, yet may not be fast enough to break the vegetation.

Mudflows

These are more rapid movements, occurring on steeper slopes, and exceeding 1 km per hr. When Nevado del Ruiz erupted in Colombia in 1985, the resultant mudflow reached the town of Armero at an estimated speed of over 40 km/hr. Mudflows are most likely to occur following periods of intensive rainfall, when both volume and weight are added to the soil giving it a higher water content than an earthflow (Case Study 15a). Mudflows may result from a combination of several factors (Figure 2.16).

Soil creep usually occurs on slopes of about 5° and produces **terracettes** (Figure 2.14). These are step-like features, often 20–50 cm in height, which develop as the vegetation is stretched and torn: they are often used and accentuated by grazing animals, especially sheep. The effects of soil creep are shown in Figure 2.15.

Solifluction

This process, meaning 'soil flow', is a slightly faster movement usually averaging between 5 cm and 1 m a year. It often takes place under periglacial conditions (Chapter 5) where vegetation cover is limited. During the

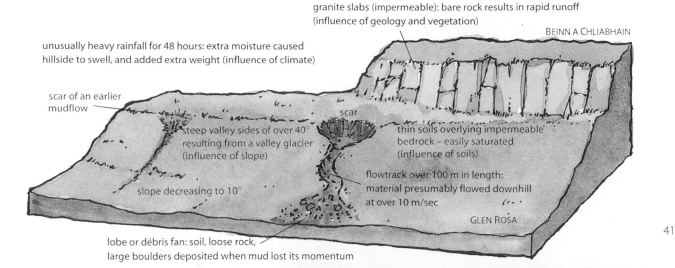

granite slabs (impermeable): bare rock results in rapid runoff (influence of geology and vegetation)

BEINN A CHLIABHAIN

unusually heavy rainfall for 48 hours: extra moisture caused hillside to swell, and added extra weight (influence of climate)

scar of an earlier mudflow

scar

steep valley sides of over 40° resulting from a valley glacier (influence of slope)

thin soils overlying impermeable bedrock – easily saturated (influence of soils)

flowtrack over 100 m in length: material presumably flowed downhill at over 10 m/sec

slope decreasing to 10°

GLEN ROSA

lobe or débris fan: soil, loose rock, large boulders deposited when mud lost its momentum

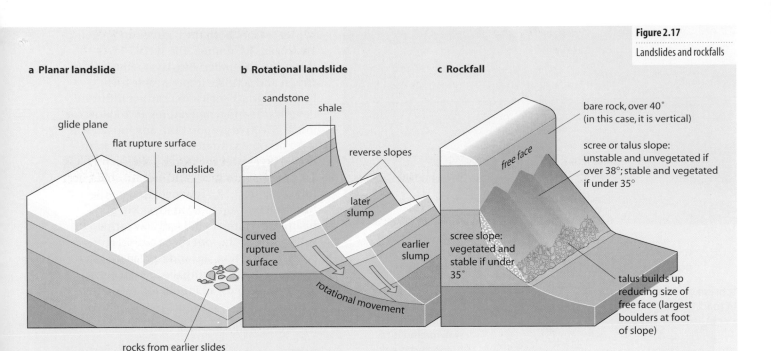

Figure 2.17
Landslides and rockfalls

a Planar landslide

glide plane

flat rupture surface

landslide

rocks from earlier slides

b Rotational landslide

sandstone

shale

reverse slopes

later slump

earlier slump

curved rupture surface

rotational movement

c Rockfall

free face

bare rock, over 40° (in this case, it is vertical)

scree or talus slope: unstable and unvegetated if over 38°; stable and vegetated if under 35°

scree slope: vegetated and stable if under 35°

talus builds up reducing size of free face (largest boulders at foot of slope)

Figure 2.18

Landslides on the Norfolk coast

Figure 2.19

Rockfalls in the crater of Vesuvius, Italy

C Rapid movements

Slides

The fundamental difference between slides and flows is that flows suffer internal derangement whilst, in contrast, slides move 'en masse' and are not affected by internal derangement. Rocks which are jointed or have bedding planes roughly parallel to the angle of slope are particularly susceptible to landslides. Slides may be planar or rotational (Figure 2.17a and b). In a planar slide, the weathered rock moves downhill leaving behind it a flat rupture surface (Figure 2.17a). Where rotational movement occurs, a process sometimes referred to as **slumping**, a curved rupture surface is produced (Figure 2.17b). Rotational movement can occur in areas of homogeneous rock, but is more likely where softer materials (clay or sands) overlie more resistant or impermeable rock (limestone or granite). Slides are common in many coastal areas of southern and eastern England. In Figure 2.18, the cliffs, composed of glacial deposits, are retreating rapidly due to frequent slides. The slumped material can be seen at the foot of the cliff.

D Very rapid movements

Rockfalls

These are spontaneous, though relatively rare, debris movements on slopes which exceed 40°. They may result from extreme physical or chemical weathering in mountains, pressure release, storm-wave action on sea cliffs, or earthquakes. Material, once broken from the surface, will either bounce or fall vertically to form scree, or talus, at the foot of a slope (Figures 2.17c and 2.19).

Places 6

The Vaiont Dam, North Italy

The Vaiont Dam was opened in northern Italy in 1960. The third-highest concrete dam in the world, it had been built in a narrow, steep-sided valley whose sides consist of alternating layers of clays and limestones, and where landslides were common. Three years later, following a period of heavy rain which saturated the clays, a mass of rock, mud, earth and vegetation slid over the limestone and into the reservoir. The dam itself withstood the landslide but a wave of water, 70 m in height, spilled over the lip creating a towering wall of water which swept down the valley and destroyed the town of Longarone, causing almost 1900 deaths.

Peruvian earthquake

In May 1970 an earthquake measuring 7.7 on the Richter scale shook parts of Peru. The shock waves loosened a mass of ice and snow near the summit of Huascaran, the country's highest peak at 6768 m. The ice and snow fell 3000 m, picking up rocks and boulders as it went. The resultant avalanche destroyed villages and hit the town of Yungay at an estimated 480 km/hr. When rescue workers eventually reached the area three days later, they found very few survivors out of a population of 20 000 and only the tops of several 30 m palm trees marking the spot where the town had previously stood.

Nevado del Ruiz, Colombia

The snow-covered volcano of Nevado del Ruiz, which had not shown signs of activity since 1595, erupted in November 1985. The snow and ice were melted by the ejection of lava, ash and hot rocks. The combined effects of the meltwater and torrential rain turned ash from previous eruptions into mud. A tidal wave of mud travelled down the Lagunillas valley at an estimated height of 30 m and at a speed of 80 km/hr. The mudflow, or lahar, hit the town of Armero just before midnight, burying it under 8 m of mud and killing 21 000 of the 22 000 inhabitants.

Q The types of mass movement can be classified according to speed of movement and the amount of water necessary to assist this movement.

a speed of movement

Extremely slow	Very slow	Slow	Moderate	Rapid	Very rapid	Extremely rapid
1 cm/year	1 m/year	1 km/year	1 km/month	1 km/hour	25 km/hour	10 m/sec

②
④
①
③
⑤

b location

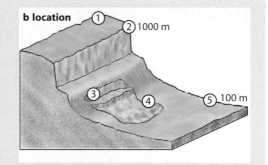

Figure 2.20

Mass movements

1 Redraw the graph (Figure 2.20a) and add the following labels in the appropriate places: earth/mudflow, solifluction, rockfall, slide, soil creep.

2 Redraw the sketch shown in Figure 2.20b and add labels for the same five movements as listed in 1. Indicate the location of the following landforms: scree (talus), terracettes, lobe.

Development of slopes

Slope development is the result of the inter-action of several factors. Rock structure and lithology, soil, climate, vegetation and human activity are probably the most significant. All are influenced by the time over which the processes operate (Places 7). Slopes are an integral part of the drainage basin system (Chapter 3) as they provide water and sediment for the river channel.

The effects of rock structure and lithology

- Areas of bare rock are vulnerable to mechanical weathering (e.g. frost shattering) and some chemical weathering processes.
- Areas of alternating harder/more resistant rocks and softer/less resistant rocks are more likely to experience movement, e.g. clays on limestones (Vaiont Dam, Places 6).
- An impervious underlying rock will cause the topsoil to become saturated more quickly, e.g. glacial deposits overlying granite.
- Steep gradients are more likely to suffer slope failure than gentler ones. In Britain, especially in lowland areas, most slopes are under 5° and few are over 40°.

- Failure is also likely on slopes where the equilibrium (balance) of the system (Framework 1, page 39), has been disturbed, e.g. a glaciated valley.
- The presence of joints, cracks and bedding planes can allow increased water content and so lead to sliding (Vaiont Dam, Places 6).
- Earthquakes (Mt Huascaran in Peru, Places 6) and volcanic eruptions (Nevado del Ruiz in Colombia, Places 6) can cause extreme slope movements.

Soil

- Thin soils tend to be more unstable. As they can support only limited vegetation, there are fewer roots to bind the soil together.
- Unconsolidated sands have lower internal cohesion than clays.
- A porous soil, e.g. sand, is less likely to become saturated than one which is impermeable, e.g. clay.
- In a non-saturated soil (Figure 2.21a), the surface tension of the water tends to draw particles together. This increases cohesion and reduces soil movement. In a saturated soil (Figure 2.21b), the pore water pressure (page 246) forces the particles apart, reducing friction and causing soil movement.

Figure 2.21

The effect of pore-water pressure and capillary action on soil movement

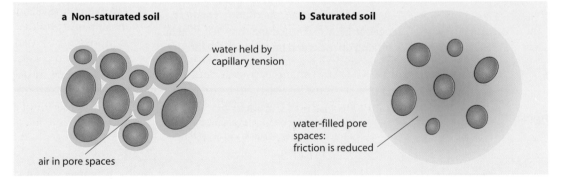

a Non-saturated soil

water held by capillary tension

air in pore spaces

b Saturated soil

water-filled pore spaces: friction is reduced

Climate

- Heavy rain and meltwater both add volume and weight to the soil.
- Heavy rain increases the erosive power of any river at the base of a slope and so, by removing material, makes that slope less stable.
- Areas with freeze–thaw or wet–dry periods are subjected to alternating expansion–contraction of the soil.
- Heavy snowfall adds weight and is thus conducive to rapid movements (e.g. avalanches, Case Study 4).

Vegetation

- A lack of vegetation means that there are fewer roots to bind the soil together.
- Sparse vegetation cover will encourage surface runoff as precipitation is not intercepted (page 51).

Human influence

- Deforestation increases (afforestation decreases) the rate of slope movement.
- Road construction or quarrying at the foot of slopes upsets the equilibrium, e.g. during the building of the M5 in the Bristol area.

- Slope development processes may be accentuated either by building on steep slopes (Hong Kong and Rio de Janeiro, Case Study 2, page 47) or by using them to deposit industrial or mining waste (Aberfan, Case Study 2).

- The vibration caused by heavy traffic can destabilise slopes (Mam Tor, Derbyshire).
- The grazing of animals and ploughing help loosen soil and to remove the protective vegetation cover.

Places 7 Slope development on Arran

Slopes, such as those of the Glen Rosa valley on the Isle of Arran, usually develop as a result of a combination of factors. The steep sides of this valley are the result of several processes; some have ceased to operate and others can still be seen working today.

The valley is a fine example of a glacial trough with its sides steepened by the action of a valley glacier and weakened by pressure release caused by the removal of surface rock by ice (Figure 2.4). When the ice melted during subsequent periglacial times, the bare, jointed granite rocks provided ideal conditions for physical weathering in the form of freeze–thaw activity and the development of scree (talus) slopes. Physical weathering still operates today,

as do chemical weathering processes, such as hydrolysis, resulting from the island's heavy rainfall. The steep slopes and the heavy rainfall contribute to a range of mass movement processes which include soil creep, solifluction, mudflows (Figure 2.16) and rockfalls. The steepness of the slopes is partly maintained by the Rosa Water (which had formed the pre-glacial valley) as it removes material and keeps the sides unstable. Finally, the slopes are modified by the movements of sheep and deer and by human activities (the tourist/walker/climber affecting vegetation, causing pollution and erosion and reducing wildlife).

Slope elements

Two models try to show the shape and form of a typical slope. The first, Figure 2.22a, is more widely used than the second (Figure 2.22b) — although, in this author's view, the first is less easily seen in the British landscape. Regardless of which model is used, confusion unfortunately arises because of the variation in nomenclature used to describe the different facets of the slope.

In reality, few slopes are likely to match up perfectly with either model, and each

individual slope is likely to show more elements than those in Figure 2.22. In the field, one way to plot the distribution of these facets to see if they form any pattern is to make a morphological map. Morphological symbols can be used to show the form of the slope (concave, convex or rectilinear) and to show changes in slope, which may be either a sharp break or a gradual change. Figure 2.23 is the result of a transect taken down a valley side in Glen Rosa, Arran.

Figure 2.22

Slope element models

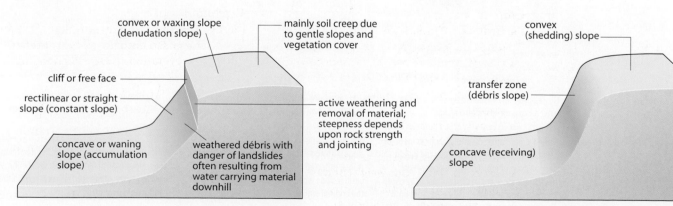

a Rectilinear

convex or waxing slope (denudation slope)

mainly soil creep due to gentle slopes and vegetation cover

cliff or free face

rectilinear or straight slope (constant slope)

concave or waning slope (accumulation slope)

weathered débris with danger of landslides often resulting from water carrying material downhill

active weathering and removal of material; steepness depends upon rock strength and jointing

b Convex–concave

convex (shedding) slope

transfer zone (débris slope)

concave (receiving) slope

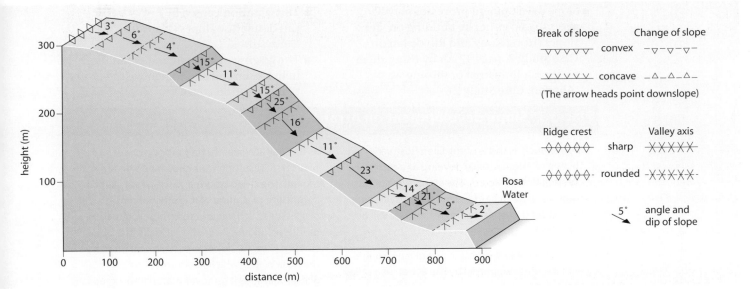

Break of slope

⊽⊽⊽⊽⊽ convex

⌄⌄⌄⌄⌄ concave

Change of slope

—▽–▽–▽– convex

—△–△–△– concave

(The arrow heads point downslope)

Ridge crest

◇◇◇◇◇ sharp

◇◇◇◇◇ rounded

Valley axis

×××× sharp

×-×-×-× rounded

5° ↘ angle and dip of slope

Figure 2.23

Morphological map drawn along a transect down a valley side in Glen Rosa, Arran

Figure 2.23

Morphological map drawn along a transect down a valley side in Glen Rosa, Arran

Figure 2.24

Slope development theories

Slope development through time

How slopes have developed over time is one of the more controversial topics in geomorphology. This is partly due to the time needed for slopes to evolve and partly due to the variety of combinations of processes acting upon slopes in various parts of the world. Slope development in differing environments has led to three divergent theories being proposed: **slope decline**, **slope replacement** and **parallel retreat**.

Figure 2.24 is a summary of these theories.

None of the theories of slope development can be universally accepted, although each may have local relevance in the context of the climate and geology (structure) of a specific area. At the same time, two different climates or processes may produce the same type of slope — e.g. cliff retreat due to sea action in a humid climate or to weathering in a semi-arid climate.

	Slope decline (W. M. Davis, 1899)	**Slope replacement** (W. Penck, 1924)	**Parallel retreat** (L. C. King, 1948, 1957)
region of study	Theory based on slopes in what was to Davis a normal climate, north-west Europe and north-east USA.	Conclusions drawn from evidence of slopes in the Alps and Andes.	Based on slopes in South Africa.
climate	Humid climates.	Tectonic areas.	Semi-arid landscapes. Sea cliffs with wave-cut platforms.
description of slope	Steepest slopes at beginning of process with a progressively decreasing angle in time to give a convex upper slope and a concave lower slope.	The maximum angle decreases as the gentler lower slopes erode back to replace the steeper ones giving a concave central portion to the slope.	The maximum angle remains constant as do all slope facets apart from the lower one which increases in concavity.
	slope decline stage 3 stage 2 stage 4 stage 1 concave curve watershed worn down convex curve peneplain By stage 4 land has been worn down into a convex-concave slope	**slope replacement** stage 3 stage 2 stage 1 A A A C B C B C B talus-scree slope B will replace slope A; slope C will eventually replace slope B	**slope retreat** stage 4 stage 2 stage 3 stage 1 convex free face concave débris slope pediment (can be removed by flash floods)
changes over time	Assumed a rapid uplift of land with an immediate onset of denudation. The uplifted land would undergo a cycle of erosion where slopes were initially made steeper by vertical erosion by rivers but later became less steep (slope decline) until the land was almost flat (peneplain).	Assumed landscape started with a straight rock slope with equal weathering overall. As scree (talus) collected at the foot of the cliff it gave a gentler slope which, as the scree grew, replaced the original one.	Assumed that slopes had two facets — a gently concave lower slope or pediment and a steeper upper slope (scarp). Weathering caused the parallel retreat of the scarp slope allowing the pediment to extend in size.

Case Study 2

People and slope failure

All slopes are affected by gravity and, consequently, by one or more of the several mass movement processes by which weathered material is transported downhill. Where slopes are gentle (about 5°), the movement of material is slow and has relatively little effect upon property or life. As slope angles increase, so too do the rate and frequency of slope movement and the risk of sudden slope failure. Slope failure, occurring in the form of either mudflows or landslides, is a natural event. When slope failure occurs in populated areas, it becomes a potentially dangerous natural hazard. However, the probability of slope failure in populated areas is often increased by thoughtless, or a total lack of, human planning.

Three examples of how slope instability and the risk of slope failure may be increased by human activity are when land is used for:

a the extraction of a natural resource or the dumping of waste material;

b urbanisation; and

c recreation.

Figure 2.25

Aberfan immediately after the mudflow

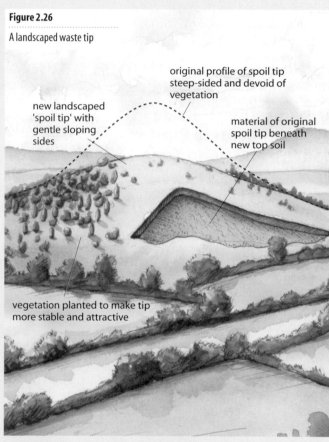

Figure 2.26

A landscaped waste tip

original profile of spoil tip steep-sided and devoid of vegetation

new landscaped 'spoil tip' with gentle sloping sides

material of original spoil tip beneath new top soil

vegetation planted to make tip more stable and attractive

Dumping of waste material: The mudflow at Aberfan

Aberfan, like many other settlements in the South Wales valleys, grew up around its colliery. However, the valley floors were rarely wide enough to store the coal waste and so it became common practice to tip it high above the town on the steep valley sides. At Aberfan, the spoil tips were on slopes of 25°, over 200 m above the town and, unknowingly, on a line of springs. Water from these springs added weight to the waste heaps which reduced their internal cohesion. Following a wet October in 1966 and a night of heavy rain, slope failure resulted in the waste material suddenly and rapidly flowing downhill. The resultant mudflow, estimated to contain over 100 000 cu m of material, engulfed part of the town which included the local junior school (Figure 2.25). The time was just after 0900 hours on 21 October, soon after lessons in the school had begun. Of the 147 deaths that morning, 116 were children and five, their teachers in the school.

Since then, the colliery complex has been removed and potentially dangerous tips in the valley have been lowered, regraded and landscaped to prevent the occurrence of a similar event (Figure 2.26). While this process has been repeated throughout the South Wales coalfield (and deep mining in that area has now ended), there is still some local concern over the stability of waste tips.

Figure 2.27

Settlements on unstable slopes in Rio de Janeiro, Brazil

Urbanisation: Hillside developments in Rio de Janeiro and Hong Kong

Many parts of the world are still experiencing rapid urbanisation (page 384). Urbanisation, especially in economically less developed countries, results from high birth rates, which cause a rapid growth in population, or migration from the surrounding rural areas, or both combined. All the best, and safest, building land has often already been used, leaving new-comers to such cities as Rio de Janeiro and Lima with no option but to settle on already-overcrowded hillsides (Figure 2.27).

In Rio, many of the hillsides are very steep and prone to severe landslides. The hot, wet climate provides the optimum conditions for chemical weathering to operate, and for the bedrock to be broken down quickly into a deep regolith. The heavy periods of rain rapidly saturate the regolith enabling it to flow over the underlying rock and down the steep hillsides.

Many recent migrants to Rio have been forced to erect temporary shelters in favelas (Places 43, page 408) on the steep hillsides as these sites are not attractive to people who can afford to pay to live in safer locations. Although individual buildings in a favela may be flimsy, their total number is so large that there is a considerable increase in weight, and therefore pressure, on the slopes. As a result, the slopes become even less stable and there is a greater risk of slope failure. Landslides have, under extreme weather conditions, caused considerable damage and loss of human life within Rio's favelas. One landslide in 1966 resulted in 279 deaths, while another in 1988 killed 277 people and left over 19 000 homeless.

Hong Kong has one of the highest population densities in the world. All suitable flat land on Hong Kong island has been built upon and further growth, either for commercial or domestic purposes, has to be on land reclaimed from the sea, on steep hillsides or on existing sites which are re-used by pulling down existing property and erecting even taller buildings. Despite Hong Kong's wealth and its high level of technology, buildings on steeper slopes have been known to collapse (Figure 2.28). The most vulnerable time is during a typhoon when the combined effects of torrential rainfall and the weight of buildings have caused slope failure and landslides. Collapsed buildings killed 64 people in 1966 and 22 in 1976.

Figure 2.28

Consequences of a landslide in Hong Kong

Recreation: Scarborough

The increasing number of people who want to take advantage of healthy exercise in scenic areas is a cause for environmental concern. Walkers, horse-riders and mountain-bike cyclists are examples of groups of people who contribute to soil erosion in favoured upland areas such as the National Parks of England and Wales. Coastal areas also have problems, especially in seaside resorts where buildings have been constructed on cliff tops in order to command the 'uninterrupted views of the sea' requested by many of their customers. However, the extra weight of buildings, especially if these are large hotels, can increase both the instability of the cliff and the risk of slope failure.

On the Yorkshire coast, for example, many properties have collapsed over the years as the cliff-line has retreated — though none, perhaps, quite as dramatically as the Holbeck Hall Hotel in Scarborough (Figure 2.29). Following several drier than average summers, the early months of 1993 were very wet. The top layer of the cliff became saturated and began to move slowly downhill, taking part of the hotel with it and rendering the remainder of the building unsafe.

Figure 2.29

Holbeck Hall Hotel, Scarborough

Formulae for chemical weathering processes referred to in text:

Oxidation $4FeO + O_2 \rightarrow 2Fe_2O_3$
(ferrous oxide + oxygen $\rightarrow$ ferric oxide)

Hydration $CaSO_4 + 2H_2O \rightarrow CaSO_4 2H_2O$
(anhydrite + water $\rightarrow$ gypsum)

Hydrolosis The formula for hydrolosis differs depending on the rock type involved in the reaction, so no single formula is applicable. The following, for the hydrolosis of feldspar/granite to kaolin, is a common example.

$K_2O, Al_2O_3, 6SiO_2 + H_2O \rightarrow Al_2O_3, 2SiO_2, 2H_2O$
(feldspar + water $\rightarrow$ kaolin)

Carbonation This process is in two stages:
$H_2O + CO_2 \rightarrow H_2CO_3$
(water + carbon dioxide $\rightarrow$ carbonic acid)
$CaCO_3 + H_2CO_3 \rightarrow Ca(HCO_3)_2$
(calcium carbonate + carbonic acid $\rightarrow$ calcium bicarbonate)

Acid rain $2SO_2 + O_2 + 2H_2O \rightarrow 2H_2SO_4$
(sulphur dioxide + oxygen + water $\rightarrow$ weak sulphuric acid)

References

Brunsden, D. and Doornkamp, J. (1977) *The Unquiet Landscape*. David & Charles.

Goudie, A. (1993) *The Nature of the Environment*. Basil Blackwell.

McCullagh, P. (1978) *Modern Concepts in Geomorphology*. Oxford University Press.

Ollier, C. (1991) *Weathering and Landforms*. Thomas Nelson.

Small, J. and Witherick, M. (1989) *A Modern Dictionary of Geography*. E. J. Arnold.

Trudgill, S. T. (1983) *Weathering and Erosion*. Butterworths.

White, I., Mottershead, D. and Harrison, S. (1992) *Environmental Systems: An Introductory Text*. Chapman & Hall.

Chapter 3

Drainage basins and rivers

"All the rivers run into the sea; yet the sea is not full; unto the place from whence the rivers come, thither they return again."

The Bible, Ecclesiastes, 1:7

A **drainage basin** is an area of land drained by a river and its tributaries. Its boundary is marked by a ridge of high land beyond which any precipitation will drain into adjacent basins. This boundary is called a **watershed**.

A drainage basin may be described as an **open system** and it forms part of the hydrological or water cycle. If a drainage basin is viewed as a system (Framework 1) then its characteristics are:

- **inputs** in the form of precipitation (rain and snow);
- **outputs** where the water is lost from the system either by the river carrying it to the sea or through **evapotranspiration** (the loss of water directly from the ground, water surfaces and vegetation).

Within this system, some of the water

- is **stored** in lakes and/or in the soil; or
- passes through a series of **transfers** or flows, e.g. infiltration, percolation, throughflow.

Figure 3.1

The drainage basin as an open system

Elements of the drainage basin system

Figure 3.1 shows the drainage basin system as it is likely to operate in a temperate humid region such as the British Isles.

Precipitation

This forms the major input into the system, though amounts vary over time and space. As a rule, the greater the intensity of a storm, the shorter its duration. Convectional thunderstorms are short, heavy and may be confined to small areas, whereas the passing of a warm sector of a depression (page 215) will give a longer period of more steady rainfall extending over all the basin.

Evapotranspiration

The two components of evapotranspiration contribute to form an output from the system. **Evaporation** is the physical process by which moisture is lost directly into the atmosphere from water surfaces and the soil due to the effects of air movement and the sun's heat. **Transpiration** is a biological process by which water is lost from a plant through the minute pores (stomata) in its leaves. Evaporation rates are affected by temperature, wind speed, humidity, hours of sunshine and other climatic factors.

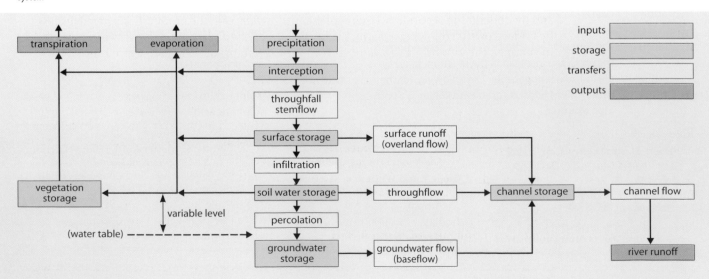

Transpiration rates depend on the time of year, the type and amount of vegetation, the availability of moisture and the length of the growing season. It is also possible to distinguish between the potential and the actual evapotranspiration of an area. For example, in deserts there is a high **potential evapotranspiration** because the amount of moisture that could be lost is greater than the amount of water actually available. On the other hand, in Britain the amount of water available for evapotranspiration nearly always exceeds the amount which actually takes place, hence the term **actual evapo-transpiration**.

Interception
The first raindrops of a rainfall event will fall on vegetation which shelters the underlying ground. This is called **interception storage**. It is greater in a woodland area or where tree crops are grown than on grass or arable land. If the precipitation is light and of short dura-tion, much of the water may never reach the ground and it may be quickly lost from the system through evaporation. Estimates sug-gest that in a woodland area up to 30 per cent of the precipitation may be lost through interception, which helps to explain why soil erosion is limited in forests. According to Newson (1975), "Interception is a dynamic process of filling and emptying a shallow store (about 2 mm in most UK trees). The emptying occurs because evaporation is very efficient for small raindrops held on tree sur-faces." In an area of deciduous trees, both interception and evapotranspiration rates will be higher in summer.

If a rainfall event persists, then water begins to reach the ground by three possible routes: dropping off the leaves, or **through-fall**; flowing down the trunk, or **stemflow**; and by undergoing **secondary interception** by undergrowth. Following a warm, dry spell in summer, the ground may be hard; at the start of a rainfall event water will then lie on the surface (**surface storage**) until the upper layers become sufficiently moistened to allow it to soak slowly downwards. If precipi-tation is very heavy initially, or if the soil becomes saturated, then excess water will flow away over the surface, a transfer known as **surface runoff** (or, in Horton's term, **over-land flow**).

Infiltration
In most environments, overland flow is relatively rare except in urban areas — which have impermeable coverings of tarmac and concrete — or during exceptionally heavy storms. Soil will gradually admit water from the surface, if the supply rate is moderate, allowing it slowly to infiltrate vertically through the pores in the soil. The maximum rate at which water can pass through the soil is called its **infiltration capacity** and is expressed in mm/hr. The rate of infiltration depends upon the amount of water already in the soil (**antecedent precipitation**), the **porosity** (Figure 8.2) and structure of the soil, the nature of the soil surface (e.g. crusted, cracked, ploughed), and the type, amount and seasonal changes in vegetation cover. Some of the water will flow laterally as **throughflow**. During drier periods, some water may be drawn up towards the surface by **capillary action**.

Percolation
As water reaches the underlying soil or rock layers, which tend to be more compact, its progress is slowed. This constant movement, called percolation, creates **groundwater storage**. Water eventually collects above an impermeable rock layer, or it may fill all pore spaces, creating a **zone of saturation**. The upper boundary of the saturated material, i.e. the upper surface of the groundwater layer, is known as the **water table**. Water may then be slowly transferred laterally as **groundwater flow** or **baseflow**. Except in areas of Carboniferous limestone, groundwa-ter levels usually respond slowly to surface storms or short periods of drought (Figure 3.5). During a lengthy dry period, some of the groundwater store will be utilised as river levels fall. In a subsequent wetter period, groundwater must be replaced before the level of the river can rise appreciably (Figure 3.3). If the water table reaches the surface, it means that the ground is saturated; excess water will then form a marsh where the land is flat, or will become surface runoff if the ground is sloping.

Channel flow
Although some rain does fall directly into the channel of a river (**channel precipitation**), most water reaches it by a combination of three transfer processes: surface runoff (over-land flow), throughflow, or groundwater flow (baseflow). Once in the river, as **channel storage**, water flows towards the sea and is lost from the drainage basin system.

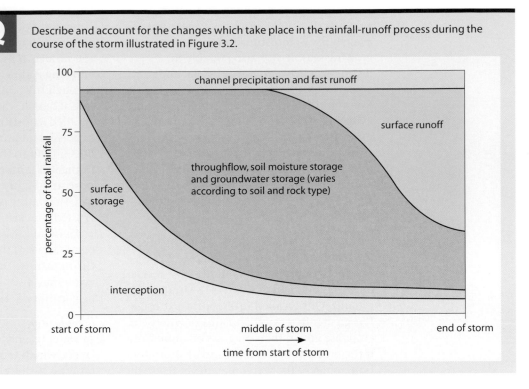

> **Q** Describe and account for the changes which take place in the rainfall-runoff process during the course of the storm illustrated in Figure 3.2.

The water balance

This shows the state of equilibrium in the drainage basin between the inputs and outputs. It can be expressed as:

$$P = Q + E \pm \text{change in storage}$$

where:

P = precipitation (measured using rain gauges);

Q = runoff (measured by discharge flumes in the river channel); and

E = evapotranspiration. (This is far more difficult to measure — how can you measure accurately transpiration from a forest?)

In Britain, the annual precipitation almost always exceeds evapotranspiration — though sometimes the situation may be reversed, as in the dry summers of 1975, 1976 and 1984. (This reversal is more likely in the south and east of England than in the north and west of Scotland.) Should evapotranspiration exceed precipitation, any surplus soil moisture will be utilised, leaving a **soil moisture deficiency**. Later, in the cooler, wetter autumn, there must be a period of **soil moisture recharge** until the soil reaches its **field capacity** (page 246).

Figure 3.3 is a graph based on an area of south-east England. During the winter, precipitation exceeds evapotranspiration, giving a **water surplus** and considerable runoff — the soils will be wet and river levels high. In summer, however, evapotranspiration exceeds precipitation and so plants and humans **utilise** water from the soil store, leaving it depleted and causing river levels to fall. By autumn, precipitation again exceeds evapotranspiration, although the first of the surplus water has to be used to **recharge** the soil store. At no time has there been a **water deficit**.

Figure 3.3

A graph illustrating the water balance

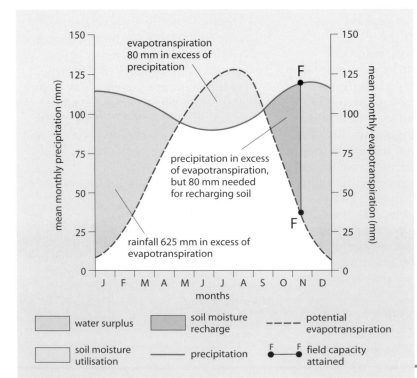

Q Figure 3.4 shows the water balance for two towns in the USA.

Figure 3.4

The water balance for two towns in the USA

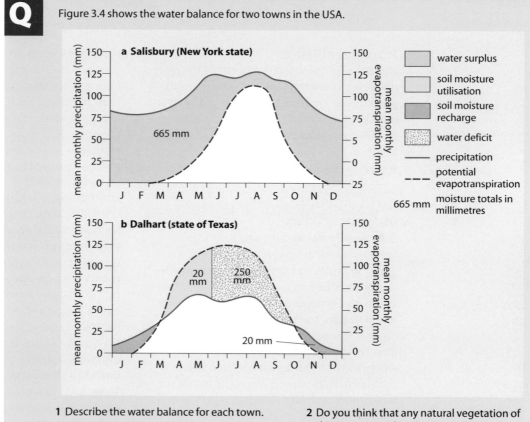

1 Describe the water balance for each town.

2 Do you think that any natural vegetation of the areas around each town is likely to be deciduous forest, grassland or semi-desert? Give your reasons.

The storm hydrograph

An important aspect of **hydrology** (the study of water) is how a drainage basin reacts to a period of rain. This is important because it can be used in predicting the flood risk and in making the necessary precautions to avoid damage to property and loss of life. The response of a river can be studied by using the **storm** or **flood hydrograph**. The hydrograph is a means of showing the discharge of a river at a given point over a short period of time. **Discharge** is the amount of water origi-nating as precipitation which reaches the channel by surface runoff, throughflow and baseflow. Discharge is therefore the water *not* stored in the drainage basin by interception, as surface storage, soil moisture storage or groundwater storage or lost through evapo-transpiration (Figure 3.1). The model of a storm hydrograph, Figure 3.5, shows how the discharge of a river responds to an individual rainfall event.

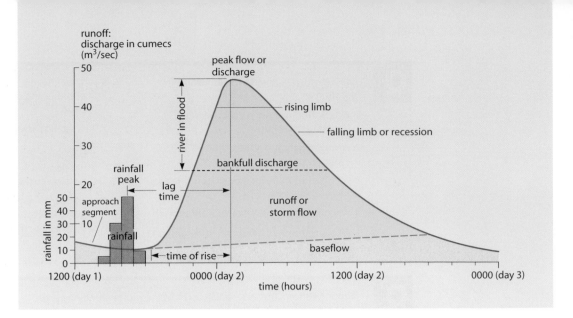

Figure 3.5

The storm hydrograph

Measuring discharge

Discharge is the velocity (speed) of the river, measured in metres (m) per second, multiplied by the cross-sectional area of the river, measured in sq m. This gives the volume in cu m/sec or **cumecs**. It can be expressed as:

$$Q = A \times V$$

where:

Q = discharge;

A = cross-sectional area;

V = velocity (Figure 3.5).

Interpreting the hydrograph

Refer to the hydrograph in Figure 3.5. The graph includes the **approach segment** which shows the discharge of the river before the storm (the antecedent flow rate). When the storm begins, the river's response is negligible for although some of the rain does fall directly into the channel, most falls elsewhere in the basin and takes time to reach the channel. However, when the initial surface runoff and, later, the throughflow eventually reach the river there is a rapid increase in discharge as indicated by the **rising limb**. The steeper the rising limb, the faster the response to rainfall — i.e. water reaches the channel more quickly. The **peak discharge** (peak flow) occurs when the river reaches its highest level. The period between maximum precipitation and peak discharge is referred to as the **lag time**. The lag time varies according to conditions within the drainage basin, e.g. soil and rock type, slope and size of the basin, drainage density, type and amount of vegetation and water already in storage. Rivers with a short lag time tend to experience a higher peak discharge and are more prone to flooding than rivers with a long lag time (Figure 3.10). The **falling** or **receding**

limb is the segment of the graph where discharge is decreasing and river levels are falling. This segment is usually less steep than the rising limb because throughflow is being released relatively slowly into the channel. By the time all the water from the storm has passed through the channel at a given location, the river will have returned to its baseflow level — unless there has been another storm within the basin. **Stormflow** is the discharge, both surface and subsurface flow, attributed to a single storm. **Baseflow** is very slow to respond to a storm, but by continually releasing groundwater it maintains the river's flow during periods of low precipitation. Indeed, baseflow is more significant over a longer period of time than an individual storm and reflects seasonal changes in precipitation, snow melt, vegetation and evapotranspiration. Finally, on the graph, **bankfull discharge** occurs when a river's water level reaches the top of its channel; any further increase in discharge will result in flooding of the surrounding land.

Controls in the drainage basin and on the storm hydrograph

In some drainage basins, river discharge increases very quickly after a storm and may give rise to frequent, and occasionally catastrophic, flooding. Following a storm, the levels of such rivers fall almost as rapidly and, after dry spells, can become very low. Rivers in other basins seem neither to flood nor to fall to very low levels. There are several factors which contribute to regulating the ways in which a river responds to precipitation.

1 Basin size, shape and relief

Size If a basin is small it is likely that rainfall will reach the main channel more rapidly than in a larger basin where the water has much further to travel. Lag time will therefore be shorter in the smaller basin.

Shape It has long been accepted that a circular basin is more likely to have a shorter lag time and a higher peak flow than an elongated basin (Figure 3.6a). All the points on the watershed of the former are approximately equidistant from the gauging station, whereas in the latter it takes longer for water from the extremities of the basin to reach the gauging station. However, Newson (1994) has pointed out that studies made in many regions of the world have shown that basin shape is less reliable as a flood indicator than basin size and slope.

Relief The slope of the basin and its valley sides also affect the hydrograph. In steep-sided upland valleys, water is likely to reach the river more quickly than in gently sloping lowland areas (Figure 3.6c).

Q

1 Using Figure 3.6a, how many hours will pass before all the floodwater passes the gauging stations at **A** and **B**?

2 Draw the probable storm hydrographs for **Basin A** and **Basin B**. Explain the shapes which you have predicted.

3 Fit the three hydrographs **P**, **Q** and **R** in Figure 3.6b with the drainage basins **X**, **Y** and **Z**.

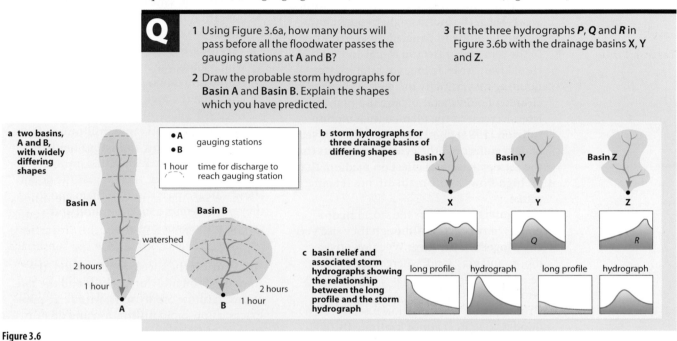

a two basins, A and B, with widely differing shapes

Basin A

watershed

Basin B

● A / ● B gauging stations

1 hour time for discharge to reach gauging station

2 hours

1 hour

A

2 hours

1 hour

B

b storm hydrographs for three drainage basins of differing shapes

Basin X   Basin Y   Basin Z

X   Y   Z

P   Q   R

c basin relief and associated storm hydrographs showing the relationship between the long profile and the storm hydrograph

long profile   hydrograph   long profile   hydrograph

Figure 3.6

Drainage basin shape

---

2 Types of precipitation

Prolonged rainfall Flooding most frequently occurs following a long period of heavy rainfall when the ground has become saturated and infiltration has been replaced by surface runoff (overland flow).

Intense storms (e.g. convectional thunderstorms) When heavy rain occurs, the rainfall intensity may be greater than the infiltration capacity of the soil (e.g. in summer in Britain, when the ground may be harder). The resulting surface runoff is likely to produce rapid rises in river levels (flash floods).

Snowfall Heavy snowfall means that water is held in storage and river levels drop. When temperatures rise rapidly (in Britain, this may be with the passage of a warm front and its associated rainfall, page 215), meltwater soon reaches the main river. It is possible that the ground will remain frozen for some time, in which case infiltration will be impeded.

3 Temperature

Extremes of temperature can restrict infiltration (very cold in winter, very hot and dry in summer) and so increase surface runoff. If evapotranspiration rates are high, then there will be less water available to flow into the main river.

4 Land use

Vegetation Vegetation may help to prevent flooding by intercepting rainfall and storing moisture on its leaves before it evaporates back into the atmosphere. Estimates suggest that tropical rainforests intercept up to 80 per cent of rainfall (30 per cent of which may later evaporate) whereas arable land may intercept only 10 per cent. Interception is less during the winter in Britain when deciduous trees have shed their leaves and crops have been harvested to expose bare earth. Plant roots, especially those of trees, reduce throughflow by taking up water from the soil.

Figure 3.7

The effect of vegetation on storm hydrographs of the River Wye and the River Severn (geology and precipitation are the same in both basins)

Flooding is more likely to occur in deforested areas — e.g. the increasingly frequent and serious flooding in Bangladesh is attributed to the removal of trees in Nepal and other Himalayan areas. In areas of afforestation, flooding may initially increase as the land is cleared of old vegetation and drained, but later decrease as the planted trees mature. Newson (1994) points out that, after 20 years of data collecting, the evidence suggests that the canopy has more effect on medium than on high flows as the main ditches remain active.

Figure 3.7 contrasts the storm hydrographs of two rivers. Although they rise very close together, the River Wye flows over moors and grassland, whereas the River Severn flows through an area of coniferous forest.

Urbanisation Urbanisation has increased flood risk. Water cannot infiltrate through tarmac and concrete, and gutters and drains carry water more quickly to the nearest river. Small streams may be either canalised so that (with friction reduced) the water flows away more quickly, or culverted, which allows only a limited amount of water to pass through at one time (Figure 3.8).

5 Rock type (geology)

Rocks which allow water to pass through them are said to be **permeable**. There are two types of permeable rock:

- **Porous**, e.g. sandstone and chalk, which contain numerous pores able to fill with and store water (Figure 8.2).
- **Pervious**, e.g. Carboniferous limestone, which allow water to flow along bedding planes and down joints within the rock, although the rock itself is impervious (Figure 8.1).

As both types permit rapid infiltration, there is little surface runoff and only a limited number of surface streams. In contrast, **impermeable rocks**, such as granite, do not allow water to pass through them and so they produce more surface runoff and a greater number of streams.

6 Soil type

This controls the speed of infiltration, the amount of soil moisture storage and the rate of throughflow. Sandy soils, with large pore spaces, allow rapid infiltration and do not encourage flooding. Clays have much smaller pore spaces; this reduces infiltration and throughflow, but encourages surface runoff and increases the risk of flooding.

7 Drainage density

This refers to the number of surface streams in a given area (page 60). The density is higher on impermeable rocks and clays, and lower on permeable rocks and sands. The higher the density, the greater is the probability of flash floods. A **flash flood** is a sudden rise of water in a river, shown on the hydrograph as a shorter lag time and a higher peak flow in relation to normal discharge.

Figure 3.8

An urban river

Figure 3.9

The effect of urbanisation on the drainage basin system

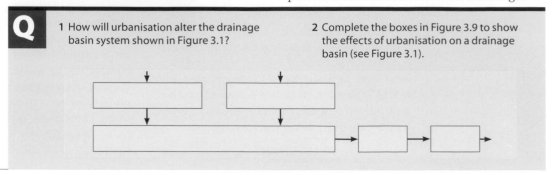

Q

1 How will urbanisation alter the drainage basin system shown in Figure 3.1?

2 Complete the boxes in Figure 3.9 to show the effects of urbanisation on a drainage basin (see Figure 3.1).

8 Tides and storm surges

High spring tides tend to prevent river flood water from escaping into the sea. Flood water therefore builds up in the lower part of the valley. If high tides coincide with gale force winds blowing onshore and a narrowing estuary, the result may be a **storm surge** (Places 15, page 132). This happened in south-east England and in the Netherlands in 1953 and prompted the construction of the Thames Barrier and the implementation of the Dutch Delta Plan.

Figure 3.10

Two storm hydrographs

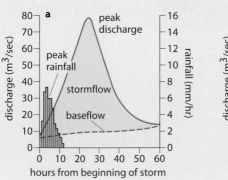

Q Figure 3.10 shows two hydrographs. Which one, **a** or **b**, is more likely to correspond to each of the pairs in the following situations?

1 a long period of steady rain and a short, torrential downpour;

2 a steep-sided and a gently sloping valley;

3 an area with a low drainage density and one with a high density;

4 an area of granite and an area of chalk;

5 an area of mature deciduous trees and an area of heathland in summer;

6 a rural area and an urban area;

7 an area where the soil is saturated and an area which is not saturated;

8 following a very cold period with some snow and a mild period in autumn;

9 a mature deciduous woodland in winter and in summer;

River régimes

The régime of a river is the term used to describe the annual variation in discharge. The average régime, which can be shown by either the mean daily or the mean monthly figures, is determined primarily by the climate of the area — e.g. the amount and distribution of rainfall, together with the rates of evapotranspiration and snowmelt.

Local geology may also be significant. There are few rivers flowing today under wholly natural conditions, especially in Britain. Most are managed, regulated systems which result from human activity.

Régimes of rivers, which are used to demonstrate seasonal variations, may be either simple, with one peak period of flow, or complex with several peaks (Places 8).

Places 8 River Don, Yorkshire

Discharge of the River Don

Figure 3.11

Rainfall and runoff for the River Don, Yorkshire (*after* Riley, Briggs and Tolley)

Figure 3.11 shows the rainfall and runoff figures for the River Don (South Yorkshire) for one year. Discharge is usually at its highest in winter when Britain receives most of its depressions and when evapotranspiration is limited due to the low temperatures. Early spring may also show a peak if the source of the river is in an upland area liable to heavy winter snowfalls — in this case, the Pennines. In contrast, river levels are lowest in summer when most of Britain receives less rainfall and when evapotranspiration rates are at their highest. There is often a correlation, or relationship, between the two variables of rainfall and runoff. This relationship can be shown by means of a scattergraph (Figure 22.6). Rainfall is plotted along the base (the **x** axis) because it is the independent variable — i.e. it does not depend on the amount of runoff. Runoff is plotted on the vertical or **y** axis because it is the dependent variable — i.e. runoff does depend upon the amount of rainfall.

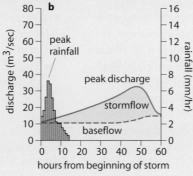

Morphometry of drainage basins

Morphometry means the 'measurement of shape or form'.

The development of morphometric techniques was a major advance in the quantitative (as opposed to the qualitative) description of drainage basins (Framework 2). Instead of studies being purely subjective, it became possible to compare and contrast different basins with precision. Much of the early work in this field was by R.E. Horton. In the mid 1940s he devised the 'Laws of drainage composition' which established a hierarchy of streams ranked according to 'order'. One of these laws, the **law of stream number**, states that within a drainage basin a constant geometric relationship exists between stream order and stream number (Figure 3.13a).

Figure 3.12 shows how one of Horton's successors, A.N. Strahler, defined streams of different order. All the initial, unbranched source tributaries he called **first order** streams. When two first order streams join they form a **second order**; when two second order streams merge they form a **third order**, and so on. Notice that it needs two stream segments of equal order to join to produce a segment of a higher order, while the order remains unchanged if a lower order segment joins a higher order segment. For example, a second order plus a second order gives a third order but if a second order stream joins a third order, the resultant stream remains as a third order. A basin may therefore be described in terms of the highest order stream within it, e.g. a 'third order basin' or a 'fourth order basin'.

If the number of segments in a stream order is plotted on a semi-log graph against the stream order, then the resultant best-fit-line will be straight (Figure 3.13a). On a semi-log graph, the vertical scale, showing the dependent variable (Figure 22.5), is divided into cycles, each of which begins and ends ten times greater than the previous cycle, e.g. a range of 1 to 10, 10 to 100, 100 to 1000, and so on. (If the horizontal scale, showing the independent variable, had also been divided into cycles instead of having an arithmetic scale, then Figure 3.13 would have been referred to as a log-log graph (Figure 18.20).) Logarithmic graphs are valuable when:

- The rate of change is of more interest than the amount of change: the steeper the line the greater the rate of change.
- There is a greater range in the data than there is space to express on an arithmetic scale (a log scale compresses values).
- There are considerably more data at one end of the range than the other.

Figure 3.13a shows a perfect negative correlation (Figure 22.6): as the independent variable (in this case the stream order) increases, then the dependent variable (the number of streams) decreases. Studies of stream ordering for most rivers in the world produce a similar straight-line relationship. For any exceptions to Horton's law of stream ordering, further studies can be made to determine which local factors alter the relationship. Relationships also exist between stream order and the mean length of streams (Figure 3.13b), and stream order and mean drainage basin area (Figure 3.13c).

Figure 3.12

Strahler's method of stream ordering

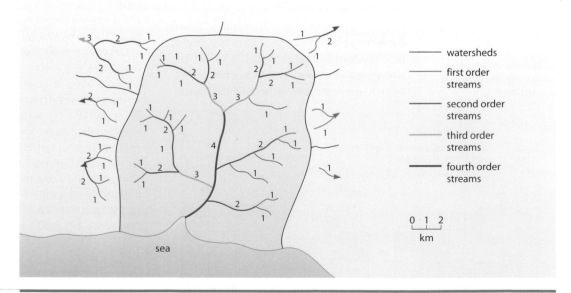

Figure 3.13

Relationships between stream order and other variables

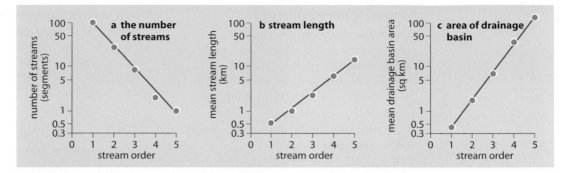

Q

1 Describe the relationships between stream order and
  a mean stream length (Figure 3.13b);
  b mean drainage basin area (Figure 3.13c).

2 In each case, is the correlation (the relationship between the independent and the dependent variables) positive or negative?

Framework 2
Quantitative techniques and statistical methods of data interpretation

As geography adjusted to a more scientific approach in the 1960s, a series of statistical techniques were adopted which could be used to quantify field data and add objectivity to the testing of hypotheses and theories. This period is often referred to as the 'Quantitative Revolution'.

At first it seemed to many, the author included, that mathematics had taken over the subject but it is now accepted that these techniques are a useful aid provided they are not seen as an end in themselves. They provide a tool which, if carefully handled and understood, gives greater precision to arguments, helps in the identification of patterns and may contribute to the discovery of relationships and possible cause – effect links. In short, by providing greater accuracy in handling data they reduce the reliance upon subjective conclusions.

It is essential to select the most appropriate techniques for the data and for the job in hand. Therefore some understanding of the statistical methods involved is important.

Statistical methods may be profitably employed in these areas.

1 **Sampling** (Framework 4, page 144) Rapid collection of the data is made possible.

2 **Correlation and regression** (Framework 14, page 573) This not only shows possible relationships between two variables but quantifies or measures the strength of those relationships.

3 **Spatial distributions** (Framework 14, page 574) Not only may this approach be used to identify patterns, but it may demonstrate how likely it is that the resultant distributions occurred by chance.

When these new techniques first appeared in schools in the 1970s, they appeared extremely daunting until it was realised that often the difficulty of the worked examples detracted from the usefulness of the technique itself. Where such techniques appear in this book, the mathematics have been simplified to show more clearly how methods may be used and to what effect. With the wider availability of calculators and computers it has become easier to take advantage of more complex calculations to test geographical hypotheses (Framework 7, page 284). Much of the 'number crunching' has now been removed by the increasing availability of statistical packages for micro-computers.

Comparing drainage basins

Horton's work has made it possible to compare different drainage basins scientifically (quantitatively) rather than relying on subjective (qualitative) descriptions by individuals. It allows people studying drainage basin morphometry in different parts of the world to use the same standards, measurements and 'language'.

Figure 3.14 shows two imaginary and adjacent basins. These can be compared in several different ways — see Table in Question 1 opposite, where the data for basin **A** have already been inserted.

The bifurcation ratio

This is the relationship between the number of streams of one order and those of the next highest order. It is obtained by dividing the number of streams in one order by the number in the next highest order, e.g. for basin A in Figure 3.14:

$$\frac{N1 \text{ (number of first order streams)}}{N2 \text{ (number of second order streams)}} = \frac{26}{6} = 4.33$$

and then finding the mean of all the ratios in the basin being studied, i.e.

$$\frac{4.33 + 3.00 + 2.00}{3} = 3.11 = \text{bifurcation ratio for basin A.}$$

The human significance of the bifurcation ratio is that as the ratio is reduced so the risk of flooding within the basin increases. It also indicates the flood risk for parts, rather than all, of the basin. Most British rivers have a bifurcation ratio of between 3 and 5.

Figure 3.14

A comparison between two drainage basins on clays and sands

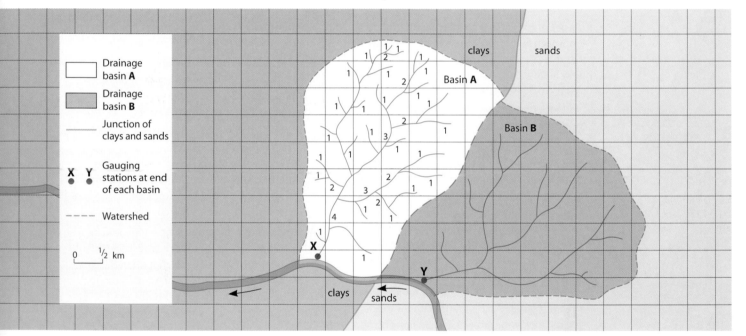

Drainage density

This is found by measuring the total length of all the streams within the basin (L) and dividing by the area of the whole basin (A). It is therefore the average length of stream within each unit area. For basin A in Figure 3.14, this will be:

$$\frac{L}{A} = \frac{22.65}{12.50} = 1.81 \text{ km per km}^2$$

In Britain most drainage densities lie between 2 and 4 km per km^2 but this varies considerably according to local conditions. A number of factors influence drainage density.

It tends to be highest in areas where the land surface is impermeable, where slopes are steep, where rainfall is heavy and prolonged, and where vegetation cover is lacking.

a Geology and soils On very permeable rocks or soils (e.g. chalk, sands) drainage densities may be under 1 km per km^2, whereas this increases to over 5 km per km^2 on highly impermeable surfaces (e.g. granite, clays). In Figure 3.14 with two adjacent drainage basins of approximately equal size, shape and probably rainfall, the difference in drainage density is likely to be due to basin **A** being on clays and basin **B** on sands.

b Land use The drainage density, especially of first order streams, is much greater in areas with little vegetation cover. The density decreases, as does the number of first order streams, if the area becomes afforested. Deserts tend to have the highest densities of first order channels, even if the channels are dry for most of the time.

c Time As a river pattern develops over a period of time, the number of tributaries will decrease as will the drainage density.

d Precipitation Densities are usually highest in areas where rainfall totals and intensity are also high.

e Relief Density is usually greater on steeper slopes than on more gentle slopes.

 1 Complete the table for drainage basin **B** in Figure 3.14.

		Basin A	Basin B
1 Number (N) of streams (segments) in each order	N_1 (first order)	26	
	N_2 (second order)	6	
	N_3 (third order)	2	
	N_4 (fourth order)	1	
2 Total number of streams	$\sum N$	35	
3 Bifurcation ratios	$\dfrac{N_1}{N_2}$	4.33	
	$\dfrac{N_2}{N_3}$	3	
	$\dfrac{N_3}{N_4}$	2	
4 Average bifurcation ratio		3.11	

		Basin A	Basin B
5 Total length (L) of streams in each order	L_1 (first order)	12	
	L_2 (second order)	7	
	L_3 (third order)	2.4	
	L_4 (fourth order)	1.25	
6 Total stream length (km)	$\sum L$	22.65	
7 Mean stream length (L) in each order(km)	$\bar{L}_1$ (first order)	0.46	
	$\bar{L}_2$ (second order)	1.17	
	$\bar{L}_3$ (third order)	1.20	
	$\bar{L}_4$ (fourth order)	1.25	
8 Total area A of drainage basin (km²)	A	12.50	
9 Drainage density	$\dfrac{\sum L}{A}$	1.81	

Figure 3.15

The response of a British drainage basin to extremes of rainfall

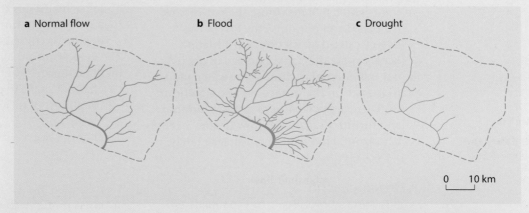

a Normal flow   b Flood   c Drought

0   10 km

Figure 3.16

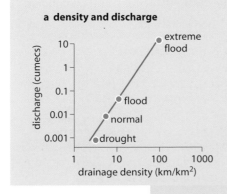
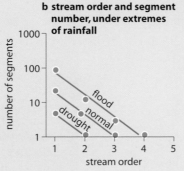

a density and discharge

b stream order and segment number, under extremes of rainfall

2 Figure 3.15 illustrates the influence of rainfall on the density of surface rivers in a drainage basin. Describe and give reasons for the changes in density following
  a periods of dry weather (drought); and
  b very wet conditions (flood).

3 Figure 3.16 shows the relationship between drainage density and discharge and between the number of segments and stream order. Describe and give reasons for the changes in each graph
  a following a period of drought; and
  b under flood conditions.

Figure 3.17

Turbulence in a river: the confluence of the Rio Amazon (red with silt) and the Rio Negro (black with plant acids). The waters remain unmixed for 20-30 km

Figure 3.18

Types of flow in a river

River form and velocity

A river will try to adopt a channel shape that best fulfils its two main functions: transporting water and sediment. It is important to understand the significance of channel shape in order to identify the controls on the flow of a river.

Types of flow

As water flows downhill under gravity, it seeks the path of least resistance — i.e. a river possesses potential energy and follows a route which will maximise the rate of flow (velocity) and minimise the loss of this energy caused by friction. Most friction occurs along the banks and bed of the river, but the internal friction of the water and air resistance on the surface are also significant.

There are two patterns of flow, **laminar** and **turbulent**. Laminar flow (Figure 3.18a) is a horizontal movement of water so rarely experienced in rivers that it is usually discounted. Such a method of flow, if it existed, would travel over sediment on the river bed without disturbing it. Turbulent flow, the dominant method, consists of a series of erratic eddies, both vertical and horizontal, in a downstream direction (Figures 3.17 and 3.18b). Turbulence varies with the velocity of the river which, in turn, depends upon the amount of energy available after friction has been overcome. It is estimated that under 'normal' conditions about 95 per cent of a river's energy is expended in order to overcome friction.

Influence of velocity on turbulence

- If the velocity is high, the amount of energy still available after friction has been overcome will be greater and so turbulence increases. This results in sediment on the bed being disturbed and carried downstream. The faster the flow of the river, the larger the quantity and size of particles which can be transported. The transported material is referred to as the river's **load**.
- When the velocity is low, there is less energy to overcome friction. Turbulence decreases and may not be visible to the human eye. Sediment on the river bed remains undisturbed. Indeed, as turbulence maintains the transport of the load, a reduction in turbulence may lead to deposition of sediment.

The velocity of a river is influenced by three main factors:

1 channel shape in cross-section;
2 roughness of the channel's bed and banks; and
3 channel slope.

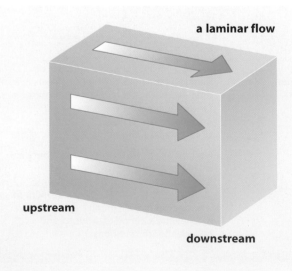

a laminar flow

upstream

downstream

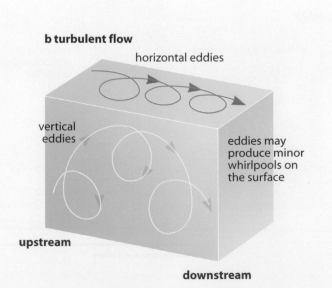

b turbulent flow

horizontal eddies

vertical eddies

eddies may produce minor whirlpools on the surface

upstream

downstream

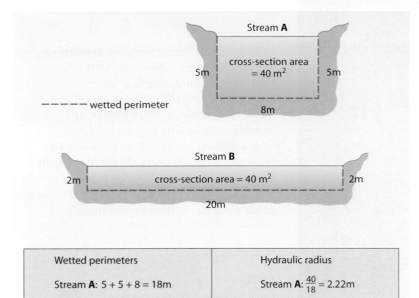

 wetted perimeter

Wetted perimeters	Hydraulic radius
Stream **A**: 5 + 5 + 8 = 18m	Stream **A**: $\frac{40}{18}$ = 2.22m
Stream **B**: 2 + 2 + 20 = 24m	Stream **B**: $\frac{40}{24}$ = 1.66m

Figure 3.19

The wetted perimeter, hydraulic radius and efficiency of two differently shaped channels with equal area

The **wetted perimeter** is the total length of the bed and bank sides in contact with the water in the channel. Figure 3.19 shows two channels with the same cross-section area but with different shapes and hydraulic radii.

Stream **A** has a larger hydraulic radius, meaning that it has a smaller amount of water in its cross-section in contact with the wetted perimeter. This creates less friction which in turn reduces energy loss and allows greater velocity. Stream **A** is said to be the more efficient of the two rivers.

Stream **B** has a smaller hydraulic radius, meaning that a larger amount of water is in contact with the wetted perimeter. This results in greater friction, more energy loss and reduced velocity. Stream **B** is less efficient than stream **A**.

The shape of the cross-section controls the point of maximum velocity in a river's channel. The point of maximum velocity is different in a river with a straight course where the channel is likely to be approximately symmetrical (Figure 3.20) compared to a meandering channel where the shape is asymmetrical (Figure 3.21).

1 Channel shape

This is best described by the term **hydraulic radius** — i.e. the ratio between the area of the cross-section of a river channel and the length of its wetted perimeter. The cross-section area is obtained by measuring the width and the mean depth of the channel.

Q

1 Account for the differences in velocity within the symmetrically shaped channel in Figure 3.20.

2 Figure 3.21 shows an asymmetrical channel.
   a Using the information given in the table, complete the data plotted on the cross-section diagram.
   b Draw **isovels** (lines joining places of equal velocity) at intervals of 0.1 m per second.

   c Shade in the part of the channel with the maximum velocity.
   d Label the area in which you consider that deposition is most likely to occur.

3 Account for the differences in isovel patterns between Figures 3.20 and 3.21.

Figure 3.21

An asymmetrical channel: velocities in the cross-section of a typically meandering river

Figure 3.20

A symmetrical channel: velocities in a straight stretch of river

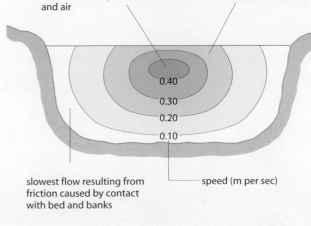

Depth (m)	Distance from Bank **X** (m)				
	0.2	0.4	0.6	0.8	1.0
0.1	0.47	0.29	0.23	0.11	0.05
0.2	0.53	0.32	0.18	0.08	
0.3	0.44	0.22	0.07		
0.4	0.32	0.10	0.01		
0.5	0.10				

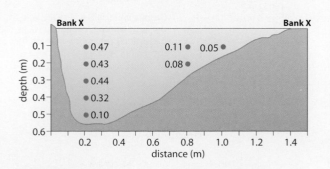

Figure 3.22

A boulder-strewn river bed in the upper Afon Glaslyn (the Aber Glaslyn pass), Snowdonia – notice the scree slopes in the background

Figure 3.23

Why a river increases in velocity towards its mouth

2 Roughness of channel bed and banks

A river flowing between banks composed of coarse material with numerous protrusions and over a bed of large, angular rocks (Figure 3.22) meets with more resistance than a river with cohesive clays and silts forming its bed and banks.

Figure 3.23 shows why the velocity of a mountain stream is less than that of a lowland river. As bank and bed roughness increase, so does turbulence. Therefore a mountain stream is likely to pick up loose material and carry it downstream.

Roughness is difficult to measure, but Manning, an engineer, calculated a **roughness coefficient** by which he interrelated the three factors affecting the velocity of a river. In his formula, known as 'Manning's N':

$$v = \frac{R^{0.67} \ S^{0.5}}{n}$$

where:

v = mean velocity of flow;
R = hydraulic radius;
S = channel slope;
n = boundary roughness.

The formula gives a useful approximation: the higher the value, the rougher the bed and banks. For example,

Bed profile	Sand and gravel	Coarse gravel	Boulders
Uniform	0.02	0.03	0.05
Undulating	0.05	0.06	0.07
Highly irregular	0.08	0.09	0.10

3 Channel slope

As more tributaries and water from surface runoff, throughflow and groundwater flow join the main river, the discharge, the channel cross-section area and the hydraulic radius will all increase. At the same time, less energy will be lost through friction and the role of bedload material will decrease. As a result, the river flows over a gradually decreasing gradient — the characteristic concave **long profile** (**thalweg**) as shown in Figure 3.24.

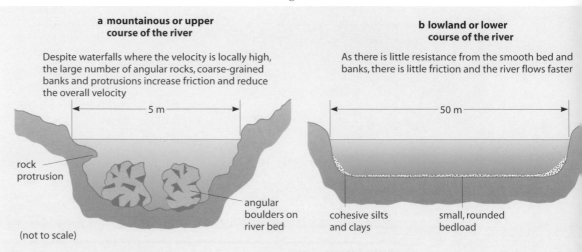

a mountainous or upper course of the river

Despite waterfalls where the velocity is locally high, the large number of angular rocks, coarse-grained banks and protrusions increase friction and reduce the overall velocity

5 m

rock protrusion

angular boulders on river bed

(not to scale)

b lowland or lower course of the river

As there is little resistance from the smooth bed and banks, there is little friction and the river flows faster

50 m

cohesive silts and clays

small, rounded bedload

Figure 3.24

The characteristic long profile
of a river

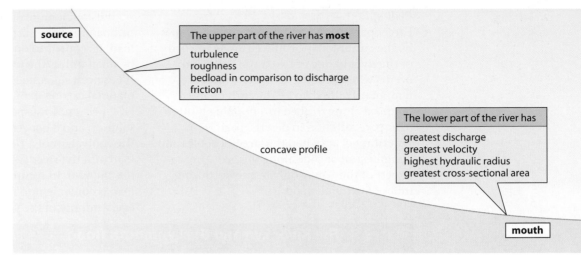

source

The upper part of the river has **most**

turbulence
roughness
bedload in comparison to discharge
friction

concave profile

The lower part of the river has

greatest discharge
greatest velocity
highest hydraulic radius
greatest cross-sectional area

mouth

In summarising this section it should be noted that:

- A river in a deep, broad channel, often with a gentle gradient and a small bedload, will have a greater velocity than a river in a shallow, narrow, rock-filled channel — even if the gradient of the latter is steeper.
- The velocity of a river increases as it nears the sea — unless, like the Colorado (see Case Study 3b) and the Nile, it flows through deserts where water is lost through evaporation or by human extraction for water supply.
- The velocity increases as the depth, width and discharge of a river all increase.
- As roughness increases, so too does turbulence and the ability of the river to pick up and transport sediment.

Transportation

Any energy remaining after the river has overcome friction can be used to transport sediment. The amount of energy available increases rapidly as the discharge, velocity and turbulence increase, until the river reaches flood levels. A river in flood has a large wetted perimeter and the extra friction is likely to cause deposition on the flood plain. A river at bankfull stage can move large quantities of soil and rock — its load — along its channel. In Britain, most material carried by a river is either sediment being redistributed from its banks or material reaching the river from mass movement on its valley sides.

The load is transported by three main processes: **suspension**, **solution** and as **bedload** (Figure 3.25).

Suspended load
Very fine particles of clay and silt are dislodged and carried by turbulence in a fast-flowing river. The greater the turbulence and velocity, the larger the quantity and size of particles which can be picked up. The material held in suspension usually forms the greatest part of the total load; it increases in amount towards the river's mouth, giving the water its brown or black colour.

Dissolved or solution load
Water flowing within a river channel contains acids (e.g. carbonic acid from precipitation). If the bedrock is readily soluble, like limestone, it is constantly dissolved in the running water and removed in solution. Except in limestone areas, the material in solution forms only a relatively small proportion of the total load.

Figure 3.25

Transportation processes in a
river or stream

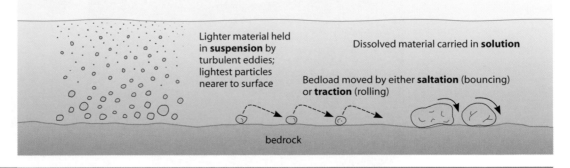

Lighter material held in **suspension** by turbulent eddies; lightest particles nearer to surface

Dissolved material carried in **solution**

Bedload moved by either **saltation** (bouncing) or **traction** (rolling)

bedrock

Bedload

Larger particles which cannot be picked up by the current may be moved along the bed of the river in one of two ways. **Saltation** occurs when pebbles, sand and gravel are temporarily lifted up by the current and bounced along the bed in a hopping motion (compare saltation in deserts, page 167). **Traction** occurs when the largest cobbles and boulders roll or slide along the bed. The largest of these may only be moved during times of extreme flood.

It is much more difficult to measure the bedload than the suspended or dissolved load. Its contribution to the total load may be small unless the river is in flood (Places 9). It has been suggested that the proportion of material carried in one year by the River Tyne is 57 per cent in suspension, 35 per cent in solution and 8 per cent as bedload. This is the equivalent of a 10-tonne lorry tipping its load into the river every 20 minutes throughout the year. In comparison, the Amazon's load is equivalent to four such lorries tipping every minute of the year!

Places 9 The River Lyn and the Lynmouth flood

One of the worst floods in living memory in Britain was that which devastated the North Devon town of Lynmouth in 1952. The West Lyn River flows through a narrow, steep-sided valley and has a steep gradient. It forms a small drainage basin on impermeable rocks on the northern flanks of Exmoor. August of that year had been very wet and left the ground saturated. On 15 August, an estimated 230 mm of rain fell in a freak storm lasting 14 hours. There was immediate surface runoff into the river causing a flash flood. Although its drainage basin is 100 times larger, the River Thames is said to have had a greater discharge only twice in the last century.

The course of the West Lyn had previously been altered at Lynmouth and its channel narrowed by the building of hotels and other tourist amenities. The already-narrow arch of the road bridge had become blocked by boulders and vegetation. During the flood, the river, too swollen to be contained within its channel, flowed down the main street of the town following the path of least resistance. The size of the bedload was enormous. Over 100 000 tonnes of boulders were left in the main street after the waters receded, and one boulder found in the basement of a hotel weighed 7.5 tonnes (the weight of a double-decker bus). Elsewhere, rocks of up to 10 cu m had been moved by traction. Measurement of the suspended load was impossible, but it did, on deposition, form an offshore delta. The flood claimed 34 lives, and destroyed 90 houses and hotels, 130 cars and 19 boats (Figure 3.26).

Apart from the information centre housing photos and newspaper extracts, the only evidence af the 1952 event is a mark, incredibly high up on a rock cliff in the gorge, which marked the maximum height of the flood and a small hydro-electric power station, which, even under low water conditions, shows that the force of the river is capable of providing enough energy to sustain the basic needs of Lynmouth.

Figure 3.26

Bedload left in the main street of Lynmouth in the aftermath of the 1952 flood

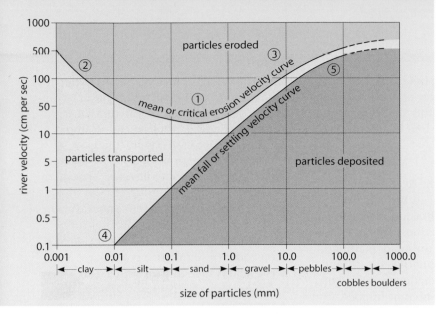

Figure 3.27

The Hjulström graph, showing the relationship between velocity and particle size. This shows the velocities necessary ('critical') for the initiation of movement (erosion); for deposition (sedimentation); and the area where transportation will continue to occur once movement has been initiated

Competence and capacity

Two further terms should be noted at this point: the competence and the capacity of a river. **Competence** is the maximum size of material which a river is capable of transporting. **Capacity** is the total load actually transported. When the velocity is low, only small particles such as clay, silt and fine sand can be picked up (Figure 3.27). As the velocity increases, larger material can be moved. Because the maximum particle mass which can be moved increases with the sixth power of the velocity, rivers in flood can move considerable amounts of material. For example, if the stream velocity increased by a factor of four, then the mass of boulders which could be moved would increase by 4^6 or 4096 times; if by a factor of five, the maximum mass it could transport would be multiplied 15 625 times.

The relationship between particle size (competence) and water velocity is shown on the Hjulström graph (Figure 3.27). The

mean, or **critical**, **erosion velocity** curve gives the approximate velocity needed to pick up and transport, in suspension, particles of various sizes. The material carried by the river (capacity) is responsible for most of the subsequent erosion. The **mean fall** or **settling**, **velocity** curve shows the velocities at which particles of a given size become too heavy to be transported and so will fall out of suspension and be deposited.

The graph shows two important points:
1. Sand can be transported at lower velocities than either finer or coarser particles. Particles of about 0.2 mm diameter can be picked up by a velocity of 20 cm per second (labelled **1** on the graph) whereas finer clay particles (**2**), because of their cohesive properties, need a velocity similar to that of pebbles (**3**) to be dislodged. During times of high discharge and velocity, the size and amount of the river's load will increase considerably, causing increased erosion within the channel.
2. The velocity required to maintain particles in suspension is less than the velocity needed to pick them up. Indeed, for very fine clays (**4**) the velocity required to maintain them is virtually nil — at which point the river must almost have stopped flowing! This means that material picked up by turbulent tributaries and lower order streams can be kept in suspension by a less turbulent, higher order main river. For coarser particles (**5**), the boundary between transportation and deposition is narrow, indicating that only a relatively small drop in velocity is needed to cause sedimentation.

Figure 3.28

Velocities at which particles of different sizes are picked up and deposited

Size of particles in the river	Velocity at which particles may be picked up (cm per sec) (mean erosion velocity curve)	Velocity at which particles may be deposited (cm per sec) (mean fall velocity curve)
clay		
silt		
sand		
gravel		
pebbles		
cobbles		
boudlers		

Use Figure 3.27 to answer the following questions:

1. a Complete the table in Figure 3.28 to show the velocity at which particles of different sizes will be picked up and deposited.
   b Does Figure 3.27 show the competence, capacity or load of a river?

2. Which process of transportation is likely to be most significant in a river flowing:
   a over limestone;
   b on glacial sands and gravels;
   c on boulder clay;
   d through a narrow gorge and then onto a flat, wide plain.

Erosion

The material carried by a river can contribute to the wearing away of its banks and, to a lesser extent and mainly in the upper course, its bed. There are four main processes of erosion.

Corrasion

Corrasion occurs when the river picks up material and rubs it along its bed and banks, wearing them away by **abrasion**, rather like sandpaper. This process is most effective during times of flood and is the major method by which the river erodes both vertically and horizontally. If there are hollows in the river bed, pebbles are likely to become trapped. Turbulent eddies in the current can swirl pebbles around to form **potholes** (Figure 3.29).

Attrition

As the bedload is moved downstream, boulders collide with other material and the impact may break the rock into smaller pieces. In time, angular rocks become increasingly rounded in appearance.

Hydraulic action

The sheer force of the water as the turbulent current hits river banks, (on the outside of a meander bend), pushes water into cracks. The air in the cracks is compressed, pressure is increased and, in time, the bank will collapse. **Cavitation** is a form of hydraulic action caused by bubbles of air collapsing. The resultant shock waves hit and slowly weaken the banks. This is the slowest and least effective erosion process.

Solution or corrosion

This occurs continuously and is independent of river discharge or velocity. It is related to the chemical composition of the water, e.g. the concentration of carbonic and humic acid.

Deposition

When the velocity of a river begins to fall, it has less energy and so no longer has the competence or capacity to carry all its load. So, starting with the largest particles, material begins to be deposited (Figure 3.27).

Deposition occurs when:
- discharge is reduced following a period of low precipitation;
- velocity is lessened on entering the sea or a lake (resulting in a delta);
- shallower water occurs on the inside of a meander (resulting in a point bar);
- the load is suddenly increased, (caused by débris from a landslide)
- the river overflows its banks so that the velocity outside the channel is reduced (resulting in a flood plain).

As the river loses energy, the following changes are likely:
- The heaviest or bedload material is deposited first. It is for this reason that the channels of mountainous streams are often filled with large boulders (Figure 3.22). Large boulders increase the size of the wetted perimeter.
- Gravel, sand and silt — transported either as bedload or in suspension — will be carried further, to be deposited over the flood plain or in the channel of the river as it nears its mouth (Figure 3.36).
- The finest particles of silt and clay, which are carried in suspension, may be deposited where the river meets the sea — either to infill an estuary or to form a delta (Figure 3.37).
- The dissolved load will not be deposited, but will be carried out to sea where it will help to maintain the saltiness of the oceans.

Figure 3.29

Potholes in the bed of the Afon Glaslyn, Snowdonia

Figure 3.30

The Afon Glaslyn, Snowdonia, showing features and processes at selected sites (O.S. 1:50 000, sheets 115 and 124, grid references 630545 to 575370)

Q Figure 3.30 is based upon an Open University television programme on the Afon Glaslyn in North Wales and aims to show the relationships between transportation, erosion and deposition.

1 Using the evidence given in Figure 3.30, say whether you agree with the hypotheses (Framework 7, page 284) that, as the river loses energy:
  a the largest material, carried as bedload, will be deposited first;

b material carried in suspension will be deposited over floodplains or in the channel of the river as it nears its mouth;
c as the competence of the river decreases, material is likely to be carried greater distances;
d material transported as dissolved load will be carried out to sea.

2 How would you further test these hypotheses in the field?

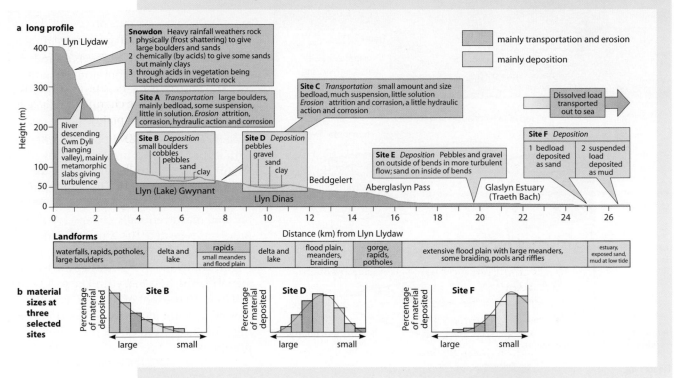

Fluvial features

As the velocity of a river increases, surplus energy becomes available which may be harnessed to transport material and cause erosion. Where the velocity decreases, an energy deficit is likely to result in depositional features.

Results of erosion

V-shaped valleys and interlocking spurs
As was seen in Figure 3.22, the channel of a river in its upper course is often choked with large, angular boulders. This bedload produces a large wetted perimeter which uses up much of the river's energy. Erosion is minimal because little energy is left to pick up and transport material. However, following periods of heavy rainfall or after rapid snowmelt, the discharge of a river may rise rapidly. As the water flows between

boulders, turbulence increases and may result either in the bedload being taken up into suspension or, as is more usual because of its size, in its being rolled or bounced along the river bed. The result is intensive **vertical erosion** which enables the river to create a steep-sided valley with a characteristic 'V' shape (Figure 3.31). The steepness of the valley sides depends upon several factors.

- **Climate**: is there sufficient rain to instigate mass movement on the valley sides and to increase discharge sufficiently for the river to generate enough energy to move its bedload?
- **Rock structure**: for example, Carboniferous limestone is a resistant, permeable rock which often produces almost vertical valley sides.
- **Vegetation**: vegetation may help to bind the soil together and thus keep the hillslope more stable.

Figure 3.31

'V'-shaped valley with interlocking spurs, small rapids and no flood plain

Interlocking spurs form because the river is forced to follow a winding course around the protrusions of the surrounding highland. As the resultant spurs interlock, the view up or down the valley is restricted (Figure 3.31).

A process characteristic at the source of a river is **headward erosion**, or **spring sapping**. Here, where throughflow reaches the surface, the river may erode back towards its watershed as it undercuts the rock, soil and vegetation.

Waterfalls

A waterfall forms when a river, after flowing over relatively hard rock, meets a band of less resistant rock or, as is common in South America and Africa, where it flows over the edge of a plateau. As the water approaches the brink of the falls, velocity increases because the water in front of it loses contact with its bed and so is unhampered by friction (Figure 3.32). The underlying softer rock is worn away as water falls onto it. In time, the harder rock may become undercut and unstable, and may eventually collapse. As this process is repeated, the waterfall retreats upstream leaving a deep, steep-sided gorge (Figures 3.32 and 3.33). At Niagara, the falls are retreating by 1 m a year. The rock, which collapses to the foot of the falls, is swirled around by turbulence, usually in times of high discharge, and carves out a deep **plunge pool**.

top of falls collapses and retreats

Iguaçu River

horizontal lavas

as waterfall retreats it leaves a gorge with vertical sides, 80 m tall

horizontal layers of resistant Triassic lavas

softer rocks being undercut, causing the overlying lavas to collapse

large, fallen, angular boulders are swirled around, forming a plunge pool

Iguaçu River flowing over rapids

Figure 3.32

Field sketch of the Iguaçu Falls at the border of Argentina and Brazil

Figure 3.33

The Iguaçu Falls

Rapids

Rapids develop where the gradient of the river bed increases without a sudden break of slope (as in a waterfall) or where the stream flows over a series of gently dipping bands of harder rock. Rapids increase the turbulence of a river and hence its erosive power.

Effects of fluvial deposition

Deposition of sediment takes place when there is a decrease in energy or an increase in capacity which makes the river less competent to transport its load. This can occur anywhere from the upper course, where large boulders may be left, to the mouth, where fine clays may be deposited.

Flood plains

Rivers have most energy when at their bankfull stage. Should the river continue to rise, then the water will cover any adjacent flat land. The land susceptible to flooding in this way is known as the **flood plain** (Figure 3.34). As the river spreads over its flood plain, there will be a sudden increase in both the wetted perimeter and the hydraulic radius. This results in an increase in friction, a corresponding decrease in velocity and the deposition of material previously held in suspension. The thin veneer of silt, deposited by each flood, increases the fertility of the land , while the successive flooding causes the flood plain to build up in height (as yet it has proved impossible to bore down to bedrock in the lower Nile valley). The flood plain may also be made up of material deposited as point bars on the inside of meanders (page 74) and can be widened by the **lateral erosion** of the meanders. The edge of the flood plain is often marked by a prominent slope known as the **bluff line** (Figure 3.35).

Levées

When a river overflows its banks, the increase in friction produced by the contact with the flood plain causes material to be deposited. The coarsest material is dropped first to form a small, natural embankment (or levée) alongside the channel (Figure 3.35). During subsequent periods of low discharge, further deposition will occur within the main channel causing the bed of the river to rise and the risk of flooding to increase. To try to contain the river, the embankments are sometimes artificially strengthened and heightened (the levée protecting St Louis from the Mississippi is 15.8 m higher than the flood plain which it is meant to protect). Some rivers, such as the Mississippi and Huang He, flow above the level of their flood plains which means that if the levées collapse there can be serious damage to property and loss of life (Case Study 3a).

Figure 3.34

Floods in Sussex, January 1994

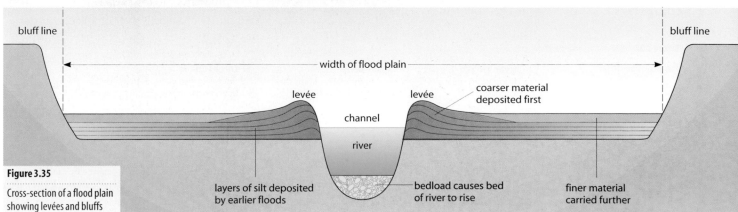

Figure 3.35

Cross-section of a flood plain showing levées and bluffs

bluff line

width of flood plain

bluff line

levée

levée

coarser material deposited first

channel

river

layers of silt deposited by earlier floods

bedload causes bed of river to rise

finer material carried further

Braiding

For short periods of the year, some rivers carry a very high load in relation to their velocity — e.g. during snowmelt periods in Alpine or Arctic areas. When a river's level falls rapidly, competence and capacity are reduced, and the channel may become choked with material, causing the river to braid — i.e. to divide into a series of diverging and converging segments (Figure 3.36).

Figure 3.36

A braided river, South Island, New Zealand

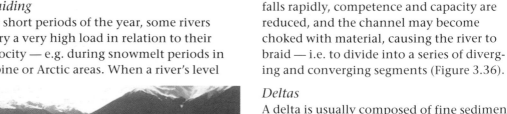

Deltas

A delta is usually composed of fine sediment which is deposited when a river loses energy and competence as it flows into an area of slow-moving water such as a lake or the sea. When rivers like the Mississippi or the Nile reach the sea, the meeting of fresh and salt water produces an electric charge which causes clay particles to coagulate and to settle on the sea bed, a process called **flocculation**. Deposits are laid on the sea bed in a three-fold sequence (Figure 3.38). The finest materials are carried furthest and form the **bottomset beds**. These will be covered by slightly coarser materials which are deposited to form a slope and make up the **foreset beds**. The upper layers, nearest to the land and composed of still coarser deposits, are the horizontal **topset beds**.

Deltas are so called because it was thought that their shape resembled that of delta, the fourth letter of the Greek alphabet (Δ). In fact, deltas vary greatly in shape but geomorphologists have grouped them into three basic forms:

- **arcuate**: having a rounded, convex outer margin, e.g. the Nile;
- **cuspate**: where the material brought down by a river is spread out evenly on either side of its channel, e.g. the Tiber;
- **bird's foot**: where the river has many **distributaries** bounded by sediment and which extend out to sea like the claws of a bird's foot, e.g. the Mississippi (Figure 3.37).

Although deltas provide some of the world's most fertile land, their flatness makes them high flood-risk areas, while the shallow and frequently changing river channels hinder navigation.

Figure 3.37

The Mississippi delta

Figure 3.38

The structure of a delta

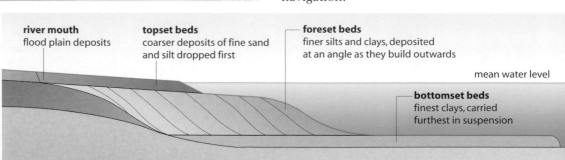

river mouth
flood plain deposits

topset beds
coarser deposits of fine sand and silt dropped first

foreset beds
finer silts and clays, deposited at an angle as they build outwards

mean water level

bottomset beds
finest clays, carried furthest in suspension

bedrock

Effects of combined erosion and deposition

Pools, riffles and meanders

Rivers rarely flow in a straight line. Indeed, testing under laboratory conditions suggests that a straight course is abnormal and unstable. How meanders begin to form is uncertain, but they appear to have their origins during times of flood and in relatively straight sections where pools and riffles develop (Figure 3.39). The usual spacing between **pools**, areas of deeper water, and

Figure 3.39

A possible sequence in the development of a meander

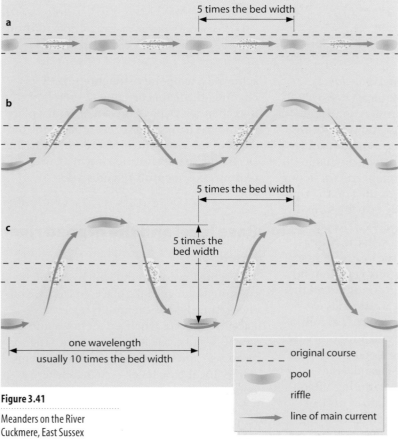

Figure 3.41

Meanders on the River Cuckmere, East Sussex

Figure 3.40

A pool and riffles in the River Gelt, Cumbria

riffles, areas of shallower water, is usually very regular being five to six times that of the bed width. The pool is an area of greater erosion where the available energy in the river builds up due to a reduction in friction. Across the riffle area, energy is dissipated. As a higher proportion of the total energy is then needed to overcome friction, the erosive capacity is decreased and, except at times of high discharge, material is deposited (Figure 3.40). The regular spacings of pools and riffles, spacings which are almost perfect in an alluvial stretch of river, are believed to result from a series of secondary flows which exist within the main flow. Secondary flows include **helicoidal flow**, a corkscrew movement, as shown in Figure 3.17b, and a series of converging and diverging lateral rotations. Helicoidal flow is believed to be responsible for moving material from the outside of one meander bend and then depositing much of it on the inside of the next bend. It is thought, therefore, that it is the secondary flows which increase the sinuosity (the curving nature) of the meander (Figure 3.41), producing a regular meander wavelength which is about ten times that of the bed width. Sinuosity is described as:

$$\frac{\text{actual channel length}}{\text{straight-line distance}}$$

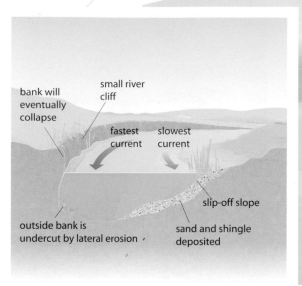

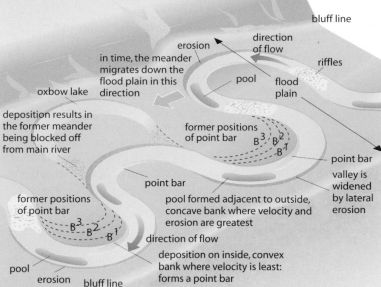

Figure 3.42

Cross-section of a meander

Figure 3.43

Meanders, point bars and oxbow lakes, showing migration of meanders and changing positions of point bars over time

Meanders, point bars and oxbow lakes

A meander has an asymmetrical cross-section (Figure 3.42) formed by erosion on the outside bend, where discharge and velocity are greatest and friction is at a minimum, and deposition on the inside, where discharge and velocity are at a minimum and friction is at its greatest. Material deposited on the convex inside of the bend may take the form of a curving **point bar** (Figure 3.43). The particles are usually graded in size, with the largest material being found on the upstream side of the feature (there is rarely any gradation up the slope itself). As erosion continues on the outer bend, the whole meander tends to migrate slowly downstream. Material forming the point bar becomes a contributory factor in the formation of the flood plain. Over time, the sinuosity of the meander may become so pronounced that, during a flood, the river cuts through the narrow neck of land in order to shorten its course. Having achieved a temporary straightening of its channel, the main current will then flow in mid channel. Deposition can now take place next to the banks and so, eventually, the old curve of the river will be abandoned, leaving a crescent-shaped feature known as an **oxbow lake** or **cutoff** (Figures 3.43 and 3.44).

Base level and the graded river

Base level

This is the lowest level to which erosion by running water can take place. In the case of rivers, this theoretical limit is sea level. Exceptions occur when a river flows into an inland sea (e.g. the River Jordan into the Dead Sea) and if there happens to be a temporary **local base level** — such as where a river flows into a lake, where a tributary joins a main river, or where there is a resistant band of rock crossing a valley.

Q 1 Complete the following table to show some differences between the processes and features (landforms) in two sections of the same river: one near to its source (upper course); the other near to its mouth (lower course).

	Upper course	Lower course
Long profile/gradient	steep	gentle
Valley width	narrow	
Valley depth		
Cross-section of valley shape		
Velocity of river		
Discharge		
Number of tributaries		

	Upper course	Lower course
Stream order		
Main processes of erosion		
Main processes of transportation		
Size of deposited material		
Shape of deposited material		
Features (landforms) formed by erosion		
Features (landforms) formed by deposition		

Figure 3.44

Meanders and an oxbow lake, Alaska, USA

Grade

The concept of **grade** is one of a river forming an open system (Framework 1, page 39) in a state of dynamic equilibrium where there is a balance between the rate of erosion and the rate of deposition. In its simplest interpretation, a graded river has a gently sloping long profile with the gradient decreasing towards its mouth (Figure 3.45a). This balance is always transitory as the slope (profile) has to adjust constantly to changes in discharge and sediment load. These can cause short-term increases in either the rate of erosion or deposition until the state of equilibrium has again been reached. This may be illustrated by two situations:

- The long profile of a river happens to contain a waterfall and a lake (Figure 3.45b). Erosion is likely to be greatest at the waterfall, while deposition occurs in the lake. In time, both features will be eliminated.
- There is a lengthy period of heavy rainfall within a river basin. As the volume of water rises and consequently the velocity

and load of the river increase, so too will the rate of erosion. Ultimately, the extra load carried by the river leads to extra deposition further down the valley or out at sea.

In a wider interpretation, grade is a balance not only in the long profile, but also in the river's cross-profile and in the roughness of its channel. In this sense, balance or grade is when all aspects of the river's channel (width, depth and gradient) are adjusted to the discharge and load of the river at a given point in time. If the volume and load change, then the river's channel morphology must adjust accordingly.

Changes in base level

There are two groups of factors which influence changes in base level:

- **Climatic**: the effects of glaciation and changes in rainfall;
- **Tectonic**: crustal uplift, following plate movement, and local volcanic activity.

As will be seen in Chapter 6, changes in base level affect coasts as well as rivers. There are two types of base level movement: positive and negative.

- **Positive change** occurs when sea level rises in relation to the land (or the land sinks in relation to the sea). This results in a decrease in the gradient of the river with a corresponding increase in deposition and potential flooding of coastal areas.
- **Negative change** occurs when sea level falls in relation to the land (or the land rises in relation to the sea). This movement causes land to emerge from the sea, steepening the gradient of the river and therefore increasing the rate of fluvial erosion. This process is called **rejuvenation**.

Figure 3.45

River profiles

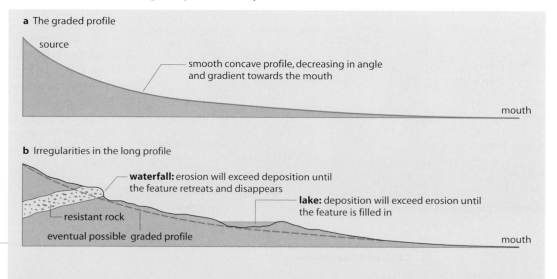

a The graded profile

source

smooth concave profile, decreasing in angle and gradient towards the mouth

mouth

b Irregularities in the long profile

waterfall: erosion will exceed deposition until the feature retreats and disappears

lake: deposition will exceed erosion until the feature is filled in

resistant rock

eventual possible graded profile

mouth

Figure 3.46

The effect of rejuvenation in the long profile

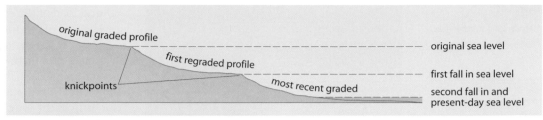

Figure 3.47

A rejuvenated river, Antalya, Turkey: the land has only recently experienced tectonic uplift and the river has had insufficient time to re-adjust to the new sea-level

profiles (Figure 3.46). Where the rise in the land (or drop in sea-level) is too rapid to allow a river sufficient time to erode vertically to the new sea level, it may have to descend as a waterfall over recently emerged sea cliffs (Figure 3.47). In time, the river will cut downwards and backwards and the waterfall will retreat upstream. The **knickpoint**, usually indicated by the presence of a waterfall, marks the maximum extent of the newly graded profile (Places 10). Should a river become completely regraded, which is unlikely because of the timescale involved, the knickpoint and all of the original graded profile will disappear.

Rejuvenation

A negative change in base level increases the potential energy of a river enabling it to revive its erosive activity; in doing so, it upsets any possible graded long profile. Beginning in its lowest reaches, next to the sea, the river will try to regrade itself.

During the Pleistocene glacial period, Britain was depressed by the weight of ice. Following deglaciation, the land slowly and intermittently rose again (**isostatic uplift**, page 112). Thus rejuvenation took place on more than one occasion, with the result that many rivers today show several partly graded

River terraces and incised meanders

River terraces are remnants of former flood plains which, following vertical erosion caused by rejuvenation, have been left high and dry above the maximum level of present-day flooding. They offer excellent sites for the location of towns (e.g. London, Figures 3.49 and 14.5). Above the present flood plain of the Thames at London are two earlier ones forming the Taplow and Boyn Hill terraces. If a river cuts rapidly into its flood plain, a pair of terraces of equal height may be seen flanking the river and creating a **valley-in-valley** feature. However, more often than not, the

Figure 3.48

The River Greta

Places 10 River Greta, Yorkshire Dales National Park

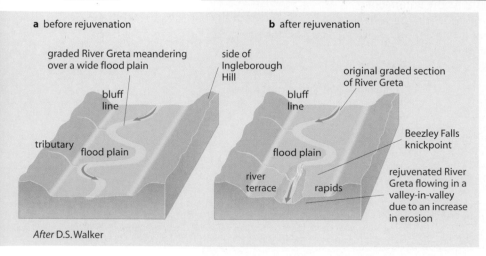

a before rejuvenation

graded River Greta meandering over a wide flood plain

bluff line

tributary

flood plain

b after rejuvenation

side of Ingleborough Hill

original graded section of River Greta

bluff line

flood plain

river terrace

rapids

Beezley Falls knickpoint

rejuvenated River Greta flowing in a valley-in-valley due to an increase in erosion

After D.S. Walker

The River Greta, in north-west Yorkshire, is a good example of a rejuvenated river. Figure 3.48a is a reconstruction to show what its valley (upstream from the town of Ingleton) might have looked like before the fall in base level. Figure 3.48b is a simplified sketch showing how the same area appears today. The Beezley Falls are a knickpoint. Above the falls, the valley has a wide, open appearance. Below the falls, the river flows over a series of rapids and smaller falls in a deep, steep-sided valley.

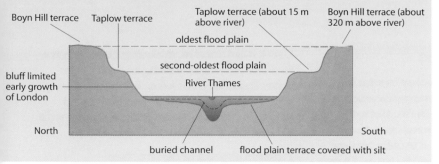

Figure 3.49

Cross-section illustrating the paired river terraces of the Thames at London

river cuts down relatively slowly, enabling it to meander at the same time. The result is that the terrace to one side of the river may be removed as the meanders migrate downstream. Figure 3.51 shows terraces, not paired, on a small stream crossing a beach on southern Arran. In this case, rejuvenation takes place twice daily as the tide ebbs and sea-level falls.

If the uplift of land (or fall in sea level)

continues for a lengthy period, the river may cut downwards to form incised meanders. There are two types of incised meander. **Entrenched meanders** have a symmetrical cross-section and result from either a very rapid incision by the river or the valley sides being resistant to erosion (the River Wear at Durham, Figures 14.7b and 3.50a). **Ingrown meanders** occur when the uplift of the land, or incision by the river, is less rapid, allowing the river time to shift laterally and to produce an asymmetrical cross-valley shape (the River Wye at Tintern Abbey, Figure 3.50b). As with meanders in the lower course of a normal river, incised meanders can also change their channels to leave an abandoned meander with a central **meander core** (Figure 3.50b).

Figure 3.50

Incised meanders and associated cross-valley profiles

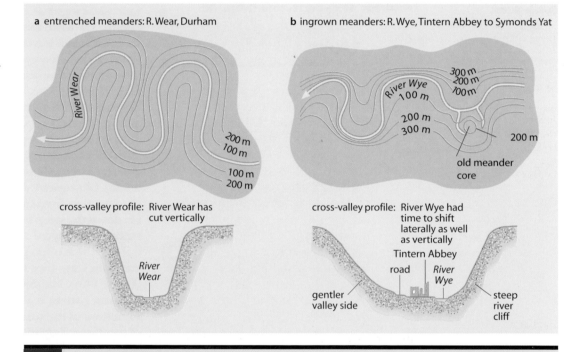

Figure 3.51

Rejuvenation on a micro scale: a small stream crossing a beach at Kildonan, Arran, has cut downwards to the level of the falling tide

Q

1 Make a sketch of Figure 3.51 and on it label: ingrown meanders, river terraces and valley-in-valley features.

2 The absence of which feature suggests that this stream has already re-graded itself?

Drainage patterns

A **drainage pattern** is the way in which a river and its tributaries arrange themselves within their drainage basin (see Horton's Laws, page 58). Most patterns evolve over a lengthy period of time and usually become adjusted to the structure of the basin. There is no widely accepted classification, partly because most patterns are descriptive.

Patterns independent of structure
Parallel This, the simplest pattern, occurs on newly uplifted land or other uniformly sloping surfaces which allow rivers and tributaries to flow downhill more or less parallel with each other — e.g. rivers flowing south-eastwards from the Aberdare Mountains in Kenya (Figure 3.52a).
Dendritic Deriving its name from the Greek word *dendron*, meaning a tree, this is a tree-like pattern in which the many tributaries (branches) converge upon the main river (trunk). It is a common pattern and develops in basins having one rock type with no variations in structure (Figure 3.52b).

Patterns dependent on structure
Radial In areas where the rocks have been lifted into a dome structure (e.g. the batholiths of Dartmoor and Arran) or where a conical volcanic cone has formed (e.g. Mount Etna), rivers radiate outwards from a central point like the spokes of a wheel (Figure 3.52c).
Trellised or rectangular In areas of alternating resistant and less resistant rock, tributaries will form and join the main river at right angles (Figure 3.52d). Sometimes each individual segment is of approximately equal length. The main river, called a **consequent river** because it is a consequence of the initial uplift or slope (compare parallel drainage), flows in the same direction as the dip of the rocks (Figure 3.53a). The tributaries which develop, mainly by headward erosion along areas of weaker rocks, are called **subsequent streams** because they form at a later date than the consequents. In time, these subsequents create wide valleys or vales (Figure 3.53b). **Obsequent streams** flow in the opposite direction from the consequent streams — i.e. down the steep scarp slope of the escarpment (Figure 3.53b). It is these obsequents which often provide the sources of water for scarp-foot springline settlements (Figure 14.6). The development of this drainage pattern is also responsible for the formation of the **scarp and vale topography** of south-east England (page 183).

Figure 3.52
Drainage patterns

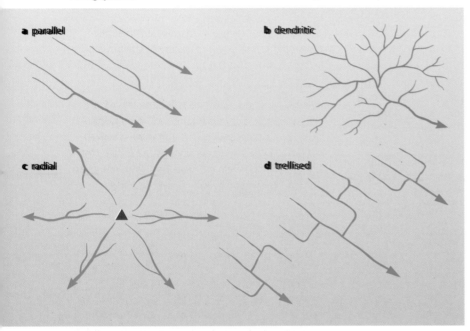

a parallel
b dendritic
c radial
d trellised

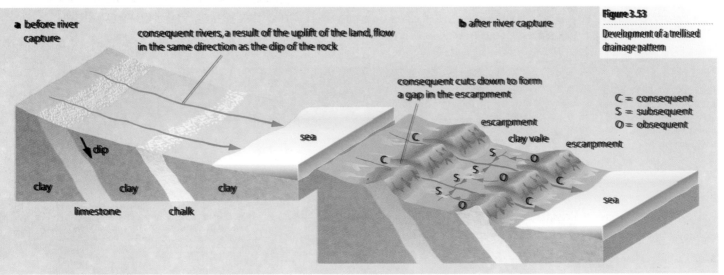

Figure 3.53
Development of a trellised drainage pattern

a before river capture
consequent rivers, a result of the uplift of the land, flow in the same direction as the dip of the rock
dip
clay   clay   clay
limestone   chalk

b after river capture
consequent cuts down to form a gap in the escarpment
sea
C
C
escarpment
clay vale
escarpment
S   O
S   O
S   O
C
sea

C = consequent
S = subsequent
O = obsequent

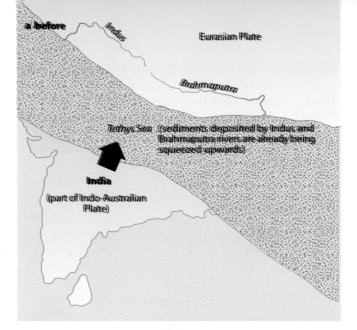

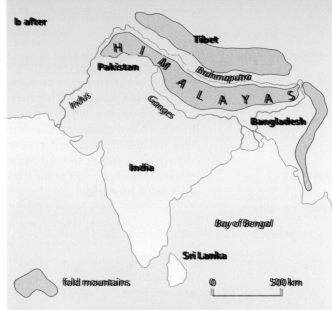

a before

Figure 3.54

Antecedent drainage
(Himalayas)

Patterns apparently unrelated to structure

Antecedent   Antecedence is when the drainage pattern developed before such structural movements as the uplift or folding of the land, and where vertical erosion by the river was able to keep pace with the later uplift. The Brahmaputra River rises in Tibet, but turns southwards to flow through a series a deep gorges in the Himalayas before reaching the Bay of Bengal (Figure 3.54). It must at one stage have flowed southwards into the Tethys Sea (Figure 1.4) which had existed before the Indo-Australian Plate moved northwards and collided with the Eurasian Plate forming the Himalayas (Figure 1.15). The Brahmaputra, with an increasing gradient and load, was able to cut downwards through the rising Himalayas to maintain its original course.

Superimposed   In several parts of the world, including the English Lake District, the drainage pattern seems to have no relationship to the present-day surface rocks. When the Lake District was uplifted into a dome, the newly formed volcanic rocks were covered by sedimentary limestones and sandstones. The radial drainage pattern which developed, together with later glacial processes, cut through and ultimately removed the surface layers of sedimentary rock to superimpose itself upon the underlying volcanic rocks.

River capture

Rivers, in attempting to adjust to structure, may capture the headwaters of their neighbours. For example, most eastward-flowing English rivers between the Humber and central Northumberland have had their courses altered by **river capture** or **piracy** (Figure 3.55).

Figure 3.56a shows a case where there are two consequent rivers with one having a greater discharge and higher erosional activity than the other. Each has a tributary (subsequents X and Y) flowing along a valley of weaker rock, but subsequent X (the tributary of the master, or larger, consequent) is likely to be the more vigorous. Subsequent X will, therefore, cut backwards by headward

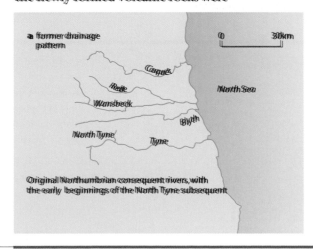

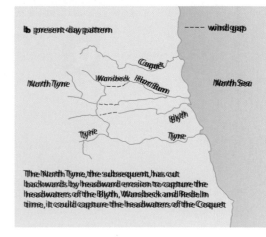

Figure 3.55

River capture, Northumberland

erosion until it reaches subsequent Y (the tributary of the weaker consequent); then, by a process known as **watershed migration** (Figure 3.56b), it will begin to enlarge its own drainage basin at the expense of the smaller river. In time, the headwaters of the minor consequent will be captured and diverted into the drainage basin of the major consequent (Figure 3.56c).

The point at which the headwaters of the minor river change direction is known as the **elbow of capture**. Below this point, a **wind gap** marks the former course of the now **beheaded consequent**. (A wind gap is a dry valley which was cut through the hills by a former river). The beheaded river is also known as a **misfit stream** as its discharge is far too low to account for the size of the valley through which it flows (Figure 3.56c).

Figure 3.56

Stages in river capture shown in plan and cross-section

a before capture (piracy) occurs

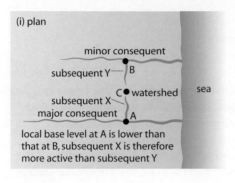

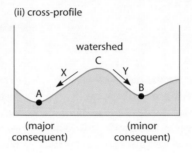

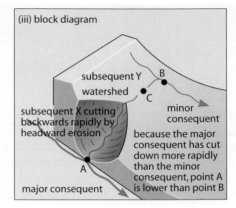

b watershed migration (recession)

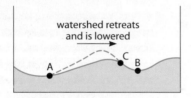

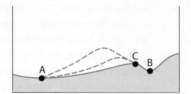

c after capture has taken place

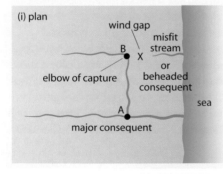

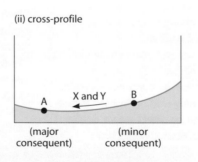

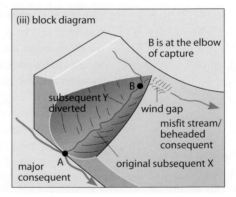

Q

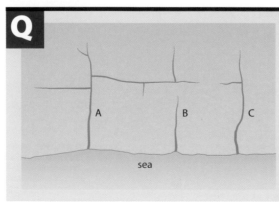

1 Redraw Figure 3.57.
   a On it, label: the master consequent, a subsequent river, an obsequent river, elbow of capture, wind gap, misfit stream, beheaded consequent.
   b Mark on the watershed for river A.

2 What changes in the drainage pattern are likely to take place in the future? Give reasons for your answer.

Figure 3.57

A drainage pattern

Drainage basins and rivers

Case Study 3

The need for river managment a. River flooding: The Mississippi, 1993

Flooding by rivers is a natural event which, because people often choose to live in flood-risk areas, becomes a hazard (page 29). To people living in the Mississippi valley "that their river should flood is as natural as sunshine in Florida or snowfall in the Rockies". Without human intervention, the Mississippi would flood virtually every year. Indeed, it has been this frequency of flooding which has, over many centuries, allowed today's river to flow for much of its course over a wide, fertile, flat, alluvial floodplain (Figures 3.58 and 3.59).

Figure 3.58

The flood hazard and the Mississippi River

"Usually, of course, the great floods occur in the lower river, in the last 1 600 km below Cairo, Illinois. This is where the plain flattens out (the river drops less than 120 m from here to its mouth) and where the Ohio and Tennessee flow into the Mississippi.

Of the water that flows past Memphis, only about 38 per cent comes from the Missouri–Mississippi network. The bulk comes from the Ohio and Tennessee, from the lush Appalachians, rather than the dry mid-West. 'We don't mind too much about the Missouri', says Donna Willett, speaking for the US Army Corps of Engineers (who have the responsibility of flood prevention). "It can rain there for weeks, and we wouldn't mind. We can handle three times the water coming down in those floods. But the Ohio, well, that's another story. When that starts rising, we start watching. . ."

Enquiry route on river flooding	Application to a specific event: the Mississippi, 1993
1 Where is the river/drainage basin located?	The Mississippi — together with its main tributaries, the Missouri and the Ohio — drains one-third of the USA and a small part of Canada (Figure 3.59).
2 What is the frequency of flooding?	Left to its own devices, flooding would be an almost annual event with late spring being the peak period
3 What is the magnitude of flooding?	Until recently, major floods occurred every 5–10 years (there were six in the 1880s) and a serious/extreme flood occurred approximately once every 40 years.
4 What are the natural causes of flooding?	Usually it results from heavy rainfall (January–May) in the Appalachian Mountains, especially if this coincides with snowmelt (Figure 3.59).
5 What are the consequences of flooding?	Initially, it was to develop the wide, alluvial flood plain. The 1927 flood caused 217 deaths; 700 000 people were evacuated; the river became up to 150 km wide (usual width 1 km); livestock and crops were lost; services were destroyed.
6 What attempts can be made to reduce the flood hazard?	Until the 1927 flood, the main policy was 'hold by levées' — by 1993, some levées were 15 m high (Figure 3.60). After 1927, new schemes included building dams and storage reservoirs (six huge dams and 105 on Missouri); afforestation to reduce/delay runoff; creating diversion spillways (e.g. Bonnet Carré floodway diverts flood water into Lake Pontchartrain and the sea); cutting through meanders to straighten and shorten the course (Figure 3.60).
7 How successful have the attempts to reduce flooding been?	In 1883, Mark Twain claimed that "You cannot tame that lawless stream". By 1973, it appeared that the river had been tamed: there was no further flooding . . . until 1993. Has human intervention made the danger worse? (page 84)

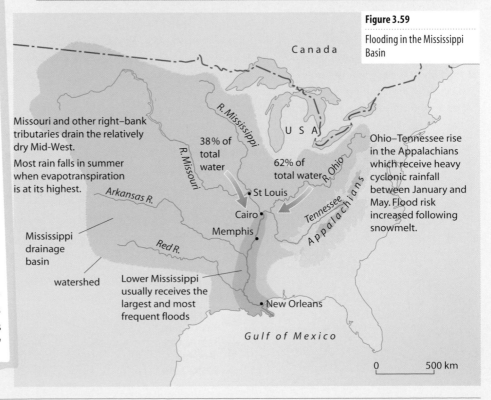

Figure 3.59

Flooding in the Mississippi Basin

Missouri and other right–bank tributaries drain the relatively dry Mid-West.

Most rain falls in summer when evapotranspiration is at its highest.

Canada

R. Mississippi

U S A

38% of total water

62% of total water

R. Ohio

St Louis

Cairo

Memphis

R. Missouri

Arkansas R.

Tennessee

Appalachians

Ohio–Tennessee rise in the Appalachians which receive heavy cyclonic rainfall between January and May. Flood risk increased following snowmelt.

Mississippi drainage basin

watershed

Red R.

Lower Mississippi usually receives the largest and most frequent floods

New Orleans

Gulf of Mexico

0   500 km

1 a) Height (metres) of levees at Memphis

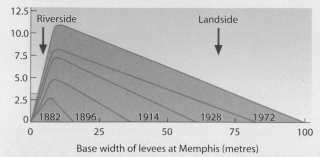

Base width of levees at Memphis (metres)

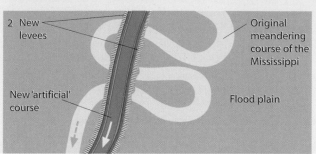

2 New levees — New 'artificial' course — Original meandering course of the Mississippi — Flood plain

b) The 1993 flood at St Louis

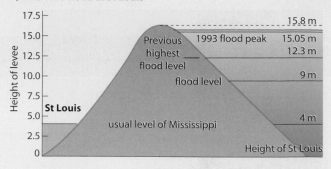

By making the course straighter and shorter, flood water could be removed from the river basin as quickly as possible. It was achieved by cutting through the narrow necks of large meanders. Between 1934 and 1945 one stretch of the river alone was reduced from 530 km to almost 230 km. By shortening the distance, the gradient and therefore the velocity, of the river increases. (But rivers try to create meanders rather than to flow naturally in straight courses.)

Figure 3.60

Two engineering schemes to try to control flooding

Engineering/planning schemes in the Mississippi basin

Prior to the 1993 flood, it was perceived that the flow of the Mississippi had been controlled. This had been achieved through a variety of flood prevention schemes (Figure 3.60).

- Levées had been heightened, in places to over 15 m, and strengthened. There were almost 3000 km of levées along the main river and its tributaries.
- By cutting through meanders, the Mississippi had been straightened and shortened: for 1750 km, it flows in artificial channels.
- Large spillways had been built to take excess water during times of flood.
- The flow of the major tributaries (Missouri, Ohio and Tennessee) had been controlled by a series of dams.

Why did the Mississippi flood in 1993?

The Mid-West was already having a wet year when record-setting spring and summer rains hit. The rain ran off the soggy ground and into rapidly rising rivers. Several parts of the central US had over 200 per cent more rain than was usual for the time of year (Figure 3.61). It was the ferocity, location and timing of the flood which took everyone by surprise. Normally, river levels are falling in mid-summer, the upper Mississippi was not perceived to be the major flood-risk area, and people believed that flooding in the basin had been controlled. Floodwater at St Louis reached an all-time high (Figure 3.61). Satellite photographs showed the extent of the flooding (Figure 3.62).

After the flood: should rivers run freer?

Since the first levée was built on the Mississippi in 1718, engineers have been channelling the river to protect farmland and towns from floodwaters. But have the levées, dams and diversion channels actually aggravated the flooding? There are two schools of thought. One advocates accepting that rivers are part of a complex ecological balance and that flooding should be allowed as a natural event (Figure 3.66). The other argues for better defences and a more effective control of rivers (Figure 3.67).

Figure 3.61

Extract from US Today, a daily newspaper

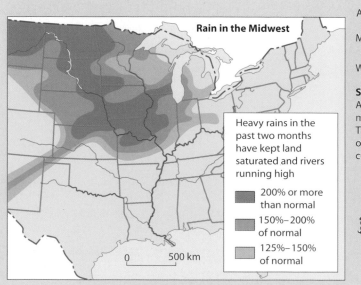

Rain in the Midwest

Heavy rains in the past two months have kept land saturated and rivers running high

- 200% or more than normal
- 150%–200% of normal
- 125%–150% of normal

0   500 km

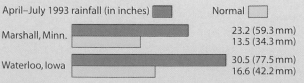

April–July 1993 rainfall (in inches) ▪   Normal ▫

Marshall, Minn.	23.2 (59.3 mm)
	13.5 (34.3 mm)
Waterloo, Iowa	30.5 (77.5 mm)
	16.6 (42.2 mm)

St Louis

Although there were some nervous moments, the city's massive 11–mile long, 52-foot floodwall protected the downtown from flooding. The river crested here August 1 at a record 49.4 feet, and the amount of water flowing past the Gateway Arch surpassed a record 1 million cubic feet per second.

flood stage: 30 feet

highest crest
Aug 1: 49.4

river level

bankfull level
9.1 m

Peak discharge
(15.05 m)

July 1   Aug 8

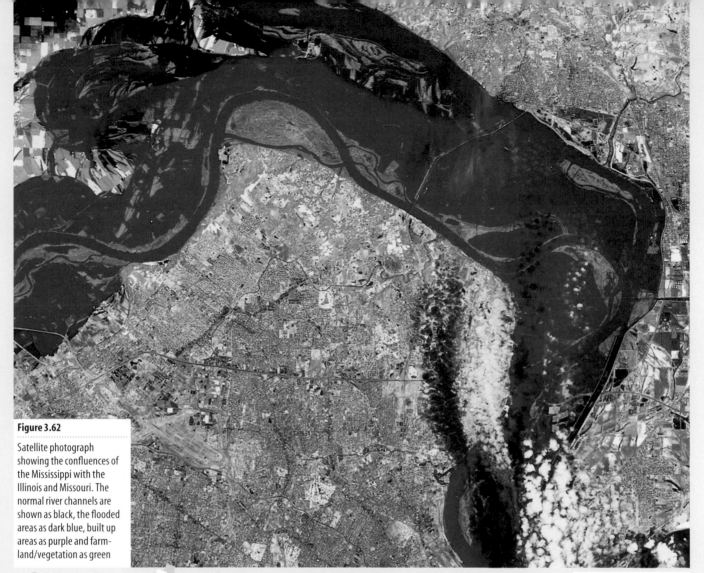

Figure 3.62

Satellite photograph showing the confluences of the Mississippi with the Illinois and Missouri. The normal river channels are shown as black, the flooded areas as dark blue, built up areas as purple and farmland/vegetation as green

Figure 3.63

The consequences of flooding in the St. Louis area

The consequences of flooding in the St Louis area (US Today 9 August 1993)

Flood of '93

Damage	Deaths	Evacuated	Houses	Crops
$10.5 billion	45	74 000	45 000	$6.5 billion

Nearly half of the counties in nine states bordering the upper reaches of the Mississippi and Missouri rivers have been declared federal disaster areas. This is the first step in becoming eligible for federal aid, including direct grants from Congress, Federal Emergency Management Agency and many other groups:

Declared disaster areas

Peak discharge: 26 June

Illinois: In the fight against flooding rivers, 17 levees were breached, including one that flooded the town of Valmeyer and 70 000 acres of surrounding farmland. One flood–related death was reported.

In Alton, the treatment plant was flooded Aug 1, cutting off water to the town's 33 000 residents. "Our levee did not breach, but the water came in through the street, the drains, anywhere there was a hole, at such a rate that pumps couldn't keep up," says Mayor Bob Towse.

Statewide property losses may top $365 million, including damage to 140 miles of roads and eight bridges. Agricultural damage is estimated at more than $610 million. An estimated 4% of the state's cropland–900 000 acres– was flooded. In addition, 15 727 people were displaced, 860 businesses closed and nearly 9000 jobs lost.

Missouri: The highest death toll–25–and the greatest property damage–$1.3 billion–of all flooded states were reported here. Statewide, 13 airports have been closed, and 25 000 residents evacuated. Flooding on 1.8 million acres of farmland has caused about $1.7 billion in crop losses.

Heroic efforts apparently saved historic Ste Genevieve, which has been battling rising waters since the start of July.

"On a floodplain unaltered by levees and asphalt, floodwaters can spread slowly, depositing rich silt, causing less erosion, and potenialy cresting at a lower levels. Undeveloped land absorbs some water and holds the excess until it can drain off." (National Geographic)

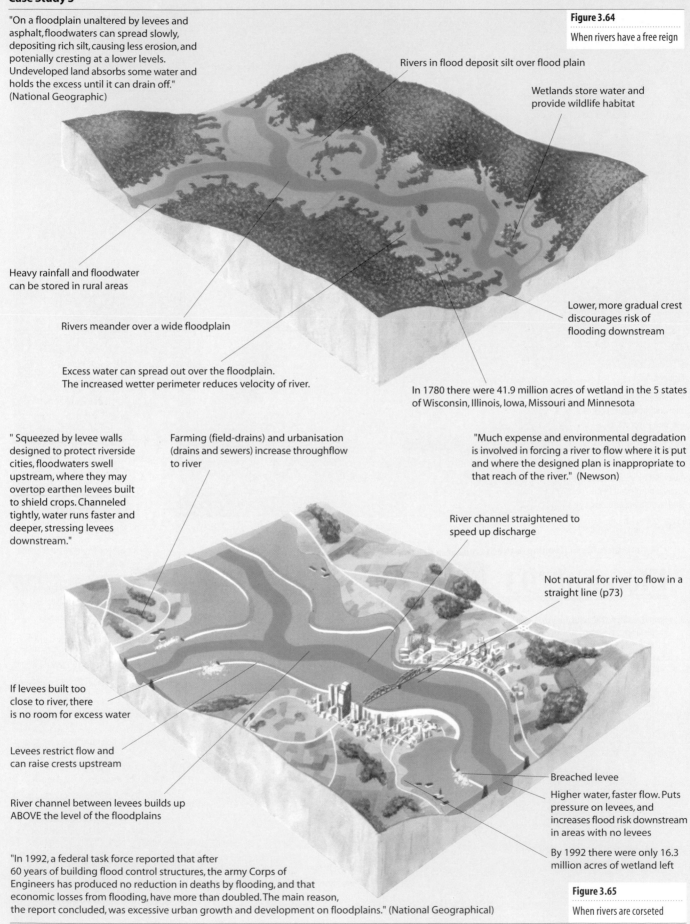

Figure 3.64

When rivers have a free reign

Rivers in flood deposit silt over flood plain

Wetlands store water and provide wildlife habitat

Heavy rainfall and floodwater can be stored in rural areas

Rivers meander over a wide floodplain

Excess water can spread out over the floodplain. The increased wetter perimeter reduces velocity of river.

Lower, more gradual crest discourages risk of flooding downstream

In 1780 there were 41.9 million acres of wetland in the 5 states of Wisconsin, Illinois, Iowa, Missouri and Minnesota

" Squeezed by levee walls designed to protect riverside cities, floodwaters swell upstream, where they may overtop earthen levees built to shield crops. Channeled tightly, water runs faster and deeper, stressing levees downstream."

Farming (field-drains) and urbanisation (drains and sewers) increase throughflow to river

"Much expense and environmental degradation is involved in forcing a river to flow where it is put and where the designed plan is inappropriate to that reach of the river." (Newson)

River channel straightened to speed up discharge

Not natural for river to flow in a straight line (p73)

If levees built too close to river, there is no room for excess water

Levees restrict flow and can raise crests upstream

River channel between levees builds up ABOVE the level of the floodplains

Breached levee

Higher water, faster flow. Puts pressure on levees, and increases flood risk downstream in areas with no levees

By 1992 there were only 16.3 million acres of wetland left

"In 1992, a federal task force reported that after 60 years of building flood control structures, the army Corps of Engineers has produced no reduction in deaths by flooding, and that economic losses from flooding, have more than doubled. The main reason, the report concluded, was excessive urban growth and development on floodplains." (National Geographical)

Figure 3.65

When rivers are corseted

Case Study 3

b. Managing a river: the Colorado

"The River Colorado (not the longest river in the U.S.) is the most dramatic, most used and closer than most basins in the U.S. to using the last drop of available water for man's needs" (National Academy of Sciences, 1968)

The Colorado is regarded by modern Americans as a symbol of the development and conflicts in the American West. The Colorado has been used as a source of water for over 2000 years by Native American peoples such as the Hohokam, Anasazi and Mogollon, who have long vanished from the region. During the last 100 years, it has changed from an uncontrolled river passing through a semi-desert, capturing the imagination of explorers such as John Wesley Powell, into one supplying the demands of 20 million people. Colorado water is now controlled by 11 major dams in 7 states and Mexico, produces 120 million kw of electricity and irrigates 800 000 hectares of farmland. Except for the occasional local flood flow, no water from the Colorado has reached the Gulf of California in the last 30 years.

Issues such as who has the right to the river water, who should control the river flow, how the water should be allocated, how it should be used, and who should pay for the schemes proposed, are of major importance to the states of the south-west. Also, in recent years, there has been an increasingly vocal conservation lobby questioning the construction of great dam projects.

What is the background to these issues?

1 The characteristics of the river

The river rises in the snowfields of the Rockies, where the annual precipitation may be as much as 2540 mm of snow during the winter — although the mountain states rarely receive more than 500 mm of rainfall a year. Temperatures throughout the Colorado basin are high during the summer, which means that there is a net loss of water through evaporation.

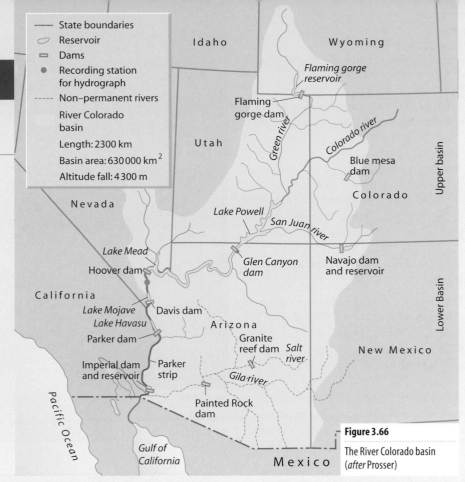

Key:
— State boundaries
⬭ Reservoir
▭ Dams
● Recording station for hydrograph
----- Non–permanent rivers
— River Colorado basin
Length: 2300 km
Basin area: 630 000 km^2
Altitude fall: 4 300 m

Figure 3.66

The River Colorado basin (*after* Prosser)

Evapotranspiration can be as high as 1830 mm per annum. The snow meltwater is essential to the maintenance of the flow of the river, which is at its highest in April and May. Annual flows vary between a recorded all-time low of 4.9 cu km (in 1917) and a maximum of 21.4 cu km (in 1977).

The lower basin had been liable to flood — with disastrous consequences in the period 1905-09, when the river burst its banks, flooding the Imperial Valley agricultural lands and re-creating the Salton Sea. Some of the highest recorded flows occurred in the El Nino floods of 1983, which affected the Parker valley region. This was in spite of the completion of the major dams upstream.

The river contains a high sediment load — Colorado means red colour — and it rarely has a clear appearance except in its mountain reaches. This means that it is unsuitable for public water supply in its natural state. The sediment has formed some of the richer alluvial soils on the Mexican border and in the delta area.

Figure 3.67

Population change in the south-west

2 Population growth in the south-west

In the past 50 years, there has been a marked change in population distribution in the US. The west and south-west have attracted people from the declining industrial areas of the north-east. The south-west is becoming more urban. California's population has increased rapidly, creating a huge demand for water. Seventy per cent of the water for San Diego is piped daily from the Colorado. 182 000 hectares of irrigated land in Imperial Valley use 20 per cent of California's allocation of water from the Colorado via the All American Canal. Arizona's towns, such as Phoenix and Tucson in the Gila river basin, have become new centres of development; their needs for water also have to be met.

	1981	1991	Percentage change
Arizona	2 718 000	3 811 000	+40
Colorado	2 890 000	3 302 000	+14.2
California	23 668 000	30 316 000	+28
Nevada	800 000	1 266 000	+58.2
New Mexico	1 303 000	1 529 000	+17.3
Utah	1 461 000	1 749 000	+19.7
Wyoming	470 000	439 000	-6.6

3 Other demands for water

In the upper-basin states of Wyoming, Colorado and Utah, the water is needed for cattle farming and for the irrigation of fodder crops, as well as for human consumption. East of the Great Divide, 283 500 ha of farmland are supplied with water via the Big Thompson and other diversion projects, which also help to meet the demands of the urban populations in Denver, Boulder and Colorado Springs.

Why is there a need to manage the basin?

The river flows through or between seven states and Mexico. The demands of the states are different, although the greatest perceived need is for hydro-electric power to supply the towns and farms of the region.

River water is used for irrigation throughout the basin. California was the first state to construct canals to carry irrigation water to the farms being developed in the Imperial Valley. This occurred in the late 19th century. At the same time, Mormon settlers moved into the remote areas along the river in Utah and into Arizona. They were farmers providing food for themselves and their small communities. In the upper basin, the cattle-ranchers of the interior of Wyoming took water from the river and its tributaries, the Green and the Grand Rivers. As farmers arrived from the east, there was conflict over who had use of the water. The question of water rights is complex, with prior appropriation (first come, first served) being normal.

The need for an agreement between the states to control the river was evident by the 1920s. Two years of discussion took place and the Colorado Compact was signed in 1922. States who first built a dam and provided irrigation water from this dam may not have their right to the water

Figure 3.68

Lake Powell behind Glen Canyon Dam

taken away by another state building upstream at a later date. This system favours California.

The Great Depression of the early 1930s followed the signing of the Colorado compact. Many farmers from the Mid-West moved towards the south-west, especially to California. An urgent need for irrigation water developed; but who should pay for it?

In an effort to provide help for the unemployed during the Great Depression, the government set up the Bureau of Reclamation which, together with the Corps of Army Engineers, was to plan and construct major engineering projects requiring federal (national) funds, throughout the west.

How is the river being managed?

There is no overall control of the river from its source to the Gulf of California.

Management is through a number of federal- and state-funded projects, some of which are multi-purpose.

The first of the large schemes was the Hoover Dam. Its primary purpose was to control the flow of the river, with the provision of hydro-electric power (HEP) being seen as a means of paying the huge costs of construction. This later changed as state and federal governments saw the benefits to be gained from the provision of cheap power in and to a relatively remote area. In the period after 1945 there was a rapid movement to develop the south-west. More great dams were planned on the river and their primary aim was to provide HEP, then flow control and irrigation.

Figure 3.69

Main projects on the Colorado River

1. Flood control.	2. Hydro-electricity	3. Irrigation	4. Urban water supply	5. Recreation
Dams: Fontanelle, Flaming Gorge, Glen Canyon, Hoover, Davis, Imperial	Fontanelle, Flaming Gorge, Glen Canyon, Power Hoover, Davis, Imperial	Hoover, Davis, Headgate, Paola Verde Imperial and Morelos Dams: water piped to farmers in Wyoming and Utah from the Green River and the Salt Valley. Central Arizona Project: water piped from the Colorado to the Gila basin. All-American Canal takes water to Imperial valley, California	Central Arizona Project; to Phoenix and Tucson. Big Thompson Project across the Great Divide to Denver, Colorado Springs and towns along the Rocky Mountain Front.	Glen Canyon: Lake Powell, boating and water sports — the largest water recreation area in the south-west. Lake Mead (Hoover Dam). Lake Havasu (Parker Dam). Rafting and boating along stretches of the river, including Grand Canyon.

Figure 3.70

The Colorado Compact allocation

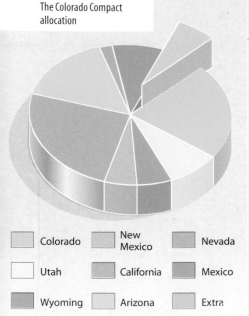

Colorado	New Mexico	Nevada
Utah	California	Mexico
Wyoming	Arizona	Extra

From its confluence with the Green River in southern Utah, the Colorado flows across deeply dissected, level, sandstone plateaux having cut a series of great gorges, or canyons, in their surfaces. The most famous of these is the Grand Canyon (Figure 7.19).

These canyons are valuable sites for storage dams. Glen Canyon Dam is 178 m high and 475 m long, and stores water in Lake Powell which extends for 290 km to the north-east (Figure 3.68). Hoover Dam, 600 km to the west, was built in Black Canyon.

It is 3.5 m taller than Glen Canyon Dam and holds more water — because of the depth of the canyon.

Environmental and conservationist movements have caused a further change in attitude to the control of the river. They believe in the preservation of the wilderness in its original state, objecting to intrusion into National Parks and Reservations. There has been successful opposition to the construction of dams in the Grand Canyon National Park and at Echo Park in the Dinosaur National Monument. The Navajo Indians have the right to the water within their tribal territories alongside the river. Many traditional Native American sites close to the river, including Rainbow Arch in Glen Canyon (Figure 3.71), are in danger from changing levels in Lake Powell. This remote area was one which the conservationists unsuccessfully tried to prevent from being flooded to form Lake Powell in 1962.

South of Hoover Dam, the river flows across a lower and wider valley. Rainfall and runoff in this region are low.

Figure 3.71

Rainbow Arch

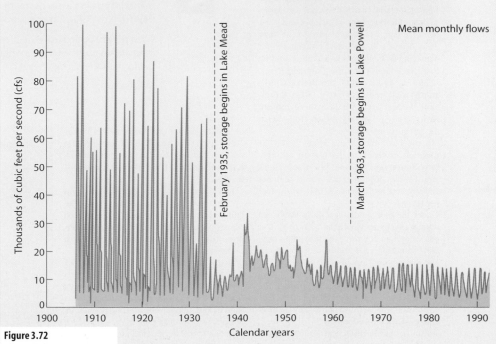

Figure 3.72

Colorado hydrograph: flow below Hoover Dam site showing changes as a result of control of flows by dams

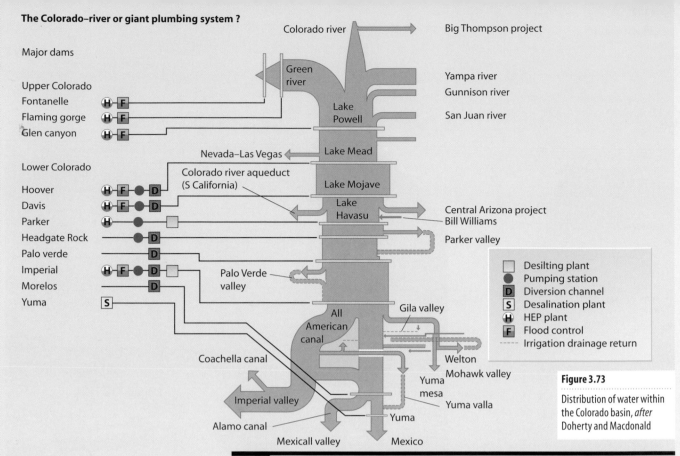

The Colorado–river or giant plumbing system ?

Major dams

Upper Colorado
Fontanelle   (H)–[F]
Flaming gorge   (H)–[F]
Glen canyon   (H)–[F]

Colorado river   Big Thompson project

Green river

Yampa river
Gunnison river
San Juan river

Lake Powell

Lake Mead   Nevada–Las Vegas

Lower Colorado

Hoover   (H)–[F]–●–[D]
Davis   (H)–[F]–●–[D]
Parker   (H)–●–[ ]
Headgate Rock   ●–[D]
Palo verde   [D]
Imperial   (H)–[F]–●–[D]–[ ]
Morelos   [D]
Yuma   [S]

Colorado river aqueduct (S California)

Lake Mojave

Lake Havasu

Central Arizona project
Bill Williams

Parker valley

Palo Verde valley

All American canal

Gila valley

Coachella canal

Welton
Mohawk valley
Yuma mesa
Yuma valla
Yuma

Imperial valley

Alamo canal

Mexicall valley   Mexico

	Legend
[ ]	Desilting plant
●	Pumping station
[D]	Diversion channel
[S]	Desalination plant
(H)	HEP plant
[F]	Flood control
-----	Irrigation drainage return

Figure 3.73

Distribution of water within the Colorado basin, *after* Doherty and Macdonald

---

Q

1 Identify the major differences in usage of water between the upper and lower basins.

2 Which sections of the basin would be most likely to suffer shortages during periods of drought?

3 What could be the impacts of the 'Irrigation drainage returns' to the main river?

Q

1 Why should the environmentalists be in favour of steady flow from the Glen Canyon Dam?

2 What may happen to the lake level and to power supplies when California suffers a heat wave?

3 Changing lake levels have an impact on recreation uses both at Lake Powell and in Grand Canyon. Explain this.

Figure 3.74

Extract from the Arizona Republic, June 1992

Flow limits imposed at Glen Canyon Dam

Something new will roll through the gates of the glen Canyon Dam beginning today: a relatively steady flow of water.

Wild fluctuations in the Colorado River will end as the US Bureau of Reclamation places interim restrictions on the dam's hydro–electric plant. The ebb and flow from the dam has been blamed for washing away beaches in the Grand Canyon and for damaging fish populations and other wildlife habitats in and along the Colorado River.

'This is good news,' the Sierra Club's south–west representative, Rob Smith, said Wednesday. 'We like these interim flows. They're largely drawn from what the scientists recommended.'

In the past, the river's water level has dropped as much as 13 feet in a single day, Smith said. In one day, water flows could be as low as 1000 cubic feet per second and as high as 31500 cubic feet per second.

Under the new guidelines, daily flows can drop or rise by no more than 8000 cubic feet per second on any day. The maximum flow will be limited to 20000 cubic feet per second, and a minimum flow of 5000 cubic feet per second must be maintained.

'We're taking this step to protect the environment. That's the important thing,' Reclamation Commissioner Dennis Underwood said Wednesday.

Although environmentalists applauded the bureau's move, operators of thepower plant were not pleased.

A spokesman for the Western Area Power Administration, the federal agency that sells the dam's power, said the interim flows will seriously affect the operation of the dam.

'The interim flows lop off about 30 per cent of our generating capacity,' said the administration's deputy area director, Ken Maxey of Salt Lake City.The new regulations may cost utilities $23 million in extra operating costs a year, he added.

It is unclear how much of the cost increases will be passed along to Arizona consumers. The Salt River Project receives some power from the Glen Canyon Dam, but much of it is traded, not purchased, and would not be affected a spokesman said.

Arizona Public Service Co. does not receive any of the dam's power. The bureau will test the new guidelines for 90 days before enacting

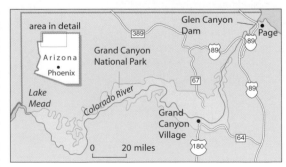

them as a long term measure. The stop gap measures will remain in place until a full–scale environmental–impact study is completed in 1993.

(The Arizona Republic, June 1992)

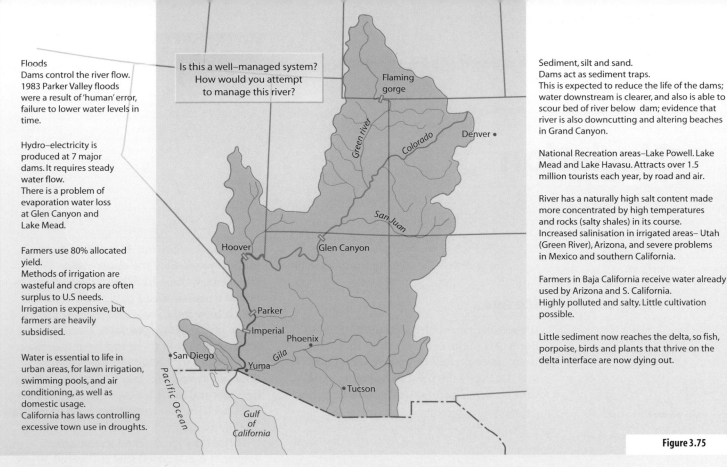

Floods
Dams control the river flow.
1983 Parker Valley floods
were a result of 'human' error,
failure to lower water levels in
time.

Hydro–electricity is
produced at 7 major
dams. It requires steady
water flow.
There is a problem of
evaporation water loss
at Glen Canyon and
Lake Mead.

Farmers use 80% allocated
yield.
Methods of irrigation are
wasteful and crops are often
surplus to U.S needs.
Irrigation is expensive, but
farmers are heavily
subsidised.

Water is essential to life in
urban areas, for lawn irrigation,
swimming pools, and air
conditioning, as well as
domestic usage.
California has laws controlling
excessive town use in droughts.

Is this a well–managed system?
How would you attempt
to manage this river?

Sediment, silt and sand.
Dams act as sediment traps.
This is expected to reduce the life of the dams;
water downstream is clearer, and also is able to
scour bed of river below dam; evidence that
river is also downcutting and altering beaches
in Grand Canyon.

National Recreation areas–Lake Powell. Lake
Mead and Lake Havasu. Attracts over 1.5
million tourists each year, by road and air.

River has a naturally high salt content made
more concentrated by high temperatures
and rocks (salty shales) in its course.
Increased salinisation in irrigated areas– Utah
(Green River), Arizona, and severe problems
in Mexico and southern California.

Farmers in Baja California receive water already
used by Arizona and S. California.
Highly polluted and salty. Little cultivation
possible.

Little sediment now reaches the delta, so fish,
porpoise, birds and plants that thrive on the
delta interface are now dying out.

Figure 3.75

How effective is the managment of the Colorado?

Baja California

"Fifty miles south of the U.S. Border in Mexico's Baja California, the great river of the West that I had followed from beginning to end was gone, the water in its bed a shallow, narrow sump of salt and pesticide-laced run-off from crop irrigation." (Jim Carrier, National Geographic, June 1991, p4)

The issue of water supply to the farmers of Baja California was partly resolved after the construction of the Yuma and Morelos Dams. A total of 1848 million cum should be supplied through the Compact. Irrigation water from districts close to the Mexican border is returned to the river, but with a high saline content so that farmers in Mexico are disadvantaged. An expensive new desalinisation plant has been constructed at Yuma by the US. Insufficient water of good quality is forcing Mexican farmers to leave 20 per cent of their land uncultivated.

A further issue is the Californian plan to line the All-American Canal carrying water to the Imperial Valley. At present, this loses 130.5 million cum a year, much of which recharges Mexican aquifers. The Mexican authorities consider that the underground water near Mexicali will be depleted, again affecting the farming community, if the canal is sealed.

The delta

The Colorado delta is now a wasteland. Until the river flow was controlled, it was home to birds such as the egret; to wild cats, especially the jaguar; to marine species, such as the totoaba, a large fish, and porpoises; and to marine grasses which used to thrive on the interface between land and sea. There are now few communities attempting to make a life in this area.

Tasks:

1 The US government has decided to appoint a Commission to report on future management of the Colorado basin. Prepare a report to the Commission from the standpoint of an environmentalist on the need for greater control over development of the river.

2 Annotate a map of the Colorado basin to show the main management issues in one colour and the solutions (or otherwise) in a second colour.

References

Gregory, K. J. and Walling, D. E. (1973) *Drainage Basin Form and Process*. E. J. Arnold.

Hacking, E. (1992) Investigating river flooding, *Geography Review*, 5(5), p. 19.

McCullagh, P. (1978) *Modern Concepts in Geomorphology*. Oxford University Press.

Newson, M. D. (1975) *Flooding and Flood Hazard in the UK*. Oxford University Press.

Newson, M. (1994) *Hydrology and the River Environment*. Oxford University Press.

Planet Earth: Floods. (1983) Time–Life Books.

Planet Earth: Rivers and Lakes. (1985) Time–Life Books.

Prosser, R. (1992) *Natural Systems and Human Responses*. Thomas Nelson.

Small, J. and Witherick, M. (1989) *A Modern Dictionary of Geography*. E.J. Arnold.

Weyman, D. (1973) *Runoff Processes and Streamflow Modelling*. Oxford University Press.

Wilson, J. (1984) *Statistics in Geography for 'A' level Students*. Schofield & Sims.

Chapter 4

Glaciation

"Great God! this is an awful place."
The South Pole, Robert Falcon Scott, *Journal,* **1912**

Ice ages

It appears that roughly every 200–250 million years in the earth's history there have been major periods of ice activity (Figure 4.1). Of these, the most recent and significant occurred during the Pleistocene period of the Quaternary era (Figure 1.1). In the 2 million years since the onset of the Quaternary, there have been fluctuations in global temperature of between 5° and 6°C which have led to cold phases (**glacials**) and warm phases (**interglacials**). Until recently it was believed that there had been four main glacials, but in the last decade evidence obtained from cores taken from ocean floor deposits and polar ice caps has led glaciologists to claim that there have been more than 20 (Figure 4.2).

When the ice reached its maximum extent, it is estimated that it covered 30 per cent of the earth's land surface (compared with some 10 per cent today). However, its effect was felt not only in polar latitudes and mountainous areas, for each time the ice advanced there was a change in the global climatic belts (Figure 4.3). Only 18 000 years ago, at the time of the maximum advance within the last glacial, ice covered Britain as far as the Bristol Channel, the Midlands and Norfolk. The southern part of Britain experienced **tundra** conditions (pages 309 – 10), as did most of France.

Climatic change

Although it is accepted that climatic fluctuations occur on a variety of timescales, as yet there is no single explanation for the onset of major ice ages or for fluctuations within each ice age. The most feasible of theories to date is that of Milutin Milankovitch, mathematician/astronomer. Between 1912 and 1941, he performed exhaustive calculations which show that the earth's position in space, its tilt and its orbit around the sun all change. These changes, he claimed, affect incoming radiation from the sun and produce three main cycles of 95, 42 and 21 thousand years (Figure 4.5). His theory, and the timescale of each cycle, has been given considerable support by evidence gained, since the mid 1970s, from

Figure 4.1

A chronology of ice ages (in bold)

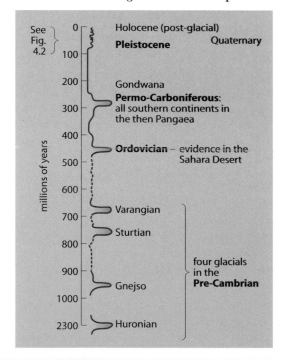

Figure 4.2

Generalised trends in mean global temperatures during the past one million years

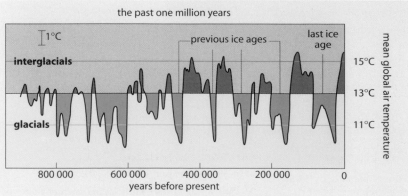

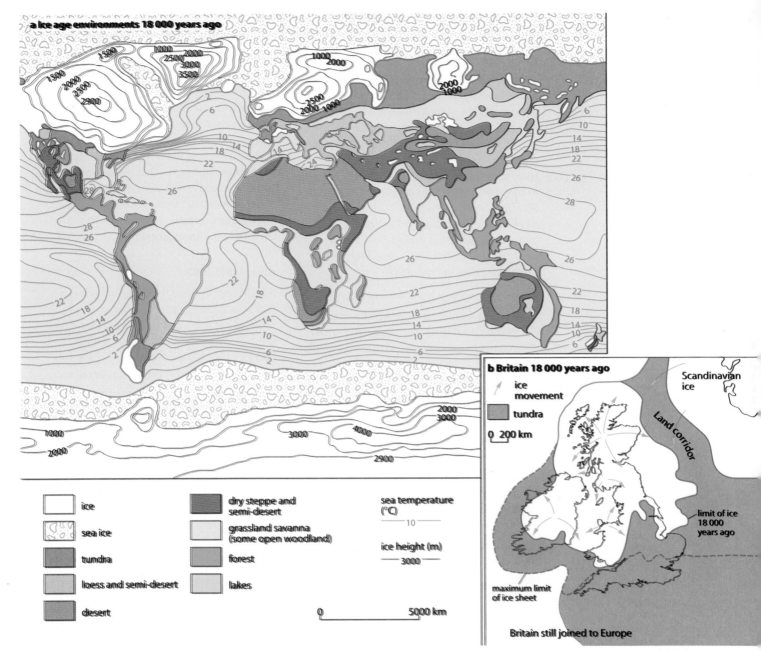

a Ice age environments 18 000 years ago

b Britain 18 000 years ago

ice movement

tundra

0 200 km

Scandinavian ice

Land corridor

limit of ice 18 000 years ago

maximum limit of ice sheet

Britain still joined to Europe

Legend:

- ice
- sea ice
- tundra
- loess and semi-desert
- desert
- dry steppe and semi-desert
- grassland savanna (some open woodland)
- forest
- lakes

sea temperature (°C)
— 10 —

ice height (m)
— 3000 —

0   5000 km

Figure 4.3

World climates and vegetation 18 000 years ago (after CLIMAP)

ocean floor cores. As yet, although the relationship appears to have been established, it is not known precisely how these celestial cycles relate to climatic change.

Other suggestions have been made as to the causes of ice ages, and it is possible that one or more of these may, at some future date, be shown to be significant (Places 11):

- Variations in sunspot activity may increase or decrease the amount of radiation received by the earth.
- Injections of volcanic dust into the atmosphere can reflect and absorb radiation from the sun (page 189 and Figure 1.41).
- Changes in atmospheric carbon dioxide

gas could accentuate the greenhouse effect (page 236). Initially extra CO_2 traps heat in the atmosphere, possibly raising world temperatures by an estimated 3°C. In time, some of this CO_2 will be absorbed by the seas, reducing the amount remaining in the atmosphere and causing a drop in world temperatures and the onset of another ice age.

- The movement of plates — either into colder latitudes or at constructive margins, where there is an increase in altitude—could lead to an overall drop in world land temperatures.
- Changes in ocean currents (page 193) or jet streams (page 211).

Antarctica, 1988

The first results of a 5-year drilling experiment by the Russians in Antarctica were announced in January 1988. It took from 1980 to 1985 to drill and carefully to extract an ice core extending downwards for 2 km. By dating the rings, which show the annual accumulation of snow (Figure 4.4), it has been estimated that ice from the bottom of the core is 160 000 years old. This should provide information about the whole of the last interglacial and glacial. The researchers hope that eventually they will reach ice 500 000 years old. Three of several experiments being undertaken are:

1 measuring dust from volcanic activity;

2 extracting bubbles of air which will contain carbon dioxide to see if the levels of that gas have changed over periods of time; and

3 comparing ratios between two types of hydrogen which may help to indicate past temperatures.

Early results seem to indicate that:

- there have been periods of considerable volcanic activity;

- the earth may well wobble on its axis causing the 21 000 year cycle suggested by Milankovitch;

- there have been earlier periods when CO_2 built up in the atmosphere with no plausible suggestions as to why this should occur. It does appear though that the last glacial began at a time when the content of CO_2 was very low.

Evidence seems to support the view that the earth is nearing the end of a brief interglacial and that, following a period of colder, drier climate (which would seriously affect food crop production), the next glacial is likely to start within the next few thousands years. This might be temporarily delayed by the increase in temperatures resulting from the extra CO_2 content of the atmosphere.

Greenland, 1992

A 3-km core through the Greenland ice cap dates back 250 000 years. During that period, snowfall has averaged 15–20 cm a year. At the same time as the snow was being compressed into ice (page 93), dust, gases and chemicals, which were present in the atmosphere, were trapped in the ice. The gases included two types of oxygen isotope, 016 and 018. The ratio between these two isotopes changes as the proportion of global water bound up in the ice changes. (The amount of 018 in the atmosphere increases as air temperatures fall and decreases as temperatures rise.) The changing ratios obtained from the Greenland core show, therefore, both short-term and long-term temperature changes over the last quarter of a million years.

Figure 4.4

Dirt bands (englacial débris) in an Icelandic glacier: the amount of ice between each dirt band represents one year's accumulation of snow

a the 95 000 year stretch

The earth's orbit stretches from being nearly circular to an elliptical shape and back again in a cycle of about 95 000 years. During the Quaternary, the major glacial–interglacial cycle was almost 100 000 years. Glacials occur when the orbit is almost circular and interglacials when it is a more elliptical shape.

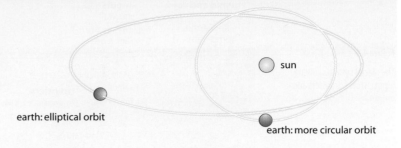

b the 42 000 year tilt

Although the tropics are set at 23.5°N and 23.5°S to equate with the angle of the earth's tilt, in reality the earth's axis varies from its plane of orbit by between 21.5° and 24.5°. When the tilt increases, summers will become hotter and winters colder, leading to conditions favouring interglacials.

$a = 21.5°$
$b = 24.5°$

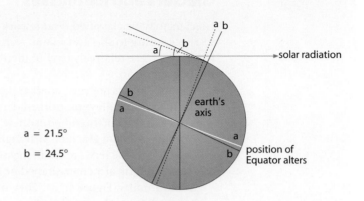

c the 21 000 year wobble

As the earth slowly wobbles in space, its axis describes a circle once in every 21 000 years.
1 At present, the orbit places the earth closest to the sun in the northern hemisphere's winter and furthest away in summer. This tends to make winters mild and summers cool. These are ideal conditions for glacials to develop.
2 The position was in reverse 12 000 years ago, and this has contributed to our present warm 'interglacial'.

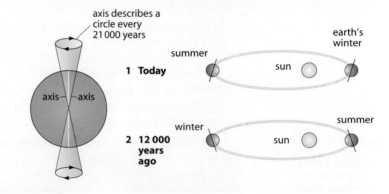

Figure 4.5

Milankovich's theories on climatic change

Snow accumulation and ice formation

As the climate gets colder, more precipitation is likely to be in the form of snow in winter and there is less time for that snow to melt in the shorter summer. If the climate continues to deteriorate, snow will lie throughout the year forming a permanent **snow line**—the level above which snow will lie all year. In the northern hemisphere, the snow line is at a lower altitude on north-facing slopes, as these receive less insolation than south-facing slopes. The snow line is also lower nearer to the poles and higher nearer to the Equator: it is at sea level in northern Greenland; at about 1500 m in southern Norway; at 3000 m in the Alps; and at 6000 m at the Equator. It is estimated that the Cairngorms in Scotland would be snow-covered all year had they been 200 m higher.

When snowflakes fall they have an open feathery appearance, trap air and have a low density. Where snow collects in hollows, it becomes compressed by the weight of subsequent falls and gradually develops into a more compact, dense form called **firn** or **névé**. Firn is compacted snow which has experienced one winter's freezing and survived a summer's melting. It is composed of randomly orientated ice crystals separated by air passages. In temperate latitudes, such as in the Alps, summer meltwater percolates into the firn only to freeze either at night or during the following winter, thus forming an increasingly dense mass. Air is progressively squeezed out and after 20–40 years the firn will have turned into solid ice. This same process may take several hundred years in Antarctica and Greenland where there is no summer melting. Once ice has formed, it may begin to flow downhill, under the force of gravity, as a **glacier**.

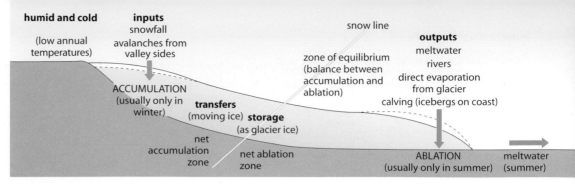

Glaciers and ice masses

Glaciers may be classified (Framework 5, page 151) according to size and shape — characteristics which are relatively easy to identify by field observation.

1 **Niche glaciers** are very small and occupy hollows and gulleys on north-facing slopes in the northern hemisphere.

2 **Corrie** or **cirque glaciers**, although larger than niche glaciers, are small masses of ice occupying armchair-shaped hollows in mountains (Figure 4.13). They often overspill from their hollows to feed valley glaciers.

3 **Valley glaciers** are larger masses of ice which move down from either an ice-field or a cirque basin source (Figure 4.7). They usually follow former river courses and are bounded by steep sides.

4 **Piedmont glaciers** are formed when valley glaciers extend onto lowland areas, spread out and merge (Figure 4.27).

5 **Ice caps** and **ice sheets** are huge areas of ice which spread outwards from central domes. Apart from exposed summits of high mountains, called **nunataks**, the whole landscape is buried. Ice sheets, which once covered much of northern Europe and North America (Figure 4.3) are now confined to Antarctica (86 per cent of present-day world ice) and Greenland (11 per cent).

Glacial systems and budgets

A glacier behaves as a system (Framework 1, page 39), with inputs, stores, transfers and outputs (Figure 4.6). Inputs are derived from snow falling directly onto the glacier or from **avalanches** along valley sides (Case Study 4). The glacier itself is water in storage. Outputs from the glacier system include evaporation, **calving** (the formation of icebergs), and meltwater streams which flow either on top of or under the ice during the summer months.

The upper part of the glacier, where inputs exceed outputs, is known as the **zone of accumulation**; while the lower part, where outputs exceed inputs, is called the **zone of ablation**. The **zone of equilibrium** is where the rates of accumulation and ablation are equal and it corresponds with the snow line (Figures 4.6 and 4.7).

The **glacier budget**, or **net balance**, is the difference between the total accumulation and the total ablation for one year. In temperate glaciers (page 96), there is likely to be a negative balance in summer when ablation exceeds accumulation, and a positive balance in winter when the reverse occurs (Figure 4.8). If the summer and winter budgets cancel each other out, the glacier appears to be stationary. It appears stationary because the **snout** — i.e. the end of the glacier — is neither advancing nor retreating, although ice from the accumulation zone is still moving down-valley into the ablation zone. Glaciers are sensitive indicators of climatic change, both short-term and long-term.

Figure 4.8

The glacial budget or net balance

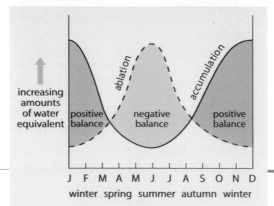

Refer to Figures 4.6, 4.7 and 4.8.

1 How does the glacier operate as a system?

2 What is meant by the following terms:
   a zone of accumulation;
   b zone of equilibrium;
   c zone of ablation?

3 Describe the shape of the two curves in Figure 4.8. Give reasons for the differences between them.

4 What is meant by the following terms:
   a positive balance in winter;
   b negative balance in summer;
   c annual balance (or glacier budget)?

5 Under what conditions might a glacier a advance; b retreat?

6 How might the following affect the net balance of a glacier:
   a a decrease in mean annual temperature;
   b a decrease in mean annual precipitation?

Figure 4.9

Processes of glacier movement

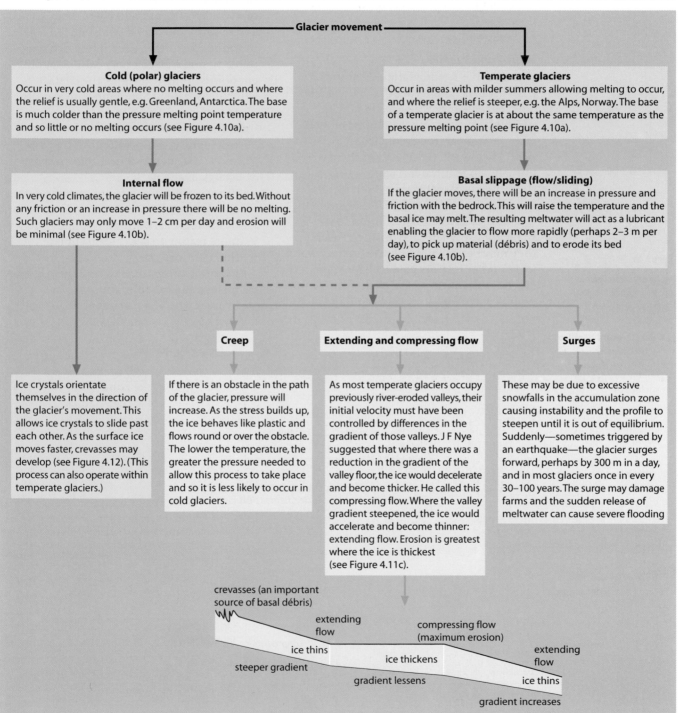

Cold (polar) glaciers
Occur in very cold areas where no melting occurs and where the relief is usually gentle, e.g. Greenland, Antarctica. The base is much colder than the pressure melting point temperature and so little or no melting occurs (see Figure 4.10a).

Temperate glaciers
Occur in areas with milder summers allowing melting to occur, and where the relief is steeper, e.g. the Alps, Norway. The base of a temperate glacier is at about the same temperature as the pressure melting point (see Figure 4.10a).

Internal flow
In very cold climates, the glacier will be frozen to its bed. Without any friction or an increase in pressure there will be no melting. Such glaciers may only move 1–2 cm per day and erosion will be minimal (see Figure 4.10b).

Basal slippage (flow/sliding)
If the glacier moves, there will be an increase in pressure and friction with the bedrock. This will raise the temperature and the basal ice may melt. The resulting meltwater will act as a lubricant enabling the glacier to flow more rapidly (perhaps 2–3 m per day), to pick up material (débris) and to erode its bed (see Figure 4.10b).

Creep

Extending and compressing flow

Surges

Ice crystals orientate themselves in the direction of the glacier's movement. This allows ice crystals to slide past each other. As the surface ice moves faster, crevasses may develop (see Figure 4.12). (This process can also operate within temperate glaciers.)

If there is an obstacle in the path of the glacier, pressure will increase. As the stress builds up, the ice behaves like plastic and flows round or over the obstacle. The lower the temperature, the greater the pressure needed to allow this process to take place and so it is less likely to occur in cold glaciers.

As most temperate glaciers occupy previously river-eroded valleys, their initial velocity must have been controlled by differences in the gradient of those valleys. J F Nye suggested that where there was a reduction in the gradient of the valley floor, the ice would decelerate and become thicker. He called this compressing flow. Where the valley gradient steepened, the ice would accelerate and become thinner: extending flow. Erosion is greatest where the ice is thickest (see Figure 4.11c).

These may be due to excessive snowfalls in the accumulation zone causing instability and the profile to steepen until it is out of equilibrium. Suddenly—sometimes triggered by an earthquake—the glacier surges forward, perhaps by 300 m in a day, and in most glaciers once in every 30–100 years. The surge may damage farms and the sudden release of meltwater can cause severe flooding

crevasses (an important source of basal débris)

extending flow

compressing flow (maximum erosion)

ice thins

ice thickens

extending flow

steeper gradient

gradient lessens

ice thins

gradient increases

Glacier movement and temperature

The character and movement of ice depend upon whether it is warm or cold, which in turn depends upon the **pressure melting point (PMP)**. The pressure melting point is the temperature at which ice is on the verge of melting. A small increase in pressure can therefore cause melting. PMP is normally 0°C on the surface of a glacier, but it can be lower within a glacier (due to an increase in pressure caused by either the weight or the movement of ice). In other words, as pressure increases, then the freezing point for water falls below 0°C.

Warm and cold ice

Warm ice has a temperature of around 0°C (PMP) throughout its depth (Figure 4.10a) and consequently is able, especially in summer, to release large amounts of melt-water. Temperatures in cold ice are permanently below 0°C (PMP) and so there is virtually no meltwater (Figure 4.10a). It is the presence of meltwater which facilitates the movement of a glacier. Temperature is therefore an alternative criterion to size or shape for use when categorising glaciers — they may be either **temperate** (mainly warm ice) or **polar** (mainly cold ice). Movement is much faster in temperate glaciers where the presence of meltwater acts as a lubricant and

reduces friction (Figure 4.10b). It can take place by one of four processes of **basal flow** (or **slipping**): **creep; extending–compressing flow;** and **surges** (Figure 4.9). Polar glaciers move less quickly as, without the presence of melt-water, they tend to be frozen to their beds. The main process here is **internal flow**, although creep and extending–compressing flow may also occur.

Both types of glacier move more rapidly on the surface and away from their valley sides (Figure 4.11a), but it is the temperate one which is the more likely to erode its bed and to carry and deposit most material as **moraine** (page 105). Recent research suggests that any single glacier may exhibit, at different points along its profile, the characteristics of both polar and temperate glaciers.

Movement is greatest:
- at the point of equilibrium — as this is where the greatest volume of ice passes and consequently where there is most energy available;
- in areas with high precipitation and ablation;
- in small glaciers, which respond more readily to short-term climatic fluctuations;
- in temperate glaciers, where there is more meltwater available; and
- in areas with steep gradients.

Figure 4.10

Comparison of temperature and velocity profiles in polar and temperate glaciers

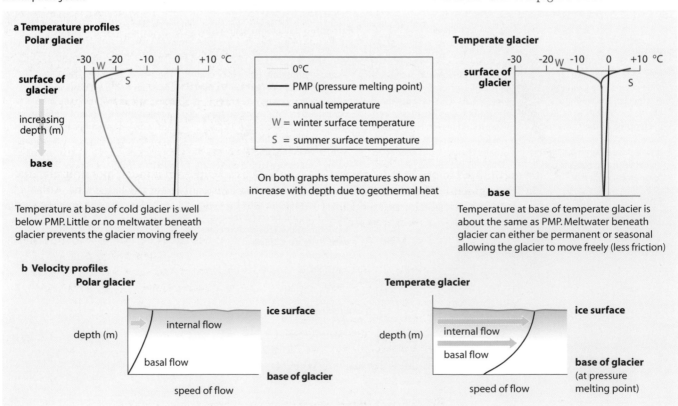

a Temperature profiles

Polar glacier

surface of glacier

increasing depth (m)

base

Temperature at base of cold glacier is well below PMP. Little or no meltwater beneath glacier prevents the glacier moving freely

—— 0°C
—— PMP (pressure melting point)
—— annual temperature
W = winter surface temperature
S = summer surface temperature

On both graphs temperatures show an increase with depth due to geothermal heat

Temperate glacier

surface of glacier

base

Temperature at base of temperate glacier is about the same as PMP. Meltwater beneath glacier can either be permanent or seasonal allowing the glacier to move freely (less friction)

b Velocity profiles

Polar glacier

depth (m)

ice surface
internal flow
basal flow
base of glacier

speed of flow

Temperate glacier

depth (m)

ice surface
internal flow
basal flow
base of glacier (at pressure melting point)

speed of flow

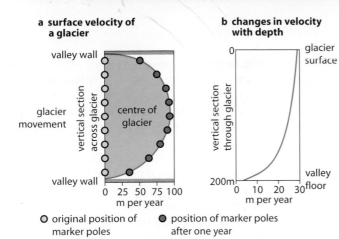

a surface velocity of a glacier

valley wall

glacier movement

vertical section across glacier

centre of glacier

valley wall

0 25 50 75 100
m per year

○ original position of marker poles

● position of marker poles after one year

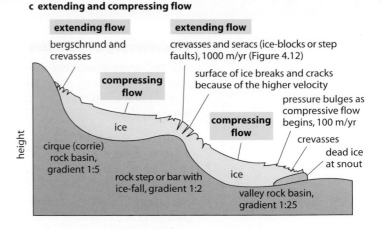

b changes in velocity with depth

0 — glacier surface

vertical section through glacier

200m — valley floor

0 10 20 30
m per year

c extending and compressing flow

extending flow
bergschrund and crevasses

compressing flow

extending flow
crevasses and seracs (ice-blocks or step faults), 1000 m/yr (Figure 4.12)

surface of ice breaks and cracks because of the higher velocity

compressing flow

pressure bulges as compressive flow begins, 100 m/yr

crevasses
dead ice at snout

height

ice

cirque (corrie) rock basin, gradient 1:5

rock step or bar with ice-fall, gradient 1:2

ice

valley rock basin, gradient 1:25

ice

Figure 4.11

Velocity and flow in a glacier

Q

1 a What is the pressure melting point?

   b Why do temperate glaciers move more quickly than cold glaciers?

   c Describe the main processes which enable/cause glaciers to move.

2 Describe, with reasons, where velocity is greatest in Figure 4.11a and 4.11b.

3 With reference to Figure 4.11c, account for:
   a the location of the areas of extending and compressing flow;
   b the formation of crevasses and seracs.

Figure 4.12

Crevasses on an icefall, Skafta Glacier, Iceland

Transportation by ice

Glaciers are capable of moving large quantities of débris. This rock débris may be transported in one of three ways:

1 **Supraglacial débris** is carried on the surface of the glacier as lateral and medial moraine (page 105). In summer, the relatively small load carried by surface melt-water streams often disappears down crevasses.

2 **Englacial débris** is material carried within the body of the glacier. It may once have been on the surface, only to be buried by later snowfalls.

3 **Subglacial débris** is moved along the floor of the valley either by the ice or by meltwater streams formed by pressure melting.

Glacial erosion

Ice that is stationary or contains little débris has limited erosive power, whereas moving ice carrying with it much morainic material can drastically alter the landscape. Although ice lacks the turbulence and velocity of water in a river, it has the 'advantage' of being able to melt and to refreeze in order to overcome obstacles in its path. Virtually all the glacial processes of erosion are physical as the climate tends to be too cold for chemical reactions to operate (Figure 2.9b).

Processes of glacial erosion

The processes associated with glacial erosion are: frost shattering, abrasion, plucking, rotational movement, and extending and compressing flow.

Frost shattering

This process (page 34) produces much loose material which may fall from the valley sides onto the edges of the glacier to form **lateral moraine**, be covered by later snowfall, or plunge down crevasses to be transported as **englacial débris**. Some of this material may be added to rock loosened by frost action as the climate deteriorated (but before glaciers formed) to form **basal débris**.

Abrasion

This is the sandpapering effect of angular material embedded in the glacier as it rubs against the valley sides and floor. It usually produces smoothened, gently sloping landforms.

Plucking

At its simplest, this process involves the glacier freezing onto rock outcrops, after which ice movement pulls away masses of rock. In reality, as the strength of the bedrock is greater than that of the ice, it would seem that only previously loosened material can be removed. Material may be continually loosened by one of three processes:

1 the relationship between local pressure and temperature (the PMP) produces sufficient meltwater for freeze–thaw activity to break up the ice-contact rock;
2 water flowing down a **bergschrund** (Figure 4.13b) or smaller crevasses will later freeze onto rock surfaces; and
3 removal of layers of bedrock by the glacier causes a release in pressure and an enlarging of joints in the underlying rocks (pressure release, page 35).

Plucking generally creates a jagged-featured landscape.

Bergschrund: a large, crevasse-like feature found near the head of a glacier

Figure 4.13

Processes in the formation of a cirque

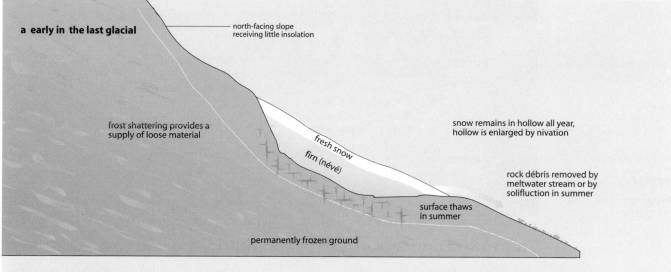

a early in the last glacial

north-facing slope receiving little insolation

frost shattering provides a supply of loose material

fresh snow
firn (névé)

snow remains in hollow all year, hollow is enlarged by nivation

rock débris removed by meltwater stream or by solifluction in summer

surface thaws in summer

permanently frozen ground

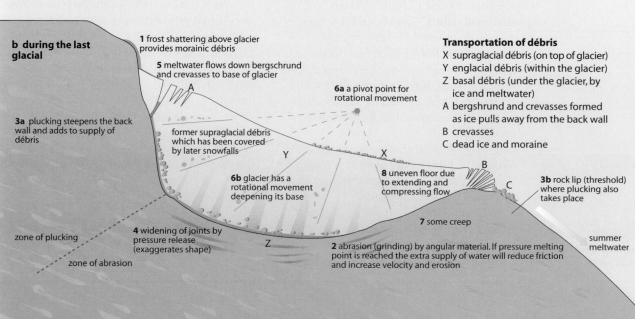

b during the last glacial

1 frost shattering above glacier provides morainic débris

5 meltwater flows down bergschrund and crevasses to base of glacier

A

6a a pivot point for rotational movement

3a plucking steepens the back wall and adds to supply of débris

former supraglacial débris which has been covered by later snowfalls

Y

X

6b glacier has a rotational movement deepening its base

8 uneven floor due to extending and compressing flow

B

C

3b rock lip (threshold) where plucking also takes place

7 some creep

zone of plucking

4 widening of joints by pressure release (exaggerates shape)

Z

2 abrasion (grinding) by angular material. If pressure melting point is reached the extra supply of water will reduce friction and increase velocity and erosion

summer meltwater

zone of abrasion

Transportation of débris
X supraglacial débris (on top of glacier)
Y englacial débris (within the glacier)
Z basal débris (under the glacier, by ice and meltwater)
A bergshrund and crevasses formed as ice pulls away from the back wall
B crevasses
C dead ice and moraine

Figure 4.14

A cirque in West Wales. The steep wall maintains its shape as freeze–thaw still operates. Broken off material forms scree which is beginning to infill the lake which itself has been dammed behind a natural rock lip

Lip orientation is the direction of an imaginary line from the centre of the back wall of the cirque to its lip

Rotational movement

This is a downhill movement of ice which, like a landslide (Figure 2.17), pivots about a point. The increase in pressure is responsible for the overdeepening of a cirque floor (Figure 4.13b).

Extending and compressing flow

Figures 4.9 and 4.11c show how this process causes differences in the rate of erosion at the base of a glacier.

Maximum erosion occurs:

- where temperatures fluctuate around 0°C, allowing frequent freeze–thaw to operate;
- in areas of jointed rocks which can be more easily frost shattered;
- where two tributary glaciers join, or the valley narrows, giving an increased depth of ice; and
- in steep mountainous regions in temperate latitudes, where the velocity of the glacier is greatest.

Landforms produced by glacial erosion

Cirques

These are amphitheatre or armchair-shaped hollows with a steep back wall and a rock basin. They are also known as corries (Scotland) and cwms (Wales).

During periglacial times (Chapter 5), before the last glacial, snow collected in hollows especially on north-facing slopes. A series of processes, collectively known as **nivation** and which included freeze–thaw, solifluction and possibly chemical weathering, operated under and around the snow patch (Figure 4.13a). These processes caused the underlying rocks to disintegrate. The resultant débris was then removed by summer meltwater streams to leave, in the enlarged hollow, an embryo cirque. It has been suggested that the overdeepening process might need several periglacials or interglacials and glacials in which to form. As the snow patch grew, its layers became increasingly compressed to form firn and, eventually, ice (page 93).

It is accepted that several processes interact to form a fully developed cirque (Figure 4.13b). Plucking is one process responsible for steepening the back wall, but this partly relies upon a supply of water for freeze–thaw and partly upon pressure release in well-jointed rocks. A rotational movement, aided by water from pressure point melting and angular subglacial débris from frost shattering, enables abrasion to over-deepen the floor of the cirque. A **rock lip** develops where erosion decreases. This may be increased in height by the deposition of morainic débris at the glacier's snout. When the climate begins to get warmer, the ice remaining in the hollow melts to leave a deep, rounded lake or tarn (Figure 4.14).

In Britain, as elsewhere in the northern hemisphere, cirques are nearly always orientated between the north-west (315°), through the north-east (where the frequency peaks) to the south-east (135°). This is largely attributable to two facts:

- Northern slopes receive least **insolation** and so glaciers remained there much longer than those facing in more southerly directions (less melting on north-facing slopes).
- Western slopes face the sea and, although still cold, the relatively warmer winds which blew from that direction were more likely to melt the snow and ice (more snow accumulated on east-facing slopes).

Of 56 cirques identified in the Snowdon area, 51 have a lip orientation of between 310° and 120° (Questions, page 100).

Framework 3 Mean, median and mode

These are all types of average (measures of dispersion, Framework 6, page 230).

1 The mean (or arithmetic average) is obtained by totalling the values in a set of data and dividing by the number of values in that set. It is expressed by the formula:

$$\bar{x} = \frac{\sum x}{n}$$

where:

$\bar{x}$ = mean, $\sum$ = the sum of, x = the value of the variable, n = the number of values in the set

The mean is reliable when the number of values in the sample is high and their range, i.e. the difference between the highest and lowest values, is low, but it becomes less reliable as the number in the sample decreases as it then becomes influenced by extreme values.

2 The median is the mid-point value of a set of data. In the sample of 15 cirques, values must be ranked in descending order. The mid-point is the orientation of the eighth cirque because there are seven values above and seven below. Had there been an even number of values then the median would have been the mean of the two middle values. The median is a less accurate measure of dispersion than the mean because widely differing sets of data can return the same median but it is less distorted by extreme values.

3 The mode is the value or class which occurs most frequently in the data. In the set of values 4, 6, 4, 2, 4 the mode would be 4. Although it is the easiest of the three 'averages' to obtain it has limited value. Some data may not have two values in the same class (e.g. 1, 2, 3, 4, 5), while others may have more than one modal value (e.g. 1, 1, 2, 4, 4).

Relationships between mean, median and mode

When data is plotted on a graph we can often make usefull observations about the shape of the curve. For example, we would expect 'A' levelresults nationally to show a few top grades, a smaller number of 'unclassifieds' and a large number of average passes. Graphically this would show a normal distribution, with all three averages at the peak. If the distribution is skewed, then by definition only the mode will lie at the peak (Figure 4.15).

Figure 4.15

Normal and skewed distributions

a the normal distribution

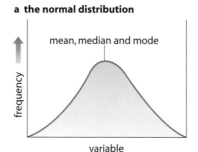

b a positively skewed distribution

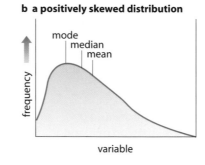

c a negatively skewed distribution

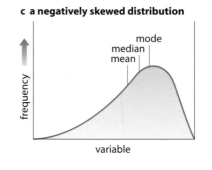

Q

The table shows the lip orientation of 15 cirques (the minimum needed for an acceptable sample) in the Glyders of Snowdonia, and 15 on the Isle of Arran.

1 What is meant by bearings of 45° and 350°?

2 For both groups of cirques, give the mean, median and mode of their orientation.

3 What do your answers show regarding the orientation of cirques in the two areas?

Cirque number	1	2	3	4	5	6	7	8	9	10	11	12	13	14	15
Cirque orientation (degrees)															
Glyders	30	60	45	55	75	50	80	50	10	15	10	35	45	50	85
Arran	05	05	10	55	15	30	95	05	185	70	120	40	30	115	110

Figure 4.16

An arête: Karakoram Mountains, Northern Pakistan

Figure 4.17

A pyramidal peak: Machhapuchare, Nepal

Figure 4.18

Glacial trough with ribbon lake: Wast Water

Arêtes and pyramidal peaks

When two adjacent cirques erode backwards or sideways towards each other, the previously rounded landscape is transformed into a narrow, rocky, steep-sided ridge called an **arête**, as in Figure 4.16 or as Striding Edge in the Lake District. If three or more cirques develop on all sides of a mountain, a **pyramidal peak**, or horn, may be formed. This feature has steep sides and several arêtes radiating from the central peak, e.g. Figure 4.17 and the Matterhorn.

Glacial troughs, rock steps, truncated spurs and hanging valleys

These features are interrelated in their formation. Valley glaciers straighten, widen and deepen preglacial valleys, turning the original 'V'-shaped, river-formed feature into the characteristic 'U' shape typical of glacial erosion — e.g. Wast Water in the Lake District (Figure 4.18). These steep-sided, flat-floored valleys are known as **glacial troughs**. The overdeepening of the valleys is credited to the movement of ice which, aided by large volumes of meltwater and subglacial débris, has a greater erosive power than that of rivers. Extending and compressing flow may overdeepen parts of the trough floor, which later may be occupied by long, narrow **ribbon lakes**, such as Wast Water, or may leave less eroded, more resistant **rock steps**.

Theories to explain pronounced overdeepening of valley floors are debated amongst glaciologists and geomorphologists. Suggested causes include: extra erosion following the confluence of two glaciers; the presence of weaker rocks; an area of rock deeply weathered in preglacial times; or a zone of well-jointed rock. Should the deepening of the trough continue below the former sea level, then during deglaciation and subsequent rises in sea level, the valley may become submerged to form a **fjord** (Figure 6.44).

Abrasion by englacial and subglacial débris and plucking along the valley sides remove the tips of preglacial interlocking spurs leaving cliff-like **truncated spurs** (Figure 4.18).

Hanging valleys result from differential erosion between a main glacier and its tributary glaciers. The floor of any tributary glacier is deepened at a slower rate so that when the glaciers melt it is left hanging high above the main valley and its river has to descend by a single or a series of waterfalls, e.g. Glen Rosa, Arran (Figure 4.19).

Figure 4.20

A roche moutonnée: Yosemite National Park, California

Figure 4.19

Hanging valley: Glen Rosa, Arran

Striations, roches moutonnées and crag and tail

These are smaller erosion features, which help to indicate the direction of ice movement. As a glacier moves across areas of exposed rock, larger fragments of angular débris embedded in the ice tend to leave a series of parallel scratches and grooves called **striations** (e.g. Central Park in New York).

A **roche moutonnée** is a mass of more resistant rock. It has a smooth, rounded upvalley or stoss slope facing the direction of ice flow, formed by abrasion, and a steep, jagged, down-valley or lee slope resulting from plucking (Figures 4.20 and 4.21).

A **crag and tail** consists of a larger mass of resistant rock or crag (e.g. the basaltic crag upon which Edinburgh Castle has been built) which protected the lee-side rocks from erosion, thus forming a gently sloping tail (e.g. the tail down which the Royal Mile extends).

 Q The left-hand column in the table is a list of features produced by glacial erosion; the second column describes these features; and the third summarises their formation. At present, neither the descriptive nor the formation points match the features. Rewrite the list using the appropriate descriptions and formation processes.

Feature	Description	Formation
Cirque	pointed peak with radiating arêtes	englacial or subglacial débris dragged over exposed rock
Arête	tributary glacier left high above the main valley	extending and compressing flow overdeepens parts of valley floor
Pyramidal peak	stepped long profile in a glacial trough (valley)	resistant rock remains after ice abrasion on the ice-direction-facing side and plucking on the lee
Glacial trough	ice-smoothened rocks with a steeper side facing down-valley	frost shattering, abrasion, plucking and rotational ice movement
Hanging valley	steep-sided, U-shaped valley	three or more cirques cut backwards
Truncated spur	small, deep, circular lake	two cirques cut back towards each other
Rock steps	an amphitheatre-shaped depression in a mountain side, with a steep back wall and a rock lip	overdeepening by abrasion by subglacial débris moving in a rotational movement; fills with water after deglaciation
Ribbon lake	steep, cliff-like valley sides	formed by extending and compressing flow in main valley or where two glaciers met
Cirque lake	narrow, knife-edged ridge	widened and deepened by valley glacier
Roche moutonnée	rocks scarred with thin parallel scratches	valley glaciers have removed the ends of interlocking spurs by abrasion
Striations	long, narrow lake in a glacial trough	ice in main valley eroded more rapidly than ice in the tributary valleys often producing a waterfall

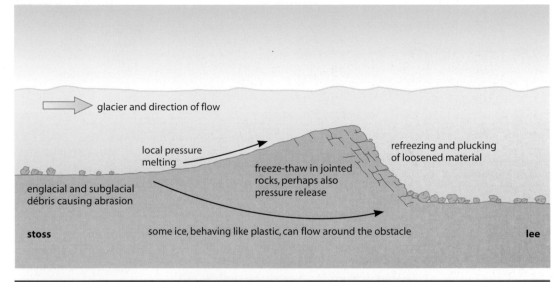

Figure 4.21

The formation of a roche moutonnée

glacier and direction of flow

local pressure melting

refreezing and plucking of loosened material

freeze-thaw in jointed rocks, perhaps also pressure release

englacial and subglacial débris causing abrasion

stoss

some ice, behaving like plastic, can flow around the obstacle

lee

Places 12 Snowdonia

Llyn Peris - ribbon lake

Nant Llanberis - glacial trough

Cwmbrwynog - arête

Snowdon - pyramidal peak

Crib y Ddysgl - arête

Y Lliwedd - arête

Bwlch Main - arête

Grib Goch - arête

Glaslyn - corrie

hanging valley

Llyn Llydaw - corrie

Cwm Dyli - hanging valley

truncated spur

waterfall

Llyn Gwynant - ribbon lake

Nant Gwynant - glacial trough

Figure 4.22

Land sketch of glacial features in Snowdonia (looking west)

Snowdonia is an example of a glaciated upland area. Although Snowdon itself has the characteristics of a pyramidal peak, the ice age was too short (by several thousand years) for the completed development of the classic pyramidal shape which makes the appearance of the Matterhorn so spectacular (Figure 4.17). What are well developed are the arêtes, such as Crib Goch and Bwlch Main, which radiate from the central peak. Between these arêtes are up to half a dozen cirques (cwms, as this is in Wales) including the eastward-facing Glaslyn and the north-eastward orientated (page 99) Llyn (lake) Llydaw. Glaslyn, which is trapped by a rock lip, is 170 m higher than Llyn Llydaw. Striations and roches moutonnées can be found in several places where the rocks are exposed on the surface. To the north and the south-east of Snowdon are the glacial troughs of Nant (valley) Llanberis and Nant Gwynant. Both valleys have the characteristic `U' shape, with steep valley sides, truncated spurs and a flat valley floor. Located on the valley floors are ribbon lakes, including Llyn Peris and Llyn Gwynant (Figure 3.30). Numerous small rivers, with their sources in hanging valleys, descend by waterfalls, as at Cwm Dyli, into the two main valleys. Although the ice has long since gone, the actions of frost and snow, together with that of rain and more recently people, continue to modify the landscape—remember that rarely does a landscape exhibit stereotyped 'textbook' features (see Figure 4.22)!

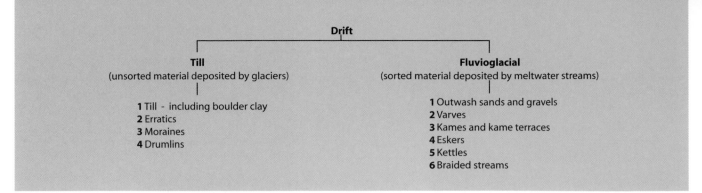

Figure 4.23

Landforms resulting from
glacial deposition

Glacial deposition

Drift is the term historically used by British geologists and glaciologists and is still in use to refer collectively to all glacial deposits (Figure 4.23). These deposits, which include boulders, gravels, sands and clays, may be subdivided into **till**, which includes all material deposited directly by the ice, and **fluvioglacial material,** which is the débris deposited by meltwater streams. Fluvioglacial material includes deposits which may have been deposited initially by the ice and which were later picked up and redeposited by meltwater — either during or after the ice age. Till consists of unsorted material, whereas fluvioglacial deposits have been sorted. Deposition occurs in upland valleys and across lowland areas. A study of drift deposits helps to explain:

- nature and extent of an ice advance;
- frequency of ice advances;
- sources and directions of ice movement; and
- postglacial chronology (including climatic changes, page 273).

Till deposits

Although the term **till** is often applied today to all materials deposited by ice, it is more accurately used to mean an unsorted mixture of rocks, clays and sands. This material was largely transported as supraglacial débris and later deposited to form moraine — either during periods of active ice movement, or at times when the glacier was in retreat. In Britain, till was commonly called **boulder clay** but — since some deposits may contain neither boulders nor clay — this term is now less fashionable. Individual stones are sub-angular — i.e. they are not rounded like river or beach material — but neither do they possess the sharp edges of rocks which have recently been broken up by frost shattering. The composition of till reflects the character of the rocks over which it has passed; East Anglia, for example, is covered by chalky till because the ice passed over a chalk escarpment.

**Lithology:
The study of the nature and composition of rocks**

Fabric till analysis is a fieldwork technique used to determine the direction and source of glacial deposits. Stones and pebbles carried by a glacier tend to become aligned with their long axes parallel to the direction of ice flow as this offers least resistance to the ice. For example, a small sample of 50 stones was taken from a moraine in Glen Rosa, Arran. As each stone was removed, its geology was examined and its orientation was carefully measured using a compass. The results allowed two conclusions to be reached:

1 The pebbles were grouped into classes of 20° and plotted onto a rose diagram (Figure 4.24). The classes were plotted as respective radii from the midpoint of the diagram and then the ends of the radii were joined up to form a star-like polygonal graph. As each stone has two orientations which must be opposites (e.g. 10° and 190°), the graph will be symmetrical. The results show that the ice must have come from the north-north-west or the south-south-east.

2 Although most of the pebbles taken in the sample were composed of local rock, some were of material not found on the island (erratics). This suggests that some of the ice must have come from the Scottish mainland.

Landforms characteristic of glacial deposition

Erratics

These are boulders picked up and carried by ice, often for many kilometres, to be deposited in areas of completely different **lithology** (Figure 4.25). By determining where the boulders originally came from, it is possible to track ice movements. For example, volcanic material from Ailsa Craig in the Firth of Clyde has been found 250 km to the south on the Lancashire plain, while some deposits on the north Norfolk coast originated from southern Norway.

Figure 4.24

Fabric till analysis: orientation of a sample of stones taken from a moraine in Glen Rosa, Arran.

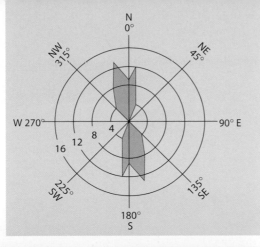

Moraine

Moraine is a type of landform which develops when the débris carried by a glacier is deposited. (It is NOT, therefore, the actual material which is being transported by the glacier). It is possible to recognise five types of moraine (Figure 4.26):

- **Lateral moraine** is formed from débris derived from frost shattering of valley sides and carried along the edges of the glacier (Figure 4.27). When the glacier melts, it leaves an embankment of material along the valley side.
- **Medial moraine** is found in the centre of a valley and results from the merging of two lateral moraines where two glaciers joined (Figure 4.27).
- **Terminal** or **end moraine** is often a high mound (or series of mounds) of material extending across a valley, or lowland area, at right angles to and marking the maximum advance of the glacier or ice sheet.
- **Recessional moraines** mark interruptions in the retreat of the ice when the glacier or ice sheet remained stationary long enough for a mound of material to build up. Recessional moraines are usually parallel to the terminal moraine.
- **Push moraines** may develop if the climate deteriorates sufficiently for the ice temporarily to advance again. Previously deposited moraine may be shunted up into a mound. It can be recognised by individual stones which have been pushed upwards from their original horizontal positions.

Figure 4.26

Types of moraine

1 cirque glacier
2 lateral moraine
3 medial moraine
4 valley glacier
5 frost shattering
6 meltwater streams
7 recessional moraine
8 push moraine
9 terminal moraine

Figure 4.27

Medial moraines , Gulkana
Glacier, Alaska

Figure 4.28

Morainic mounds on Langdale
Fells, Cumbria

using the **elongation ratio**. This is calculated by dividing the maximum length by the maximum width — drumlins are always longer than they are wide. They are usually found in **swarms** or *en echelon*.

There is much disagreement as to how drumlins are formed. Theories suggest they may be an erosion feature, or formed by deposition around a central rock, or even by meltwater. However, none of these accounts for the fact that the majority of drumlins are composed of till which, lacking a central core of rock and consisting of unsorted material, would be totally eroded by moving ice. The most widely accepted view is that they were formed when the ice became overloaded with material, thus reducing the competence of the glacier. The reduced competence may have been due to the melting of the glacier or to changes in velocity related to the pattern of extending–compressing flow. Once the material had been deposited, it may then have been moulded and streamlined by later ice movement. The most recent theory (1987) suggests that drumlins represent more competent and less deformable patches of subglacial débris.

Drumlins

These are smooth, elongated mounds of till with their long axis parallel to the direction of ice movement. Drumlins may be over 50 m in height, over 1 km in length and nearly 0.5 km in width. The steep stoss end faces the direction from which the ice came, while the lee side has a more gentle, stream-lined appearance. The highest point of the feature is near to the stoss end (Figure 4.29). The shape of drumlins can be described by

Figure 4.29

Plan of a drumlin showing
typical dimensions

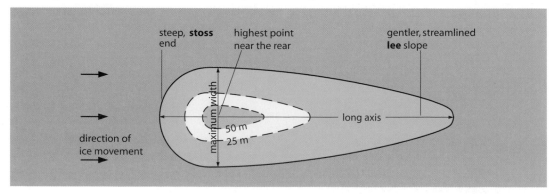

Glaciation

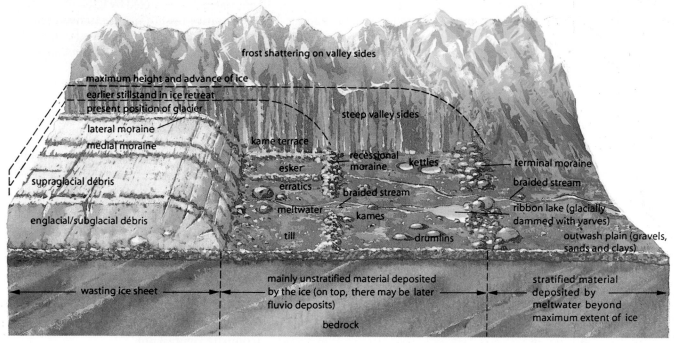

Figure 4.30

Features of lowland glaciation

Fluvioglacial landforms

Fluvioglacial landforms are those moulded by glacial meltwater and have, in the past, been considered to be mainly depositional. More recently, it has become realised that meltwater plays a far more important role in the glacial system than was previously thought, especially in temperate glaciers and in creating erosion features as well as depositional landforms. Most meltwater is derived from ablation. The discharge of glacial streams, both supraglacial and subglacial, is high during the warmer, if not warm, summer months. As the water often flows under considerable pressure, it has a high velocity and is very turbulent. It is therefore able to pick up and transport a larger amount of material than a normal river of similar size. This material can erode vertically, mainly through abrasion but partly by solution, to create subglacial valleys and large potholes, some of latter being up to 20 m in depth. Deposition occurs whenever there is a decrease in discharge and is responsible for a group of landforms (Figures 4.23 and 4.30).

Outwash plains (sandur)

These are composed of gravels, sands and, uppermost and furthest from the snout, clays. They are deposited by meltwater streams issuing from the ice either during summer or when the glacier melts. The material may originally have been deposited by the glacier and later picked up, sorted and dropped by running water beyond the maximum extent of the ice sheets. In parts of the North German Plain, deposits are up to 75 m deep. Outwash material may also be deposited on top of till following the retreat of the ice (Figure 4.30).

Varves

A varve is a distinct layer of silt lying on top of a layer of sand, deposited annually in lakes found near to glacial margins. The coarser, lighter coloured sand is deposited during late spring when meltwater streams have their peak discharge and are carrying their maximum load. As discharge decreases towards autumn when temperatures begin to drop, the finer, darker coloured silt will settle. Each band of light and dark materials represents one year's accumulation (Figure 4.31). By counting the number of varves, it is possible to date the origin of the lake; variations in the thickness of each varve will indicate warmer and colder periods (e.g. greater melting causing increased deposition).

Figure 4.31

The formation of varves in a postglacial lake

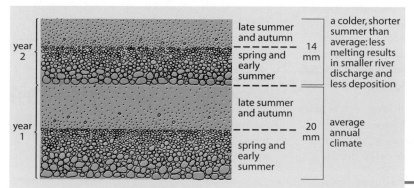

Kames and kame terraces

Kames are undulating mounds of sand and gravel deposited unevenly by meltwater, similar to a series of deltas, along the front of a stationary or slowly melting ice sheet (Figure 4.30). As the ice retreats, the unsupported kame often collapses. Kame terraces, also of sand and gravel, are flat areas found along the sides of valleys. They are deposited by meltwater streams flowing in the trough between the glacier and the valley wall. Troughs occur here because, in summer, the valley side heats up faster than the glacier ice and so the ice in contact with it melts. Kame terraces are distinguishable from lateral moraines by their sorted deposits.

Eskers

These are very long, narrow, sinuous ridges composed of sorted coarse sands and gravel. It is thought that eskers are the fossilised courses of subglacial meltwater streams. As the channel is restricted by ice walls, the hydrostatic pressure and the transported load are both considerable. As the bed of the channel builds up (there is no flood plain), material is left above the surrounding land following the retreat of the ice. Like kames, eskers usually form during times of deglaciation.

Kettles

These form from detached blocks of ice, left by the glacier as it retreats, and then partially buried by the fluvioglacial deposits left by meltwater streams. When the ice blocks melt, they leave enclosed depressions which often fill with water to form kettle-hole lakes and 'kame and kettle' topography.

Braided streams

Channels of meltwater rivers often become choked with coarse material as a result of the marked seasonal variations in discharge (compare Figure 3.36).

Places 13 Glacial landforms on Arran

Using fieldwork to answer an 'A' level question: Describe the landforms found near the snout of a former glacier

Figure 4.23 listed the types of feature formed by glacial deposition, subdividing them into those composed of unsorted material, left by the glacier, material sorted by fluvioglacial action. If the snout of a glacier had remained stationary for some time, indicating a balance between accumulation and ablation, and had then slowly retreated, several of these landforms might be visible following deglaciation. One such site studied by a sixth form was the lower Glen Rosa valley on the Isle of Arran (Figure 4.32).

The dominant feature was a mound (**A**), 14 m high, into which the Rosa Water had cut, giving a fine exposed section of the deposited material. As the mound was a long, narrow ridge-like feature extending across the valley, it was suggested that it might be either a terminal or a recessional moraine. It was concluded that the feature was ice-deposited because the material was unsorted: many of the largest boulders were high up in the exposure; also, most of the stones were sub-angular (not more rounded as might be expected in fluvioglacial deposits).

However, an observation downstream at point **B** revealed that material there was also unsorted and this, together with some large granite erratics seen earlier nearer the coast, seemed to indicate that the mound could not be a terminal moraine as it did not mark the maximum advance of the ice. When a fabric till analysis was carried out, it was noted that the average dip of the stones was about 25°, suggesting that the feature might instead have been a push moraine resulting from a minor re-advance during deglaciation. The orientation of 50 sample stones (Figure 4.24) showed that the ice must have come either from the north-north-west (probable, as this was the highland) or the south-south-east (unlikely, as the lower ground would not be the source of a glacier). An examination of the geology of the stones showed that 80 per cent were granite, and therefore were erratics carried from the upper Rosa valley; 15 per cent were schists (the local rock); and 5 per cent were other igneous rocks not found on the island. It was inferred from the presence of these other rocks that some of the ice must have originated on the Scottish mainland. Also at point **B**, an investigation of river banks showed a mass of sand and gravel with some level of sorting — as might be expected in an outwash area.

Upstream from **A** was a second mound, (**C**) filling much of the valley floor (Figure 4.33). Student suggestions as to the nature of the feature included its being a drumlin, a lateral, a medial, a recessional or even another push moraine. When measured, it was found that its

Glaciation

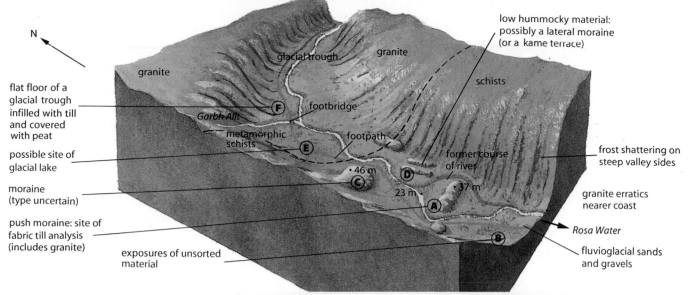

N

granite

granite

glacial trough

footbridge

Garbh Allt

metamorphic schists

footpath

46 m

23 m

37 m

flat floor of a glacial trough infilled with till and covered with peat

possible site of glacial lake

moraine (type uncertain)

push moraine: site of fabric till analysis (includes granite)

exposures of unsorted material

low hummocky material: possibly a lateral moraine (or a kame terrace)

schists

former course of river

frost shattering on steep valley sides

granite erratics nearer coast

Rosa Water

fluvioglacial sands and gravels

F **E** **C** **D** **A** **B**

Figure 4.32

Sketch to show features of glacial deposition in the lower Glen Rosa valley, Arran

Figure 4.33

Field sketch of landform at C in Figure 4.32

length was slightly greater than its width (an elongation ratio of 1.25:1) and the highest point was nearest the up-valley end; it had neither the streamlined shape nor a sufficiently high elongation ratio to be a drumlin (and there were no signs of a swarm!). It appeared to be too far from the valley side to be a lateral moraine; and as two glaciers could not have met here, neither could it have been a medial moraine. It was concluded that it *was* a moraine — perhaps formed during a stillstand in the glacier's retreat, or if the glacier lost momentum after having negotiated a bend in the glacial trough.

Across the river (**D**), was an area of low hummocky material winding along the foot of the valley side to as far as **A**. It was speculated that the feature may have been formed in one of three ways: meltwater depositing sands and gravel between the valley side and the former glacier as a kame terrace; a lateral moraine from frost shattering on the valley sides; or solifluction deposits formed as the climate grew milder and the glacier retreated. (The feature was not flat enough for a river terrace to be seriously considered).

Upstream, the valley floor was extremely flat (**E**). This could be the remains of a former glacial lake, formed when meltwater from the retreating glacier had become trapped behind the moraine at **C** and before it had had time to cut through the deposits. It was impossible to gain a profile to prove or disprove the existence of a lake. Had there been a lake, the profile might have shown a chronological sequence (since the last glacial) as depicted in Figure 4.34.

After crossing the Garbh Allt (a hanging valley), the steep-sided, flat-floored U-shape of the glacial trough through which the Rosa Water

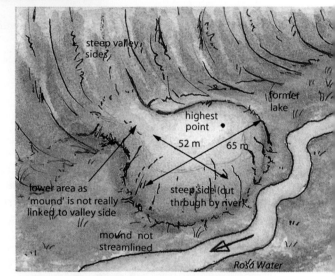

steep valley sides

former lake

highest point

52 m 65 m

lower area as 'mound' is not really linked to valley side

steep side (cut through by river)

mound not streamlined

Rosa Water

flows was visible. The flatness of the floor was probably due to the deposition of subglacial débris — although the till has since been covered by peat, a symptom of the cold, wet conditions.

Although not every feature of glacial deposition was present — there was no evidence of eskers or kettles — this small area did contain several of the landforms and deposits which might be expected at, or near to, the snout of a former glacier.

Figure 4.34

Possible profile of ground at point E in Figure 4.32

Juncus (rushes), cotton grass, heather

thick layer of peat formed under cold, wet conditions

varves and silt deposited on bed of post-glacial lake

stratified sands and gravels deposited by meltwater stream from the retreating glacier

unsorted till deposited by glacier

bedrock (schists)

The left-hand column in the table below is a list of features formed by glacial deposition; the second column describes these features; and the third summarises their formation. At present, neither the descriptive nor the formation points match the features. Rewrite the list using the appropriate description and formation processes.

Feature	Description	Formation
Till	mounds of sorted material deposited by meltwater along valley sides or as a delta at an ice front	materials deposited by ice when englacial material is too heavy to be carried by a melting glacier
Terminal moraine	small, shallow lake containing stratified material	rocks carried by glacier from their source to an area of different rock
Recessional moraines	sorted deposits of gravel, sand and clay spread over a lowland area	formed by the meeting of two lateral moraines
Lateral moraine	a long, narrow, winding ridge of sorted material	material deposited where meltwater streams are in contact with the ice
Medial moraine	unsorted, angular deposits of rock, sand and clay	meltwater spreads out over low-lying areas, decreases in velocity and deposits its load of gravels, then sands and clays
Push moraine	narrow ridge of unsorted material extending across a valley	moraine deposited during a stillstand in the ice retreat
Drumlins	unsorted material found along sides of glaciers and glaciated valleys	deposited by subglacial streams
Erratics	series of narrow ridges extending across a valley	mainly unsorted subglacial material deposited by melting glaciers
Kettles	mounds of material found in centres of glaciers and glaciated valleys	mounds of boulders deposited at maximum advance of ice
Esker	rocks which are not native to the area in which they are found	material already deposited on valley floor pushed upwards by a temporary ice re-advance
Kame	small, elongated mounds with their steep ends facing up-valley, having streamlined shapes and found in swarms	blocks of detached, dead ice left by a melting ice-sheet, later surrounded by fluvioglacial material, melt and forms small lakes
Outwash plain	unsorted material found across a valley with the stones tilted at an angle	material found along the sides of a glacier or a valley as a result of frost shattering

The rose diagram (Figure 4.35) shows the orientation between 0° and 210° of the long axes of stones obtained from a sample of till. The table shows some of the remaining orientations.

1 Complete the
   a remainder of the table;
   b star graph.

2 From which direction is the ice likely to have come? Give a reason for your answer.

3 Describe three other methods that you could use in the field to determine either the direction of flow or the source of the ice.

4 In Chapters 2, 3 and 4, mention has been made of solifluction material, a river terrace, scree, a lateral moraine and a kame terrace — all of which may be found along the foot of a valley side. How would you tackle the problem, in the field, of trying to identify the differences between these five features?

5 Describe appropriate outline fieldwork methods which might be useful in answering the following questions:
   a Describe the landforms found near to the snout of a glacier.
   b Describe and account for the main characteristics of the depositional landforms which result from the downwasting and retreat of an ice-sheet. Consider the nature of the deposits as well as their surface forms.

Figure 4.35

A rose diagram: used to show the orientation of stones in a sample of till

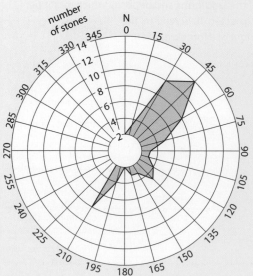

The orientation of the long axes of the stones in the till sample

Direction (°)	Number of stones
225	
240	
255	
270	3
285	3
300	
315	
330	
345	

Other effects of glaciation

Drainage diversion and proglacial lakes
Where ice sheets expand, they may divert the courses of rivers. For example, the preglacial River Thames flowed in a north-easterly direction. It was progressively diverted southwards by advancing ice (Figure 4.36).

Where ice sheets expand and dam rivers, proglacial lakes are created, e.g. Lakes Lapworth and Harrison (Figure 4.36). Before the ice age, the River Severn flowed northwards into the River Dee, but this route became blocked during the Pleistocene by Irish Sea ice. A large lake, Lapworth, was impounded against the edge of the ice until the waters rose high enough to breach the lowest point in the southern watershed. As the water overflowed through an **overspill channel**, there was rapid vertical erosion which formed what is now the Ironbridge Gorge. When the ice had completely melted, the level of this new route was lower than the original course (which was also blocked by drift), forcing the present-day River Severn to flow southwards.

Other rivers, e.g. the Warwickshire Avon (Figure 4.36) and the Yorkshire Derwent (Places 14), have also been diverted as a consequence of glacial activity. Sometimes the glacial overspill channels have been abandoned — e.g. at Fenny Compton, where the Warwickshire Avon temporarily flowed south-east into the Thames (O^1 in Figure 4.36). Proglacial lakes are also found behind eskers and recessional moraines.

Figure 4.36

Glacial diversion of drainage and proglacial lakes in England and Wales

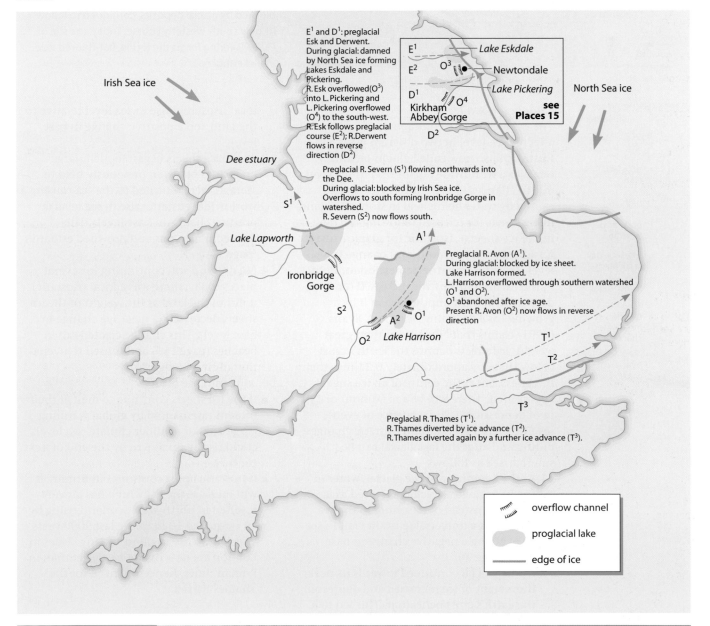

E^1 and D^1: preglacial Esk and Derwent. During glacial: damned by North Sea ice forming Lakes Eskdale and Pickering. R. Esk overflowed(O^3) into L. Pickering and L. Pickering overflowed (O^4) to the south-west. R. Esk follows preglacial course (E^2); R. Derwent flows in reverse direction (D^2)

Preglacial R. Severn (S^1) flowing northwards into the Dee. During glacial: blocked by Irish Sea ice. Overflows to south forming Ironbridge Gorge in watershed. R. Severn (S^2) now flows south.

Preglacial R. Avon (A^1). During glacial: blocked by ice sheet. Lake Harrison formed. L. Harrison overflowed through southern watershed (O^1 and O^2). O^1 abandoned after ice age. Present R. Avon (O^2) now flows in reverse direction

Preglacial R. Thames (T^1). R. Thames diverted by ice advance (T^2). R. Thames diverted again by a further ice advance (T^3).

Irish Sea ice

North Sea ice

Dee estuary

Lake Eskdale

Newtondale

Lake Pickering

see Places 15

Kirkham Abbey Gorge

Lake Lapworth

Ironbridge Gorge

Lake Harrison

overflow channel

proglacial lake

edge of ice

Figure 4.37

Proglacial lakes and overflow channels in North Yorkshire

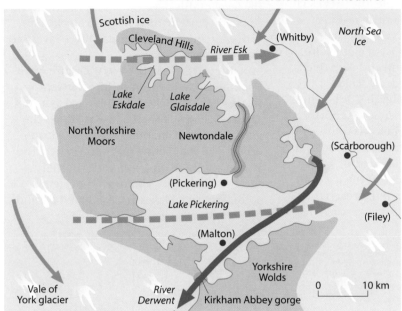

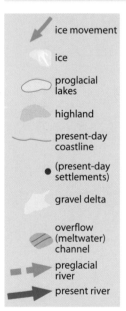

ice movement

ice

proglacial lakes

highland

present-day coastline

(present-day settlements)

gravel delta

overflow (meltwater) channel

preglacial river

present river

Lake Eskdale, a proglacial lake, formed when the North Sea ice sheet blocked the mouth of the River Esk. The level of the lake rose until its water found a new route over a low point in its southern watershed on the North Yorkshire Moors. The overflow river flowed through Lake Glaisdale before cutting the deep, narrow, steep-sided, flat-floored Newtondale valley. At the end of this valley, the river formed a delta where it flowed into another proglacial lake — Lake Pickering. Lake Pickering, also dammed by North Sea ice, found an outlet to the south-west where it formed an overflow channel — the present-day Kirkham Gorge. After the ice melted, the Esk reverted to its original course entering the sea near Whitby; Newtondale became virtually a dry valley; and the River Derwent, its eastward exit from Lake Pickering blocked by glacial deposits, continued to follow its new south-westerly course. Today, the site of Lake Pickering forms the fertile, flat-floored Vale of Pickering.

Changes in sea level

The expansion and contraction of ice sheets affected sea level in two different ways. **Eustatic** (also now called **glacio-eustatic**) refers to a worldwide fall (or rise) in sea level due to changes in the hydrological cycle caused by water being held in storage on land in ice sheets (or released following the melting of ice sheets). **Isostatic** (or **glacio-isostatic**) adjustment is a more local change in sea level resulting from the depression (or uplift) of the earth's crust by the increased (or decreased) weight imposed upon it by a growing (or a declining) ice sheet. Evans (1991) claims that "Because of their great weight, ice sheets depress the Earth's crust below them by approximately 0.3 times their thickness. So, at the centre of an ice sheet 700 m thick, there will be a maximum of 210 m of depression". The sequence of events resulting from eustatic and isostatic changes during and after the last glacial can be summarised as follows:

1 At the beginning of the glacial, water in the hydrological cycle was stored as ice on the land instead of returning to the sea. There was a universal (eustatic) fall in sea level giving a negative change in base level (page 76).

2 As the glacial continued towards its peak, the weight of ice increased and depressed the earth's crust beneath it. This led to a local (isostatic) rise in sea level relative to the land and a positive change in base level.

3 As the ice sheets began to melt, large quantities of water, previously held in storage, were returned to the sea causing a worldwide (eustatic) rise in sea level (a positive change in base level). This formed fjords, rias and drowned estuaries (page 149).

4 Finally, and still continuing in several places today, there was a local (isostatic) uplift of the land as the weight of the ice sheets decreased (a negative change in base level). This change created raised beaches (page 150) and caused rejuvenation of rivers (page 76).

Looking into the future:

■ If the ice sheets continue to melt at their present rate, caused by global warming (page 237) or a milder climate, sea levels could rise by up to 5 m by the end of next century.

■ If isostatic uplift continues in Britain, it will increase the tilt which has already resulted in north-west Scotland rising by an estimated 10 m in the last 9000 years and south-east England sinking. Tides in London are now over 4 m higher than in Roman times, hence the need for the Thames Barrier.

Case Study 4

Avalanches

Figure 4.38

An avalanche

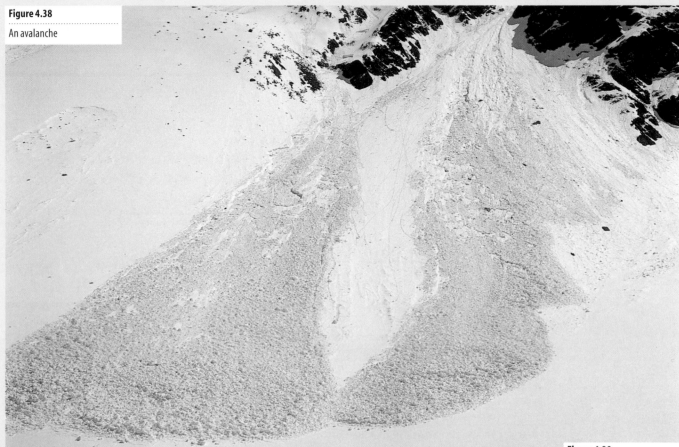

An **avalanche** is a sudden downhill movement of snow, ice and/or rock. It occurs when the weight (mass) of material is sufficient to overcome friction allowing the débris to descend at considerable speeds under the force of gravity. The average speed of descent, according to video recordings made of snow avalanches in Japan, is between 40 and 60 km per hour. It is the unpredictability of avalanches which makes them a major environmental hazard in Alpine areas (Figure 4.38).

There are several types of avalanche which makes a simple classification difficult. Figure 4.39 gives a mainly descriptive, late 19th-century classification, while Figure 4.40 gives a modern classification based more upon genetic and morphological characteristics.

Figure 4.39

A late 19th-century classification of avalanches

a Staublawinen (airborne powder snow)	Pure (completely airborne)
	Common (some contact with the ground)
b Grundlawinen (ground-hugging)	Rolling
	Sliding

a Avalanche break-away point	single point - loose snow avalanche	easier (not easy) to predict and manage
	large area or 'slab'	often localised, hardest to predict, greatest threat to off-piste skiers
b Depth	total snow depth	
	top layers of snow move over lower layers	alpine inhabitants regard this as the most dangerous
c Channel (track) width	unconfined - no channel	wide area, hard to manage
	gulley - confined to narrow track	dangerous, but easier to manage
d Nature of snow (water content)	dry snow - mainly rolling	above ground level so friction is reduced: can reach speeds of 300 km per hour - very destructive
	wet snow - mainly sliding	follows ground topography, occurs under föhn conditions (page 224), limited protection, much damage

Figure 4.40

A modern classification of avalanches

Causes

- Heavy snowfall compressing and adding weight to earlier falls.
- Steep slopes of over 25° where stability is reduced and friction is more easily overcome.
- A sudden increase in temperature, especially on south-facing slopes and under föhn wind conditions (page 224).
- Heavy rainfall falling upon snow (more likely in Scotland than in the Alps).
- Deforestation reducing slope stability. In the Alps, the recent increase in the number of trees dying due to acid rain is blamed upon the increase in road traffic.
- Vibrations, triggered by off-piste skiers, nearby traffic and, more dangerously, earth movements (Figure 4.41).
- Very long, cold and dry winters followed by heavy snowfalls in spring. Early falls will turn into ice over which the later falls will slide. (Some local people perceive this to pose the greatest avalanche risk.)

Consequences

Avalanches can block roads and railways, cut off power supplies and, under extreme conditions, destroy buildings. Also, between 1980 and 1991 there were, in Alpine Europe alone, 1210 recorded avalanche deaths. Of these, almost 50 per cent were skiers (virtually all in off-piste areas). This avalanche death rate continues to grow as the popularity of skiing increases. Figure 4.41 describes the effects of one avalanche.

Figure 4.41

An avalanche event: Ranrahirca, Peru, 10 January 1962

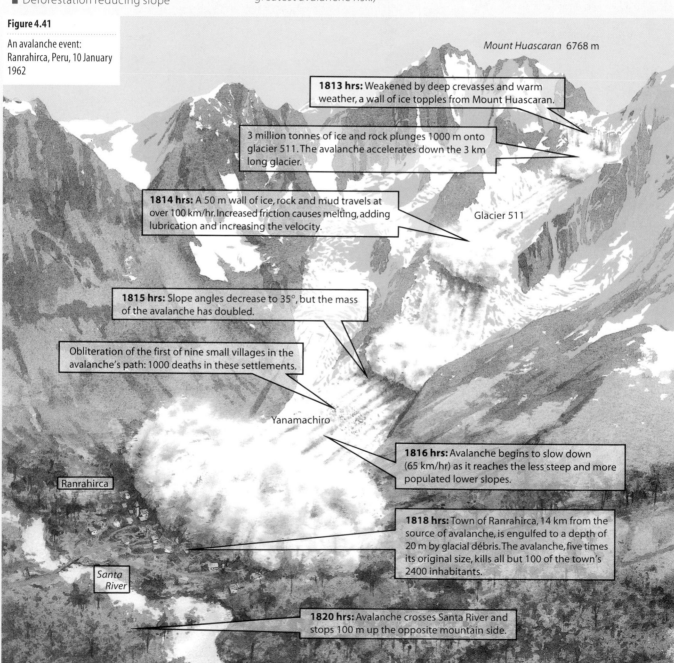

Mount Huascaran 6768 m

1813 hrs: Weakened by deep crevasses and warm weather, a wall of ice topples from Mount Huascaran.

3 million tonnes of ice and rock plunges 1000 m onto glacier 511. The avalanche accelerates down the 3 km long glacier.

1814 hrs: A 50 m wall of ice, rock and mud travels at over 100 km/hr. Increased friction causes melting, adding lubrication and increasing the velocity.

Glacier 511

1815 hrs: Slope angles decrease to 35°, but the mass of the avalanche has doubled.

Obliteration of the first of nine small villages in the avalanche's path: 1000 deaths in these settlements.

Yanamachiro

1816 hrs: Avalanche begins to slow down (65 km/hr) as it reaches the less steep and more populated lower slopes.

Ranrahirca

1818 hrs: Town of Ranrahirca, 14 km from the source of avalanche, is engulfed to a depth of 20 m by glacial débris. The avalanche, five times its original size, kills all but 100 of the town's 2400 inhabitants.

Santa River

1820 hrs: Avalanche crosses Santa River and stops 100 m up the opposite mountain side.

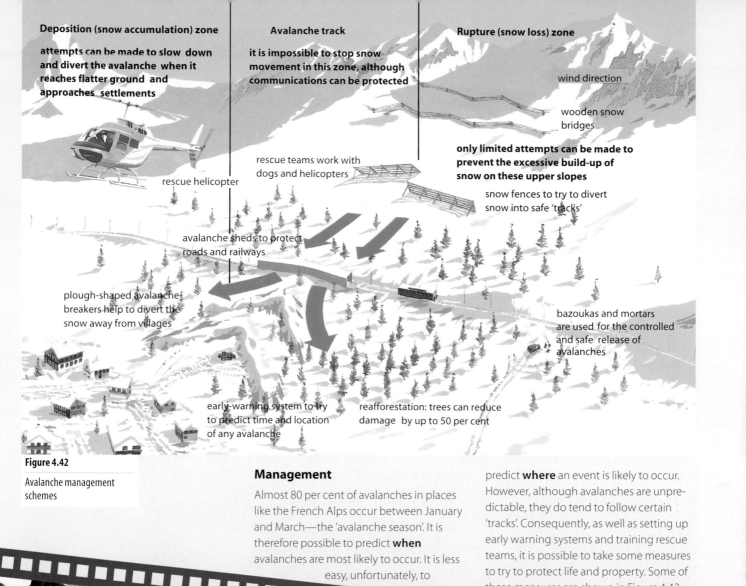

Deposition (snow accumulation) zone

attempts can be made to slow down and divert the avalanche when it reaches flatter ground and approaches settlements

Avalanche track

it is impossible to stop snow movement in this zone, although communications can be protected

Rupture (snow loss) zone

wind direction

wooden snow bridges

only limited attempts can be made to prevent the excessive build-up of snow on these upper slopes

rescue teams work with dogs and helicopters

rescue helicopter

snow fences to try to divert snow into safe 'tracks'

avalanche sheds to protect roads and railways

plough-shaped avalanche breakers help to divert the snow away from villages

bazoukas and mortars are used for the controlled and safe release of avalanches

early-warning system to try to predict time and location of any avalanche

reafforestation: trees can reduce damage by up to 50 per cent

Figure 4.42

Avalanche management schemes

Management

Almost 80 per cent of avalanches in places like the French Alps occur between January and March—the 'avalanche season'. It is therefore possible to predict **when** avalanches are most likely to occur. It is less easy, unfortunately, to predict **where** an event is likely to occur. However, although avalanches are unpredictable, they do tend to follow certain 'tracks'. Consequently, as well as setting up early warning systems and training rescue teams, it is possible to take some measures to try to protect life and property. Some of these measures are shown in Figure 4.42.

Figure 4.43

Avalanche protection and rescue schemes

References

Boyce, J. and Ferretti, J. (1984) *Fieldwork in Geography*. Cambridge University Press.

Dawson, A.G. (1992) *Ice Age Earth*. Routledge.

Evans, D. (1988) Glaciated land on the rebound, *Geography Review*.

Glacial Deposits Geography Today: A Search for Order BBC Television.

Glaciers. BBC Television/Open University.

Goudie, A. (1993) *The Nature of the Environment*. Blackwell.

McCullagh, P. (1978) *Modern Concepts in Geomorphology*. Oxford Universty Press.

Planet Earth: Glaciers (1982) Time–Life Books.

Planet Earth: Ice Ages (1983) Time–Life Books.

Prosser, R. (1993) *Natural Systems and Human Responses*. Thomas Nelson.

Valley Glaciers Geography Today: A Search for Order. BBC Television.

Chapter 5

Periglaciation

"Computer simulation models are now being used to predict . . . soil instability due to frost where buildings, roads, airfields, and other utilities are to be built in regions of frost climate."

D. Brunsden and J. Doornkamp, 1977

The term **periglacial**, strictly speaking, means 'near to or at the fringe of an ice sheet', where frost and snow have a major impact upon the landscape. However, the term is often more widely used to include any area which has a cold climate — e.g. mountains in temperate latitudes — or which has experienced severe frost action in the past — e.g. southern England during the

Quaternary ice age (Figure 4.3b). Today, the most extensive periglacial areas lie in the Arctic regions of Canada, the USA and the CIS. These areas, which have a tundra climate, soils and vegetation (pages 309–11), exhibit their own characteristic landforms.

Permafrost

Permafrost is permanently frozen ground. It occurs where soil temperatures remain below 0°C for at least two consecutive summers. Permafrost covers almost 25 per cent of the earth's land surface (Figure 5.1) although its extent changes over periods of time. Its depth and continuity also vary (Figure 5.2).

Figure 5.1

Permafrost zones of the Arctic

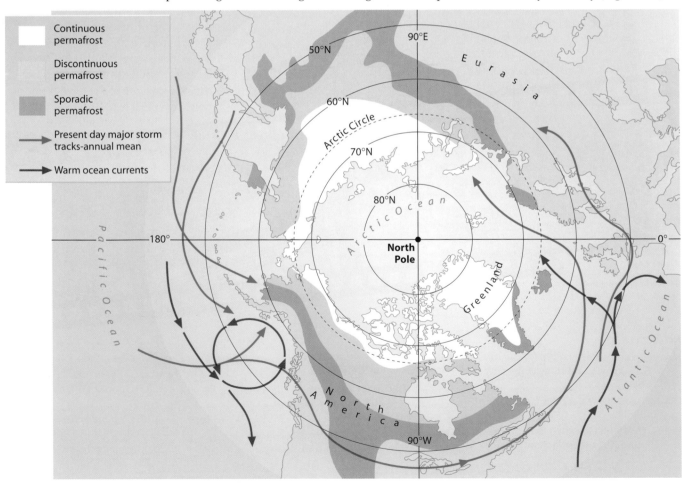

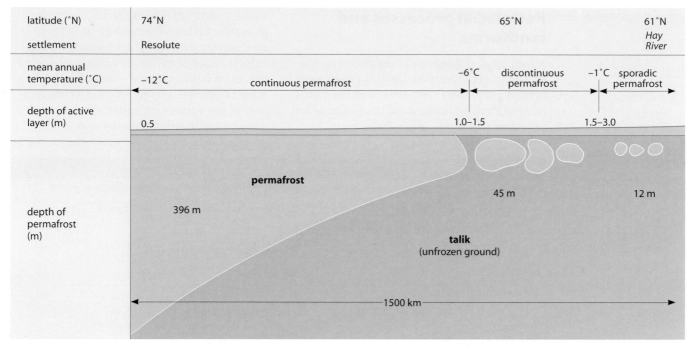

latitude (°N)	74°N		65°N	61°N
settlement	Resolute			*Hay River*
mean annual temperature (°C)	–12°C	continuous permafrost	–6°C discontinuous permafrost	–1°C sporadic permafrost
depth of active layer (m)	0.5		1.0–1.5	1.5–3.0

permafrost

396 m

45 m

12 m

talik (unfrozen ground)

|————— 1500 km —————|

depth of permafrost (m)

Figure 5.2

Transect through part of the permafrost zone in northern Canada

Figure 5.3

Soil temperatures in permafrost at Yakutsk, Siberia

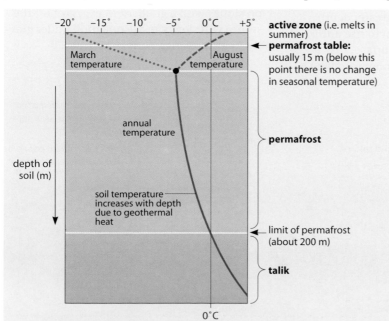

Continuous permafrost is found mainly within the Arctic Circle where the mean annual temperature is below -5°C. Here winter temperatures may fall to -50°C and summers are too cold and too short to allow anything but a superficial melting of the ground. The permafrost has been estimated to reach a depth of 700 m in northern Canada and 1500 m in Siberia. As Figure 5.1 shows, continuous permafrost extends further south in continental interiors than in coastal areas which are subject to the warming influence of the sea, e.g. the North Atlantic Drift in north-west Europe.

Discontinuous permafrost lies further south in the northern hemisphere, reaching 50°N in central Russia, and corresponds to those areas with a mean annual temperature of between -1°C and -5°C. As is shown in Figure 5.2, discontinuous permafrost consists of islands of permanently frozen ground, separated by less cold areas which lie near to rivers, lakes and the sea.

Sporadic permafrost is found where mean annual temperatures are just below freezing point and summers are several degrees above 0°C. This results in isolated areas of frozen ground (Figure 5.2).

In areas where summer temperatures rise above freezing point, the surface layer thaws to form the **active layer**. This zone, which under some local conditions can become very mobile for a few months before freezing again, can vary in depth from a few centimetres (where peat or vegetation cover protects the ground from insolation) to 5 m. The active layer is often saturated because meltwater cannot infiltrate downwards through the impermeable permafrost. Meltwater is unlikely to evaporate in the low summer temperatures or to drain downhill since most of the slopes are very gentle. The result is that permafrost regions contain many of the world's few remaining wetland environments.

The unfrozen layer beneath, or indeed any unfrozen material within, the permafrost is known as *talik*. The lower limit of the permafrost is determined by geothermal heat which causes temperatures to rise (Figure 5.3).

Periglacial processes and landforms

Most periglacial regions are sparsely populated and underdeveloped. Until the search for oil and gas in the 1960s, there had been little need to study or understand the geographical processes which operate in these areas. Although significant strides have been made in the last 30 years, there is still uncertainty as to how certain features have developed and, indeed, whether such features are still being formed today or are a legacy of a previous, even colder climate — i.e. a fossil or relict landscape. Figure 5.4 gives a classification of the various processes which operate, and the landforms which develop, in periglacial areas.

Figure 5.4

Classification of periglacial processes and landforms

	Processes	Landforms
Ground ice	Ice crystals and lenses (frost-heave)	Sorted stone polygons (stone circles and stripes: patterned ground)
	Ground contraction	Ice wedges with unsorted polygons: patterned ground
	Freezing of ground water	Pingos
Frost weathering	Frost shattering	Blockfields, talus (scree), tors (Chapter 8)
Snow	Nivation	Hollows
Meltwater	Solifluction	Solifluction Isheets, rock streams
	Streams	Braiding, dry valleys in chalk (Chapter 8)
Wind	Windblown	Loess (limon)

Ground ice

Frost-heave: ice crystals and lenses

Frost-heave includes several processes which cause either fine-grained soils such as silts and clays to expand to form small domes or individual stones within the soil to be moved to the surface (Figure 5.5). It results from the direct formation of ice — either as crystals or as lenses. The **thermal conductivity** of stones is greater than that of soil. As a result, the area under a stone becomes colder than the surrounding soil and ice crystals form. Further expansion by the ice widens the capillaries in the soil, allowing more moisture to rise and to freeze. The crystals, or the larger ice lenses which form at a greater depth, force the stones above them to rise until eventually they reach the surface. (Ask a gardener in northern Britain to explain why a plot which was left stoneless in autumn has become stone-covered by the spring following a cold winter).

During periods of thaw, meltwater leaves fine material under the uplifted stones preventing them from falling back into their original positions. In areas of repeated freezing (ideally where temperatures fall to between -4°C and -6°C) and thawing, frost-heave both lifts and sorts material to form **patterned ground** on the surface (Figure 5.6). The larger stones, with their extra weight, move outwards to form, on almost flat areas, stone circles or, more accurately, **stone polygons**. Where this process occurs on slopes with a gradient in excess of 6°, the stones will slowly move downhill under gravity to form elongated **stone stripes**.

Figure 5.5

Frost-heave and stone-sorting
a doming occurs when the ground freezes in winter and may disappear in summer when the ground thaws — the ground is chilled from above
b stones roll down into the hollows between mounds and material becomes sorted in size with the finest deposits left in the centre of the polygon and on top of the mound

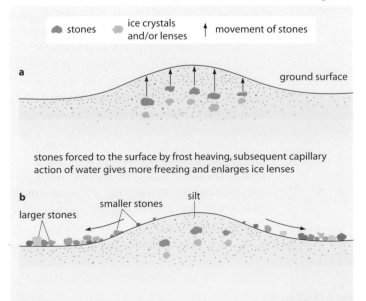

stones | ice crystals and/or lenses | ↑ movement of stones

a
ground surface

stones forced to the surface by frost heaving, subsequent capillary action of water gives more freezing and enlarges ice lenses

b
larger stones | smaller stones | silt

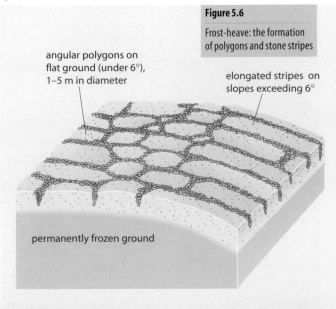

Figure 5.6

Frost-heave: the formation of polygons and stone stripes

angular polygons on flat ground (under 6°), 1–5 m in diameter

elongated stripes on slopes exceeding 6°

permanently frozen ground

Figure 5.7

Low centre, ice wedge polygons near Barrow, Alaska. The patterned ground is formed by polygons up to 30 m in diameter. The polygon boundaries mark the position of the ice wedges.

Figure 5.7

Low centre, ice wedge polygons near Barrow, Alaska. The patterned ground is formed by polygons up to 30 m in diameter. The polygon boundaries mark the position of the ice wedges.

Figure 5.8

The formation of ice wedges

Ground contraction

The refreezing of the active layer during the severe winter cold causes the soil to contract. Cracks open up which are similar in appearance to the irregularly shaped polygons found on the bed of a dried-up lake. During the following summer, these cracks fill with meltwater and, sometimes, also with water and windblown deposits. When this water refreezes, either the following winter or during cold summer nights, the cracks widen and deepen to form **ice wedges** (Figure 5.8). This process is repeated annually until the wedges, which underlie the perimeters of the polygons, grow to as much as 1 m in thickness and 3 m in depth. **Fossil wedges**, i.e. cracks filled with sands and silt left by meltwater, are a sign of earlier periglacial conditions (Figure 5.9).

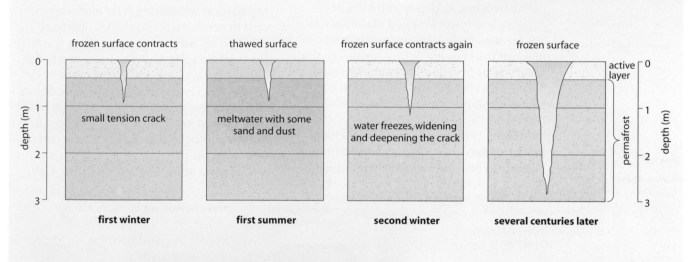

Figure 5.9

Fossil ice wedges

Processes of frost-heaving and ground contraction both produce patterned ground (Figure 5.7). However, frost-heaving results in small dome-shaped polygons with larger stones found to the outside of the circles, whereas ice contraction produces larger polygons with the centre of the circles depressed in height and containing the bigger stones. The diameter of an individual polygon can reach over 30 m.

Freezing of ground water

Pingos are dome-shaped, isolated hills which interrupt the flat tundra plains (Figure 5.10). They can have a diameter of up to 500 m and may rise 50 m in height to a summit which is often ruptured to expose an icy core. As they occur mainly in sand, they are not susceptible to frost-heaving. American geographers recognise two types of pingo (Figure 5.11a and b).

Figure 5.10

A pingo, Mackenzie Delta, Canada

Figure 5.11

Formation of pingos

Open-system pingos occur in valley bottoms and in areas of thin or discontinuous permafrost. Surface water is able to infiltrate into the upper layers of the ground where it can circulate in the unfrozen sediments before freezing. As the water freezes, it expands and forms localised masses of ice. The ice forces any overlying sediment upwards into a dome-shaped feature, in the same way as frozen milk lifts the cap off its bottle. This type of pingo, referred to as the **East Greenland type**, grows from below (Figure 5.11a).

Closed-system pingos are more characteristic of flat, low-lying areas where the permafrost is continuous. They often form on the sites of small lakes where water is trapped (en**closed**) by freezing from above and by the advance of the permafrost inwards from the lake margins. As the water freezes it will expand, forcing the ground above it to rise upwards into a dome shape. This type of pingo is known as the **Mackenzie type** as over 1400 have been recorded in the delta region of the River Mackenzie. It results from the downward growth of the permafrost (Figure 5.11b).

As the surface of a pingo is stretched, the summit may rupture and crack. Where the ice-core melts, the hill may collapse leaving a meltwater-filled hollow (Figure 5.11c). Later, a new pingo may form on the same site, and there may be a repeated cycle of formation and collapse.

Frost weathering

Mechanical weathering is far more significant in periglacial areas than is chemical weathering, with freeze–thaw being the dominant process (Figure 2.9). On relatively flat upland surfaces, e.g. the Scafell range in the Lake District and the Glyders in Snowdonia, the extensive spreads of large, angular boulders, formed *in situ* by frost action, are known as **blockfields** or **felsenmeer** (literally, a 'rock sea').

Scree, or **talus**, develops at the foot of

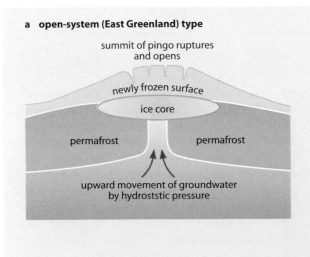

a open-system (East Greenland) type

summit of pingo ruptures and opens

newly frozen surface

ice core

permafrost

permafrost

upward movement of groundwater by hydroststic pressure

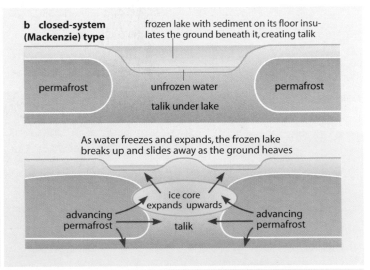

b closed-system (Mackenzie) type

frozen lake with sediment on its floor insulates the ground beneath it, creating talik

permafrost

unfrozen water

permafrost

talik under lake

As water freezes and expands, the frozen lake breaks up and slides away as the ground heaves

ice core expands upwards

advancing permafrost

talik

advancing permafrost

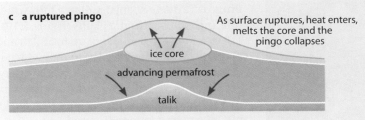

c a ruptured pingo

As surface ruptures, heat enters, melts the core and the pingo collapses

ice core

advancing permafrost

talik

hollow on site of ruptured pingo

solifluction

former dome

meltwater rampart

steep slopes, especially those composed of well-jointed rocks prone to frost action. Freeze–thaw may also turn well-jointed rocks, such as granite, into **tors** (Figure 8.14). One school of thought on tor formation suggests that these landforms result from frost shattering with the weathered débris later having been removed by solifluction. If this is the case, tors are therefore a relict (fossil) of periglacial times.

Snow

Snow is the agent of several processes which collectively are known as **nivation** (page 99). These nivation processes, sometimes referred to as 'snowpatch erosion' are believed to be responsible for enlarging hollows on hillsides (embryo cirques, Figure 4.13a). Nivation hollows are still actively forming in places like Iceland, but are relict features in southern England (as on the scarp slope of the South Downs behind Eastbourne).

Meltwater

During periods of thaw, the upper zone (active layer) melts, becomes saturated and, if on a slope, begins to move downhill under gravity by the process of solifluction (page 41). Solifluction leads to the infilling of valleys and hollows by sands and clays to form **solifluction sheets** (Figures 5.12 and 5.13a) or, if the source of the flow was a nivation hollow, a rock stream (Figure 5.15). Solifluction deposits, whether they have either in-filled valleys or have flowed over cliffs, as in southern England, are also known as **head** or, in chalky areas, **coombe** (Figure 5.13b).

The chalklands of southern England are characterised by numerous dry valleys (Figure 8.11). The most favoured of several theories put forward to explain their origin suggests that the valleys were carved out under periglacial conditions. Any water in the porous chalk at this time would have frozen, to produce permafrost, leaving the surface impermeable. Later, meltwater rivers would have flowed over this frozen ground to form 'V'-shaped valleys (page 184).

Rivers in periglacial areas have a different régime from those flowing in warmer climates. Many may stop flowing altogether during the long and very cold winter (Figure 5.14) and have a peak discharge in late spring or early summer when melting is at its maximum. With their high velocity, these rivers are capable of transporting large amounts of material when at their peak flow. Later in the year, when river levels fall rapidly, much of this material will be deposited, leaving a braided channel (Figure 3.36).

Figure 5.12

Solifluction sheet in the Cheviot Hills, Northumberland

Figure 5.13

Solifluction sheet and head

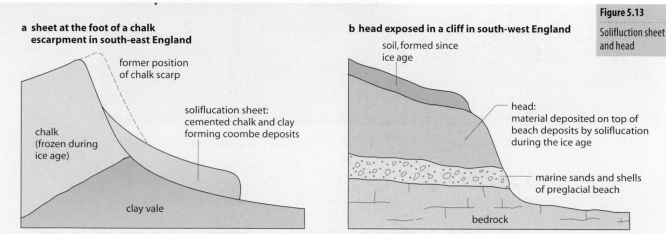

a **sheet at the foot of a chalk escarpment in south-east England**

former position of chalk scarp

chalk (frozen during ice age)

solifluction sheet: cemented chalk and clay forming coombe deposits

clay vale

b **head exposed in a cliff in south-west England**

soil, formed since ice age

head: material deposited on top of beach deposits by solifluction during the ice age

marine sands and shells of preglacial beach

bedrock

Figure 5.14

Régime of a river in a
periglacial area

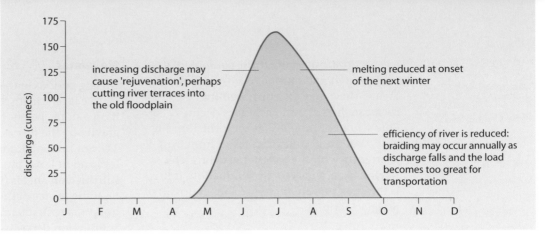

increasing discharge may
cause 'rejuvenation', perhaps
cutting river terraces into
the old floodplain

melting reduced at onset
of the next winter

efficiency of river is reduced:
braiding may occur annually as
discharge falls and the load
becomes too great for
transportation

Wind

A lack of vegetation and a plentiful supply of
fine, loose material (i.e. silt) found in glacial
environments enabled strong, cold, out-
blowing winds to pick up large amounts of
débris and to redeposit it as **loess** in areas far
beyond its source. Loess covers large areas in
a discontinuous zone extending from the
Mississippi–Missouri valley in the USA,
across France (where it is called **limon**) and
the North European Plain and into north-
west China (where in places it exceeds 300 m
in depth and forms the yellow soils of the
Huang He valley). In all areas, it gives an agriculturally productive, fine-textured, deep,
well-drained and easily worked soil which is,
however, susceptible to further erosion by
water and wind (Figure 10.33).

Figure 5.15

Land sketch showing typical
landforms found in a
periglacial area

Q

1 What is meant by periglacial?

2 Describe the processes which lead to the
formation of the landforms shown in Figure
5.15.

3 Why do gentle slopes, such as between **X**
and **Y** on Figure 5.15, develop under
periglacial conditions?

4 Under what conditions might the depth of
permafrost alter?

5 Why are many periglacial landforms in
Britain said to be 'relict'?

KEY

A blockfield
B stone polygons, garlands and stripes
C solifluction sheets/benches
D nivation hollow with snow patch
E rock stream
F débris fan
G braided stream
H ice-wedge polygons

K pingo
L tor
M talus (scree)
N čliffs with head deposits

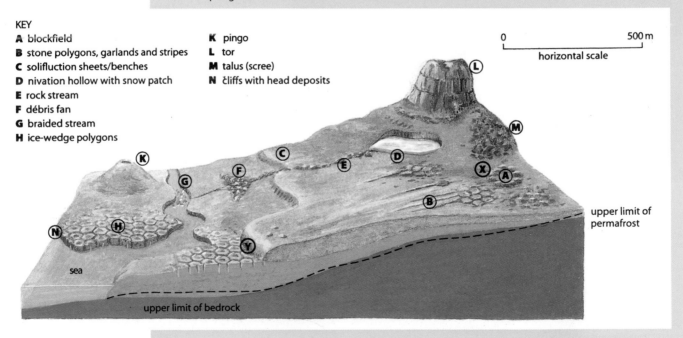

References

Brunsden, D. and Doornkamp, J. (1977) *The
Unquiet Landscape*, David and Charles.

Goudie, A. (1993) *The Nature of the Environment*.
Blackwell.

McCullagh, P. (1978) *Modern Concepts in
Geomorphology*. Oxford University Press.

Planet Earth: Ice Ages (1983) Time–Life Books.

Case Study 5

Thermokarst

Thermokarst is a landscape where the ground surface is very uneven. It develops when masses of ground ice melt. This leads to an increase in the depth of the active layer which in turn causes parts of the land surface to subside. Thermokarst is, therefore, the general name given to irregular, hummocky terrain with marshy or lake-filled hollows created by the disruption of the thermal equilibrium of the permafrost. Russian geographers claim that thermokarst only develops where:

1 There are either large amounts of ground ice (ice wedges, ice lenses and ice sheets) or unconsolidated sediments with a high ice content (where ice forms over 70 per cent of the volume);

2 The depth of seasonal or perennial thawing is either greater than the depth of ice wedges or where it reaches ice-rich rocks.

Although thermokarst has, in the past, developed due to climatic change, increasingly in the last few decades the major agent has been the human occupancy and development of periglacial areas. The thermal equilibrium, which is very delicate, may be upset by a range of human activities (Figure 5.16) which can:

a increase the thermokarst—i.e. lower the level of permafrost

■ Removal of mosses and other tundra vegetation for construction purposes means that in summer more heat reaches the soil and so the depth of thaw increases, as does the likelihood of flooding.

■ Construction of centrally heated buildings has warmed the ground underneath them, causing the buildings to subside.

■ Siting of oil, sewerage and water pipes in the active zone has increased the rate of thaw, sometimes causing fracturing of the pipes as the ground moves. Similar earth movements have caused roads and railways to lose alignment and dams and bridges to crack.

■ Drilling for oil and gas poses problems because the heat from the drills melts the permafrost. Having enlarged the drilling hole, the machinery vibrates and is less effective.

■ Road construction, even for motorways in southern England, upsets the delicate equilibrium of many slopes produced by solifluction during different climatic conditions.

b reduce the thermokarst— i.e. raise the level of permafrost

Development may also upset the thermal balance in the opposite direction. The construction of unheated buildings, for example, reduces the already low amounts of heat received during the short summer. This causes the upper surface of permafrost to rise and buildings to tilt. Similarly, early road construction increased the permafrost and several Arctic highways now run nearly a metre above the surrounding land. To offset this, all large, modern unheated buildings and airstrips are constructed on thick gravel pads to try to maintain the level of the frost table. Heat extractors are installed to try to prevent the extension of the thermokarst.

Human attempts to exploit periglacial regions commercially have only been made relatively recently in time. They have, so far, often been made without sufficient attention being paid to the seemingly universal problem of trying to balance short-term economic gain with the often longer-term environmental loss. The result has been an upset in the thermal equilibrium of many places, and destruction of the fragile environment.

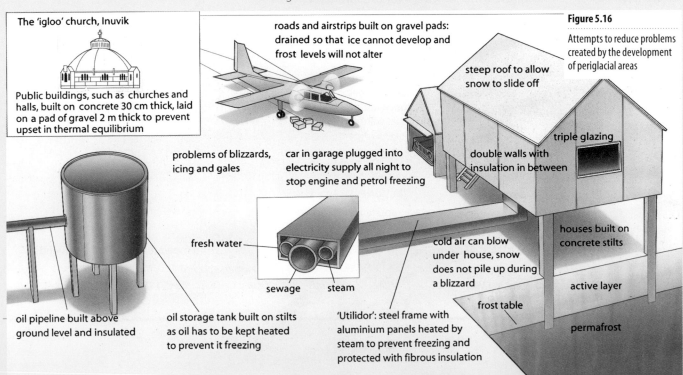

The 'igloo' church, Inuvik

Public buildings, such as churches and halls, built on concrete 30 cm thick, laid on a pad of gravel 2 m thick to prevent upset in thermal equilibrium

roads and airstrips built on gravel pads: drained so that ice cannot develop and frost levels will not alter

Figure 5.16

Attempts to reduce problems created by the development of periglacial areas

steep roof to allow snow to slide off

triple glazing

double walls with insulation in between

problems of blizzards, icing and gales

car in garage plugged into electricity supply all night to stop engine and petrol freezing

cold air can blow under house, snow does not pile up during a blizzard

houses built on concrete stilts

fresh water

sewage   steam

active layer

frost table

permafrost

oil pipeline built above ground level and insulated

oil storage tank built on stilts as oil has to be kept heated to prevent it freezing

'Utilidor': steel frame with aluminium panels heated by steam to prevent freezing and protected with fibrous insulation

Chapter 6

Coasts

"A recent estimate of the coastline of England and Wales is 2750 miles and it is very rare to find the same kind of coastal scenery for more than 10 to 15 miles together."

J. A. Steers, *The Coastline of England and Wales*, 1960

"I do not know what I may appear to the world; but to myself I seem to have been only a boy playing on the sea-shore, and diverting myself in now and then finding a smoother pebble or a prettier shell than ordinary, while the great ocean of truth lay all undiscovered before me."

Isaac Newton, *Philosophiae Naturalis Principia Mathematica*, 1687

The coast is a narrow zone where the land and the sea overlap and directly interact. Its development is affected by terrestrial, atmospheric, marine and human processes (Figure 6.1) and their interrelationships. The coast is the most varied and rapidly changing of all landforms and ecosystems.

Waves

Waves are created by the transfer of energy from the wind blowing over the surface of the sea. (An exception to this definition is those waves, **tsunamis**, which result from submarine shock waves generated by earthquake or volcanic activity.) As the strength of the wind increases, so too does **frictional drag** and the size of the waves. Waves which result from local winds and travel only short distances are known as **sea** whereas those waves formed by distant storms and travelling large distances are referred to as **swell**. The energy acquired by waves depends upon three factors: the wind velocity, the period of time during which the wind has blown, and the length of the fetch. The **fetch** is the maximum distance of open water over which the wind can blow, and so places with the greatest fetch potentially receive the highest energy waves. Parts of south-west England are exposed to the Atlantic Ocean and when the south-westerly winds blow it is possible that some waves may have originated several thousand kilometres away. Dover, by comparison, has less than 40 km of open water between it and France and consequently receives lower energy waves.

Figure 6.1

Factors affecting coasts

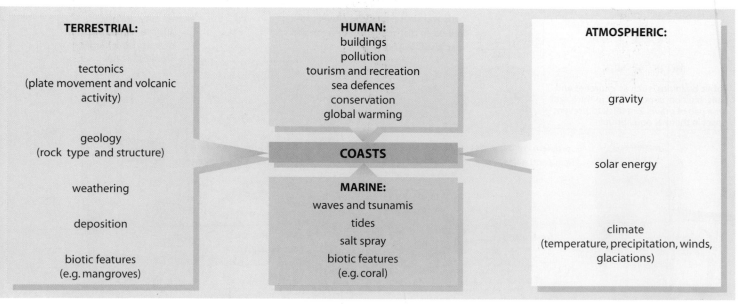

TERRESTRIAL:	HUMAN:	ATMOSPHERIC:
tectonics (plate movement and volcanic activity)	buildings pollution tourism and recreation sea defences conservation global warming	gravity
geology (rock type and structure)	**COASTS**	solar energy
weathering	MARINE:	
deposition	waves and tsunamis tides salt spray	climate (temperature, precipitation, winds, glaciations)
biotic features (e.g. mangroves)	biotic features (e.g. coral)	

Q Locate Cromer, Lowestoft and Felixstowe on a map which shows East Anglia and the North Sea. Which is the direction of the greatest fetch for each place? Which location is likely to receive the largest (highest energy) waves, and which the smallest (lowest energy)? Could the direction of greatest fetch be one factor in determining the size of the three ports?

Wave terminology

The **crest** and the **trough** are respectively the highest and lowest points of a wave (Figure 6.2).

Wave height (*H*) is the distance between the crest and the trough. The height has to be estimated when in deep water. Wave height rarely exceeds 6 m although freak, unexpected waves can be a hazard to life, property and shipping (the Land's End tragedy).

Wave period (*T*) is the time taken for a wave to travel through one wave length. This can be timed either by counting the number of crests per minute or by timing 11 waves and dividing by 10 — i.e. the number of intervals.

Wave length (*L*) is the distance between two successive crests. It can be determined by either of two formulae:

$$L = 1.56T^2$$
or $L = CT$

Wave velocity (*C*) is the speed of movement of a crest in a given period of time.

Wave steepness ($H \div L$) is the ratio of the wave height to the wave length. This ratio cannot exceed 1:7 (0.14) because at that point the wave will break. Steepness determines whether waves will build up or degrade beaches. Most waves have a steepness of between 0.005 and 0.05.

The **energy** (*E*) of a wave in deep water is expressed by the formula:

$$E \propto \text{(is proportional to) } LH^2$$

This means that even a slight increase in wave height can generate large increases in energy. It is estimated that the average pressure of a wave in winter is 11 tonnes per sq m, but this may be three times greater during a storm — it is little wonder that under such conditions sea defences may be destroyed and that wave power is a potential source of renewable energy (page 497).

Swell is characterised by waves of low height, gentle steepness, long wave length and a long period. **Sea**, with opposite characteristics, usually has higher energy waves.

Waves in deep water
Deep water is when the depth of water is greater than half the wave length:

$$(D = > \frac{L}{4})$$

The drag of the wind over the sea surface causes water and floating objects to move in an **orbital motion** (Figure 6.3). Waves are surface features (submerged submarines are unaffected by storms) and therefore the sizes of the orbits decrease rapidly with depth. Any floating object in the sea has a small net horizontal movement but a much larger vertical motion.

Waves in shallow water
As waves approach shallow water, i.e. when their depth is less than half the wavelength,

$$(D = < \frac{L}{2})$$

friction with the sea bed increases. As the base of the wave begins to slow down, the circular oscillation becomes more elliptical (Figure 6.4). As the water depth continues to decrease, so does the wave length.

Meanwhile the height and steepness of the wave increase until the upper part spills or plunges over. The point at which the wave breaks, the **plunge line**, is where the depth of water and the height of the wave are virtually equal. The body of foaming water which then rushes up the beach is called the **swash**, while any water returning down to the sea is the **backwash**.

Figure 6.2

Wave terminology

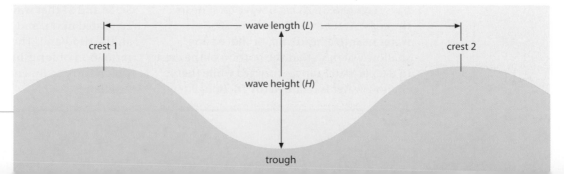

Figure 6.3

Movement of an object in deep water: the diagrams show the circular movement of a ball or piece of driftwood through five stages in the passage of one wave length (crest 1 to crest 2); although the ball moves vertically up and down and the wave moves forward horizontally, there is very little horizontal movement of the ball until the wave breaks; the movement is orbital and the size of the orbit decreases with depth

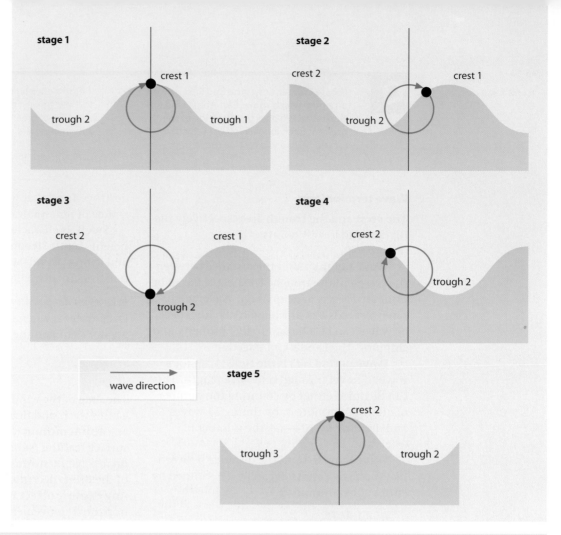

Figure 6.4

Why a wave breaks

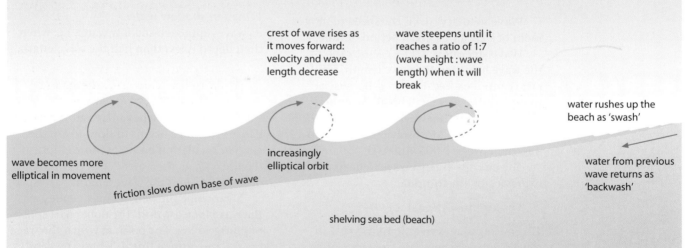

Wave refraction

Where waves approach an irregular coastline, they are refracted — i.e. they become increasingly parallel to the coastline. This is best illustrated where a headland separates two bays (Figure 6.5). As each wave crest nears the coast, it tends to drag in the shallow water near to a headland, or indeed any shallow water, so that the portion of the crest in deeper water moves forward while that in shallow water is retarded (by frictional drag) so that the wave bends. The **orthogonals** (lines drawn at right angles to wave crests) in Figure 6.5 represent four stages in the advance of a particular wave crest. It is apparent from the convergence of lines S^1, S^2, S^3 and S^4 that wave energy becomes concentrated upon, and so accentuates erosion at, the headland. The diagram also shows the formation of **longshore** (littoral) **currents**, which carry sediment away from the headland.

Figure 6.5

Wave refraction at a headland

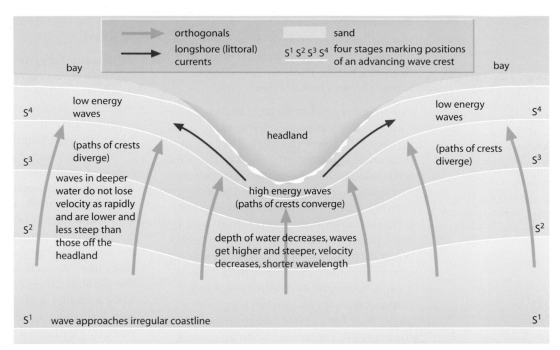

Beaches

Beaches may be divided into three sections — **backshore** (upper), **foreshore** (lower) and **nearshore** — based upon the influence of waves (Figure 6.6). A beach forms a buffer zone between the waves and the coast. If the beach proves to be an effective buffer, it will dissipate wave energy without experiencing any net change itself. Because it is composed of loose material, a beach can rapidly adapt its shape to changes in wave energy. It is, therefore, in dynamic equilibrium with its environment (Framework 1, page 39).

Beach profiles fall between two extremes: those which are wide and relatively flat; and those which are narrow and steep. The gradient of natural beaches is dependent upon the interrelationship between two main variables:

- **Wave energy** Field studies have shown a close relationship between the profile of a beach and the action of two types of wave: constructive and destructive (page 128). However, the effect of wave steepness on beach profiles is complicated by the second variable.
- **Particle size** There is also, due to differences in the relative dissipation of wave energy, a distinct relationship between beach slope and particle size. This relationship is partly due to grain size and partly to percolation rates, both of which are greater on shingle beaches than on sand (pages 129-30). Consequently, shingle beaches are steeper than sand beaches (Figure 6.6)

Figure 6.6

Wave zones and beach morphology (*after* King, 1980)

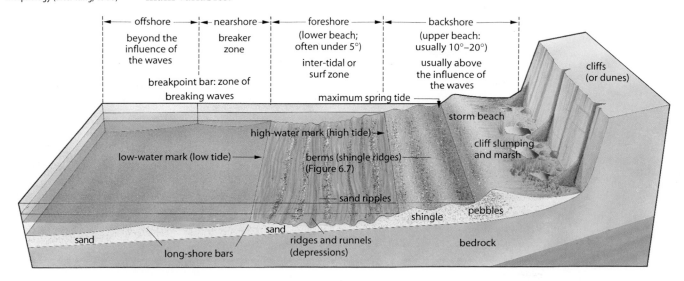

Wave energy

It is the steepness of waves that determines whether they are likely to build up or degrade a beach. There are two extreme forms: constructive and destructive.

Constructive (flat or surging) waves

Constructive waves cause sediment to build up above the low-water mark. They result from swell and are most common where the fetch is large. These waves are flat and low. The wave length is large (up to 100 m between crests) in relation to wave height (often less than 1 m). This results in a long wave period (usually only 6–8 waves breaking in each minute). Individual waves break near to the shore and, due to their flatness, are low in energy. As constructive waves commonly occur on beaches with a low angle, they have a wide area to cross and so the energy in the swash is soon dissipated, leaving a weak backwash. Consequently, sand and shingle is slowly, but constantly, moved up the beach (Figure 6.7a). This will gradually increase the gradient of the beach and leads to the formation of **berms** at its crest (Figures 6.8 and 6.9) and, especially on sandy beaches, **ridges and runnels** (Figure 6.6). Waves which approach the shore at an angle other than 90° will rapidly lose energy as they try to overcome friction on crossing the beach.

Destructive (steep or spilling) waves

Destructive waves comb material down the beach, depositing it below the low water mark. They are more common where the fetch is short. These waves are steep and high. The wave length is small (perhaps only 20 metres between crests) in relation to wave height (usually over one metre). This results in a shorter wave period (usually between 10 and 14 waves breaking in each minute). Individual waves break some distance away from the shore and, due to their steepness, are high in energy. As destructive waves are more likely to occur on beaches with a steeper angle, their energy is concentrated upon a smaller area. Although some shingle may be thrown up above the high water mark by very large waves, forming a **storm beach** (Figure 6.8), most material is carried downwards by the strong backwash to form a **longshore (breakpoint) bar** (Figures 6.6 and 6.7b). As material is constantly moved downwards, the beach profile becomes increasingly more gentle in its lower section.

For waves of a particular size and steepness, the swash and backwash adjust the beach profile until it is in equilibrium.

Figure 6.7

Differences between waves

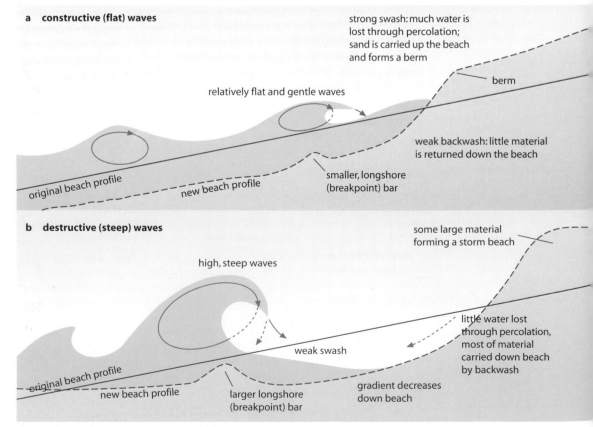

a constructive (flat) waves

relatively flat and gentle waves

strong swash: much water is lost through percolation; sand is carried up the beach and forms a berm

berm

weak backwash: little material is returned down the beach

smaller, longshore (breakpoint) bar

original beach profile

new beach profile

b destructive (steep) waves

high, steep waves

some large material forming a storm beach

weak swash

little water lost through percolation, most of material carried down beach by backwash

gradient decreases down beach

larger longshore (breakpoint) bar

original beach profile

new beach profile

Complete the following table:

	Constructive/flat waves	Destructive/steep waves
Wave height		
Wave steepness		
Wave length		
Wave period		
Frequency per minute		
High or low energy?		
Beach gradient		
Stronger swash or backwash?		

Particle size

This factor complicates the influence of wave steepness on the morphology of a beach. The fact that shingle beaches have a steeper gradient than sandy beaches is due mainly to differences in percolation rates resulting from differences in particle size — i.e. water will pass through coarse-grained shingle more rapidly than through fine-grained sand (Figure 8.2).

Figure 6.8

Storm beaches and berms: berms mark the limits of successively lower tides

S + 1 = high tide after the spring high tide
S + 2 = second high tide after the spring high tide
S + 3 = third high tide after spring high tide

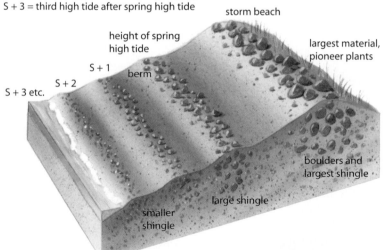

Shingle beaches

Shingle may make up the whole, or just the upper part, of the beach and, like sand, it will have been sorted by wave action. Usually, the larger the size of the shingle, the steeper the gradient of the beach — i.e. the gradient is in direct proportion to shingle size. This is an interesting hypothesis to test by experiment in the field (Framework 7, page 284).

Regardless of whether waves on shingle beaches are constructive or destructive, most of the swash rapidly percolates downwards leaving limited surface backwash. This, together with the loss of energy resulting from friction caused by the uneven surface of the shingle (compare this with the effects of bed roughness of a stream, page 64), means that under normal conditions, very little shingle is moved back down the beach. Indeed, the strong swash will probably transport material up the beach forming a berm at the spring high-tide level. Above the berm there is often a storm beach, comprised of even bigger boulders thrown there by the largest of waves, while below may be several smaller ridges, each marking the height of the successively lower high tides which follow the maximum spring tide (Figures 6.8 and 6.9).

Figure 6.9

Berms and storm beaches in north-east Anglesey, Wales

Sand beaches

Sand usually produces beaches with a gentle gradient. This is because the small particle size allows the sand to become compact when wet, severely restricting the rate of percolation. Percolation is also hindered by the storage of water in pore spaces in sand which enables most of the swash from both constructive and destructive waves to return as backwash. Little energy is lost by friction (sand presents a smoother surface than shingle) so material will be carried down the beach. The material will build up to form a longshore bar at the low-tide mark (Figure 6.6). This will cause waves to break further from the shore, giving them a wider beach over which to dissipate their energy.

The lower parts of sand beaches are sometimes crossed by shore-parallel ridges and runnels (Figure 6.6). The ridges may be broken by channels which drain the runnels at low tide.

The interrelationship between wave energy, beach material and beach profiles may be summarised by the following generalisations which refer to net movements:

- Destructive waves carry material down the beach.
- Constructive waves carry material up the beach.
- Material is carried upwards on shingle beaches.
- Material is carried downwards on sandy beaches.

Figure 6.10

Causes of tides

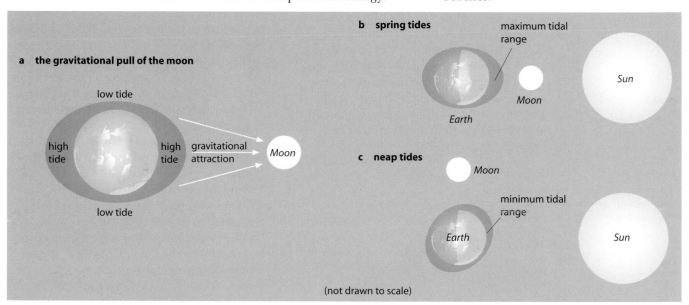

Tides

The position at which waves break over the beach, and their range, are determined by the state of the tide. It has already been seen that the levels of high tides vary (berms are formed at progressively lower levels following spring high tides; Figure 6.8). Tides are controlled by gravitational effects, mainly of the moon but partly of the sun, together with the rotation of the earth and, more locally, the geomorphology of sea basins.

The moon has the greatest influence. Although its mass is much smaller than that of the sun, this is more than compensated for by its closer proximity to the earth. The moon attracts, or pulls, water to the side of the earth nearest to it. This creates a bulge or **high tide** (Figure 6.10a), with a complementary bulge on the opposite side of the earth.

This bulge is compensated for by the intervening areas where water is repelled and which experience a **low tide**. As the moon orbits the earth, the high tides follow it.

A lunar month (the time it takes the moon to orbit the earth) is 28 days and the tidal cycle (the time between two successive high tides) is 12 hours and 25 minutes, giving two high tides, near enough, per day. The sun, with its smaller gravitational attraction, is the cause of the difference in tidal range rather than of the tides themselves. Once every 14 days (i.e. twice in a lunar month), the moon and sun are in alignment on the same side of the earth (Figure 6.10b). The increase in gravitational attraction generates the **spring tide** which produces the highest high tide, the lowest low tide and the maximum tidal range.

Midway between the spring tides are the **neap tides**, occurring when the sun, earth and moon form a right angle with the earth at the apex (Figure 6.10c). As the sun's attraction partly counterbalances that of the moon, the tidal range is at a minimum with the lowest of high tides and the highest of low tides. Spring and neap tides vary by approximately 20 per cent above and below the mean high-tide and low-tide levels.

So far, we have seen how tides might change on a uniform or totally sea-covered earth. However, the earth is not uniform and often the explanation of local tidal patterns is very complex.

The **Coriolis force** (page 208) is the effect of the earth's rotation on moving air and water. In the northern hemisphere, this effect is deflection to the right; it can have a major effect on tidal range. For example, as a rising tide moves towards Britain from the south-west, part of the water will travel northwards up the Irish Sea (Figure 6.11). As it is deflected to the right, there will be higher tides on the Welsh and English coasts than in Ireland. Later, this wave will travel outwards through the North Sea producing higher tides on the English coast than on Danish and Dutch coasts.

The morphology of the sea bed and coastline also affects tidal range. In the example of the North Sea, as the tidal wave travels south it moves into an area where both the width and the depth of the sea decrease. This results in a rapid accumulation, or funnelling, of water to give an increasingly higher tidal range — the range at Dover is several metres greater than in northern Scotland (Figure 6.11). Estuaries where incoming tides are forced into rapidly narrowing valleys will also have considerable tidal ranges — e.g. the Severn estuary with 13 m, the Rance (Brittany) with 11.6 m and the Bay of Fundy (Canada) with 15 m. It is due to these extreme tidal ranges that the Rance has the world's first tidal power station, while the Bay of Fundy and the Severn have, respectively, experimental and proposed schemes for electricity generation (page 496). Extreme narrowing of estuaries can concentrate the tidal rise so rapidly that an advancing wall of water, or **tidal bore**, may travel upriver, e.g. the Rivers Severn and Amazon. In contrast, small enclosed seas have only minimal tidal ranges, e.g. the Mediterranean with 0.01 m.

Figure 6.11

Tidal range: the effects of the Coriolis force and North Sea geomorphology

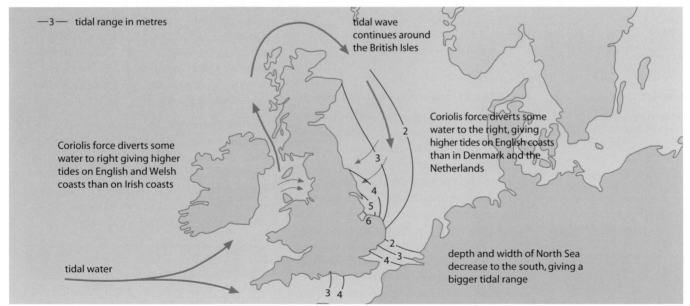

Storm surges

These are rapid rises in sea level where water is piled up against a coastline far in excess of normal conditions. Where storm surges occur in densely populated areas, they pose a major natural hazard as they can cause considerable loss of life, damage to property and disruption of everyday life. Two areas particularly prone to storm surges, the southern North Sea and the Bay of Bengal, are described in Places 15. In both instances, the causes of the storm surges are areas of deep low atmospheric pressure. In the case of the North Sea, these are depressions; in the Bay of Bengal, they are cyclones (page 219).

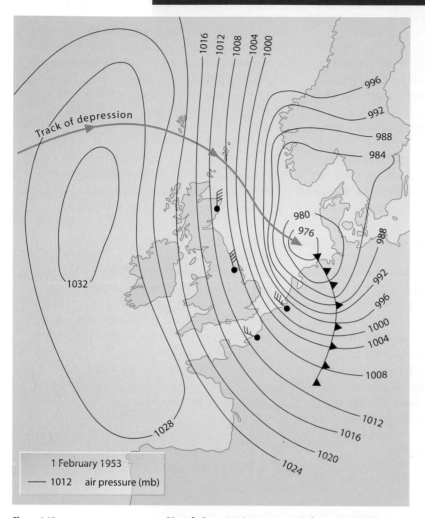

Figure 6.12

The North Sea storm surge of 1 February 1953

North Sea, 31 January–1 February 1953

A deep depression to the north of Scotland, instead of following the usual track which would have taken it over Scandinavia, turned southwards into the North Sea (Figure 6.12). As air is forced to rise in a depression (page 214), the reduced pressure tends to raise the surface of the sea area underneath it. If pressure falls by 56 millibars, as it did on this occasion, the level of the sea may rise by up to 0.5 m. The gale-force winds, travelling over the maximum fetch, produced storm waves over 6 m high. This caused water to pile up in the southern part of the North Sea. This event coincided with spring tides and with rivers discharging into the sea at flood levels. The result was a high tide, excluding the extra height of the waves, of over 2 m in Lincolnshire, over 2.5 m in the Thames estuary and over 3 m in the Netherlands. The immediate result was the drowning of 264 people in south-east England and 1835 in the Netherlands. To prevent such devastation by future surges, the Thames Barrier and the Dutch Delta Scheme have since been constructed. Both schemes needed considerable capital and technology to implement.

Bay of Bengal

The south of Bangladesh includes many flat islands formed by deposition from the Rivers Ganges and Brahmaputra. This delta region is ideal for rice growing and is home to an estimated 40m people. However during the autumn, tropical cyclones (tropical low pressure storms) funnel water northwards up the Bay of Bengal which becomes increasingly narrower and shallower towards Bangladesh. The water sometimes builds up into a surge which may exceed 4 m in height and which may be capped by waves reaching a further 4 m. The result can be a wall of water which sweeps over the defenceless islands. Three days after one such surge in 1985, the Red Cross suggested that over 40 000 people had probably been drowned, many having been washed out to sea as they slept. The only survivors were those who had climbed to the tops of palm trees and managed to cling on despite the 180 km/hr winds. The Red Cross feared outbreaks of typhoid and cholera in the area because fresh water had been contaminated. Famine was a serious threat as the rice harvest had been lost under the salty waters.

The flood hazard in Bangladesh appears to be increasing as global warming causes world sea-levels to rise and as the rivers continue to build up the delta by the deposition of silt. The World Bank has recently proposed financing a coastal protection scheme as Bangladesh itself lacks the necessary capital and technology.

Year	Height of storm surge	Death toll (estimated)
1963	5.2	22 000
1966	6.1	80 000
1970	0.1	300 000 (north-east India)
1985	5.7	40 000
1988	4.8	25 000
1990	6.3	140 000
1991	6.1	150 000

Figure 6.13

Waves breaking on Filey Brigg, Yorkshire: wave energy is absorbed by a band of rock and so the cliff behind is protected

Coastal erosion

Erosion processes operating at the coast

Wave pounding Steep waves have considerable energy. When they break as they hit the foot of cliffs or sea walls, they may generate shock-waves of up to 30 tonnes per sq m. Some sea walls in parts of eastern England need replacing within 25 years of being built due to wave pounding (Case Sudy 6).

Hydraulic pressure When a parcel of air is trapped and compressed, either in a joint in a cliff or between a breaking wave and a cliff, then the resultant increase in pressure may, over a period of time, weaken and break off pieces of rock or damage sea defences.

Abrasion/corrasion This is the wearing away of the cliffs by sand, shingle and boulders hurled against them by the waves. It is the most effective method of erosion and is most rapid on coasts exposed to storm waves.

Attrition Rocks and boulders already eroded from the cliffs are broken down into smaller and more rounded particles.

Corrosion/solution This includes the dissolving of limestones by carbonic acid in sea water (compare Figure 2.8), and the evapo-ration of salts to produce crystals which expand as they form and cause the rock to disintegrate (Figure 2.2). Salt from sea water or spray is capable of corroding several rock types. The secretions from pioneer blue-green bacteria also contribute to corrosion (page 266).

Sub-aerial Cliffs may be worn down by non-marine processes. Rain erodes cliffs either by falling directly onto them or as a result of throughflow or surface runoff from land. This water, together with the effects of weathering by wind and frost, can lead to mass movement either as soil creep on gentle slopes or slumping and landslides on steeper slopes (Figures 2.17 and 2.28).

Human activity The increase in pressure resulting from building on cliff tops and the removal of beach material which may otherwise have protected the base of the cliff both contribute to more rapid coastal erosion.

Rates of erosion may be reduced, locally, by the construction of sea defences. Human activity therefore has the effect of disturbing the equilibrium of the coast system (Case Study 6).

Factors affecting the rate of erosion

1 **Breaking point of the wave** A wave which breaks as it hits the foot of a cliff releases most energy and causes maximum erosion. If the wave hits the cliff before it breaks, then much less energy is transmitted whereas a wave breaking further offshore will have had its energy dissipated as it travelled across the beach (Figure 6.13).

2 **Wave steepness** Very steep destructive waves formed locally (sea) have more energy and erosive power than gentle constructive waves formed many kilo-metres away (swell).

3 **Depth of sea, length and direction of fetch, configuration of coastline** A steeply shelving beach creates higher and steeper waves than one with a more gentle gradient. The longer the fetch, the greater the time available for waves to collect energy from the wind. The existence of headlands with vertical cliffs tends to concentrate energy by wave refraction (page 126).

4 **Supply of beach material** Although a continual supply of material is needed to erode the cliffs, a surfeit protects the coast by absorbing wave energy.

5 **Beach width** Cliffs containing a readily available supply of sand (e.g. in northern Norfolk) will form wider beaches than those with limited amounts (e.g. Holderness). The wider the beach the greater will be the loss of energy (waves take longer to pass over it) and the slower the rate of cliff erosion.

6 **Rock resistance, structure and dip** The strength of coastal rocks influences the rate of erosion (Figure 6.14). In Britain, it is coastal areas where glacial till was deposited which are being worn back most rapidly (Places 16). When Surtsey first arose out of the sea off the south-west coast of Iceland in 1963 (Places 3, page 16), it consisted of unconsolidated volcanic ash. It was only when the ash was covered and protected by a lava flow the following year that the island's survival was seemingly guaranteed.

Figure 6.14

Rock type and average rates of erosion

Rock type	Location	Rate of erosion (m/yr)
Volcanic ash	Krakatoa	40
Glacial till	Holderness	2
Glacial till	Norfolk	1
Chalk	South-east England	0.3
Shale	North Yorkshire	0.09
Granite	South-west England	0.001

Rocks which are well-jointed or have been subject to faulting have an increased vulnerability to erosion. The steepest cliffs are usually where the rock's structure is horizontal or vertical and the gentlest where the rock dips upwards away from the sea. In the latter case, blocks may break off and slide downwards (Figure 2.17). Erosion is also rapid where rocks of different resistance overlie one another, e.g. chalk and Gault clay in Kent.

Places 16 Holderness

This stretch of coastline is retreating by an average of 2 m a year. Since Roman times, the sea has encroached by nearly 3 km, and some 50 villages mentioned in the Domesday Book of 1086 have disappeared. The storm surge of 1953 removed up to 4 m in one night.

The coast is composed of glacial till which presents little resistance to storm waves. These waves can reach 4 m in height when the wind is in the north-east — i.e. the direction of maximum fetch. The local till contains much fine material but limited amounts of sand. The finer material is washed out to sea, leaving a narrow beach which is incapable of absorbing all the energy from the breaking waves. The destructive storm waves comb material down the beach leaving the cliff face unprotected. Waves attack the base of the cliff making it unstable and, together with the weight added to the till by rainwater, may cause the cliff to collapse (Figure 6.15). The cliff-foot rubble is then available for removal by wave energy, backwash and longshore drift.

"Parts of Britain's coastline are being eroded at rates of up to 6 m a year. Each year, there are large landslips on the same scale as the one which has just destroyed the Holbeck Hall Hotel in Scarborough (Figure 2.29). Yet they rarely make the headlines because usually it is fields rather than homes that are destroyed. Scientists now warn that coastal defence structures constructed to combat erosion often increase problems further along the coast (see Case Study 6). Professor Pethick, a coastal geomorphologist at Hull University, has claimed that the best overall solution is to avoid building in the worst-affected areas, surrender land to the sea and compensate land owners. He suggests that compensation should be sorted out at EU level because large quantities of sediment from Britain's east coast are carried by currents across the North Sea to be deposited on the beaches of Belgium, the Netherlands and Germany where they act as a natural sea defence. According to Professor Pethick's estimates, the Holderness cliffs are being eroded at a rate of 2 m/yr, a loss of 1.5m cu m"

(*adapted from The Independent*, July 1993)

Figure 6.15

Holderness houses collapsing into the sea

Figure 6.16

Wave-cut notch at Flamborough Head, Yorkshire

Erosion landforms

Headlands and bays

These are most likely to be found in areas of alternating resistant and less resistant rock. Initially, the less resistant rock experiences most erosion and develops into **bays**, leaving the more resistant outcrops as **headlands**. Later, the headlands receive the highest energy waves and so become more vulnerable to erosion than the sheltered bays (Figure 6.5). The latter now experience low-energy breakers which allow sand to accumulate and so help to protect that part of the coastline.

Wave-cut platforms

Wave energy is at its maximum when a high, steep wave breaks at the foot of a cliff. This results in undercutting of the cliff to form a **wave-cut notch** (Figure 6.16). The continual undercutting causes increased stress and tension in the cliff until eventually it collapses. As these processes are repeated, the cliff retreats leaving, at its base, a gently sloping, **wave-cut platform** which has a slope angle of less than 4° (Figure 6.17). The platform, which appears relatively even when viewed from a distance, cuts across rocks regardless of their type and structure. A closer inspection of this inter-tidal feature usually reveals that it is deeply dissected by abrasion, resulting from material carried across it by tidal movements, and corrosion. As the cliff continues to retreat, the widening of the platform means that incoming waves break further out to sea and have to travel over a wider area of beach. This will dissipate their energy, reduce the rate of erosion of the headland, and limit the further extension of the platform. It has been hypothesised that wave-cut platforms cannot exceed 0.5 km in width.

Where there has been negative change in sea-level (page 75), former wave-cut platforms remain as **raised beaches** above the present influence of the sea (Figure 6.45).

Figure 6.17

Wave-cut platform at Flamborough Head, Yorkshire

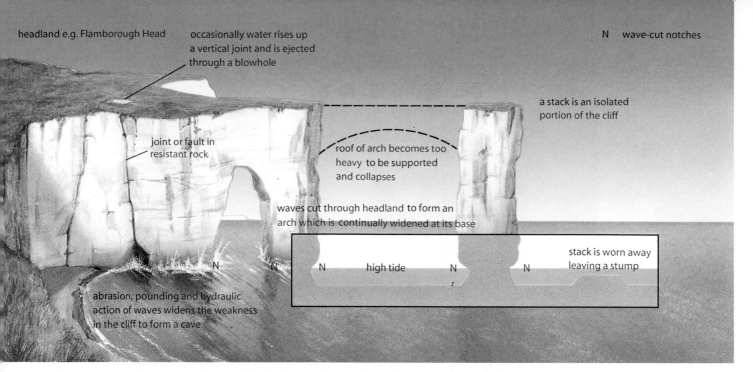

headland e.g. Flamborough Head

occasionally water rises up
a vertical joint and is ejected
through a blowhole

N   wave-cut notches

a stack is an isolated
portion of the cliff

joint or fault in
resistant rock

roof of arch becomes too
heavy to be supported
and collapses

waves cut through headland to form an
arch which is continually widened at its base

N   high tide   N   N

N   stack is worn away
leaving a stump

abrasion, pounding and hydraulic
action of waves widens the weakness
in the cliff to form a cave

Figure 6.18

The formation of caves, blowholes, arches and stacks

Figure 6.19

Old Harry near Swanage, Dorset: this shows a small wave-cut platform (centre), a cave (left foreground) and an arch within a large stack (Old Harry); a second stack (Old Harry's Wife) lies to the right

Caves, blowholes, arches and stacks

Where cliffs are of resistant rock, wave action attacks any lines of weakness such as a joint or a fault. Sometimes the sea cuts inland, along a joint, to form a narrow, steep-sided inlet called a **geo**, or to undercut part of the cliff to form a **cave**. As shown in Figure 6.18, caves are often enlarged by several combined processes of marine erosion. Erosion may be vertical, to form **blowholes**, but is more typically backwards through a headland to form **arches** and **stacks** (Figures 6.18 and 6.19).

These landforms, which often prove to be attractions to sight-seers and mountaineers, can be found at The Needles (Isle of Wight), Old Harry (near Swanage, Figure 6.19) and Flamborough Head (Yorkshire), which are all cut into chalk, and at The Old Man of Hoy (Orkneys) which is Old Red Sandstone.

Q

1 Draw annotated sketches to explain the formation of:
   a the wave-cut platform in Figure 6.17;
   b the caves, arch and stacks in Figure 6.19.

2 Why are stacks more likely to occur in jointed rocks such as chalk rather than in unconsolidated material such as glacial till?

Coasts

Transportation of beach material

Up and down the beach

As we have already seen, flat, constructive waves tend to move sand and shingle up the beach, whereas the net effect of steep, destructive waves is to comb the material downwards.

Longshore (littoral) drift

Usually wave crests are not parallel to the shore, but arrive at a slight angle. Only rarely do waves approach a beach at right angles. The wave angle is determined by wind direction, the local configuration of the coastline, and refraction at headlands and in shallow water. The oblique wave angle creates a nearshore current known as **longshore (or littoral) drift** which is capable of moving large quantities of material in a down-drift direction (Figure 6.20). On many coasts, longshore drift is predominantly in one direction — e.g. on the south coast of England, where the maximum fetch and prevailing wind are both from the south-west, there is a predominantly eastward movement of beach material. However, brief changes in wind, and therefore wave, direction can cause the movement of material to be reversed.

As material is moved along the shore, it often follows a zig-zag pattern. This is because when a wave breaks, the swash carries material up the beach at the same angle as that at which the wave approached the shore. As the swash dies away, the backwash and any material carried by it returns straight down the beach, at right angles to the waterline, under the influence of gravity. If beach material is carried a considerable distance, it becomes smaller, more rounded and better sorted.

Where beach material is being lost through longshore drift, the coastline in that locality is likely to be worn back more quickly because the buffering effect of the beach is lessened. To counteract this process, wooden breakwaters or **groynes** are built (Figure 6.21). Groynes encourage the local accumulation of sand (important in tourist resorts) but can result in a depletion of material, and therefore an increase in erosion, further along the coast (Case Study 6).

Figure 6.20

The effects of longshore drift

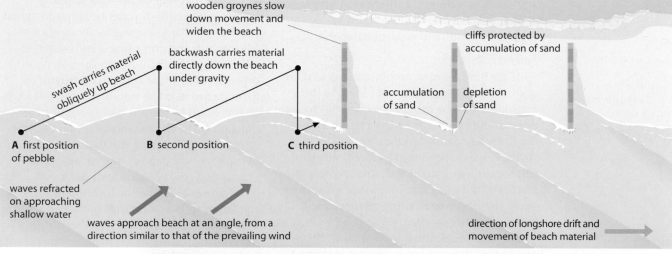

wooden groynes slow down movement and widen the beach

backwash carries material directly down the beach under gravity

swash carries material obliquely up beach

cliffs protected by accumulation of sand

accumulation of sand

depletion of sand

A first position of pebble

B second position

C third position

waves refracted on approaching shallow water

waves approach beach at an angle, from a direction similar to that of the prevailing wind

direction of longshore drift and movement of beach material

Figure 6.21

The effect of groynes on longshore drift, Southwold, Suffolk: this type of coastal management is usually undertaken at holiday resorts where sandy beaches are a major tourist attraction

Figure 6.22

A spit: Dawlish Warren at the mouth of the River Exe, Devon

Coastal deposition

Deposition occurs where the accumulation of sand and shingle exceeds its depletion. This may take place in sheltered areas with low energy waves or where rapid coastal erosion further along the coast provides an abundant supply of material. In terms of the coastal system, deposition takes place as inputs exceed outputs and the beach can be regarded as a store of eroded material.

Spits

Spits are long, narrow accumulations of sand and/or shingle with one end joined to the mainland and the other projecting out to sea or extending part way across a river estuary (Figure 6.22). Sandy spits are formed by constructive waves produced by swell; shingle spits occur mainly where destructive waves have resulted from sea; composite spits occur when the larger-sized shingle is deposited before the finer sands.

In Figure 6.23, the line **X–Y** marks the position of the original coastline. At point **A**, because the prevailing winds and maximum

Figure 6.23

Stages in the formation of a spit

fetch are from the south-west, material is carried eastwards by longshore drift. When the orientation of the old coastline began to change at **B**, some of the larger shingle and pebbles were deposited in the slacker water in the lee of the headland. As the spit continued to grow, storm waves threw some larger material above the high-water mark (**C**), making the feature more permanent; while, under normal conditions, the finer sand was carried towards the end of the spit at **D**. Many spits develop a hooked or curved end. This may be for two reasons: a change in the prevailing wind to coincide with the second-most-dominant wave direction and second-longest fetch, or wave refraction at the end of the spit carrying some material into more sheltered water.

Eventually the seaward side of the spit will retreat, while longshore drift continues to extend the feature eastwards. A series of recurved ends may form (**E**) each time there is a series of storms from the south-east giving a lengthy period of altered wind direction. Having reached its present-day position (**F**), the spit is unlikely to grow any further — partly because the faster current of the river will carry material out to sea and partly because the depth of water becomes too great for the spit to build upwards above sea level. Meanwhile, the prevailing south-westerly wind will pick up sand from the beach as it dries out at low tide and carry it inland to form **dunes** (**G**). The stability of the spit may be increased by the anchoring qualities of marram grass. At the same time, gentle, low-energy waves entering the sheltered area behind the spit deposit fine silt and mud, creating an area of **saltmarsh** (**H**).

Figure 6.26 shows the location of some of the larger spits around the coast of England and Wales. How do these relate to the direction of the maximum fetch and of the prevailing and dominant winds?

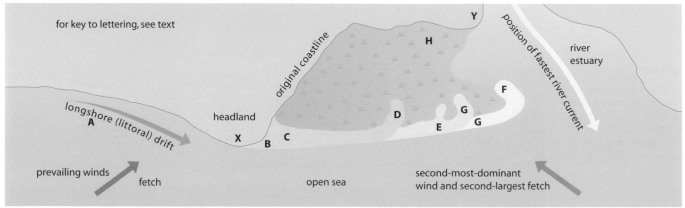

for key to lettering, see text

Y

position of fastest river current

original coastline

H

river estuary

longshore (littoral) drift

A

headland

X

B

C

D

E

G

G

F

prevailing winds

fetch

open sea

second-most-dominant wind and second-largest fetch

Figure 6.24

A tombolo: Loch Eriboll,
Highland, Scotland

Tombolos, bars and barrier islands

A **tombolo** is a beach which extends outwards to join with an offshore island (Figure 6.24). Chesil Beach, in Dorset, links the Isle of Portland to the mainland. Some 30 km long and up to 14 m high, it presents a gently smoothed face to the prevailing winds in the English Channel.

If a spit develops in a bay into which no major river flows, it may be able to build across that bay, linking two headlands, to form a **bar**. Bars straighten coastlines and trap water in lagoons on the landward side. Bars, such as that at Slapton Ley, in Devon (Figure 6.25), may also result in places where constructive waves lead to the landward migration of offshore, sea-bed material.

Barrier islands are a series of sandy islands totally detached from, but running almost parallel to, the mainland. Between the islands, which may extend for several hundred kilometres, and the mainland is a tidal lagoon (Figure 6.27). Although relatively uncommon in Britain, they are widespread globally, accounting for 13 per cent of the world's coastlines. They are easily recognisable on maps of the eastern USA (Places 17), the Gulf of Mexico, the northern Netherlands, West Africa and southern and western Australia. While their origin is uncertain, they do tend to develop on coasts with relatively high energy waves and a low tidal range. One theory suggests that they formed, below low-tide mark, as offshore bars of sand and have moved progressively landwards. An alternative theory suggests that rises in post-glacial sea-level may have partly submerged older beach ridges. In either case, the breaches between the islands seem likely to have been caused by storm waves.

Figure 6.25

A bar: Slapton Ley, Devon

Figure 6.26

Location of some major spits, tombolos and bars in England and Wales

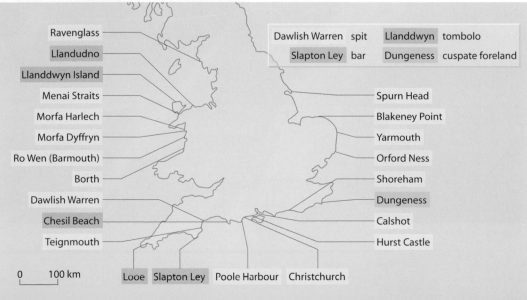

Figure 6.27

Barrier islands off North Carolina, USA, taken from the Apollo spacecraft
(X = position of Figure 6.28)

Barrier islands have a unique morphology, flora and fauna. The smooth, straight, ocean edge is characterised by wide, sandy beaches which slope gently upwards to sand dunes which are anchored by high grasses (Figure 6.28). Behind the dunes, the 'island' interior may contain shrubs and woods, deer and snakes, insects and birds. The landward side is punctuated by sheltered bays, quiet tidal lagoons, salt marshes and, towards the tropics, mangrove swamps. These wetlands provide a natural habitat for oysters, fish and birds. Although barrier islands form the interface between the land and the ocean, they seem fragile in comparison with the power which the wind and sea brings to them. It is virtually impossible for a tropical storm or hurricane to move ashore without first crossing either of the two longest stretches of barrier islands in the world: either that which extends for 2500 km from New Jersey to the southern tip of Florida (Figure 6.27); or the one stretching for 2100 km along the Gulf Coast states to Mexico.

Barrier islands are subject to a process called 'wash over'. This process, which might occur up to 40 times in some years, is when storm waves carry large quantities of sand over the island from the seaward face to the landward side. This results in the seaward side being eroded and pushed backwards. The landward marshes and mangrove swamps become suffocated, and the tidal lagoons are narrowed. From a human viewpoint, barrier islands form an essential natural defence against hurricanes and their storm-force waves.

Figure 6.28

Barrier island on Core Banks, looking south (see X on Figure 6.27 for location)

Sand dunes

Sand dunes are a dynamic landform whose equilibrium depends upon the interrelationship between mineral content (sand) and vegetation.

Longshore drift may deposit sand in the inter-tidal zone. As the tide ebbs, the sand will dry out allowing winds from the sea to move material up the beach by saltation (page 167). This process is most likely to occur when the prevailing winds come from the sea and where there is a large tidal range which exposes large expanses of sand at low tide. Sand may become trapped by seaweed and driftwood on berms or at the point of the highest spring tides. Plants begin to colonise the area (page 268) stabilising the sand and encouraging further accumulation. The regolith has a high pH value due to calcium carbonate from seashells.

Embryo dunes are the first to develop (Figure 6.29). They become stabilised by the growth of lyme and marram grass. As these grasses trap more sand, the dunes build up and become, due to the high rate of percolation, increasingly arid. Plants need either succulent leaves to store water (sand couch), or thorn-like leaves to reduce transpiration in the strong winds (prickly saltwort), or long tap-roots to reach the water table (marram grass). As more sand accumulates, the embryo dunes join to form **foredunes** which can attain a height of 5 m (Figures 6.29 and 6.30). Due to a lack of humus, their colour gives them the name **yellow dunes**. The dunes become increas-ingly grey as humus and bacteria from plants and animals are added and they gradually become more acidic and vegetation-covered. These **grey (mature) dunes** may reach a height of 10–30 m before the supply of fresh sand is cut off by their increasing distance from the beach (Figure 11.10). There may be several parallel ridges of old dunes (as at Morfa Harlech, Figure 6.31), separated by low-lying, damp slacks. Heath plants begin to dominate the area as acidity, humus and moisture content all increase (Figure 11.8). Paths cut by humans and animals expose areas of sand. As the wind funnels along these tracks, **blow-outs** may form in the now **wasting dunes**. To combat further erosion at Morfa Harlech, parts of the dunes have been fenced off and marram grass has been planted to try to re-stabilise the area and to prevent any inland migration of the dunes (Figure 10.38).

The above idealised scheme can be interrupted at any stage by storms or human use. If the supply of sand is cut off, then new embryo dunes cannot form and yellow dunes may be degraded so that it is the older, grey dunes which line the beach.

Figure 6.29

A transect across sand dunes, based on fieldwork at Morfa Harlech, North Wales

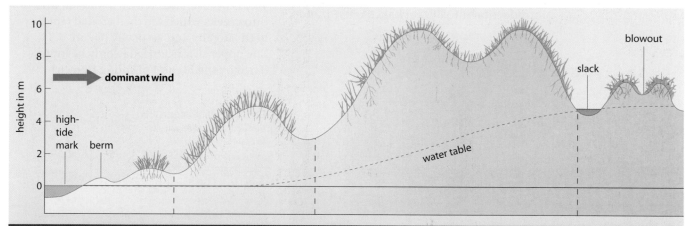

Type of dune	Embryo	Fore or yellow dunes	Grey dunes and dune ridges	Wasting dunes with blowouts
Dune height (m)	1	5	8–10	6–8
Percentage of exposed sand	80	20	less than 10	over 40 on dunes
Humus and moisture content	very little	some humus, very little moisture	humus increases inland, water content still low	high humus, brackish water in slacks
pH	over 8	slightly alkaline	increasingly acid inland: pH 6.5–7	acid: pH 5–6
Plant types	sand couch, lyme grass	marram, xerophytic species	creeping fescue, sea spurge, some marram, cotton grass and heather	heather, gorse on dunes, *Juncus* in slacks

Figure 6.30

Embryo and foredunes at Morfa Harlech, North Wales (refer also to Figures 11.8 and 11.9)

Figure 6.32

Llanhridian saltmarsh, Gower peninsula, South Wales (refer also to Figures 11.11 and 11.13)

Figure 6.31

Morfa Harlech from Harlech Castle showing foredunes, grey or wasting dunes, old cliff-line and salt marsh

Saltmarshes

Where there is sheltered water in river estuaries or behind spits, silt and mud will be deposited either by the gently rising and falling tide or by the river, thus forming a zone of **inter-tidal mud flats**. Initially, the area may only be uncovered by the sea for less than 1 hour in every 12-hour tidal cycle. Plants such as algae and *Salicornia* can tolerate this lengthy submergence and the high levels of salinity. They are able to trap more mud around them, creating a surface which remains exposed for increasingly longer periods between tides (Figure 6.32). *Spartina* grows throughout the year and since its introduction into Britain has colonised, and become dominant in, many estuaries. The landward side of the inter-tidal mud flats is marked by a small cliff (Figure 11.12), above which is the flat, **sward zone**. This zone may only be covered by the sea for less than 1 hour in each tidal cycle. Seawater collects in hollows which become increasingly saline as the water evaporates. The hollows often enlarge into **saltpans** which are devoid of vegetation except for certain algae and the occasional halophyte (page 269 and Figure 11.11). As each tide retreats, water drains into **creeks** which are then eroded rapidly both laterally and vertically (Figure 6.33). The upper sward zone may only be inundated by the highest of spring tides.

Figure 6.33

Llanhridian saltmarsh: note the sward zone, creeks and salt pan

Q

Figure 6.34a shows Blakeney spit and the adjacent part of the coastline of north Norfolk as it was in *circa* AD 1600. Figure 16.34b shows the same coast in 1979.

1 a State four coastal changes which have occurred in the area between 1600 and 1979.
   b Explain each of the changes you have identified in **1a**.

2 a With reference to Figure 6.34b, how might the shingle found at A differ from that found at B?
   b Explain the difference.

3 a With reference to Figure 6.34b, describe the changes at Old Far Point which occurred between 1954 and 1974.
   b Explain how these differences might have occurred.

4 Why have sand dunes formed at point **C** on Figure 6.34b?

5 a Where would you expect an area of saltmarsh to form?
   b Give reasons for your answer.

Figure 6.34

Blakeney Point, Norfolk

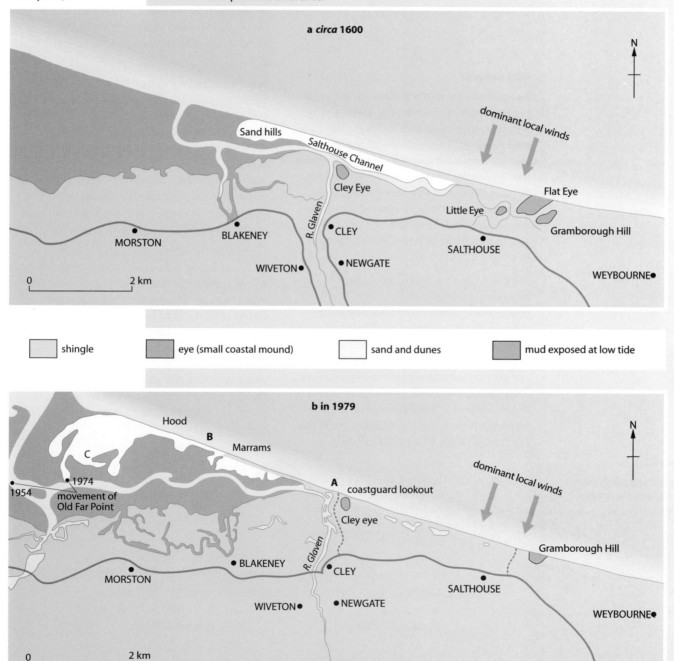

Figure 6.35

A sample population in relation to the total population

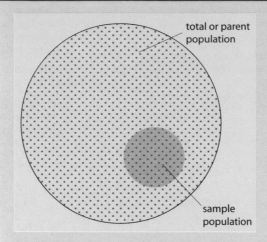

Why sample?

Geographers are part of a growing number of people who find it increasingly useful and/or necessary to use data to quantify the results of their research. The problem with this trend is that the amount of data may be very expensive, too time-consuming, or just impracticable to collect — as it would be, for example, to investigate everybody's shopping patterns in a large city; to find the number of stones on a spit; or to map the land use of all the farms in Britain.

Sampling is the method used to make statistically valid inferences when it is impossible to measure the **total population** (Figure 6.35). It is essential, therefore, to find the most accurate and practical method of obtaining a **representative sample**. If that sample can be made with the minimum of bias, then statistically significant conclusions may be drawn. However, even if every effort is made to achieve precision, it must be remembered that any sample can only be a close estimate.

Sampling basics

Most sampling procedures assume that the total population has a **normal distribution** (Figure 4.15a) which, when plotted on a graph, produces a symmetrical curve on either side of the mean value. This shows that a large proportion of the values are close to the average, with few extremes. Figure 6.36 shows a normal distribution curve and the **standard deviation** (page 231), the measure of dispersion from the mean. Where most of the values are clustered near to the mean, the standard deviation is low.

The larger the sample, the more accurate it is likely to be, and the more likely it is to resemble the parent population; it is also more likely to conform to the normal distribution curve. While the generally accepted minimum size for a sample is 30, there is no upper limit — although there is a point beyond which the extra time and cost involved in increasing the sample size do not give a significant improvement in accuracy (an example of the law of diminishing returns, page 420).

Figure 6.36 shows that, in a normal distribution, 68.27 per cent of the values in the sample occur within a range of ±1 standard deviations (SDs) from the mean; 95 per cent of the values fall within ±2 SDs; and 99 per cent within ±3 SDs. These percentages are known as **confidence limits**, or **probability levels**. Geographers usually accept the 95 per cent probability level when sampling. This means that they accept the chance that, in 5 cases out of every 100, the true mean will lie outside 2 SDs to either side of their sample mean.

Figure 6.36

A normal distribution showing standard deviations from the mean

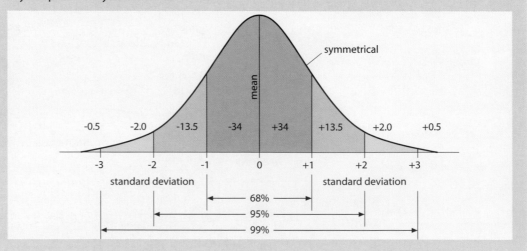

Sampling techniques

Several different methods may be used according to the demands of the required sample and the nature of the parent population. There are two major types with one refinement:

- **Random sampling** This is the most accurate method as it has no bias.
- **Systematic sampling** This method is often quicker and easier to use, although some bias or selection is involved.
- **Stratified sampling** This method is often a very useful refinement for geographers; it can be used with either a random or a systematic sample.

Random sampling

Under normal circumstances, this is the ideal type of sample because it shows no bias. Every member of the total population has an equal chance of being selected and the selection of one member does not affect the probability of selection of another member. The ideal random sample may be obtained using **random numbers**. These are often generated by computer and are available in the form of printed tables of random numbers, but if necessary they can be obtained by drawing numbers out of a hat. Random number tables usually consist of columns of pairs of digits. Numbers can be chosen by reading either along the rows or down the columns, provided only one method is used. Similarly, any number of figures may be selected — six for a grid reference, four for a grid square, three for house numbers in a long street, etc. Using the grid shown in Figure 6.37, the random number table given below yields eight 6-figure grid references: 927114; (986691 has to be excluded

because the grid does not contain these numbers); 906126; etc.

One feature of a genuine random sample is that the same number can be selected more than once — so remember that if you are pulling numbers from a hat, they should be replaced immediately after they have been read and recorded.

There are three alternative ways of using random numbers to sample areal distributions (patterns over space) (Figure 6.37).

1 **Random point** A grid is superimposed over the area of the map to be sampled. Points, or map references, are then identified using random number tables, and plotted on the map. The eight points identified earlier (in the random number table) have been plotted on Figure 6.37a. A large number of points may be needed to ensure coverage of the whole area — see Figure 6.39!

2 **Random line** Random numbers are used to obtain two end points which are then joined by a line, as in Figure 6.37b which uses the same eight random points, in the order in which they occurred in the table. Several random lines are needed to get a representative sample (e.g. lines across a city to show transects of variation in land use).

3 **Random area** Areas of constant size, e.g. grid squares or quadrats, are obtained using random numbers. By convention, the number always identifies the south-west corner of a grid square. If sample squares one-quarter the size of a grid square are used, together with the same sample points, their locations are shown on Figure 6.37c; note that the point in the north-east

Part of a random number table

9271	0143	2141	9381
1498	3796	4413	1405
6691	4294	6077	9091
9061	1148	9493	1940
2660	7126	7126	4591
3459	7585	4897	8138
6090	7962	5766	7228
2191	9271	9042	5884

Figure 6.37

Random sampling using point, line and area techniques

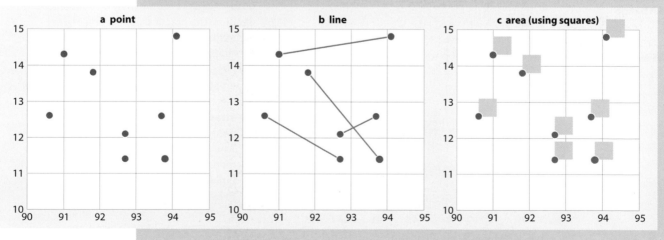

Figure 6.38

Systematic sampling using point, line and area techniques

cannot be used because part of the sample square lies outside the study area. This method can be used to sample land-use areas or the distribution of plant communities over space.

The advantages of random sampling include its ability to be used with large populations and its avoidance of bias. Careful **sample design** is needed, however, to avoid the possibility of achieving misleading results — for example, when sampling small populations, and when sampling over a large area. Also, when used in the field, it may involve considerable time and energy in visiting every point.

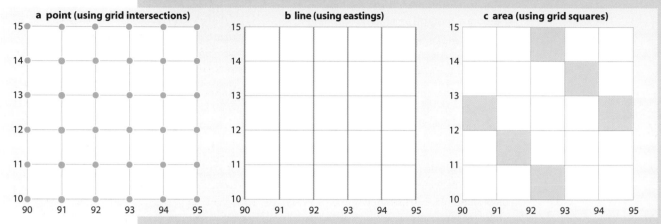

a point (using grid intersections)   **b line (using eastings)**   **c area (using grid squares)**

Systematic sampling

A systematic sample is one in which values are selected in a regular way — e.g. choosing every 10th person on a list or every 20th house in a street. This can be an easier method in terms of time and effort than random sampling. Like random sampling, it can be operated using individual points, lines or areas (Figure 6.38).

1 **Systematic point** This can show changes over distance, e.g. by sampling the land use every 100 m. It can also show change through time, e.g. by sampling from the population censuses (taken every 10 years).

2 **Systematic line** This may be used to choose a series of equally spaced transects across an area of land, e.g. a shingle spit.

3 **Systematic area** This is often used for land-use sampling, to show change with distance

or through time (if old maps or air photographs are available). Quadrats, positioned at equal intervals, are used for assessing plant distributions.

The main advantage of systematic sampling lies in its ease of use. However, its main disadvantage is that all points do not have an equal chance of selection — it may either overstress or miss an underlying pattern (Figure 6.39).

Stratified sampling

When there are significant groups of known size within the parent population, in order to ensure adequate coverage of all the sub-groups it may be advisable to stratify the sample — i.e. to divide the population into categories and sample within each. Although categorising into groups (layers or strata) may be a subjective decision, the practical application of this technique has considerable advantages for the geographer. Once the groups have been decided, they can be sampled either systematically or randomly, using point, line or area techniques.

1 **Stratified systematic sampling** This method can be useful in many situations — when interviewing people, sampling from maps, and during field work. For example, in political opinion polls, the total population to be sampled can divided (stratified) into equal age and/or socio-economic groups, e.g. 10–19, 20–29, etc. The number

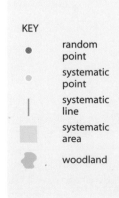

KEY
- random point
- systematic point
| systematic line
■ systematic area
▪ woodland

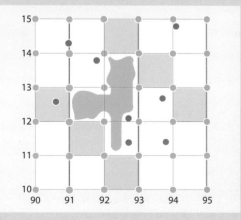

KEY

• random point

 moorland

Figure 6.40

A random point sample, stratified by area

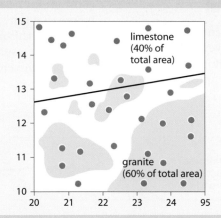

interviewed in each category should be in proportion to its known size in the parent population. This is most easily achieved by sampling at a regular interval (systematically) throughout the entire population, such that the required total sample size is obtained. For example, if a sample size of 800 is required from a total population of 8000 (i.e. a 10 per cent sample), every 10th person would be interviewed.

2 Stratified random sampling This method can be used to cover a wide range of data, both in interviewing and in geographical field work and map work. For example, Figure 6.40 shows the distribution of moorland on two contrasting rock types: granite occupies 60% of the total area and limestone 40%. To discover whether the proportion of moorland cover varies with rock type, the sampling must be in proportion to their relative extents. Thus, if a sample size of 30 points is derived using random numbers, 18 are needed within the granite area (18 is 60 per cent of 30) and 12 within the limestone area (12 is 40 per cent of 30). If it was decided to area sample, 18 quadrats would have to fall within the granite area, and 12 in the limestone.

The advantages of stratified sampling include its potential to be used either randomly or systematically, and in conjunction with point, line or area techniques. This makes it very flexible and useful, as many populations have geographical sub-groups. Care must be taken, however, to select appropriate strata.

Q

In an attempt to see the merits and problems of sampling in the field, four groups of sixth-formers were taken to a spot on the east coast of the Isle of Arran. They were told to devise an appropriate sampling technique:

a to see if beach material decreases in size towards low-water mark; and

b to assess the percentage of vegetation cover present across the spit.

Each group decided independently on a method different from the other three. These have been summarised in Figure 6.41.

1 What do you consider to be the advantages and disadvantages of each of the techniques used by the four groups in terms of the objectives of their study?

2 Which group do you consider should have obtained the most accurate results? Why?

3 Can you suggest a more appropriate method of sampling to establish **a** and **b**?

Figure 6.41

Sampling techniques on a spit as selected by sixth-formers conducting a field-work exercise

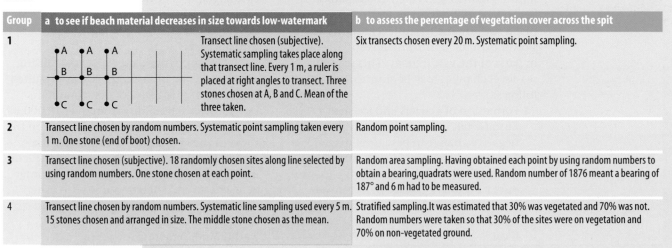

Group	a to see if beach material decreases in size towards low-watermark		b to assess the percentage of vegetation cover across the spit
1	*(diagram of transect with points A, B, C)*	Transect line chosen (subjective). Systematic sampling takes place along that transect line. Every 1 m, a ruler is placed at right angles to transect. Three stones chosen at A, B and C. Mean of the three taken.	Six transects chosen every 20 m. Systematic point sampling.
2	Transect line chosen by random numbers. Systematic point sampling taken every 1 m. One stone (end of boot) chosen.		Random point sampling.
3	Transect line chosen (subjective). 18 randomly chosen sites along line selected by using random numbers. One stone chosen at each point.		Random area sampling. Having obtained each point by using random numbers to obtain a bearing, quadrats were used. Random number of 1876 meant a bearing of 187° and 6 m had to be measured.
4	Transect line chosen by random numbers. Systematic line sampling used every 5 m. 15 stones chosen and arranged in size. The middle stone chosen as the mean.		Stratified sampling. It was estimated that 30% was vegetated and 70% was not. Random numbers were taken so that 30% of the sites were on vegetation and 70% on non-vegetated ground.

Changes in sea-level

Although the daily movement of the tide alters the level at which waves break onto the foreshore, the average position of sea-level in relation to the land has remained relatively constant for nearly 6000 years (Figure 6.42). Before that time there had been several major changes in this mean level, the most dramatic being a result of the Quaternary ice age and plate movements.

Figure 6.42

Eustatic changes in sea-level since 18 000 BC

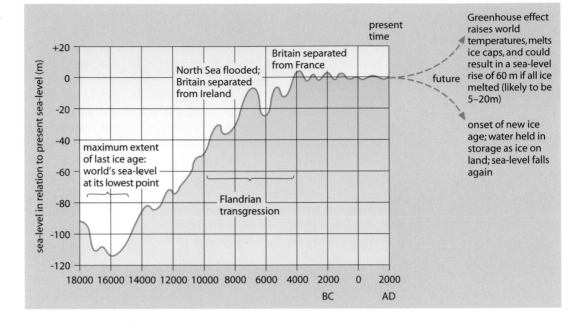

During times of maximum glaciation, large volumes of water were stored on the land as ice, probably three times more than today. This modification of the **hydrological cycle** meant that there was a worldwide, or **eustatic** (glacio-eustatic, page 112), fall in sea-level of an estimated 100–150 m.

As ice accumulated, its weight began to depress those parts of the crust lying beneath it. This caused a local, or **isostatic** (glacio-isostatic, page 112), change in sea-level.

The world's sea-level was at its minimum 18 000 years ago when the ice was at its maximum (Figure 6.42). Later, as temperatures began to rise and ice caps melted, there was first a eustatic rise in sea-level followed by a slower isostatic uplift which is still operative in parts of the world today. This sequence of sea-level changes may be summarised as follows:

1 Formation of glaciers and ice sheets. Eustatic fall in sea-level gives rise to a negative change in base level (page 75).

2 Continued growth of ice sheets. Isostatic depression of the land under the ice produces a positive change in base level.

3 Ice sheets begin to melt. Eustatic rise in sea-level with a positive change in base level.

4 Continued decline of ice sheets and glaciers. Isostatic uplift of the land under former ice sheets results in a negative change in base level.

During this deglaciation, there may have been a continuing, albeit small, eustatic rise in sea-level but this has been less rapid than the isostatic uplift so that base level appears to be falling. Measurements suggest that parts of north-west Scotland are still rising by 4 mm a year and some northern areas of the Gulf of Bothnia (Scandinavia) by 20 mm a year. The uplift in northern Britain is causing the British Isles to tilt and the land in south-east England to be depressed. This process is of utmost importance to the future natural development and human management of British coasts.

Tectonic changes, especially at collision and destructive plate margins, have resulted in the uplift (**orogeny**) of new mountain ranges. Sea shells dating from over 18 000 BP (before present) have been found high in the rocks of the Himalayas, the Alps and the Andes. Local tilting (**epeirogeny**) of the land in the Mediterranean has led to the submergence of several ancient ports, with others now stranded above the present-day sea-level. In south-east England and the Netherlands, epeirogeny has increased the risk of marine flooding. The emergence of

Figure 6.43

A ria: Salcombe, Devon

the Mid-Atlantic Ridge (page 16) could have resulted in a eustatic rise in sea-level of several hundred metres, while the effects of ocean floor spreading could have led to a reversed eustatic change since a 1 per cent increase in the area of the oceans could produce a fall of 40 m in ocean levels. Local volcanic and earthquake activity has also altered the balance between land and sea-levels.

Landforms created by sea-level changes

Changes in sea-level have affected:
- the shape of coastlines and the formation of new features by increased erosion or deposition;
- the balance between erosion and deposition by rivers (page 75) resulting in the drowning of lower sections of valleys or in the rejuvenation of rivers; and
- the migration of plants, animals and people.

Landforms resulting from submergence
Eustatic rises in sea-level following the decay of the ice sheets led to the drowning of many low-lying coastal areas.

Figure 6.44

Geiranger Fjord, Norway

During the ice age, those rivers which continued to flow were able to cut their valleys downwards to the lower base level. When the ice melted and sea-level subsequently rose, the lower parts of the main valley and its tributaries were drowned to produce sheltered, winding inlets, called **rias**. Rias are characteristic of south-west Britain and the north-west of France and Spain (Figure 6.43).

Whereas rias formed when river valleys flowed at right angles to the sea, **Dalmatian coasts** are found where river valleys had formed parallel to the coastline.

Fjords are located where glaciers were able to erode below sea-level. When the ice melted, the valleys were flooded by the rising sea to form long, deep, narrow inlets with precipitous sides and hanging valleys: fjords are drowned glacial troughs (Figure 6.44). Glaciers appear to have followed lines of weakness, such as a pre-glacial river valley or a major fault line. Unlike a ria, which gets progressively deeper towards the sea, a fjord has a relatively shallow seaward entrance, known as a **threshold**, whose origin is debatable.

Lowland areas which have been glaciated and drowned are called **fjards**, e.g. Strangford Lough, Northern Ireland.

Landforms resulting from emergence
Following the global rise in sea-level, and still occurring in several parts of the world today, came the isostatic uplift of land as the weight of the ice sheets decreased. Landforms created as a result of land rising relative to the sea include raised beaches and erosion surfaces.

Raised beaches As the land rose, former wave-cut platforms and their beaches were raised above the reach of the waves. Raised beaches are characteristic of the west coast of Scotland (Figure 6.45). They are recognised by a line of degraded cliffs fronted by what was originally a wave-cut platform. Within the old cliff-line may be relict landforms such as wave-cut notches, caves, arches and stacks (Figure 6.46). The presence of such features indicates that isostatic uplift could not have been constant. It has been estimated that it would have taken an unchanging sea-level up to 2000 years to cut each wave-cut platform. (This evidence has been used to show that the climate did not ameliorate steadily following the ice age.)

Figure 6.46

The abandoned cliff-line at King's Cave, Arran, with its 8-m raised beach

Figure 6.45

Raised beaches on the Isle of Arran: the lower relates to the younger 8-m beach; the upper to the older 30-m beach

Early workers in the field claimed that there were three levels of raised beach on the west coast of Scotland, found at 25, 50 and 100 ft above the present sea-level. These are now referred to as the 8-m, 15-m and 30-m raised beaches. However, this description is now considered too simplistic, since it has been accepted that places nearest to the centre of the ice depression have risen the most and that the amount of uplift decreases with distance from that point. Thus, for example, the much-quoted '8-m raised beach' on Arran in fact lies at heights of 4–6 m. Where the raised beach is extensive, there is a considerable difference in height between the old cliff on its landward side and the more recent cliff to the seaward side — e.g. the 30-m beach in south-east Arran rises from 24 to 38 m.

It is now more acceptable to estimate the time at which a raised beach was formed by carbon-dating sea shells found in former beach deposits, rather than by referring solely to its height above sea-level (i.e. to indicate a 'late glacial raised beach' rather than a '100-ft/30-m beach'). Figure 6.47 is a labelled transect showing raised beaches on the west coast of Arran.

Erosion surfaces In Dyfed, the Gower peninsula (South Wales) and Cornwall, flat planation surfaces dominate the scenery. Where their general level is between 45 and 200 m, the surfaces are thought to have been cut during the Pleistocene period when sea-levels were higher — hence the alternative name of **marine platforms** (Figure 6.48).

Rock structure

Concordant coasts and **discordant coasts** are located where the natural relief is determined by rock structure (geology). They form where the geology consists of alternate bands of resistant and less resistant rock which form hill ridges and valleys (Chapter 8). Concordant coasts occur where the rock structure is parallel to the coast, as at Lulworth Cove, Dorset (Figure 6.49). Should there be local tectonic movements, a eustatic

Figure 6.48

Erosion surfaces (marine peneplanation) at St David's, Dyfed, South Wales

Figure 6.47

Diagrammatic transect across raised beaches in west Arran

(not to scale)

30-m upper raised beach

abandoned sandstone cliffs

cave, 25 m deep, with rounded stones formed by earlier storm waves

present wave-cut platform covered in pebbles and boulders

small cliff 2 m high

former high-tide level

wave-cut notch

present high-tide level

lower raised beach 15 m wide, cave 4 m above present high-tide level

present storm beach

Figure 6.49

Lulworth coast, Dorset, a concordant (Pacific) coastline

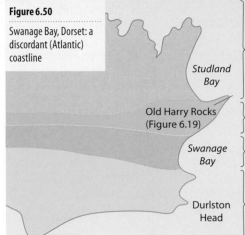

Figure 6.50

Swanage Bay, Dorset: a discordant (Atlantic) coastline

Studland Bay — low-lying area (inland) and bays (coast) form on less resistant sands and clays

Old Harry Rocks (Figure 6.19) — ridge (inland) and cliffs (coast) develop on the more resistant chalk

Swanage Bay — vale (inland) and bay (coast) form on less resistant clay

Durlston Head — ridge (inland) and headland with cliffs (coast) develop on the more resistant limestone

rise in sea-level, or a breaching of the coastal ridge, then summits of the ridge may be left as islands and separated from the mainland by drowned valleys. These can be seen on atlas maps showing the former Yugoslavia (Dalmatian coast) or San Francisco and southern Chile (Pacific coasts). Discordant coasts occur where the coast 'cuts across' the rock structure, as in Swanage Bay, Dorset (Figure 6.50). Here the ridges end as cliffs at headlands, while the valleys form bays.

Framework 5 Classification

Why classify?

Geographers frequently utilise classifications, e.g. types of climate, soil and vegetation, forms and hierarchy of settlement, and types of landform. This is done to try to create a sense of order by grouping together into classes features which have similar, if not identical, characteristics into identifiable categories. For example, no two stretches of coastline will be exactly the same, yet by describing Blakeney Point as a spit (page 143) it may be assumed that its appearance and the processes leading to its formation are similar to those outlined in Figure 6.23, even if there are local differences in detail.

How to classify

When determining the basis for any classification, care must be taken to ensure that:

- only meaningful data and measures are used;
- within each group or category, there should be the maximum number of similarities;
- between each group, there should be the maximum number of differences;
- there should be no exceptions, i.e. all the features should fit into one group or another; and
- there should be no duplication, i.e. each feature should fit into one category only.

As classifications are used for convenience and to assist understanding, they should be easy to use. They should not be oversimplified (too generalised), or too complex (unwieldy); but should be appropriate to the purpose for which they are to be used.

No classification is likely to be perfect and several approaches may be possible.

An example

The following landforms have already been referred to in this book:

> arch; braided river; corrie; delta; esker; hanging valley; knickpoint; moraine; raised beach; rapids; spit; wave-cut platform.

Can you think of at least three different ways in which they may be categorised? Some possibilities are listed below:

a Perhaps the simplest classification is a two-fold division based upon whether they result from erosion; or from deposition.

b They could be reclassified into two different categories: those formed under a previous climate (i.e. relict features); and those still being formed today.

c The most obvious may be a three-fold division into coastal; glacial; and fluvial landforms.

d A more complex classification would result from combining either **a** and **b**; or **a** and **c**; to give six groups.

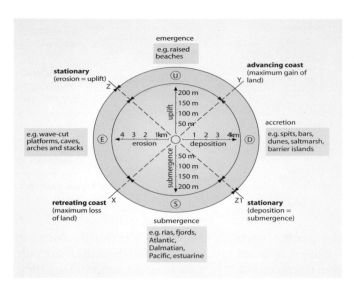

Figure 6.51

Valentin's classification of coasts, showing the conditions under which the maximum advance or retreat of the coast will occur and how it is possible for a coast to be in equilibrium (stationary).

Classification of coasts

Considering the wide variety of coastlines, it should not be surprising that several different attempts have been made to classify coasts, each based upon a range of criteria.

Changes in sea level (Johnson, 1919)

This, the earliest method, divided coasts into four categories:

1 **Emergent**: those resulting from a fall in sea-level and/or uplift of the land. Coastlines formed in this way are basically straight in appearance (e.g. barrier islands).

2 **Submergent**: those showing a recent rise in sea-level and/or a fall in the land surface. This type is characterised by an indented coastline (e.g. rias and fjords).

3 **Stable** or **neutral**: **a** those showing no signs of changes in sea-level or in the land; **b** those produced by non-marine processes (e.g. a river delta, a volcanic island).

4 **Compound**: those with a mixture of at least two of the three previously listed groups.

This classification came to be regarded as too simplistic when new evidence accumulated to show that there had been several changes in sea-level (local and global) and that many coasts exhibited signs both of submergence and emergence.

Relationship between coastal and other processes of erosion and deposition (Shepard, 1963)

This has been shown to be overgeneralised, having only two categories. These are:

1 **Primary**: where the influence of the sea has been minimal, e.g. fjords (ice), rias and deltas (rivers), and islands (volcanic or coral).

2 **Secondary**: where marine processes have been dominant, e.g. stacks and spits.

Advancing and retreating coasts (Valentin, 1952)

This method also assumes that all coasts can be fitted into a two-fold classification (Figure 6.51).

1 **Advancing**: where marine deposition or the uplift of the land is dominant.

2 **Retreating**: where marine erosion or the submergence of the land is more significant.

Tectonic coasts (Inman and Nordstrom, 1971)

This classification was based upon plate tectonic theory. They recognised four types of coastal plate boundaries:

1 **Diverging plates** (the Red Sea).

2 **Converging plates** (the island arcs of Japan and the Philippines).

3 **Major transform faults** (California).

4 **Stable plate boundaries** (India and Australia).

Energy-produced coastlines (Davis, 1980)

This, the most recent attempt to categorise coasts, is gaining credibility because it relates directly to the amount of energy expended by different types of wave upon a particular stretch of coastline.

1 **High-energy environments**: where destructive storm waves breaking on shingle beaches are most typical.

2 **Low-energy environments**: where constructive swell waves breaking upon sandy beaches are more frequent.

3 **Protected environments**: where wave action is limited in small, sheltered, sea areas.

 1 **a** Choose one system of coastal classification. Describe and explain the principles upon which the system is based.

b Describe some of the problems of applying your chosen classification to cover all coastal processes and/or landforms.

2 If you get an opportunity to study a stretch of coastline, discuss the problems you found in trying to fit it into any one coastal classification with which you are familiar.

Future sea-level rise and coastal stability

We have already seen (page 148) that over very long periods of geological time (tens of millions of years) sea-level has been controlled by the movement of ocean plates and the uplift of continents, and that over shorter periods (the last million years) sea-level has been controlled by the volume of ice on the land. Since the 'Little Ice Age' in the 17th century, when glaciers in the Alps and arctic areas advanced, the world has been slowly warming (Figure 11.8). This seems to explain why sea-level has risen about 15–25 cm in the last century and, accepting that global warming also results from the release of 'greenhouse' gases, why it is expected to rise at an increasing rate in the next 100 years (Figure 6.52). Sea-levels are expected to rise due to:

- the thermal expansion of the oceans (warmer water is less dense and so occupies a greater volume;
- the melting of smaller alpine glaciers releasing water at present held in storage. (The polar ice caps are not forecast to melt significantly as warmer world temperatures are expected to increase evaporation from the oceans which, in turn, is likely to increase snowfall in Greenland and Antarctica.)

The present rate of sea-level rise is between 1 and 2 mm per year (i.e. 0.1 and 0.2 m per 100 years). This rate is expected to rise to over 5 mm/year (0.5 m/100 years) in the next two or three decades. The extreme scenario, based upon the continued increase in the rate of global warming, is a rise in excess of 10 mm/year (1 m/100 years).

Natural coastal ecosystems do appear to have the ability to adjust to sea-level changes. Salt marshes and mangrove swamps can trap mud in sufficient quantities to enable them to grow upwards by 10 mm per year, while both coral and sand also seem capable of making any necessary response.

Coastal geomorphologists suggest, however, that the present rate of sea-level rise is insufficient by itself to cause significant increases in coastal erosion in Britain. However, these observations refer to natural coasts. Coastlines fixed by and affected by human activity (Case Study 6) pose very different problems. A rise of 1 m over the next century could inundate 25 per cent of Bangladesh and 30 per cent of Egypt's arable land (Figure 6.53). Many of the world's major ports and conurbations would also be threatened, including London, Tokyo, Bangkok, Calcutta and Singapore. Several low-lying ocean states, such as the Maldives, could become almost totally submerged. Storm surges are likely to become an even greater threat in places such as the Bay of Bengal.

Figure 6.52

Projections of future sea-level rise as a result of global warming: the extreme values cover the 95 per cent probability range (after Clayton, 1992)

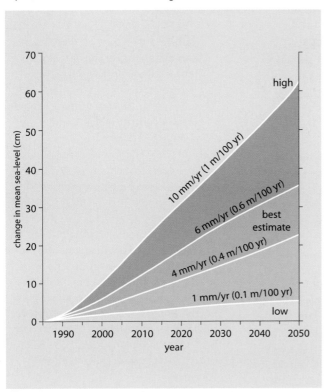

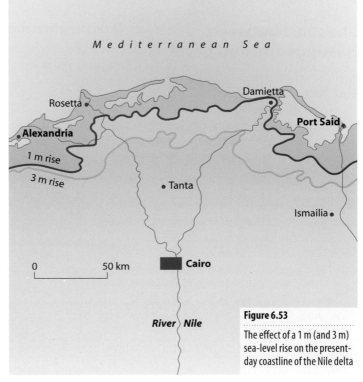

Figure 6.53

The effect of a 1 m (and 3 m) sea-level rise on the present-day coastline of the Nile delta

Case Study 6

The management of coastal erosion and flooding on the East Sussex coast

The need for coastal management

People have always settled on the coast and have had to live with the twin problems of coastal flooding and erosion. Over the last 10 000 years, the sea-level has fluctuated due to worldwide eustatic change and local isostatic change (page 112) as ice sheets retreated at the end of the Pleistocene period. Over the last 500 years, many ports and resorts have grown up along the coast.

The traditional response to coastal erosion and flooding, from Victorian times onwards, has been to build hardened defences such as sea walls, which absorb rather than dissipate energy. This policy was continued after large-scale coastal flooding in south-east England in January 1953 (Places 15, page 132). Many of these defences have now reached the end of their practical life and, with sea-level currently rising between 3 and 5 mm a year in south-east England, coastal management is an important issue. Deciding the best management approach is not simple, and there are many factors to take into account — some of which are shown in Figure 6.54.

This Case Study explores the problems of coastal management near Hastings in East Sussex. It will be seen that human interventions may cause new problems — often unintentionally — further along the coast.

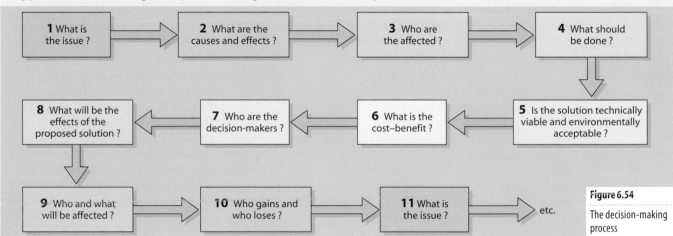

Figure 6.54

The decision-making process

The coast between Hastings and Pett Level

Hastings is a large, coastal resort town in East Sussex, southern England. The cliffed coastline extends approximately 10 km eastwards from Hastings to the cliff-top coastal village of Fairlight and as far as Cliff End (Figure 6.55). Beyond Cliff End, artificial sea defences protect the low-lying Pett Level marsh, behind which the cliffs of the former coastline now lie several kilometres inland (Figure 6.55). The geology along this section of the coast consists of unconsolidated Tertiary deposits of sands, gravels and sandstones. Natural rates of erosion are 0.5–1.0 m per year, with high-energy waves from the Atlantic having a fetch of 5000 km (page 124).

Coastal works at Hastings

Coastal land use changes over the last 100 years at Hastings have had a profound effect on this whole stretch of coastline.

- The cliff face has been armoured and protected to prevent rock falls and cliff retreat, with the result that the natural input of sediment into the coastal system has been stopped.
- Groynes were built to the west in Victorian times to encourage the beach to accrete (build up), thus providing protection for the town.
- The harbour (Figure 6.56), built in 1896, accelerated this beach building by providing an effective sediment trap for the longshore drift, which flows in an easterly direction. This has had the benefit of creating a well built-up beach up to 300 m wide on the updrift (western) side of the harbour arm (Figure 6.57). (Any surplus material is carried eastwards by the longshore drift, well beyond Hastings.) An effective barrier to the largest storm waves, the beach makes sea walls and other hard, engineered defences unnecessary.
- A £1 000 000 improvement and extension to the harbour was built in the 1970s to provide a larger area for fishing boats. This has increased the sediment trap effect.
- The town has built outwards over 150 years and the buildings, the promenade and the wide beach ensure that waves no longer break behind the beach at the former cliff base, as the whole sea-front is now artificial.

However, these changes at Hastings have been to the detriment of areas downdrift — i.e. towards the east at Fairlight and Pett Level.

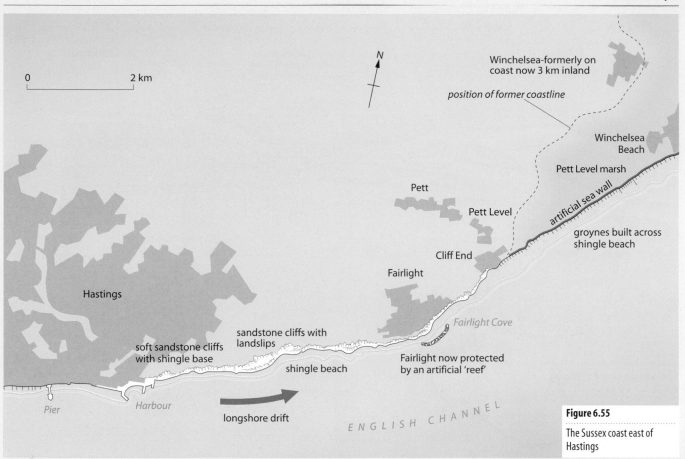

Figure 6.55

The Sussex coast east of Hastings

Figure 6.56

Hastings harbour at low tide

Figure 6.57

The western beach, Hastings

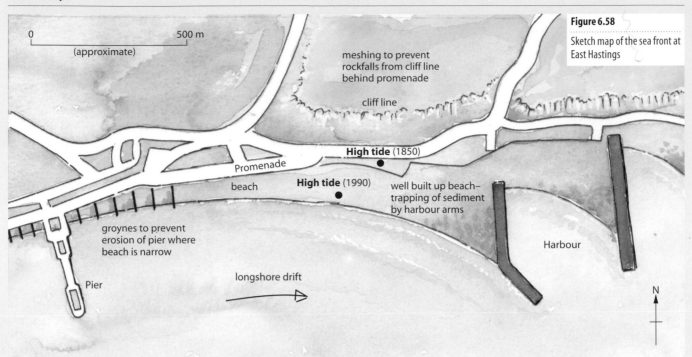

Figure 6.58

Sketch map of the sea front at East Hastings

meshing to prevent rockfalls from cliff line behind promenade

cliff line

High tide (1850)

Promenade

beach

High tide (1990)

well built up beach– trapping of sediment by harbour arms

Harbour

groynes to prevent erosion of pier where beach is narrow

Pier

longshore drift

N

0   500 m
(approximate)

The effects on Fairlight

The coastal village of Fairlight has experienced erosion of its 30 m high, near-vertical cliffs and the loss of land and homes. The main cause has been basal undercutting followed by mass movements resulting from sub-aerial processes (page 42) in the form of slides and slumping (Figure 6.59); this has led to rapid cliff retreat (compare Places 16, page 134). High-energy waves at the cliff foot ensured that eroded material was quickly dispersed, thus encouraging the continual movement of slides above the tidal zone. Throughout the 1970s, rates of retreat were 0.5–1.0 m a year. However, there were extreme events, such as the loss of 66 m in one night at Fairlight Cove in December 1979: heavy rainfall produced saturated groundwater conditions, mobilising landslides that had been initiated by winter storms. In 1989, houses at the cliff top were abandoned and the adjacent 'Sea View' road was closed to traffic (Figure 6.60).

Figure 6.59

The cliffs at Fairlight village looking west, and showing evidence of mass movement

Figure 6.60

Condemned houses on Fairlight clifftop

Plan view

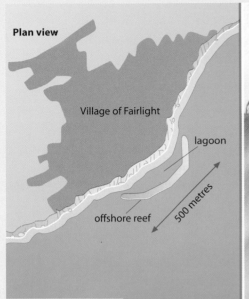

Village of Fairlight

lagoon

offshore reef

500 metres

Figure 6.61

Coastal erosion and the sea defences at Fairlight

Cross-section
not to scale

Fairlight   "Sea View" road

unconsolodated
sands and clays

cliffs

30 m

still water lagoon
reduces wave energy
at the cliff foot.

reef of quarried
material

20 m

waves break on reef

10 m

sea

Figure 6.62

Houses on the clifftop at Fairlight, showing the artificial reef below

Figure 6.63

The artificial reef at Fairlight

It was believed that the 1970s extension of the harbour arm at Hastings had reduced the sediment input into the coastal system west of Fairlight and had exacerbated cliff erosion by reducing the width of the beach. With little sand in the environment, the base of the cliffs was thus exposed to higher wave energies with the result that the rate of cliff retreat had increased to more than 1 m per year by the late 1980s. More and more properties were coming under threat of cliff fall (Figure 6.62). Therefore, in 1990, an offshore reef was constructed in the inter-tidal zone. This 500 m long structure of quarried stone encouraged waves to break further from the shore. It also created a lagoon in front of the beach at the base of the cliffs (see Figure 6.63), which greatly reduced wave energy at the cliff foot. Material was able to accumulate at the base of the cliff, thus stabilising and anchoring any subsequent mass movement. Erosion has been reduced by 90 per cent, although cliff retreat still continues at a slow rate due to weathering of the cliff face. However, the reef itself has acted as a sediment trap, to the detriment of Pett Level which lies further to the east.

The problems created at Pett Level

Pett Level is a flat, low-lying marshy area. Once part of the sea, it has been successively drained by people since medieval times in order to provide low-quality agricultural land. The village of Winchelsea, 3 km inland, was originally on the coast (Figure 6.55). Cliff End is a small coastal village to the west of the marsh and the whole area is sparsely populated, with rough pasture for sheep grazing and some income from summer tourism. Due to the contraction of sediment since the marsh was drained, most of Pett Level is at or slightly below sea-level. Flooding is obviously a danger and the effect of a high tide coupled with a storm is obvious! Rising sea-levels can only make the problem worse.

In 1940, during the Second World War when every piece of agricultural land was vitally important, a large earthen embankment was constructed to protect the marsh from flooding. This embankment is maintained by the National Rivers Authority (NRA) and it has to be protected from erosion by hardened defences such as groynes, sections of wooden revetment (Figure 6.64) and a sea wall near Cliff End where there is a caravan park (Figure 6.65). On the seaward side, the embankment is topped with concrete stepping to prevent erosion (Figure 6.66). Groynes encourage sediment to build up and minimise undermining of the embankment. Any breach of the sea wall would be catastrophic for the low-lying area behind. Additionally, the NRA has a policy of continually 'nourishing' the beach with quarried cobbles and shingle, particularly during times of winter storms when waves have the greatest energy and erosive potential.

The embankment protecting Pett Level from flooding has experienced increased erosion since 1987. The beach has required additional nourishment by the NRA (Figure 6.67) to ensure that it does not become diminished and less resistant to high-energy storm events. The sediment trap of the harbour works at Hastings and the reef at Fairlight, which itself interferes with longshore drift to some extent, mean that constant nourishment and maintenance of the Pett Level defences are required to prevent damage. How long this will be possible or economically feasible, given the additional factor of rising sea-levels due to global warming, is uncertain.

Figure 6.64

Protective embankment and wooden groynes at Pett Level

Figure 6.65

The flood wall at Pett Level caravan park

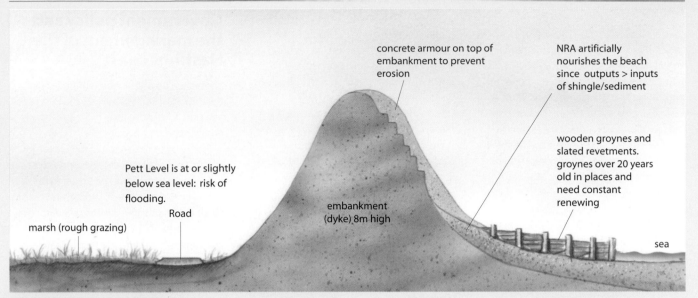

concrete armour on top of embankment to prevent erosion

NRA artificially nourishes the beach since outputs > inputs of shingle/sediment

wooden groynes and slated revetments. groynes over 20 years old in places and need constant renewing

Pett Level is at or slightly below sea level: risk of flooding.

Road

embankment (dyke) 8m high

marsh (rough grazing)

sea

Coastal management strategies for the future: the decision-making debate

It can reasonably be argued that coastal towns are very valuable and therefore need to be defended. Can the same argument be applied to small coastal villages or to marginal land located next to the sea? Coastal management is overseen for the government by the Ministry of Agriculture, Fisheries and Food (MAFF). Most of the £200 million spent each year is devolved through the NRA which actually carries out most of the work in England and Wales. Cost–benefit analysis has nearly always been applied to projects since the 1953 floods: i.e. what is the value of the land and property to be protected and how much will the protection cost? Is there significant benefit for the cost of protection? If the value of the land, including buildings, roads and services was greater than the cost of the defences, then the government provided a 70 per cent grant towards the cost of protection, the remainder coming from local authority funds. Even under this 'traditional' thinking, remote rural settlements, such as the farming community of Mappleton on the Holderness coast, would have difficulty justifying coastal protection. Agricultural land values in north-east England are relatively low: the financial value of land experiencing rapid erosion is virtually nil! However, such traditional thinking is now changing and even areas that might be considered worth protecting are finding that funds are less forthcoming.

The conservation organisation English Nature has argued for some time for a more long-term and environmentally acceptable approach to coastal management. Previous thinking concentrated on the prevention of a problem — for instance, a sea wall to stop erosion. It is now suggested that we should consider avoidance strategies (moving inland) and prediction. An increased understanding of climatology, coastal geomorphology and systems modelling should help us to predict outcomes and then to plan coastal management in such a way that our actions do not make matters worse.

Traditional approaches to coastal management are becoming increasingly discredited. For example, solid defences such as sea walls do not dissipate the energy of incoming waves, they simply change their direction. The waves flow back down the slope of the beach, scouring away beach material in the process. The beach is lowered and the base of the wall undermined, leading to its eventual collapse. Furthermore, sea walls prevent erosion of the land behind — beneficial in some respects, but cutting off access to a useful supply of sediment for the beach and downdrift areas. Sea walls are also very costly to build and maintain.

English Nature has suggested that some coastal areas should be allowed to erode and/or flood. The policy of managed retreat has been advocated. By allowing a strip of land, usually low-value agricultural land, adjacent to the seashore to flood, a natural salt marsh is created. This provides a conservation and recreational resource, and is cheaper than building and maintaining hard defences. Furthermore, English Nature has suggested that managed retreat makes increasingly more sense in the context of rising sea-levels: protection is provided for up to 50 years and managed retreat is aesthetically acceptable (although farmers and landowners would disagree).

The government currently faces several problems in dealing with coastal management:

- Sea-levels are rising more rapidly than at any time in living memory, and hence coastal flooding and erosion will become acute problems.
- Many existing defences built after the 1953 floods are now reaching the end of their lives and are in a poor state of repair.
- Funds are limited and there are many demands made upon them.

The policy of managed retreat has attracted positive comment from the government. In October 1993, the Minister announced the new MAFF thinking on coastal management, in the form of a policy document. The government believes that before permission, let alone funding for any coastal management scheme is given, the following three criteria must be fulfilled:

- It must be justifiable in cost–benefit terms.
- It must be technically viable: it must actually work and must have minimal effects elsewhere.
- It must be 'green' or 'environmentally friendly': it must look nice, not cause environmental degradation and must fit in with current conservation thinking.

Some rural areas of the East Anglian coast (where nearly half the NRA's annual coastal management budget is spent) might now find themselves 'undefendable'! The Country Landowners' Association and many farmers are outraged by the implications. However, given the complexity of the issue and the constant and increasing threat of rising sea-levels, perhaps it is inevitable that some groups or individuals will be disappointed. It is very difficult to decide the social and cultural value of a place: governments tend to be more concerned with financial matters and 'value for money' for the taxpayer. Rural land values at the coast are falling (we have agricultural 'overproduction' in Europe; page 448) and, even when the productivity of farmland is taken into account, it seems that some areas will be left to their fate — e.g. the more sparsely populated sections of the Holderness coast.

Government policy and the management of the Hastings coast

Should we be prepared to continue defending Hastings, Fairlight and Pett Level at all costs? Will the recent changes in government policy have any effect?

Professor Keith Clayton of the University of East Anglia, a coastal geomorphologist, believes that we should be prepared to allow some coastal areas to flood and erode. Far better to compensate the farmer for the loss of land and accept the free supply of sand and sediment to the beach, than to defend at all costs, with expensive hard defences that cause damage elsewhere.

Figure 6.68

Pett Level house on stilts: protection against flooding

As coastal towns such as Hastings form an important local economic resource, there would appear little likelihood of them being 'abandoned to the sea' — unless they become indefensible due to extreme sea-level rise. (Hastings, on a clifftop site, would not be vulnerable to flooding.) Although Fairlight is a fairly affluent retirement village, it is small and so several clifftop properties have already been abandoned — one built as recently as 1983! Although the creation of the reef has arrested cliff undercutting, retreat will continue to occur at a slow rate due to cliff face weathering. New houses are still being built at Fairlight, though in parts of the village several hundred metres back from the cliff top. It seems that Fairlight is safe for the forseeable future but the long-term effects on the reef of rising sea-levels will need to be considered.

Pett Level is most likely to be affected by the change in government policy. The erosion of the embankment has become an increasing problem and, with rising sea-levels, it is likely that it will continue to be so. Would Pett Level be a candidate for managed retreat? Should residents be forced to move or find their own solutions (Figure 6.68) to the problem of flooding? What would a cost–benefit analysis suggest should be done? Some difficult decisions are required within the next few years and we may see a rather different picture of coastal management in Britain as an increasing number of coastal protection schemes, many of which cost millions of pounds, are failing in their battles with the forces of nature (Figure 6.69).

Figure 6.69

"Moving against the tide"

Over the last two centuries, structurally engineered "hard" defences — sea walls, bank and groynes — have gone up all around England's coast to safeguard people and property from flooding and erosion, and to reclaim land for development and agriculture. The cost of building and maintaining these defences has been high. This year alone, the Ministry of Agriculture, Food and Fisheries (MAFF) and the coastal authorities — National Rivers Authority (NRA), local authorities and the Internal Drainage Boards (IDBS) — will spend £300 million on flood and coastal defences in England and Wales. And with present sea defences unable to cope with rising sea level, costs look set to skyrocket.

The Government has finally acknowledged that replacing or improving engineered sea defences is not necessarily the long-term answer to its coastal defence needs — either on financial or environmental grounds. In its latest flood and coastal defence policy statement, published late last year, MAFF said that the goal of protecting the coastline can best be achieved by adapting and supplementing natural coastal processes.

Claire Hutchings, *Geographical Magazine*, March 1994

References

Clayton, K. (1992) *Coastal Geomorphology*. Thomas Nelson.

Coastal Dunes: Geography Today. ITV.

Coastal Management: Geography 16–19 Project (1985) Longman.

Defence Against the Sea: Place and People. ITV.

Estuary Management and Planning: Geography 16–19 Project (1985) Longman.

Goudie, A. (1993) *The Nature of the Environment*. Blackwell.

McCullagh, P. (1978) *Modern Concepts in Geomorphology*. Oxford University Press.

Pethick, J. (1984) *An Introduction to Coastal Geomorphology*. Edward Arnold.

Planet Earth: Edge of the Sea. (1985) Time–Life Books.

Science in Geography, Books 1–4 (1974) Oxford University Press.

Waves and Beaches: Geography Today. ITV.

Wilson, J. (1984) *Statistics in Geography for 'A'-level Students*. Schofield & Sims.

Chapter 7

Deserts

"Little by little the sky was darkened by the mixing dust, and the wind felt over the earth, loosened the dust and carried it away"

J. Steinbeck, *The Grapes of Wrath*, 1939

What is a desert?

"The deserts of the world, which occur in every continent including Antarctica, are areas where there is a great deficit of moisture, predominantly because rainfall levels are low. In some deserts this situation is in part the result of high temperatures, which mean that evaporation rates are high. It is the shortage of moisture which determines many of the characteristics of the soils, the vegetation, the landforms, the animals, and the activities of humans" (Goudie and Watson, 1990).

A desert environment has conventionally been described in terms of its deficiencies — water, soils, vegetation and population. Deserts include those parts of the world which produce the smallest amount of organic matter and have the lowest net primary production (NPP, page 284). In reality, many desert areas have potentially fertile soils, evidenced by successful irrigation schemes; all have some plant and animal life, even if special adaptations are necessary for their survival; and some are populated by humans, occasionally only seasonally by nomads but elsewhere permanently — e.g. in large cities like Cairo and Karachi.

The traditional definition of a desert is an area receiving less than 250 mm of rain per year. While very few areas receive no rain at all (Places 18), amounts of precipitation are usually small and occurrences are both infrequent and unreliable. Climatologists have sometimes tried to differentiate between cold deserts where for at least one month a year the mean temperature is below 6°C, and hot deserts. Several geomorphologists have used this to distinguish the landforms found in the hot sub-tropical deserts — i.e. our usual mental image of a desert — from those found in colder latitudes, e.g. the Gobi Desert and the tundra.

Modern attempts to define deserts are more scientific and are specifically linked to the water balance (page 52). This approach is based on the relationship between the input of water as precipitation (*P*), the output of moisture resulting from evapotranspiration (*E*) and changes in water held in storage in the ground. In parts of the world where there is little precipitation annually or where there is a seasonal drought, the **actual evapotranspiration** (*AE*) is compared with **potential evapotranspiration** (*PE*) — the amount of water loss that would occur if sufficient moisture was always available to the vegetation cover. C. W. Thornthwaite (1931) was the first to define an **aridity index** using this relationship (Figure 7.1).

Figure 7.1

The index of aridity

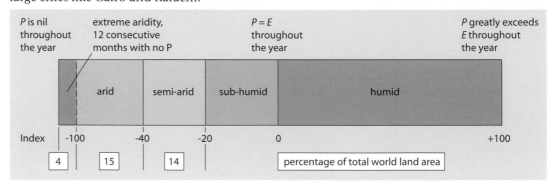

Location and causes of deserts

On the basis of climatic characteristics, including Thornwaite's aridity index, one-third of the world's land surface can be classified as desert — i.e. arid and semi-arid. Alarmingly, this figure, and therefore the extent of deserts, may be increasing (Case Study 7).

As shown in Figure 7.2, the majority of deserts lie in the centre or on the west coast of continents between 15° and 30° north and south of the Equator. This is the zone of sub-tropical high pressure where air is subsiding (the descending limb of the Hadley cell, Figure 9.35). On page 209, there is an explanation of how warm, tropical air is forced to rise at the Equator, producing convectional rain, and how later that air, once cooled and stripped of its moisture, descends at approximately 30° north and south of the Equator. As this air descends it is compressed, warmed and produces an area of permanent high pressure. As the air warms, it can hold an increasing amount of water vapour which causes the lower atmosphere to become very dry. The low relative humidity, combined with the fact that there is little surface water for evaporation, gives clear skies.

A second cause of deserts is the rain shadow effect produced by tall mountain ranges. As the prevailing winds in the sub-tropics are the trade winds, blowing from the north-east in the northern hemisphere and the south-east in the southern, then any barrier, such as the Andes, prevents moisture from reaching the western slopes. Where plate movements have pushed up mountain ranges in the east of a continent, the **rain shadow effect** creates a much larger extent of desert (e.g. 82 per cent of the land area of Australia) than when the mountains are to the west, as in South America.

Aridity is increased as the trade winds blow towards the Equator, becoming warmer and therefore drier. Where the trade winds blow from the sea, any moisture which they might have held will be precipitated on eastern coasts leaving little moisture for mid-continental areas. The three major deserts in the northern hemisphere which lie beyond the sub-tropical high pressure zone (the Gobi and Turkestan in Asia and the Great Basins of the USA) are mid-continental regions far removed from any rain-bearing winds, and surrounded by protective mountains.

A third combination of circumstances giving rise to deserts is also shown in Figure 7.2. Several deserts lie along western coasts where the ocean water is cold. In each case, the prevailing winds blow parallel to the coastline and, due to the earth's rotation, they tend to push surface water seaward at right angles to the wind direction. The Coriolis force (page 208) pushes air and water coming from the south towards the left in the southern hemisphere and water from the north to the right in the northern hemisphere. Consequently, very cold water is drawn upwards to the ocean surface, a process called **upwelling**, to replace that driven out to sea. Any air which then crosses this cold water is cooled and its capacity to hold moisture is diminished. Where these cooled winds from the sea blow on to a warm land surface, advection fogs form (page 205 and Places 18).

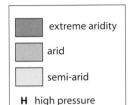

- ▣ extreme aridity
- ▢ arid
- ▢ semi-arid

H high pressure
R rainshadow
M mid-continent
U upwelling of cold water

1 Australia (e.g. Simpson Great Sandy) **H R M**
2 Gobi **M**
3 Thar **H**
4 Iran **H M**
5 Turkestan **M**
6 Arabia **H**
7 Somalia **H**
8 Kalahari **HM**
9 Namib **HU**
10 Sahara **H** (**U** in west, **M** in centre)
11 Patagonia **R**
12 Monte **H**
13 Atacama **HRU**
14 Sonora **H**
15 Mojave **HR** (**U** on coasts)
16 Great Basins **R**

Figure 7.2

Arid lands of the world

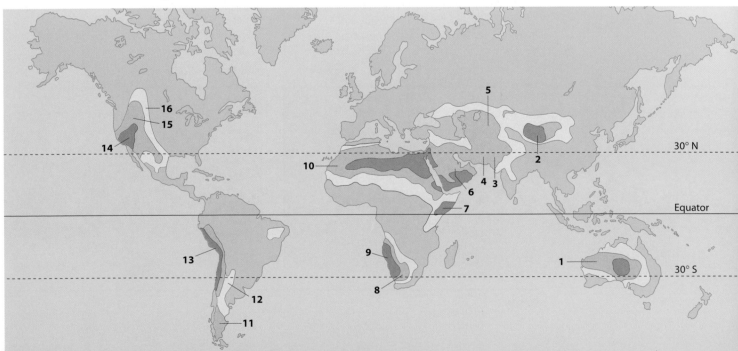

Figure 7.3

Welwischia Mirabilis, Namibia

The prevailing winds in the Atacama, which lies in the sub-tropical high pressure belt, blow northwards along the South American coast. These winds, and the northward-flowing Humboldt (Peruvian) current over which they blow, are pushed westwards (to the left) and out to sea by the Coriolis force as they approach the Equator. This allows the upwelling of cold water from the deep Peru–Chile sea trench (Figure 1.12) that provides the rich nutrients to nourish the plankton which form the basis of Peru's fishing industry. The upwellings also cool the air above them which then drifts inland and over the warmer desert. The meeting of warm and cold air produces advection fogs (page 205) which provide sufficient moisture for a limited vegetation cover. Inland parts of the Atacama are alleged to be the only truly rainless desert in the world, but even here the occasional rainfall event does occur.

Desert landscapes: what does a desert look like?

Deserts provide a classic example of how easy it is to portray or to accept an inaccurate mental picture of different places (or people) in the world. What is your image of a desert? Is it a landscape of sand dunes similar to those shown in Figures 7.15–7.18, perhaps with a camel or palm tree somewhere in the background? Large areas of dunes, known as **erg**, do exist — but they cover only about one-quarter of the world's deserts. Most deserts consist either of bare rock, known as **hammada** (Figure 7.4) or stone-covered plains, called **reg** (Figure 7.5). Deserts contain a great diversity of landscapes. This diversity is due to geological factors (tectonics and rock type) as well as to climate (temperature, rainfall and wind) and resultant weathering processes.

Figure 7.4

A rocky (hammada) desert, Wadi Rum, Jordan

Figure 7.5

A stony (reg) desert, Sahara

Arid processes and landforms

In their attempts to understand the development of arid landforms, geographers have come up against three main difficulties:

- How should the nature of the weathering processes be assessed? Desert weathering was initially assumed to be largely mechanical and to result from extreme diurnal ranges in temperature. More recently, the realisation that water is present in all deserts in some form or other has led to the view that chemical weathering is far more significant than had previously been thought. Latest opinions seem to suggest that the major processes, e.g. exfoliation and salt weathering, may involve a combination of both mechanical and chemical weathering.
- What is the relative importance of wind and water as agents of erosion, transportation and deposition in deserts?
- How important have been the effects of climatic change on desert landforms? During some phases of the Quaternary, and previously when continental plates were in different latitudes, the climate of present arid areas was much wetter than it is today. How many of the landforms that we see now are, therefore, relict and how many are still in the process of being formed?

Mechanical weathering

Traditionally, weathering in deserts was attributed to mechanical processes resulting from extremes of temperature. Deserts, especially those away from the coast, are usually cloudless and are characterised by daily extremes of temperature. The lack of cloud cover can allow day temperatures to exceed 40°C for much of the year; while at night, rapid radiation often causes temperatures to fall to zero. Although in some colder, more mountainous deserts, frost shattering is a common process, it was believed that the major process in most deserts was **insolation weathering** (page 35). Insolation weathering occurs when, during the day, the direct rays of the sun heat up the surface layers of the rock. These surface layers, lacking any protective vegetation cover, may reach 80°C. The different types and colours of minerals in most rocks, especially igneous, heat up and cool down at different rates, causing internal stress and fracturing. This process was thought to cause the surface layers of exposed rock to peel off — **exfoliation** — or individual grains to break away — **granular disintegration**. Where surface layers do peel away, newly exposed surfaces experience pressure release (page 35). This is believed to be a contributory process in the formation of rounded exfoliation domes such as Ayers Rock in Australia (Figure 7.6).

Doubts about insolation weathering began when it was noted that the 4500-year-old ancient monuments in Egypt showed little evidence of exfoliation, and that monuments in Upper Egypt, where the climate is extremely arid, showed markedly fewer signs of decay than those located in Lower Egypt, where there is a limited rainfall. D. T. Griggs (1936) conducted a series of laboratory experiments in which he subjected granite blocks to extremes of temperature in excess of 100°C. After the equivalent of almost 250 years of diurnal temperature change, he found no discernible difference in the rock. Later, he subjected the granite to the same temperature extremes while at the same time spraying it with water. Within the equivalent of two and a half years of diurnal temperature change, he found the rock beginning to crack. His conclusions, and those of later geomorphologists, suggest that some of the weathering previously attributed to insolation can now be ascribed to chemical changes caused by moisture. Although rainfall in deserts may be limited, the rapid loss of temperature at night frequently produces dew (175 nights a year in Israel's Negev) and the mingling of warm and cold air on coasts (e.g. of the Atacama) causes advection fog. There is sufficient moisture, therefore, to combine with certain minerals to cause the rock to swell (hydration) and the outer layers to peel off (exfoliation). At present, it would appear that the case for insolation weathering is neither proven nor disproven and that it may be a consequence of either mechanical weathering, or chemical weathering, or both.

Figure 7.6

An exfoliation dome: Ayers Rock, Australia (compare Figure 2.3)

The second mechanical process in desert environments, **salt weathering**, is more readily accepted although the action of salt can cause chemical, as well as physical, changes in the rock. Salts in rainwater, or salts brought to the surface by capillary action, form crystals as the moisture is readily evaporated in the high temperatures and low relative humidities. Further evaporation causes the salt crystals to expand and mechanically to break off pieces of the rock upon which they have formed (page 35). Subsequent rainfall, dew or fog may be absorbed by salt minerals causing them to swell (hydration) or chemically to change their crystal structure. Where salts accumulate near or on the surface, particles may become cemented together to form **duricrusts**. These hard crusts are classified according to the nature of their chemical composition. (Students with a special interest in geology or chemistry may wish to research the meaning of the terms: **calcretes**, **silcretes** and **gypcretes**.) Another form of crust, **desert varnish**, is a hard, dark glazed surface found on exposed rocks which have been coated by a film composed largely of oxides of iron and manganese (Figure 7.7) and, possibly, bacterial action. It is hoped that the dating of desert varnish may help to establish a chronology of climatic changes in arid and semi-arid environments.

Figure 7.7

Carvings in desert varnish, Wadi Rum, Jordan

The importance of wind and water

Geomorphologists, working in Africa at the end of the last century, believed the wind to be responsible for most desert landforms. Later fieldwork, carried out mainly in the higher and wetter semi-arid regions of North America, recognised and emphasised the importance of running water and, in doing so, de-emphasised the role of wind. Today, it is more widely accepted that both wind and water play a significant, but locally varying, part in the development of the different types of desert landscape.

Aeolian (wind) processes

Transport

The movement of particles is determined by several factors. Aeolian movement is greatest where winds are strong (usually over 20 km/hr), turbulent, come from a constant direction and blow steadily for a lengthy period of time. Of considerable importance, too, is the nature of the regolith. It is more likely to be moved if there is no vegetation to bind it together or to absorb some of the wind's energy; if it is dry and unconsolidated; if particles are small enough to be transported; and if material has been loosened by farming practices. While such conditions do occur locally in temperate latitudes, e.g. coastal dunes, summits of mountains and during dry summers in arable areas, the optimum conditions for transport by wind are in arid and semi-arid environments.

Wind can move material by three processes: suspension, saltation and surface creep. The effectiveness of each method is related to particle size (Figure 7.8).

Suspension Where material is very fine, i.e. less than 0.15 mm in diameter, it can be picked up by the wind, raised to considerable heights and carried great distances. There have been occasions, though perhaps recorded only once a decade, when red dust from the Sahara has been carried northwards and deposited as 'red rain' over parts of Britain. Visibility in deserts is sometimes reduced to

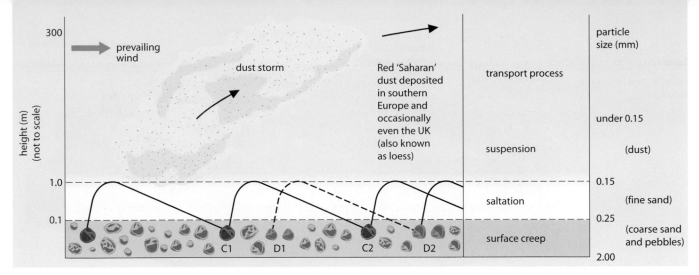

height (m) (not to scale)

300

prevailing wind

dust storm

Red 'Saharan' dust deposited in southern Europe and occasionally even the UK (also known as loess)

1.0

0.1

C1   D1   C2   D2

particle size (mm)

transport process

under 0.15

suspension   (dust)

0.15

saltation   (fine sand)

0.25

surface creep   (coarse sand and pebbles)

2.00

Figure 7.8

Processes of wind transportation

Figure 7.9

A desert pavement in Jordan, created by deflation

less than 1000 m and this is called a **dust storm**. The number of recorded dust storms on the margins of the Sahara has increased rapidly in the last 25 years as the drought of that region has intensified. In Mauritania, there was an average of only 5 days a year with dust storms during the early part of the 1960s compared with an average of 58 days/yr over a similar period in the early 1980s.

Saltation When wind speeds exceed the threshold velocity (the speed required to initiate grain movement), fine and coarse-grained sand particles are lifted. They may rise almost vertically for several centimetres before returning to the ground in a relatively flat trajectory of less than 12° (Figure 7.8). As the wind continues to blow, the sand particles bounce along, leapfrogging over one another. Even in the worst storms, sand grains are rarely lifted higher than 2 m above the ground.

Surface creep Every time a sand particle, transported by saltation, lands, it may dislodge and push forward larger particles (more than 0.25 mm in diameter) which are too heavy to be uplifted. This constant bombardment gradually moves small stones and pebbles over the desert surface.

Erosion

There are two main processes of wind erosion: deflation and abrasion.

Deflation This is the progressive removal of fine material by the wind leaving pebble-strewn desert pavements or reg (Figures 7.9 and 7.10). Over much of the Sahara, and especially in Sinai in Egypt, vast areas of monotonous, flat and colourless pavement are the product of an earlier, wetter climate. Pebbles were transported by water from the surrounding highlands and deposited with sand, clay and silt on the lowland plains. Later, the lighter particles were removed by the wind causing the remaining pebbles to settle and to interlock like cobblestones.

Elsewhere in the desert, dew may collect in hollows and material may be loosened by chemical weathering and then removed by wind to leave **closed depressions** or **deflation hollows**. Closed depressions are numerous and vary in size from a few metres across to the extensive Qattara depression in Egypt which reaches a depth of 134 m below sea-level. Closed depressions may also have a tectonic (the south-west of the USA) or a solution origin (limestone areas in Morocco). The Dust Bowl, formed in the American Mid-West in the 1930s, was a consequence of deflation following a severe drought in a region where inappropriate farming techniques had been introduced. Vast quantities of valuable topsoil were blown away, some of which was deposited as far away as Washington, DC.

Figure 7.10

The process of deflation

deflation: silt and sand removed by wind leaving stones

land surface is lowered

leaving desert pavement: a coarse mosaic of stones resembling a cobbled street which protects against further erosion

167

Figure 7.11

Appearance and formation of yardangs

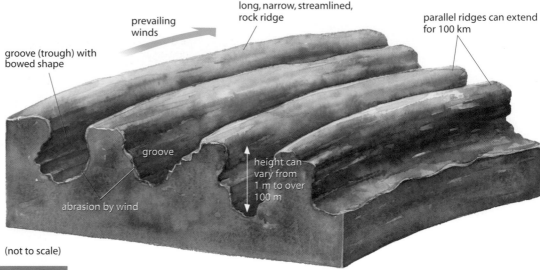

prevailing winds

long, narrow, streamlined, rock ridge

parallel ridges can extend for 100 km

groove (trough) with bowed shape

groove

abrasion by wind

height can vary from 1 m to over 100 m

(not to scale)

Figure 7.12

Landshore yardangs, Western desert, Egypt

Figure 7.13

The movement of crescent-shaped barchan

Abrasion is a sandblasting action effected by materials as they are moved by saltation. This process smooths, pits, polishes and wears away rock close to the ground. Since sand particles cannot be lifted very high, the zone of maximum erosion tends to be within 1 m of the earth's surface. Abrasion produces a number of distinctive landforms which include ventifacts, yardangs and zeugen.

Ventifacts are individual rocks with sharp edges and, due to abrasion, smooth sides. The white rock in the foreground of Figure 7.9 has a long axis of 25cm.

Yardangs are extensive ridges of rock, separated by grooves (troughs), with an alignment similar to that of the prevailing winds (Figure 7.11 and 7.12). In parts of the Sahara, Arabian and Atacama Deserts, they are large enough to be visible on air photographs and satellite imagery.

Zeugen are tabular masses of resistant rock separated by trenches where the wind has cut vertically through the cap into underlying softer rock.

Deposition

Dunes develop when sand grains, moved by saltation and surface creep, are deposited. Although large areas of dunes, known as ergs, cover about 25 per cent of arid regions, they are mainly confined to the Sahara and Arabian Deserts, and are virtually absent in North America. Much of the early fieldwork on dunes was carried out by R. A. Bagnold in North Africa in the 1920s. He noted that some, but by no means all, dunes formed around an obstacle — a rock, a bush, a small hill or even a dead camel; and most dunes were located on surfaces which were even and sandy and not on those which were irregular and rocky. He concentrated on two types of dune: the **barchan** and the **seif**. The barchan is a small, crescent-shaped dune, about 30 m high, which is moved by the wind (Figures 7.13 and 7.15). The seif, named after an Arab curved sword, is much larger (100 km in length and 200 m in height) and more common — although the process of its formation is more complex than initially thought by Bagnold. Textbooks often overemphasise these two dunes, especially the barchan which is a relatively uncommon feature.

a in plan

prevailing wind

gentle, slightly concave slope

X

Y

maximum height 30 m

horn moves faster than centre of dune as there is less sand to move

horn

barchans migrate, moving forwards by up to 30 m/yr

b in profile

prevailing wind

saltation and surface creep on gentle slope

eddying helps to maintain steep slopes

A

X

B

Y

A steep, upper, slip slope of coarse grains and with continual sand avalanches due to unconsolidated material (unlike a river, coarse grains are at the top)

B gentle, basal apron with sand ripples: the finer grains, as on a beach, give a gentler gradient than coarser grains

Figure 7.14

Classification of sand dunes (*after* Goudie)

While Bagnold had to travel the desert in specially converted cars, modern geographers derive their picture of desert landforms from aerial photographs and Landsat images. These new techniques have helped to identify several types of dune, and the modern classification, still based on morphology, contains several additional types (Figure 7.14). Dune morphology depends upon the supply of sand, wind direction, availability of vegetation and the nature of the ground surface.

Type of dune	Description	Supply of sand	Wind direction and speed	Vegetation cover	Speed of dune movement
barchan 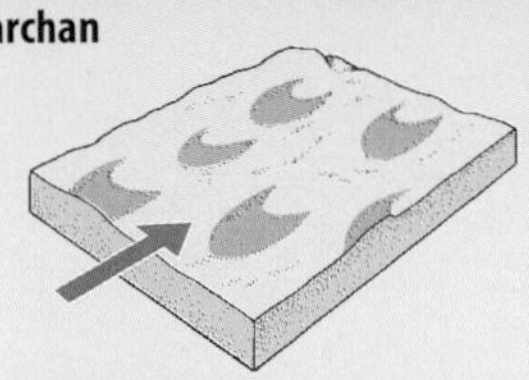	individual dunes, crescent shape with horns pointing downwind (Figures 7.13 and 7.15)	limited	constant direction, right angles to dune	none	highly mobile
barchanoid ridges 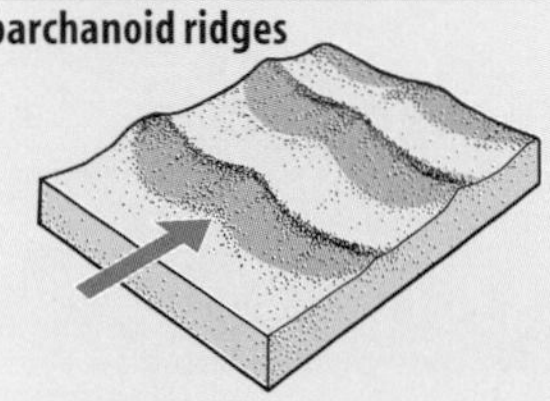	asymmetrical, orientated at right angles to wind, rows of barchans forming parallel ridges	limited	constant direction, right angles to dune	none	mobile
transverse 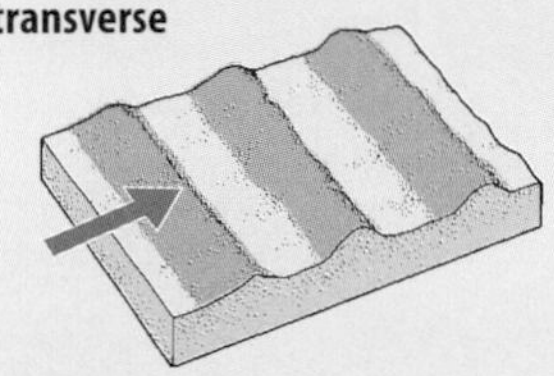	orientated at right angles to wind but lacking barchanoid structure, resemble ocean waves (Figure 7.16)	abundant (thick) sand cover	steady winds (trades), constant direction but with reducing speeds, right angles to dune	vegetation stabilises sand	sand checked by barriers, limited mobility
dome 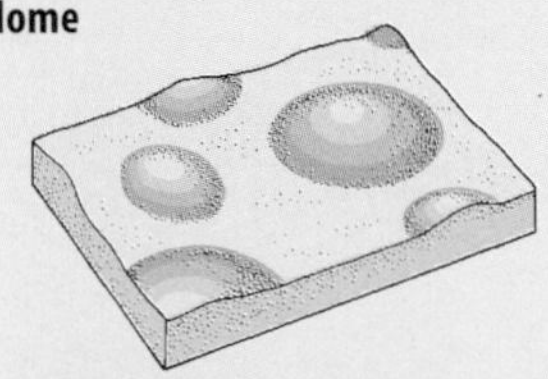	dome shaped (height restricted by wind)	appreciable amounts of coarse sand	strong winds limit height of dune	none	virtually no movement
seif (linear) 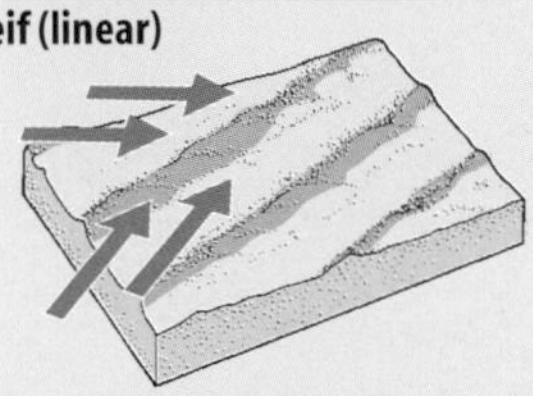	longitudinal, parallel dunes with slip faces on either side, can extend for many km (Figure 7.17)	large	persistent, steady winds (trades), with slight seasonal or diurnal changes in direction	none	regular (even) surface, virtually no movement
parabolic 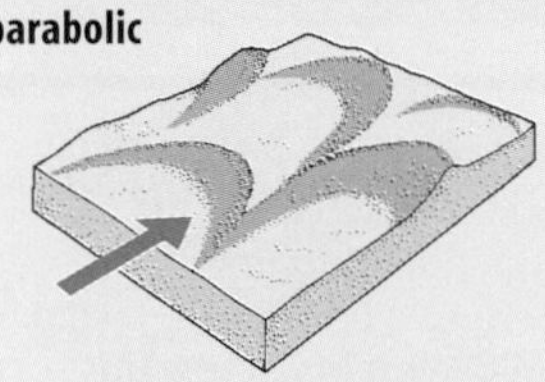	hairpin-shaped with noses pointing downwind, a type of blow-out (eroded) dune where middle section has moved forward, may occur in clusters	limited	constant direction	where present, can anchor sand	highly mobile (by blow-outs in nose of dune)
star 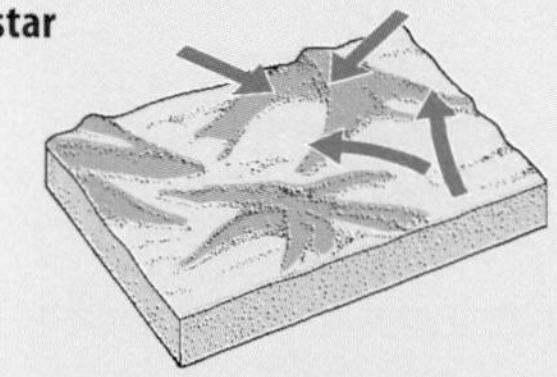	complex dune with a star (starfish) shape (compare arêtes radiating from central peak) (Figure 7.18).	limited	effective winds blow from several directions	none	virtually no movement
reversing 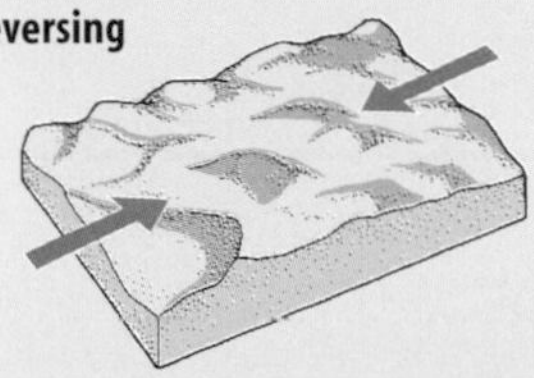	undulating, haphazard shape	limited	winds of equal strength and duration from opposite directions	none	virtually no movement

Figure 7.15

Barchan dunes near Luderitz, Namibia

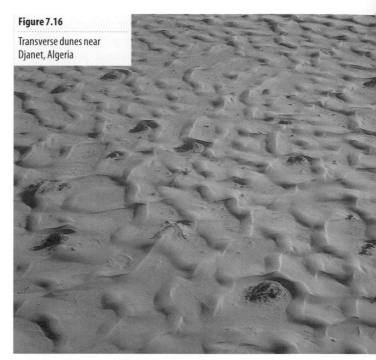

Figure 7.16

Transverse dunes near Djanet, Algeria

Figure 7.18

Star dunes, Sossusvlei, Namibia

Figure 7.17

Seif (linear) dunes, Sossusvlei, Namibia

 Q

1 "Traditionally, weathering in deserts was attributed to mechanical processes resulting from extremes of temperature."
   a Describe how weathering can result from extremes of temperature.
   b Why did doubts arise concerning the effectiveness of this type of weathering?

   c i What other processes of mechanical weathering operate in deserts?
   ii What landforms result from mechanical weathering in deserts?
   d Under what conditions does chemical weathering operate in deserts?

2 a Describe three methods by which the wind can transport material.
   b Read the introduction of John Steinbeck's *The Grapes of Wrath*.. Why did the American Mid-West become a 'dust bowl'?

3 How important do you consider wind to be as an agent of erosion in deserts?

4 a What are the major factors affecting the morphology of sand dunes?
   b What are the main types of sand dune?

The effects of water

It has already been noted that, in arid areas, moisture must be present for processes of chemical weathering to operate. We have also seen that often rainfall is low, irregular and infrequent with long-term fluctuations. Although most desert rainfall occurs in low-intensity storms, the occasional sudden, more isolated, heavy downpour, does occur. There are records of several extreme desert rainfall events, each equivalent to the 3-monthly mean rainfall of London. The impact of water is, therefore, very significant in shaping desert landscapes.

Rivers in arid environments fall into three main categories:

Exogenous Exogenous rivers are those like the Colorado, Nile, Indus, Tigris and Euphrates, which rise in mountains beyond the desert margins. These rivers continue to flow throughout the year even if their discharge is reduced by evaporation when they cross the arid land. (The last four rivers mentioned provided the location for some of the earliest urban settlements, page 360). The Colorado has, for over 300 km of its course, cut down vertically to form the Grand Canyon. The canyon, which in places is almost 2000 m (over 1 mile) deep, has steep sides partly due to rock structure and partly

due to insufficient rainfall to degrade them (Figure 7.19 and Case Study 20).

Endoreic Endoreic drainage occurs where rivers terminate in inland lakes. Examples are the River Jordan into the Dead Sea and the Bear into the Great Salt Lake.

Ephemeral Ephemeral streams, which are more typical of desert areas, flow intermittently, or seasonally, after rain storms. Although often shortlived, these streams can generate high levels of discharge due to several local characteristics. First, the torrential nature of the rain exceeds the infiltration capacity of the ground and so most of the water drains away as surface runoff (overland flow, page 51). Secondly, the high temperatures and the frequent presence of duricrust combine to give a hard, impermeable surface which inhibits infiltration. Thirdly, the lack of vegetation means that no moisture is lost or delayed through interception and the rain is able to hit the ground with maximum force. Fourthly, fine particles are displaced by rainsplash action and, by infilling surface pore spaces, further reduce the infiltration capacity of the soil. It is as a result of these minimal infiltration rates that slopes of less than 2° can, even under quite modest storm conditions, experience extensive overland flow.

Studies in Kenya, Israel and Arizona suggest that surface runoff is likely to occur within 10 minutes of the start of a downpour (Figure 7.21). This may initially be in the form of a **sheet flood** where the water flows evenly over the land and is not confined to channels. Much of the sand, gravel and pebbles covering the desert floor is thought to have been deposited by this process; yet, as the event has rarely been witnessed, it is assumed that deposition by sheet floods occurred mainly during earlier wetter periods called **pluvials**.

Very soon, the collective runoff becomes concentrated into deep, steep-sided ravines known as **wadis** (Figure 7.20) or **arroyos**. Normally dry, wadis may be subjected to irregular **flash floods** (Places 19). The average occurrence of these floods is once a year in the semi-arid margins of the Sahara and once a decade in the extremely arid interior. This infrequency of floods compared to the great number and size of wadis, suggests that they were created when storms were more frequent and severe — i.e. they are a relict feature.

Figure 7.19

The Grand Canyon, Arizona, USA

Figure 7.20

A wadi near Qumran, Israel: in the background is the Dead Sea.

Camping in a wadi is something which experienced desert travellers avoid. It is possible to be swept away by a flash flood which occurs virtually without warning — there may have been no rain at your location, and perhaps nothing more ominous than a distant rumble of thunder. Indeed, the first warning may be the roar of an approaching wall of water. One minute the bed of the wadi is dry, baked hard under the sun and littered with weathered débris from the previous flood or from the steep valley sides (Figure 7.20), while the next minute it is a raging torrent.

The energy of the flood enables large boulders to be moved by traction and enormous amounts of coarse material to be taken into suspension — some witnesses have claimed it is more like a mudflow. Friction from the roughness of the bed, the large amounts of sediment and the high rates of evaporation soon cause a reduction in the stream's velocity. Deposition then occurs, choking the channel, followed by braiding as the water seeks new outlets. Within hours, the floor of the wadi is dry again.

The rapid runoff does not replenish groundwater supplies and without the groundwater contribution to base flow, characteristic of humid climates, rivers cease to flow. At the mouth of the wadi, where the water can spread out and energy is dissipated, material is deposited to form an **alluvial fan** or **cone** (Figure 7.22). If several wadis cut through a highland close to each other, their semi-circular fans may merge to form a **bahada (bajada)** which is an almost continuous deposit of sand and gravel.

Figure 7.21

Typical storm hydrograph for a flash flood in a wadi

Q The typical storm hydrograph for a wadi (Figure 7.21) contrasts with that of a British river (Figure 3.5).

Account for the differences in lag time, the steep rising *and* falling limbs, the high peak discharge and the short duration of flow.

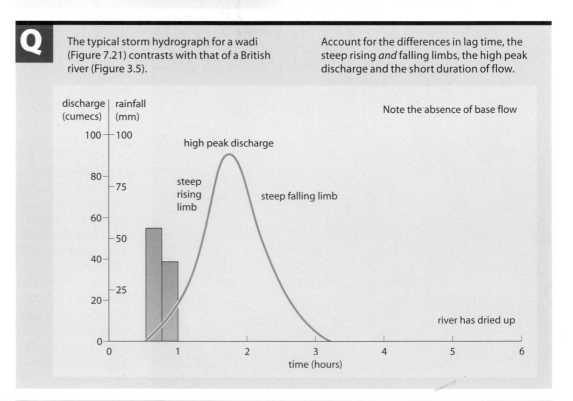

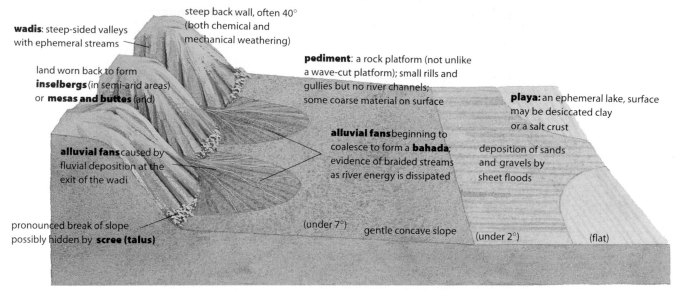

wadis: steep-sided valleys with ephemeral streams

steep back wall, often 40° (both chemical and mechanical weathering)

land worn back to form **inselbergs** (in semi-arid areas) or **mesas and buttes** (arid)

pediment: a rock platform (not unlike a wave-cut platform); small rills and gullies but no river channels; some coarse material on surface

playa: an ephemeral lake, surface may be desiccated clay or a salt crust

alluvial fans caused by fluvial deposition at the exit of the wadi

alluvial fans beginning to coalesce to form a **bahada**; evidence of braided streams as river energy is dissipated

deposition of sands and gravels by sheet floods

pronounced break of slope possibly hidden by **scree (talus)**

(under 7°)   gentle concave slope

(under 2°)   (flat)

Figure 7.22

Pediments and playas

Pediments and playas

Stretching from the foot of the highlands, there is often a gently sloping area either of bare rock or of rock covered in a thin veil of débris (Figures 7.22 and 7.23). This is known as a **pediment**. There is often an abrupt break of slope at the junction of the highland area and the pediment. Two main theories suggest the origin of the pediment, one involving water. This theory proposes that weathered material from cliff faces, or débris from alluvial fans, was carried during pluvials by sheet floods. The sediment planed the lowlands before being deposited, leaving a gently concave slope of less than 7° (Figure 7.22). The alternative theory involves the parallel retreat of slopes resulting from weathering (King's hypothesis, Figure 2.24).

Playas are often found at the lowest point of the pediment. They are shallow, ephemeral, saline lakes formed after rain storms. As the rain water rapidly evaporates, flat layers of either clay, silt or salt are left. Where the dried-out surface consists of clay, large **desiccation cracks**, up to 5 m deep, are formed. When the surface is salt-covered, it produces the 'flattest landform on land'.

Rogers Lake, in the Mojave Desert, California, has been used for spacecraft landings, while the Bonneville salt flats in Utah have been the location for land-speed record attempts.

Occasionally, isolated, flat-topped remnants of former highlands, known as **mesas**, rise sheer from the pediment. Some mesas, in Arizona, have summits large enough to have been used as village sites by the Hopi Indians. **Buttes** are smaller versions of mesas. The most spectacular mesas and buttes lie in Monument Valley National Park in Arizona (Figure 7.24).

Figure 7.23

Pediment at foot of highlands, Wadi Rum, Jordan

Figure 7.24

Mesas and buttes in the Monument National Park, Arizona, USA

Relationship between wind and water

Some desert areas are dominated by wind, others by water. Areas where wind appears to be the dominant geomorphological agent are known as **aeolian domains**. The effectiveness of the wind increases where, and when, amounts of rainfall decrease. As rainfall decreases, so too does any vegetation cover. This allows the wind to transport material unhindered, and rates of erosion (abrasion and deflation) and deposition (dunes) to increase. **Fluvial domains** are those where water processes are dominant or, as evidence increasingly suggests, have been dominant in the past. Vegetation, which stabilises material, increases as rainfall increases or where coastal fog and dew are a regular occurrence.

Evidence also suggests that wind and water can interact in arid environments and that landforms produced by each do co-exist within the same locality. However, the balance between their relative importance has often altered, mainly due to climatic change either over lengthy periods of time (e.g. the 18 000 years since the time of maximum glaciation) or during shorter fluctuations (e.g. since the mid 1960s in the Sahel). At present, and especially in Africa, the decrease in rainfall in the semi-arid desert fringes means that the rôle of water is probably declining, while that of the wind is increasing.

Climatic change

There have already been references to pluvials within the Sahara Desert (pages 165 and 171). Prior to the Quaternary era, these may have occurred when the African plate lay further to the south and the Sahara was in a latitude equivalent to that of the present-day savannas. In the Quaternary era, the advance of the ice sheets resulted in a shift in windbelts which caused changes in precipitation patterns, temperatures and evaporation rates. At the time of maximum glaciation (18 000 BP) desert conditions appear to have been more extensive than they are today (Figure 7.25). Since then, as suggested by radio-carbon dating (page 232), there have been frequent, relatively short-lived pluvials, the last occurring about 9000 BP. Evidence for a once-wetter Sahara is given in Figure 7.26.

Herodotus, a historian living in ancient Greece, described the Garamantes civilisation which flourished in the Ahaggar Mountains 3000 years ago. This people, who recorded their exploits in cave paintings at Tassili des Ajjers, hunted elephants, giraffes, rhinos and antelope. Twenty centuries ago, North Africa was the 'granary of the Roman Empire'. Wadis are too large and deep and alluvial cones too widespread to have been formed by today's occasional storms, while sheet floods are too infrequent to have moved so much material over pediments. Radiating from the Ahaggar and Tibesti Mountains, aerial photographs and satellite imagery have revealed many dry valleys which once must have held permanent rivers (Figure 16.46). Lakes were also once much larger and deeper. Around Lake Chad, shorelines 50 m above the present level are visible and research suggests that lake levels might once have been over 100 m higher. (Lake Bonneville in the USA is only one-tenth of its former maximum size and, like Lake Chad, is drying up rapidly.) Small crocodiles found in the Tibesti must have been trapped in the slightly wetter uplands as the desert advanced. Also, pollen analysis has shown that oak and cedar forests abounded in the same region 10 000 years ago. Groundwater in the Nubian sandstone has been dated, by radio-isotope methods, to be over 25 000 years old and may have accumulated at about the same time as fossil laterite soils (page 297).

Figure 7.25

Extent of sand dunes in Africa

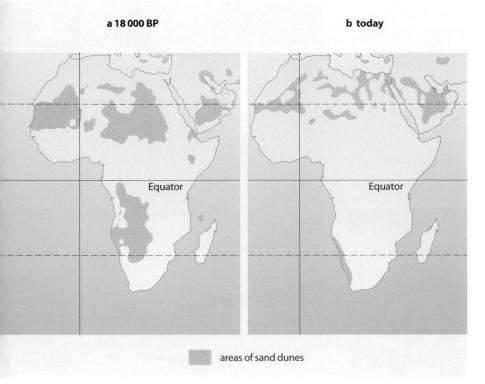

a 18 000 BP

b today

Equator

Equator

▨ areas of sand dunes

Deserts

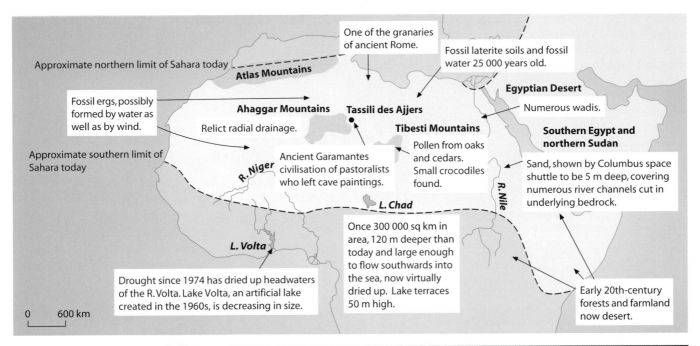

Figure 7.26

Evidence of pluvials in the Sahara

Q 1 Assess the relative importance of water and wind in the development of landforms in arid areas.

2 Outline the evidence which has led to the conclusion that hot arid areas have in the past experienced more humid conditions than those which prevail at the present time.

Case Study 7

Desertification: fact or fiction?

"Literally, desertification means the making of a desert, but although the word has been in use for more than 40 years, few can agree on exactly what is means: there are more than 100 different definitions of the term" (Middleton, 1993). The divergence of definitions is due largely to the uncertainty as to the causes of desertification. Goudie says that "the question has been asked

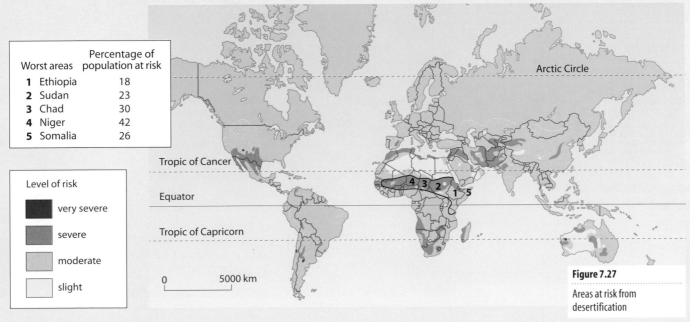

Worst areas	Percentage of population at risk
1 Ethiopia	18
2 Sudan	23
3 Chad	30
4 Niger	42
5 Somalia	26

Level of risk

- very severe
- severe
- moderate
- slight

Figure 7.27

Areas at risk from desertification

whether this process is caused by temporary drought periods of high magnitude, is due to longer-term climatic change towards aridity, is caused by man-induced climatic change, or is the result of human action through man's degradation of the biological environments in arid zones. Most people now believe that it is produced by a combination of increasing human and animal populations, which cause the effects of drought years to become progressively more severe so that the vegetation is placed under increasing stress".

Those places perceived to be at **greatest risk** from desertification are shown in Figure 7.27. (This map and the four levels of risk were devised at the 1977 Nairobi Conference on Desertification.) It is generally agreed that the desert is encroaching into semi-arid, desert margins, especially in the Sahel—a broad belt of land on the southern side of the Sahara (2–4 in Figure 7.27).

Some of the main interrelationships between the believed causes of desertification are shown in Figure 7.28.

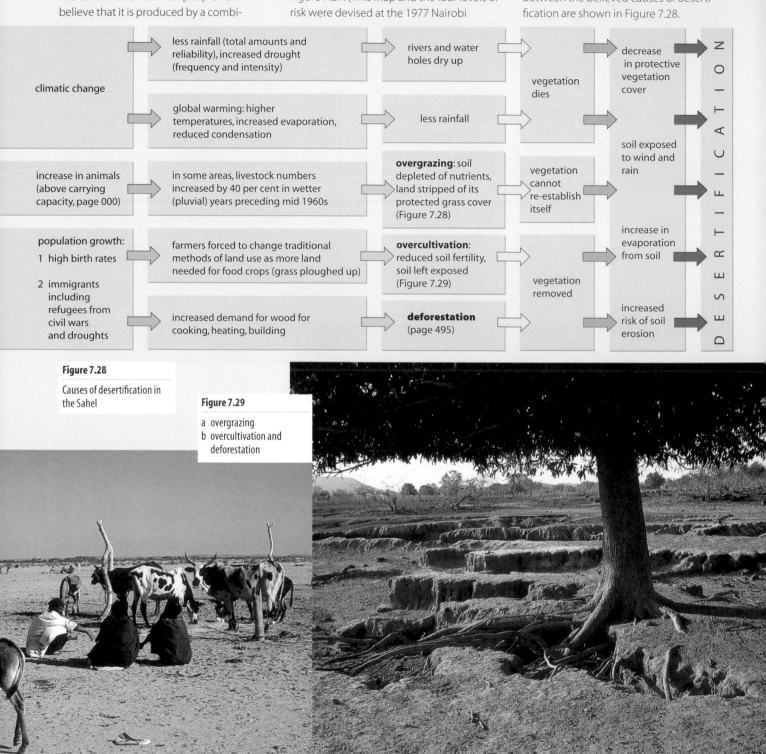

Figure 7.28

Causes of desertification in the Sahel

Figure 7.29

a overgrazing
b overcultivation and deforestation

In 1975, Hugh Lamprey a bush pilot and environmentalist, claimed that, since his previous study 17 years earlier, the desert in the Sudan had advanced southwards by 90–100 km. In 1982 and at the height of one of Africa's worst-ever recorded droughts, UNEP (United Nations Environmental Programme), claimed that the Sahara was advancing southwards by 6–10 km a year and that, globally, 21 million hectares of once-productive soil were being reduced each year to zero productivity, that 850 million people were being affected, and that 35 per cent of the world's land surface was at risk. This latter figure

was again quoted by UNEP at the 1992 Rio Earth Summit.

Recent scientific studies, made by using satellite imagery and through more detailed fieldwork, have thrown considerable doubt upon both the causes and the effects of desertification. Three examples are given in Figure 7.30.

However, such doubts do not mean that the risk of desertification has disappeared. They do suggest that we should be wary of overgeneralisation, and that our assessment of risk should be based on detailed evidence.

The semi-arid lands are fragile environments. Their boundaries are subject to change resulting from variations in rainfall, from variations in human use, and perhaps also from global warming. It is often difficult to separate natural causes from human ones, and short-term fluctuations from long-term trends. Regardless of whether desertification is already a major environmental hazard or whether it poses a future environmental risk, one thing is certain: any increase in desertification is likely to be the result of people's mismanagement of their natural resources.

Researchers at the University of Lund in Sweden have carried out field surveys and examined satellite pictures of the Sudan in an effort to confirm Lamprey's findings. Earlier this year (1993) in the journal Ambio, Ulf Hellden, head of the university's remote sensing laboratory and working for UNEP, reported "no major shifts in the northern cultivation limit, no major sand dune transformation, no major changes in vegetation cover" beyond the dramatic but short-term effects of variable rainfall.

A belt of sand dunes that Lamprey said formed the advancing front of the Sahara in the Northern Kordofan province of Sudan showed no sign of movement between 1962 and 1984. Nor was there any evidence of patches of desert growing round boreholes or villages - a phenomenon frequently claimed to be the result of overgrazing by herds of cattle (page 440). Hellden claims that "As long as the effects of desertification in Africa and elsewhere are not documented according to scientific standards, there is a risk that the desertification issue will become a political and development fiction rather than a scientific fact".

Amid the shimmering sands of the desert margins, dogma often stands in for reality. One such dogma holds that desert margins have a fixed 'carrying capacity' of humans and animals. According to a recent executive director of UNEP, "when the number is exceeded, the whole piece of land will quickly degenerate. Population pressure is definitely one of the major causes of desertification".

True? New research suggests not. Long-term studies show that in the desert margins of two of Africa's largest and fastest-growing countries - Nigeria and Kenya - the opposite has happened. Rapidly-increasing populations emerge from these studies as saviours of the landscape, rescuing it from rampant soil erosion, and protecting trees and conserving water.

Ridley Nelson, a researcher for the World Bank (which has spent heavily on planting trees to halt the deserts), says that "advancing sand in not a major global problem. Claims that the Sahara is expanding at some horrendous rate are unacceptible to-day". Andrew Warren and Clive Agnew of the ecology and conservation unit at University College, London, agree that "active sand dunes seldom threaten valuable land".

Nelson also questions UNEP's figures concerning the rate of, and the areas affected by, desertification. He says that "they are meaningless. They come largely from answers to a questionnaire sent out to governments in 1982, at the height of a major African drought. Nobody knows how these governments arrived at their figures."

Figure 7.30

Recent views on desertification

References

Desert Processes . Open University, BBC Television.

Goudie, A. (1993) *The Nature of the Environment.* Blackwell.

Goudie, A. and Watson, A. (1990) *Desert Morphology.* Thomas Nelson.

McCullagh, P. (1978) *Modern Concepts in Geomorphology.* Oxford University Press.

Middleton, N. (1993) The desertification debate, *Geography Review*, September, p. 30.

Planet Earth: Arid Lands (1984) Time–Life Books.

Thomas, D.S.G. (1989) *Arid-zone Geomorphology.* Belhaven Press.

Chapter 8

Rock types and landforms

"At first sight it may appear that rock type is the dominant influence on most landscapes ...
As geomorphologists, we are more concerned with the ways in which the characteristics of rocks respond to the processes or erosion and weathering than with the detailed study of rocks themselves."

Roy Collard, *The Physical Geography of Landscape*, 1988

Lithology refers to the physical characteristics of a rock

Previous chapters have demonstrated how landscapes at both local and global scales have developed from a combination of processes. Plate tectonics, weathering and the action of moving water, ice and wind both create and destroy landforms. Yet these processes, however important they are at present or have been in the past, are insufficient to explain the many different and dramatic changes of scenery which can occur within short distances, especially in the British Isles.

Each individual rock type is capable of producing its own characteristic scenery. Landforms are greatly influenced by a rock type's vulnerability to weathering, its permeability and its structure.

To show how these three factors affect different rocks and to explain their resultant landforms and potential economic use, four rock types have been selected as exemplars. Carboniferous limestone and chalk (both sedimentary rocks), and granite and basalt (both igneous) have been chosen because, arguably, these produce some of the most distinctive types of landform and scenery.

Lithology and geomorphology

Vulnerability to weathering

Mechanical weathering in Britain occurs more readily in rocks which are jointed. Water can penetrate either down the **joints** or along the **bedding planes** (Figure 8.1) of Carboniferous limestone, or into cracks resulting from pressure release or contraction on cooling within granite and basalt (pages 26 and 35). Subsequent freezing and thawing along these lines of weakness causes frost shattering (page 34).

Chemical weathering is a major influence in limestone and granite landforms. Limestone, composed mostly of calcium carbonate, is slowly dissolved by the carbonic acid in rainwater, i.e. the process of carbonation (page 37). Granite consists of quartz, feldspar and mica. It is susceptible to hydration, where water is incorporated into the rock structure causing it to swell and crumble (page 36), and to hydrolysis, when the feldspar is chemically changed into clay (page 37). Quartz, in comparison with other minerals, is one of the least prone to chemical weathering.

Mottershead has emphasised that "the mechanical resistance of rocks depends on the strength of the individual component minerals and the bonds between them, and that chemical resistance depends on the individual chemical resistances of the component minerals. Mechanical strength decreases if just one of these component minerals becomes chemically altered".

Figure 8.1

Bedding planes with joints and angle of dip

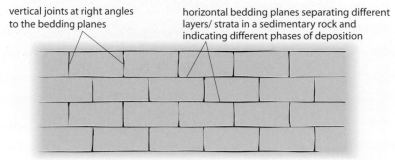

a massively bedded Carboniferous limestone

vertical joints at right angles to the bedding planes

horizontal bedding planes separating different layers/ strata in a sedimentary rock and indicating different phases of deposition

b thinly bedded chalk

the angle of dip is the difference between the actual inclination of the rock and the horizontal

horizontal

10° dip

joints still at right angles to bedding planes

gently dipping bedding planes

Permeability

Permeability is the rate at which water may be stored within a rock or is able to pass through it. Permeability can be divided into two types:

Primary permeability or porosity This depends upon the texture of the rock and the size, shape and arrangement of its mineral particles. The areas between the particles are called **pore spaces** and their size and alignment determine how much water can be absorbed by the rock. Porosity is usually greatest in rocks which are coarse-grained, such as gravels, sands, sandstone and oolitic limestone, and usually lowest in those which are fine-grained, such as clay and granite. (It is possible to have fine-grained sandstone and coarse-grained granite.)

The **infiltration capacity** of sands is estimated to average 200 mm/hour, whereas in clay it is only 5 mm/hour. Pore spaces are larger where the grains are rounded rather than angular and compacted (Figure 8.2). Porosity can be given as an index value based upon the percentage of the total volume of the rock which is taken up by pore space — e.g. clay 20 per cent, gravel 50 per cent. When all the pore spaces are filled with water, the rock is said to be **saturated**. The water table marks the upper limit of saturation (Figure 8.9). Permeable rocks which store water are called **aquifers**.

Secondary permeability or perviousness This occurs in rocks which have joints and fissures along which water can flow. The most pervious rocks are those where the joints have been widened by solution, e.g. Carboniferous limestone, or by cooling, e.g. basalt. A rock may be pervious because of its structure, though water may not be able to pass through the rock mass itself.

Where rocks are porous or pervious, water rapidly passes downwards to become groundwater, leaving the surface dry and without evident drainage — chalk and limestone regions have few surface streams. **Impermeable rocks** neither absorb water nor allow it to pass through them, e.g. granite. These rocks will have, therefore, a higher drainage density (page 56).

Infiltration capacity is the maximum rate at which water percolates into the ground

Figure 8.2

Pore spaces and infiltration capacity

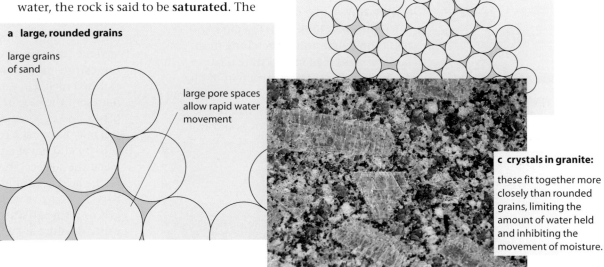

a large, rounded grains

large grains of sand

large pore spaces allow rapid water movement

b small, rounded grains

although there are more pore spaces, they are much smaller: water clings to grains (surface tension) preventing the passage of moisture (Figure 10.13)

c crystals in granite:

these fit together more closely than rounded grains, limiting the amount of water held and inhibiting the movement of moisture.

Structure

Resistance to erosion depends upon whether the rock is massive and stratified, folded or faulted. Usually the more massive the rock and the fewer its joints and bedding planes, the more resistant it is to weathering and erosion. Conversely, the softer, more jointed and less compact the rock, the more vulnerable it is to denudation processes. Usually, more resistant rocks remain as upland areas (granite), while those which are less resistant form lowlands (clay).

However, there are exceptions. Chalk, which is relatively soft and may be well-jointed, forms rolling hills because it allows water to pass through it and so fluvial activity is limited. Carboniferous or Mountain limestone, having joints and bedding planes, produces jagged karst scenery because although it is pervious it has a very low porosity.

Limestone

Limestone is a rock consisting of at least 80 per cent calcium carbonate. In Britain, most limestone was formed during four geological periods each of which experienced different conditions. The following list begins with the oldest rocks. Use an atlas to find their location.

Carboniferous limestone This is hard, grey, crystalline and well-jointed. It contains many fossils, including corals, crinoids and brachiopods. These indicate that the rock was formed on the bed of a warm, clear sea and adds to the evidence that the British Isles once lay in warmer latitudes. Carboniferous limestone has developed its own unique landscape, known as **karst**, which in Britain is seen most clearly in the Peak District and Yorkshire Dales National Parks.

Magnesian limestone This is distinctive because it contains a higher proportion of magnesium carbonate. In Britain, it extends in a belt from the mouth of the River Tyne to Nottingham. In the Alps, it is known as **dolomite**.

Jurassic (oolitic) limestone This forms a narrow band extending southwards from the North Yorkshire Moors to the Dorset coast. Its scenery is similar to that typical of chalk.

Cretaceous chalk This is a pure, soft, well-jointed limestone. Stretching from Flamborough Head in Yorkshire (Figures 6.16 and 6.17), it forms the escarpment of the Lincoln Wolds, the East Anglian Heights and the North and South Downs, before ending up as the 'White Cliffs' at Dover and at Beachy Head, the Needles and Swanage (Figure 6.19). Cretaceous chalk is assumed to be the remains of small marine organisms which lived in clear, shallow seas.

The most distinctive of the limestone landforms are found in Carboniferous limestone and chalk.

Carboniferous limestone

This rock develops its own particular type of scenery primarily because of three characteristics. First, it is found in thick beds separated by almost horizontal bedding planes and with joints at right angles (Figure 8.1). Secondly, it is pervious but not porous, meaning that water can pass along the bedding planes and down joints but not through the rock itself. Thirdly, calcium carbonate is soluble. Carbonic acid in rainwater together with humic acid from moorland plants, dissolve the limestone and widen any weaknesses in the rock — i.e. the bedding planes and joints. Acid rain also speeds up carbonation and solution, page 205. As there is minimum surface drainage and little breakdown of bedrock to form soil, the vegetation cover tends to be thin or absent. In winter, this allows frost shattering to produce scree at the foot of steep cliffs.

It is possible to classify carboniferous limestone landforms into four types:

1 **Surface features caused by solution** Limestone pavements are flat areas of exposed rock. They are flat because they represent the base of a dissolved bedding plane, and exposed because the surface soil may have been removed by glacial activity and never replaced. Where joints reach the surface, they may be widened by the acid rainwater (carbonation, page 37) to leave deep gashes called **grikes**. Some grikes at Malham in northwest Yorkshire are 0.5 m wide and up to 2 m deep. Between the grikes are flat-topped yet dissected blocks referred to as **clints** (Figure 2.8). In time, the grikes widen and the clints are weathered down until a lower bedding plane is exposed and the process of solution–carbonation is repeated.

2 **Drainage features** Rivers which have their source on surrounding impermeable rocks, such as the shales and grits of northern England, may disappear down **swallow holes** or **sinks** as soon as they reach the limestone (Figure 8.3). The

Figure 8.3

A stream disappearing down a swallow hole near Alum Pot, Ingleborough

Figure 8.4

The resurgence at the foot of Malham Cove

Figure 8.6

Stalactites, stalagmites and pillars, Carlsbad Caverns, New Mexico

streams flow underground finding a pathway down enlarged joints, forming **potholes**, and along bedding planes. Where solution is more active, underground **caves** may form. Corrosion often widens the caverns until parts of the roof collapse, providing the river with angular material ideal for corrasion. Heavy rainfall very quickly infiltrates downwards so caverns and linking passages may become water-filled within minutes. The resultant turbulent flow can transport large stones and the floodwater may prove fatal to cavers and potholers. Rivers make their way downwards, often leaving caverns abandoned as the water finds a lower level, until they reach underlying impermeable rock. A **resurgence** occurs where the river reappears on the surface, often at the junction of permeable and impermeable rocks (Figure 8.4).

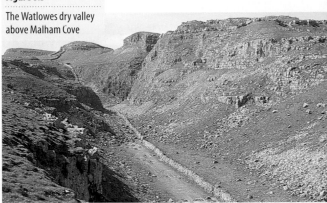

Figure 8.5

The Watlowes dry valley above Malham Cove

3 **Surface features resulting from underground drainage** Steep-sided valleys are likely to have been formed as rivers flowed over the surface of the limestone, probably during periglacial times when permafrost acted as an impermeable layer. When the rivers were able to revert to their subterranean passages, the surface valleys were left dry (Figure 8.5). Should a series of underground caverns become too large to support the rock above them, the limestone will collapse to form a **gorge**. If the area above an individual cave collapses, a small surface depression called a **doline** is formed. In the former Yugoslavia, where the term 'karst' originated, huge depressions called **poljes** may have formed in a similar way. Poljes may be up to 400 sq km in area.

4 **Underground depositional features** Groundwater may become saturated with calcium bicarbonate which is formed by the chemical reaction between carbonic acid in rainwater and calcium carbonate in the rock. However, when this 'hard' water reaches a cave, much of the carbon dioxide bubbles out of solution back into the air — i.e. the process of carbonation in reverse. Aided by the loss of some moisture by evaporation, calcium carbonate (calcite) crystals are subsequently precipitated. Water dripping from the ceiling of the cave initially forms pendant soda straws which, over a very long period of time, may grow into icicle-shaped **stalactites** (Figure 8.6). Experiments in Yorkshire caves suggest that stalactites grow at about 7.5 mm per year. As water drips on to the floor, further deposits of calcium carbonate form the more rounded, cone-shaped **stalagmites** which may, in time, join the stalactites to give **pillars**. Elsewhere, water trickles down rocks leaving a layer or curtain of calcite crystals called a **flowstone**.

Figure 8.7

The karst towers of Guilin, south China

The limestones which outcrop near Guilin have formed a unique karst landscape. The massively bedded, crystalline rock, which in places is 300 m thick, has been slowly pushed upwards from its sea-bed origin by the same tectonic movements which formed the Himalayas and the Tibetan Plateau far to the west. The heavy summer monsoon rain, sometimes exceeding 2000 mm, has led to rapid fluvial erosion by such rivers as the Li Jiang (Li River). The availability of water together with the high sub-tropical temperatures (Guilin is 25°N) encourage highly active chemical weathering (solution–carbonation).

The result has been the formation of a landscape which for centuries has inspired Chinese artists and, recently, has attracted growing numbers of tourists. To either side of the river are natural domes and towers, some of which rise almost vertically 150 m from surrounding paddy fields (Figures 8.7 and 16.12), giving the valley its gorge-like profile. Caves, visible on the sides of the towers, were formed by underground tributaries to the Li Jiang when the main river was flowing at levels considerably higher than those of today.

Figure 8.8

Characteristic features of Carboniferous limestone scenery

Q

Figure 8.8 is a model showing typical landforms found in an area of Carboniferous limestone or karst. The features have been labelled A–Z.

1 Using the following list, match each term with the appropriate letter: abandoned cave; bedding planes; cavern; clints and grikes; dolines; dry valley; gorge; impermeable rock (twice); joints; limestone pavement; limestone plateau; permeable rock; pillar; pothole; resurgence; scar; scree; stalactites; stalagmites; surface river (four times); swallow hole; underground river.

2 Explain the particular location of each feature.

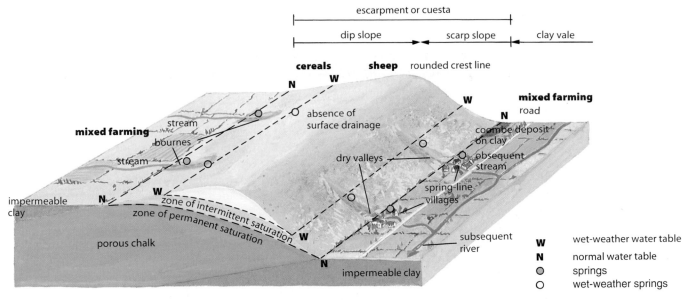

escarpment or cuesta

dip slope · scarp slope · clay vale

cereals · sheep · rounded crest line

N · W

mixed farming

stream · bournes · stream

absence of surface drainage

mixed farming · road

coombe deposit on clay

dry valleys

obsequent stream

spring-line villages

impermeable clay · N · W

zone of intermittent saturation

zone of permanent saturation

porous chalk

W · N

subsequent river

impermeable clay

W · wet-weather water table
N · normal water table
○ · springs
○ · wet-weather springs

Figure 8.9

Scarp and vale scenery: an idealised section through a chalk escarpment in south-east England

Speleology is the scientific study of caves

Economic value of Carboniferous limestone
Human settlement on this type of rock is usually limited and dispersed (page 368) due to limited natural resources, especially the lack of water and good soil. Villages such as Castleton (Derbyshire) and Malham (Yorkshire) have grown up near to a resurgence.

Limestone is often quarried for the cement and steel industries or as ornamental stone, but the resultant scars have led to considerable controversy (Case Study 8). The conflict is between the economic advantages of extracting a valuable raw material and providing local jobs, versus the visual eyesore, noise, dust and extra traffic resulting from the operations — e.g. the Hope valley, Derbyshire.

Farming is hindered by the dry, thin, poorly developed soils for, although most upland limestone areas of Britain receive high rainfall totals, water soon flows underground. The rock does not readily weather into soil-forming particles, such as clay or sand, but is dissolved and the residue is then leached (page 239). On hard limestones, rendzina soils may develop (page 253). These soils are unsuitable for ploughing and their covering of short, coarse, springy grasses favours only sheep grazing. In the absence of hedges and trees, drystone walls were com-

monly built as field boundaries. The scenery attracts walkers and school parties, while underground features lure cavers, potholers and **speleologists**.

Chalk

Chalk, in contrast to Carboniferous limestone scenery, consists of gently rolling hills with rounded crest lines. Typically, chalk has steep, rather than gorge-like, dry valleys and is rarely exposed on the surface (Figure 8.9).

The most distinctive feature of chalk is probably the **escarpment**, or **cuesta**, e.g. the North Downs and South Downs (Figure 8.10). Here the chalk, a pure form of limestone, was gently tilted by the earth movements associated with the collision of the African and Eurasian Plates. Subsequent erosion has left a steep scarp slope and a gentle dip slope. In south-east England, clay vales are found at the foot of the escarpment (Figure 3.53b).

Although chalk — like Carboniferous limestone — has very little surface drainage, its surface is covered in numerous dry valleys (Figure 8.11). Given that chalk can absorb and allow rainwater to percolate through it, how could these valleys have formed?

Figure 8.10

South Downs chalk escarpment, Poynings, Sussex

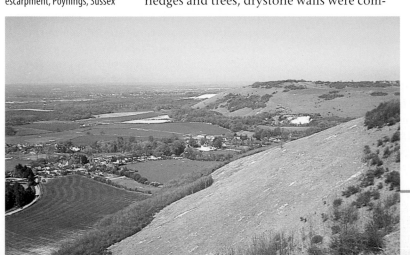

Figure 8.11

A dry valley in chalk: Devil's Dyke, South Downs, Sussex

Goudie lists 16 different hypotheses that have been put forward regarding the origins of dry valleys. These he has grouped into three categories:

Uniformitarian These hypotheses assume that there have been no major changes in climate or sea-level and that 'normal' — i.e. fluvial — processes of erosion have operated without interruption. A typical scenario would be that the drainage system developed on impermeable rock overlying the chalk; and subsequently became superimposed upon it (page 79).

Marine These hypotheses are related to changes in sea or base level (page 76). One, which has a measure of support, suggests that when sea-levels rose eustatically at the end of the last ice age (page 112), water tables and springs would also have risen. Later, when the base level fell, so too did the water table and spring line, causing valleys to become dry.

Palaeoclimatic This group of hypotheses, based on climatic changes during and since the ice age, is the most widely accepted. One hypothesis claims that under periglacial conditions any water in the pore spaces would have been frozen, causing the chalk to behave as an impermeable rock (page 121). As temperatures were low, most precipitation would fall as snow. Any meltwater would have to flow over the surface, forming valleys which are now relict landforms (Figure 8.11). An alternative hypothesis stems from occasions when places receive excessive amounts of rainfall and streams temporarily reappear in dry valleys. Climatologists have shown that there have been times since the ice age when rainfall was considerably greater than it is today. Figure 8.9 shows the normal water table with its associated spring line. If there is a wetter than average winter, or longer period, when moisture loss through evaporation is at its minimum, then the level of permanent saturation will rise. Notice that the wet-weather water table causes a rise in the spring line and so seasonal rivers, or **bournes**, will flow in the normally dry valleys. Remember also that there will be a considerable lag time between the peak rainfall and the time that the bournes will begin to flow (throughflow rather than surface run-off on chalk). The springs are the source of obsequent streams (Figure 3.53b).

The presence of coombe deposits, resulting from solifluction (page 121), also links chalk landforms with periglacial conditions.

Economic value of chalk
The main commercial use of chalk is in the production of cement, but there are objections on environmental grounds to both quarries and the processing works. Settlement tends to be in the form of nucleated villages strung out in lines along the foot of an escarpment, originally to take advantage of the assured water supply from the springs (Figures 8.9 and 8.10). Water-storing chalk aquifers have long been used as a natural, underground reservoir by inhabitants of London, although recent increases in demand have exceeded the rate at which this artesian water has been replaced. The result has been the lowering of the water table.

Chalk weathers into a thin, dry, calcareous soil with a high pH. Until this century, the springy turf of the Downs was mainly used to graze sheep and to train race horses. Horse racing is still important locally, as at Epsom and Newmarket, but much of the land has been ploughed and converted to the growing of wheat and barley. In places, the chalk is covered by a residual deposit of **clay-with-flints** which may have been an insoluble component of the chalk or may have been left from a former overlying rock. This soil is less porous and more acidic than the calcareous soil and several such areas are covered by beech trees — or were, before the violent storm of October 1987 (Places 23, page 217). Flint has been used as a building material and was the major source for Stone Age tools and weapons.

Granite

Granite was formed when magma was intruded into the earth's crust. Initially, as on Dartmoor and in northern Arran, the magma created deep-seated, dome-shaped batholiths (page 26), since which time the rock has been exposed by various processes of weathering and erosion. Having been formed at a depth and under pressure, the rate of cooling was slow and this enabled large crystals of quartz, mica and feldspar to form. As the granite continued to cool, it contracted and a series of cracks were created vertically and horizontally, at irregular intervals. These cracks may have been further enlarged, millions of years

Rock types and landforms

Figure 8.12

Jointing in granite, Goat Fell, Arran

whitish clay called **kaolinite**. Where the change occurs at a greater depth (perhaps due to hydrothermal action), it produces **kaolin**. Quartz, which is not affected by chemical weathering, remains as loose crystals (Figure 2.7).

The most distinctive granite landform is the **tor** (Figure 8.13). There are two major theories concerning its formation, based on physical and chemical weathering respectively. Both, however, suggest the removal of material by solifluction and hence lead to the opinion that tors are relict features.

The first hypothesis suggests that blocks of exposed granite were broken up, sub-aerially, by frost shattering during periglacial times. The weathered material was then moved downhill by solifluction to leave the more resistant rock upstanding on hill summits and valley sides.

The second, proposed by D. L. Linton and more widely accepted, suggests that joints in the granite were widened by sub-surface chemical weathering (Figure 8.14). Linton suggested that deep weathering occurred during the warm Pliocene period (Figure 1.1) when rainwater penetrated the still-unexposed granite. As the joints widened, roughly rectangular blocks or core-stones were formed. The weathered rock is believed to have been removed by solifluction during periglacial times to leave outcrops of granite tors, separated by shallow depressions. The spacing of the joints is believed to be critical in tor formation: large, resistant core-stones have been left where joints were spaced far apart; where they were closely packed and weathering was more active, clay-filled depressions have developed. The rounded nature of the core-stones (Figure 8.14) is caused by **spheroidal weathering**, a form of exfoliation (page 35).

later, by pressure release as overlying rocks were removed (Figure 8.12).

The coarse-grained crystals render the rock non-porous but, although many texts quote granite as an example of an impermeable rock, water can find its way along the many cracks making some areas permeable. Despite this, most granite areas usually have a high drainage density (though on parts of Dartmoor the density is less than on some surrounding rocks) and, as they occur in upland parts of Britain which have a high rainfall, they are often covered by marshy terrain.

Although a hard rock, granite is susceptible to both physical and chemical weathering. The joints, which can hold water, are widened by frost shattering while the different rates of expansion and cooling of the various minerals within the rock cause granular disintegration (page 35). The feldspar and, to a lesser extent, mica can be changed chemically by hydrolysis (page 37). This means that calcium, potassium, sodium, magnesium and, if the pH is less than 5.0, iron and aluminium, are released from the chemical structure. Where the feldspar is changed near to the surface it forms a

Figure 8.13

Hound Tor, Dartmoor

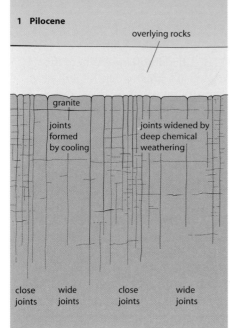
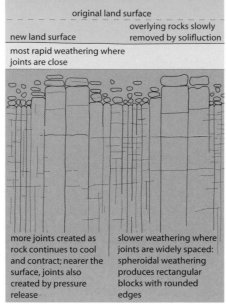
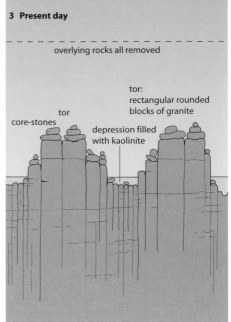

1 Pilocene

overlying rocks

granite

joints formed by cooling

joints widened by deep chemical weathering

close joints · wide joints · close joints · wide joints

2 Pleistocene

original land surface

new land surface · overlying rocks slowly removed by solifluction

most rapid weathering where joints are close

more joints created as rock continues to cool and contract; nearer the surface, joints also created by pressure release

slower weathering where joints are widely spaced: spheroidal weathering produces rectangular blocks with rounded edges

3 Present day

overlying rocks all removed

tor: rectangular rounded blocks of granite

tor

core-stones

depression filled with kaolinite

Figure 8.14

The formation of tors (*after* D.L. Linton)

Economic value of granite

As a raw material, granite can be used for building purposes; Aberdeen, for example, is known as 'the granite city'. Kaolin, or china clay, is used in the manufacture of pottery. Peat, which overlies wide areas of granite bedrock, is an acidic soil which is often severely gleyed (page 251) and saturated with water, forming blanket bogs. The resultant heather-covered moorland is often unsuitable for farming but provides ideal terrain for grouse, and army training. With so much surface water and heavy rainfall, granite areas provide ideal sites for reservoirs. Tors, such as Hound Tor on Dartmoor (Figure 8.13), may become tourist attractions, but granite environments tend to be inhospitable for settlement.

Basalt

Unlike granite, basalt formed on the earth's surface, usually at constructive plate margins. The basic lava, on exposure to the air, cooled and solidified very rapidly. The rapid cooling produced small, fine-grained crystals and large cooling cracks which, at places like the Giant's Causeway in Northern Ireland (Figure 1.25) and Fingal's Cave on the Isle of Staffa, are characterised by perfectly shaped hexagonal, columnar jointing. Basalt can be extruded from either fissures or a central vent (page 23). When extruded from fissures, the lava often covers large areas of land, hence the term flood basalts, to produce flat plateaux such as the Deccan Plateau in India and the Drakensbergs in South Africa. Successive eruptions often build upwards to give, sometimes aided by later erosion, stepped hillsides beneath flat, tabular summits (e.g. the Drakensbergs, Lanzarote and Antrim). When extruded from a central vent, the viscous lava produces gently sloping shield volcanoes (Figure 1.19b). Shield volcanoes can reach considerable heights — Mauna Loa (Hawaii) rises over 9000 m from the Pacific seabed making it, from base to summit, the highest mountain on earth.

Economic value of basalt

Basaltic landforms can sometimes be monotonous, such as places covered in flood basalts, and sometimes scenic and spectacular, as the Giant's Causeway, the Hawaiian volcanoes and the Iguaçu Falls in Brazil (Figures 3.32 and 3.33). Basaltic lava can weather relatively quickly into a deep, fertile soil as on the Deccan in India and in the coffee-growing region of south-east Brazil. It can also be used for road foundations.

Q

The following questions are intended to be answered as essays.

1 Discuss the relationship between climate and the processes of rock weathering in Carboniferous limestone, chalk, granite and basalt.

2 With reference to Carboniferous limestone, chalk and clay, granite and basalt, discuss the hypothesis that rock type and rock structure are dominant controlling factors in the formation of landforms in these areas.

Case Study 8

Quarrying in National Parks

Figure 8.15

Limestone quarries in Derbyshire

Figure 8.16

Limestone can reduce emission of SO_2 from power stations

Advantages	Disadvantages
Provides jobs for local people, often in rural areas with little alternative employment	Noise from blasting, even if only done at controlled times, and heavy traffic
Income from employment and rates paid by quarry firms help local communities (shops and other services	Dust from explosions and movement of traffic)
Extra jobs may reduce rural depopulation	Visual pollution of scars caused by quarries and of ugly buildings
Provides raw materials for other industries, local and national	Spoil heaps (waste tips)
Can help national economy (reducing costs of imports/increasing value of exports)	Run-off from quarry can include sediment which can reach rivers
Can lead to improvements in local communications (e.g. wider roads)	Heavy lorries on narrow rural roads, especially if roads are in scenic/tourist areas

Figure 8.17

Advantages and disadvantages of quarrying

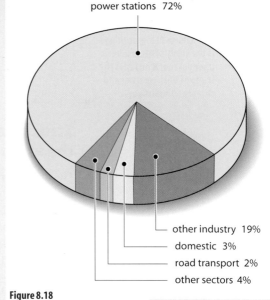

Figure 8.18

Sulphur dioxide emissions by sector in UK, 1990

Quarrying can be an important source of local employment and, especially in economically less developed countries, of national wealth. However, any social and economic gain can easily be offset by environmental loss. In Britain, it is perhaps the extraction of limestone which causes most public concern, possibly because many of the quarries are in National Parks (Figure 8.15). The advantages and disadvantages of quarrying are summarised in Figure 8.17.

A major environmental problem is the release into the atmosphere of sulphur dioxide (SO_2) from thermal power stations producing electricity (Figures 8.16 and 8.18). A major cause of acid rain, SO_2 also poses a threat to health. Two possible methods of reducing SO_2 emissions from power stations are to use 'fluidised beds' (where coal and limestone are burned together so that sulphur 'sticks' to the limestone) or 'desulphurisation plants' (where SO_2 is sprayed with water turning it into sulphuric acid, which is then neutralised by adding lime). Unfortunately, either process means an increase in demand for, and therefore in the quarrying of, limestone. Thus, the reduction of one environmental problem (air pollution) is likely to add to that of another environmental issue (quarrying).

References

Goudie, A. (1993) *The Nature of the Environment*. Blackwell.

Monkhouse, F. J. and Small, J. (1978) *A Dictionary of the Natural Environment*. Edward Arnold.

Ollier, C. (1991) *Weathering and Landforms*. Thomas Nelson.

Planet Earth: Underground Worlds (1982) Time–Life Books.

Chapter 9

Weather and climate

"When two Englishmen meet, their first talk is of the weather."

Samuel Johnson, *The Idler*

The distinction between climate and weather is one of scale. Weather refers to the state of the atmosphere at a local level, usually on a short time scale of minutes to months. It emphasises aspects of the atmosphere that affect human activity — such as sunshine, cloud, wind, rainfall and temperature. Climate is concerned with the long-term behaviour of the atmosphere in a specific area. Climatic characteristics are represented by data on temperature, pressure, wind, precipitation, humidity, etc. which are used to calculate daily, monthly and yearly averages (Framework 6, page 230) and to build up global patterns (Chapter 12).

The science of meteorology is the study of atmospheric phenomena; it includes the study of both weather and climate (climatology).

Structure and composition of the atmosphere

The atmosphere is an envelope of transparent, odourless gases held to the earth by gravitational attraction. While the furthest limit of the atmosphere is said by international convention to be at 1000 km, most of the atmosphere, and therefore our climate and weather, is concentrated within 16 km of the earth's surface at the Equator and 8 km at the poles. Fifty per cent of atmospheric mass is within 5.6 km of sea level and 99 per cent is within 40 km. Atmospheric pressure decreases rapidly with height but, as recordings made by radiosondes, weather balloons and more recently weather satellites have shown, temperature changes are more complex. Changes in temperature mean that the atmosphere can be conveniently divided into four distinctive layers (Figure 9.1); moving outwards from the earth's surface, these are:

1 **Troposphere** Temperatures in the troposphere decrease by 6.4°C with every 1000 m increase in altitude (environmental lapse rate, page 198). This is because the earth's surface is warmed by incoming solar radiation which in turn heats the air next to it by conduction. Pressure falls as the effect of gravity decreases, although wind speeds usually increase with height. The layer is unstable and contains most of the atmosphere's water vapour, cloud, dust and pollution. The tropopause, which forms the upper limit to the earth's climate and weather, is marked by an isothermal layer where temperatures remain constant despite any increase in height.

2 **Stratosphere** The stratosphere is characterised by a steady increase in temperature (temperature inversion, page 200) caused by a concentration of **ozone** (O_3). This gas absorbs incoming **ultra-violet (UV) radiation** from the sun. Winds are light in the lower parts, but increase with height; pressure continues to fall and the air is dry. The stratosphere, like the two layers above it, acts as a protective shield against meteorites which usually burn out as they enter the earth's gravitational field. The **stratopause** is another isothermal layer where temperatures do not change with increasing height.

3 **Mesosphere** Temperatures fall rapidly as there is no water vapour, cloud, dust or ozone to absorb incoming radiation. This layer experiences the atmosphere's lowest temperatures (–90°C) and strongest winds (nearly 3000 km/hr). The **mesopause**, like the tropopause and stratopause, shows no change in temperature.

4 **Thermosphere** Temperatures rise rapidly with height, perhaps to reach 1500°C. This is due to an increasing proportion of atomic oxygen in the atmosphere which, like ozone, absorbs incoming ultra-violet radiation.

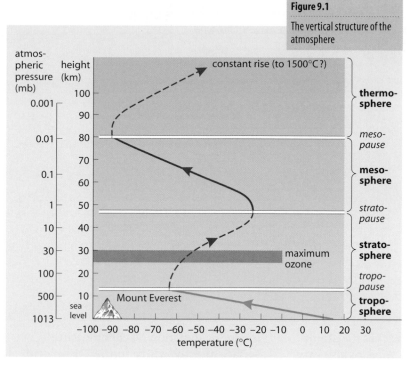

Figure 9.1

The vertical structure of the atmosphere

changes in temperature with height

←—— fall

——— constant

–►-- rise

Energy in the atmosphere

The sun is the earth's prime source of energy. The earth receives energy as incoming **short-wave** solar radiation (also referred to as **insolation**). It is this energy which controls our planet's climate and weather and which, when converted by photosynthesis in green plants, supports all forms of life. The amount of incoming radiation received by the earth is determined by four astronomical factors (Figure 9.3): the solar constant, the distance from the sun, the altitude of the sun in the sky and the length of night and day. Figure 9.3 is theoretical in that it assumes there is no atmosphere around the earth. In reality, much insolation is absorbed, reflected and scattered as it passes through the atmosphere (Figure 9.4).

Absorption of incoming radiation is mainly by ozone, water vapour, carbon dioxide and particles of ice and dust. Clouds and, to a lesser extent, the earth's surface **reflect** considerable amounts of radiation back into space. The ratio between incoming radiation and the amount reflected, expressed as a percentage, is known as the **albedo**. The albedo varies with cloud type from 30–40 per cent in thin clouds, to 50–70 per cent in thicker stratus and 90 per cent in cumulo-nimbus

Atmospheric gases

The various gases which combine to form the atmosphere are listed in Figure 9.2. Of these, nitrogen and oxygen together make up 99 per cent by volume. Of the others, water vapour (lower atmosphere), ozone (O_3) (upper atmosphere) and carbon dioxide (CO_2) have an importance far beyond their seemingly small amounts. It is the depletion of O_3 (Places 21) and the increase in CO_2 (Case Study 9) which are causing concern to scientists.

Figure 9.2

The composition of the atmosphere

Gas		Percentage by volume	Importance for weather and climate	Other functions/source
Permanent gases:	nitrogen	78.09	} Mainly passive	Needed for plant growth.
	oxygen	20.95		Produced by photosynthesis; reduced by deforestation.
Variable gases:	water vapour	0.20–4.0	Source of cloud formation and precipitation, reflects/absorbs incoming radiation. Keeps global temperatures constant. Provides majority of natural 'greenhouse effect'.	Essential for life on earth. Can be stored as ice/snow.
	carbon dioxide	0.03	Absorbs long-wave radiation from earth and so contributes to 'greenhouse effect'. Its increase due to human activity is a major cause of global warming.	Used by plants for photosynthesis; increased by burning fossil fuels and by deforestation.
	ozone	0.00006	Absorbs incoming ultra-violet radiation.	Reduced/destroyed by chlorofluorocarbons (CFCs).
Inert gases:	argon	0.93		
	helium, neon, krypton	trace		
Non-gaseous:	dust	trace	Absorbs/reflects incoming radiation. Forms condensation nuclei necessary for cloud formation.	Volcanic dust, meteoritic dust, soil erosion by wind.
Pollutants:	sulphur dioxide, nitrogen oxide, methane	trace	Affects radiation. Causes acid rain.	From industry, power stations and car exhausts.

(Note: the figures refer to dry air and so the variable amount of water vapour is not usually taken into consideration)

189

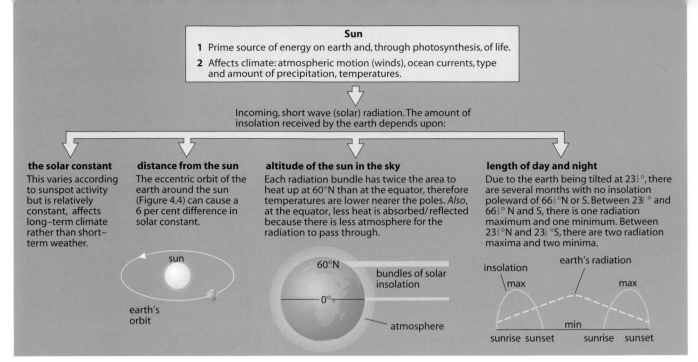

Sun

1 Prime source of energy on earth and, through photosynthesis, of life.

2 Affects climate: atmospheric motion (winds), ocean currents, type and amount of precipitation, temperatures.

Incoming, short wave (solar) radiation. The amount of insolation received by the earth depends upon:

the solar constant

This varies according to sunspot activity but is relatively constant, affects long–term climate rather than short–term weather.

distance from the sun

The eccentric orbit of the earth around the sun (Figure 4.4) can cause a 6 per cent difference in solar constant.

sun

earth's orbit

altitude of the sun in the sky

Each radiation bundle has twice the area to heat up at 60°N than at the equator, therefore temperatures are lower nearer the poles. *Also*, at the equator, less heat is absorbed/reflected because there is less atmosphere for the radiation to pass through.

60°N

bundles of solar insolation

0°

atmosphere

length of day and night

Due to the earth being tilted at 23½°, there are several months with no insolation poleward of 66½°N or S. Between 23½ ° and 66½° N and S, there is one radiation maximum and one minimum. Between 23½°N and 23½ °S, there are two radiation maxima and two minima.

insolation

earth's radiation

max

max

min

sunrise sunset   sunrise   sunset

Figure 9.3

Incoming radiation received by the earth (assuming that there is no atmosphere)

(when only 10 per cent reaches the atmosphere below cloud level). Albedos also vary over different land surfaces, from less than 10 per cent over oceans and dark soil, to 15 per cent over coniferous forest and urban areas, 25 per cent over grasslands and deciduous forest, 40 per cent over light-coloured deserts and 85 per cent over reflecting fresh snow. Where deforestation and overgrazing occur, the albedo increases. This reduces the possibility of cloud formation and precipitation and increases the risk of **desertification** (Case Study 7). **Scattering** occurs when incoming radiation is diverted by molecules of gas. It takes place in all directions and some of the radiation will reach the earth's surface as **diffuse** radiation.

As a result of absorption, reflection and scattering only about 24 per cent of incoming radiation reaches the earth's surface directly, with a further 21 per cent arriving at ground level as diffuse radiation (Figure 9.4). Incoming radiation is converted into heat energy when it reaches the earth's surface. As the ground warms, it radiates energy back into the atmosphere where 94 per cent is absorbed (only 6 per cent is lost to space) mainly by water vapour and carbon dioxide — the greenhouse effect (Case Study 9). Without the natural greenhouse effect, which traps so much of the outgoing radiation, world temperatures would be 33°C colder than they are at present and life on earth would be impossible. (During the ice age, it was only 4°C cooler.) This outgoing (terrestrial) radiation is **long-wave** or **infra-red** radiation.

Figure 9.4

The solar energy cascade

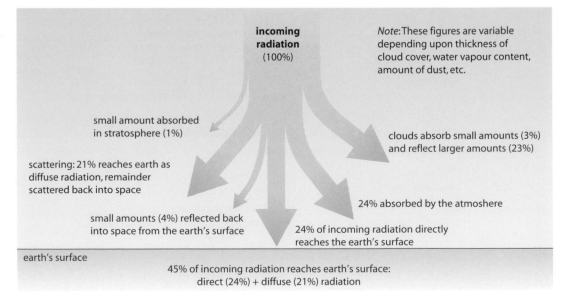

incoming radiation (100%)

Note: These figures are variable depending upon thickness of cloud cover, water vapour content, amount of dust, etc.

small amount absorbed in stratosphere (1%)

clouds absorb small amounts (3%) and reflect larger amounts (23%)

scattering: 21% reaches earth as diffuse radiation, remainder scattered back into space

24% absorbed by the atmoshere

small amounts (4%) reflected back into space from the earth's surface

24% of incoming radiation directly reaches the earth's surface

earth's surface

45% of incoming radiation reaches earth's surface: direct (24%) + diffuse (21%) radiation

The major concentration of ozone is in the stratosphere 25–30 km above sea level (Figure 9.1). Ozone acts as a shield, protecting the earth from the damaging effects of ultra-violet (UV) radiation from the sun. However, there is serious concern as this shield appears to be breaking down. Each spring over Antarctica, an area the size of the USA is showing increasing signs of ozone depletion. By 1993, ozone had been reduced to between one-half and two-thirds of its 1970 amount, with most of this depletion occurring after 1990. Since 1989, depletions have also been recorded over the Arctic. An increase in UV radiation has been linked to an increase in skin cancer (fair skin is at greater risk than dark skin) and cataracts, and a suppressed immunity to disease. This damage is believed to be caused by humans releasing into the atmosphere a family of chemicals which contain chlorine (chlorofluorocarbons or CFCs). Chlorine breaks down ozone. Although supersonic aircraft have also been blamed, the major culprits include propellants used in aerosols such as hair-spray, deodorants and fly-killer, refrigerator coolant and manufacturing processes which produce foam packaging. Scientists claim that a 1 per cent depletion in ozone causes a 5 per cent increase in cases of skin cancer and that the depletion rate has been 6 per cent since 1970. Following a world conference of 120 nations in 1989, the EU decided to ban all use of CFCs by 2000 AD.

In contrast, vehicle exhaust systems generate dangerous quantities of ozone close to the earth's surface, causing damage to plant tissues in crops and trees and affecting the health of humans and animals. For example, high asthma levels, particularly in children in London, were linked with pollution from traffic during the hot summer of 1994. The problem became so acute that health warnings were broadcast.

Q

1 Name the two warm layers in the atmosphere? Why are they warm?

2 Compare the characteristics of the troposphere and the stratosphere under the headings: **a** temperature; **b** pressure; **c** water vapour.

3 How do changes in the amount of ozone, water vapour and carbon dioxide in the atmosphere affect the amount of radiation absorbed?

4 Suggest measures which could and should be taken to restrict the potential damage caused by:
   a increasing amounts of ozone and carbon dioxide in the lower atmosphere; and
   b the depletion of ozone in the stratosphere.

Figure 9.5

The heat budget

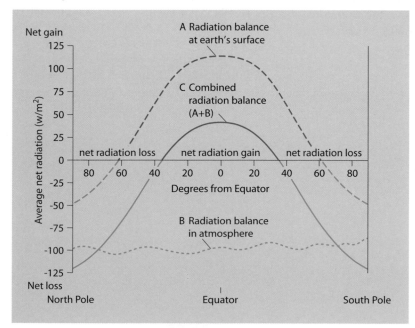

The heat budget

Since the earth is neither warming up nor cooling down, there must be a balance between incoming insolation and outgoing terrestrial radiation. Figure 9.5 shows that:

- There is a net gain in radiation everywhere on the earth's surface (curve A) except in polar latitudes which have high albedo surfaces.
- There is a net loss in radiation throughout the atmosphere (curve B).
- After balancing the incoming and outgoing radiation, there is a net surplus between 35°S and 40°N (the difference in latitude is due to the larger land masses of the northern hemisphere) and a net deficit to the poleward sides of those latitudes (curve C).

This means that there is a **positive heat balance** within the tropics and a **negative heat balance** both at high latitudes (polar regions) and high altitudes. Two major **transfers** of heat, therefore, take place to prevent tropical areas from overheating (Figure 9.6).

1 **Horizontal heat transfers** Heat is transferred away from the tropics, thus preventing the Equator from becoming increasingly hotter and the poles increasingly colder. Winds (air movements including jet streams, page 211; hurricanes, page 219; and depressions, page 214) are responsible for 80 per cent of this heat transfer, and ocean currents for 20 per cent (page 198).

2 **Vertical heat transfers** Heat is also transferred vertically, thus preventing the earth's surface from getting hotter and the atmosphere colder. This is achieved through **radiation**, **conduction**, **convection** and the transfer of **latent heat**.

Variations in the radiation balance occur at a number of spatial and temporal scales. Regional differences may be due to the uneven distribution of land and sea, altitude and the direction of prevailing winds. Local variations may result from **aspect** and amounts of cloud cover. Seasonal and diurnal variations are related to the altitude of the sun and the length of night and day.

Global factors affecting insolation

Factors which influence the amount of insolation received at any point, and therefore its radiation balance and heat budget, vary considerably over time and space.

Long-term factors
These are relatively constant at a given point.

- **Height above sea level** The atmosphere is not warmed directly by the sun, but by heat radiated from the earth's surface and distributed by conduction and convection. As the height of mountains increases, they present a decreasing area of land surface from which to heat the surrounding air. In addition, as the density or pressure of the air decreases, so too does its ability to hold heat (Figure 9.1). This is because the molecules in the air which receive and retain heat become fewer and more widely spaced as height increases.

- **Altitude of the sun** As the angle of the sun in the sky decreases, the land area heated by a given ray and the depth of atmosphere through which that ray has to pass both increase. Consequently, the amount of insolation lost through absorption, scattering and reflection also increases. Places in lower latitudes therefore have higher temperatures than those in higher latitudes.

- **Land and sea** Land and sea differ in their ability to absorb, transfer and radiate heat energy. The sea is more transparent than the land, and is capable of absorbing heat down to a depth of 10 metres. It can then transfer this heat to greater depths through the movements of waves and currents. The sea also has a greater **specific heat capacity** than that of land. Specific heat capacity is the amount of energy required to raise the temperature of 1kg of a substance by 1°C, expressed in kilojoules per kg per °C. The specific heat capacity of water is 4.2 kj/kg/°C, that of soil is 2.1 kj/kg/°C and that of sand 0.84 kj/kg/°C. (Expressed in kilocalories,

Figure 9.6

Heat transfers in the atmosphere

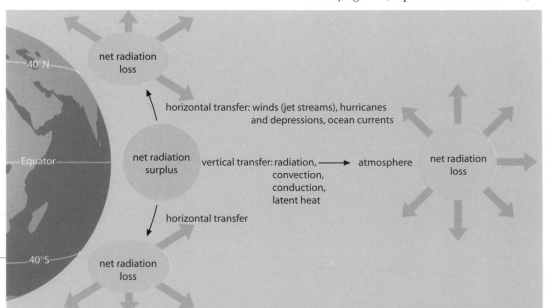

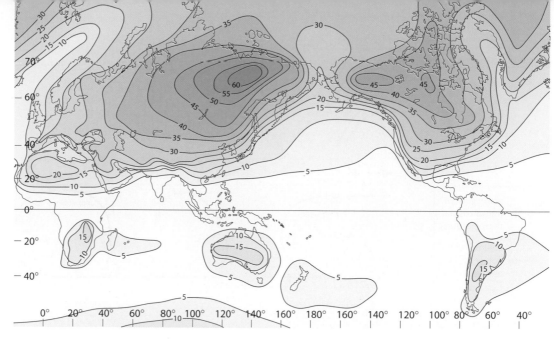

Figure 9.7

Mean annual ranges in global temperature (°C)

The joule (1000 joules = 1 kilojoule) is a unit of energy or work. It is equivalent to 4.2 kilocalories. The rate at which energy is transferred or used is measured in watts.

the specific heat capacity of water is 1.0, that of land is 0.5 and that of sand 0.2.) This means that water requires twice as much energy as soil and five times more than sand to raise an equivalent mass to the same temperature. During summer, therefore, the sea heats up more slowly than the land. In winter, the reverse is the case and land surfaces lose heat energy more rapidly than water. The oceans act as efficient 'thermal reservoirs'. This explains why coastal environments have a lower annual range of temperature than locations at the centres of continents (Figure 9.7).

■ **Prevailing winds** The temperature of the wind is determined by its area of origin and by the characteristics of the surface over which it subsequently blows (Figure 9.8). A wind blowing from the sea tends to be warmer in winter and cooler in summer than a corresponding wind coming from the land.

■ **Ocean currents** These are a major component in the process of horizontal transfer of heat energy. Warm currents carry water polewards and raise the air temperature of the maritime environments where they flow. Cold currents carry water towards the Equator and so lower the temperatures of coastal areas (Figure 9.9).

Figure 9.10 shows the difference between the mean January temperature of a place and the mean January temperatures of other places with the same latitude; this difference is known as a **temperature anomaly**. (The term 'temperature anomaly' is used specifically to describe temperature differences from a mean. It should not be confused with the more general definition of 'anomaly' which refers to something which does not fit into a general pattern.) For example, Stornoway (Figure 9.10) has a mean January temperature of 4°C which is 20°C higher than the average for other locations lying at 58°N. Such anomalies result primarily from the uneven heating and cooling rates of land and sea and are intensified by the horizontal transfer of energy by ocean currents and prevailing winds. Remember that the sun appears overhead in the southern hemisphere at this time of year and isotherms have been reduced to sea level — i.e. temperatures are adjusted to eliminate some of the effects of relief, thus emphasising the influence of prevailing winds, ocean currents and continentality.

Figure 9.8

Simplified diagram showing the effect of prevailing winds on land and sea temperatures

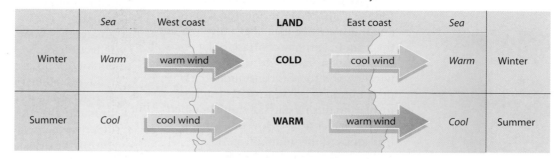

Figure 9.9

Major ocean currents

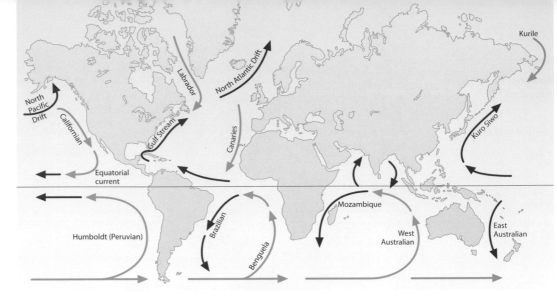

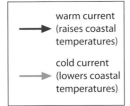

warm current
(raises coastal
temperatures)

cold current
(lowers coastal
temperatures)

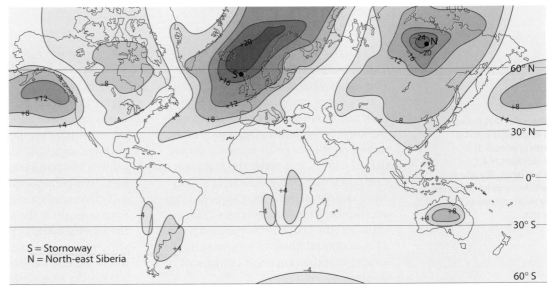

S = Stornoway
N = North-east Siberia

Figure 9.10

Temperature anomalies for
January (*after* D.C. Money)

Short-term factors

- **Seasonal changes** At the spring and autumn equinoxes (21 March and 22 September) when the sun is directly over the Equator, insolation is distributed equally between both hemispheres. At the summer and winter solstices (21 June and 22 December) when, due to the earth's tilt, the sun is overhead at the tropics, the hemisphere experiencing 'summer' will receive maximum insolation.

- **Length of day and night** Insolation is only received during daylight hours and reaches its peak at noon. There are no seasonal variations at the Equator, where day and night are of equal length throughout the year. In extreme contrast, polar areas receive no insolation during part of the winter when there is continuous darkness, but may receive up to 24 hours of insolation during parts of summer when the sun never sinks below the horizon (the lands of the midnight sun).

Local influences on insolation

- **Aspect** Hillsides alter the angle at which the sun's rays hit the ground (Places 22). In the northern hemisphere, north-facing slopes, being in shadow for most or all of the year, are cooler than those facing south. The steeper the south-facing slope, the higher the angle of the sun's rays to it and therefore the higher will be the temperature. North- and south-facing slopes are referred to, respectively, as the **adret** and **ubac**.

- **Cloud cover** The presence of cloud reduces both incoming and outgoing radiation. The thicker the cloud, the greater the amount of absorption, reflection and scattering of insolation, and of terrestrial radiation. Clouds may reduce daytime temperatures, but they also act as an insulating blanket to retain heat at night. This means that tropical deserts, where skies are clear, are warmer during the day and cooler at night than humid equatorial regions with a greater cloud cover. The world's greatest diurnal ranges of temperature are therefore found in tropical deserts.

- **Urbanisation** This alters the albedo and creates urban 'heat islands' (page 225).

Atmospheric moisture

Water is a liquid compound which is converted by heat into vapour (gas) and by cold into a solid (ice). Its presence serves three essential purposes. First, it maintains life on earth: flora, in the form of natural vegetation (biomes) and crops; and fauna — i.e. all living creatures including humans. Secondly, water in the atmosphere, mainly as a gas, absorbs, reflects and scatters insolation to keep our planet at a habitable temperature. Thirdly, atmospheric moisture is of vital significance as a means of transferring surplus energy from tropical areas either horizontally to polar latitudes or vertically into the atmosphere to balance the heat budget (Figure 9.5).

Despite this need for water, its existence in a form readily available to plants, animals and humans is limited. It has been estimated that 97.2 per cent of the world's water is in the oceans and seas; in this form, it is only useful to plants tolerant of saline conditions (**halophytes**) and to the populations of a few rich countries who can afford desalinisation plants (the Gulf oil states).

Places 22 An alpine valley

Many alpine valleys in Switzerland and Austria have an east–west orientation which means that their valley sides face either north or south. South-facing slopes are much warmer and drier than those facing north (Figure 9.11). The south-facing slopes have more plant species, a higher tree-line, a greater land use with alpine pastures at higher altitudes and fruit and hay lower down; also, they usually provide the best sites for settlement. In contrast, north-facing slopes are snow-covered for a much longer period; they are less suited to farming; the tree-line is lower and they tend to be left forested. However, on the valley floors, as severe frosts are likely to occur during times of temperature inversion (page 200), sensitive plants and crops do not flourish.

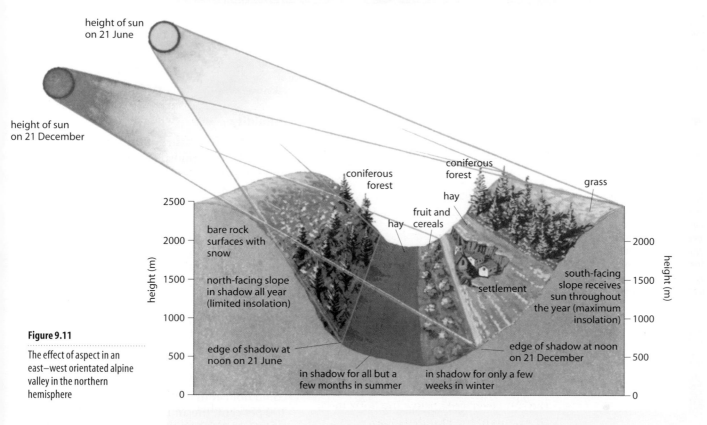

Figure 9.11

The effect of aspect in an east–west orientated alpine valley in the northern hemisphere

Figure 9.12

The hydrological cycle
(compare with Figure 3.1)

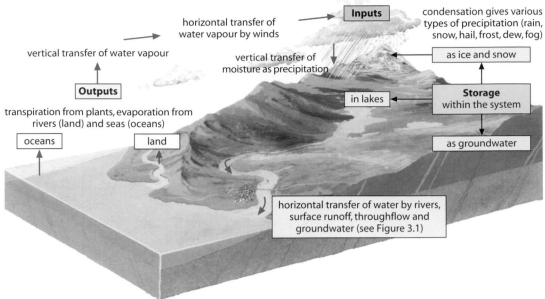

Approximately 2.1 per cent of water in the hydrosphere is held in storage as polar ice and snow. Only 0.7 per cent is fresh water found either in lakes and rivers (0.1 per cent), as soil moisture and groundwater (0.6 per cent) or in the atmosphere (0.001 per cent). At any given time, the atmosphere only holds, on average, sufficient moisture to give every place on the earth 2.5 cm (about 10 days' supply) of rain. There must therefore be a constant recycling of water between the oceans, atmosphere and land. This recycling is achieved through the **hydrological cycle** (Figure 9.12).

Figure 9.13

The world's water balance

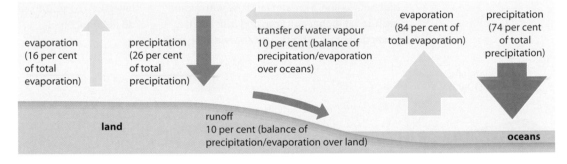

Figure 9.14

Air temperatures and absolute humidity for saturated air

Absolute humidity is the mass of water vapour in a given volume of air measured in grams per cubic metre (g/cu m). Specific humidity is similar but is expressed in grams of water per kilogram of air.

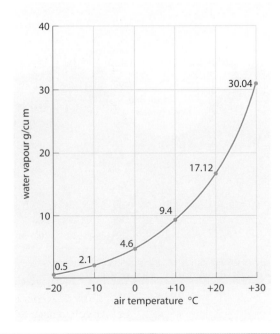

Humidity

Humidity is a measure of the water vapour content in the atmosphere and depends upon the temperature of the air. At any given temperature, there is a limit to the amount of moisture that the air can hold. When this limit is reached, the air is said to be **saturated**. Cold air can hold only relatively small quantities of vapour before becoming saturated but this amount increases rapidly as temperatures rise (Figure 9.14). This means that the amount of precipitation obtained from warm air is generally greater than that from cold air. **Relative humidity** (RH) is the amount of water vapour in the air at a given temperature expressed as a percentage of the maximum amount of vapour that the air could hold at that temperature. If the RH is 100 per cent, the air is saturated. If it lies between 80 and 99 per cent, the air is said to be 'moist' and the weather is humid or clammy. When the RH drops to 50 per cent, the air is 'dry'— figures as low as 10 per cent have been recorded over hot deserts.

If unsaturated air is cooled and atmospheric pressure remains constant, a critical temperature will be reached when the air becomes saturated (i.e. RH = 100 per cent). This is known as the **dew point**. Any further cooling will result in the condensation of excess vapour, either into water droplets where condensation nuclei are present, or into ice crystals if the air temperature is below 0°C. This is shown in the following worked example.

1 The early morning air temperature was 10°C. Although the air could have held 100 units of water at that temperature, at the time of the reading it held only 90. This meant that the RH was 90 per cent.

2 During the day, the air temperature rose to 12°C. As the air warmed it became capable of holding more water vapour, up to 120 units. Owing to evaporation, the reading reached a maximum of 108 units which meant that the RH remained at 90 per cent — i.e. $(108 \div 120) \times 100$.

3 In the early evening, the temperature fell to 10°C at which point, as stated above, it could hold only 100 units. However, the air at that time contained 108 units so, as the temperature fell, dew point was reached and the 8 excess units of water were lost through condensation.

Condensation

This is the process by which water vapour in the atmosphere is changed into a liquid or, if the temperature is below 0°C, a solid. It usually results from air being cooled until it is saturated. Cooling may be achieved by:

1 **Radiation** (contact) **cooling** This typically occurs on calm, clear evenings. The ground loses heat rapidly through terrestrial radiation and the air in contact with it is then cooled by conduction. If the air is moist, some vapour will condense to form radiation fog, dew, or — if the temperature is below freezing point — hoar frost (page 204).

2 **Advection cooling** This results from warm, moist air moving over a cooler land or sea surface. Advection fogs (Places 18, page 164) in California and the Atacama Desert (page 164) are formed when warm air from the land drifts over cold offshore ocean currents (Figure 9.9).

As both radiation and advection involve horizontal rather than vertical movements of air, the amount of condensation created is limited.

3 **Orographic** and **frontal uplift** Warm, moist air is forced to rise either as it crosses a mountain barrier (orographic ascent) or when it meets a colder, denser mass of air at a front (page 212).

4 **Convective** or **adiabatic cooling** This is when air is warmed during the daytime and rises in pockets as **thermals** (Figure 9.15). As the air expands, it uses energy and so loses heat and the temperature drops. Because air is cooled by the reduction of pressure with height rather than by a loss of heat to the surrounding air, it is said to be adiabatically cooled (see lapse rates, page 198).

As both orographic and adiabatic cooling involve vertical movements of air, they are more effective mechanisms of condensation.

Condensation does not occur readily in clean air. Indeed, if air is absolutely pure, it can be cooled below its dew point to become **supersaturated** with an RH in excess of 100 per cent. Laboratory tests have shown that clean, saturated air can be cooled to −40°C before condensation or, in this case, **sublimation**. However, air is rarely pure and usually contains large numbers of condensation nuclei. These microscopic particles, referred to as **hygroscopic nuclei** because they attract water, include volcanic dust (heavy rain always accompanies volcanic eruptions); dust from windblown soil; smoke and sulphuric acid originating from urban and industrial areas; and salt from sea spray. Hygroscopic nuclei are most numerous over cities, where there may be up to 1 million per cu cm, and least common over oceans (only 10 per cu cm). Where large concentrations are found, condensation can occur with an RH as low as 75 per cent — as in the smogs of Los Angeles (page 205 and Case Study 15a).

Figure 9.15

Convective cooling

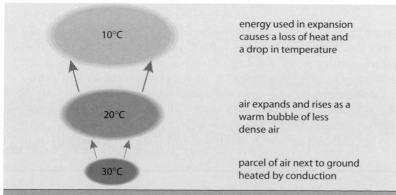

energy used in expansion causes a loss of heat and a drop in temperature

air expands and rises as a warm bubble of less dense air

parcel of air next to ground heated by conduction

ground heated by insolation

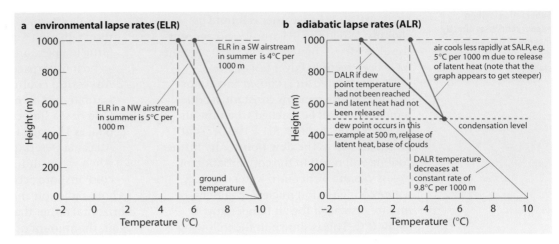

a environmental lapse rates (ELR)

ELR in a SW airstream in summer is 4°C per 1000 m

ELR in a NW airstream in summer is 5°C per 1000 m

ground temperature

b adiabatic lapse rates (ALR)

air cools less rapidly at SALR, e.g. 5°C per 1000 m due to release of latent heat (note that the graph appears to get steeper)

DALR if dew point temperature had not been reached and latent heat had not been released

dew point occurs in this example at 500 m, release of latent heat, base of clouds

condensation level

DALR temperature decreases at constant rate of 9.8°C per 1000 m

Lapse rates

The **environmental lapse rate** (ELR) is the decrease in temperature usually expected with an increase in height through the troposphere (Figure 9.1). The ELR is approximately 6.5°C per 1000 m, but varies according to local air conditions. It may vary due to several factors: **height** — ELR is lower nearer ground level; **time** — it is lower in winter or during a rainy season; over different **surfaces** — it is lower over continental areas; and between different **air masses** (Figure 9.16a).

The **adiabatic lapse rate** (ALR) describes what happens when a parcel of air rises and the decrease in pressure is accompanied by an associated increase in volume and a decrease in temperature (Figure 9.15). Conversely, descending air will be subject to an increase in pressure causing a rise in temperature. In either case, there is negligible mixing with the surrounding air. There are two adiabatic lapse rates.

1 If the upward movement of air does not lead to condensation, the energy used by expansion will cause the temperature of the parcel of air to fall at the **dry adiabatic lapse rate** (DALR on Figure 9.16b). The DALR, which is the rate at which an unsaturated parcel of air cools as it rises or warms as it descends, remains constant at 9.8°C per 1000 m (i.e. approximately 1°C per 100 m).

2 When the upward movement is sufficiently prolonged to enable the air to cool to its dew point temperature, condensation occurs and the loss in temperature with height is then partly compensated by the release of latent heat (Figure 9.16b). Saturated air, which therefore cools at a slower rate than unsaturated air, loses heat at the **saturated adiabatic lapse rate** (SALR). The SALR can vary because the warmer the air the more moisture it can hold, and so the greater the amount of latent heat released following condensation. The SALR may be as low as 4°C per 1000 m and as high as 9°C per 1000 m. It averages about 5.4°C per 1000 m (i.e. approximately 0.5°C per 100 m). Should temperatures fall below 0°C, then the air will cool at the **freezing adiabatic lapse rate** (FALR). This is the same as the DALR as very little moisture is present at low temperatures.

Air stability and instability

Parcels of warm air which rise through the lower atmosphere cool adiabatically. The rate and maintenance of any vertical uplift depend upon the temperature–density balance between the rising parcel and the surrounding air. In a simplified form, this balance is the relationship between the environmental lapse rate and the dry and saturated adiabatic lapse rates.

Stability

The state of **stability** is when a rising parcel of unsaturated air cools more rapidly than the air surrounding it. This is shown diagrammatically when the ELR lies to the right of the DALR, as in Figure 9.17. In this example the ELR is 6°C per 1000 m and the DALR is 9.8°C per 1000 m. By the time the rising air has reached 1000 m, it has cooled to 10.2°C which leaves it colder and denser than the surrounding air which has only cooled to 14°C. If there is nothing to force the parcel of air to rise, e.g. mountains or fronts, it will sink back to its starting point.

Latent heat is the amount of heat energy needed to change the state of a substance without affecting its temperature. When ice changes into water or water into vapour, heat is taken up to help with the processes of melting and evaporation. This absorption of heat results in the cooling of the atmosphere. When the processes work in reverse — i.e. vapour condenses into water or water freezes into ice — heat energy is released and the atmosphere is warmed.

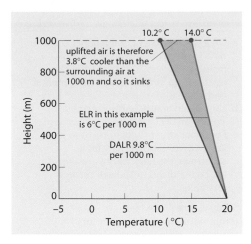

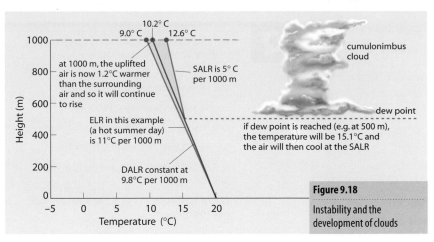

cumulonimbus cloud

dew point

if dew point is reached (e.g. at 500 m), the temperature will be 15.1°C and the air will then cool at the SALR

Figure 9.18

Instability and the development of clouds

Figure 9.17

Stability: changes in lapse rates and air temperature with height

The air is described as stable because dew point may not have been reached and the only clouds which might have developed would be shallow, flat-topped cumulus which do not produce precipitation (Figure 9.21). Stability is often linked with anticyclones (page 218) when any convection currents are suppressed by sinking air to give dry, sunny conditions.

Instability

Conditions of **instability** arise in Britain on hot days. Localised heating of the ground warms the adjacent air by conduction, creating a higher lapse rate. The resultant parcel of rising unsaturated air cools less rapidly than the surrounding air. In this case, as shown in Figure 9.18, the ELR lies to the left of the DALR. The rising air remains warmer and lighter than the surrounding air. Should it be sufficiently moist and if dew point is reached, then the upward movement may be accelerated to produce towering cumulus or cumulo-nimbus type cloud (Figure 9.21). Thunderstorms are likely

(Figure 9.22) and the saturated air, following the release of latent heat, will cool at the SALR.

Conditional instability

This type of instability occurs when the ELR is lower than the DALR but higher than the SALR. In Britain, it is the most common of the three conditions. The rising air is stable in its lower layers and, being cooler than the surrounding air, would normally sink back again. However, if the mechanism which initially triggered the uplift remains, then the air will be cooled to its dew point. Beyond this point, cooling takes place at the slower SALR and the parcel may become warmer than the surrounding air (Figure 9.19). It will now continue to rise freely, even if the uplifting mechanism is removed, as it is now in an unstable state. Instability is conditional upon the air being forced to rise in the first place, and later becoming saturated so that condensation occurs. The associated weather is usually fine in areas at altitudes below condensation level, but cloudy and showery in those above.

Figure 9.19

Conditional instability

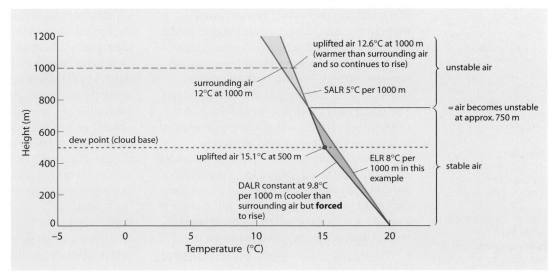

Q

1 Assume that the following conditions occur in three airstreams up to 2000 m above sea-level:

sea-level air temperature is always 10°C;

condensation is always at 500 m;

DALR is 10°C per 1000 m (1°C per 100 m);

SALR is 5°C per 1000 m (0.5°C per 100 m);

ELR of Airstream 1 is 4°C per 1000 m (0.4°C per 100 m);

ELR of Airstream 2 is 12°C per 1000 m (1.2°C per 100 m); and

ELR of Airstream 3 is 8°C per 1000 m (0.8°C per 100 m).

Figure 9.20

Environmental lapse rates (ELRs) and adiabatic lapse rates (ALRs)

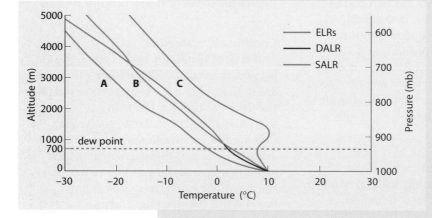

a On a single sheet of graph paper, plot the data for each of the three airstreams from sea-level to 2000 m to show the relationship between the environmental and adiabatic lapse rates.

b Label each graph 'stable', 'unstable', or 'conditionally unstable' as appropriate.

c For the airstream which you have labelled 'conditionally unstable', determine the height above sea-level at which the displaced air will become unstable.

2 Study Figure 9.20 which is a temperature–height diagram, or tephigram, showing relationships between adiabatic and environmental lapse rates.

a For the zone below 2000 m, describe the temperature conditions for the environmental situation shown by curve **C**.

b State three different mechanisms which are likely to trigger the upward movement of a parcel of air from ground level.

c Given the information in Figure 9.20, for each of the environmental lapse rate curves **A**, **B** and **C**, state which of the three atmospheric conditions of stability, instability and conditional stability is demonstrated.
Justify each of your answers.

d For the curve **B**, assuming that the parcel of air is lifted mechanically to a height of over 1000 m, calculate the lower and upper limits of cloud formation.

e State the weather conditions normally associated with stability and instability.

Temperature inversions

As the lapse rate exercises have shown, the temperature of the air usually decreases with altitude, but there are certain conditions when the reverse occurs. **Temperature inversions**, where warmer air overlies colder air, may occur at three levels in the atmosphere. Figure 9.1 showed that temperatures increase with altitude in both the stratosphere and the thermosphere. Inversions can also occur near ground level and high in the troposphere. High-level inversions are found in depressions where warm air overrides cold air at the warm front or is undercut by colder air at the cold front (page 213). Low-level, or ground, inversions usually occur under anticyclonic conditions (page 218) when there is a rapid loss of heat from the ground due to radiation at night, or when warm air is advected over a cold surface (line **C**, Figure

9.20). Under these conditions, fog and frost (page 204) may form in valleys and hollows.

Clouds

Clouds form when air cools to dew point and vapour condenses into water droplets and/or ice crystals. There are many different types of cloud, but they are often difficult to distinguish as their form constantly changes. The general classification of clouds, used in the *International Cloud Atlas*, was proposed by Luke Howard in 1803. His was a descriptive classification, based upon cloud shape and height (Figure 9.21). He used four Latin words: **cirrus** (a lock of curly hair); **cumulus** (a heap or pile); **stratus** (a layer); and **nimbus** (rain-bearing). He also compiled composite names using these four terms, such as cumulo-nimbus, cirrostratus; and added the prefix 'alto-' for middle-level clouds.

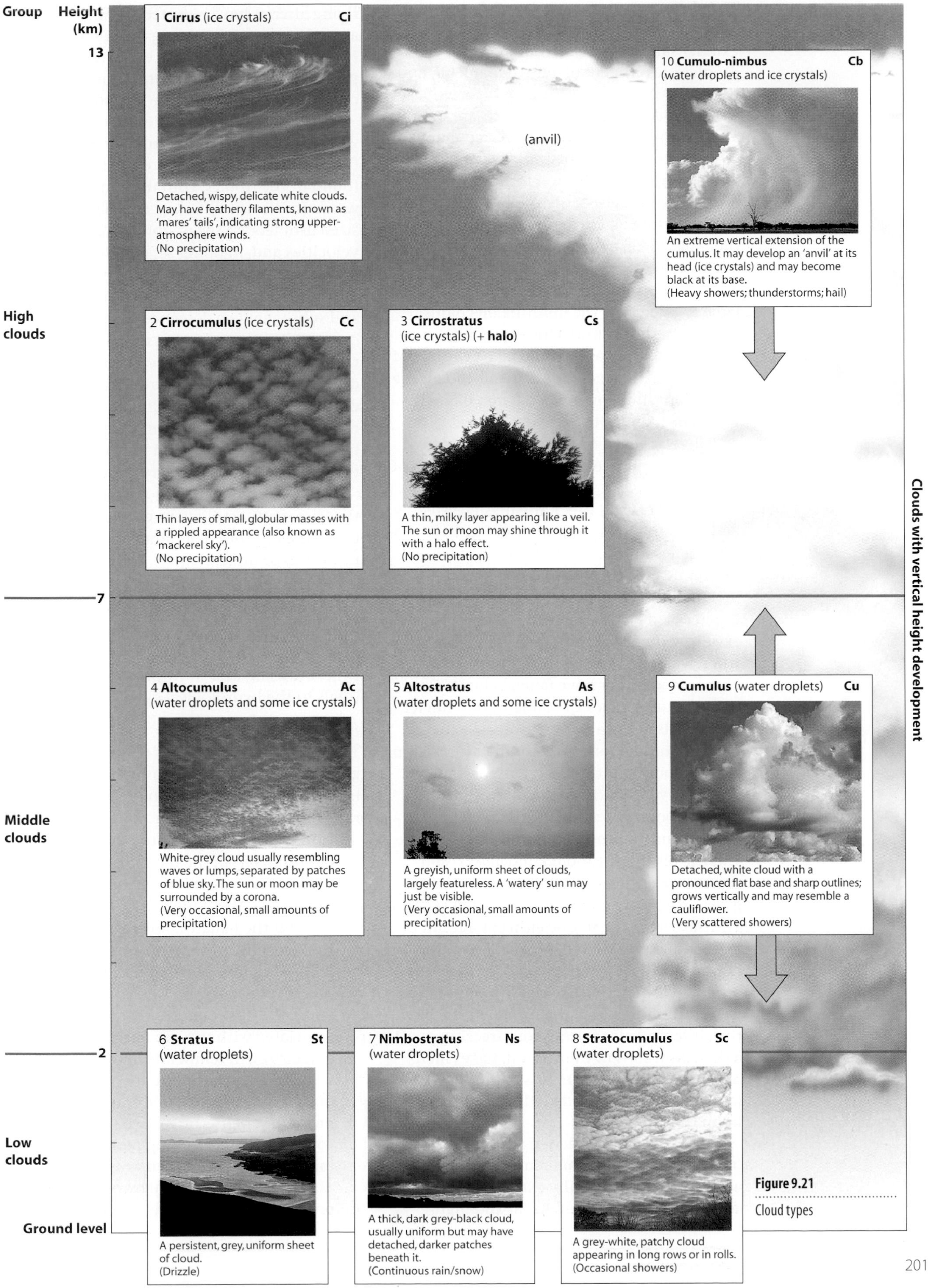

Group | **Height (km)**

13

1 Cirrus (ice crystals) **Ci**

Detached, wispy, delicate white clouds. May have feathery filaments, known as 'mares' tails', indicating strong upper-atmosphere winds.
(No precipitation)

(anvil)

10 Cumulo-nimbus **Cb**
(water droplets and ice crystals)

An extreme vertical extension of the cumulus. It may develop an 'anvil' at its head (ice crystals) and may become black at its base.
(Heavy showers; thunderstorms; hail)

High clouds

2 Cirrocumulus (ice crystals) **Cc**

Thin layers of small, globular masses with a rippled appearance (also known as 'mackerel sky').
(No precipitation)

3 Cirrostratus **Cs**
(ice crystals) (+ **halo**)

A thin, milky layer appearing like a veil. The sun or moon may shine through it with a halo effect.
(No precipitation)

7

4 Altocumulus **Ac**
(water droplets and some ice crystals)

White-grey cloud usually resembling waves or lumps, separated by patches of blue sky. The sun or moon may be surrounded by a corona.
(Very occasional, small amounts of precipitation)

5 Altostratus **As**
(water droplets and some ice crystals)

A greyish, uniform sheet of clouds, largely featureless. A 'watery' sun may just be visible.
(Very occasional, small amounts of precipitation)

9 Cumulus (water droplets) **Cu**

Detached, white cloud with a pronounced flat base and sharp outlines; grows vertically and may resemble a cauliflower.
(Very scattered showers)

Middle clouds

2

6 Stratus **St**
(water droplets)

A persistent, grey, uniform sheet of cloud.
(Drizzle)

7 Nimbostratus **Ns**
(water droplets)

A thick, dark grey-black cloud, usually uniform but may have detached, darker patches beneath it.
(Continuous rain/snow)

8 Stratocumulus **Sc**
(water droplets)

A grey-white, patchy cloud appearing in long rows or in rolls.
(Occasional showers)

Low clouds

Ground level

Clouds with vertical height development

Figure 9.21

Cloud types

Precipitation

Condensation produces minute water droplets, less than 0.05 mm in diameter, and, if the dew point temperature is below freezing, ice crystals. The droplets are so tiny and weigh so little that they are kept buoyant by the rising air currents which created them. So although condensation forms clouds, clouds do not necessarily produce precipitation. As rising air currents are often strong, there has to be a process within the clouds which enables the small water droplets and/or ice crystals to become sufficiently large to overcome the uplifting mechanism and fall to the ground. One experiment suggested that whereas it would take a droplet of 0.01 mm diameter five days to fall 100 m, the equivalent time for one of 6.0 mm diameter would be 1.8 minutes.

There are currently two main theories which attempt to explain the rapid growth of water droplets.

The **ice crystal mechanism**, suggested by Bergeron and later supported by Findeisen, is often referred to as the Bergeron–Findeisen mechanism. It appears that when the temperature of air is between -5°C and -25°C, supercooled water droplets and ice crystals exist together. Super-cooling takes place when water remains in the atmosphere after temperatures have fallen below 0°C — usually due to a lack of condensation nuclei. Ice crystals are in a minority because the **freezing nuclei** necessary for their formation are less abundant than condensation nuclei. The relative humidity of air is ten times greater above an ice surface than over water. This means that the water droplets evaporate and the resultant vapour condenses (sublimates) back onto the ice crystals which then grow into hexagonal-shaped snowflakes. The flakes grow in size — either as a result of further condensation or by fusion as their numerous edges interlock on collision with other flakes. They also increase in number as ice splinters break off and form new nuclei. If the air temperature rises above freezing point as the snow falls to the ground, flakes melt into raindrops. Experiments to produce rainfall artificially by cloud-seeding are based upon this process.

The Bergeron–Findeisen theory is supported by evidence from temperate latitudes where rainclouds usually extend vertically above the freezing level. Radar and high-flying aircraft have reported snow at high altitudes when it is raining at sea-level. However, as clouds rarely reach freezing point in the tropics, the formation of ice crystals is unlikely in those latitudes.

The **collision and coalescence** process was suggested by Longmuir. 'Warm' clouds (i.e. those containing no ice crystals), as found in the tropics, contain numerous water droplets of differing sizes. Different sized droplets are swept upwards at different velocities and, in doing so, collide with other droplets. It is thought that the larger the droplet, the greater the chance of collision and subsequent coalescence with smaller droplets. When coalescing droplets reach a radius of 3 mm, their motion causes them to disintegrate to form a fresh supply of droplets. The thicker the cloud (cumulonimbus), the greater the time the droplets have in which to grow and the faster they will fall, usually as thundery showers.

Latest opinions suggest that these two theories may complement each other, but that a major process of raindrop enlargement has yet to be understood.

Types of precipitation

Although the definition of precipitation includes sleet, hail, dew, hoar frost, fog and rime, only rain and snow provide significant totals in the hydrological cycle.

Rainfall

There are three main types of rainfall, distinguished by the mechanisms which cause the initial uplift of the air. Each mechanism rarely operates in isolation.

1 **Convergent** and **cyclonic** (**frontal**) rainfall results from the meeting of two air streams in areas of low pressure. Within the tropics, the trade winds, blowing towards the Equator, meet at the intertropical convergence zone or ITCZ (page 209). The air is forced to rise and, in conjunction with convection currents, produces the heavy afternoon thunderstorms associated with the equatorial climate (page 292). In temperate latitudes, depressions form at the boundary of two air masses. At the associated fronts, warm, moist, less dense air is forced to rise over colder, denser air, giving periods of prolonged and sometimes intense rainfall. This is often augmented by orographic precipitation.

Weather and climate

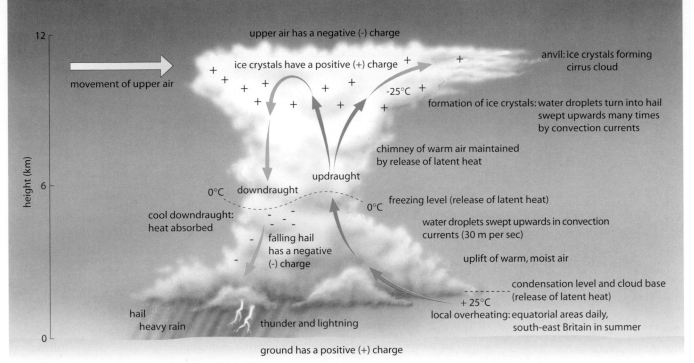

Figure labels (clockwise from top): upper air has a negative (-) charge · movement of upper air · ice crystals have a positive (+) charge · -25°C · anvil: ice crystals forming cirrus cloud · formation of ice crystals: water droplets turn into hail swept upwards many times by convection currents · chimney of warm air maintained by release of latent heat · updraught · freezing level (release of latent heat) · 0°C · water droplets swept upwards in convection currents (30 m per sec) · uplift of warm, moist air · condensation level and cloud base (release of latent heat) · +25°C · local overheating: equatorial areas daily, south-east Britain in summer · downdraught · 0°C · cool downdraught: heat absorbed · falling hail has a negative (-) charge · hail · heavy rain · thunder and lightning · ground has a positive (+) charge

height (km): 12 / 6 / 0

Figure 9.22

Convectional rainfall: the development of a thunderstorm

Drizzle is defined as water droplets under 0.5 mm in diameter.

2 **Orographic** or **relief** rainfall results when near-saturated, warm maritime air is forced to rise where confronted by a coastal mountain barrier. Mountains reduce the water-holding capacity of rising air by enforced cooling and can increase the amounts of cyclonic rainfall by retarding the speed of depression movement. Mountains also tend to cause air streams to converge and funnel through valleys. Rainfall totals increase where mountains are parallel to the coast, as is the Canadian Coast Range, and where winds have crossed warm off-shore ocean currents, as they do before reaching the British Isles. As air descends on the leeward side of a mountain range, it becomes compressed and warmed and condensation ceases, creating a **rain shadow** effect where little rain falls.

3 **Convectional** rainfall occurs when the ground surface is locally overheated and the adjacent air, heated by conduction, expands and rises. During its ascent, the air mass remains warmer than the surrounding environmental air and it is likely to become unstable (page 199) with towering cumulo-nimbus clouds forming. These unstable conditions, possibly augmented by frontal or orographic uplift, force the air to rise in a 'chimney' (Figure 9.22). The updraught is maintained by energy released as latent heat at both condensation and freezing levels. The cloud summit is characterised by ice crystals in an anvil shape, the top of the cloud

being flattened by upper-air movements. When the ice crystals and frozen water droplets, i.e. hail, become large enough, they fall in a downdraught. The air through which they fall remains cool as heat is absorbed by evaporation. The downdraught reduces the warm air supply to the 'chimney' and therefore limits the lifespan of the storm. Such storms are usually accompanied by thunder and lightning. One of several theories put forward to account for lightning suggests that the ice crystals in the upper cloud create a positive charge while the earth has a negative charge. **Lightning** is the visible discharge of electricity between clouds or between clouds and the ground. Convection is one of the processes by which surplus heat and energy from the earth's surface are transferred vertically to the atmosphere in order to maintain the heat balance (page 192).

Snow, sleet, glazed frost and hail

Snow forms under similar conditions to rain (Bergeron–Findeisen process) except that as dew point temperatures are under 0°C, then the vapour condenses directly into a solid (sublimation). Ice crystals will form if hygroscopic or freezing nuclei are present and these may aggregate to give snowflakes. As warm air holds more moisture than cold, snowfalls are heaviest when the air temperature is just below freezing. As temperatures drop, it becomes 'too cold for snow'. Figure 9.23 shows the typical conditions under which snow might fall in Britain.

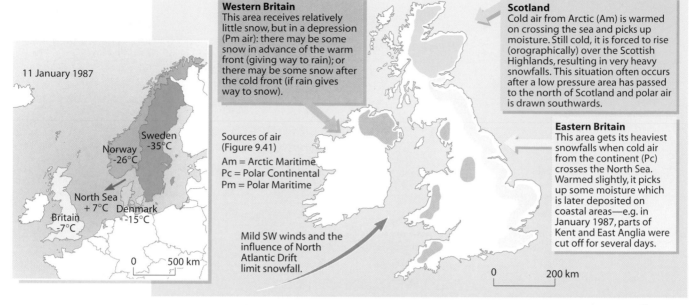

Western Britain
This area receives relatively little snow, but in a depression (Pm air): there may be some snow in advance of the warm front (giving way to rain); or there may be some snow after the cold front (if rain gives way to snow).

Scotland
Cold air from Arctic (Am) is warmed on crossing the sea and picks up moisture. Still cold, it is forced to rise (orographically) over the Scottish Highlands, resulting in very heavy snowfalls. This situation often occurs after a low pressure area has passed to the north of Scotland and polar air is drawn southwards.

Eastern Britain
This area gets its heaviest snowfalls when cold air from the continent (Pc) crosses the North Sea. Warmed slightly, it picks up some moisture which is later deposited on coastal areas—e.g. in January 1987, parts of Kent and East Anglia were cut off for several days.

11 January 1987

Sweden −35°C
Norway −26°C
North Sea +7°C
Denmark −15°C
Britain −7°C

Sources of air (Figure 9.41)
Am = Arctic Maritime
Pc = Polar Continental
Pm = Polar Maritime

Mild SW winds and the influence of North Atlantic Drift limit snowfall.

0   500 km

0   200 km

Figure 9.23

Causes of uneven snowfall patterns across Britain

Sleet is a mixture of ice and snow formed when the upper air temperature is below freezing, allowing snowflakes to form, and the lower air temperature is around 2 to 4°C, which allows their partial melting.

Glazed frost is the reverse of sleet and occurs when water droplets form in the upper air but turn to ice on contact with a freezing surface. When glazed frost forms on roads, it is known as 'black ice'.

Hail is made up of frozen raindrops which exceed 5 mm in diameter. It usually forms in cumulo-nimbus clouds, resulting from the uplift of air by convection currents, or at a cold front. It is more common in areas with warm summers where there is sufficient heat to trigger off the uplift of air, and less common in colder climates. Hail frequently proves a serious climatic hazard in cereal-growing areas such as the American Prairies.

Dew, hoar frost, fog and rime
Dew, hoar frost and radiation fog all form under calm, clear, anticyclonic conditions when there is rapid terrestrial radiation at night. Dew point is reached as the air cools by conduction and moisture in the air, or transpired from plants, condenses. If dew point is above freezing, **dew** will form; if it is below freezing, **hoar frost** develops. Frost may also be frozen dew. Dew and hoar frost usually occur within 1 m of ground level.

If the lower air is relatively warm, moist and contains hygroscopic nuclei, and if the ground cools rapidly, **radiation fog** may form. Where visibility is more than 1 km it is mist, if less than 1 km, fog. In order for radiation fog to develop, a gentle wind is needed to stir the cold air adjacent to the ground so that cooling affects a greater thickness of air. Radiation fogs usually occur in valleys, are densest around sunrise, and consist of droplets which are sufficiently small to remain buoyant in the air. Fog is likely to thicken if temperature inversions take place (Figures 9.24 and 9.25) — i.e. when cold surface air is trapped by overlying warmer, less dense air. It is under such conditions, in urban and industrial areas, that smoke and other pollutants released into the air are retained as smog (Figures 9.26 and 15.50).

Figure 9.24

Temperature inversion: radiation fog in a valley, Iceland

Figure 9.25

A low-level temperature inversion

less-warm air
warmer air
temperature inversion
trapped   pollutants
ELR
cold air

Height (m)
300
200
100
0

-20 -15 -10 -5   0   5   10
Temperature (°C)

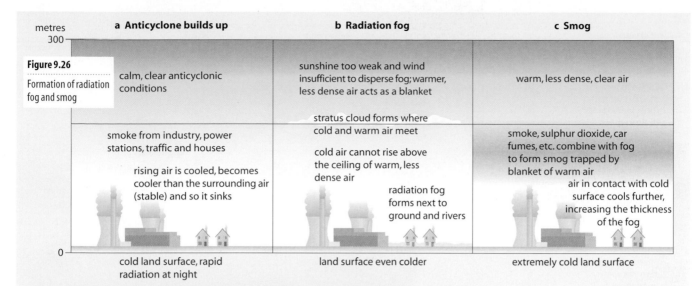

	a Anticyclone builds up	b Radiation fog	c Smog

Figure 9.26

Formation of radiation fog and smog

metres 300 — 0

a Anticyclone builds up

calm, clear anticyclonic conditions

smoke from industry, power stations, traffic and houses

rising air is cooled, becomes cooler than the surrounding air (stable) and so it sinks

cold land surface, rapid radiation at night

b Radiation fog

sunshine too weak and wind insufficient to disperse fog; warmer, less dense air acts as a blanket

stratus cloud forms where cold and warm air meet

cold air cannot rise above the ceiling of warm, less dense air

radiation fog forms next to ground and rivers

land surface even colder

c Smog

warm, less dense, clear air

smoke, sulphur dioxide, car fumes, etc. combine with fog to form smog trapped by blanket of warm air

air in contact with cold surface cools further, increasing the thickness of the fog

extremely cold land surface

Figure 9.27

Rime frost, Woodlands St Mary, Berkshire

Advection fog forms when warm air passes over or meets with cold air to give rapid cooling. In the coastal Atacama Desert (page 164), sufficient droplets fall to the ground as 'fog-drip' to enable some vegetation growth.

Rime (Figure 9.27) occurs when supercooled droplets of water, often in the form of fog, come into contact with, and freeze upon, solid objects such as telegraph poles and trees.

Acid rain

This is an umbrella term for the presence in rainfall of a series of pollutants which are produced mainly by the burning of fossil fuels. Coal-fired power stations, heavy industry and vehicle exhausts emit sulphur dioxide and nitrogen oxides. These are carried by prevailing winds across seas and national frontiers to be deposited either directly onto the earth's surface as dry deposition or to be converted into acids (sulphuric and nitric acid) which fall to the ground in rain as wet deposition. Clean rainwater has a pH value of between 5 and 6 (Figure 10.18). Today, rainfall over most of north-west Europe has a pH value of between 4 and 5; the lowest ever recorded being 2.4.

The effects of acid rain include an increase in levels of water acidity. This has caused the death of fish and plant life in many Scandinavian rivers and lakes and the pollution of fresh water supplies. Forests are being destroyed as important soil nutrients (calcium and potassium) are washed away, to be replaced by manganese and aluminium which are harmful to root growth. In time, trees shed their needles and leaves as they become less resistant to drought, frost and disease. Forecasts suggest that 90 per cent of Germany's trees may have died by early in the next century, with consequential effects on forest wildlife habitats. Acid rain has been linked with a decline in human health as seen by the increasing incidence of Alzheimer's disease (which may result from higher concentrations of aluminium), bronchitis and lung cancer. As soils become more acidic, crop yields are likely to fall. Chemical weathering is eroding buildings as far afield as St Paul's Cathedral, the Acropolis and the Taj Mahal, while 'black snow' falls annually in the Cairngorms.

 Q Describe carefully the factors which are thought to cause acid rain and the effects it has on the environment. How might the problems of acid rain be overcome and what are the difficulties involved in implementing proposed solutions? (You will need to undertake research to answer this question fully.)

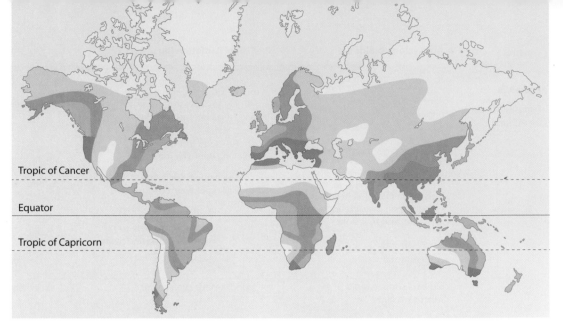

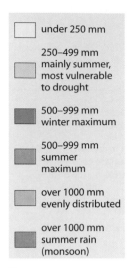

under 250 mm

250–499 mm
mainly summer,
most vulnerable
to drought

500–999 mm
winter maximum

500–999 mm
summer
maximum

over 1000 mm
evenly distributed

over 1000 mm
summer rain
(monsoon)

Figure 9.28

World precipitation: mean
annual totals and seasonal
distribution

World precipitation: distribution and reliability

Geographers are interested in describing distributions and in identifying and accounting for any resultant patterns. Where precipitation is concerned, geographers have, in the past, concentrated upon long-term distributions which show either mean annual amounts or seasonal variations. Long-term fluctuations vary considerably across the globe but, nevertheless, a map showing world precipitation does show identifiable patterns (Figure 9.28).

Equatorial areas have high annual rainfall totals due to the continuous uplift of air resulting from the convergence of the trade winds and strong convectional currents (page 209). The presence of the ITCZ ensures that rain falls throughout the year; it is the passage of the overhead sun, twice a year, which accounts for many places receiving a double rainfall peak. Further away from the Equator, rainfall totals decrease and the length of the dry season increases. These tropical areas, especially those away from the coast, experience heavy convectional rainfall in summer, when the sun is overhead, followed by a dry winter. Latitudes adjacent to the two tropics receive minimal amounts of rainfall as they correspond to areas of high pressure caused by subsiding, and therefore warming, air (Figure 7.2 and page 209).

To the poleward side of this arid zone, rainfall quantities begin to increase again and the length of the dry season decreases. These temperate latitudes receive large amounts of rainfall, spread evenly throughout the year, due to cyclonic conditions and local orographic effects. Towards the polar areas, where cold air descends to give stable conditions, precipitation totals decrease and rain gives way to snow. Between 30° and 40° north and south (in the west of continents) the Mediterranean climate is characterised by winter rain and summer drought. This general latitudinal zoning of rainfall is interrupted locally by the apparent movement of the overhead sun, the presence of mountain ranges or ocean currents, the monsoon, and continentality (distance from the sea).

More recently, geographers have become increasingly concerned with shorter-term variations. In many parts of the world, economic development and lifestyles are more closely linked to the duration, intensity and reliability of rainfall rather than to annual amounts. Precipitation is more valuable when it falls during the growing season (Canadian Prairies) and less effective if it occurs when evapotranspiration rates are at their highest (Sahel countries). In the same way, lengthy episodes of steady rainfall as experienced in Britain provide a more beneficial water supply than storms of a short and intensive duration which occur in tropical semi-arid climates. This is because moisture penetrates the soil more gradually and the risks of soil erosion, flooding and water shortages are reduced.

Of utmost importance is the reliability of rainfall. There appears to be a strong positive correlation (page 572) between rainfall totals and rainfall reliability — i.e. as rainfall totals increase, so too does rainfall reliability. In Britain and the Amazon Basin, rainfall is reliable with relatively little variation in annual totals from year to year (Figure 9.29).

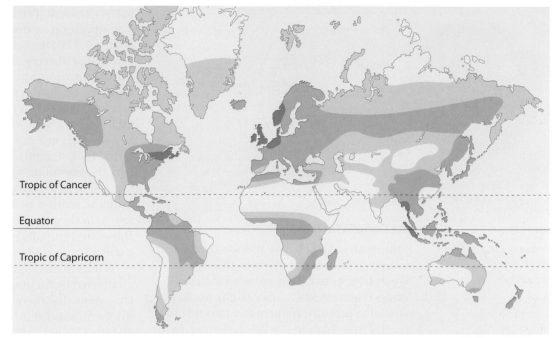

Figure 9.29

World rainfall reliability

Percentage departure from the mean

- over 30
- 21–30
- 11–20
- 10 and under

Elsewhere, especially in monsoon or tropical continental climates, there is a pronounced wet and dry season. Consequently, if the rains fail one year, the result can be disastrous for crops, and possibly also for animals and people. The most vulnerable areas, such as north-east Brazil and the Sahel countries, lie near to desert margins (Figure 9.29). Here, where even a small variation of 10 per cent below the mean can be critical, many places often experience a variation in excess of 30 per cent.

Atmospheric motion

Figure 9.30

The two basic pressure systems which affect Britain

The movement of air in the atmospheric system may be vertical (i.e. rising or subsiding) or horizontal; in the latter case it is commonly known as wind. Winds result from differences in air pressure which in turn may be caused by differences in temperature and the force exerted by gravity as pressure decreases rapidly with height (Figure 9.1). An increase in temperature causes air to heat, expand, become less dense and rise, creating an area of low pressure below. Conversely, a drop in temperature produces an area of high pressure. Differences in pressure are shown on maps by **isobars** which are lines joining places of equal pressure. To draw isobars, pressure readings are normally reduced to represent pressure at sea-level. Pressure is measured in millibars (mb) and it is usual for isobars to be drawn at 4-millibar intervals. Average pressure at sea-level is 1013 mb. However, the isobar pattern is usually more important in terms of explaining the weather than the actual figures. The closer together the isobars, the greater the difference in pressure — the **pressure gradient** — and the stronger the wind. Wind is nature's way of balancing out differences in pressure as well as temperature and humidity.

Figure 9.30 shows the two basic pressure systems which affect the British Isles. In addition to the differences in pressure, wind speed and wind direction, the diagrams also show that winds blow neither directly at right angles to the isobars along the pressure gradient, nor parallel to them. This is due to the effects of the Coriolis force and of friction, which are additional to the forces exerted by the pressure gradient and gravity.

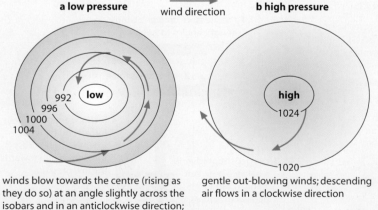

a low pressure

wind direction

992 | low
996
1000
1004

winds blow towards the centre (rising as they do so) at an angle slightly across the isobars and in an anticlockwise direction; winds are usually strong due to the steep pressure gradient

b high pressure

high
1024

1020

gentle out-blowing winds; descending air flows in a clockwise direction

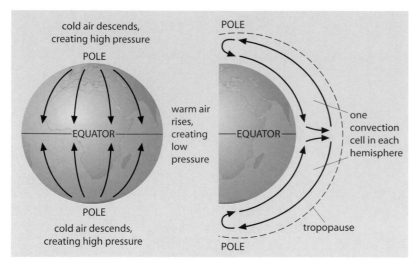

Figure 9.31

Air movement on a rotation-free earth

The Coriolis force

If the earth did not rotate and was composed entirely of either land or water, there would be one large convection cell in each hemisphere (Figure 9.31). Surface winds would be parallel to pressure gradients and would blow directly from high to low pressure areas. In reality, the earth does rotate and the distribution of land and sea is uneven. Consequently, more than one cell is created (Figure 9.35) as rising air, warmed at the equator, loses heat to space — there is less cloud cover to retain it — and as it travels further from its source of heat. A further consequence is that moving air appears to be deflected to the right in the northern hemisphere and to the left in the southern hemisphere. This is a result of the Coriolis force.

Imagine that Person A stands in the centre of a large rotating disc and throws a ball to Person B, standing on the edge of that moving disc. As Person A watches, the ball appears to take a curved path away from Person B — due to the fact that, while the ball is in transit, Person B has been moved to a new position by the rotation of the disc (Figure 9.32). Similarly, the earth's rotation through 360° every 24 hours means that a wind blowing in a northerly direction in the northern hemisphere appears to have been diverted to the right on a curved trajectory by 15° of longitude for every hour (though to an astronaut in a space shuttle, the path would look straight). This helps to explain why the prevailing winds blowing from the tropical high pressure zone approach Britain from the south-west rather than from the south. In theory, if the Coriolis force acted alone, the resultant wind would blow in a circle.

Winds in the upper troposphere, unaffected by friction with the earth's surface, show that there is a balance between the forces exerted by the pressure gradient and the Coriolis deflection. The result is the **geostrophic wind** which blows parallel to isobars (Figure 9.33). The existence of the geostrophic wind was recognised in 1857 by a Dutchman, Buys Ballot, whose law states that 'if you stand, in the northern hemisphere, with your back to the wind, low pressure is always to your left and high pressure to your right'.

Friction, caused by the earth's surface, upsets the balance between the pressure gradient and the Coriolis force by reducing the effect of the latter. As the pressure gradient becomes relatively more important when friction is reduced with altitude, the wind blows across isobars towards the low pressure (Figure 9.30). Deviation from the geostrophic wind is less pronounced over water because its surface is smoother than that of land.

Figure 9.32

The Coriolis force in the northern hemisphere

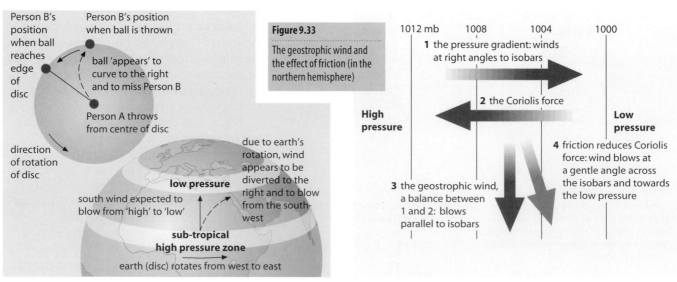

Figure 9.33

The geostrophic wind and the effect of friction (in the northern hemisphere)

A hierarchy of atmospheric motion

An appreciation of the movement of air is fundamental to an understanding of the workings of the atmosphere and its effects on our weather and climate. The extent to which atmospheric motion influences local weather and climate depends on winds at a variety of scales and their interaction in a hierarchy of patterns. One such hierarchy, which is useful in studying the influence of atmospheric motion, was suggested by B. W. Atkinson in 1988. Although defining four levels, he stressed that there were important interrelationships between each (Figure 9.34).

Figure 9.34

A hierarchy of atmospheric motion systems (after Atkinson, 1988)

Scale	Characteristic horizontal size (km)	Systems
1 Planetary	5000–10 000	Rossby waves, ITCZ
2 Synoptic (macro)	1000–5000	Monsoons, hurricanes, depressions, anticyclones
3 Meso-scale	10–1000	Land and sea breezes, mountain and valley winds, föhn, thunderstorms
4 Small (micro)	0.1–10	Smoke plumes, urban turbulence

Planetary scale: atmospheric circulation

It has already been shown that there is a surplus of energy at the Equator and a deficit in the outer atmosphere and nearer to the poles (Figure 9.6). Therefore, theoretically, surplus energy should be transferred to areas with a deficiency by means of a single convective cell (Figure 9.31). This would be the case for a non-rotating earth, a concept first advanced by Halley (1686) and expanded by Hadley (1735). The discovery of three cells was made by Ferrel (1856) and refined by Rossby (1941). Despite many modern advances using **radiosonde** readings, satellite imagery and computer modelling, this tricellular model still forms the basis of our understanding of the general circulation of the atmosphere.

The tricellular model

The meeting of the trade winds in the equatorial region forms the Inter-tropical Convergence Zone or ITCZ. The trade winds, which pick up latent heat as they cross warm, tropical oceans, are forced to rise by violent convection currents. The unstable, warm, moist air is rapidly cooled adiabatically to produce the towering cumulo-nimbus clouds, frequent afternoon thunderstorms and low pressure characteristic of the equa-torial climate (page 292). It is these strong upward currents which form the 'powerhouse of the general global circulation' and which turn latent heat first into sensible heat and later into potential energy. At ground level, the ITCZ experiences only very gentle, variable winds known as the **doldrums**.

As rising air cools to the temperature of the surrounding environmental air, uplift ceases and it begins to move away from the Equator. Further cooling, increasing density, and diversion by the Coriolis force cause the air to slow down and to subside, forming the descending limb of the **Hadley cell** (Figures 9.35 and 9.36). In looking at the northern hemisphere (the southern is its mirror image), it can be seen that the air subsides at about 30°N of the Equator to create the

Figure 9.35

Tricellular model showing atmospheric circulation in the northern hemisphere

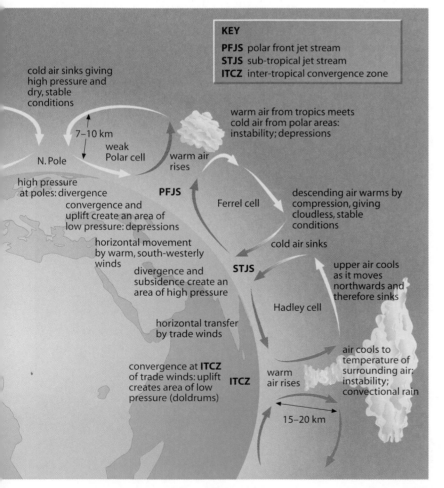

KEY
PFJS polar front jet stream
STJS sub-tropical jet stream
ITCZ inter-tropical convergence zone

cold air sinks giving high pressure and dry, stable conditions

7–10 km

weak Polar cell

N. Pole

high pressure at poles: divergence

convergence and uplift create an area of low pressure: depressions

horizontal movement by warm, south-westerly winds

divergence and subsidence create an area of high pressure

horizontal transfer by trade winds

convergence at **ITCZ** of trade winds: uplift creates area of low pressure (doldrums)

warm air from tropics meets cold air from polar areas: instability; depressions

warm air rises

PFJS

Ferrel cell

descending air warms by compression, giving cloudless, stable conditions

cold air sinks

STJS

Hadley cell

upper air cools as it moves northwards and therefore sinks

ITCZ

warm air rises

air cools to temperature of surrounding air; instability; convectional rain

15–20 km

Figure 9.36

Tricellular model to show atmospheric circulation in the northern hemisphere and the altitude of the tropopause

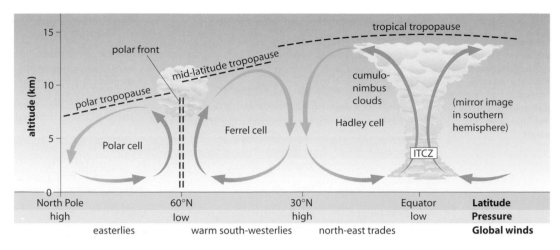

Figure 9.37

Image taken by the Meteosat geosynchronous satellite, 25.9.83: notice the clouds resulting from uplift at the ITCZ (not a continuous belt), the clear skies over the Sahara, the polar front over the north Atlantic and a depression over Britain

subtropical high pressure belt with its clear skies and dry, stable conditions (Figure 9.37). On reaching the earth's surface, the cell is completed as some of the air is returned to the Equator as the north-east trade winds.

The remaining air is diverted polewards, forming the warm south-westerlies which collect moisture when they cross sea areas. These warm winds meet cold Arctic air at the polar front (about 60°N) and are uplifted to form an area of low pressure and the rising limb of the **Ferrel** and **Polar cells** (Figures 9.35 and 9.36). The resultant unstable conditions produce the heavy cyclonic rainfall associated with mid-latitude **depressions**. Depressions are another mechanism by which surplus heat is transferred. While

some of this rising air eventually returns to the tropics, some travels towards the poles where, having lost its heat, it descends to form another stable area of high pressure. Air returning to the polar front does so as the cold easterlies.

This overall pattern is affected by the apparent movement of the overhead sun to the north and south of the Equator. This movement causes the seasonal shift of the heat Equator, the ITCZ, the equatorial low pressure zone and global wind and rainfall belts. Any variation in the characteristics of the ITCZ — i.e. its location or width — can have drastic consequences for the surrounding climates, as seen in the Sahel droughts of the early 1970s and most of the 1980s.

Rossby waves and jet streams

Evidence of strong winds in the upper troposphere first came when World War I Zeppelins were blown off-course and several inter-war balloons were observed travelling at speeds in excess of 200 km/hr. Pilots in World War II, flying at heights above 8 km, found eastward flights much faster and their return westward journeys much slower than expected, while north–south flights tended to be blown off-course. The explanation was found to be a belt of upper-air westerlies, the **Rossby waves**, which often follow a meandering path (Figure 9.38a). The number of meanders, or waves, varies seasonally with usually four to six in summer and three in winter. These waves form a complete pattern around the globe (Figure 9.38b and c).

Further investigation has shown that the velocity of these upper westerlies is not internally uniform. Within them are narrow bands of extremely fast-moving air known as **jet streams**. Jet streams, which help in the rapid transfer of energy, can exceed speeds of 230 km/hr, which is sufficient to carry a balloon, or ash from a volcano, around the earth within a week. Of five recognisable jet streams, two are particularly significant with a third having seasonal importance.

The **polar front jet stream** (PFJS, Figure 9.35) varies between latitudes 40° and 60° in both hemispheres and forms the division between the Ferrel and Polar cells — i.e. the boundary between warm tropical and cold polar air. The PFJS varies in extent, location and intensity and is mainly responsible for giving fine or wet weather on the earth's surface. Where, in the northern hemisphere, the jet stream moves south (Figure 9.39a), it brings with it cold air which descends in a clockwise direction to give dry, stable conditions associated with areas of high pressure (**anticyclones**, page 218). When the now-warmed jet stream backs northwards, it takes with it warm air which rises in an anticlockwise direction to give the strong winds and heavy rainfall associated with areas of low pressure (**depressions**, page 214). As the usual path of the PFJS over Britain is oblique — i.e. towards the north-east — this accounts for our frequent wet and windy weather. Occasionally, this path may be temporarily altered by a stationary or **blocking anticyclone** (Figures 9.39b and 9.47) which may produce extremes of climate such as the hot, dry summers of 1976 and 1989 or the cold January of 1987.

The **subtropical jet stream** or STJS occurs about 25° to 30° from the Equator and forms the boundary between the Hadley and Ferrel cells (Figures 9.35 and 9.36). This meanders less than the PFJS, has lower wind velocities, but follows a similar west–east path.

The **easterly equatorial jet stream** is more seasonal, being associated with the summer monsoon of the Indian sub-continent (page 222).

Figure 9.38

Rossby waves and jet streams (northern hemisphere)

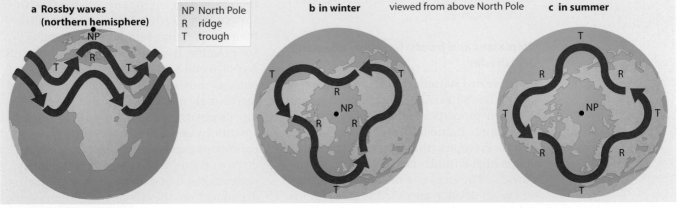

Figure 9.39

The polar front jet stream (PFJS) (northern hemisphere)

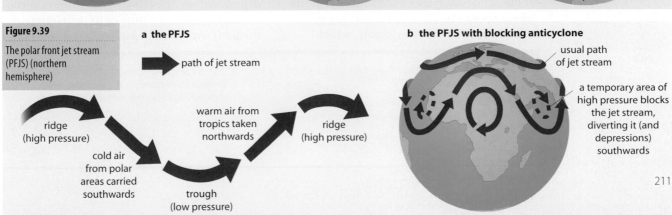

211

Q

With reference to the polar front jet stream (PFJS):

1 Explain how the PFJS contributes to the poleward transfer of energy (the heat budget).

2 Using the labels given below, complete the flow chart (Figure 9.40) to explain how the jet stream creates areas of high and low pressure in mid-latitudes:

Figure 9.40

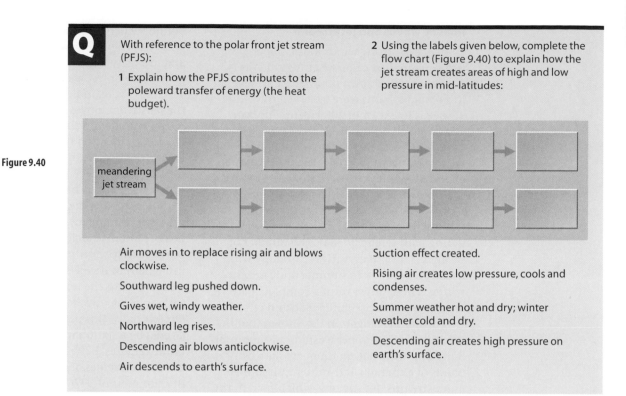

Air moves in to replace rising air and blows clockwise.

Southward leg pushed down.

Gives wet, windy weather.

Northward leg rises.

Descending air blows anticlockwise.

Air descends to earth's surface.

Suction effect created.

Rising air creates low pressure, cools and condenses.

Summer weather hot and dry; winter weather cold and dry.

Descending air creates high pressure on earth's surface.

Macro-scale: synoptic systems

The concept of air masses is important because air masses help to categorise world climate types (Chapter 12). In regions where one air mass is dominant all year, there is little seasonal variation in weather — for example, at the tropics and at the poles. Areas such as the British Isles, where air masses constantly interchange, experience much greater seasonal and diurnal variation in their weather.

Air masses and fronts: how they affect the British Isles

If air remains stationary in an area for several days, it tends to assume the temperature and humidity properties of that area. Stationary air is mainly found in the high pressure belts of the sub-tropics (the Azores and the Sahara) and in high latitudes (Siberia and northern Canada). The areas in which homogeneous air masses develop are called **source regions**. Air masses can be classified according to the **latitude** in which they originate, which determines their temperature — Arctic (*A*), Polar (*P*), or Tropical (*T*) — and the nature of the **surface** over which they develop, which affects their moisture content — maritime (*m*), or continental (*c*).

The five major air masses which affect the British Isles at various times of the year (*Am*, *Pm*, *Pc*, *Tm* and *Tc*) are derived by combining these characteristics of latitude and humidity (Figure 9.41). When air masses move from their source region they are modified by the surface over which they pass and this alters their temperature, humidity and stability. For example, tropical air moving northwards is cooled and becomes more stable, while polar air moving south becomes warmer and increasingly unstable. Each air mass therefore brings its own characteristic weather conditions to the British Isles. The general conditions expected with each air mass are given in Figure 9.42. However, it should be remembered that each air mass is unique and dependent on: the climatic conditions in the source region at the time of its development; the path which it subsequently follows; the

Q

1 What is meant by the terms 'air mass' and 'source regions'?

2 Under what conditions will tropical air become more stable and polar air less stable?

3 Give a reasoned account of the effects of various air masses on the weather of the British Isles in both summer and winter.

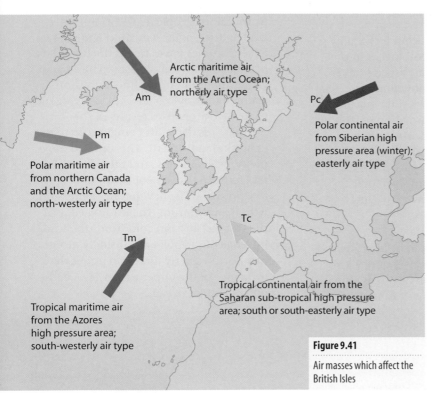

Polar maritime air from northern Canada and the Arctic Ocean; north-westerly air type

Arctic maritime air from the Arctic Ocean; northerly air type

Am

Pm

Pc

Polar continental air from Siberian high pressure area (winter); easterly air type

Tc

Tm

Tropical maritime air from the Azores high pressure area; south-westerly air type

Tropical continental air from the Saharan sub-tropical high pressure area; south or south-easterly air type

Figure 9.41

Air masses which affect the British Isles

season in which it occurs; and, since it has a three-dimensional form, the vertical characteristics of the atmosphere at the time.

When two air masses meet, they do not mix readily due to differences in temperature and density. The point at which they meet is called a **front**. A **warm front** is found where warm air is advancing and being forced to override cold air. A **cold front** occurs when advancing cold air undercuts a body of warm air. In both cases, the rising air cools and usually produces clouds, easily seen on satellite weather photograph (Figure 9.48); these clouds often generate precipitation. Fronts may be several hundred kilometres wide and they extend at relatively gentle gradients up into the atmosphere. The most notable type of front, the **polar front**, occurs when warm, moist, *Tm* air meets colder, drier, *Pm* air. It is at the polar front that depressions form.

Figure 9.42

Air masses and British weather

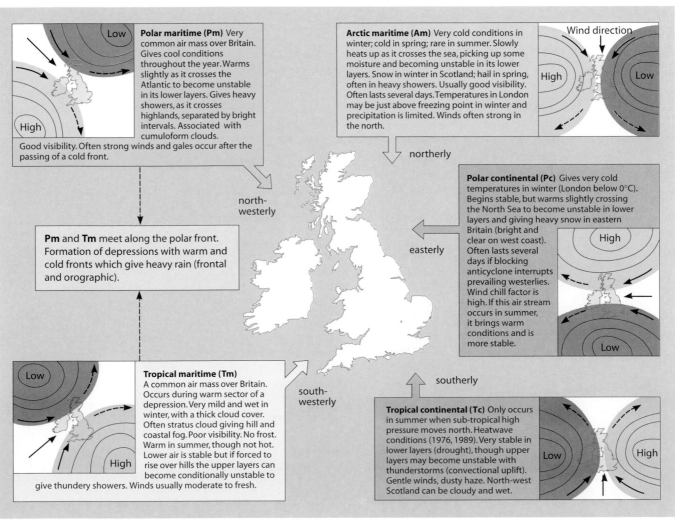

Polar maritime (Pm) Very common air mass over Britain. Gives cool conditions throughout the year. Warms slightly as it crosses the Atlantic to become unstable in its lower layers. Gives heavy showers, as it crosses highlands, separated by bright intervals. Associated with cumuloform clouds. Good visibility. Often strong winds and gales occur after the passing of a cold front.

Arctic maritime (Am) Very cold conditions in winter; cold in spring; rare in summer. Slowly heats up as it crosses the sea, picking up some moisture and becoming unstable in its lower layers. Snow in winter in Scotland; hail in spring, often in heavy showers. Usually good visibility. Often lasts several days. Temperatures in London may be just above freezing point in winter and precipitation is limited. Winds often strong in the north.

Wind direction

northerly

Pm and **Tm** meet along the polar front. Formation of depressions with warm and cold fronts which give heavy rain (frontal and orographic).

north-westerly

easterly

Polar continental (Pc) Gives very cold temperatures in winter (London below 0°C). Begins stable, but warms slightly crossing the North Sea to become unstable in lower layers and giving heavy snow in eastern Britain (bright and clear on west coast). Often lasts several days if blocking anticyclone interrupts prevailing westerlies. Wind chill factor is high. If this air stream occurs in summer, it brings warm conditions and is more stable.

Tropical maritime (Tm) A common air mass over Britain. Occurs during warm sector of a depression. Very mild and wet in winter, with a thick cloud cover. Often stratus cloud giving hill and coastal fog. Poor visibility. No frost. Warm in summer, though not hot. Lower air is stable but if forced to rise over hills the upper layers can become conditionally unstable to give thundery showers. Winds usually moderate to fresh.

south-westerly

southerly

Tropical continental (Tc) Only occurs in summer when sub-tropical high pressure moves north. Heatwave conditions (1976, 1989). Very stable in lower layers (drought), though upper layers may become unstable with thunderstorms (convectional uplift). Gentle winds, dusty haze. North-west Scotland can be cloudy and wet.

Depressions

The polar front theory was put forward by a group of Norwegian meteorologists in the early 1920s. Although some aspects have been refined since the innovation of radiosonde readings and satellite imagery, the basic model for the formation of frontal depressions remains valid. The following account describes a 'typical' or 'model' depression (Framework 9, page 326). It should be remembered, however, that individual depressions may vary widely from this model.

Depressions follow a life-cycle in which three main stages can be identified: embryo, maturity and decay (Figure 9.43).

1 The embryo depression begins as a small wave on the polar front. It is here that warm, moist, tropical (*Tm*) air meets colder, drier, polar (*Pm*) air (Figure 9.43a). Recent studies have shown the boundary between the two air masses to be a zone rather than the simple linear division claimed in early models. The convergence of the two air masses results in the warmer, less dense air being forced to rise in a spiral movement. This upward movement results in 'less' air at the earth's surface, creating an area of below-average or low pressure. The developing depression, with its warm front (the leading edge of the tropical air) and cold front (the leading edge of the polar air), usually moves in a north-easterly direction under the influence of the upper westerlies, i.e. the polar front jet stream.

Figure 9.43

The life-cycle of a depression

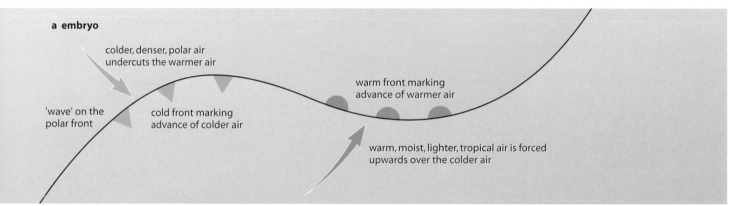

a embryo

colder, denser, polar air undercuts the warmer air

'wave' on the polar front

cold front marking advance of colder air

warm front marking advance of warmer air

warm, moist, lighter, tropical air is forced upwards over the colder air

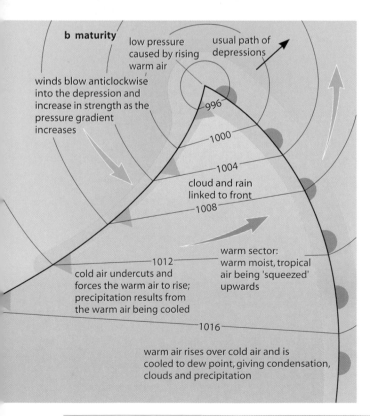

b maturity

low pressure caused by rising warm air

usual path of depressions

winds blow anticlockwise into the depression and increase in strength as the pressure gradient increases

996
1000
1004

cloud and rain linked to front

1008

warm sector: warm moist, tropical air being 'squeezed' upwards

1012

cold air undercuts and forces the warm air to rise; precipitation results from the warm air being cooled

1016

warm air rises over cold air and is cooled to dew point, giving condensation, clouds and precipitation

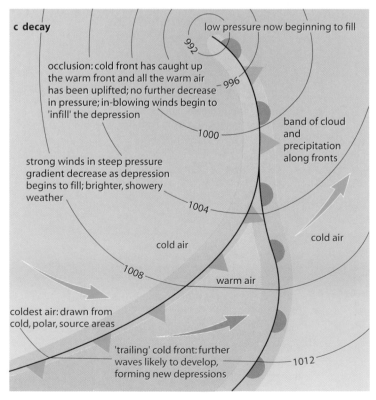

c decay

low pressure now beginning to fill

992

occlusion: cold front has caught up the warm front and all the warm air has been uplifted; no further decrease in pressure; in-blowing winds begin to 'infill' the depression

996

1000

band of cloud and precipitation along fronts

strong winds in steep pressure gradient decrease as depression begins to fill; brighter, showery weather

1004

cold air

cold air

1008

warm air

coldest air: drawn from cold, polar, source areas

'trailing' cold front: further waves likely to develop, forming new depressions

1012

2 A mature depression is recognised by the increasing amplitude of the initial wave (Figure 9.43b). Pressure continues to fall as more warm air, in the warm sector, is forced to rise. As pressure falls and the pressure gradient steepens, the inward-blowing winds increase in strength. Due to the Coriolis force (page 208), these anticlockwise-blowing winds come from the south-west. As the relatively warm air of the warm sector continues to rise along the warm front, it eventually cools to dew point. Some of its vapour will condense to release large amounts of latent heat, and clouds will develop. Continued uplift and cooling will cause precipitation as the clouds become both thicker and lower.

Satellite photographs have shown that there is likely to be a band of 'meso-scale precipitation' extending several hundred kilometres in length and up to 150 km in width along, and just in front of, a warm front. As temperatures rise and the uplift of air decreases within the warm sector, there is less chance of precipitation and the low cloud may break to give some sunshine. The cold front moves faster and has a steeper gradient than the warm front (Figure 9.44).

Progressive undercutting by cold air at the rear of the warm sector gives a second episode of precipitation — although with a greater intensity and a shorter duration than at the warm front. This band of

Figure 9.44

Weather associated with the passing of a typical mid-latitude depression

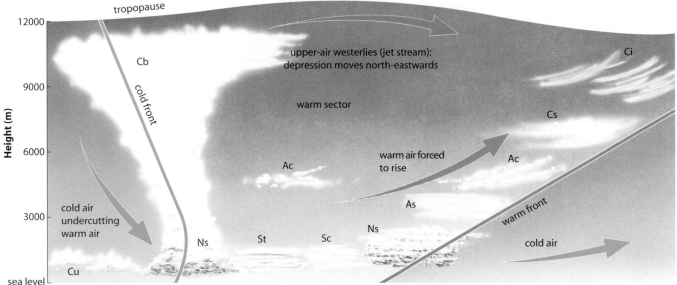

read from right to left (i.e. from 1 to 5)

	5. Behind the cold front	**4. Passing of the cold front**	**3. Warm sector**	**2. Passing of the warm front**	**1. Approach of depression**
Pressure	rise continues more slowly	sudden rise	steady	fall ceases	steady fall
Wind direction	NW	veers from SW to NW	SW	veers from SSE to SW	SSE
Wind speed	squally; speed slowly decreases (e.g. force 3–6)	very strong to gale force (e.g. force 6–8)	decreases (e.g. force 2–4)	strong (e.g. force 5–6)	slowly increases (e.g. force 1–3)
Temperature (e.g. winter)	cold (e.g. 3°C)	sudden decrease	warm/mild (e.g. 10°C)	sudden rise	cool (e.g. 6°C)
Relative humidity	rapid fall	high during precipitation	steady and high	high during precipitation	slow rise
Cloud	decreasing; in succession, Cb and Cu	very thick and towering Cb	low or may clear; St. Sc. Ac	low and thick Ns	high and thin; in succession, Ci, Cs, Ac, As
Precipitation	heavy showers	short period of heavy rain or hail	drizzle or stops raining	continuous rainfall, steady and quite heavy	none
Visibility	very good; poor in showers	poor	often poor	decreases rapidly	good but beginning to decrease

meso-scale precipitation may be only 10–50 km in width. Although the air behind the cold front is colder than that in advance of the warm front (having originated in and travelled through more northerly latitudes), it becomes unstable forming cumulo-nimbus clouds and heavy showers. Winds often reach their maximum strength at the cold front and change to a more north-westerly direction after its passage (Figure 9.44).

3 The depression begins to decay when the cold front catches up the warm front to form an **occlusion** or **occluded front** (Figure 9.43c). By this stage, the *Tm* air will have been squeezed upwards leaving no warm sector at ground level. As the uplift of air is reduced, so too are (or will be) the amount of condensation, the release of latent heat and the amount and pattern of precipitation — there may be only one episode of rain. Cloud cover begins to decrease, pressure rises and wind speeds decrease as the colder air replaces the uplifted air and 'infills' the depression.

Figure 9.45

The passage over a British weather station of a non-occluded depression in winter

Q

1 Reference has already been made to Figure 9.44, a model showing the weather associated with the passing of a typical mid-latitude depression. Now study Figure 9.45 (**a** to **e**) which plots various observations made at a British weather station during the passage of a non-occluded depression in winter.

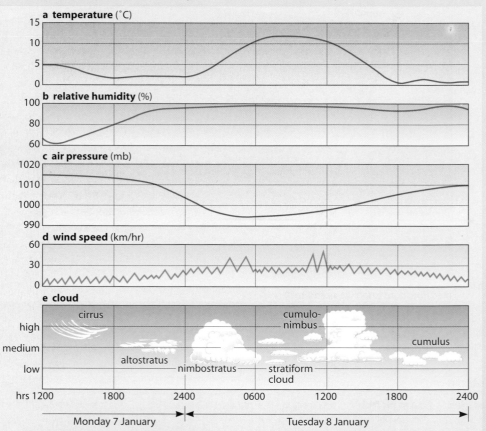

a With reference to Figure 9.45 **a**, **b** and **c** only, state the day and time when the warm front and the cold front passed over the weather station. Briefly justify the answers you have given.

b State the relationships between variations in wind speed (Figure 9.45**d**) and the passage of the warm and cold fronts. Account for these relationships.

c Describe the sequence of types and amounts of precipitation you would expect to occur at the weather station during the period from noon on 7 January to midnight 8 January.

d Explain how one other weather element, not shown in Figure 9.45, might be used to monitor the passage of the depression.

2 Read the account of 'the great storm in south-east England, 16 October 1987' in Places 23.

a In what ways was this storm *not* a typical depression?

b Why is it difficult accurately to forecast an event such as this?

c What could be done to reduce the damage and to predict the occurrence of a storm of similar intensity?

Places 23 The great storm in south-east England, 16 October 1987

This storm, the worst to affect south-east England since 1703, developed so rapidly that its severity was not predicted in advance weather forecasts.

11 October: High winds and heavy rain forecast for the end of the week.

15 October 1200 hrs: Depression expected to move along the English Channel with fresh to strong winds.
2130 hrs: TV weather forecast: strong winds gusting to 50 km/hr.

16 October 0030 hrs: Radio weather forecast: warning of severe gales.
0130 hrs: Police and fire services alerted about extreme winds.
0500 hrs: Winds reach 94 km/hr at Heathrow and 100 km/hr on parts of the south coast.
0800 hrs: Centre of depression reaches the North Sea.

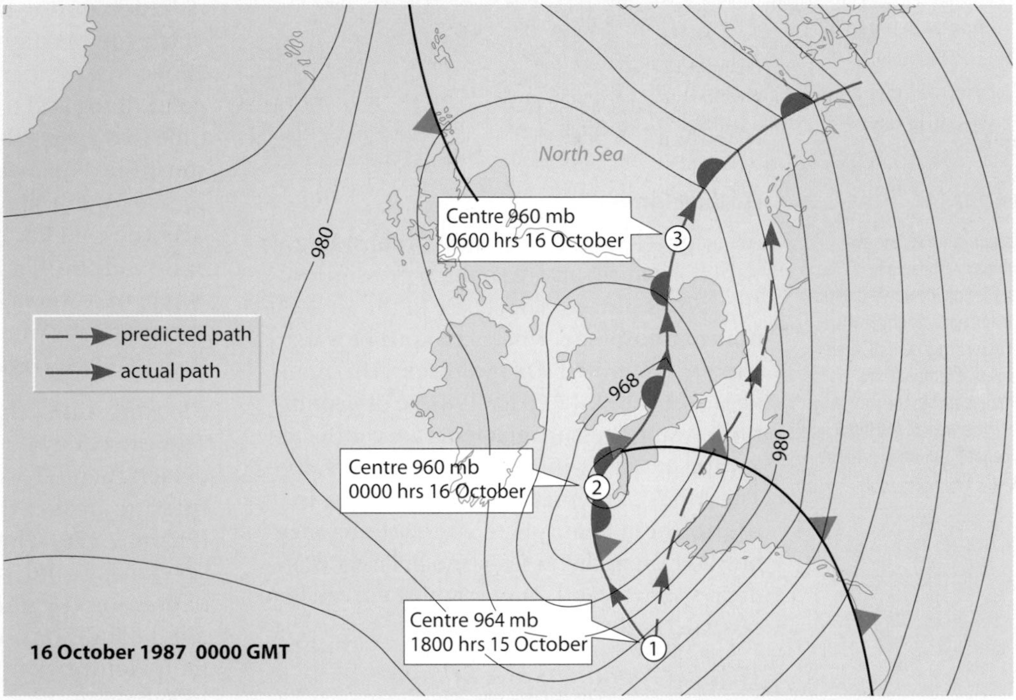

Figure 9.46
The great storm over south-east England, 16 October 1987

The storm began as a small wave on a cold front in the Bay of Biscay, where the few weather ships give only limited information. It was caused by contact between very warm air from Africa and cold air from the North Atlantic. It appeared to be a 'typical' depression until, at about 1800 hrs on 15 October, it unexpectedly deepened giving a central pressure reading of 958 millibars and an exceptionally steep pressure gradient. The exact cause of this is unknown but it was believed to result from a combination of an exceptionally strong jet stream (initiated on 13 October by air spiralling upwards along the east coast of North America in Hurricane Floyd) and extreme warming over the Bay of Biscay (see hurricanes, page 219). Together, these could have caused an excessive release of latent heat energy which North American meteorologists compare with the effect of detonating a 'bomb'. It was this unpredicted deepening, combined with the change of direction from the English Channel towards the Midlands, which caught experts by surprise.

Although the storm passed within a few hours, and luckily during the night when most people were asleep, it left a trail of death and destruction. There were 16 deaths; several houses collapsed and many others lost walls, windows and roofs; an estimated 15 million trees were blown over, blocking railways and roads; one-third of the trees in Kew Gardens were destroyed; power lines were cut and, in some remote areas, not restored for several days; few commuters managed to reach London the next day; a ferry was blown ashore at Folkestone; and insurance claims set an all-time record.

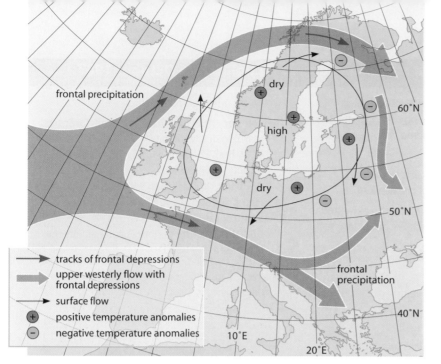

frontal precipitation

dry

dry

high

60°N

50°N

40°N

frontal
precipitation

10°E

20°E

— tracks of frontal depressions

⟹ upper westerly flow with
frontal depressions

→ surface flow

(+) positive temperature anomalies

(−) negative temperature anomalies

Figure 9.47

A blocking anticyclone over
Scandinavia: the upper
westerlies divide upwind of the
block and flow around it with
their associated rainfall; there
are positive temperature
anomalies within the southerly
flow to the west of the block
and negative anomalies to the
east

Anticyclones

An anticyclone is a large mass of subsiding air
which produces an area of high pressure on
the earth's surface. The source of the air is the
upper atmosphere, where amounts of water
vapour are limited. On its descent, the air
warms at the DALR (page 198), so dry conditions result. Pressure gradients are gentle, resulting in weak winds or calms (Figure 9.30b).
The winds blow outwards and clockwise in
the northern hemisphere. Anticyclones may
be 3000 km in diameter — much larger than
depressions — and, once established, can give
several days of settled weather.

Weather conditions over Britain
Summer Intense insolation gives hot,
sunny days without cloud or rain. Rapid

radiation at night, under clear skies, can lead
to temperature inversions and the formation
of dew and mist, although these rapidly clear
the following morning. Coastal areas may
experience advection fogs and land and sea
breezes, while highlands have mountain and
valley winds (pages 223–24). If the air has its
source over North Africa — that is, if it is a *Tc*
air mass — then heatwave conditions tend to
result. Often, after several days,
thermals increase to give thunderstorms.

Winter Although the sinking air again
gives cloudless skies, there is little incoming
radiation during the day due to the low angle
of the sun. At night, the absence of clouds
allows low temperatures and the development of fog and frost. These may take a long
time to disperse the next day in the weak
sunshine. Polar continental (*Pc*) air, with its
source in central Asia and a slow movement
over the cold European landmass, is cold, dry
and stable until it reaches the North Sea
where its lower layers acquire some warmth
and moisture, often leading to snowfalls on
the east coast (Figure 9.23).

Blocking anticyclones
These occur when cells of high pressure
detach themselves from the major high
pressure areas of the sub-tropics or poles
(Figure 9.39b). Once created, they last for
several days and 'block' eastward-moving
depressions (Figure 9.47) to create anomalous conditions such as extremes of
temperature, rainfall and sunshine — as
in Britain in the summer of 1989 and the
winter of 1987.

Figure 9.48

Depression over the North
Atlantic

Q 1 Redraw the outline of north-west Europe
shown in Figure 9.47. Locate a high
pressure area over the Irish Sea. Mark on
the two possible tracks likely to be followed
by any depression. Describe how these
tracks will affect Britain's weather.

2 Figures 9.48 and 9.49 show satellite
photographs of the British Isles: the first
under a depression, the second under an
anticyclone. With reference to these
photographs and information given in this
section, draw up a list to show the
differences between a depression and an
anticyclone using the following headings:

- cloud type
- cloud amount
- precipitation
- local winds
- speed of movement

- time of life-cycle
- surface pressure
- wind speed
- wind direction

- vertical air movement
- stability
- relative humidity

Name, in each case, the most likely type(s) of
air mass.

Figure 9.49

Anticyclone over the
British Isles

Tropical cyclones

Tropical cyclones are systems of intensive low pressure known locally as **hurricanes**, **typhoons**, **cyclones** and **willy-willies** (Figure 9.50). As yet, there is still insufficient conclusive evidence as to the process of their formation, although knowledge has been considerably improved recently due to air-flights through and over individual systems and the use of weather satellites. Tropical cyclones tend to develop

- over warm tropical oceans, where sea temperatures exceed 26°C and where there is a considerable depth of warm water;
- in autumn, when sea temperatures are at their highest;
- in the trade wind belt, where the surface winds warm as they blow towards the Equator;
- between latitudes 5° and 20° north or south of the Equator (nearer to the equator the Coriolis force is insufficient to enable the feature to 'spin').

Once formed, they move westwards — often on erratic, unpredictable courses — swinging poleward on reaching land, where their energy is rapidly dissipated (Figure 9.50). They are another mechanism by which surplus energy is transferred away from the tropics (Figure 9.6).

Hurricanes

Hurricanes are the tropical cyclone of the Atlantic. They form after the ITCZ has moved to its most northerly extent enabling air to converge at low levels, and can have a diameter of up to 650 km. Unlike depressions, hurricanes occur when temperatures, pressure and humidity are uniform over a wide area in the lower troposphere for a lengthy period, and anticyclonic conditions exist in the upper troposphere. These conditions are essential for the development, near the earth's surface, of the intense low pressure and strong winds. To enable the hurricane to move, there must be a continuous source of heat to maintain the rising air currents. There must also be a large supply of moisture to provide the latent heat, released by condensation, to drive the storm and to provide the heavy rainfall. It is estimated that in a single day a hurricane can release an amount of energy equivalent to that released by 500 000 atomic bombs the size of that dropped on Hiroshima in World War II. Only when the storm has reached maturity does the central **eye** develop. This is an area of subsiding air, some 30–50 km in diameter, with light winds, clear skies and anomalous high temperatures (Figure 9.51). The descending air increases instability by warming and exaggerates the storm's intensity.

The hurricane rapidly declines once the source of heat is removed — i.e. when it moves over colder water or a land surface; these increase friction and cannot supply sufficient moisture. The average lifespan of a tropical cyclone is 7–14 days. The characteristic weather conditions associated with the passage of a typical hurricane are shown diagrammatically in Figure 9.51, and from space in Figure 9.52.

Figure 9.50

Global location and mean frequency of tropical cyclones

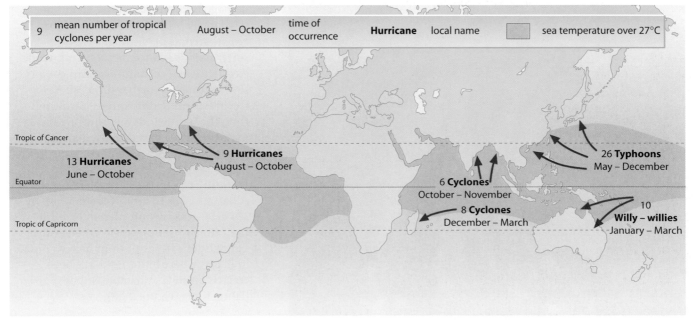

Vertical movement	updraughts increasing →		spiral uplift		subsiding air	spiral uplift		updraughts decreasing →	
Clouds	few Cu	Cu	Cu and some Cb	giant CB and Ci	none	giant Cb and Ci	Cu and some Cb	Cu	small Cu
Precipitation	none	showers	heavy showers	torrential rain 250 mm/day	none	torrential rain 250 mm/day	heavy showers	showers	none
Wind speed	gentle	fresh, gusty	locally very strong	hurricane force, 160 km/hr	calm	hurricane force 160 km/hr	locally very strong	fresh gusty	gentle
Wind direction	NNW	NW	WNW	WNW	calm	SSE	SSE	SE	ESE
Temperatures (plus examples)	high (30°C)	still high (30°C)	falling (26°C)	low (24°C)	high (32°C)	low (24°C)	rising (26°C)	high (28°C)	high (30°C)
Pressure	average, 1012 mb	steady, 1010 mb	slowly falling, 1006 mb	rapid fall	low, 960 mb	rapid rise	slowly rising, 1004 mb	steady, 1010 mb	average, 1012 mb

Figure 9.51

Weather associated with the passage of a hurricane or tropical cyclone

Figure 9.52

Satellite image of Hurricane Gladys, 12.10.68

Tropical cyclones are a major natural hazard which often cause considerable loss of life and damage to property and crops (Places 24). There are four main causes of damage:

1 **High winds** which often exceed 160 km/hr and, in extreme cases, 300 km/hr. Whole villages may be destroyed in economically less developed countries (of which there are many in the tropical cyclone belt), while even reinforced buildings in the south-east USA may be damaged. Countries whose economies rely largely on the production of a single crop (bananas in Nicaragua) may suffer serious economic problems, while electricity and communications can be severed causing further difficulties.

2 **Ocean storm (tidal) surges**, resulting from the high winds and low pressure, may inundate coastal areas, many of which are densely populated (Bangladesh, Places 15, page 132).

3 **Flooding** can be caused either by a storm (tidal) surge or the intensive rainfall. In 1974, 800 000 people died in Honduras as their flimsy homes were washed away.

4 **Landslides** can result from heavy rainfall in places where buildings have been erected on steep, unstable slopes (Hong Kong, Figure 2.28).

Bangladesh

November 1970

A tropical cyclone moved northwards up the narrowing, shallowing Bay of Bengal. Winds of over 200 km/hr and a storm surge 8 m high hit the densely populated Ganges delta. Over 4 million people were affected; 300 000 people died and 1 million were left homeless; half a million cows and oxen were drowned; two-thirds of the fishing fleet was lost and 80 per cent of the rice crop ruined.

May 1985

Three days after a tropical cyclone hit the coastal islands of Bangladesh, countless bodies could still be seen floating in the sea while hundreds of survivors, on bamboo rafts and floating roof tops, were awaiting rescue from the flood waters. The Red Cross estimated that the tidal surge, 9 m in height and penetrating 150 km inland across the flat delta region, may have claimed the lives of 40 000 people. An official source feared that on the island of Hatia alone, 6000 people, many in their sleep, were washed out to sea and that the only survivors were the few who managed to cling to the tops of palm trees in the 180 km/hr winds. Links had still to be established with several of the more remote islands. The Red Cross were fearing, in the short term, an outbreak of typhoid and cholera, as fresh water supplies had been contaminated; and, in the long term, a food shortage, as the rice crop had been lost and it would take next year's monsoon rains to wash the salt out of the soil. Thousands of animals and most of the coastal fishing fleet were also believed to have been lost.

The Caribbean: Hurricane Hugo

September 1989

"The most violent hurricane for ten years, and the eighth of the 1989 season, was, last night, cutting a swathe across the Caribbean. Hurricane Hugo, as the new storm has been christened, was reported to have a diameter of 1000 km. Already 12 people have been killed, 80 injured, and another 3000 made homeless on the French paradise island of Guadeloupe. However, due to electricity cables and telephone wires being cut, it could be days before the full toll is known. Some experts fear that, as the storm continues its path towards Puerto Rico and Florida, the final devastation could be as bad as in 1979 when Hurricane David left 1200 dead across the Caribbean and Florida.

The Royal Navy frigate *Alacrity* sailed to the rescue of more than 12 000 people made homeless on Montserrat, where six people were killed. The navy provided food, water and medical aid. Clearing-up operations continued on nearby Antigua where a 4 m tidal surge caused considerable damage.

Hurricane Hugo struck again when winds of 200 km/hr hit Puerto Rico. The island had been on full alert for two days against flash floods and mud slides. Fears that a tidal wave would rush into coastal valleys and swamp villages had already led to the evacuation of thousands of people. After depositing 400 mm of rain in 24 hours, Hugo has now changed course and is heading for the Bahamas. Hurricane expert teams between Florida and New England are on stand-by should Hugo change direction again and turn towards the American mainland."

Figure 9.53

Path taken by Hurricane Hugo

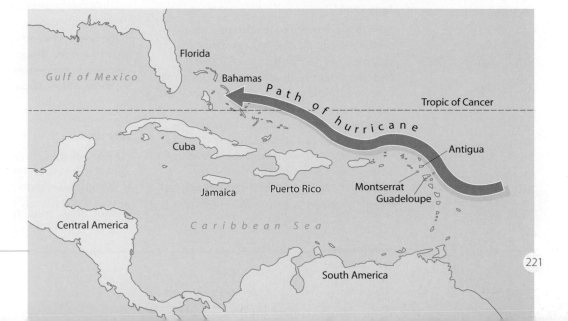

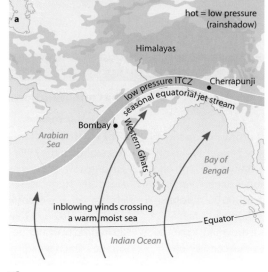

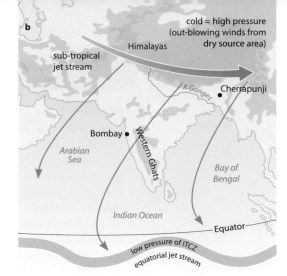

Figure 9.54

The monsoon in the Indian
sub-continent

The word 'monsoon' is
derived from the Arabic for
a season, but the term is
more commonly used in
meteorology to denote a
seasonal reversal of wind
direction.

The monsoon

The major monsoon occurs in south-east
Asia and results from three factors:

1 The extreme heating and cooling of large
   land masses in relation to the smaller heat
   changes over adjacent sea areas. This in
   turn affects pressure and winds.
2 The northward movement of the ITCZ
   during the northern hemisphere summer.
3 The uplift of the Himalayas which, some
   6 million years ago, became sufficiently
   high to interfere with the general circulation of the atmosphere.

The south-west or summer monsoon

As the overhead sun appears to move
northwards to the Tropic of Cancer in June,
it draws with it the convergence zone associated with the ITCZ (Figure 9.54a). The
increase in insolation over northern India,
Pakistan and central Asia means that heated
air rises, creating a large area of low pressure.
Consequently, warm moist *Em* (equatorial
maritime) and *Tm* air, from over the Indian
Ocean, is drawn first northwards and then,

because of the Coriolis force, is diverted
north-eastwards. The air is humid, unstable
and conducive to rainfall. Amounts of precipitation are most substantial on India's
west coast, where the air rises over the
Western Ghats, and in the Himalayas:
Bombay has 2000 mm and Cherrapunji
13000 mm in four summer months. The
advent of monsoon storms allows the
planting of rice. Rainfall totals are accentuated as the air rises by both orographic and
convectional uplift and the 'wet' monsoon is
maintained by the release of substantial
amounts of latent heat.

The north-east or winter monsoon

During the northern winter, the overhead
sun, the ITCZ and the sub-tropical jet stream
all move southwards (Figure 9.54b). At the
same time, central Asia experiences intense
cooling which allows a large high pressure
system to develop. Airstreams which move
outwards from this high pressure area will be
dry because their source area is semi-desert.
They become even drier as they cross the

Places 25   New Delhi, June and July 1994

June 1994

"Rain brought welcome relief to the Indian
capital yesterday, a day after 18 people
collapsed and died on the streets in the
blistering heat, pushing the summer death toll
in northern India to nearly 350. Heavy showers
cooled the furnace-like city, reeling under a
three-week heat wave that has kept daytime
temperatures at an almost constant 45°C and
which had, the previous day, experienced its
hottest day in 50 years when the mercury
soared to 47.6°C. It was the first pre-monsoon
rain of the season to lash New Delhi, and

children celebrated by soaking themselves in
the rain, with many elderly citizens joining them
in the belief that monsoon rains help cure
blisters and skin diseases caused by extreme
heat. More thunderstorms are expected by the
weekend which should mark the onset of the
summer monsoon."

July 1994

"The July death toll from relentless monsoon
rains across India and Pakistan rose to more
than 590 as a several waves of severe storms
passed across the sub-continent."

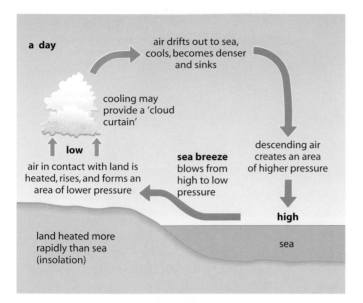

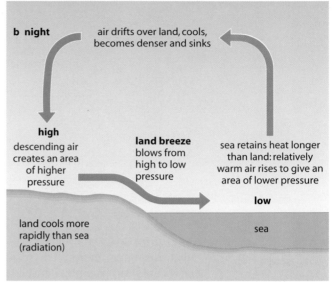

Figure 9.55

Land and sea breezes in Britain

Himalayas and adiabatically warmer as they descend to the Indo-Gangetic plain. Bombay receives less than 100 mm of rain during these eight months.

The monsoon, which in reality is much more complex than the model described above, affects the lives of one-quarter of the world's population. Unfortunately, monsoon rainfall, especially in the Indian sub-continent, is unreliable (Figure 9.29). If the rains fail, then drought and famine ensue: 1987 was the ninth year in a decade when the monsoon failed in north-west India. If, conversely, there is excessive rainfall then large areas of land experience extreme flooding (Bangladesh in 1987 and 1988).

Meso-scale: local winds

Of the three meso-scale circulations described here, two — **land and sea breezes** and **mountain and valley winds** — are caused by local temperature differences; the third — the **föhn** — results from pressure differences either side of a mountain range.

The land and sea breeze

This is an example, on a diurnal timescale, of a circulation system resulting from differential heating and cooling between land surfaces and adjacent sea areas. The resultant pressure differences, although small and localised, produce gentle breezes which affect coastal areas during calm, clear anticyclonic conditions. When the land heats up rapidly each morning, lower pressure forms and a gentle breeze begins to blow from the sea to the land (Figure 9.55a). By

early afternoon, this breeze has strengthened sufficiently to bring a freshness which, in the tropics particularly, is much appreciated by tourists at the beach resorts. Yet by sunset, the air and sea are both calm again.

Although the circulation cell rarely rises above 500 m in height or reaches more than 20 km inland in Britain, the sea breeze is capable of lowering coastal temperatures by 15°C and can produce advection fogs such as the 'sea-fret' of eastern Britain.

At night, when the sea retains heat longer than the land, there is a reversal of the pressure gradient and therefore of wind direction (Figure 9.55b). The land breeze, the gentler of the two, begins just after sunset and dies away by sunrise.

The mountain and valley wind

This wind is likely to blow in mountainous areas during times of calm, clear, settled weather. During the morning, valley sides are heated by the sun — especially if they are steep, south-facing (in the northern hemisphere) and lacking in vegetation cover. The air in contact with these slopes will heat, expand and rise (Figure 9.56a), creating a pressure gradient. By 1400 hours, the time of maximum heating, a strong uphill or **anabatic wind** blows up the valley and the valley sides — ideal conditions for hang-gliding! The air becomes conditionally unstable (Figure 9.19), often producing cumulus cloud and, under very warm conditions, cumulonimbus with the possibility of thunderstorms on the mountain ridges. A compensatory sinking of air leaves the centre of the valley cloud-free.

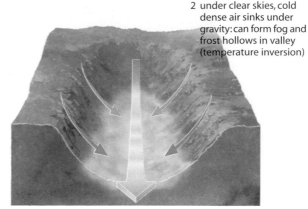

a day (anabatic flow)

updraughts may produce cloud on hills

descending air gives clear skies

3 winds less strong if valley sides face north (less heating)

2 wind blows up valley sides

1 wind blows up-valley

b night (katabatic flow)

2 under clear skies, cold dense air sinks under gravity: can form fog and frost hollows in valley (temperature inversion)

1 wind blows down-valley

Figure 9.56

Mountain and valley winds

During the clear evening, the valley loses heat through radiation. The surrounding air now cools and becomes denser. It begins to drain, under gravity, down the valley sides and along the valley floor as a mountain wind or **katabatic wind** (Figure 9.56b). This gives rise to a temperature inversion (Figure 9.25) and, if the air is moist enough, in winter may create fog (Figure 9.24) or a **frost hollow**. Maximum wind speeds are generated just before dawn, normally the coldest time of the day. Katabatic winds are usually gentle in Britain, but are much stronger if they blow over glaciers or permanently snow-covered slopes. In Antarctica, they may reach hurricane force.

The föhn

The **föhn** is a strong, warm and dry wind which blows periodically to the lee of a mountain range. It occurs in the Alps when a depression passes to the north of the mountains and draws in warm, moist air from the Mediterranean. As the air rises (Figure 9.57), it cools at the DALR of 1°C per 100 m (page 198). If, as in Figure 9.57, condensation occurs at 1000 m, there will be a release of latent heat and the rising air will cool more slowly at the SALR of 0.5°C per 100 m. This means that when the air reaches 3000 m it will have a temperature of 0°C instead of the -10°C had latent heat not been released. Having crossed the Alps, the descending air is compressed and warmed at the DALR so that, if the land dropped sufficiently, the air would reach sea-level at 30°C. This is 10°C warmer than when it left the Mediterranean. Temperatures may rise by 20°C within an hour and relative humidity can fall to 10 per cent.

This wind, also known as the **chinook** on the Prairies, has considerable effects on human activity. In spring, when it is most likely to blow, it lives up to its Red Indian name of 'snow-eater' by melting snow and enabling wheat to be sown; and in

Figure 9.57

The föhn

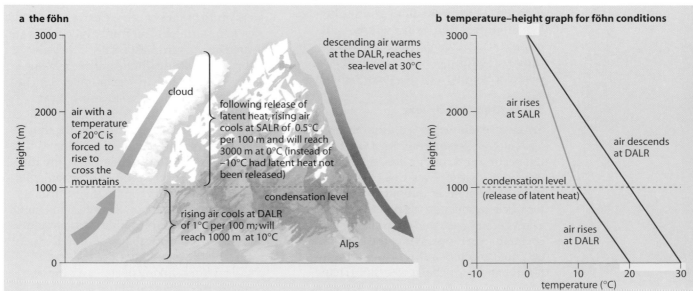

a the föhn

air with a temperature of 20°C is forced to rise to cross the mountains

cloud

following release of latent heat, rising air cools at SALR of 0.5°C per 100 m and will reach 3000 m at 0°C (instead of –10°C had latent heat not been released)

descending air warms at the DALR, reaches sea-level at 30°C

condensation level

rising air cools at DALR of 1°C per 100 m; will reach 1000 m at 10°C

Alps

b temperature–height graph for föhn conditions

air rises at SALR

air descends at DALR

condensation level (release of latent heat)

air rises at DALR

Figure 9.58

Distribution of minimum
temperatures over London,
14 May 1959 *(after* Chandler)

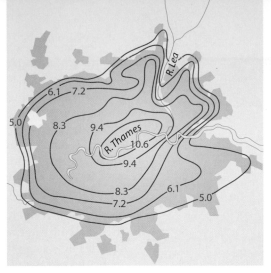

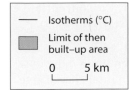

— Isotherms (°C)

▨ Limit of then
built–up area

0   5 km

Switzerland it clears the Alpine pastures of
snow. Conversely, its warmth can cause
avalanches, forest fires and the premature
budding of trees.

Microclimates

Microclimatology is the study of climate
over a small area. It includes changes
resulting from the construction of large
urban centres as well as those existing
naturally between different types of land
surface, e.g. forests and lakes.

Urban climates

Large cities and conurbations experience
climatic conditions which differ from those
of the surrounding countryside. They generate more dust and condensation nuclei than
natural environments; they create heat; they
alter the chemical composition and the
moisture content of the air above them; and
they affect both the albedo and the flow of
air. Urban areas therefore have distinctive
climates.

Temperature
Although tower blocks cast more shadow,
normal building materials tend to be non-
reflective and so absorb heat during the daytime. Dark-coloured roofs, concrete or brick
walls and tarmac roads all have a high
thermal capacity which means that they are
capable of storing heat during the day and
releasing it slowly during the night. Further
heat is obtained from car fumes, factories,
power stations, central heating and people
themselves. The term **urban heat island**
acknowledges that, under calm conditions,
temperatures are warmest in the more
built-up city centre and decrease towards
the suburbs and open countryside (Figure
9.58). In urban areas:

- day temperatures are, on average, 0.6°C
  warmer;
- night temperatures may be 3° or 4°C
  warmer as dust and cloud act like a blanket to reduce radiation and buildings give
  out heat like storage radiators;
- the mean winter temperature is 1° to 2°C
  warmer (rural areas are even
  colder when snow-covered as this
  increases their albedo);
- the mean summer temperature may be
  5°C warmer;
- the mean annual temperature is warmer
  by between 0.6°C in Chicago and 1.3°C
  in London compared with that of the
  surrounding area.

Note how, in Figure 9.58, temperatures not
only decrease towards London's boundary
but also beside the Rivers Thames and Lea.
The urban heat island is why large cities have
less snow, fewer frosts, earlier budding and
flowering of plants and a greater need, in
summer, for air conditioning than neighbouring rural areas.

Sunlight
Despite having higher mean temperatures,
cities receive less sunshine and more cloud
than their rural counterparts. Dust and other
particles may absorb and reflect as much as
50 per cent of insolation in winter, when the
sun is low in the sky and has to pass through
more atmosphere, and 5 per cent in summer.
High-rise buildings also block out light.

Wind
Wind velocity is reduced by buildings which
create friction and act as windbreaks. Urban
mean annual velocities may be up to 30 per
cent lower than in rural areas and periods of
calm may be 10–20 per cent more frequent.
In contrast, high-rise buildings, such as the
skyscrapers of New York and Hong Kong
(Figure 20.8), form 'canyons' through which
wind may be channelled. These winds may
be strong enough to cause tall buildings to
sway and pedestrians to be blown over and
troubled by dust and litter. The heat island
effect may cause local thermals and reduce
the wind chill factor. It also tends to generate
con-siderable small-scale turbulence and
eddies. In 19th-century Britain, the most
sought-after houses were usually on the western and south-western sides of cities, to be
up-wind of industrial smoke and pollution
(Mann's model, page 389).

Location	Visibility less than 40 m (very dense fog)	Visibility less than 1000 m (less dense fog)
Kingsway (central London)	19	940
Kew (middle suburbs)	79	633
London Airport (outer suburbs)	46	562
South-east England (mean of 7 stations)	20	494

Relative humidity

Relative humidity is lower in urban areas where the warmer air can hold more moisture and where the lack of vegetation and water surface limits evapotranspiration.

Cloud

Urban areas appear to receive thicker and up to 10 per cent more frequent cloud cover than rural areas. This may result from convection currents generated by the higher temperatures and the presence of a larger number of condensation nuclei.

Precipitation

The mean annual precipitation total and the number of days with less than 5 mm of rainfall are both about 10 per cent greater in major urban areas. Reasons for this are the same as for cloud formation. Strong thermals increase the likelihood of thunder by 25 per cent and the occurrence of hail by up to 400 per cent. The higher urban temperatures may turn the snow of rural areas into sleet and limit, by up to 15 per cent, the number of days with snow lying on the ground. On the other hand, the frequency, length and intensity of fog, especially under anticyclonic conditions, is much greater — there may up to 100 per cent more in winter and 25 per cent more in summer, caused by the concentration of condensation nuclei (Figure 9.59).

Atmospheric composition

There may be three to seven times more dust particles over a city than in rural areas. Large quantities of gaseous and solid impurities are emitted into urban skies by the burning of fossil fuels, by industrial processes and from car exhausts. Urban areas may have up to 200 times more sulphur dioxide and ten times more nitrogen oxide (the major components of acid rain) than rural areas, as well as ten times more hydrocarbons and twice as much carbon dioxide. These pollutants tend to increase cloud cover and precipitation, cause smog (page 205), give higher temperatures and reduce sunlight.

Forest and lake microclimates

Different land surfaces produce distinctive local climates. Figure 9.60 summarises and compares some of the characteristics of microclimates found in forests and around lakes. As with urban climates, research and further information are still needed to confirm some of the statements.

Figure 9.60

Microclimates of forests and water surfaces

Microclimate feature	Forest (coniferous and deciduous)	Water surface (lake, river)
Incoming radiation and albedo	Much incoming radiation is absorbed and trapped. Albedo for coniferous forest is 15%; deciduous 25% in summer and 35% in winter; and desert scrub 40%.	Less insolation absorbed and trapped. Albedo may be over 60% — i.e. higher than over seas/oceans. Higher on calm days.
Temperature	Small diurnal range due to blanket effect of canopy. Forest floor is protected from direct sunlight. Some heat lost by evapotranspiration.	Small diurnal range because water has a higher specific heat capacity. Cooler summers and milder winters. Lakesides have a longer growing season.
Relative humidity	Higher during daytime and in summer, especially in deciduous forest. Amount of evapotranspiration depends on length of day, leaf surface area, wind speed, etc.	Very high, especially in summer when evaporation rates are also high.
Precipitation	Heavy rain can be caused by high evapotranspiration rates, e.g. in tropical rain forests. On average, 30–35% of rain is intercepted: more in deciduous woodland in summer than in winter.	Air is humid. If forced to rise, air can be unstable and produce cloud and rain. Amounts may not be great due to fewer condensation nuclei. Fogs form in calm weather.
Wind speed and direction	Trees reduce wind speeds, especially at ground level. (They are often planted as windbreaks.) Trees can produce eddies.	Wind may be strong due to reduced friction. Large lakes (e.g. L. Victoria) can create land and sea breezes (page 223).

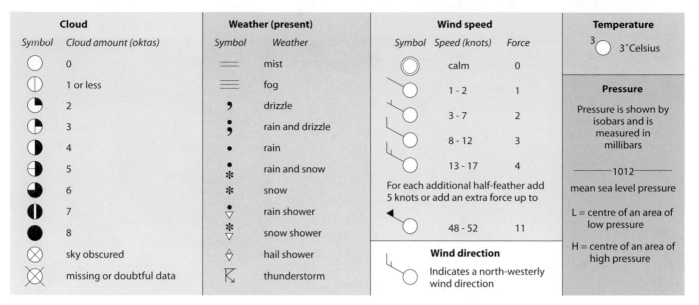

Cloud		Weather (present)		Wind speed			Temperature
Symbol	**Cloud amount (oktas)**	**Symbol**	**Weather**	**Symbol**	**Speed (knots)**	**Force**	

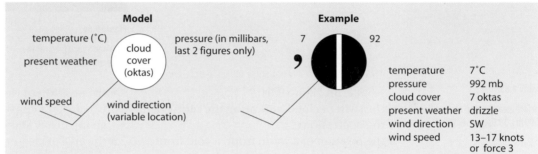

Figure 9.61

Weather symbols for cloud, precipitation, wind speed, temperature, pressure and wind direction

Figure 9.62

A weather station model and an example of an actual weather situation

Weather maps and forecasting in Britain

A weather map or **synoptic chart** shows the weather for a particular area at one specific time. It is the result of the collection and collation of a considerable amount of data at numerous weather stations, i.e. from a number of sample points (Framework 4, page 144). These data are then refined, usually as quickly as possible and now using computers, and are plotted using internationally accepted weather symbols. A selection of these symbols is shown in Figure 9.61. Weather maps are produced for different purposes and at various scales.

1 The daily weather map, as seen on television or in a national newspaper, aims to give a clear, but highly simplified, impression of the weather.

2 At a higher level, a synoptic map shows selected meteorological characteristics for specific **weather stations**. The **station model** in Figure 9.62 shows six elements: temperature, pressure, cloud cover, present weather (e.g. type of precipitation), wind direction and wind speed.

3 At the highest level, the Meteorological Office produces maps showing finite detail, e.g. amounts of various types of cloud at low, medium and high levels, dew point temperatures, barometric tendency (i.e. trends of pressure change), etc. The role of the weather forecaster is to try to determine the speed and direction of movement of various air masses and any associated fronts, and to try to predict the type of weather these movements will bring. Forecasters now make considerable use of **satellite images** (Figure 9.64). Satellite images are photos, taken by weather satellites as they continually orbit the earth. These photos, which are relayed back to earth, are invaluable in the prediction of short-term weather trends. Although forecasting is increasingly assisted by information from satellites, radar and computers, which show upper air as well as surface air conditions in a 3-dimensional model, the complexity and unpredictability of the atmosphere can still catch the forecaster by surprise (Places 23, page 217). Part of this problem is related to the fact that meteorological information is a sample (Framework 4, page 144) rather than a total picture of the atmosphere and so there is always a risk of the anomaly becoming the reality.

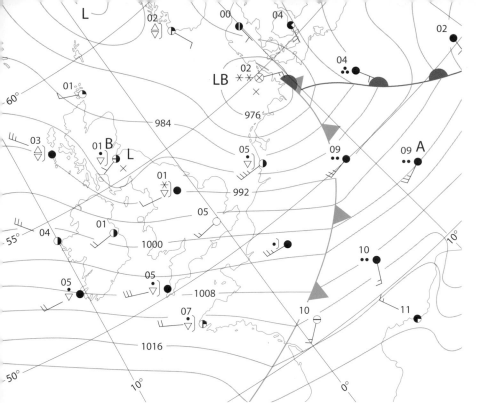

Figure 9.63

Surface weather map at 0000 hrs, 19 December 1986

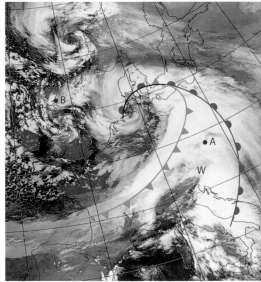

Figure 9.64

Infra-red image taken at the same time as Figure 9.63

Notice, on Figure 9.64:
- the swirl of cloud indicating the centre of the depression (L);
- the position of a warm front, a cold front and an occlusion (these have been super-imposed onto the infra-red image);

- the position of the warm sector (*W*) and two areas of convective cumulus clouds giving heavy showers (*C1 and C2*); and
- the area of clearer skies associated with the higher pressure over Spain (S).

Figure 9.65

Weather map for 0600 hrs, 25 February 1969

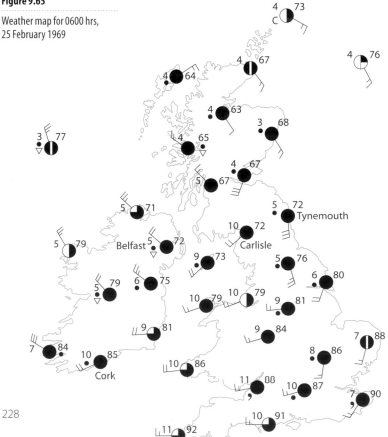

Q

1 With reference to Figures 9.63 and 9.64
 a How do you account for the large amount of cloud over Denmark, central Germany and central France?
 b Describe and suggest reasons for the differences in weather between central Germany (Station A) and Edinburgh (Station B)?

2 a On a copy of the map in Figure 9.65, use the information given to insert the probable position of a warm front, a cold front and an occlusion. You should consider differences in temperature, wind direction, wind speed and present weather. You may find Figures 9.43 and 9.44 useful.
 b Plot the likely pattern of isobars at 4 mb intervals onto the map shown in **a**. (If you work backwards from 1012 mb, the first isobar you should plot will be for 992 mb).

3 With reference to Figure 9.65
 a Describe and account for the differences in the weather between Tynemouth, Carlisle and Belfast.
 b Assuming that the low pressure system in Figure 9.65 is moving north-east, draw two station models to show the weather likely to be experienced by Tynemouth and Carlisle at 1200 hrs on 25 February 1969.

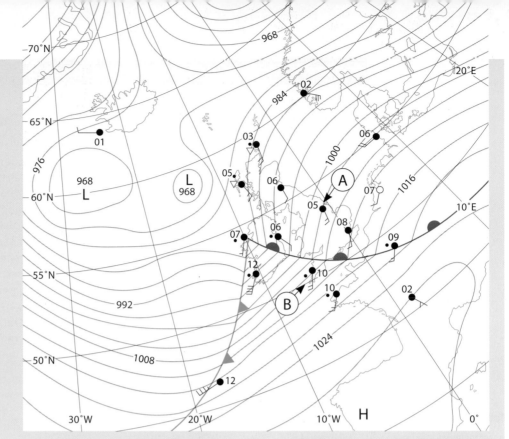

Figure 9.66

Weather map for 1200 hrs, 2 January 1984

4 Figures 9.66 and 9.67 are weather maps of the North Atlantic and western Europe for the period 2nd and 3rd January 1984.

   a Describe the weather being experienced at location A (Doncaster) on 2nd January, as shown on Figure 9.66.

   b Using Figure 9.66 describe and suggest reasons for the differences in weather between location A (Doncaster) and location B (Plymouth) on 2nd January.

   c By 3rd January (Figure 9.67) location B (Plymouth) has experienced major changes in weather. Describe and suggest reasons for the changes that are evident between 2nd and 3rd January at Plymouth.

   d Suggest why variations in weather may be experienced across the British Isles in the conditions prevailing on 3rd January (Figure 9.67).

Figure 9.67

Weather map for 1200 hrs, 3 January 1984

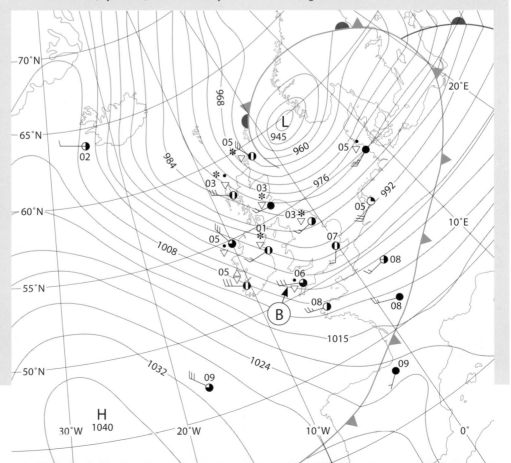

Throughout this chapter on weather and climate, mean climatic figures have been quoted. To build up these pictures of global, regional and local climate patterns, statistics have been obtained by averaging readings, usually for temperature and precipitation, over a 35-year timescale. However, these averages themselves are often not as significant as the **range** or the degree to which they vary from, or are dispersed about, the mean.

For example, two tropical weather stations may have equal annual rainfall totals when measured over 35 years. Station A may lie on the equator and experience reliable rainfall with little variation from one year to the next. Station B may experience a monsoon climate where in some years the rains may fail entirely while in others they cause flooding.

The measure of dispersion from the mean can be obtained by using any one of three statistical techniques:

- the range;
- the interquartile range; or
- the standard deviation.

These techniques are included here because meteorological data both require and benefit from their use, but they may be applied to most branches of geography where there is a danger that the mean, taken alone, may be misleading (the problems of overgeneralisation are discussed in Framework 8, page 321). Again, it must be stressed that use of a quantitative technique does not guarantee objective interpretation of data: great care must be taken to ensure that an appropriate method of manipulating the data is chosen.

It has already been seen how it is possible, given a data set, to calculate the mean and the median (Framework 3, page 99). However, neither statistic gives any idea of the spread, or range, of that data. As the example of the two tropical weather stations shows, mean values on their own give only part of the full picture. The spread of the data around the mean should also be considered.

Range

This very simple method involves calculating the difference between the highest and lowest value of the sample population, e.g. the annual range in temperature for London is 14°C (July 18°C, January 4°C). The range emphasises the extreme values and ignores the distribution of the remainder.

Interquartile range

The interquartile range consists of the middle 50 per cent of the values in a distribution; 25 per cent each side of the median (middle value). This calculation is useful because it shows how closely the values are grouped around the median (Figure 9.68). It is easy to calculate; it is unaffected by extreme values; and it is a useful way of comparing sets of similar data.

The example in Figure 9.68 gives temperatures for 19 weather stations in the British Isles at 0600 on 14 January 1979. These temperatures have been ranked in the table.

Figure 9.68

Finding the interquartile range

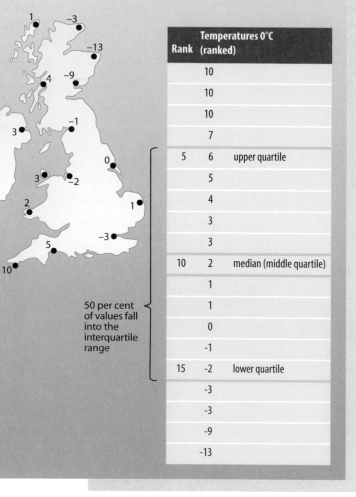

Rank	Temperatures 0°C (ranked)	
	10	
	10	
	10	
	7	
5	6	upper quartile
	5	
	4	
	3	
	3	
10	2	median (middle quartile)
	1	
	1	
	0	
	-1	
15	-2	lower quartile
	-3	
	-3	
	-9	
	-13	

50 per cent of values fall into the interquartile range

The **upper quartile** (UC) is obtained by using the formula:

$$UQ = \left(\frac{n+1}{4}\right) \quad \text{i.e.} \quad \left(\frac{19+1}{4}\right) = 5$$

This means that the UQ is the fifth figure from the top of the ranking order, i.e. 6°C. The **lower quartile** (LQ) is found by using a slightly different formula:

$$LQ = \left(\frac{n+1}{4}\right) \times 3 \quad \text{i.e.} \quad \left(\frac{19+1}{4}\right) \times 3 = 15$$

This shows the LQ to be the 15th figure in the ranking order, i.e. -2°C. You will notice that the middle quartile is the same as the median. The **interquartile range** is the difference between the upper and lower quartiles, i.e. 6°C - -2°C = 8°C.

Another measure of dispersion, the **quartile deviation**, is obtained by dividing the interquartile range by two, i.e. 8°C ÷ 2 = 4°C

The smaller the interquartile range, or quartile deviation, the greater the grouping around the median and the smaller the dispersion or spread.

Standard deviation

This is the most commonly used method of measuring dispersion and although it may involve lengthy calculations it can be used with the arithmetic mean and it removes extreme values. The formula for the standard deviation is:

$$\sigma = \sqrt{\frac{\Sigma(x - \bar{x})^2}{n}}$$

where: σ = standard deviation;

  x = each value in the data set;

  $\bar{x}$ = mean of all values in the data set; and

  n = number of values in the data set.

Let us suppose that the minimum temperatures for 10 weather stations in Britain on a winter's day were, in °C, 5, 8, 3, 2, 7, 9, 8, 2, 2 and 4. The standard deviation of this data set is worked out in Figure 9.69, proceeding as follows:

1 Find the mean ($\bar{x}$).

2 Subtract the mean from each value in the set:
$x - \bar{x}$.

3 Calculate the square of each value in **2**, to remove any minus signs: $(x - \bar{x})^2$.

4 Add together all the values obtained in **3**:
$\Sigma(x - \bar{x})^2$.

5 Divide the sum of the values in **4** by n:
$$\frac{\Sigma(x - \bar{x})^2}{n}$$

6 Take the square root of **5** to obtain the standard deviation:
$$\sqrt{\frac{\Sigma(x - \bar{x})^2}{n}}$$

The resulting standard deviation of $\sigma = 2.65$ is a low value indicating that the data are closely grouped around the mean.

Figure 9.69

Finding the standard deviation

Minimum temperatures for 10 weather stations in Britain on a winter's day

The mean of 5, 8, 3, 2, 7, 9, 8, 2, 2, 4 is: $\bar{x} = \dfrac{50}{10} = 5$

Weather station	Temperature at each station (x)	x - x̄	(x × x̄)²
1	5	5 - 5 = 0	0
2	8	8 - 5 = 3	9
3	3	3 - 5 = -2	4
4	2	2 - 5 = -3	9
5	7	7 - 5 = 2	4
6	9	9 - 5 = 4	16
7	8	8 - 5 = 3	9
8	2	2 - 5 = -3	9
9	2	2 - 5 = -3	9
10	4	4 - 5 = -1	1

$\Sigma(x - \bar{x})^2 = 70$

$$\sigma = \sqrt{\frac{70}{10}}$$

$$\sigma = \sqrt{7} \quad \therefore \text{ standard deviation } = 2.65$$

	Annual precipitation totals (mm)	
Year	Salina Cruz	Stornoway
1	1665	877
2	699	1082
3	550	1203
4	1188	963
5	1040	1241
6	886	1194
7	1091	1072
8	578	900
9	762	1146
10	701	1094
11	798	1098
12	1040	1318
13	911	791
14	2356	1035
15	1681	1151

Q Figure 9.70 gives the annual precipitation totals for Stornoway (north-west Scotland) and Salina Cruz (Mexico) over a period of 15 years.

1 For each station calculate:
  a the precipitation range;
  b the upper and lower quartiles of the precipitation data;
  c the interquartile range of the precipitation data;
  d the quartile deviation;
  e the mean annual precipitation total;
  f the standard deviation of the annual precipitation over the 15-year period.

2 Compare carefully the results for the two stations, calculated in a–f above. How do you account for these differences?

Figure 9.70

Weather and climate

Climatic change

Climates have changed and still are constantly changing at all scales, from local to global, and over varying time-spans. There have been, however, surges of change over time which meteorologists and earth scientists are continually trying to clarify and explain.

Evidence of past climatic changes

- **Rocks** are found today which were formed under climatic conditions and in environments which no longer exist (Figure 1.1). In Britain, for example, coal was formed under hot, wet tropical conditions; sandstones were laid down during arid times; various limestones accumulated on the floors of warm seas; and glacial deposits were left behind by retreating ice sheets.
- **Fossil landscapes** exist, produced by certain geomorphological processes which no longer operate. Examples include glacially eroded highland in north and west Britain (Chapter 4), granite tors on Dartmoor (page 185) and wadis in deserts (page 172).
- Evidence exists of **changes in sea-level** (Arran, page 150) and changes in lake levels (Sahara, Figure 7.26).
- **Pollen analysis** shows which plants were dominant at a given time. Each plant species has a distinctively shaped pollen grain. If these grains land in an oxygen-free environment, such as a peat bog, they resist decay. Although pollen can be transported considerable distances by the wind and by wildlife, it is assumed that grains trapped in peat form a representative sample of the vegetation which was growing in the surrounding area at a given time; also, that this vegetation was a response to the climatic conditions prevailing at that time. Vertical sections made through peat show changes in pollen (i.e. vegetation), and these changes can be used as evidence of climatic change (the vegetation–climatic timescale in Figures 11.18 and 11.19).
- **Dendrochronology** or tree-ring dating is the technique of obtaining a core from a tree-trunk and using it to determine the age of the tree. Tree growth is rapid in spring, slower by the autumn and, in temperate latitudes, stops in winter. Each year's growth is shown by a single ring.

However, when the year is warm and wet, the ring will be larger because the tree grows more quickly than when the year is cold and dry. Tree-rings therefore reflect climatic changes. Recent work in Europe has shown that tree growth is greatest under intense cyclonic activity and is more a response to moisture than to temperature. Tree-ring timescales are being established by using the remains of oak trees, some nearly 10 000 years old, found in river terraces in south-central Europe. Bristlecone pines, still alive after 5000 years, give a very accurate measure in California.

- **Chemical methods** include the study of oxygen and carbon isotopes. An isotope is one of two or more forms of an element which differ from each other in atomic weight (i.e. they have the same number of protons in the nucleus, but a different number of neutrons). For example, two isotopes in oxygen are O-16 and O-18. The O-16 isotope, which is slightly lighter, vaporises more readily; whereas O-18, being heavier, condenses more easily. During warm, dry periods, the evaporation of O-16 will leave water enriched with O-18 which, if it freezes into polar ice, will be preserved as a later record (Places 11, page 92). Colder, wetter periods will be indicated by ice with a higher level of O-16. The most accurate form of dating is based on C-14, a radioactive isotope of carbon. Carbon is taken in by plants during the carbon cycle (Figure 11.25). Carbon-14 decays radioactively at a known rate and can be compared with C-12, which does not decay. Using C-12 and C-14 from a dead plant, scientists can determine the date of death to a standard error of ± 5 per cent. This method can accurately date organic matter up to 50 000 years old (page 273).
- **Historical records** of climatic change include:

- cave paintings of elephants in the central Sahara (page 175);
- vines growing successfully in southern England between AD 1000 and 1300;
- graves for human burial in Greenland which were dug to a depth of 2 m in the 13th century, but only 1 m in the 14th and could not be dug at all in the 15th due to the extension of permafrost;

- fairs held on the frozen River Thames in Tudor times;
- the measurement of recent advances and retreats of Alpine glaciers.

Causes of climatic change

Several theories, covering varying timescales, have been advanced to try to explain climatic change. At present there is no unanimous consensus of opinion as to its causes: climatic change may be explained by one of these theories, several in combination, or by a theory yet to be propounded. The suggested theories include:

1 **Variations in solar energy** Although it was initially believed that solar energy output did not vary over time (hence the term 'solar constant' in Figure 9.3), increasing evidence suggests that sunspot activity, which occurs in cycles, may significantly affect our climate — times of high annual temperatures on earth appear to correspond to periods of maximum sunspot activity.

2 **Astronomical relationships between the sun and the earth** There is increasing evidence supporting Milankovitch's cycles of change in the earth's orbit, tilt and wobble (Figure 4.4). This would account for changes in the amounts of solar radiation reaching the earth's surface.

3 **Changes in oceanic circulation** Changes in oceanic circulation affect the exchange of heat between the oceans and the atmosphere. This can have both long-term effects on world climate (where currents at the onset of the Quaternary ice age flowed in opposite directions to those at the end of the ice age) and short-term effects (El Nino, Case Study 9).

4 **Meteorites** A major extinction event, which included the dinosaurs, took place about 65 million years ago. This event was believed to have been caused by one or more meteors colliding with the earth. This seems to have caused a reduction in incoming radiation, a depletion of the ozone layer, a lowering of global temperatures and an increase in acid rain.

5 **Volcanic activity** It has been accepted for some time that volcanic activity has influenced climate in the past and continues to do so. World temperatures are lowered after any large eruption (Krakatoa, Places 27, page 267; and Pinatubo, Case Study 1, page 31), and after a period of several eruptions, when the increase in dust particles absorbs and scatters more of the incoming radiation (Figure 9.4). Evidence suggests that these major eruptions may temporarily offset the greenhouse effect. Precipitation also increases due to the greater number of hygroscopic nuclei (dust particles) in the atmosphere (page 197).

6 **Plate tectonics** Plate movements have led to redistributions of land masses and to long-term effects on climate. These effects may result from a land mass 'drifting' into different latitudes (British Isles, page 20); or from the sea bed being pushed upwards to form fold mountains (page 18). The presence of fold mountains can lead to a colder climate (a suggested cause of the Quaternary ice age, page 91) and can act as a barrier to atmospheric circulation — the Asian monsoon was established by the creation of the Tibetan Plateau (page 222).

7 **Composition of the atmosphere** Gases in the atmosphere can be increased and altered following volcanic eruptions. At present, increasing concern is being expressed at the build up of CO_2 gas in the atmosphere and the resultant greenhouse effect (see Case Study 9), together with the use of aerosols (Places 21, page 191) which are blamed for the depletion of ozone in the upper atmosphere.

Climatic change in Britain

Britain's climate has undergone changes in the longest term (page 20 and Figure 1.1); during and since the onset of the Quaternary (Figures 4.2 and 11.18); and in the more recent short term. Following the 'little ice age' which lasted from about AD 1540 to 1700, temperatures generally increased to reach a peak in about 1940. After that time, there was a tendency for summers to become cooler and wetter, springs to be later, autumns milder and winters more unpredictable. However, since the onset of the 1980s there appears to have been a considerably warming of our climate, with seven of the ten warmest years this century occurring during that decade — a fact which lends support to the theory of global warming (Case Study 9).

Case Study 9

Short-term and long-term climatic changes

Short-term change: El Nino

The oceans, as we have seen, have a considerable heat storage capacity which makes them a major influence on world climates. If ocean temperatures change, this will have a considerable effect upon weather patterns in adjacent land masses. Interactions between the ocean and the atmosphere have become, recently, a major area of scientific study.

The most important and interesting example of the ocean–atmosphere interrelationship is provided by the El Nino event which occurs, periodically, in the Pacific Ocean. Under normal circumstances, the surface temperature of the western Pacific Ocean exceeds 28°C. This creates a low pressure system over south-east Asia which draws in warm, moist air and which is responsible for the heavy rainfall in that part of the world. In contrast, the eastern parts of the Pacific Ocean are much cooler due to an upwelling of cold, nutrient-rich water. They are also drier (the Atacama desert in Peru) as prevailing winds blow from the land. Figure 9.71 shows the resultant Walker Circulation Cell, named after the person who put forward the model, and its associated weather.

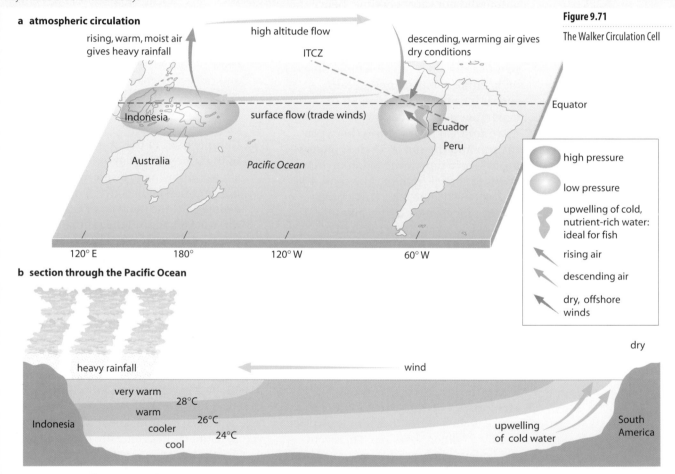

a atmospheric circulation

rising, warm, moist air gives heavy rainfall

high altitude flow

ITCZ

descending, warming air gives dry conditions

Indonesia

surface flow (trade winds)

Ecuador

Peru

Australia

Pacific Ocean

Equator

120° E   180°   120° W   60° W

high pressure

low pressure

upwelling of cold, nutrient-rich water: ideal for fish

rising air

descending air

dry, offshore winds

b section through the Pacific Ocean

heavy rainfall

wind

dry

very warm   28°C

warm   26°C

cooler   24°C

cool

Indonesia

upwelling of cold water

South America

Figure 9.71

The Walker Circulation Cell

An El Nino event, usually referred to as an El Nino Southern Oscillation (ENSO), occurs periodically—perhaps every 4–7 years. It is called 'El Nino" which means 'little child' because, in those years when it does occur, it appears just after Christmas. In contrast to normal conditions, surface water temperatures in excess of 28°C extend much further eastwards, allowing the ITCZ to migrate southwards and causing the trades to weaken in strength, and even to be reversed (Figure 9.72). This reversal affects the surface sea temperatures of the eastern Pacific and the atmosphere and air masses which pass over it. The heaviest rainfall is now over the eastern Pacific, with south-east Asia much drier and, during extreme El Nino events, even exposed to drought.

The upwelling of cold water off South America is reduced, allowing water temperatures to rise by 2–6°C. However, as the warmer water lacks oxygen, nutrients and plankton, fish life is drastically reduced. Evidence suggests that, with less marine life to take it up, concentrations of CO_2 in the atmosphere increase, contributing to the greenhouse effect.

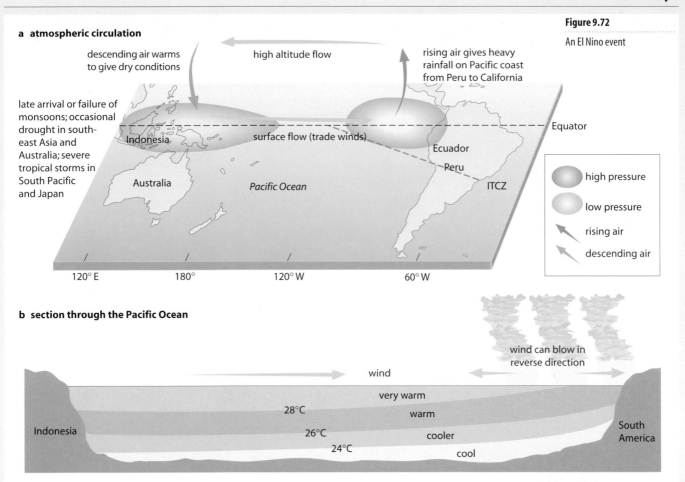

Figure 9.72

An El Nino event

a atmospheric circulation

descending air warms to give dry conditions

high altitude flow

rising air gives heavy rainfall on Pacific coast from Peru to California

late arrival or failure of monsoons; occasional drought in south-east Asia and Australia; severe tropical storms in South Pacific and Japan

Indonesia

surface flow (trade winds)

Equator

Ecuador

Peru

Australia

Pacific Ocean

ITCZ

120° E   180°   120° W   60° W

	high pressure
	low pressure
	rising air
	descending air

b section through the Pacific Ocean

wind can blow in reverse direction

wind

very warm

28°C

warm

Indonesia

26°C

cooler

South America

24°C

cool

There is increasing evidence that the ENSO has a significant effect on climates in places far beyond the Pacific margins. For example,

- the warmest El Nino event was in 1982–83 when sea temperatures rose by nearly 6°C. This was followed, in 1983-84, by severe drought in the Sahel and southern Africa;
- El Nino events correspond to extremely cold conditions in northern North America, and to stormy conditions, floods and high waves, in California;
- Britain and north-west Europe experience exceptionally wet and mild winters, possibly due to the ENSO affecting the Rossby waves and altering the course of the jet stream.

Although the cause of ENSO has to be proved, there is a fascinating link between the event and major volcanic eruptions within the tropical Pacific Ocean (more precisely, between 10°S and 25°N). The links, or perhaps just coincidences, are shown in Figure 9.73.

At present, global observations are being made, questions asked and models tested to see if indeed there are relationships between ENSO and atmospheric circulation on the one hand, and between volcanic eruptions and El Nino events on the other (Figure 9.74). Some answers should appear in the next decade. You, as a geographer, should 'Watch this space!'

Eruption	Year	El Nino event
Agung (Indonesia)	1963	1965
Mayon (Philippines)	1968	1968
El Chinon (Mexico)	1982	1982–83
Nevado del Ruiz (Colombia)	1985	1986
Pinatubo (Philippines)	1991	1992–93

Figure 9.73

Volcanic eruptions and El Nino events

Figure 9.74

"In all these connections or coincidences there are strings of 'might', 'possibly' and 'maybe'. The interactions of the solid Earth, its atmosphere, its oceans and its biosphere, and the incoming and outgoing radiation are poorly understood. The exchanges among them are not only complex but are generally investigated by scientists from many differing fields who may not understand the problems of another discipline. It is like blind men describing the elephant, with the additional problem that each blind man speaks a different language."

Decker and Decker, *Mountains of Fire*, 1991

Long-term change: The Greenhouse Effect

The earth is warmed during the day by incoming, short-wave radiation (insolation) from the sun and cooled at night by outgoing, longer-wave, infrared radiation. As, over a lengthy period of time, the earth is neither warming up nor cooling down, there must be a balance between incoming and outgoing radiation (page 191). While incoming radiation is able to pass through the atmosphere (which is 99 per cent nitrogen and oxygen, Figure 9.2), some of the outgoing radiation is trapped by a blanket of trace gases. Because they trap heat as in a greenhouse, these are referred to as greenhouse gases. Without these natural greenhouse gases, the earth's average temperature would be 33°C colder than it is today—far too cold for life in any form. (During the last ice age, temperatures were only 4°C cooler.) Water vapour provides the majority of the natural greenhouse effect, with lesser contributions from carbon dioxide, methane, nitrous oxide and ozone.

During the last 150 years there has been, with the exception of water vapour which remains a constant in the system, a rise in greenhouse gas concentrations. This has been due largely to the increase in world population and a corresponding growth in human activity, especially agricultural and industrial (Figure 9.75).

By adding these gases to the atmosphere, we are increasing its ability to trap heat (Figure 9.76). Most scientists now accept that the greenhouse effect is causing global warming. World temperatures, which have risen by 0.5°C since the middle of the last century, increased rapidly in the 1980s when seven of this century's warmest years were recorded (Figure 9.77). Predictions suggest that there could be a further rise of between 1.5°C and 4.5°C by the year 2100.

Figure 9.75

The major greenhouse gases

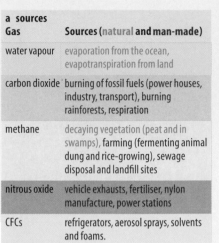

a sources

Gas	Sources (natural and man-made)
water vapour	evaporation from the ocean, evapotranspiration from land
carbon dioxide	burning of fossil fuels (power houses, industry, transport), burning rainforests, respiration
methane	decaying vegetation (peat and in swamps), farming (fermenting animal dung and rice-growing), sewage disposal and landfill sites
nitrous oxide	vehicle exhausts, fertiliser, nylon manufacture, power stations
CFCs	refrigerators, aerosol sprays, solvents and foams.

b rate of increase

carbon dioxide 72%

CFCs 13%

methane 10%

nitrous oxide 5%

Figure 9.76

The radiation balance and the greenhouse effect

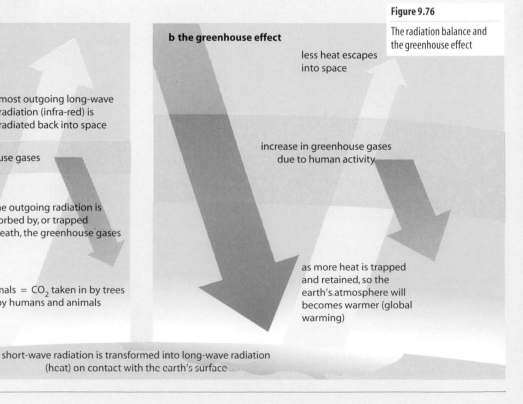

a the radiation balance

incoming short-wave radiation (ultraviolet) passes directly through the natural greenhouse gases

most outgoing long-wave radiation (infra-red) is radiated back into space

natural greenhouse gases

some outgoing radiation is absorbed by, or trapped beneath, the greenhouse gases

previously a balance:
CO_2 given off by humans and animals = CO_2 taken in by trees
O_2 given out by trees = O_2 used by humans and animals

b the greenhouse effect

less heat escapes into space

increase in greenhouse gases due to human activity

as more heat is trapped and retained, so the earth's atmosphere will becomes warmer (global warming)

short-wave radiation is transformed into long-wave radiation (heat) on contact with the earth's surface

While most scientists now agree that global warming is taking place, there is considerable uncertainty, and disagreement, as to its effect on the earth's climate. As the atmosphere gets warmer, then so too will water in the oceans. As sea water warms, it will expand, causing a eustatic rise in its level by a predicted 0.25–1.0 metre by the year 2100. This alone could partly submerge low-lying coral islands, such as the Maldives, and increase the flood risk in countries with large river deltas, such as Egypt and Bangladesh (page 153). More debatable is the effect of a warmer earth on the polar ice-caps. Initially, it was assumed that they would melt causing an appreciable rise in sea-level. Recent evidence (June 1994) suggests that while temperatures in the Antarctic have risen appreciably, there have been no observed signs of the ice sheets melting. It is believed that while warming will tend to melt glaciers and the edges of ice sheets, the increased snowfall expected in a wetter climate (caused by

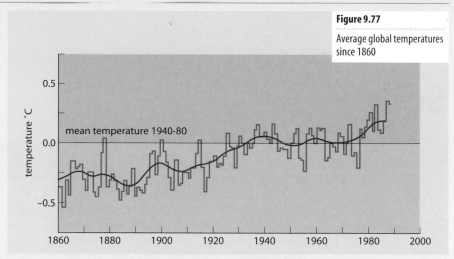

Figure 9.77

Average global temperatures since 1860

higher rates of evaporation) would increase their volume. If ice sheets did melt, which is by no means certain, then polar areas would experience the greatest warming as a reduction in snow cover would alter their albedo. The distribution of precipitation across the world is also likely to change. Computer predictions suggest that while some parts of the world will become wetter and agriculturally more productive, others, especially those around 40°N, will become drier with a less reliable rainfall (Figure 9.78). As these latitudes contain many important cereal-growing regions, there could be an increased risk of global food shortage.

Figure 9.78

Weather forecast for the year 2100: assuming, by then, CO_2 content in the atmosphere has doubled

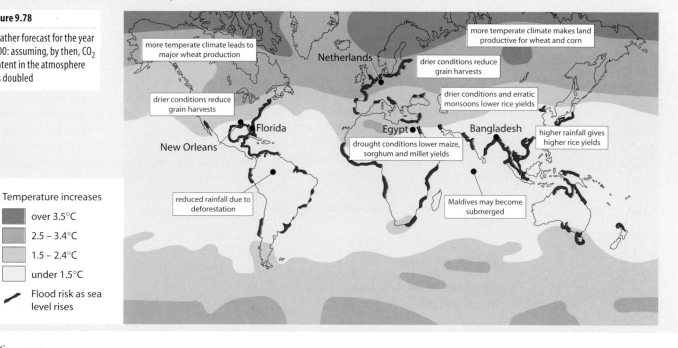

Temperature increases

- over 3.5°C
- 2.5 – 3.4°C
- 1.5 – 2.4°C
- under 1.5°C
- Flood risk as sea level rises

References

Barry, R. G. and Chorley, R. J. (1989) *Atmosphere, Weather and Climate.* Methuen.

Chandler, T. J. (1981) *Modern Meteorology and Climatology.* Thomas Nelson.

Decker, R. W. and Decker, B. B. (1991) *Mountains of Fire.* Cambridge University Press.

Goudie, A. (1993) *The Nature of the Environment.* Blackwell.

Gribbin, J. (1978) *The Climatic Threat.* Collins Fontana.

Money, D. C. (1985) *Climate, Soils and Vegetation.* University Tutorial Press.

Musk, L. (1988) *Weather Systems.* Cambridge University Press.

O'Hare, G. and Sweeney, J. (1986) *The Atmospheric System.* Oliver & Boyd.

Pickering, K. and Owen, L. (1994) *Global Environmental Issues.* Routledge.

Planet Earth: Storms (1981) Time–Life Books.

Planet Earth: The Atmosphere (1983) Time–Life Books.

Prosser, R. (1992) *Natural Systems and Human Responses.* Thomas Nelson.

Wilson, J. (1984) *Statistics in Geography for 'A' level Students.* Schofield & Sims.

Soils

Soil forms the thin surface layer of the earth's crust. It provides the foundation for plant and, consequently, animal life on land. The most widely accepted scientific definition is that by J. Joffe (1949) who stated that:

> "the soil is a natural body of animal, mineral and organic constituents differentiated into horizons [Figure 10.5] of variable depth which differ from the material below in morphology, physical make up, chemical properties and composition, and biological characteristics."

A simpler definition is that soil results from the interrelationships between, and interaction of, several physical, chemical and biological processes all of which vary according to different natural environments.

The study of soil, its origins, characteristics and utilisation (**pedology**) is a science in itself.

Soil formation

The first stage in the formation of soil is the weathering of parent rock to give a layer of loose, broken material known as **regolith**. Regolith may also be derived from the deposition of alluvium, drift, loess and volcanic material. The second stage, the formation of **true soil** or **topsoil**, results from the addition of water, gases (air), living organisms (biota) and decayed organic matter (humus).

Pedologists have identified five main factors involved in soil formation. As all of these are closely interconnected and interdependent, their relationship may be summarised as follows:

soil = f(parent material + climate
    + topography + organisms + time)
where: f = function of

Parent material

When a soil develops from an underlying rock, its supply of minerals is largely dependent on that parent rock. The minerals are susceptible to different rates and processes of weathering — see the example of granite, Figure 10.1. Parent rock contributes to control of the depth, texture, drainage (permeability) and quality (nutrient content) of a soil and also influences its colour. In parts of Britain, parent rock may be the major factor in determining the soil type (limestone or granite).

Climate

Climate determines the type of soil at a global scale. The distribution of world soil types corresponds closely to patterns of climate and vegetation. Climate affects the rate of weathering of the parent rock, with the most rapid breakdown being in hot, humid environments.

Figure 10.1

The influence of a parent rock, granite, on soil formation

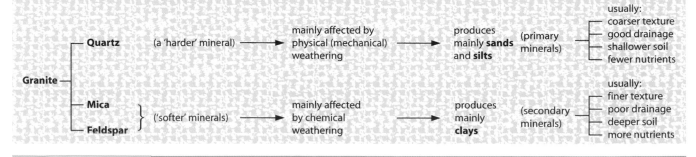

Precipitation affects the type of vegetation which grows in an area which in turn provides humus; for example, more humus is found in tropical rainforests than in the tundra. Rainfall totals and intensity are also important. Where rainfall is heavy, the downward movement of water through the soil transports mineral salts with it, a process known as **leaching**. Where rainfall is light or where evapotranspiration exceeds precipitation, water and mineral salts are drawn upwards towards the surface by the process of **capillary action**. Leaching tends to produce acidic soils and capillary action alkaline soils (page 248).

Temperatures determine the length of the growing season and affect the supply of humus. The speed of vegetation decay is fastest in hot, wet climates as temperatures also influence the activity and number of soil organisms and the rate of evaporation — i.e. whether leaching or capillary action is dominant.

Topography (relief)
As the height of the land increases, so too do amounts of precipitation, cloud cover and wind, while temperatures and the length of the growing season both decrease. Aspect is an important local factor (page 194) with south-facing slopes in the northern hemisphere being warmer and drier than those facing north. The angle of slope affects drainage and soil depth. The steeper the slope, the faster the rate of throughflow and surface runoff of water which may accelerate mass movement and increase the risk of soil erosion. Soils on steep slopes are likely to be thin, poorly developed and relatively dry. The more gentle the slope, the slower the rate of movement of water through the soil

and the greater the likelihood of water-logging and the formation of peat. There is little risk of soil erosion but the increased rate of weathering, due to the extra water, and the receipt of material moved downslope tend to produce deep soils. A **catena** is where soils are related to the topography of a hill-side and is a sequence of soil types down a slope. The catena (Figure 10.2) is described in more detail on page 256.

Organisms (biota)
Plants, bacteria, fungi and animals all interact in the **nutrient cycle** (page 278). Plants take up mineral nutrients from the soil and return them to it after they die. This re-cycling of plant nutrients (Figure 12.7) is achieved by the activity of micro-organisms, such as bacteria and fungi, which assist in the decomposition and decay of dead vegetation. At the same time, macro-organisms, which include worms and termites, mix and aerate the soil. Human activity is increasingly affecting soil development through the addition of fertiliser, the breaking up of horizons by ploughing, draining or irrigating land, and by unwittingly accelerating or deliberately controlling soil erosion.

Time
Soils usually take a long time to form, perhaps up to 400 years for 10 mm and, under extreme conditions, 1000 years for 1 mm. It can take 3000–12 000 years to produce a sufficient depth of mature soil for farming. Young soils tend to retain many characteristics of the parent material from which they are derived. With time, they acquire new characteristics resulting from the addition of organic matter and the activity of organisms. **Horizons** or layers develop as soils reach a state of equilibrium with their environment.

Figure 10.2

A catena: the relationship between soil and slope (not drawn to scale)

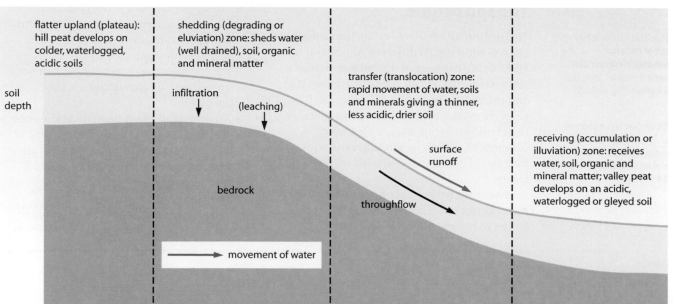

flatter upland (plateau): hill peat develops on colder, waterlogged, acidic soils

shedding (degrading or eluviation) zone: sheds water (well drained), soil, organic and mineral matter

transfer (translocation) zone: rapid movement of water, soils and minerals giving a thinner, less acidic, drier soil

soil depth

infiltration

(leaching)

surface runoff

receiving (accumulation or illuviation) zone: receives water, soil, organic and mineral matter; valley peat develops on an acidic, waterlogged or gleyed soil

bedrock

throughflow

movement of water

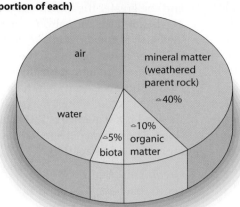

pore space containing **air** and/or **water = 45%**
(can be **45% water**, or **45% air**, but is more
usually a proportion of each)

mineral matter + organic matter
+ biota **= 55%**

air

mineral matter
(weathered
parent rock)

≃40%

water

≃5%
biota

≃10%
organic
matter

In northern Britain, upland soils must be less
than 10 000 years old, as that was the time of
the last glaciation when any existing soil
cover was removed by ice. The time taken for
a mature soil to develop depends primarily
upon parent material and climate. Soils
develop more rapidly on rocks which weather into sandy material than on those producing clays; and in hot, wet climates rather
than in colder and/or drier environments.

A mature soil consists of four components: mineral matter, organic matter
including biota (page 246), water and air.
The relative proportions of these components in a 'normal' soil, by volume, is given
in Figure 10.3.

Figure 10.3

Relative proportions, by
volume, of components in a
'normal' soil (*after* Courtney
and Trudgill)

Q

1 Describe briefly the relative importance of
the main factors affecting the formation of
soil as shown in Figure 10.4.

Parent material	Time		Human influences
permeability mineral content		**Soil**	

Climate

weathering
precipitation
temperature

Organisms (biota)

organic matter
nutrient cycle/recycling
mixing and aeration

Topography (relief)

altitude
aspect
slope angles

Figure 10.4

Factors affecting the formation
of soil

2 Although air and water together account
for some 45 per cent of the total volume of
soil, under what conditions may water
account for
a 45 per cent;
b zero?

3 Why is the proportion of mineral matter
much greater in soils in colder, drier areas
than in warmer, wetter areas?

4 Why do the number of biota and the rate of
organic matter production both increase in
soils as rainfall and temperatures increase?

5 How would you expect the proportions of
the four components in a peat soil to differ
from those given in Figure 10.3?

The soil profile

The **soil profile** is a vertical section through
the soil showing its different horizons (Figure
10.5). It is a product of the balance between
soil system inputs and outputs (Figure 10.6)
and the redistribution of, and chemical
changes in, the various soil constituents.
Different soil profiles are described in
Chapter 12, but an idealised profile is given
here to aid familiarisation with several new
terms.

The three major soil horizons, which may
be subdivided, are referred to by specific
letters to indicate their genetic origin:

- The upper layer, or **A horizon**, is where
  biological activity and humus content are
  at their maximum. It is also the zone
  which is most affected by the leaching of
  soluble materials and by the downward
  movement, or **eluviation**, or clay
  particles.
- Beneath this, the **B horizon** is the zone of
  accumulation, or **illuviation**, where clays
  and other materials removed from the *A*
  horizon are redeposited. The *A* and *B*
  horizons together make up the true soil.
- The **C horizon** consists mainly of
  recently weathered material (regolith)
  resting on the bedrock.

**Eluviation is the washing
out of material — i.e. the
removal of organic and
mineral matter from the
A horizon (Figure 10.5)**

**Illuviation is the process
of inwashing — i.e. the
redeposition of organic
and mineral matter in the
B horizon**

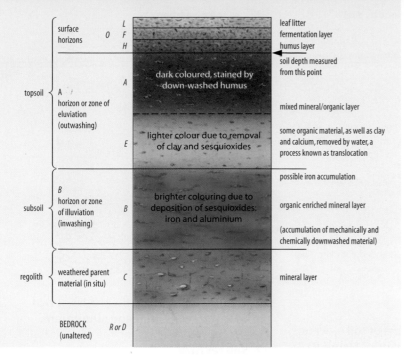

Figure 10.5

An idealised soil profile in Britain

Labels in figure:

surface horizons — O — L, F, H — leaf litter / fermentation layer / humus layer — soil depth measured from this point

topsoil — A — horizon or zone of eluviation (outwashing) — A — dark coloured, stained by down-washed humus — mixed mineral/organic layer

E — lighter colour due to removal of clay and sesquioxides — some organic material, as well as clay and calcium, removed by water, a process known as translocation

subsoil — B — horizon or zone of illuviation (inwashing) — B — brighter colouring due to deposition of sesquioxides: iron and aluminium — possible iron accumulation / organic enriched mineral layer / (accumulation of mechanically and chemically downwashed material)

regolith — weathered parent material (in situ) — C — mineral layer

BEDROCK (unaltered) — R or D

While this threefold division is useful and convenient, it is, as will be seen later, over-simplified. Several examples show this:

- Humus may be mixed throughout the depth of the soil, or it may form a distinct layer. Where humus is mixed with the soil to give a crumbly, black, nutrient-rich layer it is known as **mull**. Where humus is slow to decompose, as in cold, wet upland areas, it produces a fibrous, acidic and nutrient-deficient surface horizon known as **mor** (peat moorlands).
- The junctions of horizons may not always be clear.
- All horizons need not always be present.
- The depth of soil and of each horizon vary at different sites. Local conditions produce soils with characteristic horizons

differing from the basic *A*, *B*, *C* pattern: for example, a waterlogged soil, suffering from a shortage of oxygen, develops a gleyed (*G*) horizon (page 254).

The soil system

Figure 10.6 is a model showing the soil as an open system where materials and energy are gained and lost at its boundaries. The system comprises inputs, stores, outputs and recycling or feedback loops (Framework 1, page 39).

Inputs include:

- water from the atmosphere or through flow from higher up the slope;
- gases from the atmosphere and the respiration of soil animals;
- mineral nutrients from weathered parent material, which are needed as plant food;
- organic matter and nutrients from decaying plants and animals; and
- solar energy and heat.

Outputs include:

- water lost to the atmosphere through evapotranspiration;
- nutrients taken up by plants as food;
- nutrients lost through leaching and throughflow; and
- loss of soil particles through soil creep and erosion.

Recycling

Some of the nutrients taken up and stored by plants may later be returned to the soil via leaf litter during the following autumn or when plants die and decompose.

Figure 10.6

The 'open' soil system

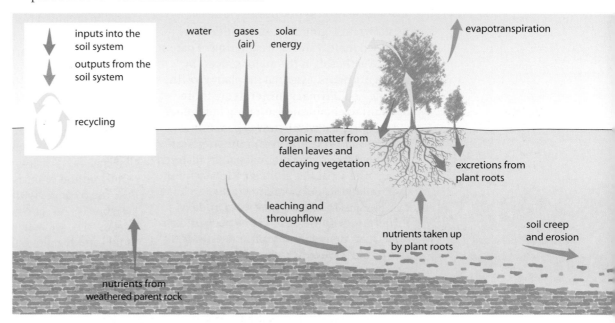

Labels in figure: inputs into the soil system / outputs from the soil system / recycling / water / gases (air) / solar energy / evapotranspiration / organic matter from fallen leaves and decaying vegetation / excretions from plant roots / leaching and throughflow / nutrients taken up by plant roots / soil creep and erosion / nutrients from weathered parent rock

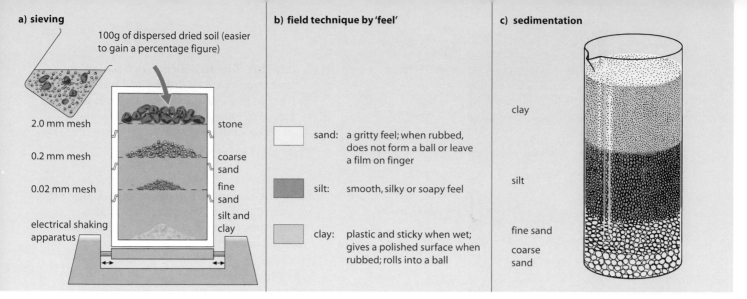

a) sieving

100g of dispersed dried soil (easier to gain a percentage figure)

2.0 mm mesh — stone

0.2 mm mesh — coarse sand

0.02 mm mesh — fine sand

silt and clay

electrical shaking apparatus

b) field technique by 'feel'

sand: a gritty feel; when rubbed, does not form a ball or leave a film on finger

silt: smooth, silky or soapy feel

clay: plastic and sticky when wet; gives a polished surface when rubbed; rolls into a ball

c) sedimentation

clay

silt

fine sand

coarse sand

Figure 10.7

Measuring soil texture (*after* Courtney and Trudgill)

Soil properties

The four major components of soil — water, air, mineral and organic matter (Figure 10.3) — are all closely interlinked. The resultant interrelationships produce a series of 'properties', ten of which are listed and described below.

1 mineral (inorganic) matter;
2 texture;
3 structure;
4 organic matter (humus);
5 moisture;
6 air;
7 organisms (biota);
8 nutrients;
9 acidity (pH value);
10 temperature.

It is necessary to understand the workings of these properties to appreciate how a particular soil can best be managed.

1 Mineral (inorganic) matter

As shown in Figure 10.1, soil minerals are obtained mainly by the weathering of parent rock. Weathering is the major process by which nutrients, essential for plant growth, are released. **Primary minerals** are those resistant to chemical weathering but vulnerable to physical weathering. These minerals, such as quartz (sands), retain the chemical characteristics of the original parent material after being separated from it by frost action, as in upland Britain, or temperature extremes, as found in deserts. **Secondary minerals** are those which have been broken down and altered by various processes of chemical weathering: oxidation, carbonation, hydrolysis and hydration (Chapter 2). Chemical weathering forms clays, of which several varieties exist, and **sesquioxides**

(Figure 10.5) which are the oxides of two primary minerals, iron and aluminium.

2 Soil texture

The term 'texture' refers to the degree of coarseness or fineness of the mineral matter in the soil. It is determined by the proportion of **sand**, **silt** and **clay** particles. Particles larger than sand are grouped together and described as stones. In the field, it is possible to decide whether a soil sample is mainly sand, silt or clay by its 'feel'. As shown in Figure 10.7b, a sandy soil feels gritty and lacks cohesion; a silty soil has a smoother, soaplike feel as well as having some cohesion; and a clay soil is sticky and plastic when wet and may, being very cohesive, be rolled into various shapes.

This method gives a quick guide to the texture, but it lacks the precision needed to determine the proportion of particles in a given soil with any accuracy. This precision may be obtained from either of two laboratory measurements, both of which are dependent upon particle size. The Soil Survey of Great Britain accepts the International scale, which gives the following diameter sizes:

coarse sand	between 2.0 and 0.2 mm
fine sand	between 0.2 and 0.02 mm
silt	between 0.02 and 0.002 mm
clay	less than 0.002 mm

One method of measuring texture involves the use of sieves with differing meshes (Figure 10.7a). The sample must be dry and needs to be well-shaken. A mesh of 0.2 mm, for example, allows fine sand, silt and clay particles to pass through it, while trapping the coarse sand. The weight of particles remaining in each sieve is expressed as a percentage of the total sample.

Figure 10.8

The texture of different soil types

In the second method, sedimentation (Figure 10.7c), a weighed sample is placed in a beaker of water, thoroughly shaken and then allowed to settle. According to

Stoke's Law, "the settling rate of a particle is proportional to the diameter of that particle". Consequently, the larger, coarser, sand grains settle quickly at the bottom of the beaker and the finer, clay particles settle last, closer to the surface (compare Figure 3.27). The Soil Survey tends to use both methods because sieving is less accurate in measuring the finer material and sediment-ation is less accurate with coarser particles.

The results of sieving and sedimentation are usually plotted either as a pie chart (Figure 10.8) or as a triangular graph (Figure 10.9). As the proportions of sand, silt and clay vary considerably, it is traditional to have 12 texture categories (Figure 10.9).

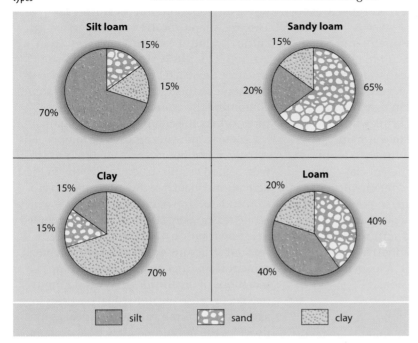

The importance of texture

As texture controls the size and spacing of soil pores, it directly affects the soil water content, water flow and extent of aeration. Clay soils tend to hold more water and are less well drained and aerated than sandy soils (page 245).

Texture also controls the availability and retention of nutrients within the soil. Nutrients stick to — i.e. are **adsorbed** onto — clay particles and are less easily leached by infiltration or throughflow than in sandy soils (page 247).

Plant roots can penetrate coarser soils more easily than finer soils, and 'lighter' sandy soils are easier to plough for arable farming than 'heavier' clays.

Texture greatly influences soil structure (see following section).

How does texture affect farming?

The following comments are generalised as it must be remembered that soils vary enormously.

Sandy soils, being well-drained and aerated, are easy to cultivate and permit crop roots (e.g. carrots) to penetrate. However, they are vulnerable to drought because they lack humus. They also need considerable amounts of fertiliser because nutrients and organic matter are often leached out and not replaced.

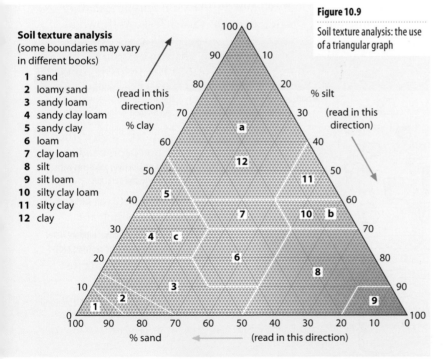

Soil texture analysis
(some boundaries may vary in different books)

1 sand
2 loamy sand
3 sandy loam
4 sandy clay loam
5 sandy clay
6 loam
7 clay loam
8 silt
9 silt loam
10 silty clay loam
11 silty clay
12 clay

Figure 10.9

Soil texture analysis: the use of a triangular graph

Figure 10.10

Five soil samples

Sample	Clay (%)	Silt (%)	Sand (%)
d	61	-26	13
e	33	7	60
f	-8	79	13
g	5	5	90
h	34	36	30

Q

1 Using Figure 10.9, give the percentage constituents of soils **a**, **b** and **c**.

2 On a sheet of triangular graph paper, plot and match correct names to the samples in Figure 10.10.

3 What type of weathering is likely to produce **a** sand; **b** clay? What is the connection between weathering and soil texture?

4 a What is meant by primary minerals and secondary minerals?

 b Name one type of parent rock likely to give soil **1** on the graph, and one type likely to give soil **12**. State your reasons.

Silty soils also tend to lack mineral and organic nutrients. The smaller pore size means that more moisture is retained than in sands but heavy rain tends to 'seal' or cement the surface, increasing the risk of sheetwash and erosion.

Clay soils are rich in nutrients and organic matter but they are difficult to plough and, after heavy rain, are prone to waterlogging and may become gleyed (pages 251and 254). Plant roots find difficulty in penetration. Clays expand when wet, shrink when dry and take the longest time to warm up.

The ideal soil for agriculture is a **loam** (Figure 10.8). This has sufficient clay (20 per cent) to hold moisture and retain nutrients; sufficient sand (40 per cent) to prevent waterlogging, to be well-aerated and to be light enough to work; and sufficient silt (40 per cent) to act as an adhesive, holding the sand and clay together. A loam is likely to be least susceptible to erosion.

3 Soil structure

It is the aggregation of individual particles which gives the soil its structure. In undisturbed soils, these aggregates form different shapes known as **peds**. It is the shape and alignment of the peds which, combined with particle size/texture, determine the size and number of the pore spaces through which water, air, roots and soil organisms can pass. The size, shape, location and suggested agricultural value of each ped type is given in Figure 10.11. While six different types of structure are listed here, it should be noted that some authorities only give five — some combining the crumb and granular structures, others the columnar and prismatic.

There is some uncertainty as to how peds actually form, but it is accepted that soils with a good crumb structure give the highest agricultural yield, are more resistant to erosion and develop best under grasses — which is one reason why fallow should be included in a crop rotation. Sandy soils have the weakest structures as they lack the chemical cement from calcium carbonate (or the secretions of organisms), clays and humus needed to cause the individual particles to aggregate. A crumb structure is ideal as it provides the optimum balance between air, water and nutrients.

Figure 10.11

Soil structures

Notes
1 Some soils may be structureless, e.g. sands.
2 Some soils may have more than one structure (ped) in a horizon.
3 Each horizon is likely to have its own distinctive ped.

Type of structure (ped)	Size of structure (mm)	Description of peds	Shape of peds	Location (horizon: texture)	Agricultural value
crumb	1–5	small individual particles similar to breadcrumbs; porous		A horizon: loam soil	the most productive; well aerated and drained — good for roots
granular	1–5	small individual particles; usually non-porous		A horizon: clay soil	fairly productive; problems with drainage and aeration
platy	1–10	vertical axis much shorter than horizontal, like overlapping plates; restrict flow of water	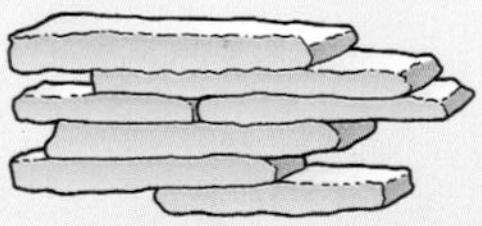	B horizon: silts and clays, or when compacted by farm tractors	the least productive; hinders water and air movement; restricts roots
blocky	10–75	irregular shape with horizontal and vertical axes about equal; may be rounded or angular but closely fitting	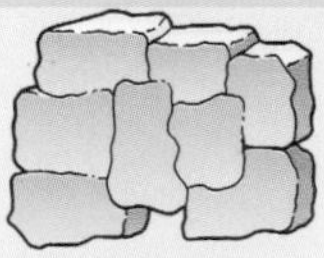	B horizon: clay-loam soils	productive: usually well drained and aerated
prismatic	20–100	vertical axis much larger than horizontal; angular caps and sides to columns	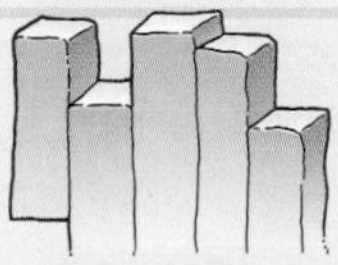	B and C horizons: often limestones or clays	usually quite productive: formed by wetting and drying; adequate water movement and root development
columnar	20–100	vertical axis much larger than horizontal; rounded caps and sides to columns	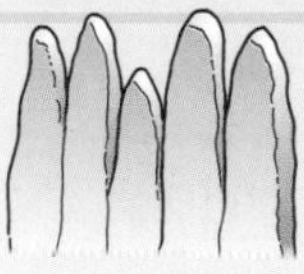	B and C horizons: alkaline and desert soils	quite productive (if water available)

1 Distinguish between the terms soil 'texture' and soil 'structure'.

2 How does soil texture influence soil structure?

3 Which types of soil structure (peds) are most beneficial to farmers?

4 Match the six structures named in Figure 10.11 to the letters **a** to **e** used in Figure 10.12.

Figure 10.12

Differences in soil structure
(*after* Courtney and Trudgill)

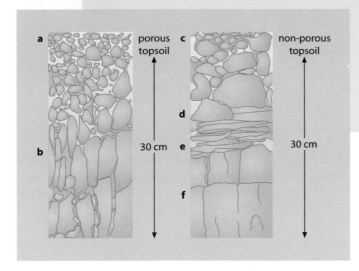

4 Organic matter

Organic matter, or humus, is derived mainly from decaying plants and animals or from the secretions of living organisms. Fallen leaves and decaying grasses and roots are the main sources of humus. As bacteria and fungi break down the organic matter, three distinct layers may be seen in the soil profile (Figure 10.5).

1 *L* or **leaf litter** layer: plant remains are still visible.

2 *F* or **fermentation** layer: decay is most rapid, although some plant remains are still visible.

3 *H* or **humus** layer: the process of decay has been completed and no plant remains are visible.

Humus gives the soil a black or dark brown colour. The highest amounts of humus are found in areas of temperate grassland forming the **chernozems**, or black earths (page 303), of the North American Prairies, the Russian Steppes and the Argentinian Pampas. In tropical rainforests, heavy rainfall soon leaches out any humus remaining after the considerable uptake and storage by plants from the soil. In drier climates, there may be insufficient vegetation to give an adequate supply.

Humus is a major source of nutrients and it combines with clays to form the **clay–humus complex** (page 247). The clay–humus complex is essential for a fertile soil as it provides it with a high water- and nutrient-holding capacity. Humus acts as a cement, binding the soil particles together and thus reducing the risk of erosion by improved cohesion.

5 Soil moisture

Soil moisture is important because it affects the upward and downward movement of water and nutrients. It helps in the development of horizons; it supplies water for living plants and organisms; it provides a solvent for plant nutrients; it controls soil temperature; and it determines the incidence of erosion. The amount of water in a soil at a given time can be expressed as:

$$W \propto R - (E + T + D)$$
$$\text{(input)} - \text{(outputs)}$$

where: W = water in the soil;
R = rainfall/precipitation;
T = transpiration;
$\propto$ = proportional to;
E = evaporation;
D = drainage.

Drainage depends upon the balance between the **water retention capacity** (water storage in a soil) and the infiltration rate. This is controlled by the soil's texture and structure. It has already been shown how texture and structure affect the size and distribution of pore spaces. Clays have numerous small pores (**micropores**) which may retain water for lengthy periods, but which also restrict infiltration rates (page 51). Sands have fewer but much larger **macropores** which permit water to pass through more quickly (a rapid infiltration rate), but have a low water retention capacity. A loam provides a more balanced supply of water, in the micropores, and air, in the macropores.

The presence of moisture in the soil does not necessarily mean that it is available for plant use. Plants growing in clays may still suffer from water stress even though clay has a high water-holding capacity. Soil water can be classified according to the tension at which it is held, and it is measured in atmospheres of pressure (atm). Following a heavy storm or a lengthy episode of rain or snowmelt, all the pore spaces may be filled,

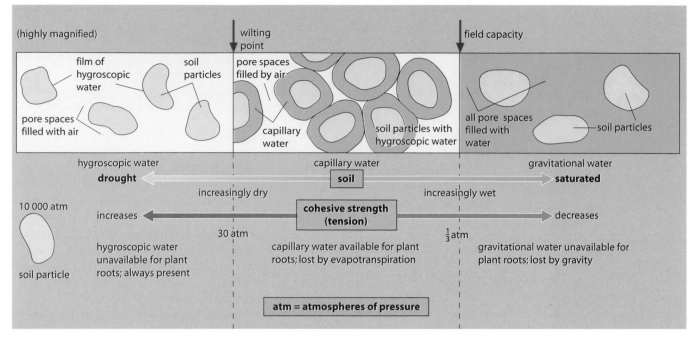

Figure 10.13

Availability of soil moisture for plant use

with the result that the soil becomes saturated. When infiltration ceases, water with a low cohesive strength (low surface tension) drains away rapidly under gravity. This is called **gravitational**, or **free**, **water** and is not available for use by plants. Once this excess water has drained away, the remaining moisture that the soil can hold is said to be its **field capacity** (Figure 10.13).

Moisture at field capacity is held either as **hygroscopic water** or as **capillary water**. Hygroscopic water is always present, regardless of how dry a soil becomes, but it is unavailable for plant use. It is found as a thin film around the soil particles to which it sticks due to the strength of its surface tension. Capillary water is attracted to, and forms a film around, the hygroscopic water, but has a lower cohesive strength. It is capillary water which is freely available to plant roots. However, this water can be lost to the soil by evapotranspiration. When a plant loses more water through transpiration than it can take up through its roots it is said to suffer **water stress** and it begins to wilt. At **wilting point**, photosynthesis (page 274) is reduced but, provided water can be obtained relatively soon or if the plant is adapted to drought conditions, this need not be fatal. Figure 10.13 shows the different water-holding characteristics of soil.

6 Air

Air fills the pore spaces left unoccupied by soil moisture. It is essential for plant growth and living organisms. Compared with atmospheric air, air in the soil contains more carbon dioxide, released by plants and soil

biota, and more water vapour; but less oxygen, as this is consumed by bacteria. Biota need oxygen and give off carbon dioxide. These gases are exchanged through the process of diffusion.

7 Soil organisms (biota)

Soil organisms include bacteria, fungi, and earthworms. They are more active and plentiful in warmer, well-drained and aerated soils (mull) than they are in colder, more acidic and less well-drained and aerated soils (mor) (page 241).

Organisms are responsible for three important soil processes:

- **Decomposition**: detritivores, such as earthworms, mites, woodlice and slugs, begin this process by burying leaf litter (detritus), which hastens its decay, and eating some of it. Their faeces (wormcasts, etc) increase the surface area of detritus upon which fungi and bacteria can act. Fungi and bacteria secrete enzymes which break down the organic compounds in the detritus. This releases nutrient ions essential for plant growth (soil nutrients, below), into the soil while some organic compounds remain as humus.
- **Fixation**: by this process, bacteria can transform nitrogen in the air into nitrate which is an essential nutrient for plant growth.
- **Development of structure:** fungi help to bind individual soil particles together to give a crumb structure, while burrowing animals create passageways which help the circulation of air and water and facilitate root penetration.

8 Soil nutrients

Nutrient is the term given to chemical elements found in the soil which are essential for plant growth and the maintenance of the fertility of a soil. There are four **primary elements** (carbon, C; hydrogen, H; nitrogen, N; and oxygen, O) which are needed in considerable quantities; five **secondary elements** (calcium, Ca; magnesium, Mg; potassium, K; sulphur, S; and phosphorus, P) needed in smaller amounts; and several trace elements (e.g. molybdenum, Mo) required in minute amounts. Ca helps the growth of roots and new shoots; Mg is a component of chlorophyll; K and Na help in the formation of starches and oils; and Mo activates enzymes. Of these N, K and Ca are most likely to become deficient in areas of cultivation.

Nutrients may be obtained from:

1 Nutrients in solution may originate in rain water. Although these are readily available to plants, they may be leached out of the soil in gravitational water.

Secondary elements are especially vulnerable.

2 Fertiliser may be added artificially.

3 Minerals may be released from the parent rock by weathering or from decaying organic matter by soil organisms. These are soluble and dissolve into the soil solution, producing positively (+) charged ions called **cations**.

4 Nutrients attached to the clay–humus complex provide the major reserves. The particles of clay and humus develop negatively (-) charged ions on their surface known as **anions**. The negatively charged clay and humus particles attract the positively charged minerals in the soil solution and the cations are adsorbed (i.e. they become attached) onto the particles (Figure 10.14). The resultant double layer of negative and positive ions is known as the **Gouy Layer** (Figure 10.15).

Cation exchange usually takes place between soil particles and soil solution, but it also occurs between soil particles and plant roots. It is an important process as it allows nutrients, adsorbed to particles of clay and humus, to become detached so that they can be absorbed by plant roots (Figure 10.17). Cations of Ca^{++}, Mg^{++}, K^+ and Na^+ are released from the clay–humus particles and replace an equal number of H^+ cations which were initially found either in the soil solution or attached to plant roots. As well as providing nutrients for plant roots, the cation exchange releases hydrogen which in turn increases the acidity of the soil (see next section). Acidity accelerates the rate of weathering of the parent rock, releasing more minerals to replace those used by the plants or lost through leaching in gravitational water.

Different soils have different **cation exchange capacities** (CEC) — i.e. different abilities to retain cations for plant use. As Figure 10.16 shows, sands have a much lower ability to retain plant nutrients than does humus and so will be less fertile.

Figure 10.14

The Gouy layer: a film of cations (+) around a clay–humus particle (-)

clay–humus particle (or micelle) with negative charge

Gouy layer
(-) anions
(+) cations

Figure 10.15

Cation and anion concentration around a clay–humus particle (*after* Courtney and Trudgill)

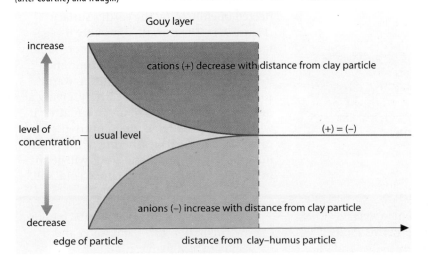

Gouy layer

increase

cations (+) decrease with distance from clay particle

level of concentration — usual level

(+) = (-)

decrease

anions (-) increase with distance from clay particle

edge of particle

distance from clay–humus particle

	CEC (measured in mille-equivalents per 100 g of soil)
sandy soils	1–5
clays	3–50
humus	150–400

Figure 10.16

Soil cation exchange capacities (CECs)

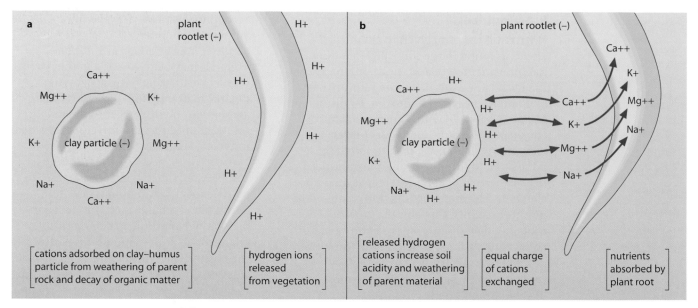

Figure 10.17

The process of cation exchange (*after* Courtney and Trudgill)

9 Acidity (pH)

As mentioned in the previous section, soil contains positively charged hydrogen cations. **Acidity** or **alkalinity** is a measure of the degree of concentration of these cations It is measured on the pH scale (Figure 10.18), which is logarithmic (compare the Richter scale, Figure 1.3). This means that a reading of 6 is 10 times more acidic than a reading of 7 (which is neutral), and 100 times more acid than one of 8 (which is alkaline). Most British soils are slightly acidic, although in upland Britain acidity increases as the heavier rainfall leaches out elements such as calcium faster than they can be replaced by weathering. Acid soils therefore tend to need constant liming if they are to be farmed successfully.

A slightly acid soil is the optimum for farming in Britain as this helps to release secondary elements. However, if a soil becomes too acidic it releases iron and aluminium which, in excess, may become toxic and poisonous to plants and organisms. Increased acidity makes organic matter more soluble and therefore vulnerable to leaching; and it discourages living organisms, thus reducing the rate of breakdown of plant litter and causing the formation of peat.

In areas where there is a balance between precipitation and evapotranspiration, soils are often neutral, as in the American Prairies (page 303); while in areas with a water deficiency, as in deserts (page 299), soils are more alkaline.

10 Soil temperature

Incoming radiation can be absorbed, reflected or emitted by the earth's surface (Figure 9.4). The topsoil, especially if vegetation cover is limited, heats up more rapidly than the subsoil during the daytime and loses heat more rapidly at night. A 'warm' soil will have greater biota activity, giving a more rapid breakdown of organic matter; it will be more likely to contain nutrients because the chemical weathering of the parent material will be faster; and seeds will germinate more readily in it than in a 'cold' soil.

Figure 10.18

The pH scale showing acidity and alkalinity

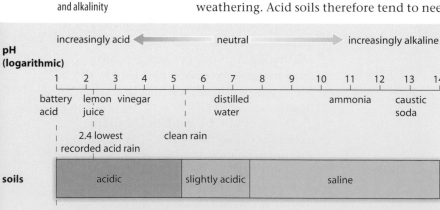

Q	With reference to the ten properties listed above, describe what makes the optimum	soil for farming in two contrasting environments of your choice.

Begin by reading a book which describes in detail how to dig a soil pit and how to describe and explain the resultant profile (e.g. Courtney and Trudgill, 1984, or O'Hare, 1987; see References at end of chapter).

First, make sure you obtain permission to dig a pit. The site must be carefully chosen. You will need to find an undisturbed soil — so avoid digging near to hedges, trees, footpaths or on recently ploughed land. Ideally, make the surface of the pit 0.7 sq m, and the depth 1 m (unless you hit bedrock first). Carefully lay the turf and soil on plastic sheets. Clear one face of the pit, preferably one facing south as this will get the maximum light, to get a 'clean' profile so that you can complete your recording sheet. (The one in Figure 10.19 is a very detailed example.) Sometimes you will not be able to take all the readings due to problems such as lack of clarity, time and equipment; sometimes some details will not be relevant to a particular enquiry.

Make a detailed field sketch before replacing the soil and turf. You may have to complete several tasks in the laboratory before writing up your description. You can gather information from a soil without needing to know how it formed or what type it is. Remember, it is unlikely that your answer will exactly fit a model profile. It may show the characteristics of a podsol (Figure 12.39) if you live in the north of Britain (cooler, wetter uplands) or of a brown forest earth (Figure 12.34) if you live in the south — but do not **force** your profile to fit a model.

Figure 10.19

Soil recording sheets

a soil site

Recorded by		Date		Locality		Six-figure grid reference
Parent rock (geological map)	Altitude (estimated from Ordnance Survey map)	Angle of slope (abney level)	Aspect (bearing or compass point)	Relief (uniform, concave or convex slope, terrace)
Exposure (exposed, sheltered)	Drainage (shedding or receiving site, flood plain, terrace, boggy)	Natural vegetation or type of farming (tree species, ground vegetation, crops, animals)	Previous few days' weather (warm, cold, wet, dry)	Other local details (remember your labelled field sketch)

b soil profile

Horizon	Depth of horizon (cm)	Lower boundary of horizon	Colour	Texture	Stoni-ness	Structure (peds)	Consist-ency	pH	Moisture content	Porosity	Organic matter	Roots	Carbon-ates	Soil biota and/or animals
How to read	measure from base of humus layer	sharp, abrupt, clear, indistinct, gradual, irregular, smooth, broken	use Munsell colour chart	percent-age clay, silt or sand 'feel' sieves sedimen-tation	size of stones, number of stones, shape of stones	structure-less	loose, friable, firm, hard, plastic, sticky, soft	litmus paper or soil-testing kit	weigh sample, evaporate water, reweigh sample, or: use a moisture meter	time taken for a beakerful of water to infiltrate	type estimate percent-age	weigh, burn sample (and roots), reweigh sample, calculate percent-age	add dilute hydro-chloric acid; if it effer-vesces, sample is over 1% carbonate	number, types
A														
B														
C														

Figure 10.20

Soil-forming processes

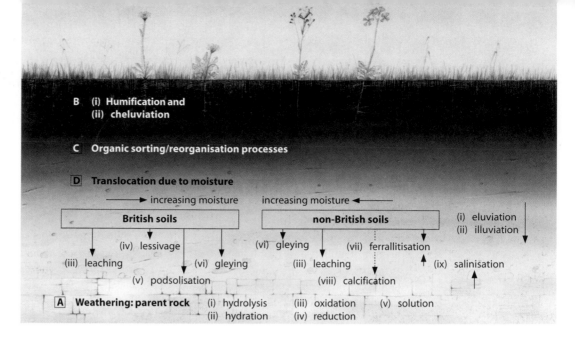

B (i) **Humification and**
(ii) **cheluviation**

C **Organic sorting/reorganisation processes**

D **Translocation due to moisture**

→ increasing moisture   increasing moisture ←

| **British soils** | | **non-British soils** | (i) eluviation
(ii) illuviation |

(iv) lessivage   (vi) gleying   (vii) ferrallitisation

(iii) leaching   (vi) gleying   (iii) leaching   (ix) salinisation

(v) podsolisation   (viii) calcification

A **Weathering: parent rock**   (i) hydrolysis   (iii) oxidation   (v) solution
(ii) hydration   (iv) reduction

Processes of soil formation

Numerous processes are involved in the formation of soil and the creation of the profiles, structures and other features described above. Soil-forming processes depend on all the five factors described on page 238. Some of the more important processes are shown in Figure 10.20.

A Weathering

As described on page 242, weathering produces primary and secondary minerals as well as determining the rates of release of nutrients and the soil depth, texture and drainage. In systems terms, this means that minerals are released as inputs into the soil system from the bedrock store and transferred into the soil store (Figure 10.6).

B Humification and cheluviation

Humification is the process by which organic matter is decomposed to form humus (section 4, page 245) — a task performed by soil organisms. Humification is most active in the *H* horizon of the soil profile (Figure 10.5) where it can result in mull (pH 5.5–6.5), mor (pH 3.5–4.5) (page 241), or the intermediate moder (pH 4.5–5.5).

As organic matter decomposes, it releases nutrients and organic acids. These acids — known as **chelating agents** — attack clays and rock, releasing iron and aluminium. The dissolved minerals are transported downwards under the influence of the chelating agents: the process of **cheluviation**.

C Organic sorting

Several processes operate within the soil to reorganise mineral and organic matter into horizons; to contribute to the aggregation of particles and to the formation of peds. Earthworm activity is a significant factor in sorting material into different particle sizes.

D Translocation of soil materials

Translocation is the movement of soil components in any form (solution, suspension) or direction (downward, upward). It usually takes place in association with soil moisture.

Q

1 a List the five processes of chemical weathering shown in Figure 10.20.
  b Refer back to Chapter 2 and describe each process, explaining how it helps in the formation of soil.

2 Which process of chemical weathering is not listed in Figure 10.20? Suggest a reason for its omission.

Note: Among other functions, hydrolysis is responsible for the formation of the (+) cations of Ca, Na, K and Mg and several (-) anions referred to under cation exchange.

British soils

In Britain, there is usually a soil moisture budget surplus due to an excess of precipitation over evapotranspiration (water balance, Figure 3.3) or, locally, to poor drainage. This excess leads to the three translocation processes of leaching, podsolisation and gleying, to which some pedologists add a fourth — lessivage.

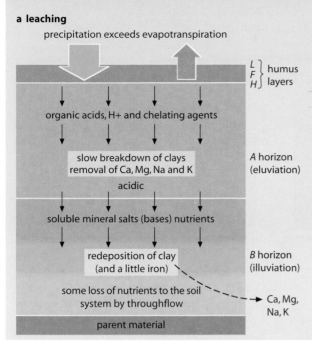

a leaching

precipitation exceeds evapotranspiration

$\begin{rcases} L \\ F \\ H \end{rcases}$ humus layers

organic acids, H+ and chelating agents

slow breakdown of clays
removal of Ca, Mg, Na and K
acidic

A horizon (eluviation)

soluble mineral salts (bases) nutrients

redeposition of clay
(and a little iron)

B horizon (illuviation)

some loss of nutrients to the soil
system by throughflow → Ca, Mg, Na, K

parent material

b podsolisation

precipitation greatly exceeds evapotranspiration

mor humus

$\begin{rcases} L \\ F \\ H \end{rcases}$ humus layers

many organic acids, H+ and chelating agents

rapid breakdown of clays
leaching of bases: Ca, Mg, Na and K
pH under 4.5 releases Fe and Al

A horizon (eluviation)

white/grey acidic horizon

Ca Mg Na K Fe Al Si N

hard pan

redeposition in sequence of organic
matter, Fe, Al, and clay

B horizon (illuviation)

serious loss of bases to the
soil system by throughflow → Ca, Mg, Na, K

parent material

Figure 10.21

The processes of leaching and podsolisation (see also Figure 12.39)

Leaching

Leaching is the removal of soluble material in solution. Where precipitation exceeds evapotranspiration and soil drainage is good, rain water — containing oxygen, carbonic acid and organic acids, collected as it passes through the surface vegetation — causes chemical weathering, the breakdown of clays and the dissolving of soluble salts (bases). Ca and Mg are eluviated from the *A* horizon, making it increasingly acid as they are replaced by hydrogen ions, and are subsequently illuviated in the underlying *B* horizon (Figure 10.21a).

Podsolisation

This is a more intense form of leaching. It is most common in cool climates where precipitation is greatly in excess of evapotranspiration and where soils are well drained or sandy. Podsolisation is also defined as the removal of sesquioxides (and chelates) under extreme leaching. As the surface vegetation is often coniferous forest, heathland or moors, rain percolating through it becomes progressively more acidic and may reach a pH of below 4.5 (Figure 10.18). This in turn dissolves an increasing amount and number of bases (Ca, Mg, Na and K), silica and, ultimately, the sesquioxides of iron and aluminium (Figure 10.21b). The resultant **podsol soil** (Figure 12.39) therefore has two distinct horizons: the bleached *A* horizon, drained of coloured minerals by leaching; and the reddish-brown *B* horizon where the sesquioxides have been illuviated. Often the iron deposits form a **hard pan** which is a characteristic of a podsol.

Gleying

This occurs when the output of water from the soil system is restricted, giving **anaerobic** or **waterlogged** conditions (page 254). This is most likely to occur on gentle slopes, in depressions where the underlying rock is impermeable, or following periods of heavy rain. Under such conditions the pore spaces fill with stagnant water which becomes de-oxygenised. The reddish coloured oxidised iron, iron III (Fe^{+++} or ferric iron), is chemically reduced to form iron II (Fe^{++} or ferrous iron) which is grey-blue in colour. Occasionally, pockets of air re-oxygenise the iron II to give scatterings of red mottles. Although many British soils show some evidence of gleying, the conditions develop most extensively on moorland plateaux.

Lessivage

This is a particular type of leaching resulting from clay particles being carried downwards in suspension. This process can lead to the breakdown of peds.

Courtney and Trudgill have summarised the relationship between leaching, podsolisation and gleying, and precipitation and drainage (Figure 10.22).

Non-British soils

Leaching occurs in most areas, such as the tropical rainforest, where precipitation exceeds evapotranspiration. It is the rapid translocation of bases that makes these soils infertile once the protective forest cover has been removed. **Gleying** occurs worldwide where environmental conditions are similar to those described for its operation in Britain.

Figure 10.22

Soil-forming processes and the water balance (Figure 3.3) (*after* Courtney and Trudgill)

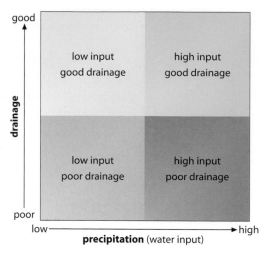

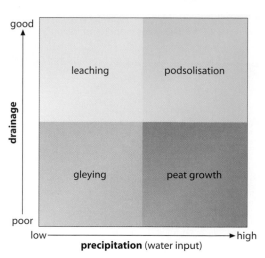

Ferrallitisation

This is the process by which parent rock is changed into a soil consisting of clays (kaolinite) and sesquioxides (hydrated oxides of iron and aluminium). In humid tropical areas, with constantly high temperatures and rainfall for all or most of the year, there is rapid chemical weathering. This initially produces clays which later break down to form silica — which is removed by leaching — and the sesquioxides of iron and aluminium — which remain, giving the characteristic red colour of many tropical soils (Figure 10.23a). Ferrallitisation is the reverse of podsolisation where the silica remains and the iron and aluminium are removed. In tropical rainforests, with rain throughout the year, **ferrallitic** soils develop (page 294). In savanna areas, with alternating dry and wet seasons, **ferruginous** soils form (page 297).

Calcification

Calcification is a process typical of low rainfall areas where precipitation is either equal to, or slightly higher than, evapotranspiration. Although there may be some leaching, it is insufficient to remove all the calcium which then accumulates, in relatively small amounts, in the *B* horizon (Figure 10.23b; and chernozems, page 303).

Salinisation

This occurs when potential evapotranspiration is greater than precipitation in places where the water table is near to the surface. It is therefore found locally in dry climates and is not a characteristic of desert soils. As moisture is evaporated from the surface, salts are drawn upwards in solution by capillary action. Further evaporation results in the deposition of salt as a hard crust (Figure 10.23c). Salinisation has become a critical problem in many irrigated areas, such as California, page 458.

Figure 10.23

Soil-forming processes

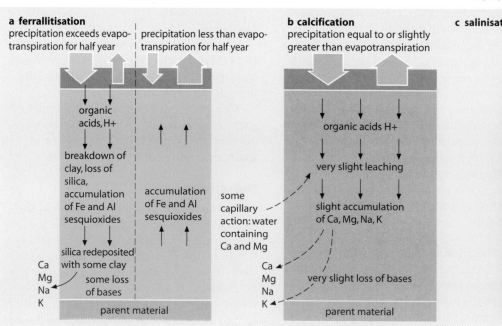

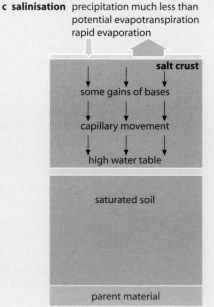

Zonal, azonal and intrazonal soils

Zonal soils

Zonal soils are mature soils. They result from the maximum effects of climate and living matter (vegetation) upon parent rock in areas where there are no extremes of weathering, relief or drainage and where the landscape and climate have been stable for a long time. Consequently, zonal soils have had time to develop distinctive profiles and, usually, clear horizons. A description of the major zonal soils, and how their formation can be linked to climate and vegetation, is given in Chapter 12 and Figure 12.2.

Azonal soils

Azonal soils, in contrast to zonal soils, have a more recent origin and are said to be immature as they occur where soil-forming processes have had insufficient time to operate fully. As a consequence, these soils show the characteristics of their origin (i.e. parent rock, agent of deposition), do not have well defined horizons, and are not associated with specific climatic–vegetational zones. Azonal soils, in Britain, include **scree** (weathering); **alluvium** ((fluvial); **till** (glacial); **sands** and **gravels** (fluvio-glacial); **sand dunes** (aeolian and marine); **salt marsh** (marine) and **volcanic** (tectonic) **soils**.

Intrazonal soils

Intrazonal soils reflect the dominance of a single local factor, such as parent rock or extremes of drainage. As they are not related to general climatic controls, they are not found in zones. They can be divided into:

- **Calcimorphic** or **calcareous** soils develop upon a limestone parent rock, (rendzina and terra rossa, Figure 10.24).
- **Hydromorphic** soils are those having a constantly high water content (gleyed soils and peat).
- **Halomorphic** soils have high levels of soluble salts which render them saline.

Calcimorphic

1 **Rendzina** The rendzina (Figure 10.25) develops where limestones or chalk are the parent material and where grasses form the surface vegetation (the English Downs). The grasses produce a leaf litter rich in bases. This encourages considerable activity by organisms which help with the rapid recycling of nutrients. The *A* horizon therefore consists of a black/dark brown mull humus. Due to the continual release of calcium from the parent rock and a lack of hydrogen cations, the soil is alkaline with a pH of between 7.0 and 8.0. The calcium-saturated clays, with a crumb or blocky structure, tend to limit the movement of water and so there is little leaching. Consequently there is no *B* horizon. The underlying limestones, affected by chemical weathering, leave very little insoluble residue and this, together with the permeable nature of the bedrock, results in a thin soil with limited moisture reserves.

Figure 10.24

Calcimorphic soils: terra rossa and rendzina

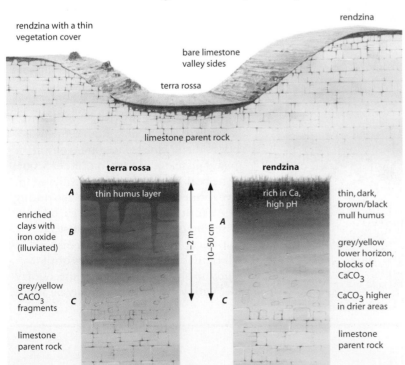

rendzina with a thin vegetation cover

rendzina

bare limestone valley sides

terra rossa

limestone parent rock

terra rossa

A — thin humus layer

enriched clays with iron oxide (illuviated) — B

grey/yellow CACO₃ fragments — C

limestone parent rock

1–2 m

rendzina

rich in Ca, high pH

thin, dark, brown/black mull humus

A

grey/yellow lower horizon, blocks of CaCO₃

10–50 cm

CaCO₃ higher in drier areas

C

limestone parent rock

Figure 10.25

A rendzina, Grude Imotsk polje, Yugoslavia

253

Figure 10.26

Gleying: a hydromorphic soil

Figure 10.27

Peat: the Sutherland Flow country, Scotland

2 Terra rossa As its name suggests, terra rossa (Figure 10.24) is a red-coloured soil (it has been called a 'red rendzina'). It is found in areas of heavy, even if seasonal, rainfall where the calcium carbonate parent rock is chemically weathered (carbonation) and silicates are leached out of the soil to leave a residual deposit rich in iron hydroxides. It usually occurs in depressions within the limestone and in Mediterranean areas where the vegetation is garrigue (Figure 12.24).

Hydromorphic

1 Gley soils Gleying (Figure 10.26) occurs in saturated soils when the pore spaces become filled with water to the exclusion of air. The lack of oxygen leads to anaerobic conditions (page 251) and the reduction (chemical weathering) of iron compounds from a ferric (Fe^{+++}) to a ferrous (Fe^{++}) form. The resultant soil has a grey/blue colour with scatterings of red mottles. Because gleying is a result of poor drainage and is almost independent of climate, it can occur in any of the zonal soils. Pedologists often differentiate between **surface gleys**, caused by slow infiltration rates through the topsoil, and **groundwater gleys**, resulting from a seasonal rise in the water table or the presence of an impermeable parent rock.

2 Peat Where a soil is waterlogged and the climate is too cold and/or wet for organisms to break down vegetation completely, layers of peat accumulate (Figure 10.27). These conditions mean that litter input (supply) is greater than the rate of decomposition by organisms whose activity rates are slowed down by the low temperatures and the anaerobic conditions. Peat is regarded as a soil in its own right when the layer of poorly decomposed material exceeds 38 cm in depth. Peat can be divided according to its location and acidity. **Blanket peat** is very acidic; it covers large areas of wet upland plateaux in Britain (Kinder Scout in the Peak District); and it is believed to have formed 5000–8000 years ago during the Atlantic climatic phase (Figure 11.18). **Raised bogs**, also composed of acidic peat, occur in lowlands with a heavy rainfall. Here the peat accumulates until it builds up above the surrounding countryside. **Valley**, or **basin, peat** may be almost neutral or only slightly acidic if water has drained off surrounding calcareous uplands (the Somerset Levels and the Fens); otherwise, it too will be acid (Rannoch Moor in Scotland). Fen peat is a high quality agricultural soil.

Halomorphic

Halomorphic soils contain high levels of soluble salts and have developed through the process of salinisation (page 252 and Figure 16.56). They are most likely to occur in hot, dry climates where, in the absence of leaching, mineral salts are brought to the surface by capillary action and where the parent rock or groundwater contains sodium chloride (common salt) or other salts. The water, on reaching the surface, evaporates to leave a thick crust (e.g. Bonneville salt flats in Utah, page 173) in which only salt-resistant plants (halophytes, page 269) can grow.

Soil classification

Although the grouping together of different soil types under the headings zonal, azonal and intrazonal is probably the simplest and most convenient method of classifying soils, it is too generalised to make it scientifically valuable. The need for geographers to classify material has already been discussed (Framework 5, page 151). By classifying soils, predictive statements can be made which enable farmers, among others, to appreciate the capabilities, limitations and management requirements of a particular soil. As with other classification systems, soils must be grouped so that there is minimum diversity within each group and maximum diversity between groups. What constitutes a working basis is complicated by the fact that on a **global scale** different soil-forming processes operate from those on a **regional scale** (Cumbria) or **local scale** (within a valley). In addition, it seems that more criteria have been used to try to classify soils than, for example, to try to classify coasts, climates or settlements. The uniqueness of each soil has led to the development of increasingly large and more complex classifications with their often unfamiliar terminology.

- The earliest classification, published in the 1880s in Russia, was based upon the major soil types found across that country. Still widely used in accounting for the distribution of soils on a **global scale** (Chapter 12), it takes climate as being the most significant single factor in soil development. The resultant types, which include the Russian terms 'podsol' and 'chernozem', are recognised by their distinctive 'zones' or profiles. This classification tended to ignore local factors such as parent rock and changes in relief.
- In 1940, the Soil Survey of Great Britain adopted a scheme based on six main soil groups found **regionally** in Britain. These groups (Figure 10.28) form a classification most appropriate to the British Isles. However it cannot be applied effectively on a global scale, nor does it have sufficient detail to be useful at a more local level. It was revised in the 1970s to take into account factors other than climate, e.g. **local** properties such as parent rock, composition and genesis of the soil, drainage and relief. This classification has 7 main groups and 35 subgroups (Figure 10.28).
- In 1975, the United States Department of Agriculture (USDA) published its '7th Approximation' classification (so-called because it was the seventh revision). The system was based upon descriptions of soil profiles and contains 10 classes and 47 subtypes (Figure 10.28).
- Also in 1975, the United Nations Food and Agriculture Organisation (FAO) produced its own classification; this is the most widely used at present. Seeking greater accuracy, it is more detailed and consists of 26 major groups and over 240 subgroups.

Figure 10.28

Some major soil classifications

1940 British classification	1970s British classification		USDA '7th approximation'
	Main group	Number of subgroups	
podsol	lithomorphic	7	entisols
brown earth	brown soils	8	vertisols
calcareous/rendzina	podsolic	4	inceptisols
gley	pelosols	3	aridosols
organic/peat	gley	9	mollisols
undifferentiated alluvium	manmade	2	spodosols
	peat	2	alfisols
			ultisols
			oxisols
			histosols

The soil catena

A catena (Latin for 'chain') is a sequence of soils down a slope where each facet is different from, but linked to, its adjacent facets (Figure 10.2). Catenas therefore illustrate the way in which soils can change down a slope where there are no marked changes in climate or parent rock. Each catena is an example of a small-scale, open system involving inputs, processes and outputs. The slope itself is in a delicate state of dynamic equilibrium (Figure 2.12) with the soils and landforms being in a state of flux and where the ratio of erosion and deposition varies between the different slope facets. It takes a considerable period of time for catenary relationships to become established and therefore the best catenas can be found in places with a stable environment — such as in parts of Africa, where there have been relatively few recent changes in the landscape and climate.

Q

1 With reference to any soil classification which you have studied, describe the basis for this classification and briefly list its advantages and disadvantages.

2 With reference to Figure 10.2 and Chapter 10, draw a soil catena and on it locate:
 a the shedding, transfer and receiving slopes;
 b the area likely to have: the highest pH, most leaching, basin (valley) peat, blanket peat, most organic material, the lowest pH, maximum throughflow, brown earths, gleying, most soil organisms.
 In each case, give a reason for your answer.
 c Your answers to **b** were based on the bedrock being impermeable. What difference would it make if the bedrock was limestone (calcareous)?

3 A sixth-form field course to the Isle of Arran included taking various measurements down a valley side in Glen Rosa. The valley, at this point, consisted of metamorphic schists (Figure 4.32). The results are given in Figure 10.29.
 a Account for the differences in (i) acidity; (ii) moisture; and (iii) soil depth at different points on the slope.

b Draw a simple soil profile to show the probable conditions at Sites **A**, **F** and **K**.

4 Figure 10.30 shows the relief and surface geology of an upland edge and adjacent lowland in eastern England. The upland receives about 730 mm of precipitation per annum and mean monthly temperatures vary from 2.7°C in January to 15.5°C in July and August. Six sites, **A–F** on Figure 10.30, have their soil profile characteristics shown on the accompanying graphs.
 a Describe and explain the characteristics of the soil profile at site **F**.
 b Suggest why, from the evidence in Figure 10.30, the soil at site **E** is less prone to waterlogging than that at site **F**.
 c Describe and explain the changes in soil profile characteristics that occur downhill from site **B**, through site **C**, to site **D**.
 d Comment upon the advantages and limitations for arable farming of the area above 120 m containing site **A**.

5 Account for the main characteristics of **each** of the soil profiles shown on Figure 10.32. (The site of each soil profile is shown on Figure 10.31).

Figure 10.29

Readings taken on a soil catena, Isle of Arran

Site	Altitude (m)	Slope angle (°)	Depth of soil (cm)	pH	Moisture content (moisture meter)
A	320	2	170	4.4	6.0
B	310	10	110	3.8	5.2
C	280	16	45	4.4	1.5
D	240	19	42	4.7	2.0
E	195	28	24	5.0	2.2
F	165	27	18	5.6	2.0
G	125	30	28	5.9	2.5
H	95	33	20	5.7	1.5
I	80	20	21	5.8	1.5
J	55	17	70	4.5	3.5
K	40 (river)	11	70	4.8	3.8

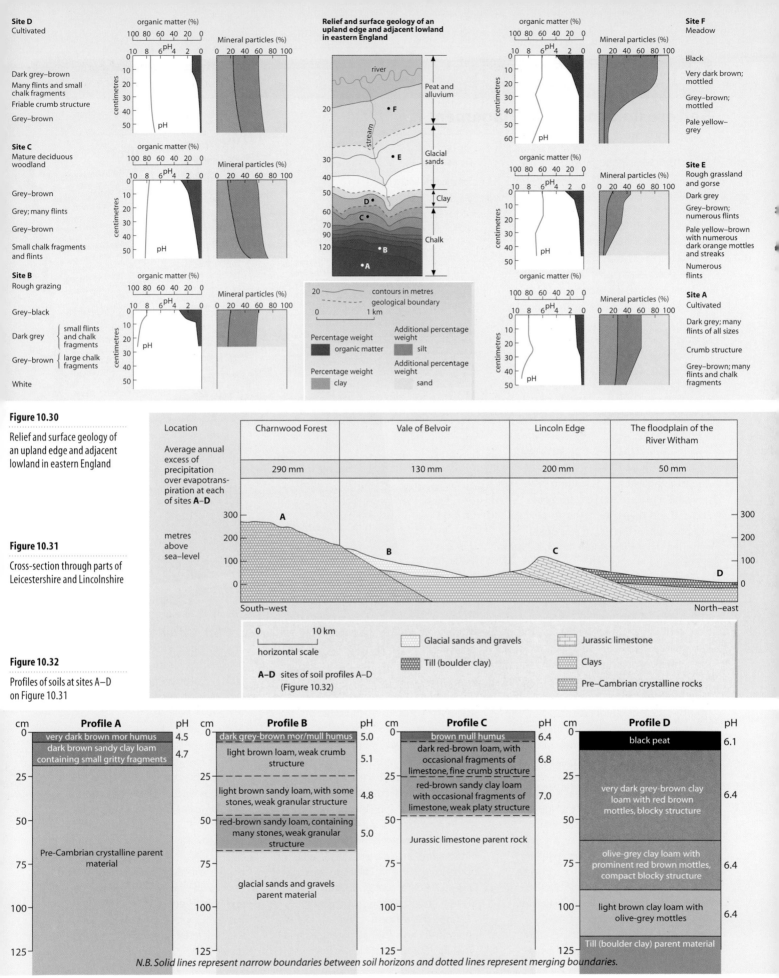

Figure 10.30

Relief and surface geology of an upland edge and adjacent lowland in eastern England

Figure 10.31

Cross-section through parts of Leicestershire and Lincolnshire

Figure 10.32

Profiles of soils at sites A–D on Figure 10.31

Case Study 10

Soil erosion and soil management

As we have seen (page 239), soil usually takes thousands of years to become sufficiently deep and mature for economic use. During that time, there is always some natural loss through leaching, mass movement and erosion by either water or wind. Normally there is an equilibrium, however fragile, between the rate at which soil forms and that at which it is eroded or degraded. That natural balance is being disturbed by human mismanagement with increasing frequency and with serious consequences.

Recent estimates suggest that 7 per cent of the world's topsoil is lost each year. The World Resources Institute claims that Burkina Faso loses 35 tonnes of soil per hectare per year. Other comparable figures are: Ethiopia 42; Nepal 70; Deccan Plateau (India) 100; and the loess plateau of North China 251 (Figure 10.34). Soil removed during a single rain or dust storm may never be replaced. The Soil Survey of England and Wales claims that 44 per cent of arable soils in the UK, an area once considered not to be under threat, are now at risk (Figure 10.33).

Soil degradation

Degradation is the result of human failures to understand and manage the soil. The major cause of soil erosion is the removal of the natural vegetation cover, leaving the ground exposed to the elements. The most serious of such removals is deforestation. In countries such as Ethiopia (Places 61, page 472) the loss of trees, resulting from population growth and the extra need for farmland and fuelwood, means that the heavy rains, when they do occur, are no longer intercepted by the vegetation. Rainsplash (the direct impact of raindrops, Figure 2.12) loosens the topsoil and prepares it for removal by sheetwash (overland flow). Water flowing over the surface has little time to infiltrate into the soil or recharge the soil moisture store (page 51). Where the water evaporates, a hard crust may form, making the surface less porous and increasing the amount of surface runoff. More topsoil tends to be carried away

Figure 10.33

Soil erosion in Britain

Soil erosion sweeps shires

"Soil erosion, the scourge of Third World countries, is hitting Britain. Official figures show that nearly half of Britain's topsoil is in danger of being blown and washed away by wind and rain.

This emerging crisis adds weight to Prince Charles's recent attack on conventional agricultural methods and to his support for organic farming. He cited 'large-scale soil erosion' as one of the 'unacceptable' side-effects of modern agriculture.

One-third of the world's arable land is expected to turn to dust by the year 2000. Around 80 per cent of Africa's topsoil is in danger and India loses some 12 billion tons of soil every year, while the American Mid-West seems to be eroding as fast as in the Dust Bowl era of the Thirties.

However, until recently, Britain was thought to be unaffected, thanks to its stable soils and climate, and so the problem has largely been ignored by the government. Several reports now show that this attitude is misguided.

The Government's own Soil Survey of England and Wales reported that 44 per cent of the country's arable soil was at risk, while a recent European Parliament report said that five million acres (2 million hectares) of the UK were threatened by erosion. In addition, a survey of farmers has revealed that one-third believe that some of the fields on their farms are affected.

It now seems certain that vast amounts of soil are being lost. Detailed studies have recorded annual losses of 80 tons from every acre in Norfolk, 73 tons an acre in West Sussex and 60 tons an acre in Shropshire."

The Observer, 29 January 1989

where there is little vegetation because there are neither plant roots nor organic matter to bind it together. Small channels or rills may be formed which, in time, may develop into large gulleys making the land useless for agriculture (Figure 10.34).

Even where the soil is not actually washed away, heavy rain may accelerate leaching and remove nutrients and organic matter at a rate faster than that at which they can be replaced by the weathering of bedrock or vegetation breakdown (as in the Amazon basin, Figure 12.7 and page 441). The loss of trees also reduces the rate of transpiration and therefore the amount of moisture in the air. There are fears that large-scale deforestation will turn areas at present under rain-forest into deserts.

Although the North American Prairies and the African savannas were grassland when the European settlers first arrived, it is now believed that these areas too were once forested and were cleared, mainly by firing, by the local Indians and African (Case Study 12). The burning of vegetation initially provides nutrients for the soil but once these have been leached by the rain or utilised by crops there is little replacement of organic material. Where the grasslands have been ploughed up for cereal cropping, the breakdown of soil structure (peds) has often led to their drying out and becoming easy prey to wind erosion (Figure 10.33). Large quantities of topsoil were blown away to create the American Dust Bowl in the 1930s, while a similar fate has more recently been experienced by many of the Sahel countries. In Britain, the removal of hedges to create larger fields — easier for modern machinery — has led to accelerated soil erosion by wind (page 457)

Loess plateau of North China

This region experiences the most rapid soil loss in the world. During and following the ice age, Arctic winds transported large amounts of loess and deposited this fine, yellow material to a depth of 200 m in the Huang He basin. Following the removal of the subsequent vegetation cover of trees and grasses to allow cereal farming (especially under the directions of Chairman Mao), the unconsolidated material has been washed away by the heavy summer monsoon rains at the rate of 1 cm per year. It is estimated that 1.6 bn tonnes of soil reach the Huang He River during each annual summer flood. This material, the most carried by any river in the world, has given the Huang He its name

Figure 10.34

Loess in China

— i.e. the 'Yellow River'. A further problem is that 6 cm of silt settles annually on the river's bed so that the river now flows 10 m above its flood plain. Should the large flood banks be breached, the river can drown thousands of people (over 1 million in the 1939 flood) and ruin all crops. The Huang He is also known as 'China's Sorrow'.

Ploughing land for crops can have adverse effects on soils. Deep ploughing destroys the soil structure by breaking up peds and burying organic material too deep for plant use. It also loosens the topsoil for future wind and water erosion. The weight of farm machinery can compact the soil surface or produce platy peds, both of which reduce infiltration capacity and inhibit aeration of the soil. Ploughing up-and down-hill creates furrows which increase the rate of surface runoff and the process of gullying.

Overgrazing, especially on the African savannas, also accelerates soil erosion (as by the Rendille of Kenya, page 440). Many African tribes have long measured their wealth in terms of the numbers, rather than the quality, of their animal herds. As the human populations of these areas continue to expand rapidly, so too do the numbers of herbivorous animals needed to support them. This almost inevitably leads to overgrazing and the reduction of grass cover (Case Study 7). When new shoots appear after the rains, they are eaten immediately by cattle, sheep, goats and camels. The arrival of the rains causes erosion; the failure of the rains results in animal deaths.

Burkina Faso

As the size of cattle and goat herds has grown, the already scant dry scrub savanna vegetation on the southern fringes of the Sahara has been totally removed over increasingly large areas. As the Sahara 'advances', the herders are forced to move southwards into moister environments where they compete for land with sedentary farmers who are already struggling to produce sufficient food for their own increasing numbers. This disruption of equilibrium further reduces the land carrying capacity (page 354) — i.e. the number of people that the soil and climate of an area can permanently support when the land is planted with staple crops. These farmers have long been aware that three years' cropping had to be followed by at least eight fallow years in order for grass and trees to re-establish themselves and organic matter to be replenished. The arrival of the herders has brought a land shortage resulting in crops being grown on the same plots every year and the nutrient-deficient soil, typical of most of tropical Africa, is rapidly becoming even less productive. This overcropping, a problem in many of the world's subsistence areas, uses up humus and other nutrients, weakens soil structures and leaves the surface exposed and thus susceptible to accelerated erosion.

Figure 10.35

Soils and land use in Burkina Faso

Figure 10.36

Eutrophication in a British river

In places where there is a rapid population growth, land which was previously allowed a fallow resting period now has to be cultivated each year (Figure 10.35) — as are other areas which were previously considered to be too marginal for crops. Monoculture, the cultivation of the same crop each year on the same piece of land, repeatedly uses up the same soil nutrients.

In many parts of the world where livestock are kept and firewood is at a premium, dung has to be used as a fuel instead of being applied to the land. In parts of Ethiopia, the sale of dung — mixed with straw and dried into 'cakes' — is often the only source of income for rural dwellers. If this dung were to be applied to the fields, rather than sold to the towns, harvests could be increased by over 20 per cent. However, the concern for most farmers is survival today rather than planning for tomorrow.

Water is essential for a productive soil. The first known civilisations, which grew up in river valleys (Figure 14.1), relied upon irrigation, as do many areas of the modern world. Unfortunately, irrigation in a hot, dry climate tends to lead to salinisation with dissolved salts being brought, by capillary action, into the root zone of agricultural trees and crops (Figure 16.56). Salinisation is also a problem in coastal areas, such as

Bangladesh, which are subject to marine flooding. Wells, sunk in dry climates, use up reserves of groundwater which may have taken many centuries to accumulate and which cannot be replaced quickly (fossil water stores, page 174). The resultant lowering of the water table makes it harder for plant roots to obtain moisture. The sinking of wells in sub-Saharan Africa, following the drought of the early 1980s, has unintentionally created difficulties. The presence of an assured water supply has attracted numerous migrants and their animals and this has accelerated the destruction of the remaining trees and exacerbated the problems of overgrazing. Even well-intentioned aid projects may therefore be environmentally damaging.

Fertilisers and pesticides are not always beneficial to a soil if applied repeatedly over lengthy periods. Chemical fertilisers do not add humus and so fail to improve or maintain soil structure. There is considerable concern over the leaching of nitrate fertilisers into streams and underground water supplies, especially if these supplies are used for domestic purposes. Where nitrates reach rivers they enrich the water and encourage the rapid growth of algae and other aquatic plants which use up oxygen, through the process of eutrophication, to leave insufficient for plant life (Figures 10.36

and 16.53). The use of pesticides (including insecticides and fungicides) can increase yields by up to 100 per cent by killing off insect pests. However, their excessive and random use also kills vital soil organisms. Organic matter then decomposes more slowly and the release of nutrients is retarded. Chemical pesticides are blamed for a 60–80 per cent reduction in the 800 species of fauna found in the Paris basin as well as for the decline in Britain's bee population.

Soil management

Fertility refers to the ability of a soil to provide for the unconstrained or optimum growth of plants. The capacity to produce high or low yields depends upon the nutrient content, structure, texture, drainage, acidity and organic content of a particular soil as well as the relief, climate and farming techniques. For ideal growth, plants must have access to nine primary and secondary elements and several trace elements (page 247). Under normal recycling (Figure 10.6), these nutrients will be returned to the soil as the vegetation dies and decomposes. When a crop is harvested there is less organic material left to be recycled. As nutrients are taken out of the soil system and not replaced, there will be an increasing shortage of macro-nutrients, particularly nitrogen, calcium, phosphorus and potassium. Where this occurs, and when other nutrients are dissolved and leached from the soil, fertiliser is essential if yields are to be maintained.

Soils need to be managed carefully if they are to produce maximum agricultural yields and cause least environmental damage.

If the most serious cause of erosion is the removal of vegetation cover, the best way to protect the soil is likely to be the addition of vegetation. Afforestation provides a long-term solution because, once the trees have grown, their leaves intercept rainfall while their roots help to bind the soil together and reduce surface runoff (Figure 10.37). The growing of ground-cover crops reduces rainsplash and surface runoff, and can protect newly ploughed land from exposure to climatic extremes. Marram grass anchors sand (Figure 10.38), while gulleys can be seeded and planted with brushwood. Certain crops and plants, especially leguminous species such as peas, beans, clover and gorse, are capable of fixing atmospheric nitrogen in the soil, thus improving its quality. Trees can also be planted to act as windbreaks and shelterbelts. This reduces the risk of wind erosion as well as providing habitats for wildlife.

Soil can also be managed by improving

Figure 10.37

Afforestation in Nepal

Afforestation in Nepal

Deforestation in this Himalayan country resulted in serious soil erosion within its own borders and was blamed, in some quarters, for the increased incidence of river flooding in Bangladesh to the south-east.

In the 1970s, a district forestry officer met a group of village leaders from one small valley. The villagers agreed to stall-feed their cattle and buffalo when they were not needed for work.

This meant that the animals ate what they were given rather than anything and everything, as they did in the forest where they had previously been allowed to graze. Their manure could now be collected and spread on the land instead of being deposited on the forest floor.

Using dung as a fertiliser, tree seedlings were grown in nurseries, and then transplanted. Hardy pines were planted, but a deciduous undergrowth was encouraged.

As well as reducing erosion and stabilising slopes, the trees provided an extra source of income and began to re-establish both an economic equilibrium and a more stable environmental balance.

farming methods. Most arable areas benefit from a rotation of crops, including grasses, which improve soil structures and reduce the likelihood of soil-borne diseases which may develop under monoculture. Many tropical soils need a recovery period of 5–15 years under shrub or forest for each 3–6 years under crops. In areas where slopes reach up to 12°, ploughing should follow along the contours to prevent excessive erosion. On even steeper slopes, terracing helps to slow down runoff, giving water more time to infiltrate and thus reducing its erosive ability (Figure 16.31).

Figure 10.38

Planting marram grass to 'fix' (stabilise) sand dunes

Strip cropping can involve either the planting of crops in strips along the contours or the intercropping of different crops in the same field. Both methods are illustrated in Figure 10.39. The crops may differ in height, time of harvest and use of nutrients. In tropical areas where lateritic soils have developed, deep ploughing may be able to break up the hard pan.

Where evapotranspiration exceeds precipitation, dry farming can be adopted. This entails covering the soil with a mulch of straw and/or weeds to reduce moisture loss and limit erosion. In the Sahel countries, the drastic depopulation of cattle following the droughts of the 1980s has given herders a chance to restock with smaller (reducing overgrazing), better quality (giving more meat and milk) herds so that incomes do not fall and the soils are given time to recover.

The addition of humus helps to bind loose soil and so reduces its vulnerability to erosion. Soil structure and texture may be improved, theoretically, by adding lime to acid soils, which improves their drainage and therefore makes them warmer; by adding humus, clay or peat to sands, to give body and to improve their water-holding capacity; and by adding sand to heavy clays, so improving drainage and aeration and making them lighter to work.

Figure 10.39

Strip farming along contours in the southern USA

In practice, such methods are rarely used due to the expense involved.

Chemical (inorganic) fertilisers help to replenish deficient nutrients, especially nitrogen, potassium and phosphorus. However, their use is expensive, especially to farmers in economically less developed countries, and can cause environmental damage. Many farmers in poorer countries cannot afford such fertilisers and have to rely upon organic fertiliser. Animal dung and straw left after the cereal harvest are mixed together and spread over the ground. This improves soil structure and, as it decays, returns nutrients to the soil. Where crop rotations are practised, grasses add organic matter and legumes provide nitrogen. In Britain and North America, a growing number of farmers are turning to organic farming for environmental reasons (Figure 10.40).

Many soils suffer from either a shortage or a surfeit of water. In irrigated areas, water must be continually flushed through the system to prevent salinisation. In areas of heavy and/or seasonal rainfall, dams may be built to control flooding and to store surplus water. The drainage of waterlogged soils can be improved by adding field drains.

In sub-Saharan Africa, there is a need to construct more wells not just to provide a more reliable water supply but also to prevent overcrowding at the few existing ones (Places 51, page 439). In several Sahelian countries, stones have been used to build small dams which trap water for long enough for some to infiltrate into the ground; they also collect the soil carried away by surface run-off (Figures 10.41 and 16.65). In North China, 'check dams', 6 m high and built from loess, have been constructed by communal effort. Silt, washed from the hills, is trapped behind the dams, the surplus water is drained away through pipes, and new farmland is created (Figure 10.42).

Organic farming in Washington State, USA

Two adjacent farms in the American state of Washington share similar soil properties, relief and climate. Since 1948 one of the farms has been managed organically, the other conventionally. By 1990, the organic farm had, in comparison with the conventional farm:

- A much greater mass of microbes and enzyme activity. Microbes help in the breakdown of organic matter into humus and in the release of nutrients into a form usable by plants. They also stabilise soil structures, fix nitrogen and break down some pesticides.
- Nearly two-thirds more organic matter on the surface of the soil. Organic matter improves soil structure and increases the moisture retention capacity of the soil, the cation exchange capacity, and the amounts of available nitrogen and potassium.
- A topsoil 16 cm thicker, mainly due to the conventional farm losing 32.4 tonnes of soil per hectare to water erosion compared to only 8.3 tonnes on the organic farm.
- A soil that was much easier to work (plough), in which plants germinated and grew more readily, and which sustained higher crop yields.

Figure 10.40

Organic farming in Washington State, USA

Stone lines in Burkina Faso

Figure 10.41

Stone lines in Burkina Faso

This project, begun by Oxfam in 1979, aimed to introduce water-harvesting techniques for tree planting. It met with resistance from local people who were reluctant to divert land and labour from food production, or to risk wasting dry-season water needed for drinking.

Attention was therefore diverted to improving food production by using the traditional local technique of placing lines of stones across slopes to reduce runoff (Figures 10.41 and 16.65). When aligned with the contours, these lines dammed rainfall giving it time to infiltrate. Unfortunately, most slopes were so gentle, under 2°, that local farmers could not determine the contours. A device costing less than £3 solved the problem. A calibrated transparent hose, 15 m long, is fixed at each end to the tops of stakes of equal length and filled with water. When the water level is equal at both ends of the hose, the bottom of the stakes must be on the same contour. The lines can be made during the dry season when labour is not needed for farming. Although they take up only 1 or 2 per cent of cropland, they can increase yields by over 50 per cent. They also help to replenish falling water tables and can regenerate the barren, crusted earth because soil, organic matter and seeds collect on the upslope side of the stone lines and plants begin to grow again.

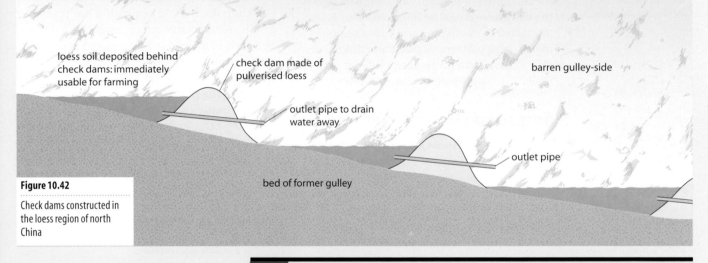

loess soil deposited behind check dams: immediately usable for farming

check dam made of pulverised loess

barren gulley-side

outlet pipe to drain water away

outlet pipe

bed of former gulley

Figure 10.42

Check dams constructed in the loess region of north China

Q

1 Describe, with reference to specific examples, how human failures to understand and manage the soil have led to an increase in soil erosion.

2 Describe, with reference to specific examples, how careful soil management can improve agricultural yields and cause minimal environmental damage.

References

Bradshaw, M. (1977) *Earth, the Living Planet*. Hodder & Stoughton.

Bridges, E. M. (1970) *World Soils*. Cambridge University Press.

Briggs, D. (1977) *Soils*. Butterworth.

Clarke, G. R. (1971) *The Study of Soil in the Field*. Oxford University Press.

Courtney, F. M. and Trudgill, S. T. (1984) *The Soil*. Edward Arnold.

Dawson, A. C. (1992) *Ice Age Earth*. Routledge.

Goudie, A. (1993) *The Nature of the Environment*. Blackwell.

Knapp, B. (1979) *Soil Processes*. Allen & Unwin.

Money, D. C. (1978) *Climate, Soils and Vegetation*. University Tutorial Press.

O'Hare, G. (1987) *Soils, Vegetation and Ecosystems*. Oliver & Boyd.

Pickering, K. and Owen, L. (1994) *Global Environmental Issues*. Routledge.

Prosser, R. (1992) *Natural Systems and Human Responses*. Thomas Nelson.

The Living Planet (1987) BBC Television/Earthscan.

Timberlake, L. (1987) *Only One Earth*. BBC/Earthscan.

Chapter 11

Biogeography

> *"The Earth's green cover is a prerequisite for the rest of life. Plants alone, through the alchemy of photosynthesis, can use sunlight energy, and convert it to the chemical energy animals need for survival."*
>
> James Lovelock, *The Gaia Atlas of Planet Management*, 1985

Biogeography may be defined as the study of the distribution of plants and animals over the earth's surface. The biogeographer is interested in describing and explaining meaningful patterns of plant and animal distributions in a given area, either at a particular time or through a time-period.

Seres and climax vegetation

A **sere** is a stage in a sequence of events by which the vegetation of an area develops over a period of time. The first plants to colonise an area and develop in it are called the **pioneer community** (or **species**). A **prisere** is the complete chain of successive seres beginning with a pioneer community and ending with a **climax vegetation** (Figure 11.1a). F. E. Clements suggested, in 1916, that for each climatic zone only one type of climax vegetation could evolve. He referred to this as the **climatic climax vegetation**; we now know it as the **monoclimax concept**. The climatic climax occurs when the vegetation has become in harmony or equilibrium with the local environment — i.e. when the natural vegetation has reached a delicate but stable balance with the climate and soils of an area (Chapter 12). Each successive seral community usually shows an increase in the number of species and the height of the plants.

Each individual sere is referred to by one or more of the larger species within that community — the so-called **dominant species**. The dominant species may be the **largest** plant or tree in the community which exerts the maximum influence on the local environment or habitat, or the most **numerous** species in the community. In parts of the

Figure 11.1

A seral progression, with possible interruptions

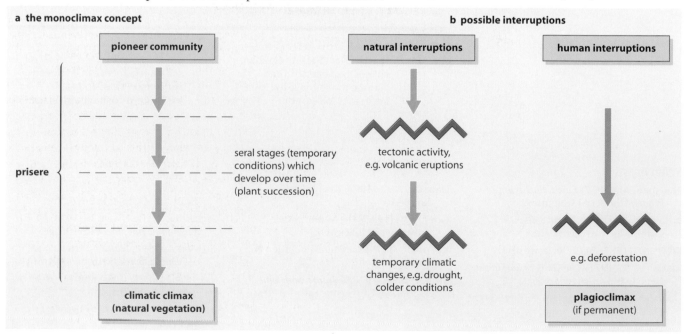

a the monoclimax concept

| pioneer community |

prisere

seral stages (temporary conditions) which develop over time (plant succession)

climatic climax (natural vegetation)

b possible interruptions

| natural interruptions |

tectonic activity, e.g. volcanic eruptions

temporary climatic changes, e.g. drought, colder conditions

| human interruptions |

e.g. deforestation

plagioclimax (if permanent)

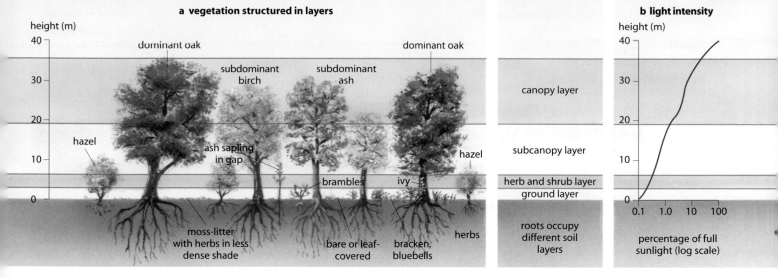

a vegetation structured in layers

height (m)

dominant oak
subdominant birch
subdominant ash
dominant oak

hazel
ash sapling in gap
brambles
ivy
hazel

moss-litter with herbs in less dense shade
bare or leaf-covered
bracken, bluebells
herbs

b light intensity

height (m)

canopy layer

subcanopy layer

herb and shrub layer
ground layer

roots occupy different soil layers

percentage of full sunlight (log scale)

Figure 11.2

Vegetation structure and light intensity typical of a temperate deciduous woodland (*after* O'Hare)

world where the climatic climax is forest — i.e. areas with higher rainfall — the plant community tends to be structured in layers (Figures 11.2 and 12.4). It can take several thousand years to reach a climatic climax.

There are, however, very few parts of today's world with a climatic climax. This is partly because few physical environments remain stable sufficiently long for the climax to be reached: most are affected by tectonic or temporary climatic changes (an area becomes warmer, colder, wetter or drier). More recently, however, instability has resulted from such human activities as deforestation, the ploughing of grass-land and acid rain. Where human activity has permanently arrested and altered the natural vegetation, the resultant com-munity is said to be a **plagioclimax** (Figure 11.1b) — examples include Britain's heather moorlands, the African savannas (page 296) and the temperate grasslands (page 302).

While it is still accepted that climate exerts a major influence upon vegetation, the

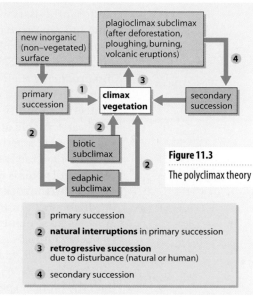

Figure 11.3

The polyclimax theory

1 primary succession
2 **natural interruptions** in primary succession
3 **retrogressive succession** due to disturbance (natural or human)
4 secondary succession

linear monoclimax concept has been replaced by the **polyclimax theory**. This theory acknowledges the importance not only of climate, but of several (poly) local factors including drainage, parent rock, relief and microclimate. The polyclimax theory, therefore, relates the climax vegetation to a variety of factors. Figure 11.3 shows how the climax vegetation may result from a **primary** or a **secondary succession**. A primary succession occurs on a new or previously sterile land surface, or in water. Figure 11.4 shows how the four more commonly accepted non-vegetated environments in Britain develop until they all reach the same climax vegetation: the oak woodland. A secondary succession occurs in an area which was once vegetated but, for various reasons, has been since laid bare. A **subclimax** occurs when the vegetation is prevented from reaching its climax due to interruptions by local factors such as soils and human interference.

Figure 11.4

Primary successions

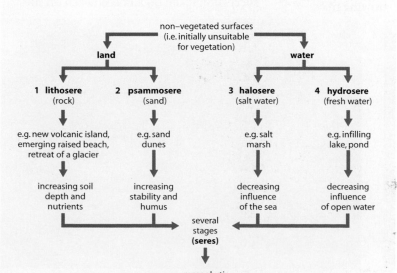

non-vegetated surfaces (i.e. initially unsuitable for vegetation)

land
water

1 **lithosere** (rock)
2 **psammosere** (sand)
3 **halosere** (salt water)
4 **hydrosere** (fresh water)

e.g. new volcanic island, emerging raised beach, retreat of a glacier
e.g. sand dunes
e.g. salt marsh
e.g. infilling lake, pond

increasing soil depth and nutrients
increasing stability and humus
decreasing influence of the sea
decreasing influence of open water

several stages (**seres**)

mesophytic (transitional: adapted to neither very dry conditions [xerophytic] nor very wet conditions [hydrophytic]) e.g. oak climax

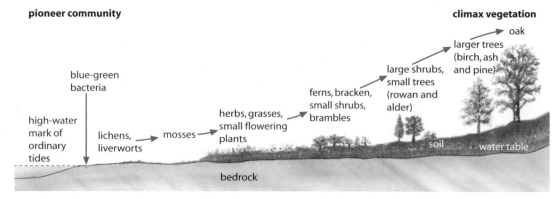

Figure 11.5

Fieldsketch of a lithosere on a newly emerging rocky coastline (raised beach), Arran

Four basic seres forming a primary succession

1 Lithosere

Areas of bare rock will initially be colonised by blue-green bacteria and single-celled photosynthesisers which have no root systems and can survive where there are few mineral nutrients. Blue-green bacteria are autotrophs (page 275), photosynthesising and producing their own organic food source. Lichens and mosses also make up the pioneer community (Figure 11.5). These plants are capable of living in areas lacking soil, devoid of a permanent supply of water and experiencing extremes of temperature. Lichen and various forms of weathering help to break up the rock to form a veneer of soil in which more advanced plant life can then grow. As these plants die, they are converted by bacteria into humus which develops a richer soil. Seeds, mainly of grasses, then colonise the area. As these plants are taller than the pioneer species, they will replace the lichen and mosses as the dominants — although the lichens and mosses will still continue to grow in the community. As the plant succession evolves over a period of time, the grasses will give way as dominants to fast-growing shrubs, which in turn will be replaced by relatively fast-growing trees

(rowan). These will eventually face competition from slower-growing trees (ash) and, finally, the oak which forms the climax vegetation. It should be noted that although each stage of the succession is marked by a new dominant, many of the earlier species remain growing although some are shaded out.

Figure 11.5 shows an idealised primary succession across a newly emerging rocky coastline. It excludes the increasing number of species found at each stage of the seral succession. The species are determined by local differences in rainfall, temperature and sunlight, bedrock and soil type, aspect and relief. Lithoseres can develop on bare rock exposed by a retreating glacier (page 273), on ash or lava following a volcanic eruption on land (Krakatoa, Places 27) or forming a new island (Surtsey, Places 3, page 16), or, as in Figure 11.5, on land emerging from the sea as a result of isostatic uplift following the melting of an ice cap (page 148).

Over time, the area shown to have the pioneer community passes through several stages until the climatic climax is reached— assuming that the land continues to rise, that there is no significant change in the local climate, and that there is no human interference. Figure 11.6 shows two stages in the succession taken on a raised beach on the

Figure 11.6

Primary succession on a lithosere, Arran
a Lichen, mosses and grasses on a rocky coastline
b Bracken and deciduous trees behind a rocky beach

Biogeography

east coast of Arran. Figure 11.6a shows lichen, favouring a south-facing aspect on gently dipping rocks, and mosses, growing in darker north-facing hollows. Beyond, where soil has begun to form and where the water table is high, grasses and bog myrtle have entered the succession. Figure 11.6b was taken where the soil depth and amount of humus have increased and the water table is lower, as indicated by the presence of bracken. The reeds to the right are found in a hollow where the water table is nearer to the surface. In the middle distance are small deciduous trees with, behind them, taller oaks indicating a climax vegetation.

Places 27 Krakatoa

In August 1883, a series of volcanic eruptions reduced the island of Krakatoa to one-third of its previous size and left a layer of ash over 50 m deep. No vegetation or animal life was left on the island or in the surrounding sea. Yet within three years (Figure 11.7), 26 species had reappeared and, in 1933, 271 plant and 720 insect species, together with several reptiles, were recorded. The first recolonisers arrived in three ways. Most were seeds blown from surrounding islands by the wind, while others drifted in from the sea or were carried by birds. However, in this example, the concept of plant succession, put forward later by F. E. Clements in 1916, seems open to dispute, as many of the plants which recolonised Krakatoa arrived there by chance — for example, a piece of driftwood with a particular seed type just happened to be washed ashore, whereas it could just as easily have missed the island altogether.

Figure 11.7

Primary succession, Krakatoa: vegetation distribution according to height above sea level, 1983

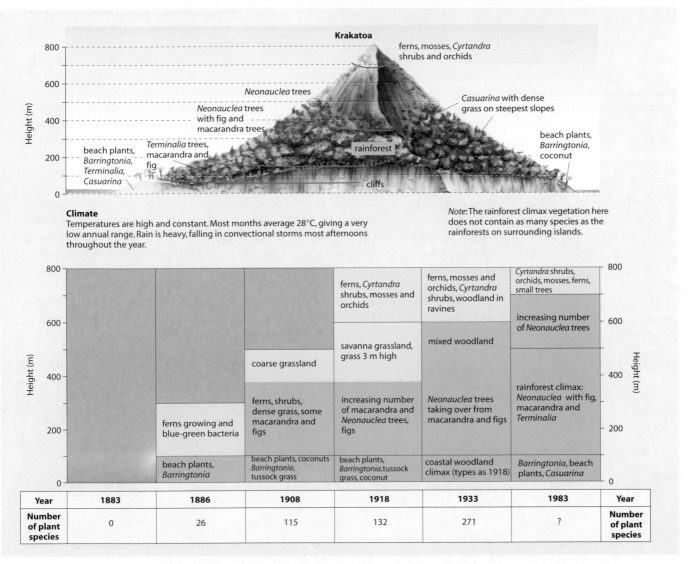

Climate
Temperatures are high and constant. Most months average 28°C, giving a very low annual range. Rain is heavy, falling in convectional storms most afternoons throughout the year.

Note: The rainforest climax vegetation here does not contain as many species as the rainforests on surrounding islands.

Year	1883	1886	1908	1918	1933	1983	Year
Number of plant species	0	26	115	132	271	?	Number of plant species

Q 1 With reference to Chapter 1, explain why a volcanic eruption occurred on Krakatoa.

2 Using Figure 11.7, describe carefully and give reasons for the secondary succession on Krakatoa following the eruption in 1883.

3 With reference to Figure 11.7, say to what extent you consider that a climax vegetation has again been reached on the island. Give reasons for your answer.

Figure 11.9

Primary succession on a psammosere: colonisation of foredunes (compare Figures 6.30 and 6.31)

Figure 11.8

Transect across sand dunes to show a psammosere, Morfa Harlech, north Wales

2 Psammoseres

A psammosere succession develops on sand and is best illustrated by taking a transect across coastal dunes (Figure 11.8). The first plants to colonise, indeed to initiate dune formation, are usually lyme grass, sea couch grass and marram grass. Sea couch grass grows on berms around the tidal high-water mark and is often responsible for the formation of embryo dunes (Figure 6.29). On the yellow fore-dunes, which are arid, being above the highest of tides and experiencing rapid percolation by rainwater, marram grass becomes equally important.

The main dune ridge, which is extremely arid and exposed to wind, is likely to be vegetated exclusively by marram grass. Marram has adapted to these harsh conditions by having leaves which can fold to reduce surface area, which are shiny and which can be aligned to the wind direction: three factors capable of limiting evapotranspiration. Marram also has long roots to tap underground water supplies and is able to grow upwards as fast as sand deposition can cover it. Grey dunes, behind the main ridge, have lost their supply of sand and are sheltered from the prevailing wind. Their greater humus content, from the decomposition of earlier marram grass, enables the soil to hold more moisture. Although marram is still present, it faces increasing competition from small, flowering plants and herbs such as sea spurge (with succulent leaves to store water) and heather.

The older ridges, further from the water, have both more and taller species. Dune slacks may form in hollows between the ridges if the water table reaches the surface. Plants such as creeping willow, yellow iris, reeds and rushes and shrubs are indicators of a deeper and wetter soil. On the landward side of the dunes, perhaps 400 m from the beach, are small deciduous trees including ash and hawthorn and, as the soil is sandy, pine plantations. Furthest inland comes the oak climax. Figure 11.8 shows a psammosere based on sand dunes at Morfa Harlech, north Wales. Figures 11.9 and 6.30 show marram and lyme grass forming the yellow fore-dunes, with gorse and heather on the greyer dunes behind. Figures 11.10 and 6.31 show vegetation on the inland ridges.

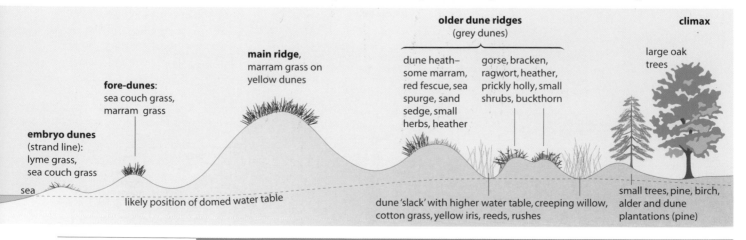

older dune ridges (grey dunes)   **climax**

main ridge, marram grass on yellow dunes

dune heath– some marram, red fescue, sea spurge, sand sedge, small herbs, heather

gorse, bracken, ragwort, heather, prickly holly, small shrubs, buckthorn

large oak trees

fore-dunes: sea couch grass, marram grass

embryo dunes (strand line): lyme grass, sea couch grass

sea

likely position of domed water table

dune 'slack' with higher water table, creeping willow, cotton grass, yellow iris, reeds, rushes

small trees, pine, birch, alder and dune plantations (pine)

Figure 11.11

A saltpan on the Suffolk coast, covered only by the highest of tides

Figure 11.10

Vegetation across a grey dune ridge and a dune slack, Braunton Burrows, Devon

Figure 11.12

Transect showing a primary succession in a halosere, Llanhridian Marsh, Gower Peninsula, south Wales

3 Haloseres

In river estuaries, large amounts of silt are deposited by the ebbing tide and inflowing rivers. The earliest plant colonisers are green algae and eel grass which can tolerate submergence by the tide for most of the 12-hour cycle and which trap mud, causing it to accumulate. Two other colonisers are *Salicornia* and *Spartina townsendii* which are **halophytes** — i.e. plants which can tolerate saline conditions. They grow on the inter-tidal mud flats (Figure 6.32), with a maximum of 4 hours exposure to the air in every 12 hours. *Spartina* has long roots enabling it to trap more mud than the initial colonisers of algae and *Salicornia*, and so, in most places, it has become the dominant vegetation. The inter-tidal flats receive new sediment daily, are waterlogged to the exclusion of oxygen, and have a high pH value.

The sward zone, in contrast, is inhabited by plants which can only tolerate a maximum of 4 hours submergence in every 12 hours. Here the dominant species are sea lavender, sea aster and grasses, including the 'bowling green turf' of the Solway Firth. However, although the vegetation here tends to form a thick mat, it is not continuous. Hollows may remain where the sea water becomes trapped leaving, after evaporation, saltpans in which the salinity is too great for plants (Figure 11.11). As the tide ebbs, water draining off the land may be concentrated into creeks (Figure 6.33). The upper sward zone is only covered by spring tides and here *Juncus* and other rushes grow. Further inland, non-halophytic grasses and shrubs enter the succession, to be followed by small trees and ultimately by the climax oak vegetation. Figure 11.12 is a transect based on the salt marshes on the north coast of the Gower Peninsula in south Wales. Figure 11.13 shows several stages in the halosere succession.

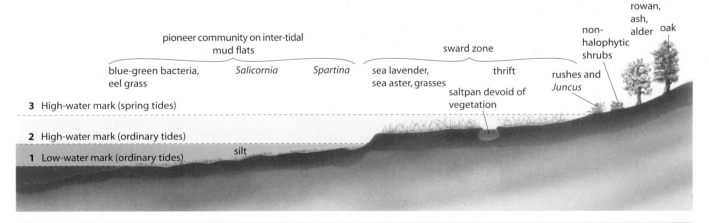

pioneer community on inter-tidal mud flats

sward zone

rowan, ash, alder oak

non-halophytic shrubs

blue-green bacteria, eel grass — *Salicornia* — *Spartina* — sea lavender, sea aster, grasses — thrift — saltpan devoid of vegetation — rushes and *Juncus*

3 High-water mark (spring tides)

2 High-water mark (ordinary tides)

1 Low-water mark (ordinary tides)

silt

Figure 11.13

Primary succession in a halosere

4 Hydroseres

Lakes and ponds originate as clear water which contains few plant nutrients. Any sediment carried into the lake will enrich its water with nutrients and begin to infill it. The earliest colonisers will probably be algae and mosses whose spores have been blown onto the water surface by the wind. These grow to form vegetation rafts which provide a habitat for bacteria and insects. Next will be water-loving plants which may either grow on the surface, e.g. water lilies and pondweed, or be totally submerged (Figure 11.14). Bacteria recycle the nutrients from

Figure 11.14

Idealised primary succession in a hydrosere

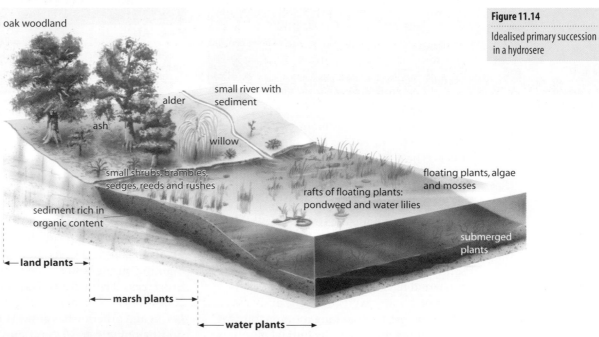

oak woodland

alder

small river with sediment

ash

willow

small shrubs, brambles, sedges, reeds and rushes

floating plants, algae and mosses

sediment rich in organic content

rafts of floating plants: pondweed and water lilies

submerged plants

←— land plants —→

←— marsh plants —→

←— water plants —→

Figure 11.15

Primary succession in a hydrosere at the head of a reservoir in Cumbria

the pioneer community, and marsh plants such as bulrushes, sedges and reeds begin to encroach into the lake. As these marsh plants grow outwards into the lake and further sediment builds upwards at the expense of the water, small trees will take root forming a marshy thicket. In time, the lake is likely to contract in size, to become deoxygenised by the decaying vegetation and eventually to disappear and be replaced by the oak climax vegetation. This primary succession is shown in Figure 11.14. Figure 11.15 shows land plants encroaching at the head of a reservoir, while Figure 11.16 illustrates the water, marsh and land plant succession in and around a small lake.

Incidentally, you do not have to be an expert botanist to recognise the plants named in these primary successions; you just need access to a good plant recognition book!

Figure 11.16

Plant succession in a small lake, Sussex

Figure 11.17

Vegetation succession in a lake (hydrosere) in lowland England

Q Study Figure 11.17a, which shows the vegetation succession in a lake (a hydrosere) in lowland England, and Figure 11.17b, which plots the number of species in the various communities over time. The lake covers an area of 6 sq km and has a maximum depth of 5 m. Describe and suggest reasons for the changes in vegetation.

a the main stages of the hydrosere

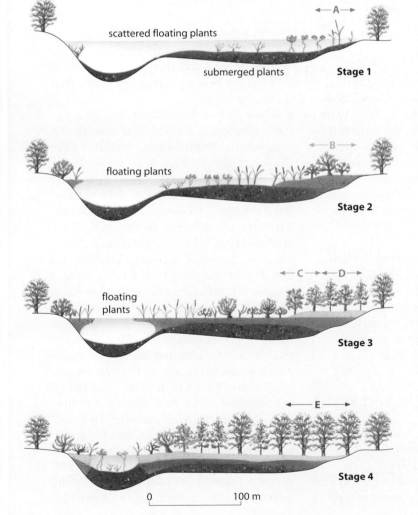

scattered floating plants

submerged plants

Stage 1

floating plants

Stage 2

floating plants

Stage 3

Stage 4

0   100 m

b changes in vegetation communities at each stage

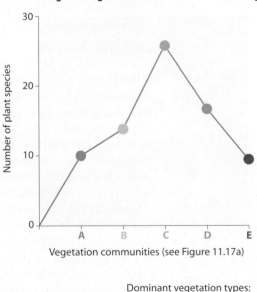

Dominant vegetation types:

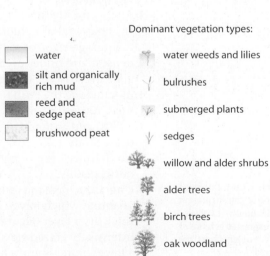

water	water weeds and lilies
silt and organically rich mud	bulrushes
reed and sedge peat	submerged plants
brushwood peat	sedges
	willow and alder shrubs
	alder trees
	birch trees
	oak woodland

Secondary succession

A climatic climax occurs when there is stability in transfers of material and energy in the ecosystem (page 274) between the plant cover and the physical environment. However, there are several factors which can arrest the plant succession before it has achieved this dynamic equilibrium, or which may alter the climax after it has been reached. These include:

- a mudflow or landslide (Places 28);
- deforestation or afforestation;
- overgrazing by animals or the ploughing up of grasslands;
- burning grasslands, moorlands, forests and heaths;
- draining wetlands;
- disease (e.g. Dutch elm); and
- changes in climate.

Places 28 Mudflow on Arran

The mudflow shown in Figure 2.15 occurred in October 1981 and completely covered all the existing vegetation. Twelve months later it was estimated that 20 per cent of the flow had been recolonised, a figure which had grown to 40 per cent in 1984 and 70 per cent in 1988. Had this been a primary succession, lichens and mosses would have formed the pioneer community and they would probably have covered only a small area. The pioneer plants would probably also have been randomly distributed and, even after seven years, the species would have been few in number and small in height.

Instead, by 1988, much of the flow had already been recolonised. It could be seen that most of the plants were found near the edges of the flow and were not randomly distributed, and there were already several species including grasses, heather, bog myrtle and mosses, some of which exceeded 50 cm in height.

These observations suggested a secondary succession with plants from the surrounding climax area having invaded the flow, mainly due to the dispersal of their seeds by the wind.

The effect of fire

The severity of a fire and its effect on the ecosystem depends largely upon the climatic conditions at the time. The fire is likely to be hottest in dry weather and, in the northern hemisphere, on sunny south-facing slopes where the vegetation is driest. The spread of a fire is fastest when the wind is strong and blowing uphill. The extent of disruption also depends upon the type and the state of the vegetation. The following is a list of examples in rank order of severity.

1 Areas with Mediterranean climate where the chaparral of California and the maquis–garrigue of southern Europe are densest and tinder-dry in late summer after the seasonal drought. Recent examples include the bush fires in south-east Australia and the south of France (late 1980s) and around Los Angeles in 1993 (Case Study 15a).
2 Coniferous forests where the leaf litter burns readily.
3 Ungrazed grasslands and, especially, the savannas which have a low biomass but a thick litter layer (Figure 11.28).
4 Intensively grazed grasslands which have a lower biomass and a limited litter layer.

5 Deciduous woodlands which, despite the presence of a thick litter layer, are often slow to burn.

Following a fire, the blackened soil has a lower albedo and absorbs heat more readily and, without its protective vegetation cover, the soil is more vulnerable to erosion. Ash initially increases considerably the quantity of inorganic nutrients in the soil and bacterial activity is accelerated. Any seedlings left in the soil will grow rapidly as there is now plenty of light, no smothering layer of leaf litter, plenty of nutrients, a warmer soil and, at first, less competition from other species. Heaths and moors which have been fired are conspicuous by their greener, more vigorous growth. A fire climax community contains plants with seeds which have a thick protective coat and which may germinate because of the heat of the fire. The community may have a high proportion of species which can sprout quickly after the fire — plants which are protected by thick, insulating bark (cork oak in the chaparral and baobab in the savannas (Figure 12.14)), or which have underground tubers or rhizomes insulated by the soil. It has been suggested that the grasslands of the American

Biomass is the total mass of living organisms present in a community at a given time; expressed in terms of oven-dry weight (mass) per unit area.

Prairies and the African savannas are not climatic climax vegetation, but are the result of firing by indigenous Indian and African tribes.

Vegetation changes in the Holocene

The Holocene is the most recent of the geological periods (Figures 1.1 and 11.18). The last glacial advance in Britain ended about 18 000 BP. Although the extreme south of England remained covered with hardy tundra plants, most of northern Britain was left as bare rock or glacial till. Had the climate gradually and constantly ameliorated, a primary succession would have taken place, from south to north, as previously described for a lithosere. However, we know that there have been several major fluctuations in climate during those 18 000 years which have resulted in significant changes in the climax vegetation (Figure 11.18).

There are several techniques for determining vegetation change, including pollen analysis, dendrochronology, radiocarbon dating and historical evidence (page 232). Each plant has a characteristic pollen grain. Where pollen is blown by the wind onto peat bogs, such as at Tregaron in west Wales, the grains are trapped by the peat. As more peat accumulates over the years, the pollen of successively later times indicates which were the dominant and subdominant plants of the period (Figures 11.18 and 11.19). As each plant grows best within certain defined temperature and precipitation limits, it is possible to determine when the climate either improved (ameliorated) or deteriorated. Dendrochronology — dating by means of the annual growth rings of trees — has shown that the bristlecone pine of California can be dated back some 5000 years. Few British trees go back even as far as the first written records. Radio-carbon dating is based on changing amounts of radioactivity in the atmosphere and in plants. Notice in Figure 11.18, which links climatic and vegetation changes, how forests increase as the climate ameliorates, and how heathland and peat moors take over when the climate deteriorates.

Figure 11.18

Climatic and vegetation changes in Britain during the Holocene

	Date	Phase/period	Climate	Vegetation	Cultures
BC	pre-15 000	final glaciation	glacial	none in northern Britain; tundra in southern England	none
	15 000–12 000	periglacial	cold, 6°C in summer	tundra	Palaeolithic
	12 000–10 000	Allerød	warming slowly to 12°C in summer	tundra with hardy trees, e.g. willow and birch	Palaeolithic
	10 000–8 000	pre-Boreal	glacial advance: colder, 4°C in summer	Arctic/Alpine plants, tundra	Mesolithic
	8 000–6 000	Boreal	continental: winters colder and drier, summers warmer than today	forests: juniper first then pine and birch and finally oak, elm and lime	Mesolithic
	6 000–3 000	Atlantic	maritime: warm summers, 20°C; mild winters, 5°C; wet	our 'optimum' climate and vegetation: oak, alder, hazel, elm and lime (too cold for lime today); peat on moors	beginning of Neolithic; first deforestation about 3500 BC
	3 000–500	sub-Boreal	continental: warmer and drier	elm and lime declined; birch flourished; peat bogs dried out	Neolithic period, settled agriculture; beginning of Bronze Age
	500–0	sub-Atlantic	maritime: cooler, stormy and wet	peat bogs re-formed; decline in forests due to climate and farming	settled agriculture
AD	0–1000	historical times	improvement: warmer and drier	clearances for farming	Roman occupation during early part
	1000–1550		decline: much cooler and wetter	further clearances: little climax vegetation left; medieval farming	
	1550–1700		'little ice age': colder than today		
	post–1700		gradual improvement	recently some afforestation: coniferous trees	Agrarian and Industrial Revolutions

Q Figure 11.19 shows how the surface of Britain has changed in the last 12 000 years. Describe the conditions in

a 10 000 BC;

b 7000 BC;

c 4000 BC;

d 1000 BP; and

e the present.

In each case, give reasons for your answer.

Ecology and ecosystems

The term **ecology**, which comes from the Greek word *oikos* meaning 'home', refers to the study of the interrelationships between organisms and their habitats. An organism's home or **habitat** lies in the biosphere—i.e. the surface zone of the earth and its adjacent atmosphere in which all organic life exists. The scale of each home varies from small **micro-habitats**, such as under a stone or a leaf, to **biomes**, which include tropical rainforests and deserts (Figure 11.20). Fundamental to the four ecological units listed in Figure 11.20 is the concept of the **environment**. The environment is a collective term to include all the conditions in which an organism lives. It can be divided into:

a the physical, non-living or **abiotic environment**, which includes temperature, water, light, humidity, wind, carbon dioxide, oxygen, pH, rocks and nutrients in the soil; and

b the living or **biotic environment**, which comprises all organisms: plants, animals, humans, bacteria and fungi.

The ecosystem

An ecosystem is a natural unit in which the life-cycles of plants, animals and other organisms are linked to each other and to the non-living constituents of the environment to form a natural system (Framework 1, page 39). The **community** consists of all the different species within a habitat or ecosystem. The **population** is all the individuals of a particular species in a habitat.

An ecosystem depends on two basic processes: **energy flows** and **material cycling**. As the flow of energy is only in one direction and because it crosses the system boundaries, this aspect of the ecosystem behaves as an **open** system. Nutrients, which are constantly recycled for future use, are circulated in a series of **closed** systems.

1 Energy flows

The sun is the primary source of energy for all living things on earth. As energy is retained only briefly in the biosphere before being returned to space, ecosystems have to rely upon a continual supply. The sun provides *heat energy* which cannot be captured by plants or animals but which warms up the communities and their non-living surroundings. The sun is also a source of *light energy* which can be captured by green plants and transformed into chemical energy through the process of **photosynthesis**. Without photosynthesis, there would be no life on earth. Light, chlorophyll, warmth, water and carbon dioxide are required for this process to operate. Carbon dioxide, which is absorbed through stomata in the

Figure 11.20

A hierarchical structure of ecological units

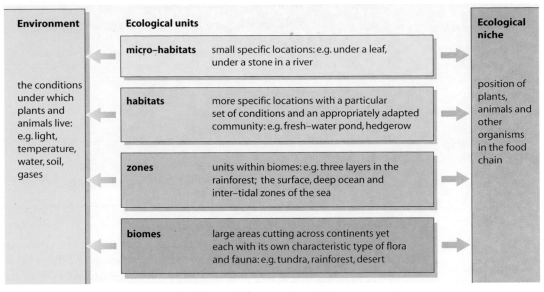

leaves of higher plants, reacts indirectly with water taken up by the roots when temperatures are suitably high, to form carbohydrate. The energy needed for this comes from sunlight which is 'trapped' by chlorophyll. Oxygen is a by-product of the process. The carbohydrate is then available as food for the plant.

Food chains and trophic levels

A food chain arises when energy, trapped in the carbon compounds initially produced by plants through photosynthesis, is transferred through an ecosystem. Each link in the chain feeds on and obtains energy from the one preceding it, and in turn is consumed by and provides energy for the following link (Figure 11.21).

Figure 11.21

Three examples of food chains through four trophic levels

Level 1	Level 2	Level 3	Level 4
grass	worm	blackbird	hawk
leaf	caterpillar	shrew	badger
phytoplankton	zooplankton	fish	human

There are usually, but not always, four links in the chain. Each link or stage is known as a **trophic** or **energy level** (Figure 11.22). In order for the first link in the chain to develop, the non-living environment has to receive both energy from the sun and the other factors (water, CO_2, etc.) needed for photosynthesis.

The **first trophic level** is occupied by the **producers** or **autotrophs** ('self-feeders') which include green plants capable of producing their own food by photosynthesis. All other levels are occupied by **consumers** or **heterotrophs** ('other-feeders'). These include animals which obtain their energy either by eating green plants directly or by eating animals which have previously eaten green plants. The **second trophic level** is where **herbivores**, the primary consumers, eat the producers. The **third trophic level** is where smaller **carnivores** (meat-eaters) act as secondary consumers feeding upon the herbivores. The **fourth trophic level** is occupied by the larger carnivores, the tertiary consumers. Also known as **omnivores** (or diversivores), this group — which includes humans — eats both plants and animals and so has two sources of food. Figure 11.22 shows the main trophic or feeding levels in a food chain. **Detritovores**, such as bacteria and fungi, are consumers which operate at all trophic levels.

Figure 11.22

Trophic levels

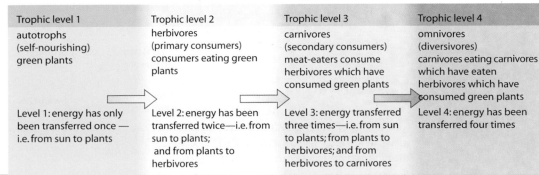

Figure 11.23

Energy flows in the ecosystem

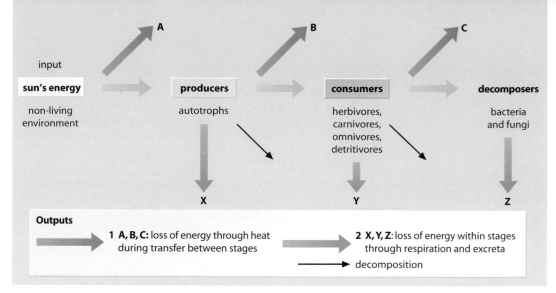

However, no transfer of energy is 100 per cent efficient and, as Figure 11.23 shows, energy is lost through respiration, by the decay of dead organisms and in excreta within each unit of the food chain, and also as heat given off when energy is passed from one trophic level to the next. Consequently, at each higher level, fewer organisms can be supported than at the previous level, even though their individual size generally increases. Simple food chains are rare; there is usually a variety of plants and animals at each level forming a more complicated **food web**. This range of species is necessary since a sole species occupying a particular trophic level in a simple food chain could be 'consumed' and this would adversely affect the organisms in the succeeding stages.

The progressive loss of energy through the food chain imposes a natural limit on the total mass of living matter (the **biomass**) and on the number of organisms that can exist at each level. It is convenient to show these changes in the form of a pyramid (Figure 11.24). A pyramid of organism numbers is of limited value for comparing ecosystems for two reasons. First, it is difficult to count the

numbers of grasses or algae per unit area. Secondly, it does not take into account the relative sizes of organisms — a bacterium would count the same as a whale! A pyramid of biomass takes into account the difference in size between organisms, but cannot be used to compare masses at different trophic levels in the same ecosystem or at similar trophic levels in different ecosystems. This is because biomass will have accumulated over different periods of time.

Humans are found at the end of a food chain and human population is dependent upon the length of the chain (and therefore the amount of energy lost). In other words, in a shorter food chain, less energy will have been lost by the time it reaches humans and so the land can support a higher density of population. In a longer food chain, more energy will have been lost by the time the food is consumed by humans which means that the carrying capacity (page 356) is lower and fewer people can be supported by a given area of land — as in western Europe, where most of the population are accustomed to animal products as well as crops.

Figure 11.24

The trophic pyramid

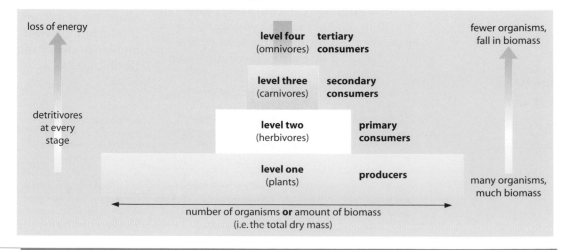

2 Material cycling

Chemicals needed to produce organic material are circulated around the ecosystem and are continually recycled. Various chemicals can be absorbed by plants either as gases from the atmosphere or as soluble salts from the soil. Each cycle consists, at its simplest, of plants taking up chemical nutrients which, once they have been used, are passed on to the herbivores and then the carnivores which feed upon them. As organisms at each of these trophic levels die, they decompose and nutrients are returned to the system. Two of these cycles, the carbon and nitrogen cycles, are illustrated in Figures 11.25 and 11.26. In each case, the most basic cycle is given (diagram **a**); followed by a more detailed example, although still not in its total complexity (diagram **b**).

Figure 11.25

The carbon cycle (*after* M. B. V. Roberts)

Figure 11.26

The nitrogen cycle (*after* M. B. V. Roberts)

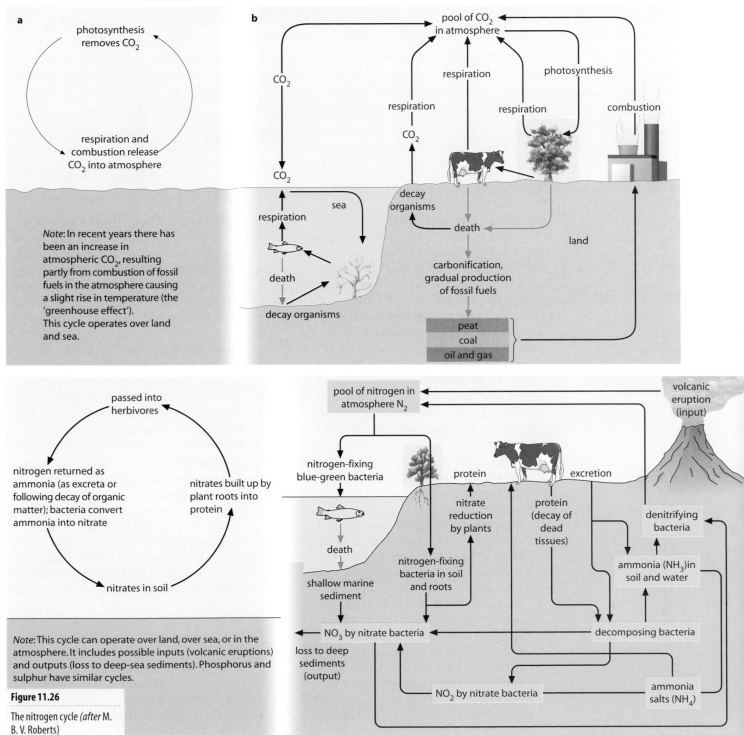

Note: In recent years there has been an increase in atmospheric CO_2, resulting partly from combustion of fossil fuels in the atmosphere causing a slight rise in temperature (the 'greenhouse effect'). This cycle operates over land and sea.

Note: This cycle can operate over land, over sea, or in the atmosphere. It includes possible inputs (volcanic eruptions) and outputs (loss to deep-sea sediments). Phosphorus and sulphur have similar cycles.

Figure 11.27

A model of the mineral
nutrient cycle
(*after* Gersmehl)

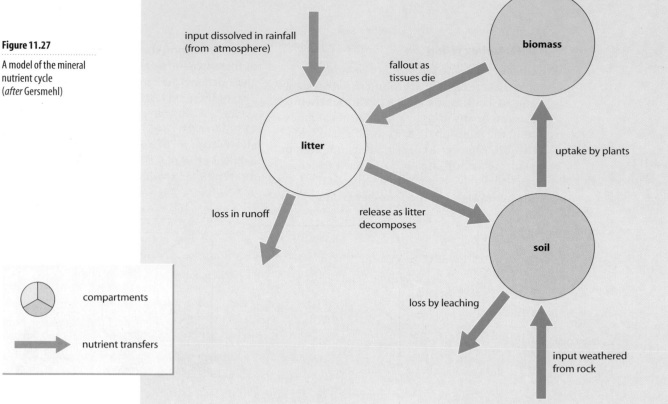

Figure 11.27 A model of the mineral nutrient cycle (after Gersmehl)

Model of the mineral nutrient cycle

This model, developed by P. F. Gersmehl in 1976, attempts to show the differences between ecosystems in terms of nutrients stored in, and transferred between, three compartments (Figure 11.27): **litter** — the surface layer of vegetation which may eventually become humus; **biomass** — the total mass of living organisms, mainly plant tissue, per unit area; and **soil**.

Figure 11.28 shows the mineral nutrient cycles for three biomes: **a** the coniferous forest (taiga); **b** the temperate grassland (prairies and steppes); and **c** the tropical rainforest (selvas):

a Taiga Litter is the largest store of mineral nutrients in the taiga. Although forest, the biomass is relatively low because the coniferous trees form only one layer, have little undergrowth, contain a limited variety of species and have needle-like leaves. The soil contains few nutrients because, following their loss through leaching and as surface runoff (after snowmelt when the ground is still frozen), replacement is slow: the low temperatures restrict the rate of chemical weathering of parent rock. The layer of needles is often thick, but their thick cuticles and the low temperatures discourage the action of the decomposers (page 246). The breakdown of litter into humus is thus very slow. These factors account for the low fertility potential of the podsol soils of the taiga (pages 307–8).

b Steppes/prairies Soil is the largest store of mineral nutrients in the temperate grasslands. The biomass store is small due to the climate which provides insufficient moisture to support trees and temperatures cold enough to reduce the growing season to approximately six months. Indeed, much of the biomass is found beneath the surface as rhizomes and roots. The grass dies back in winter and nutrients are returned rapidly to the soil. The soil retains most of these nutrients because the rainfall is insufficient for effective leaching and the climate is conducive to both chemical and physical weathering which release further nutrients from the parent rock. The presence of bacteria also speeds up the return of nutrients from the litter to the soil. These factors help to account for the high fertility potential of the black chernozem soils associated with temperate grasslands (pages 302–3).

c Selvas The tropical rainforests have, of all the major environments, the highest rates of transfer—an annual rate ten times greater than that of the taiga. The biomass is the largest store of mineral nutrients in the tropical rainforests. High annual temperatures, the heavy, evenly distributed rainfall and the year-long growing season all contribute to the tall, dense and rapid growth of vegetation. The biomass is composed of several layers of plants and countless different species. The many plant roots take up vast amounts of nutrients. In comparison, the litter store is limited, despite the continuous fall of leaves, because the hot, wet climate provides the ideal environment for bacterial action (both in numbers and type) and the decomposition of dead vegetation. In areas where the forest is cleared, the heavy rain soon removes the nutrients from the soil by leaching or surface runoff. The litter layer is rapidly reduced so that the soil has to rely upon the replacement of nutrients from the bedrock, a process which is usually rapid as climatic conditions are optimal for chemical weathering (Figure 2.9b). Initially nutrients such as phosphorus may increase if the forest is burnt, but deforestation usually leads to a rapid decline in the fertility of most ferrallitic soils (pages 292–94).

Q

1 Use Figure 11.28 to answer the following questions:
  a Explain the differences between the tropical rainforest (selvas) and the coniferous forest (taiga) ecosystems in terms of their major nutrient stores.
  b Explain the differences between the tropical rainforest (selvas) and temperate grassland (prairie) ecosystems in terms of their three major nutrient stores.
  c Why are the transfer flows in the taiga ecosystem so small?
  d Why are the transfer flows in the tropical rainforest ecosystem so high?

Figure 11.28

Mineral nutrient cycle in three different environments *(after Gersmehl)*

a taiga (northern coniferous forest)

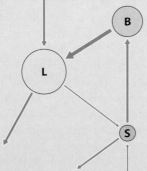

b steppe and prairie (mid–latitude continental grassland)

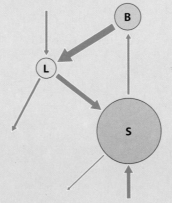

c selvas (tropical rainforest)

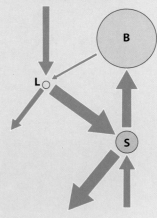

B biomass **L** litter **S** soil

 compartments (circle size proportional to amount stored)

 nutrient transfers (arrows are proportional to amount of flow)

Figure 11.29

Characteristics of four ecosystems

2 Study Figure 11.29 which shows the characteristics of four ecosystems, and then answer the following questions.
  a Draw an accurate diagram to show the amount of nutrients stored in each compartment and the quantities of nutrient transfer within the tropical grassland (savanna) ecosystem.
  b For either the forest cycles (A in Figure 11.29) or the grassland cycles (B in Figure 11.29) explain the similarities and differences between the cycles.
  c For each of the four ecosystems, explain how human activities can interrupt or modify the nutrient cycle. Refer in detail to examples you have studied and, if possible, attempt to show, as an annotated diagram, the altered mineral nutrient cycle.

		Nutrient storage			Annual nutrient transfer		
	Ecosystem type	Stored in biomass	Stored in litter	Stored in soil	Soil to biomass	Biomass to litter	Litter to soil
Forest cycles **A**	Equatorial rainforest	11 081	178	352	2 028	1 540	4 480
	Coniferous forest	3 350	2 100	142	178	145	86
Grassland cycles B	Tropical savanna	978	300	502	319	312	266
	Temperate prairie/steppe	540	370	5 000	422	426	290
All measurements in kilogrammes per hectare							

Most of the eastern coast of Africa is protected by coral reefs (Case Study 17). Coral, who live in clear, warm, shallow tropical waters, are small organisms which exude lime. For centuries, coast-dwellers have hacked out blocks of dead coral to build their houses and mosques. In 1954, the Bamburi Portland Cement Company built a factory 10 km north of Mombasa, Kenya, to produce cement, and began the open-cast extraction of coral (page 473). Cement was essential to Kenya partly to help in the internal development of the country and partly as a vital export earner. By 1971, over 25 million tonnes of coral had been quarried, leaving a sterile wasteland covering 3.5 sq km. On that land there were no plants, no wildlife, no soil: it was a dead ecosystem. The Swiss-owned

multinational cement company then appointed Dr Rene Haller to restore the environment from what he himself described as "a lunar landscape filled with saline pools" (Figure 11.30).

After trying 26 different types of tree, Dr Haller found the key to be the *Casuarina* tree (Figure 11.31). This pioneer tree grew by 3 m a year, flourished in the sterile coral rubble, and was able to withstand both the high salinity and the high ground temperatures (up to 40°C). The constant fall of the needle-type leaves provided a habitat for red-legged millipedes which, together with the *Casuarina's* ability to 'fix' atmospheric nitrogen, helped with the formation of the first soil and provided the base for a new ecosystem. The soil was collected and more trees were planted. Over the next few years, indigenous herbs, grasses and tree species, as well as beetles, spiders and small animals were introduced into the young forest, each with its own function (niche) in the developing ecosystem. The depth of the ponds and lakes was increased until they reached the ground watertable so that a fresh-water habitat was created for fish (initially the local tilapia which are tolerant of saline water), crocodiles and hippopotami. Hippopotami excrement stimulated the growth of algae which oxygenated the water, preventing eutrophication. After only 20 years, the soil depth had reached 20 cm and the rainforest, with over 220 tree species, had become sufficiently restored to be home for over 180 recorded species of bird. The ecosystem was completed with the introduction of grazing animals (herbivores) such as the buffalo, oryx, antelope and giraffe. The re-creation of the rainforest (Figure 11.32a and b) had been completed without the use of artificial fertiliser and insecticides as Dr Haller considered these to be incompatible with his concept of a complex, balanced ecosystem.

The project has not only been an environmental success, it has also become a sustainable commercial venture with income derived from, for example, the sale of timber, bananas, vegetables, crocodiles and honey. The main source of the economy is the integrated aquaculture system (Figure 11.33) with, at its centre, the tilapia fish farm. The nutrients in the effluent water are used as fertiliser in the adjacent fruit plantation and for biogas to

Figure 11.30

The Bamburi Quarry

Figure 11.31

Casuarina trees planted in coral rubble, Bamburi Quarry

Figure 11.32

The re-created rainforest ecosystem, Bamburi Quarry

operate the pumps. From here, the water is led through a rice field into settlement ponds, where 'Nile cabbage' is grown for use in clearing the fish ponds. A crocodile farm is attached to the water system as crocodile waste, rich in phosphate and nitrogen, is a valuable fertiliser. The crocodiles are part of a planned food chain. Not only are they fed on surplus tilapia, but their eggs are eaten by monitor lizards who help to control the snake population which in turn controls the rodent population. Tourism has become a recent source of income. Baobab Farm, the name of the restored area, is open to school parties each morning and visitors in the afternoon. In 1992 it received over 100 000 visitors, making it the largest attraction in the Mombasa area. In brief, the once-barren quarry is now an ecologically and economically sound enterprise.

Dr Haller also believes that his intensive aquaculture and agroforestry techniques, geared to maximum yield of food, fuel and income from minimum acreage and inputs, offer significant hope for small-scale African farmers short of fertile land on a continent with an explosive population growth and ravaged by environmental and human-created disasters. He suggests that these methods could easily be adapted by Africans since their genesis lies in tribal techniques taught to him by local farmers.

Figure 11.33

The Baobab Farm integrated aquaculture system

Figure 11.34

From the Bamburi Quarry Nature Trail leaflet

"In 1971 the Bamburi Portland Cement Company embarked on a unique project to re-create a living environment in the vast lunar landscape of its quarry. To-day an extraordinary change is evident. Behind the glamorous façade of the tropical paradise regained, a completely balanced and commercially viable aquaculture complex is an important part of the established ecosystem. In a kind of giant jig-saw, casuarina trees, millipedes and bacteria work together, to provide the fertile basis for reclamation. Of particular interest is the surprising variety of wildlife playing a part in the natural balance of the whole system. The living creation of the Bamburi Quarry Nature Trail holds more than just a visual experience, a walk around will enable you to piece together the giant jig-saw puzzle of wasteland rehabilitation ..."

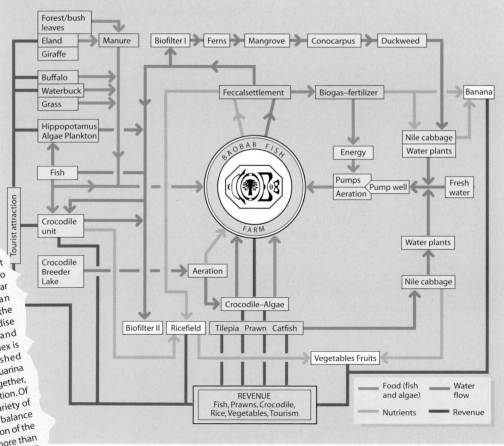

Biomes

A biome is a large global ecosystem. Each biome obtains its name from the dominant type of vegetation found within it (temperate grassland, coniferous forest, etc.). Each contains climax communities of plants and animals and can be closely linked to zonal soil types and animal communities. Climate has usually been the major controlling factor in the location and distribution of biomes, but economic development has transformed many of these natural systems. A biome can extend across a large part of a continent while its characteristics may be found in several continents (deserts and tropical rainforests). Although some authorities suggest that it is 'old fashioned' to link together climate, vegetation and soils in a 'natural region', the concept is still useful and convenient as a framework of study and as a valid hypothesis for investigation. Four main factors, **climatic**, **topographic**, **edaphic** and **biotic**, interrelate to produce and control each biome.

1 Climatic factors

- **Precipitation** largely determines the vegetation type, e.g. forest, grassland or desert. The annual amount of precipitation is usually less important than its effectiveness for plant growth—for example: How long is any dry season? Does the area receive steady, beneficial rain or short, heavy and destructive downpours? Is rainfall concentrated in summer when evapotranspiration rates are higher? Is the rainfall reliable? Does most rain fall during the growing season? Is there sufficient moisture for photosynthesis? Heavy rainfall throughout the year enables forests to grow. These may be tropical rainforests, where the plants need a constant and heavy supply of water, or coniferous forests, where trees can grow due to the lower rates of evapotranspiration. Many other parts of the world receive seasonal rainfall. Rainfall is more effective, as in places with a Mediterranean climate, when it falls in winter rather than in summer as this coincides with the time of year when evapotranspiration rates are at their lowest. However, as Mediterranean areas receive little summer rainfall, trees and shrubs growing there have to be **xerophytic** (drought-resistant) in order to survive. Rain is less effective when it falls in the summer because much of the moisture is lost through surface runoff and evapotranspiration. Effective precipitation is insufficient for trees and so savanna grasses grow in tropical latitudes and prairie grasses in more temperate areas. Places where rainfall is limited throughout the year have either a desert biome, where **ephemerals** (plants with very short life-cycles, Figure 12.19) dominate the vegetation, or a tundra biome, where precipitation falling as snow and the low temperatures combine to discourage plant growth.

- **Temperature** has a major influence on the flora — i.e. whether the forest is tropical or coniferous, or the grassland is temperate (prairie) or tropical (savanna). Where mean monthly temperatures remain above 21°C for the year and there is a continuous growing and rainy season, broad-leaved evergreen trees tend to dominate (tropical rainforests). Places where there is a resting period in tree growth, either in hot climates with a dry season or cool climates with a short growing season, are more likely to have coniferous trees as their dominant vegetation. Grasses, which include most cereals, need a minimum mean monthly temperature of 6°C in order to grow. Many plants prefer temperatures between 10°C, which is the minimum for effective photosynthesis, and 35°C. The higher the temperature, the sooner wilting point will be reached and the greater the need for water to combat losses through evapotranspiration. The lower the temperature, the fewer the number of soil organisms and the slower the breakdown of humus and recycling of nutrients needed for plant growth (Figure 12.7).

Figure 11.35

Wind-distorted tree, Mauritius

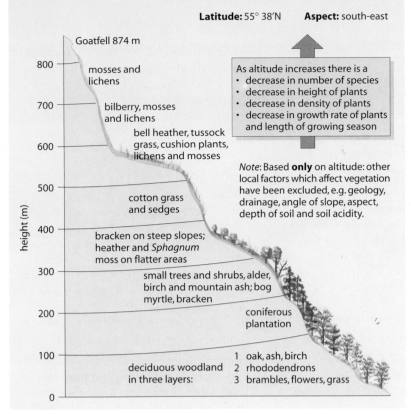

Latitude: 55° 38'N   **Aspect:** south-east

Goatfell 874 m

800 — mosses and lichens

700 — bilberry, mosses and lichens

bell heather, tussock grass, cushion plants, lichens and mosses

600 —

500 — cotton grass and sedges

400 — bracken on steep slopes; heather and *Sphagnum* moss on flatter areas

300 — small trees and shrubs, alder, birch and mountain ash; bog myrtle, bracken

coniferous plantation

200 —

100 — deciduous woodland in three layers: 1 oak, ash, birch / 2 rhododendrons / 3 brambles, flowers, grass

0 —

height (m)

As altitude increases there is a
• decrease in number of species
• decrease in height of plants
• decrease in density of plants
• decrease in growth rate of plants and length of growing season

Note: Based **only** on altitude: other local factors which affect vegetation have been excluded, e.g. geology, drainage, angle of slope, aspect, depth of soil and soil acidity.

Figure 11.36

Mapping the effect of altitude on vegetation, Goatfell, Arran

- **Light intensity** affects the process of photosynthesis. Tropical ecosystems, receiving most incoming radiation, have higher energy inputs than do ecosystems nearer to the poles. Where the amount of light decreases, as on the floor of the tropical rainforests or with increasing depth in the oceans, plant life decreases. Quality of light also affects plant growth, e.g. the increase in ultraviolet light on mountains reduces the number of species found there.
- **Winds** increase the rate of evapotranspiration and the wind-chill factor. Trees are liable to 'bend' if exposed to strong, prevailing winds (Figure 11.35).

2 Topographic factors
- As **altitude** increases, there are fewer species; they grow less tall; and they provide a less dense cover (Figure 11.36). Relief may provide protection against heavy rain (rainshadow) and wind.
- **Slope angle** influences soil depth, acidity (pH) and drainage. Steeper slopes usually have thinner soils, are less water-logged and less acidic than gentler slopes (soil catena, page 239).
- **Aspect** (the direction in which a slope faces) affects sunlight, temperatures and moisture. South-facing slopes in the northern hemisphere are more favourable to plant growth than those facing north

because they are brighter, warmer and drier (Places 22, page 195).

3 Edaphic (soil) factors
In Britain, there is considerable local variation in vegetation due to differences in soil and underlying parent rock, e.g. grass on chalk, conifers on sand, and deciduous trees on clay. Plant growth is affected by soil texture, structure, acidity, organic content, depth, water and oxygen content, and nutrients (Chapter 10).

4 Biotic factors
Biotic factors include the element of competition: between plants for light, root space and water, and between animals. Competition increases with density of vegetation. Natural selection is an important biotic factor. The composition of seral communities and the degree of reliance upon other plants and animals either for food (parasites) or energy (heterotrophs feeding on autotrophs) are also biotic factors. Today, there are very few areas of climax vegetation or biomes left in the world as most have either been altered by human activity or even entirely replaced by human-created environments. The landscape has been altered by subsidence from mining, urbanisation, the construction of reservoirs and roads, exhaustion of soils, deforestation and afforestation, fires, the clearing land for farming and the effects of tourism. The ecological balance has been upset by the use of fertilisers and pesticides, and the grazing of domestic animals.

The spatial pattern of world biomes

Figure 11.37 shows the distribution of the world's major biomes. When looking at maps of biomes in an atlas (they usually come under the heading 'Vegetation'), remember that all vegetation maps are very generalised (Framework 8, Scale and generalisation; page 322). Vegetation maps do not show local variations, transition zones or, except in extreme cases, the influence of relief. Nor is there any universal consensus among geographers and biogeographers as to the precise number of biomes. Bradshaw has suggested 16 land biomes and 5 marine; Simmons describes 13 (11 land biomes plus islands and seas); O'Hare accepts 11; while Goudie (in common with most examination syllabuses) restricts his list, as does this text, to 8 land biomes.

Biogeography

Figure 11.37

World biomes

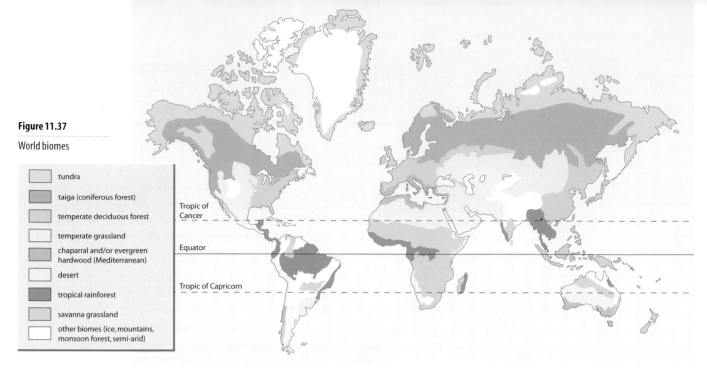

tundra

taiga (coniferous forest)

temperate deciduous forest

temperate grassland

chaparral and/or evergreen hardwood (Mediterranean)

desert

tropical rainforest

savanna grassland

other biomes (ice, mountains, monsoon forest, semi-arid)

Tropic of Cancer

Equator

Tropic of Capricorn

The eight major biomes, as shown in Figure 11.37, can be determined using a variety of criteria; two examples are discussed below and summarised in Figure 11.38.

1 **The traditional method** This links the type and global distribution of vegetation with that of the major world climatic types and zonal soils. This method was based on the understanding that it is climate which exerts the major influence and control over both vegetation and soils. The interactions between climate, soils and vegetation are described and explained in Chapter 12.

2 **The modern method** This is based upon differentiating between relative rates of producing organic matter—i.e. the speed at which vegetation grows. The rate at which organic matter grows is known as the **net primary production** or NPP, expressed in grams of dry organic matter per square metre per year (g/m²/yr). As shown in Figure 11.39, it is the tropical rainforests, with their large biomass resulting from constant high temperatures, heavy rainfall and year-round growing season, which produce the greatest amount of organic matter annually. The tundra (too cold) and the deserts (too dry) produce the least. It may be noted that the NPP for arable land is 650, lakes and rivers 400 and oceans 125.

Figure 11.39

Net primary production of eight major biomes

Net primary production (NPP)
(g/m²/yr)

- Tropical rainforests: 2200
- Deciduous forests: 1200
- Tropical grasslands: 900
- Coniferous forests: 800
- Mediterranean: 700
- Temperate grasslands: 600
- Tundra: 140
- Deserts: 90

1 Traditional method (vegetation, climate and soils subjectively linked)			**2 Modern method** (scientifically based on net primary production)		
Tropical	1	Rainforests	**High energy**	1	Rainforests
	2	Tropical grasslands		2	Deciduous forest
	3	Deserts	**Average energy**	3	Tropical grasslands
Warm temperate	4	Mediterranean		4	Coniferous forest
Cool temperate	5	Deciduous forest		5	Mediterranean
	6	Temperate grasslands		6	Temperate grasslands
Cold	7	Coniferous forest	**Low energy**	7	Tundra
	8	Tundra		8	Deserts
					(*after* I. Simmonds)

Figure 11.38

Two methods of defining the major biomes

Since the 1960s, geographers have felt an increasing need to adopt a more scientific approach to their studies. This stemmed from a number of changes which were taking place in attitudes to the study of geography and to science in a broader sense:

- The increasing scale and complexity of the subject's material and the data available.
- The rapid development of theory, often using computer modelling, from which predictions could be made.
- The realisation that, despite great care, all human observers have their own,

subjective, opinions which influence an assessment or conclusion (i.e. scientific objectivity could not be guaranteed).

The scientific approach to geography involved a series of logical steps, already practised in the physical sciences, which enabled conclusions to be drawn from precise and unbiased data (Framework 6, page 230). This approach is summarised in the flow diagram (Figure 11.40).

During a sixth-form field week on the Isle of Arran, one day was set aside for hypothesis testing. This involved seeking possible relationships between several variables on Goatfell (Figure 11.36). The hypotheses included:

- Vegetation density decreases as altitude increases.
- Soil acidity increases as altitude increases.
- Soil acidity increases as the angle of slope increases.
- Soil moisture increases as the angle of slope increases.
- Depth of soil increases as altitude decreases.
- Height of vegetation increases as altitude decreases.
- Number of species increases as altitude increases.
- Soil temperature increases as altitude decreases.

Data collection required the taking of readings at a minimum of 15 sites from sea-level to the top of Goatfell. It is important that the selection of sites is made without introducing bias (see Framework 4, page 144).

Data analysis may include drawing a scattergraph to investigate the possibility of any correlation between the two variables; calculating the strength of the relationship between the variables by using the Spearman rank correlation coefficient (page 573), and then testing the result to see how likely it is that the correlation occurred by chance (page 574).

It should then be possible to determine whether the original hypothesis is acceptable as an explanation of the data, or not. If it is rejected, then a new hypothesis should be formulated.

Figure 11.40

Hypothesis testing

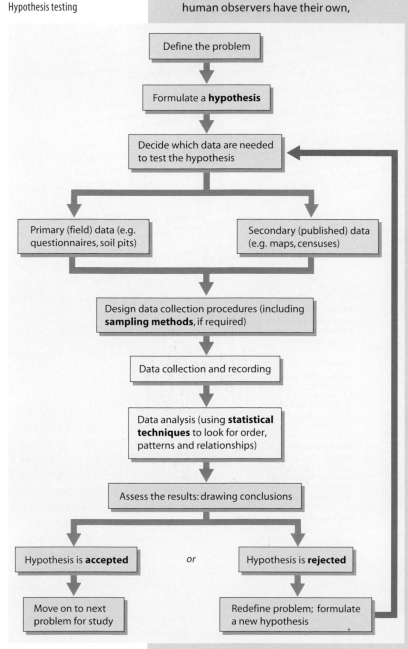

The forests of south-west Australia

Western Australia is ten times the size of UK, and about 2 per cent of the state was forested before white settlement began in 1829. The forested area stretches from Gingin, 75 km north of Perth, to Walpole, 400 km to the south (Figure 11.42). The Darling and Stirling ranges form the edge of the Darling Plateau and consist mainly of ancient igneous and metamorphic rocks. A number of river valleys cut into the plateau edge. These have broad flat valley floors.

The climate of this region is Mediterranean in type, with most rainfall in winter from May to October (700 mm); rainfall is highest (1100 mm) on the western edge of the plateau and decreases rapidly to the east. Temperatures are high in the summer (18-27°C) and cool in the winter (7-15°C). Snow has been known to fall in the Stirling Range!

These conditions allowed the development of high forests, unique to Western Australia, of hardwood trees: varieties of Eucalyptus known as karri, jarrah and marri. The great karri trees which grow to over 80 m in height are found in the south-west where the soils have a higher

Figure 11.41

HERITAGE FORESTS FACE THE AXE

IRREPLACEABLE FORESTS TO BE CLEAR FELLED FOR WOODCHIPS

In March 1994, the Australian Heritage Commission (AHC) officially included 40 areas of WA's world-unique native forests on the interim list of the Registers of the National Estate, the highest national recognition of the ecological, aesthetic, scientific or cultural value of an area. Once an area has been interim-listed, it is considered to be on the Register and entitled to protection. The Federal Minister for the environment is legally bound to prevent logging in these areas until an examination has shown that there are no "prudent and feasible alternative log sources."

In spite of this protection, CALM (Department of Conservation and Land Management) plans to clear fell many interim-listed forests such as Rocky, Sharpe

and Hawke mainly to produce export woodchips. Some of the listed areas are already being clear felled, roaded and burnt with the full knowledge of the AHC and the Federal government.

In addition, there is supposed to be a moratorium on logging in all high-conservation-value forests. Now that at least some of the best of WA's remaining native forests have been given official recognition, each of these agencies must back up their self-congratulations with action.

The only action they can reasonably take is to stop all roading and logging in WA's heritage forests immediately.

(*The Real Forest News*, April 1994, published by Western Australian Forest Alliance, Perth, WA.)

PRESSURES ON THE FOREST

Agricultural clearing
Takes place up to 500m to allow sheep rearing; wheat grown on well drained soils to east of tall forest area; forest now half extent of 165 years ago.

Settlement
Small towns expanding; most densely populated area of state outside Perth; infrastructure damages forest.

Commercial logging
280 000 cu metres p.a. sawn timber for building; timber for woodchips; industry originally used waste offcuts and damaged timber; 150 000 tonnes jarrah sent to Kemerton for charcoal in silicon manufacture; clear felling now extensive; greatest pressures in the south, but jarrah forest ecosystem under threat.

Deforestation
Leads to soil erosion, higher water table and salinisation.

Quarrying
Limestone, sand, gravel.

Mining
Bauxite, gold, tin and tantalite; 800 ha of forest lost each year; little rehabilitation.

Dieback
Soil-borne fungal disease *Phytophthora cinnamomi* affects 14% of forests; spreading because of winter logging and other human disturbance.

Prescribed burning
Frequent burns in spring reduce flora species and damage food supply for breeding birds; jarrah forest not adapted to short intervals between burns.

Pest infestations
Jarrah leaf miner, gumleaf skeletoniser, affects 620 00 ha; thinning forest canopy (due to logging and spring burning) stimulates young foliage, attracts pests.

Loss of habitat
Affects flora and flauna: 26 species of plants and animals in jarrah forests lost or in need of protection; 5 fauna species extinct in karri forests.

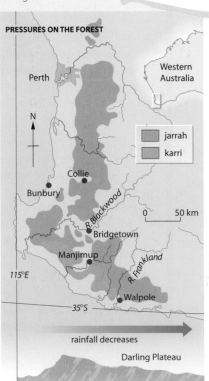

Figure 11.42

South-western Australia

moisture content and are more fertile. The quality of the forest deteriorates to the east with a variety of eucalypts reflecting lower rainfall. The jarrah forest is more extensive and has a very high species diversity (Figure 11.43). The forests provide important wildlife habitats for birds and animals, many of whom live in the hollows of the trees.

Since the coming of the white settlers, 165 years ago, half the tall forest cover has been removed (nearly 2 million hectares). Much of the early clearance was for agriculture, with pastures of clover and grasses for sheep and cattle replacing the 500 year old trees. The timber provided a valuable secondary source of income for farmers.

The commercial value of the tall forests was soon realised and large-scale lumbering began. In recent years, the clearance of timber has been rapid and extensive. The WA government controls over 1 346 000 ha of State forests, the Department of Conservation and Land Management (CALM) wishes to increase the annual cut to over 1 500 000 m3 — a 68 per cent increase over the 1987 figure. This is to allow the large timber companies to produce woodchips, saw logs, poles and 'residue' which is sent to the silicon smelter in large quantities and used as industrial charcoal. The timber mills provide employment which is important in a rural area where sheep farming has been affected by low world prices for wool.

Conservationists are attempting to stop this rapid logging increase in the virgin forests. The rate of loss of forests with many unique wild species of small mammals, birds and flora, raises questions of sustainability of the forests. It is known that already some species are extinct. At the present rate of removal, all 'old' forest will have disappeared by 2030.

The forests are an attraction to tourists, providing a further source of income for the region.

Figure 11.43
Jarrah forest

Figure 11.44
The status of native forest in south-west Australia

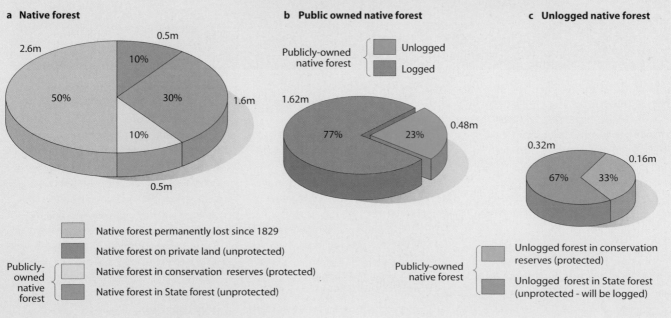

a Native forest

0.5m
2.6m
10%
50%
30%   1.6m
10%
0.5m

b Public owned native forest

Publicly-owned native forest { Unlogged / Logged

1.62m
77%   23%   0.48m

c Unlogged native forest

0.32m
0.16m
67%   33%

Native forest permanently lost since 1829
Native forest on private land (unprotected)
Publicly-owned native forest { Native forest in conservation reserves (protected)
Native forest in State forest (unprotected)

Publicly-owned native forest { Unlogged forest in conservation reserves (protected)
Unlogged forest in State forest (unprotected - will be logged)

Figure 11.45

Clear felling of karri, near Bridgetown

CALM has total responsibility for logging and regeneration of cut areas within State forests. CALM calls for tenders for cutting and hauling, then sells the logs to sawmillers and woodchippers. The chief market for West Australian timber is in the form of woodchips, sold to Japan. Since 1976 over 10 million tonnes of karri have been exported as woodchips, through the port of Bunbury.

The main method of removal is by clear felling, in compartments or coupes which vary from 200 ha in karri forests, to 800 ha in the jarrah/marri forests. Every tree in a coupe is felled, then the logged area is burned. This is mainly 'old-growth': native forest which has not been touched previously. These trees are the largest in the forest.

"It takes a double trailer truck to carry just one log. These must be coming from the last of the best virgin karri. A regular traveller on the road passes six trucks in a half-hour journey." (M. Frith, Bridgetown WA, June 1994)

There is a programme of regeneration in the karri forests. Karri seedlings, which grow more rapidly than jarrah, are planted by hand — sometimes in the jarrah forests, in an attempt to extend the range of the karri. However in dry years, like 1994, many die of drought. Also, in some of the clear-felled karri areas, three karri trees per hectare are left — to drop their seeds — then they too are felled.

Clear felling has now been reintroduced to the jarrah forests. It was abandoned in the 1940s as the jarrah trees did not re-generate successfully. The timber now being taken by this method is largely for charcoal and for chipboard. Much of the jarrah forest is mixed jarrah/marri. The marri is used widely for woodchips. With the increased demand for this product, much of the jarrah forest is being felled either by selection logging (removing all saleable trees) or by clear felling.

Impacts of deforestation

- **Salinisation** The granitic rocks of the Darling Plateau weather to a laterite (page 297) This thin layer of soil contains a high proportion of soluble salts such as calcium, magnesium and sodium. The location of the region ensures that there is also a proportion of sea salt at or near the surface which has accumulated over millions of years from the westerly winds. The presence of the high forest, with its dense vegetation and deep roots, allowed the salts to be kept at depth. Where rainfall is lower than 1100 mm, clearing of the vegetation has led to increased groundwater with a consequent mobility of salts and rise in the water table. Surface streams such as the Blackwood have become seriously salinised as a direct result of deforestation.

- **Eutrophication** In the western forest areas, with over 1100 mm of rain, agricultural and horticultural land uses and housing developments have caused eutrophication (page 456) of coastal wetlands and inlets. Why?

- **Visual and physical degradation of the landscape** This is especially bad in clear-fell areas. Where slopes are steep, tree removal reduces transpiration, increases infiltration and removes nutrients already loosened by burning. In the rainy season, soil is removed downslope with problems of rill and sheet erosion. The sediment load in streams and rivers increases, causing flooding.

- **Loss of native flora and fauna** Western Australia is famous for its wild flowers. Groundcover in the forests is being removed on an extensive scale. In the jarrah forest, 27 native species are gazetted as rare. These include the chuditch, a rare native cat, the Western ringtailed possum and water rat. Birds such as the whipbird are disappearing with the destruction of habitat through frequent fires.

What can be done?

1 CALM have a management programme which includes replanting logged areas with young karri and jarrah trees. They do not replant all clearances.

2 Straight rows of fast-growing timber such as pines and eucalypts are planted for quick returns on logged land. This plantation timber, with all the trees of similar age, height and type, gives an unnatural uniformity to the landscape.

3 Landholders are being encouraged to practise agroforestry. Trees are planted on agricultural land in belts separated by grass pasture which can be used for sheep grazing. In addition, hay crops can be sown. Little fertiliser needs to be used. The amount of timber that can be produced is twice the yield of normal plantations and in 15 years could produce as much timber as is being taken from the native forests now.

4 The number of conservation reserves could be increased with the creation of National Parks, conservation parks and nature reserves. There would need to be strong measures to ensure that these were adequately protected — it is already possible to mine within a National Park. There are outline plans to have 436 000 ha of native eucalypt forest in conservation reserves by 1998. In practice, however, felling is taking place at an increased rate within some of the areas to be designated.

Discussion

This forest destruction may not be on the scale of the Amazon deforestation, but to the people of south-west Australia it is as damaging. Jim Frith, a farmer and agroforester from near Bridgetown who has seen the rapid destruction of the forests near his farm in the last 30 years, said "We are witnessing a tragedy of gigantic proportions — the destruction of an ecosystem."

What do you think could and should be done? Do you consider that preservation of an ecosystem for future generations is necessary? Remember that in Western Europe there are only the smallest remnants of the original native mixed forests left. Should the production of paper have greater importance than protection of trees and wildlife?

If you were a young person looking for work in a town such as Bunbury, WA, your ideas might be different from those of farmers within the forest areas.

It is easy to become emotive on a topic such as this. One needs to gather the facts and draw conclusions by looking into both sides of the question.

References

Bradshaw, M. (1977) *Earth, the Living Planet*. Hodder & Stoughton.

Coastal Dunes: Geography Today (1982) BBC Television.

Goudie, A. (1993) *The Nature of the Environment*. Blackwell.

Nature at Work: Introducing Ecology (1978) British Natural History Museum Publications.

O'Hare, G. (1988) *Soils, Vegetation and Ecosystems*. Oliver & Boyd.

Pickering, K. and Owen, L. (1994) *Global Environmental Issues*. Routledge.

Roberts, M. (1986) *Biology: A Functional Approach*. Nelson.

Simmons, I. (1982) *Biogeographical Processes*. Allen & Unwin.

Chapter 12

World climate, soils and vegetation

"There was ... an instant in the distant past when the living things, the rocks, the air and the oceans merged to form the new entity, Gaia."

James Lovelock, *The Ages of Gaia*, 1989

Although it is possible to study climatic phenomena in isolation (Chapter 9), an understanding of the development of soils (Chapter 10) and vegetation (Chapter 11) necessitates an appreciation of the interrelationships between all three (Figure 12.1a). This chapter attempts to show how the integration and interaction of climate, soils and vegetation give the world its major ecosystems, or biomes, and how these have often been modified, in part or almost totally, by human activity (Figure 12.1b).

The purpose of grouping together different soil types, the difficulties in trying to achieve this, and a suggested classification, were described on page 255. The major vegetation and fauna zones were listed on page 254 and their global locations shown in Figure 11.37. In a similar way, geographers seek to classify climates (Framework 5, page 151).

Classification of climates

By studying the weather — the atmospheric conditions prevailing at a given time or times in a specific place or area — it is possible to make generalisations about the climate of that place or area — i.e. the average, or 'normal' conditions over a period of time (usually 35 years). Any area may experience short-term departures from its 'normal' climate but, at the same time, it may have long-term similarities with regions in other parts of the world.

In seeking a sense of order, the geographer tries to group together those parts of the world which have similar measurable climatic characteristics (temperature, rainfall distribution, winds, etc.) and to identify and to explain similarities and differences in spatial and temporal distributions and patterns. Areas may then be compared on a global scale to help to identify and to explain distributions of soil, vegetation and crops.

Bases for classification

The early Greeks divided the world into three zones based upon a simple temperature description: torrid (tropical), temperate and frigid (polar); they ignored precipitation.

In 1918, **Köppen** advanced the first modern classification of climate. To support his claim that natural vegetation boundaries were determined by climate, he selected as his basis what he considered were appropriate temperature and seasonal precipitation values. His resultant classification is still used today, although a modification by **Trewartha**, with 23 climatic regions, has become more widely accepted.

Thornthwaite, in the 1930s and 1940s,

Figure 12.1

Relationship between climate, vegetation and soils

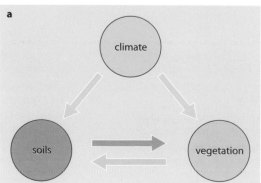

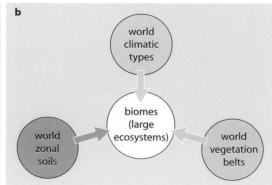

suggested and later modified a classification with a more quantitative basis. He introduced the term 'effectiveness of precipitation' (his P/E index) which he obtained by dividing the mean monthly precipitation of a place by its mean monthly evapotranspiration and taking the sum of the 12 months. The difficulty was, and still is, in obtaining accurate evapotranspiration figures. (How can you measure transpiration loss from a forest?) This classification resulted in 32 climatic regions.

In Britain, in the 1930s, **Miller** proposed a relatively simple classification based upon five latitudinal temperature zones which he determined by using just three temperature figures: 21°C (the limit for growth of coconut palms); 10°C (the minimum for tree growth); and 6°C (the minimum for grasses and cereals). He then subdivided these zones longitudinally according to seasonal distributions of precipitation. The advantages of this classification include its ease of use and convenience; its close relationship to vegetation zones and also, as these are a response on a global scale to climate and vegetation, to zonal soils.

All classifications have weaknesses: none is perfect. They do not show transition zones between climates and often the division lines are purely arbitrary; they do not allow for mesoscale variation (the Lake District and London do *not* have exactly the same climate) or microscale (local) variation; they can be criticised for being either too simplistic (Miller) or too complex (Thornthwaite); they ignore human influence and climatic change. Most tend to be based upon temperature and precipitation figures, and neglect recent studies in heat and water budgets, airmass movement and the transfer of energy. All suffer from the fact that some areas still lack the necessary climatic data to enable them to be categorised accurately.

However, climatic classifications such as those named above are rarely used today. Instead, as we saw in Chapter 11, the relationship between climate, vegetation and soils can best be described and understood through the study of ecosystems, especially the largest ecosystems: the biomes (Figure 12.1b). Figure 12.2 lists eight of the more important biomes and shows, simplistically, the links between climate, vegetation and soils. These links will be described in more detail and explained in the remainder of this chapter, using knowledge and understanding gained from Chapters 9, 10 and 11.

Figure 12.2

World biomes: the relationship between climate, vegetation and soils at the global scale

Climate type		Text reference number	Climatic characteristics	Biome (based on NPP)	Soil (zonal type)	
arctic		8	very cold all year	tundra	tundra	
cold		7	cold all year	coniferous forest (taiga)	podsols	
cool temperate	western margin	6	rain all year, winter maximum	temperate deciduous forest	brown earths	
	continental	5	summer rainfall maximum	temperate grassland	chernozems	prairie
						chestnut
warm temperate	western margins: Mediterranean	4	winter rain	Mediterranean	Mediterranean	
	eastern margins: monsoon	4A	some rain all year, summer maximum	tropical deciduous forest		
tropical	desert	3	little rain	desert (xerophytes)	red yellow desert	
	continental	2	summer rain	tropical grassland (savanna)	ferruginous	
	monsoon	1B		jungle		
	tropical eastern margins	1A	rain all year	rainforest	ferrallitic	
	equatorial	1				

1 Explain why geographers find it useful and necessary to classify climates.

2 Use your library to find out the details of **one** of the classifications listed above.
   a Outline the distinguishing characteristics of the classification which you selected and explain why that particular classification is useful.
   b Indicate what you consider to be the main strengths and limitations of the classification which you described in **a**.

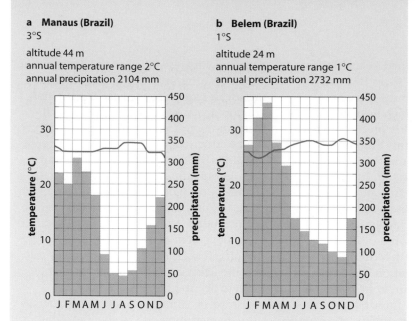

a Manaus (Brazil)
3°S

altitude 44 m
annual temperature range 2°C
annual precipitation 2104 mm

b Belem (Brazil)
1°S

altitude 24 m
annual temperature range 1°C
annual precipitation 2732 mm

Figure 12.3

Climate graphs for equatorial regions

Figure 12.4

Emergents rising above the rainforest canopy, Borneo

Figure 12.5

Buttress roots in the rainforest, Queensland, Australia

1 Tropical rainforests

The rainforest biome is located in the tropics and principally within the equatorial climate belt, 5° either side of the Equator. It includes the Amazon and Zaire basins and the coastal lands of Ecuador, West Africa and extreme south-east Asia (Figure 11.37).

Equatorial climate

Temperatures are high and constant throughout the year because the sun is always high in the sky. The annual temperature range is under 3°C inland (Manaus) and 1°C on the coast (Belem, Figure 12.3). Mean monthly temperatures, ranging from 26°C to 28°C, reflect the lack of seasonal change. Slightly higher temperatures may occur during any 'drier' season. Insolation is evenly distributed throughout the year with each day having approximately 12 hours of daylight and 12 hours of darkness. The diurnal temperature range is also small, about 10°C. Evening temperatures rarely fall below 22°C while, due to the presence of afternoon cloud, daytime temperatures rarely rise above 32°C. It is the high humidity, with its sticky, unhealthy heat, which is least appreciated by Europeans.

Annual rainfall totals usually exceed 2000 mm (Belem, 2732 mm) and most afternoons have a heavy shower (Belem has 243 rainy days per year). This is due to the convergence of the trade winds at the ITCZ and the subsequent enforced ascent of warm, moist, unstable air in strong convection currents. Evapotranspiration is rapid from the many rivers, swamps and trees. Most storms are violent with the heavy rain, accompanied by thunder and lightning, falling from cumulo-nimbus clouds. Some areas may have a drier season when the ITCZ moves a few degrees away from the Equator at the winter and summer solstices (Belem) and others have double maxima when the sun is directly overhead at the spring and autumn equinoxes. The high day-time humidity needs only a little night-time radiation to give condensation in the form of dew. The winds at ground level at the ITCZ are light and variable (doldrums) allowing land and sea breezes to develop in coastal areas (Figure 9.55).

Rainforest vegetation

It is estimated that the rainforests provide 40 per cent of the net primary production of terrestrial energy (NPP, page 284). This is a result of high solar radiation, an all-year growing season, heavy rainfall, a constant moisture budget surplus, the rapid decay of leaf litter and the recycling of nutrients.

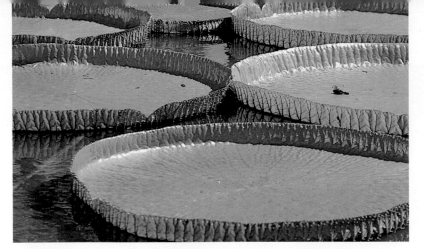

Figure 12.6 photograph

Figure 12.6

Rainforest vegetation has to adapt to the wet environment: water lilies, *Victoria regia*, native to the Amazon Basin

Figure 12.7

The nutrient cycle

Vegetation consists of trees of many different species. In Amazonia, there may be over 300 species in 1 sq km, including rosewood, mahogany, ebony, greenheart, palm and rubber. The trees, which are mainly hardwoods, have an evergreen appearance for, although deciduous, they can shed their leaves at any time during the continuous growing season. The tallest trees, **emergents**, may reach up to 50 m in height and form the habitat for numerous birds and insects. Below the emergents are three layers, all competing for sunlight (Figure 12.4).

The top layer, or **canopy**, forms an almost continuous cover which absorbs over 70 per cent of the light and intercepts 80 per cent of the rainfall. The crowns of these trees merge some 30 m above ground level. They shade the underlying species, protect the soil from erosion, and provide a habitat for most of the birds, animals and insects of the rainforest.

The second layer, or **undercanopy**, consists of trees growing up to 20 m (similar in height to deciduous trees in Britain). The lowest, or **shrub layer**, consists of shrubs and small trees which are adapted to living in the shade of their taller neighbours.

The climate is at the optimum for photosynthesis. The trees grow tall to try to reach the sunlight, and the tallest have buttress roots which emerge over 3 m above ground level to give support (Figure 12.5). The trunks are usually slender and branchless. Some, like the cacao, have flowers growing on them, and their bark is thin as there is no need for protection against adverse climatic conditions. Tree trunks also provide support for lianas, vine-like plants, which can grow to 200 m in length. Lianas climb up the trunk and along branches before plunging back down to the forest floor. Leaves are dark green, smooth and often have **drip tips** to shed excess water.

Epiphytes — plants which do not have their roots in the soil — grow on trunks, branches and even on the leaves of trees and shrubs. Epiphytes simply 'hang on' to the tree: they derive no nourishment from the host and are *not* parasites. Less than 5 per cent of insolation reaches the forest floor with the result that undergrowth is thin except in areas where trees may have been felled by shifting cultivators or where a giant emergent has fallen, dragging with it several of the top canopy trees. Vegetation is also dense along the many river banks, again because sunlight can penetrate the canopy.

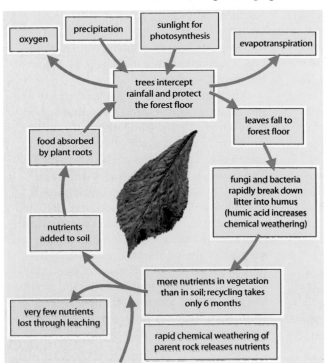

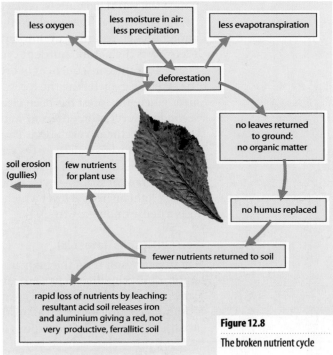

Figure 12.8

The broken nutrient cycle

Figure 12.9

Ferrallitic soil profile

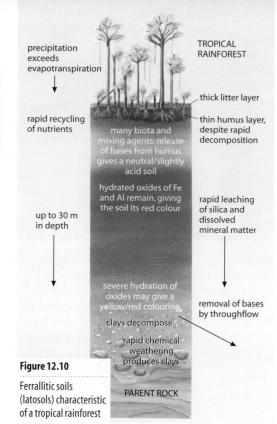

Figure 12.10

Ferrallitic soils (latosols) characteristic of a tropical rainforest

Alongside the Amazon, many trees spend several months of the year growing in water as the river and its tributaries rise over 15 m in the rainy season. Huge water lilies with leaves exceeding 2 m in width are found in flooded areas adjacent to rivers (Figure 12.6). Mangrove swamps occur in coastal areas.

Although ground animals are relatively few in number, the rainforests of Brazil alone are said to be the habitat for 2000 species of birds, 600 species of insects and mosquitoes, and 1500 species of fish.

The productivity of this biome, upon which the world depends to replace much of its used oxygen, is due largely to the rapid and unbroken recycling of nutrients. Figure 12.7 shows the natural nutrient cycle and Figure 12.8 the consequences of breaking the sytem, e.g. by felling the forest. In areas where the forest has been cleared, the secondary succession differs from that of the original climax vegetation. The new dominants are less tall; the trees are less stratified; there are fewer species and many are intolerant of shade — even though there is more light at ground level which encourages a dense undergrowth.

Ferrallitic soils (latosols)

These soils result from the high annual temperature and rainfall which cause rapid chemical weathering of bedrock and create the optimum conditions for breaking down the luxuriant vegetation. Continuous leaf fall within the forest gives a thick litter layer but the underlying humus is thin due to the rapid decomposition of organic matter by intensive biota activity. There is a rapid recycling of nutrients within the humus cycle (Figure 12.7) but the release of bases prevents the soil from becoming too acidic. The soils have a relatively low nutrient status which is maintained only by rapid and continuous replacement from the lush vegetation. Despite some protection from the tree canopy, the heavy rainfall causes severe leaching. Silica is more soluble than the sesquioxides and is removed by ferrallitisation (page 252) leaving iron and aluminium which give the soil its deep red colour (Figure 12.9).

The continual leaching and abundance of mixing agents inhibit the formation of horizons (Figure 12.10). The lower parts of the profile may have a more yellowish-red tint due to the extreme hydration of iron oxides. The clay-rich soils are also very deep, often up to 20 m, due to the rapid breakdown of parent material. Ferrallitic soils have a loose structure and, if exposed to heavy rainfall, are easily gullied and eroded. Despite their depth, the soils of the rainforest are not agriculturally productive. Once the source of nutrients (the trees) has been removed, the soil rapidly loses its fertility and local farmers, often shifting cultivators, have to move to clear new plots (Places 52, page 441).

Kano (Nigeria)
12°N

altitude 630m
annual temperature range 8° C
annual precipitation 920 mm

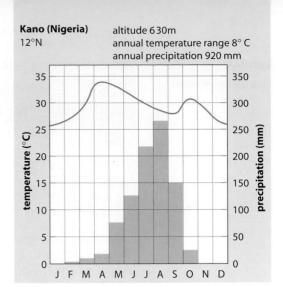

Figure 12.11

Climate graph for the tropical grassland biome

1A Tropical eastern margins

Located within the tropics, the eastern coasts of central America, Brazil, Madagascar and Queensland (Australia) receive rain throughout the year. The rain is brought by the trade winds which blow across warm, offshore ocean currents (Figure 9.9) before being forced to rise by coastal mountains. Temperatures are generally very high, although there is a slightly cooler season when the overhead sun appears to have migrated into the opposite hemisphere. The resultant vegetation and soil types are, therefore, similar to those found in the equatorial belt — i.e. rainforest and ferrallitic.

2   Tropical grasslands

These are mainly located between latitudes 5° and 15° north and south of the Equator and within central parts of continents — i.e. the Llanos (Venezuela), the Campos (Brazilian Highlands), most of central Africa surrounding the Zaire basin, and parts of Mexico and northern Australia (Figure 11.37).

Tropical continental climate

Although temperatures are high throughout the year, there is a short, slightly cooler season (in comparison with the equatorial)

when the sun is overhead at the tropic in the opposite hemisphere (Figure 12.11). The annual range is also slightly greater (Kano 8°C) due to the sun's slightly reduced angle in the sky for part of the year, the greater distance from the sea, and the less complete cloud and vegetation cover. Temperatures may drop slightly at the onset of the rainy season. For most of the year, cloud amount is limited allowing diurnal temperatures to exceed 25°C.

The main characteristic of this climate is the alternating wet and dry seasons. The wet season occurs when the sun moves overhead bringing with it the heat equator, the ITCZ, and the equatorial low pressure belt (Figure 12.12). Heavy convectional storms can give 80 per cent of the annual rainfall total in four or five months. The dry season corresponds with the moving away of the ITCZ leaving the area with the strong, steady trade winds. The trades are dry because they are warming as they blow towards the Equator and they will have shed any moisture on distant eastern coasts. Places nearer to the desert margins tend to experience dry, stable conditions (the subtropical high pressure) caused by the migration of the descending limb of the Hadley cell (page 209). Humidity is also low during this season.

Tropical or savanna grassland vegetation

The tropical grasslands are estimated to have a mean NPP of 900 g/m²/yr (page 285). This is considerably less than the rainforest partly because of the smaller number of trees, species and layers and partly because, although grasslands have the potential to return organic matter back to the soil, the rate of decomposition is reduced during the winter drought leaving considerable amounts left stored in the litter.

Figure 12.12

Causes of seasonal rainfall in the tropical continental (savanna) climate

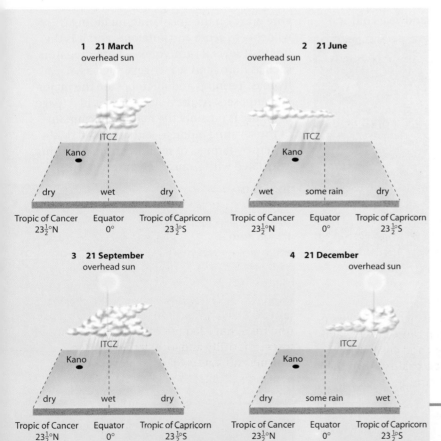

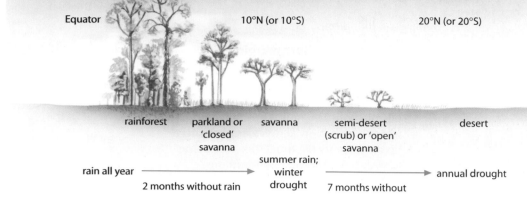

Figure 12.13

Transect across the savanna

Equator   10°N (or 10°S)   20°N (or 20°S)

rainforest   parkland or 'closed' savanna   savanna   semi-desert (scrub) or 'open' savanna   desert

rain all year ⟶ 2 months without rain ⟶ summer rain; winter drought ⟶ 7 months without ⟶ annual drought

As shown in Figure 12.13, the savanna includes a series of transitions between the rainforest and the desert. At one extreme, the 'closed' savanna is mainly trees with areas of grasses; at the other, the 'open' savanna is vegetated only by scattered tufts of grass. The trees are deciduous and, like those in Britain, lose their leaves to reduce transpiration, but, unlike in Britain, this is due to the winter drought rather than to cold. Trees are xerophytic or drought-resistant. Even when leaves do appear, they are small, waxy and sometimes thorn-like. Roots are long and extend to tap any underground water. Trunks are gnarled and the bark is usually thick to reduce moisture loss.

The baobab tree (also known as the 'upside-down tree') has a trunk of up to 10 m in diameter in which it stores water. Its root-like branches hold only a minimum number of tiny leaves in order to restrict transpiration (Figure 12.14). Some baobabs are estimated to be several thousand years old and,

like other savanna trees, are pyrophytic — i.e. their trunks are resistant to the many local fires. Acacias, with their crowns flattened by the trade winds (Figure 12.15), provide welcome though limited shade — as do the eucalyptus in Australia. Savanna trees reach 6–12 m in height. Many have 'Y'-shaped, branching trunks, ideal for the leopard to rest in after its meal! The number of trees increases near to rivers and waterholes. Grasses grow in tufts and tend to have inward-curving blades and silvery spikes. After the onset of the summer rains, they grow very quickly to over 3 m in height: elephant grass reaches 5 m. As the sun dries up the vegetation, it becomes yellow in colour. By early winter, the straw-like grass has died down, leaving seeds dormant on the surface until the following season's rain. By the end of winter, only the roots remain and the surface is exposed to wind and rain.

Over 40 different species of large herbivore graze on the grasslands, including wildebeest, zebra and antelope, and it is the home of several carnivores — both predators, such as lions, and scavengers, such as hyenas. Termites and microbes are the major decomposers. As previously mentioned (page 272), fire is possibly the major determinant of the savanna biome — either caused deliberately by farmers or resulting from lightning associated with summer electrical storms.

Figure 12.14

A baobab tree

Figure 12.15

Savanna grassland in Maasai Mara, Kenya, during the rainy season

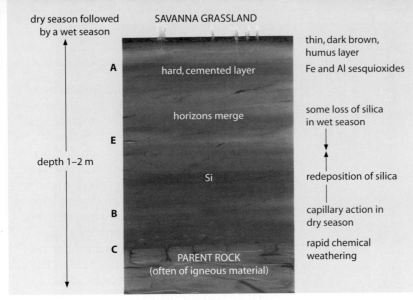

dry season followed by a wet season

SAVANNA GRASSLAND

A — hard, cemented layer

thin, dark brown, humus layer
Fe and Al sesquioxides

horizons merge

some loss of silica in wet season

E

depth 1–2 m

Si

redeposition of silica

B

capillary action in dry season

C — PARENT ROCK (often of igneous material)

rapid chemical weathering

Figure 12.16

Ferruginous soil profile

It is the fringes of the savannas, those bordering the deserts, which are at greatest risk of desertification (Case Study 7, page 175). As more trees are removed for fuel and over-grazing reduces the productivity of grasslands, the heavy rain forms gulleys and wind blows away the surface soil. Where the savanna is not farmed, there are usually more trees suggesting that grass may not be the natural climatic climax vegetation.

Ferruginous soils

As savanna grasses die back during the dry season, they provide organic matter which is readily broken down to give a thin, dark-brown layer of humus (Figure 12.16). At the same time, capillary action brings bases in solution towards the surface. During the following wet season, rapid leaching and the relatively high pH remove silica from the upper profile, leaving behind the red-coloured oxides of iron and aluminium. The seasonal change between capillary action and leaching often produces a cemented layer, or **laterite**, just below the surface, which impedes drainage, plant root penetration and ploughing. The parent material weathers into a clay which tends to become sticky and plastic during the wet season, but which can become 'brick-hard' during the drought. The term 'laterite' is derived from the Latin for 'brick' and this deposit is indeed still used as a building material.

Ferruginous soils contain few nutrients and so are not particularly suitable for agriculture. Indeed, grasslands are better suited to animals than to arable. When a lateritic crust forms near to the surface it often causes the soil above to dry out and become highly vulnerable to erosion by wind and, when the rains return, water.

3 Hot deserts

The hot deserts of the Atacama and Kalahari-Namib and those in Mexico and Australia, are all located in the trade wind belt, between 15° and 30° north or south of the Equator, and on the west coasts of continents where there are cold, offshore, ocean currents (Figures 7.2 and 11.37). The exception is the extensive Sahara-Arabian-Thar desert which owes its existence to the size of the Afro-Asian landmass.

Climate

Desert temperatures are characterised by their extremes. The annual range is often 20–30°C and the diurnal range over 50°C (Figure 12.17). Daytimes, especially in summer, receive high levels of insolation from the overhead sun, intensified by the lack of cloud cover and the bare rock or sand ground surface. In contrast, nights may be extremely cold with temperatures likely to fall below 0°C. Coastal areas, however, have much colder monthly temperatures (Arica in the Atacama has a warmest month of only 22°C) due to the presence of offshore, cold, ocean currents (Figure 9.9).

Although all deserts suffer an acute water shortage, none are truly dry. Aridity and extreme aridity have been defined by using Thornthwaite's P/E index (Figure 7.1), and four of the main causes of deserts were described on page 163. Amounts of moisture

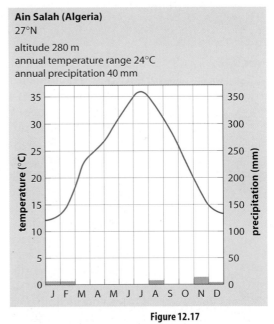

Ain Salah (Algeria)
27°N

altitude 280 m
annual temperature range 24°C
annual precipitation 40 mm

Figure 12.17

Climate graph for a hot desert biome

Figure 12.18

Saguaros cacti in the Arizona desert

Figure 12.19

Ephemerals in flower following a desert rainstorm, Saudi Arabia

Desert vegetation

Deserts have the lowest organic productivity levels of any biome (Figure 11.39). The average NPP is 90 g/m^2/yr, most of which occurs underground away from the direct heat of the sun. Vegetation has to have a high tolerance to the moisture budget deficit, intense heat and, often, salinity. Few areas are totally devoid of vegetation, although desert plants are few in species, have simple structures, no stratification by height and provide a low density cover. However, plants must be xerophytic because the lack of water hinders the ability of roots to absorb nutrients and of any green parts of the plants to photosynthesise.

Many plants are **succulents** — i.e. they can store water in their tissues. Many succulents have fleshy stems and some have swollen leaves. Cacti (Figure 12.18) absorb large amounts of water during the infrequent periods of rain. Their stems swell up, only to contract later as moisture is slowly lost through transpiration. Transpiration takes place from the stems, but is reduced by the stomata closing during the day and opening nocturnally. The stems also have a thick, waxy cuticle. Australian eucalyptids have thick, protective bark for the same purpose.

Most plants, for example cactus and thornbush, have small, spiky or waxy leaves to reduce transpiration and to deter animals. Roots are either very long to tap groundwater supplies — those of the acacia exceed 15 m — or they spread out over wide areas near to the surface to take the maximum advantage of any rain or dew, like those of the creosote bush. Bushes are, therefore, widely spaced to avoid competition for water. Some plants have bulbous roots for storing water. Seeds, which usually have a thick case protecting a pulpy centre, can lie dormant for months or several years until the next rainfall.

Following a storm, the desert blooms (Figure 12.19). Many plants are **ephemerals** and can complete their life-cycles in two or three weeks. Others, like the saltbush, are **halophytic** and can survive in salty depressions; yet others, like the date-palm, survive where the water table is near enough to the surface to form oases. Due to the lack of grass and the limited number of green plants, there are very few food chains: desert biomes have a low capacity to sustain life. There is insufficient plant food to support an abundance of animal life. Food chains (page 275) are simple, often just a single linear sequence

are usually small and precipitation is extremely unreliable. Death Valley, California, averages 40 mm a year, yet rain may fall only once every two or three years. Whereas mean annual totals vary by less than 20 per cent a year in north-west Europe, the equivalent figure for the Sahel is 80–150 per cent (page 207). Rain, when it does fall, produces rapid surface runoff which, together with low infiltration and high evaporation rates, minimises its effectiveness for vegetation. The Atacama, an almost rainless desert, has some vegetation as moisture is available in the form of advection fog (Places 18, page 164). The subsiding air, forming the descending limb of the Hadley cell, creates high pressure and produces the trade winds which are strong, persistent and likely to cause localised dust storms.

(in contrast to the interlocking webs characteristic of, for example, forests). This is why the desert ecosystem is 'fragile': organisms do not have the alternative sources of food which are available in more complex ecosystems. Many animals are small and nocturnal (the camel is an exception) and burrow into the sand during the heat of the day. Reptiles are more adaptable, but bird life is limited. The desert fringes form a delicately balanced ecosystem which is being disturbed by human activity and population growth which are, together, increasing the risk of desertification (Case study 7, page 175).

Grey desert soils

In desert areas, the climate is too dry and the vegetation too sparse for any significant chemical weathering of bedrock or the accumulation of organic material. Any moisture in the soil is likely to be drawn upwards by capillary action causing salts and bases to be deposited near to the surface. Soils are therefore alkaline, with a high concentration of magnesium, sodium and calcium. The grey colour results from the lack of moisture which, with the high pH value, restricts hydrolysis and the release of red-coloured iron. Soils are usually thin and lack horizons and structure (Figure 12.20).

Desert soils are unproductive mainly because of the lack of moisture and humus, but potentially they are not particularly infertile. Areas under irrigation are capable of producing high quality crops — although this farming technique is being threatened by salinisation (Figures 10.23c and 16.56).

4 Mediterranean (warm temperate, western margins)

This type of biome is found on the west coasts of continents between 30° and 40° north and south of the Equator — i.e. in Mediterranean Europe (which is the only area where the climate penetrates far inland), California, central Chile, Cape Province (South Africa) and parts of southern Australia (Figure 11.37).

Climate

The climate is noted for its hot, dry summers and warm, wet winters (Figure 12.21). Summers in southern Europe are hot. The sun is high in the sky, though never directly overhead, and there is little cloud. Winters are mild, partly because the sun's angle is still quite high but mainly due to the moderating influence of the sea. Other 'Mediterranean' areas are less warm in summer and have a smaller annual range due to cold, offshore currents. (Compare San Francisco, 8°C in January and 15°C in July, with Malta). Diurnal temperature ranges are often high due to the fact that many days, even in winter, are cloudless.

As the ITCZ moves northwards in the northern summer, the subtropical high pressure areas migrate with it to affect these latitudes. The trade winds bring arid conditions with the length of the dry season increasing towards the desert margins. In winter, the ITCZ, and subsequently the subtropical jet stream (page 211), move southwards allowing the westerlies, which blow from the sea,

Figure 12.20

Grey desert soils: a characteristic profile

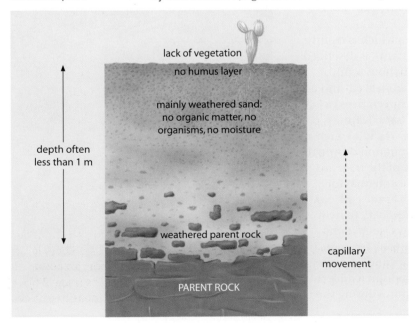

lack of vegetation

no humus layer

mainly weathered sand: no organic matter, no organisms, no moisture

depth often less than 1 m

weathered parent rock

capillary movement

PARENT ROCK

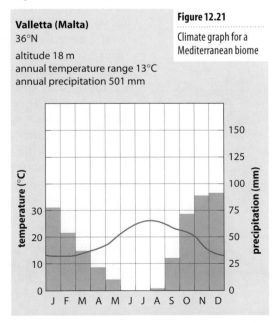

Valletta (Malta)
36°N

altitude 18 m
annual temperature range 13°C
annual precipitation 501 mm

Figure 12.21

Climate graph for a Mediterranean biome

Figure 12.22

Mediterranean winds

to bring moisture. Most areas are backed by coastal mountains and so the combined effects of orographic and frontal precipitation give high seasonal totals. Areas with adjacent, cold, offshore currents experience advection fogs (California). The Mediterranean Sea region is noted for its local winds (Figure 12.22). The **sirocco** and **khamsin** are two of the hot, dry winds which blow from the Sahara and can raise temperatures to over 40°C. The **mistral** is a cold wind which originates over the Alps and is funnelled at considerable speed down the Rhône valley.

Vegetation

The NPP of Mediterranean ecosystems is about 700 g/m²/yr (Figure 11.39). It is limited by the summer drought and has probably been reduced considerably over the centuries by human activity. Indeed, human activity, together with frequent fires, has left very little of any original climatic climax vegetation. The climax vegetation was believed to have been, in Europe at least, open woodland comprising a mixture of broad-leaved, evergreen trees (e.g. cork oak and holm oak) and conifers (e.g. aleppo pines, cypresses and cedars). The sequoia, or giant redwood, is native in California.

The present vegetation, which is mainly **xerophytic** (drought-resistant), is described as 'woodland and sclerophyllous scrub'. **Sclerophyllous** means 'hard-leaved' and is used to describe those evergreen trees or shrubs which have small, hard, leathery, waxy or even thorn-like leaves and which are efficient at reducing transpiration during the dry summer season. Many of the trees are evergreen, maximising the potential for photosynthesis. Trees such as the cork oak have thick and often gnarled bark to help reduce transpiration. Others, such as the olive and eucalyptus, have long tap roots to reach ground-water supplies and, in some cases, may have bulbous roots in which to store water. High temperatures during the

dry summer limit the amount and quality of grass. Citrus fruits, although not indigenous, are suited to the climate as their thick skins preserve moisture. Most trees only grow from 3 to 5 m in height. They provide little shade, as they grow widely spaced, and are **pyrophytic** (fire-resistant, page 272).

Where the natural woodland has been replaced, and in areas too dry for tree growth, a scrub vegetation has developed. The scrub is known as chaparral in California, maquis or garrigue in Europe, and mallee in Australia. In Mediterranean Europe, the type of scrub depends upon the underlying parent rock. **Maquis** (Figure 12.23), which is taller, denser and more tangled, grows in areas of impermeable rock (granite). It consists of shrubs, such as heathers and broom, which reach a height of 3 m. **Garrigue** (Figure 12.24) grows on drier and more permeable

Figure 12.23

Maquis vegetation

Figure 12.24

Garrigue vegetation

rocks (limestone). It is less high and less dense than maquis. Apart from gorse, with its prickles, the more common plants include aromatic shrubs such as thyme, lavender and rosemary.

The limited leaf litter tends to decompose slowly during the dry summer, even though temperatures are high enough for year-round bacterial activity. Wildlife and climax vegetation have retreated as human activity has advanced. Arguably, the Mediterranean regions of Europe and California (together with the temperate deciduous forests) form the biome most altered by human activity.

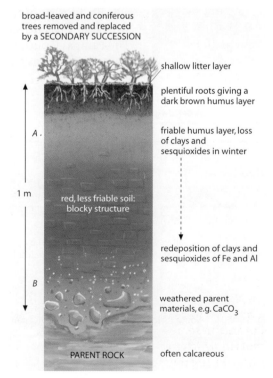

Figure 12.25

Mediterranean soils: a characteristic profile

broad-leaved and coniferous
trees removed and replaced
by a SECONDARY SUCCESSION

shallow litter layer

plentiful roots giving a
dark brown humus layer

A.

friable humus layer, loss
of clays and
sesquioxides in winter

1 m

red, less friable soil:
blocky structure

redeposition of clays and
sesquioxides of Fe and Al

B

weathered parent
materials, e.g. CaCO₃

PARENT ROCK

often calcareous

Soils

Mediterranean soils are transitional between brown earths on the wetter margins and grey desert soils at the drier fringes. Initially formed under broad-leaved and coniferous woodland, the soil is partly a relict feature from a previously forested landscape.

There are often sufficient roots and decaying plant material to provide a significant humus layer. Winter rains cause the leaching of bases, sesquioxides of iron and aluminium and the translocation of clays. The *B* horizon is therefore clay-enriched and may be coloured a bright red by the redeposition of iron and aluminium. The soils, which are often thin, are less acid than the brown earths as there is less leaching in the dry season and calcium is often released, especially in limestone areas (Figure 12.25).

In many Mediterranean areas, parent rock is locally a more important factor in soil formation than climate. This leads to the development of intrazonal soils such as rendzina and terra rossa (Figures 10.24 and 10.25).

4A Eastern margin climates in Asia (monsoon)

South-east and eastern Asia are dominated by the monsoon (page 222). Temperature figures and rainfall distributions are similar to those of places having a tropical continental climate with a very warm and dry season from November to May and a hot and very wet season from June to October (Places 25, page 222). The major difference between the two climates is that monsoon areas receive appreciably higher annual amounts of rain. The natural vegetation is jungle (tropical deciduous forest) and the dominant soil type is ferralitic. Both vegetation and soils, therefore, share many similarities with the tropical rainforest.

5 Temperate grasslands

The temperate grassland biome lies in the centre of continents approximately between latitudes 40° and 60° north of the Equator. The two main areas are the North American Prairies and the Russian Steppes (Figure 11.37).

Cool temperate continental climate

The annual range of temperature is high as there is no moderating influence from the sea (38°C at Saskatoon, Figure 12.26). The land warms up rapidly in summer to give

Saskatoon (Saskatchewan, Canada)

52°N

altitude 145 m

annual temperature range 38°C

annual precipitation 352 mm

Figure 12.26

Climate graph for a temperate continental biome

Figure 12.27

Tufted grasses on the North American Prairies, Montana

maximum mean monthly readings of around 20°C. However, the rapid radiation of heat from mid-continental areas in winter means there are several months when the temperature remains below freezing point. The clear skies also result in a large diurnal temperature range.

In Russia, precipitation decreases rapidly towards the east as distance from the sea — and therefore the rain-bearing winds — increases; in North America, however, totals are lowest to the west which is directly in the rainshadow of the Rockies. Annual amounts in both areas only average 500 mm and there is a threat of drought, as experienced in North America in 1988. Although, fortuitously, 75 per cent of precipitation falls during the summer growing season, it can occur in the form of harmful thunderstorms and hailshowers. The ground can be snow-covered for several months between October and April. Overall, there is a close balance between precipitation and evapotranspiration. In winter, both areas are open to cold blasts of arctic air although the chinook may bring temporary warmer spells to the Prairies (page 224).

Temperate grassland vegetation

This type of vegetation lies to the south of the coniferous forest belt in the dry interiors of North America and Russia. Temperate grasslands are, however, also found sporadically in parts of the southern hemisphere where they usually lie between 30° and 40°S. The Pampas (South America) and the Canterbury Plains (New Zealand) are towards

Figure 12.28

Land uses of the temperate grassland, a changed biome

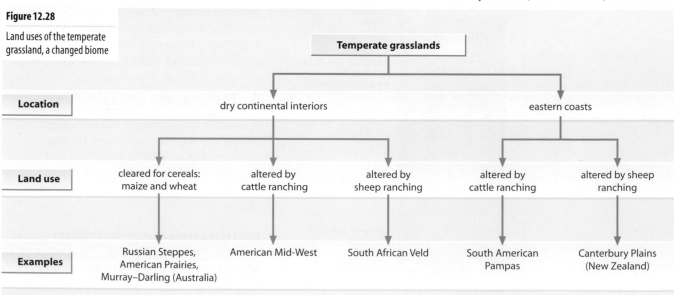

	Temperate grasslands				
Location	dry continental interiors			eastern coasts	
Land use	cleared for cereals: maize and wheat	altered by cattle ranching	altered by sheep ranching	altered by cattle ranching	altered by sheep ranching
Examples	Russian Steppes, American Prairies, Murray–Darling (Australia)	American Mid-West	South African Veld	South American Pampas	Canterbury Plains (New Zealand)

Figure 12.29

A chernozem (black earth) soil

the eastern coast, while the Murray–Darling basin (Australia) and the veld (South Africa) are further inland. The NPP of 600 g/m²/yr is considerably less than that of the tropical grasslands because the vegetation grows neither as rapidly nor as tall. Whatever the original climax vegetation of the biome may have been, the ecosystem has been significantly altered by fire and human exploitation to leave, today, grama and buffalo grass as the dominants. There are two main types of grass. Feather grasses grow to 50 cm and form a relatively even coverage, whereas tufted (tussock) grasses, reaching up to 2 m, are found in more compact clumps (Figure 12.27). The grass forms a tightly knit sod which may have restricted tree growth and certainly made early ploughing difficult. The deep roots, which often extend to a depth of 2 m in order to reach the water table, help to bind the soil together and so reduce erosion. Most of the organic material is in the grass roots and it is the roots and rhizomes which provide the largest store of nutrients (Case Study 12b).

During autumn, the grasses die down to form a turf mat in which seeds lie dormant until the snowmelt, rains and warmer temperatures of the following spring. Growth in early summer is rapid and the grasses produce narrow, inward-curving blades to limit transpiration. By the end of summer, their blue-green colour may have turned more parched. Herbaceous plants and some trees (willow) grow along water courses. In response to the windy climate, many prairie and steppe farms are protected by trees

planted as windbreaks. The decay of grasses in summer causes a rapid accumulation of humus in the soil, making the area ideal for cereals or, in drier areas, for cattle ranching (Figures 12.27 and 12.28).

The temperate grasslands are a resilient ecosystem. The grasses provide food for burrowing animals such as rabbits and gophers, and large herbivores such as antelopes, bison and kangaroos. These, in turn, may be consumed by carnivores (wolves and coyotes) or by predatory birds (hawks and eagles).

Chernozems or black earths

The thick grass cover provides a plentiful supply of mull humus which forms a black, crumbly topsoil (Figure 12.29). While the abundance of biota, especially earthworms, causes the rapid decay of organic matter during the warm summer, decomposition is arrested in any drier spells and during the long, cold winter. Humus is therefore retained near to the surface. There is effective recycling as the grasses take up and return nutrients to the soil. The late spring snowmelt and early summer storms cause some leaching (Figure 12.30), but bases such as potassium and nitrogen are moved downwards only slowly. In later summer, this is compensated by the upward movement of capillary water bringing bases nearer the surface and maintaining a neutral or slightly alkaline soil (pH 7–7.5). The grasses have an extensive root system which gives a deep (up to 1 m) dark brown to black A horizon.

The alternating dry and wet seasons immobilise iron and aluminium sesquioxides and clay within aggregates (peds) in the upper horizon and this, together with the large number of mixing agents, limits the formation of a recognisable B horizon. The subsoil, often of loess origin (page 122), is usually porous and this, together with the capillary moisture movement in summer, means that it remains dry. This upward movement of moisture causes calcium carbonate to be deposited, often in the form of nodules, in the upper C horizon.

Chernozems are regarded as the optimum soil for agriculture as they are deep, rich in organic matter, retain moisture, and have an ideal crumb structure with well-formed peds. After intensive ploughing, chernozems may require the addition of potassium and nitrates.

Figure 12.30

Characteristic profile of a chernozem (black earth)

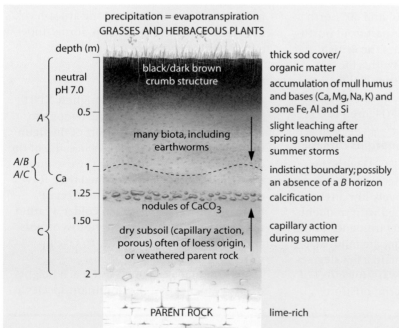

precipitation = evapotranspiration
GRASSES AND HERBACEOUS PLANTS

depth (m)

neutral pH 7.0

black/dark brown crumb structure — thick sod cover/organic matter

0.5 — accumulation of mull humus and bases (Ca, Mg, Na, K) and some Fe, Al and Si

A

many biota, including earthworms — slight leaching after spring snowmelt and summer storms

A/B
A/C — Ca — 1 — indistinct boundary; possibly an absence of a B horizon

1.25 — nodules of CaCO₃ — calcification

1.50

C — dry subsoil (capillary action, porous) often of loess origin, or weathered parent rock — capillary action during summer

2

PARENT ROCK — lime-rich

Chestnut soils

These are found in juxtaposition with the chernozems, but where the climate is drier so that evapotranspiration slightly exceeds precipitation and the resultant vegetation is sparser and more xerophytic. As the root system is less dense, both the amount and the depth of organic matter decrease and the colour becomes a lighter brown than in chernozems. Chestnut soils are more alkaline, due to increased capillary action, and suffer from more frequent summer droughts. Deposits of calcium carbonate are found near to the surface and the soil is generally shallower than a chernozem. Chestnut soils are agriculturally productive if aided by irrigation, but mismanagement can quickly lead to their exhaustion and erosion.

Prairie soils

These lie to the wetter margins of the chernozems and form a transition between them and the brown forest earths. As precipitation exceeds evapotranspiration, there is an absence of capillary action and the soil lacks the accumulation of calcium carbonate associated with chernozems. The *A/B* horizons tend to merge as there is limited leaching and strong biota activity. Decaying grasses provide much organic material and the soils are ideal for cereal crops.

6 Temperate deciduous forests

Temperate deciduous forests are located on the west coasts of continents between approximately latitudes 40° and 60° north and south of the Equator. Apart from northwest Europe (which includes the British Isles), other areas covered by this biome include the north-west of the USA, British Columbia, southern Chile, Tasmania and South Island, New Zealand (Figure 11.37).

Cool temperate western margins climate

Summers are cool (Figure 12.31) with the warmest month between 15°C and 17°C. This is a result of the relatively low angle of the sun in the sky, combined with frequent cloud cover and the cooling influence of the sea. Winters, in comparison, are mild. Mean monthly temperatures remain a few degrees above freezing due to the warming effect of the sea, the presence of warm, offshore,

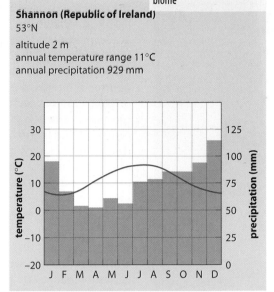

Figure 12.31

Climate graph for cool temperate western-margin biome

Shannon (Republic of Ireland)
53°N

altitude 2 m
annual temperature range 11°C
annual precipitation 929 mm

ocean currents and the insulating cloud cover. Diurnal temperature ranges are low; autumns are warmer than springs; and seasonal temperature variations depend on prevailing air masses (page 212).

This climatic zone lies at the confluence of the Ferrel and Polar cells (Figures 9.34 and 9.35), where tropical and polar air converge at the Polar Front. Warmer tropical air is forced to rise, creating an area of low pressure and forming depressions with their associated fronts. The prevailing south-westerlies, laden with vapour after crossing warm, offshore currents, give heavy orographic and frontal rain. Precipitation, often exceeding 2000 mm annually, falls throughout the year but with a winter maximum when depressions are more frequent and intense. Although snow is common in the mountains, it rarely lies for long at sea-level. Fog, most common in autumn, forms under anticyclonic conditions (page 218).

Deciduous forests

Although having the second-highest NPP of all biomes (1200 g/m/²/yr), the temperate deciduous forest falls well short of the figure for tropical rainforests, mainly because of the dormant winter season when the deciduous trees in temperate latitudes shed their leaves (Figure 11.39). Leaf fall has the effect of reducing transpiration when colder weather reduces the effectiveness of photosynthesis and when roots find it harder to take up water and nutrients.

In Britain, oaks, which can reach heights of 30–40 m, became the dominant species as

Figure 12.32

Broad-leaved deciduous (oak) woodland in Surrey, England

the climax vegetation developed through a series of several primary successions (Figure 11.4). Other trees, such as the elm (common before its population was diminished by Dutch elm disease), beech, sycamore, ash and chestnut, grow a little less high. They all develop large crowns and have broad but thin leaves (Figure 12.32). Unlike the rainforests, the temperate deciduous forests contain relatively few species. The maximum number of species per sq km in southern Britain is eight and some woodlands, such as beech, may only have a single dominant. The trees have a growing season of 6–8 months in which to bud, leaf, flower and fruit, and may only grow by about 50 cm a year.

Most woodlands show some stratification (Figure 11.2). Beneath the canopy is a lower shrub layer varying between 5 m (holly, hazel and hawthorn) and 20 m (ash and birch). This layer can be quite dense because the open mosaic of branches of the taller trees allows more light to penetrate than in the rainforests. The forest floor, if the shrub layer is not too dense, is often covered in a thick undergrowth of brambles, grass, bracken and ferns. Many flowering plants (bluebells) bloom early in the year before the taller trees have developed their full foliage. Epiphytes, which include mosses, lichens and algae, often grow on tree trunks.

The forest floor has a reasonably thick leaf litter which is readily broken down by the numerous mixing agents living in the relatively warm soil. There is a rapid recycling of nutrients, although some are lost through leaching. The leaching of humus and nutrients and the mixing by biota produce a brown-coloured soil. Soil type contributes to determine the dominant tree with oaks and elms preferring loams; beech the more acid gravels and the drier chalk; ash the lime-rich soils; and willows and alder wetter soils. There is a well-developed food chain in these forests with many autotrophs, herbivores (rabbits, deer and mice) and carnivores (foxes).

Most of Britain's natural primary deciduous woodland has been cleared for farming, for use as fuel and in building, and for urban development. Deciduous trees give way to coniferous towards polar latitudes and where there is an increase in either altitude or steepness of slope.

Brown earths

The considerable leaf litter which accumulates in autumn decomposes relatively quickly in the presence of organisms and the less acidic mull humus. Organic matter is incorporated into the *A* horizon by the action of earthworms, giving it a dark brown colour (Figure 12.33). Precipitation exceeds evapotranspiration sufficiently to cause leaching but not enough for podsolisation (Figure 10.21b). Bases, especially calcium and magnesium, are absent in the upper horizons and there is some loss of clay and sesquioxides (Figure 12.34).

The horizons merge more gradually than in a podsol, assisted by increased biota activity. The colour becomes increasingly reddish-brown with depth due to the redepo-

Figure 12.33

A brown earth

sition of iron and aluminium. There is no hard pan, so the brown earths tend to be free-draining. There is considerable recycling as the deciduous trees take up many nutrients from the soil only to return them later through fallen leaves. The soil is deeper and more fertile than the podsol and tree roots may penetrate and break up the bedrock. Due to the relatively high clay content throughout the profile, most areas of brown earth are potentially fertile through they may benefit from liming.

Figure 12.34

Characteristic profile of a brown earth

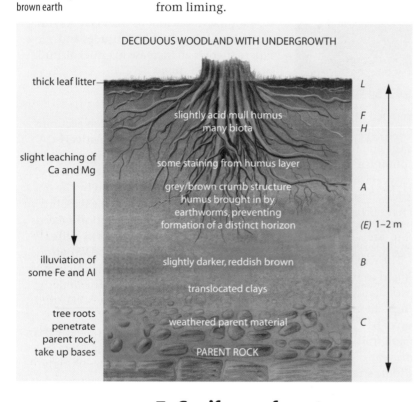

DECIDUOUS WOODLAND WITH UNDERGROWTH

thick leaf litter — L

slightly acid mull humus — F
many biota — H

slight leaching of Ca and Mg

some staining from humus layer

grey/brown crumb structure — A
humus brought in by earthworms, preventing formation of a distinct horizon — (E) 1–2 m

illuviation of some Fe and Al

slightly darker, reddish brown — B

translocated clays

tree roots penetrate parent rock, take up bases

weathered parent material — C

PARENT ROCK

7 Coniferous forests

The coniferous forest, or taiga, biome occurs in cold climates to the poleward side of 60°N in Eurasia and North America as well as at high altitudes in more temperate latitudes and in southern Chile (Figure 11.37).

Cold climates

Winters are long and cold. Minimum mean monthly temperatures may be as low as -30°C (-28°C at Dawson, Figure 12.35): there is little moderating influence from the sea and no insolation as, at this time of year, the sun never rises in places north of the Arctic Circle. Strong winds mean there is a high wind-chill factor (frostbite is a hazard to humans); any moisture is rapidly evaporated (or frozen); and snow is frequently blown about in blizzards. Summers are short, but

Figure 12.35

Graph for a cold climate biome

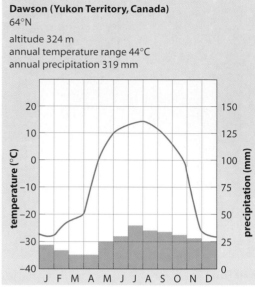

Dawson (Yukon Territory, Canada)
64°N

altitude 324 m
annual temperature range 44°C
annual precipitation 319 mm

the long hours of daylight and clear skies mean that they are relatively warm. Precipitation is light throughout the year because the air can hold only limited amounts of moisture and most places are a long way from the sea. The slight summer maximum is caused by isolated convectional rainstorms.

Coniferous forest or taiga

The coniferous forest has an average NPP of 800 $g/m^2/yr$ (Figure 11.39). The coniferous trees have developed distinctive adaptations which enable them to tolerate long, cold winters; cool summers with a short growing season; limited precipitation; and podsolic soils. The size of the dominant trees and the fact that they are evergreen — giving them the potential for year-round photosynthesis —

Figure 12.36

Coniferous forest, Gifford Pinchot National Park, Washington State, USA.

Figure 12.37

Forest floor in a coniferous forest, Cumbria, England

Figure 12.38

The coniferous forest and its transition zones

result in their relatively high NPP. The trees, which are softwoods, rarely number more than two or three species per sq km. Often there may be extensive stands of a single species, such as spruce, fir or pine. In colder areas, like Siberia, the larch tends to dominate. Although larches are cone-bearing, the European larch is deciduous and sheds its leaves in winter. All trees in the taiga, some of which attain a height of 40 m, are adapted to living in a harsh environment (Figure 12.36).

Conditions for photosynthesis become favourable in spring as incoming radiation increases and water becomes available through snowmelt (days in winter are long and dark and soil moisture is frozen). The needle-like leaves are small and the thick cuticles help to reduce transpiration during times of strong winds and during the winter when moisture is in a form unavailable for absorption by tree roots. Cones shield the seeds and thick, resinous bark protects the trunk from the extreme cold of winter and the threat of summer forest fires. The conical shape of the tree and its downward-sloping, springy branches allow the winter snows to slide off without breaking the branches. The conical shape also gives some stability against strong winds as the tree roots are usually shallow. There is usually only one layer of vegetation in the coniferous forest. The amount of ground cover is limited, partly due to the lack of sunlight reaching the forest floor and partly to the deep, acidic layer of non-decomposed needles. Plants which can survive on the forest floor include mosses, lichens and wood sorrel (Figure 12.37).

The cold climate and soils discourage earthworms and bacteria. Needles therefore decompose very slowly — less rapidly than they accumulate — to give an acid mor humus (page 241). Most of the nutrients are held in the litter (Figure 11.28a). Although precipitation is limited, evapotranspiration rates are also low with the result that leaching occurs and the few nutrients which are returned to the podsol soil are soon lost. Conifers require few nutrients, taking only 225 kg of plant nutrient annually from each hectare compared with the 430 kg taken by deciduous trees. The limited food supply means that animal life is not abundant. The dark woods are not favoured by bird life, although deer, wolves, brown bears, moose, elk and beavers are found in certain areas.

In North America and Eurasia, the coniferous forest merges into the tundra on its northern fringes (Figure 12.38). The tree line, the point above which trees are unable to grow, is often clearly marked in mountainous areas (Figure 12.36). South of the taiga lie either the deciduous forest or the temperate grassland biomes (Figure 11.37), depending upon whether the location is coastal or inland.

Podsols

Podsols develop in areas where precipitation exceeds evapotranspiration; where

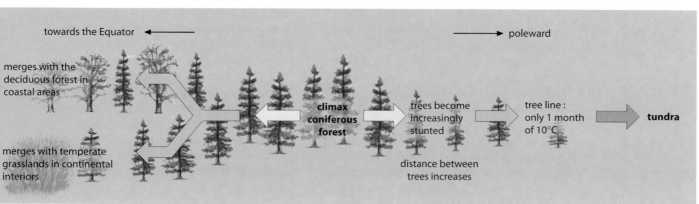

Figure 12.39

The characteristics of a podsol

a the typical podsol profile (compare Figure 10.19)

b readings from a soil pit in Delamere Forest, Cheshire

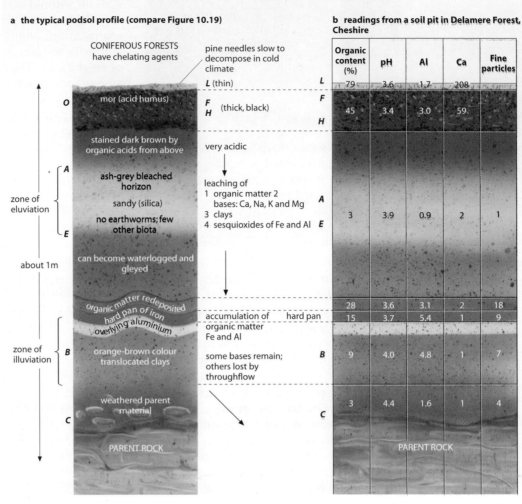

CONIFEROUS FORESTS have chelating agents

pine needles slow to decompose in cold climate

L (thin)

O — mor (acid humus)

F
H (thick, black)

stained dark brown by organic acids from above — very acidic

A — ash-grey bleached horizon

zone of eluviation

sandy (silica)

leaching of
1 organic matter 2 bases: Ca, Na, K and Mg
3 clays
4 sesquioxides of Fe and Al

no earthworms; few other biota

E

about 1m — can become waterlogged and gleyed

organic matter redeposited
hard pan of iron overlying aluminium — accumulation of organic matter Fe and Al — hard pan

zone of illuviation

B — orange-brown colour translocated clays — some bases remain; others lost by throughflow

weathered parent material

C

PARENT ROCK

	Organic content (%)	pH	Al	Ca	Fine particles
L	79	3.6	1.7	208	
F H	45	3.4	3.0	59	
A E	3	3.9	0.9	2	1
	28	3.6	3.1	2	18
	15	3.7	5.4	1	9
B	9	4.0	4.8	1	7
C	3	4.4	1.6	1	4

PARENT ROCK

coniferous forest or heathland provides the vegetation cover; and where soils are sandy. Podsols usually, but not exclusively, form in cool climates. Podsolisation is a soil-forming process which also operates within tropical ferrallitic (page 294) and ferruginous soils (page 297).

Pine needles, with their thick cuticles, provide only a thin leaf litter and inhibit the formation of humus. Any humus formed is very acid (mor) and provides chelating agents and humic acid which help to make the iron, aluminium and silica minerals more soluble. The cold climate discourages organisms and the soil is too acidic for earthworms. Consequently, well-defined horizons develop due to the slow decomposition of leaf litter and the lack of mixing agents. The downward percolation of water through the soil, especially following snowmelt, causes the leaching of bases, the translocation of clays and organic matter, and the eluviation of the sesquioxides of clay and aluminium. This leaves a narrow, ash-grey, bleached A horizon (podsol is Russian for 'ash-like')

composed mainly of quartz sand and silica (Figure 12.39). There is some dispute among pedologists as to whether the translocated materials are moved by physical, chemical or biological processes or a combination of all three.

The dark-coloured organic matter is redeposited at the top of the B horizon. Beneath this the sesquioxides first of iron and then of aluminium are deposited as a rust-coloured, hard pan. Where this hard pan, which often has a convoluted shape, becomes marked, it acts as an impermeable layer restricting the downward movement of moisture and the penetration of plant roots. This can cause some waterlogging in the E horizon to give a gleyed podsol. The lower B horizon, composed mainly of redeposited clays, has an orange-brown colour and overlies weathered parent material. Any throughflow from this horizon is likely to contain bases in solution. Although these soils are not naturally fertile, they can be improved by the addition of lime and fertiliser.

8 The tundra

The tundra, which lies to the north of the taiga, includes the extreme northern parts of Alaska, Canada and Russia, together with all of Greenland (Figure 11.37). The ground, apart from the top few centimetres in summer, remains permanently frozen (the permafrost, Chapter 5).

Arctic climate

Summers may have lengthy periods of continuous daylight but, with the angle of the sun so low in the sky, temperatures struggle to rise above freezing-point (Barrow 3°C, Figure 12.40) and the growing season is exceptionally short. Nearer the poles, the climate is one of perpetual frost. Although winters are long, dark and severe, and the sea freezes, the water has a moderating effect on temperatures keeping them slightly higher than inland places further south (Siberia). Precipitation, which falls as snow, is light — indeed, Barrow with 110 mm would be classified as a desert if temperatures were warm enough for plant growth.

Tundra vegetation

The tundra ecosystem is one with very low organic productivity. The NPP of only 140 g/m²/yr is the second-lowest of the major land biomes (Figure 11.39). In Finnish, tundra means a 'barren or treeless land', which accurately describes its winter appearance, and in Russian a 'marshy plain', which it is in summer. Any vegetation must have a high degree of tolerance of extreme cold and of moisture-deficient conditions — the latter because water is unavailable for most of the year when it is stored as ice or snow. There are fewer species of plants in the tundra than in any other biome. All are very slow- and low-growing, compact and rounded to gain protection against the wind (plants as well as humans are affected by wind-chill), and most have to complete their life-cycles within 50–60 days. There is no stratification of vegetation by height.

The five main dominants, each with its specialised local habitat, are lichens, mosses, grasses, cushion plants and low shrubs (Figure 12.42). Most have small leaves to limit transpiration and short roots to avoid the permafrost. Lichens are pioneer plants in

Figure 12.40

Climate graph for an Arctic biome

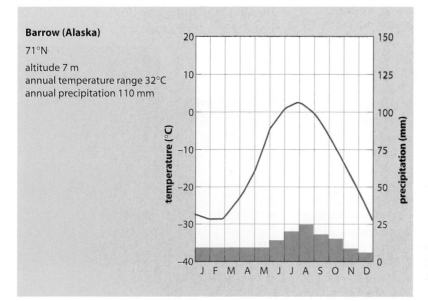

Barrow (Alaska)

71°N

altitude 7 m
annual temperature range 32°C
annual precipitation 110 mm

Figure 12.41

Waterlogged tundra in the summer season

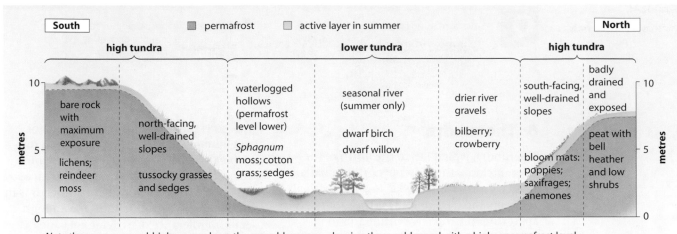

South — permafrost — active layer in summer — North

high tundra — lower tundra — high tundra

metres: 10, 5, 0

bare rock with maximum exposure

lichens; reindeer moss

north-facing, well-drained slopes

tussocky grasses and sedges

waterlogged hollows (permafrost level lower)

Sphagnum moss; cotton grass; sedges

seasonal river (summer only)

dwarf birch
dwarf willow

drier river gravels

bilberry; crowberry

south-facing, well-drained slopes

bloom mats: poppies; saxifrages; anemones

badly drained and exposed

peat with bell heather and low shrubs

metres: 10, 5, 0

Note: the more exposed, higher areas have the snow blown away leaving them colder and with a higher permafrost level.

Figure 12.42

The importance of site factors to vegetation in the tundra

areas where the ice is retreating and they can help date the chronology of an area following deglaciation (page 266). Much of the tundra is waterlogged in summer (Figure 12.41) due to the impermeable permafrost preventing infiltration. Relief is gentle and evaporation rates are low. In such areas, mosses, cotton grass and sedges thrive. On south-facing slopes and in better-drained soils, cushion plants provide a mass of colour in summer (Figure 12.43). These 'bloom mats' include arctic poppies, anemones, pink saxifrages and gentians. Where decaying vegetation accumulates (there is little bacterial action to decompose dead plants), the resultant peat is likely to be covered in heather; whereas, on drier gravels, berried plants (e.g. bilberry and crowberry) are the dominants. Adjacent to the seasonal snowmelt rivers, dwarf willows and stunted birch grow, but only to a maximum of about 30 cm and even so their crowns are often distorted and misshapen by the wind. In winter, the whole biome is covered in snow which acts as insulation for the plants.

The lack of nitrogen-fixing plants limits fertility and the cold, wet conditions inhibit the breakdown of plant material. Photosynthesis is hindered by the lack of sunlight and water for most of the year, though the presence of autotrophs, such as lichens and mosses, does provide the basis for a food chain longer than might be expected. Herbivores such as reindeer, caribou and musk-ox survive because plants like reindeer moss have a high sugar content. However, these animals have to migrate in winter to find pasture which is not covered by snow. The major carnivores are wolves and arctic fox; owls are also found.

The tundra is an extremely fragile ecosystem in a delicate balance. Once it is disturbed by human activity, such as tourism or oil exploration and extraction, it may take many years before it becomes re-established.

Tundra soils

The limited plant growth of this biome only produces a small amount of litter and, as there are few soil biota in the cold soil, organic matter decomposes only very slowly to give a thin layer of peat — a very acidic humus or mor. Where water does manage to percolate downwards, usually as meltwater in late spring, the humic acid within it (pH less than 4.5) releases iron. Underlying the soil, often at a depth of less than 50 cm, is permafrost. This, acting as an impermeable layer, severely restricts moisture percolation and causes extreme waterlogging and gleying (Figure 12.44). Few mixing agents can survive in the cold, wet, tundra soils which are thin and have no developed horizons. Where bedrock is near to the surface, the parent

Figure 12.43

'Bloom mats' in Prudhoe Bay, Alaska

Figure 12.44

Profiles typical of tundra soils

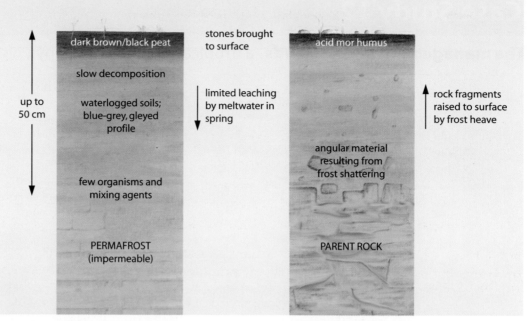

material is physically weathered by freeze–thaw action. The shattered angular fragments are raised to the surface by frost

heave, preventing the formation of horizons and creating a range of periglacial landforms (Figure 5.15).

Q

1 a Explain what is meant by the term 'climatic climax vegetation'.

b Why are there relatively few areas in the world today with a climatic climax vegetation?

2 a Describe the composition and structure of one of the following: boreal forest; tropical rainforest; temperate deciduous forest.

b Show how the characteristics of the vegetation of your chosen area have enabled it to become dominant.

3 For any one biome, describe and explain the relationship between climate, soils and vegetation.

4 Draw a fully labelled diagram to show the vertical structure and composition of **a** a mature, deciduous woodland in lowland Britain; and **b** a tropical rainforest.

5 Describe and account for the differences in net primary production (NPP) between the eight major biomes.

6 a Describe the distribution of vegetation types within an area of Mediterranean climate and show how this distribution is related to variations in relief, climate, soils and human activities.

b Describe the vegetation characteristics of the tropical grasslands and the Mediterranean lands. Discuss the extent to which these characteristics are a response to seasonal variations in climate.

7 a How has human development been responsible for the degradation of climatic climax vegetation communities and soils in the tropical rainforest?

b Explain why 'secondary vegetation' commonly results in areas which have been seriously affected by human activity.

8 a Discuss the view that many tropical grasslands and temperate grasslands are not composed of climatic climax communities.

b Why does the use of **either** the tropical grasslands **or** the temperate grasslands need careful management?

9 Explain what is meant by the following terms:

ephemerals

xerophytic

halophytic

net primary production (NPP)

stratification

biomes

epiphytes

dominants

food chain

photosynthesis

sclerophyllous

plagio-climax

Case Study 12

The management of grasslands a. Tropical grasslands in Kenya

Figure 12.45

Temperate grassland on the Loita Plain, Kenya

The main expanses of tropical grassland in Kenya lie within the Rift Valley and on the adjacent plains of the Mara (an extension of the Serengeti) and Loita (Figure 12.47). Their appearance is one of open savanna (Figures 12.13 and 12.45) with small acacia and evergreen trees (Figure 12.15). There is evidence, however, that the original climax vegetation was forest, but that this has been altered by fires, started both naturally and by humans (page 272), by overgrazing (Figure 12.46) and by climatic change.

The climate is very warm and dry for most of the year with, usually, a short season (three months) of fairly reliable and abundant rainfall and an even shorter period known as the 'little rains' (Figure 12.48). Both periods of rainfall follow soon after the ITCZ and the associated overhead sun have passed over the Equator (Figure 12.12). The annual water balance shows a deficit (Figure 3.3) so that, although there is some leaching during the rainy season, for most of the year capillary action occurs.

Figure 12.46

Scattered trees and over-grazed land in the central Rift Valley, Kenya

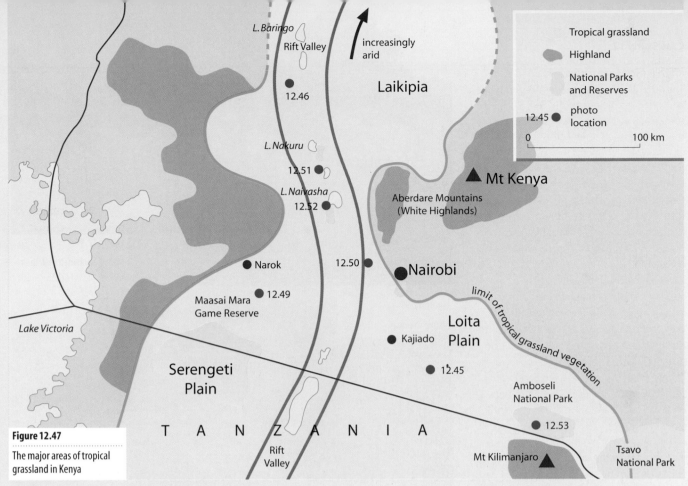

Figure 12.47

The major areas of tropical grassland in Kenya

Figure 12.48

Climate graph and water balance for Nairobi (Note - Nairobi has a higher altitude than the surrounding grass-lands and so is slightly cooler and wetter)

Nairobi (Kenya)

latitude 1°S

altitude 1820 m

annual temperature range 3°C

annual precipitation 958 mm

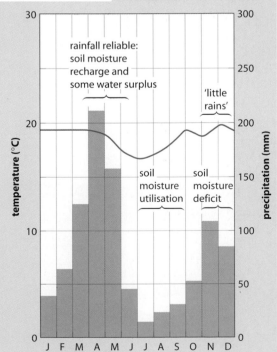

This has resulted in the development of ferruginous soils with, in places, a lateritic crust (page 297). Water supply is therefore a major management problem in this part of Kenya.

Water is obtained from springs at the foot of Mount Kilimanjaro — the mountain itself is in Tanzania — which are fed by melting snow; from several of the Rift Valley lakes (not all, as some are highly saline); from rivers (many of which are seasonal); and from waterholes. Even so, there have been, in the last 100 years alone, several major droughts when the carrying capacity of the region was exceeded. The carrying capacity (page 356) is the maximum number of a population (people, animals, plants, etc.) which can be supported by the resources of the environment in which they live — e.g. the greatest number of cattle that can be fed adequately on the available amount of grassland.

Human pressure on the natural resources

Maasai pastoralists

Maasai are defined as 'people who speak the Maa language'. Their ancestors were Nilotic, coming from southern Sudan during the first millenium A.D. They kept cattle and grew sorghum and millet. The present Maasai may have been the latest of several migration waves. Latest evidence suggests that they may have only been in Kenya for 300 years. Over time, they specialised more in cattle and came to see themselves, and be seen by others, historically and ethnically as 'people of cattle'. Figure 12.49 is a stereotype photo of the Maasai, dressed in their red cloaks and with their humped zebu cattle. While all Maasai are Maa speakers, not all Maa speakers are Maasai — nor, today, are all Maasai pastoralists! The Maasai became semi-nomadic, moving seasonally with their cattle in search of water and pasture (two wet seasons and two dry seasons meant four moves a year; Figure 12.48). Herds had to be large enough to provide sufficient milk and meat for their owners and to reproduce themselves over time, including the ability to recover from drought and disease.

Case Study 12

Kikuyu (Bantu) farmers

The Kikuyu were one of several Bantu tribes which arrived in Kenya, from the south, some 2000 years ago. They became subsistence farmers who grew crops on the higher land which bounded the eastern side of the Rift Valley. The Kikuyu and Maasai often lived a complementary life-style. For the Maasai, Kikuyu in the highlands were a secure source of foodstuffs and a place of refuge during times of drought and cattle disease. For the Kikuyu, Maasai provided a constant supply of cattle products and wives. The division between them only appeared in early colonial times when the Maasai were forcibly moved from places like Laikipia (Figure 12.47) southwards onto the newly created Maasai reservation (the districts of Narok and Kajiado). The vacuum left was filled by newly arrived European settlers, and Kikuyu (their rising numbers were causing a land shortage in the highlands). Figure 12.50 shows numerous, small, Kikuyu shambas (smallholdings) on the eastern edge of the Rift Valley to the northwest of Nairobi.

Colonial (European) settlers

While many Europeans settled in the so-called 'White Highlands', others developed huge estates within the Rift Valley. The most famous was Lord Delamere from Cheshire. He introduced, in turn, Australian sheep (they died as the local grass was mineral-deficient); British sheep and clover (the sheep died as African bees did not pollinate British clover); British cattle (wiped out by local diseases); wheat (which was more successful unless trampled by wild animals); and, finally and successfully,

Figure 12.49

Maasai herdsmen with their cattle

Figure 12.50

Kikuyu shambas

drought-resistant beef cattle. The present Delamere estate (Figure 12.51) covers 22 600 hectares (divided into 180-hectare paddocks); it has 10 900 long-horned Boran cattle (the carrying capacity is 12 000) crossed with 300 Friesian bulls; and

280 permanent workers. Although the estate is managed by 'whites', the stockmen are Maasai. More recently, multinational firms have set up large flower farms (Figure 12.52) and vegetable (especially peas and beans) farms in and near the Rift Valley. The closeness to Nairobi airport means that these perishable products can be sold in European markets, out of season, the day after being picked.

Figure 12.52

Flower-growing estate near to Lake Naivasha, Kenya

Figure 12.51

Commercial cattle ranching, Delamere Estate, Kenya

Population growth and urbanisation

Kenya, an economically less developed country, has one of the world's fastest-growing population rates. This means increased pressure on the land, especially the grasslands, to grow more subsistence crops to feed the growing domestic market; more cash crops to earn needed money from increased exports; and more land lost to urban growth.

Maasai in the mid 1990s

The traditional Maasai way of life and their grassland habitat are under constant threat. Figure 12.54 summarises, but does insufficient justice to, some of the present day problems. Change, as in many societies, is being forced upon the Maasai. While many values and traditions are still known and held, the basis of their economy — the concept of land as territory — has been so transformed that the survival of the herding system is in jeopardy. For some years, many Maasai have tried either to buy individual ranches (IRs) or to amalgamate to create group ranches (GRs), a practice which seems to fail at times of severe drought. The Maasai are also having to come to terms with a sedentary rather than a semi-nomadic life-style. Intermediate Technology (IT), a British Development Group (Places 70), has been working with Maasai to improve the standard of housing. In response to the main complaint of Maasai women, IT have helped to design a watertight cement skin which can be laid over an old mud roof (it was the women's job to apply more dung and mud onto a leaking roof during a wet night), and have improved ventilation within the house (where all cooking is done). The government have laid a pipeline from Kilimanjaro to Kajaido, to ensure a more reliable water supply. The quality of Maasai herds has improved with some cattle being sold for meat in Nairobi. The improvement to herds has been aided by IT who have helped train local villagers to become 'vets' (wasaidizi) capable of vaccinating animals and looking after common diseases. Some Maasai have begun to grow crops, while others have begun to benefit from tourism. In Amboseli National Park, the Maasai are allowed to sell artefacts from their own shop. They can retain all the income which had, previously, gone to the government.

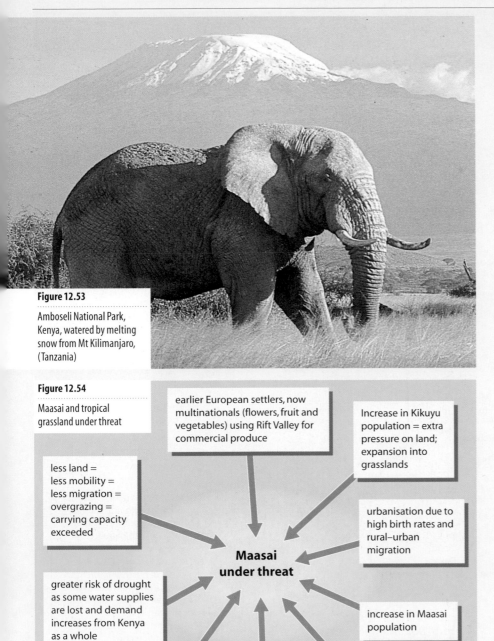

Figure 12.53

Amboseli National Park, Kenya, watered by melting snow from Mt Kilimanjaro, (Tanzania)

Figure 12.54

Maasai and tropical grassland under threat

earlier European settlers, now multinationals (flowers, fruit and vegetables) using Rift Valley for commercial produce

Increase in Kikuyu population = extra pressure on land; expansion into grasslands

less land = less mobility = less migration = overgrazing = carrying capacity exceeded

urbanisation due to high birth rates and rural–urban migration

Maasai under threat

increase in Maasai population

greater risk of drought as some water supplies are lost and demand increases from Kenya as a whole

threat to plough up areas around Narok to grow wheat for Kenya's growing population

tourism: the latest threat to the Maasai's traditional way of life, society and culture

Maasai moved out of National Parks and Reserves: loss of land and water supplies

National parks and reserves

The passing of the National Parks Ordinance in 1945 meant that specific areas were set aside either exclusively for wildlife (no permanent settlement in National Parks other than at tourist lodges) or where other types of land use were permitted only at the discretion of local councils. While wildlife has become a major source of income for Kenya, it has meant less land being available for crops and, to the Maasai, denial of access to important resources of dry-season water and pasture (Amboseli; Figure 12.53). Maasai herds were heavily depleted during the droughts of 1952 and 1972–76.

Case Study 12

b. The temperate grasslands in North America: the Prairies

Early travellers such as the Spaniard Coronado in the 16th century, who rode into Kansas from Mexico, and later French trappers and explorers in Canada, reported vast extents of waist-high, green grasses sometimes so tall that men on horseback stood in their stirrups to see where they were going. The plains seemed so vast that no limit could be found. Nineteenth-century settlers moving westwards across the Mississippi–Missouri in their wagons or drawing their handcarts, must have wondered if they would ever see woods, forests and mountains again. Today, the extent of the interior grasslands of North America is well known.

The Native Americans, who used the ecosystem, did little to alter the grassland which remained in its original state of natural balance until the late 19th century.

"It is a wild garden. Each week from April through September, about a dozen new kinds of flowers come into bloom. Once the layer of dead grass gets too thick, though, it starts to choke off the smaller grasses and wild flowers. Meantime, woody plants — they like shade and moisture — can gain a foothold in the sod and spread. If you go long enough without fire, much of this countryside will be covered with trees" quoted in Chadwick, 1993, p.113).

Grasses have a network of roots which

can extend to considerable depth to absorb water and obtain nutrients. Root systems may make up over 80per cent of the vegetative biomass in the prairie. These, together with the smaller herbs, have helped to develop a thick sod close to the surface. Plants can survive from year to year because they die back to the ground and lie dormant during the cold winters (page 303).

Soils grade in colour and fertility from brown in the western short-grass prairie (Figure 12.56) through chestnut in the mixed grass zone to the fertile black chernozems (millisols) of the tall-grass eastern zone (Figure 12.57). The chernozems have a high humus content (page 303).

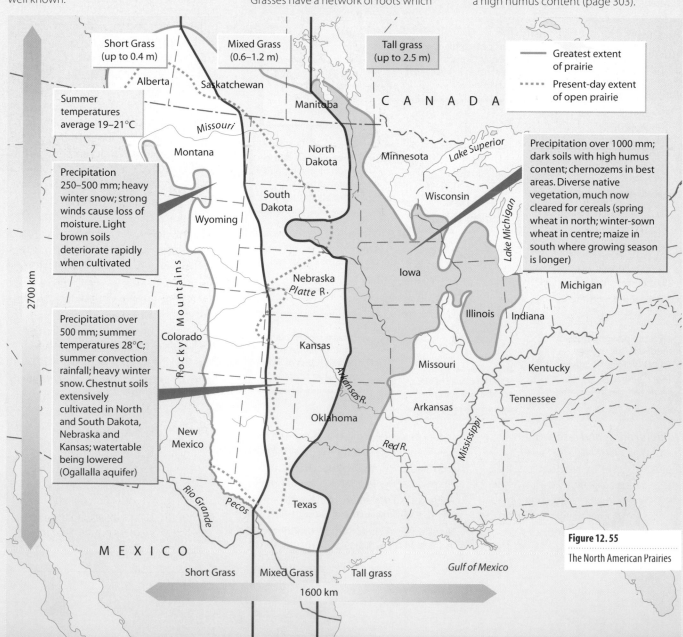

Figure 12.55

The North American Prairies

Figure 12.56

Short-grass prairie

Decaying humus releases minerals slowly for the grasses. The soils are kept light and aerated which helps to prevent compaction under heavy rain (summer convection storms) and the weight of heavy animals (bison and humans). The presence of humus also helps to conserve moisture. Due to frequent droughts, the vegetation has developed protective mechanisms, such as leaves which curl up to prevent evaporation loss, and well-developed root fibres which can obtain moisture from deep in the soil.

The rapid spring growth and early maturity of grass allows it to produce seeds early. It then becomes semi-dormant until autumn and can survive heat and drought. Late-growing species may not be able to compete and this has led to an extension of short grass into the mixed-grass zone during a succession of long dry periods.

In the 17th century, there were estimated to be 60–70 million bison roaming the grasslands with 50 million antelope, plus grizzlies, wolves and prairie dogs, together with many species of birds — hawks, lark buntings, etc. — and insects and reptiles such as snakes and lizards. Today there are few of the larger mammals left except in wilderness refuges such as National Parks.

Nutrient cycling within the prairie ecosystem

In Figure 11.28b, the small litter store reflects the relatively small amount of vegetative matter and low leaf fall. Litter decomposes into humus and nutrients are released to the soil, giving it good crumb structure. Moderate rainfall reduces loss from runoff. The large soil storage is a result of the weathering of rock and the presence of deep, rich chernozems (in the eastern and central prairies) which have accumulated a high proportion of humus (organic matter) in the temperate continental conditions. There is little or no leaching because the rainfall is exceeded by evaporation in the summer months.

How and why may this ecosystem change?

1 *Natural conditions*
- **Unpredictable rainfall** and **drought** have been a major factor in change in the North American prairies. Low rainfall in the 1930s allied to bad farming practices led to the creation of the American Dust Bowl; reduced the natural fodder for animals; and permitted an eastwards extension of the short-grass prairies into the eastern tall grass.

■ **Lightning** is a frequent cause of fire in the grasslands in summer. This destroys the surface vegetation; kills small animals; and damages the food supply. In the years following serious fires, lower bird numbers have been recorded, as many nest on the ground.

■ **Bison herds** have been effective in change. They are heavy grazers and reduce the coarser medium grasses, leaving short grasses. In the spring and early summer when mosquitoes hatch they plague the bison causing them to roll on the ground to reduce the itching! This forms depressions in the prairie surface. These bare soils may be re-colonised later by seeds carried by birds.

2 Human activity

■ **Hunting** The earliest inhabitants were the Native Americans who hunted animals for food, using fire and traps to kill unselectively. With the coming of the Europeans and the introduction of the horse and the rifle, they were able to kill large numbers of bison almost to extinction.

■ **Trapping** Fur-traders moving west from the Great Lakes and the Mississippi in the 18th and early 19th centuries helped to reduce the population of bison and elk which were slaughtered for winter food and the hides tanned and sold. The herds were forced into remote areas such as the high basins of Montana and Wyoming. Wolves and mountain lions further reduced them. It is only in the last 20 years that the bison has been increasing again as a result of careful conservation in National Parks and refuges such as the Houck Ranch in South Dakota.

■ **Cereal farming** The prairie sod proved difficult to remove and cultivate using wooden and iron ploughs, and so much of the grassland was used initially for cattle ranching. After the 1840s, when the steel plough was developed, breaking the sod for cultivation became possible. Cereal crops were successfully introduced. Overcultivation in the 1930s led to extensive soil erosion. Soil conservation methods on a large scale have helped to reduce the area which has

been permanently damaged. Droughts between 1950 and 1985 damaged the soils even more: "some plowed fields had lost a metre of their soil. As much as 25 tonnes of soil per hectare had been blown away in two counties" (23 February 1977).

■ **Cattle ranching** became the main farming activity on the western prairies (Figure 12.58). The huge herds used the range lands formerly occupied by the bison. Serious problems of overgrazing have occurred in the lower rainfall regions. Irrigation is normal in Alberta, Montana, the western parts of the Dakotas and Nebraska. This produces fodder crops for cattle who have to be fed outside during the long cold winters. It is not without its problems, for irrigation is lowering the watertable.

■ **Intensive cattle production** is now taking place on huge feedlots close to railheads such as Denver. Young cattle are fattened in stockyards on grain and silage bought in from farmers on the prairie. The animals do not lose condition by being left to roam freely on the open range.

Figure 12.57

Long-grass prairie

Figure 12.58

Cattle farming in the Prairies

How and why may this ecosystem change? What can be done to protect this ecosystem?

There are conflicting views about conservation of the grasslands. A proposal to designate 320 000 acres of prairies in Kansas as a National Park has been strongly opposed by local cattle-ranchers. However, conservationists believe that cattle ranching (Figure 12.59) has already damaged the natural balance of the area. Unless cattle ranching and controlled grazing are held in check, they argue, the prairie will become increasingly degraded.

Cattle Ranchers believe that	Conservationists believe that
controlled burning to renew pasture should be allowed (cf Native American custom)	there is too much burning; the prairie does not recover if burns are too frequent
overgrazing can be avoided by careful pasture management	new information gained from research will help both graziers and conservation
soil and water conservation are already practised	the prairie needs restoration to maintain its ecosystem
tourists, picnic sites and more roads will damage the environment	the prairie has already been damaged by cattle grazing

Figure 12.59

References

Bradshaw, M. (1977) *Earth, The Living Planet*. Hodder & Stoughton.

Courtney, F. and Trudgill, S. T. (1984) *The Soil*. Hodder & Stoughton.

Goudie, A. (1993) *The Nature of the Environment*. Blackwell.

King, T. J. (1989) *Ecology*, 2nd edn. Thomas Nelson.

Money, D. C. (1978) *Climate, Soils and Vegetation*. University Tutorial Press.

Monkhouse, F. J. and Small, J. (1989) *A Dictionary of the Natural Environment*. E. J. Arnold.

O'Hare, G. (1987) *Soils, Vegetation and Ecosystems*. Oliver & Boyd.

Prosser, R. (1992) *Natural Systems and Human Responses*. Thomas Nelson.

Roberts, M. (1986) *Biology: A Functional Approach*, 4th edn. Thomas Nelson.

Simmons, I. (1982) *Biogeographical Processes*. Allen & Unwin.

Spear, T. and Waller, R. (Eds) (1993) *Being Maasai*. James Curry.

Chapter 13

Population

"There is a real danger that in the year 2000 a large part of the world's population will still be living in poverty. The world may become over-populated and will certainly be overcrowded."

Willy Brandt, *North–South: A Programme for Survival,* 1980

Population distribution describes the way in which people are spread out across the earth's surface.

Population density describes the number of people living in a given area.

Figure 13.1

World distribution of population, 1990

In demography, the study of human population, it is important to remember that the situation is dynamic, not static. Population numbers, distributions, structures and movements constantly change in time, in space and at different levels (the micro-, meso- and macro-scales in the population system).

Distribution and density

The distribution of population over the world's surface is uneven and there are considerable variations in density.

Population distributions are often shown by means of a dot map, where each dot represents a given number of people. For example, in Figure 13.1 this method effectively shows the concentration of people in the Nile valley in Egypt, where 99 per cent of the country's population live on 4 per cent of the total land area. However, Figure 13.1 is also misleading because it suggests, incorrectly, that areas away from the Nile are totally uninhabited. In fact, parts are populated, but have insufficient numbers to warrant a symbol. When drawing a dot map, therefore, it is important to select the best possible dot value; and when using one, it is necessary to bear in mind its limitations.

Population densities are often shown by means of a choropleth map, of which Figure 13.2 is an example. Densities are obtained by

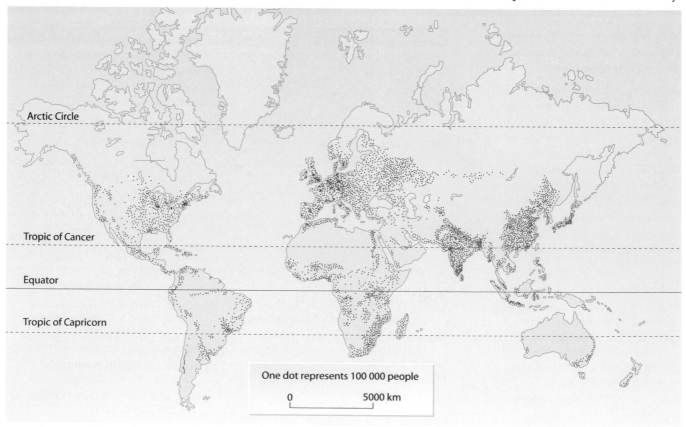

Arctic Circle

Tropic of Cancer

Equator

Tropic of Capricorn

One dot represents 100 000 people

0   5000 km

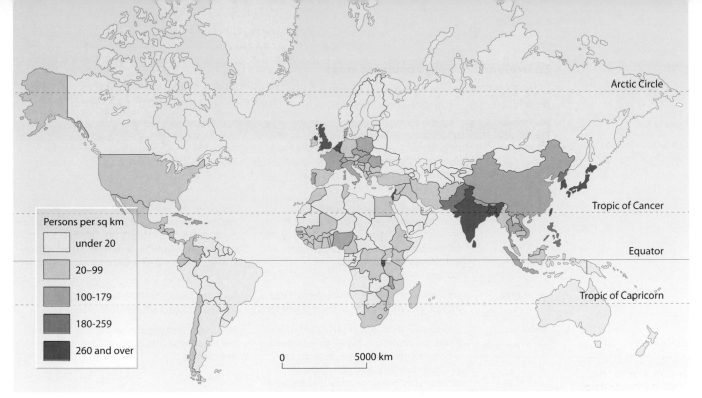

Figure 13.2

World density of population, 1990

Persons per sq km

☐	under 20
☐	20–99
☐	100-179
☐	180-259
☐	260 and over

0 ——— 5000 km

Figure 13.3

The uninhabitable Earth: how valuable are the world's soils for food production?

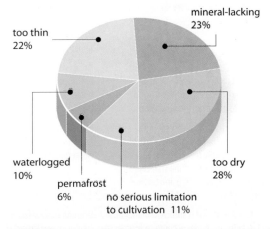

mineral-lacking 23%

too thin 22%

waterlogged 10%

permafrost 6%

no serious limitation to cultivation 11%

too dry 28%

dividing the total population of a country (or administrative area) by the total area of that country (or area). The densities are then grouped into classes, each of which is coloured lighter or darker to reflect lesser or greater density. Although these maps are easy to read, they hide concentrations of population within each unit area. Figure 13.2, for example, gives the impression that the population of Egypt is equally distributed across the country; it also suggests that there is an abrupt change in population density at the national boundary. A poorly designed system of colouring or shading can make quite small spatial differences seem large — or make huge differences look smaller.

Figures 13.1 and 13.2 both show that there are parts of the world which are sparsely populated and others which are densely populated. One useful generalisation that may be made — remembering the pitfalls of generalisation (Framework 8, page 322) — is that, at the global scale, this distribution is affected mainly by physical opportunities and constraints; whereas, at regional and local scales, it is more likely to be influenced by economic, political and social factors.

Land accounts for about 30 per cent of the earth's surface (70.9 per cent is water). Of the land area, only about 11 per cent presents no serious limitations to settlement and agriculture (Figure 13.3). Much of the remainder is desert, snow and ice, high or steep-sided mountains, and forest. Usually there are several reasons why an area is sparsely or densely populated.

Q

1 Study Figure 13.4. Certain types of environment have traditionally supported relatively low overall population densities. Is this still true today or have constraining factors been modified or overcome?

2 Read Figure 13.4 and Framework 8, page 322.

a Comment critically on the accuracy and value of the listed factors which, individually or collectively, have been offered as explanations of why certain areas have become densely populated.

b Why do geographers make generalisations like that in **a** above?

Sparsely and densely populated areas

Figure 13.4 lists some of the many factors which operate at the global scale and which may lead to an area being sparsely or densely populated. Compare these factors with the patterns shown in Figures 13.1 and 13.2.

Figure 13.4

Major factors affecting population density

Factors	Sparsely populated areas	Densely populated areas
Physical	Rugged mountains where temperature and pressure decrease with height; active volcanoes (the Andes); high plateaux (Tibet) and worn down shield lands (the Canadian Shield, Figure 1.9).	Flat, lowland plains are attractive to settlement (the Netherlands and Bangladesh) as are areas surrounding some volcanoes (Mt Etna).
Climate	Areas receiving very low annual rainfall (the Sahara Desert); areas having a long seasonal drought or unreliable, irregular rainfall (the Sahel countries); areas suffering high humidity (the Amazon basin); very cold areas, with a short growing season (northern Canada, page 306).	Areas where the rainfall is reliable and evenly distributed throughout the year; with no temperature extremes and a lengthy growing season (north-west Europe); where sunshine (the Costa del Sol) or snow (the Alps) is sufficient to attract tourists; and areas with a monsoon climate (south-east Asia).
Vegetation	Areas such as the coniferous forests of northern Eurasia and northern Canada, and the rainforests of the tropics.	Areas of grassland tend to have higher population densities than places with dense forest or desert.
Soils	The frozen soils of the Arctic (the permafrost in Siberia); the thin soils of mountains (Nepal); the leached soils of the tropical rainforest (the Amazon basin); also, increasingly large areas are experiencing severe soil erosion resulting from deforestation and overgrazing (the Sahel).	Deep, humus-filled soils (the Paris Basin) and, especially, river-deposited silt (the Ganges and Nile deltas) both favour farming.
Water supplies	Many areas lack a permanent supply of clean fresh water: mainly due either to insufficient, irregular rainfall or to a lack of money and technology to build reservoirs and wells or lay pipelines (Ethiopia).	Population is more likely to increase in areas with a reliable water supply. This may result from either a reliable, evenly distributed rainfall (northern England) or where there is the wealth and technology to build reservoirs and to provide clean water (California). Places with heavy seasonal rainfall (the monsoon lands of south-east Asia, page 222) also support many people.
Diseases and pests	These may limit the areas in which people can live or may seriously curtail the lives of those who do populate such areas (malaria in central Africa).	Some areas were initially relatively disease- and pest-free; others had the capital and medical expertise to eradicate those which were a problem (the formerly malarial Pontine Marshes, near Rome).
Resources	Areas devoid of minerals and easily obtainable sources of energy rarely attract people or industry (Paraguay).	Areas having, or formerly having, large mineral deposits and/or energy supplies (the Ruhr) often have major concentrations of population; these resources often led to the development of large-scale industry (the Pittsburgh region, USA).
Communications	Areas where it is difficult to construct and maintain transport systems tend to be sparsely populated, e.g. mountains (Bolivia), deserts (the Sahara) and forests (the Amazon basin and northern Canada).	Areas where it is easier to construct canals, railways, roads and airports have attracted settlements (the North European Plain), as have large natural ports which have been developed for trade (Singapore, page 548).
Economic	Areas with less developed, subsistence economies usually need large areas of land to support relatively few people (although this is not applicable to south-east Asia). Such areas tend to fall into three belts: tundra (the Lapps), desert fringes (the Rendille, page 440) and tropical rainforests (shifting cultivators, page 441).	Regions with intensive farming or industry can support large numbers of people on a small area of land (as in the Netherlands).
Political	Areas where the state fails to invest sufficient money or to encourage development — either economically or socially (the interior of Brazil).	Decisions may affect population distribution — e.g. by creating new cities, such as Brasilia; by opening up 'pioneer' lands for development, as in Israel.

Framework 8 Scale and generalisation

The study of an environment, whether natural or altered by human activity, involves the study of numerous different and interacting processes. The relative importance of each process may vary according to the scale of the study — i.e. global or **macro-scale**; intermediate or **meso-scale**; and local or **micro-scale**. It may also vary according to the time scale chosen — i.e. whether processes are studied through **geological time**, **historical time**, or **recent time**.

In the study of soils (Chapters 10 and 12), it is climate which tends to impose the greatest influence upon the formation and distribution of the major global (zonal) types (the podsol and chernozem). At the regional level, rock type may be the major influencing factor (Mediterranean areas with their terra rossas and rendzinas). Within a small area, such as a river valley with homogeneous climate and rock type, relief may be dominant (the catena, pages 239 and 256).

In the study of erosion, time is a major variable: a stretch of coastline may be eroded by the sea during a period of several decades or centuries; footpath erosion may occur during a single summer.

A common problem with spatial and time scales, as with models (Framework 9, page 328), is that a chosen level of detail may become inappropriate to all or part of the probelm

under study: it may become either too large and generalised, or too small and complex. For example, population distributions and densities may be studied at a variety of spatial and time scales. At the world scale (Figures 13.1 and 13.2), the pattern shown is so general and deterministic that it may lead the student into an over-simplified understanding of the processes which produced the apparent distribution and/or density. Such generalised patterns usually break down into something more complex when studied at a more local level or over a period of time.

Although it may often be easier to identify and account for distributions, densities, anomalies and changes at the national level, it is more difficult in the case of a country the size of Brazil (Figure 13.5) than it is for a smaller country such as Uruguay. It is often only when looking at a smaller region (Figure 13.6) or an urban area (Figure 13.7), perhaps over a relatively short time period, that the complexities of the various processes can be readily understood.

Places 30 Population densities at the national level

Even a quick look at the population density map of Brazil (Figure 13.5) shows a relatively simple general pattern. Over 90 per cent of Brazilians live in a discontinuous strip about 500 km wide, adjacent to the east coast. This strip accounts for less than 25 per cent of the country's total area. The density declines very rapidly towards the north-west, where several remote areas are almost entirely lacking in permanent settlement.

The area marked **1A** on Figure 13.5 is the dry north-east (the Sertao). Here the long and frequent water balance deficit (drought), high

temperatures and poor soils combine to make the area unsuitable for growing high-yield crops or rearing good-quality animals. The Sertao also lacks known mineral or energy reserves; communications are poor; and the basic services of health, education, clean water and electricity are lacking. Although birth rates are exceptionally high (many mothers have more than ten babies), there is a rapid outward migration to the urban areas (page 342), a high infant mortality rate and a short life expectancy (page 330).

Area **1B** is the tropical rainforest, drained by the River Amazon and its tributaries. Here the climate is hot, wet and humid; rivers flood annually; and there is a high incidence of disease. In the past, the forest has proved difficult to clear, but once the protective trees have gone, soils are rapidly leached and become infertile. Land communications are difficult to build and maintain. The area has suffered, as has **1A**, from a lack of federal investment and can only support subsistence economies.

Figure 13.5

Population density in Brazil: the national scale

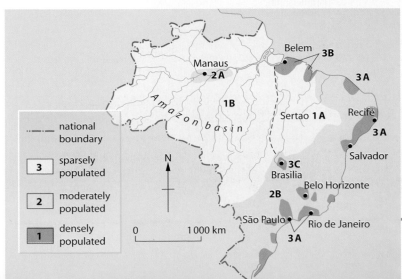

There are, however, two anomalies in Amazonia. The first is a zone along the River Amazon centred on Manaus (**2A** on Figure 13.5). Originally a Portuguese trading post, Manaus has had two growth periods. The first was associated with the rubber boom at the turn of this century, while the second began in the 1980s with the development of tourism and the granting of its new status as a free port (Places 73, page 548). The second anomaly has followed the recent exploitation of several minerals (iron ore at Carajas and bauxite at Trombetas) and energy resources (hydro-electricity at Tucuri; Places 77, page 577).

The more easterly parts of the Brazilian Plateau are moderately populated (area **2B**). The climate is cooler and it is considerably healthier than on the coast and in the rainforest. The soil, in parts, is a rich terra rossa (page 254) which here is a weathered volcanic soil ideal for the growing of coffee. Several precious minerals have been found. However, rainfall is irregular with a long winter drought; communications are still limited; and federal investment has been insufficient to stimulate much population growth.

Except where the highland reaches the sea, the eastern parts of the plateau around São Paulo and Belo Horizonte and the east coast have the highest population densities (area **3A**). Although the coastal area is often hot and humid, the water supply is good. Several natural harbours proved ideal for ports and this encouraged trade and the growth of industry. Salvador, the first capital, was the centre of the slave trade. Rio de Janeiro became the second capital, developing as an economic, cultural and administrative centre. More recently, it has received increasing numbers of tourists from overseas and migrants from the north of Brazil.

The fastest-growing city in the world, other than Mexico City, is São Paulo. The cooler climate and terra rossa soils initially led to the growth of commercial farming based on coffee. Access to minerals such as iron ore and to energy supplies later made it a major industrial centre. The São Paulo region has had high levels of federal investment, leading to the development of a good communications network and the provision of modern services.

Area **3B** is a recent growth pole (page 519) based on the discovery and exploitation of vast deposits of iron ore and bauxite, the construction of hydro-electric power stations and the advantages of access along the coastal strip and Amazon corridor. **3C** is the new federal capital, Brasilia, built in the early 1960s to try to redress the imbalance in population density and wealth between the south-east of the country and the interior.

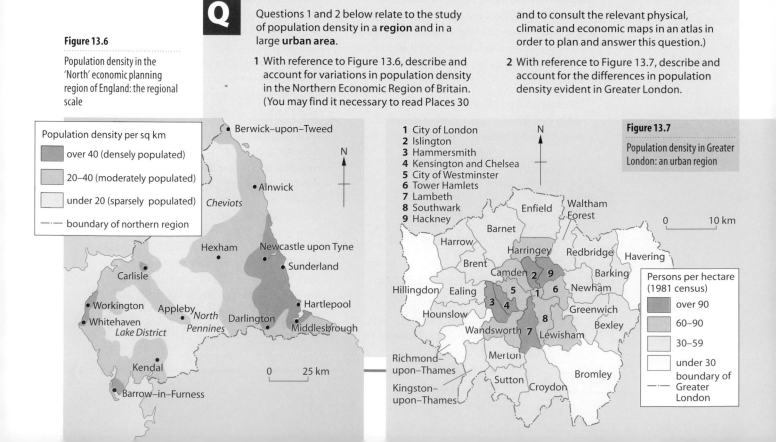

Figure 13.6

Population density in the 'North' economic planning region of England: the regional scale

Q Questions 1 and 2 below relate to the study of population density in a **region** and in a large **urban area**.

1 With reference to Figure 13.6, describe and account for variations in population density in the Northern Economic Region of Britain. (You may find it necessary to read Places 30

and to consult the relevant physical, climatic and economic maps in an atlas in order to plan and answer this question.)

2 With reference to Figure 13.7, describe and account for the differences in population density evident in Greater London.

Population density per sq km

- over 40 (densely populated)
- 20–40 (moderately populated)
- under 20 (sparsely populated)
- —·— boundary of northern region

1 City of London
2 Islington
3 Hammersmith
4 Kensington and Chelsea
5 City of Westminster
6 Tower Hamlets
7 Lambeth
8 Southwark
9 Hackney

Figure 13.7

Population density in Greater London: an urban region

Persons per hectare (1981 census)

- over 90
- 60–90
- 30–59
- under 30
- —·— boundary of Greater London

Lorenz curves

Lorenz curves are used to show inequalities in distributions. Population, industry and land use are three topics of interest to the geographer which show unequal distributions over a given area. Figure 13.8 illustrates the unevenness of population distribution over the world. The diagonal line represents a perfectly even distribution, while the concave curve (it may be convex in other examples) illustrates the degree of concentration of population within the various continents. The greater the concavity of the slope, the greater the inequality of population distribution (or industry, land use, etc.).

Figure 13.8

A Lorenz curve: the distribution of world population in 1960

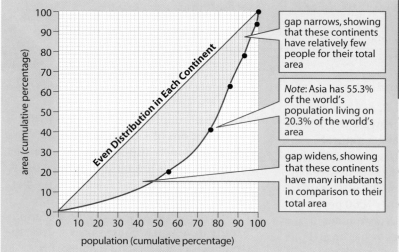

Continents ranked in descending order of population (1960)	Population (%)	Population (cumulative %)	Area (%)	Area (cumulative %)
Asia	55.3	55.3	20.3	20.3
Europe and USSR	21.2	76.5	20.1	40.4
Africa	9.3	85.8	22.3	62.7
Latin America	7.1	92.9	15.2	77.9
North America	6.6	99.5	15.8	93.7
Oceania	0.5	100.0	6.3	100.0

Q

Figure 13.9 gives the estimated population totals for each continent for 2000 A.D.

1 Work out the percentage of world population predicted to be living in each continent in 2000 A.D.

2 Calculate the cumulative percentage, ranking the continents in descending order of population size.

3 Complete a Lorenz curve using Figure 13.8 for guidance.

4 Describe any differences you notice between your graph and Figure 13.8.

Figure 13.9

Estimated world population for the year 2000

Continent	Estimated total population, 2000 A.D. (millions)	Population (%)	Population (cumulative %)	Area (%)	Area (cumulative %)
Asia	3548.1			20.3	20.3
Africa	871.8			22.3	42.6
Europe and CIS	827.2			20.1	62.7
Latin America	546.7			15.2	77.9
North America	297.1			15.8	93.7
Oceania	30.1			6.3	100.0
World	6121.8				

Population changes in time

It has already been stated (page 320) that populations are dynamic — i.e. their numbers, distributions, structure and movement (migration) constantly change over time and space. Population change is another example of an open system (Framework 1, page 39) with inputs, processes and outputs (Figure 13.10).

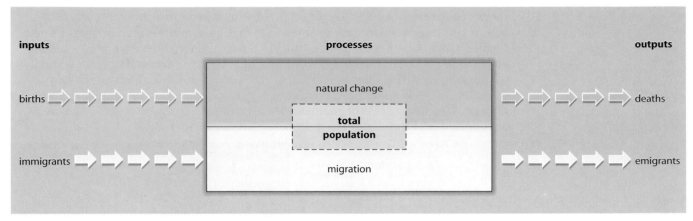

inputs | processes | outputs

births ⇒ ⇒ ⇒ ⇒ ⇒ ⇒

natural change

total population

migration

deaths

immigrants ⇒ ⇒ ⇒ ⇒ ⇒

emigrants

Figure 13.10

Population change as an open system

Birth rates, death rates and natural increase

The total population of an area is the balance between two forces of change: **natural increase** and **migration** (Figure 13.10). The natural increase is the difference between birth rates and death rates. The **birth rate** is the number of live births per 1000 people per year and the **death rate** is the number of deaths per 1000 people per year. Throughout history, until the last few years in a small number of the economically most developed countries, birth rates have nearly always exceeded death rates. Exceptions have followed major outbreaks of disease (the bubonic plague) or wars (as in Rwanda). Any natural change in the population, either an increase or a decrease, is usually expressed as a percentage and referred to as the **annual growth rate**. Population change is also affected by migration. Although migration does not affect world population totals, it does affect the way people are distributed across the world. Migration leads to *either* an increase in the population — when the number of immigrants exceeds the number of emigrants (as in Hong Kong and Zaire) — *or* a decrease in population — when the number of emigrants exceeds the number of immigrants (as in the former Yugoslavia and Rwanda).

The demographic transition model

The demographic transition model describes a sequence of changes over a period of time in the relationship between birth and death rates and overall population change. The model, based on population changes in several industrialised countries in western Europe and North America, suggests that *all* countries pass through similar demographic transition stages or **population cycles** — or will do, given time. Figure 13.11 illustrates

the model and gives reasons for the changes at each transition stage. It also gives examples of countries which appear to 'fit' the descriptions of each stage.

Like all models, the demographic transition model has its limitations (Framework 9, page 328). It failed to consider, or to predict, several factors and events:

1 Birth rates in several of the most economically developed countries have, since the model was put forward, fallen below death rates (Germany, Sweden). This has caused, for the first time, a population decline which suggests that perhaps the model should have a fifth stage added to it.

2 The model, being Eurocentric, assumed that in time all countries would pass through the same four stages. It now seems unlikely, however, that many of the economically less developed countries, especially in Africa, will ever become industrialised.

3 The model assumed that the fall in the death rate in Stage 2 was the consequence of industrialisation. Initially, the death rate in many British cities rose, due to the insanitary conditions which resulted from rapid urban growth, and it only began to fall after advances were made in medicine. The delayed fall in the death rate in many developing countries has been due mainly to their inability to afford medical facilities. In many countries, the fall in the birth rate in Stage 3 has been *less* rapid than the model suggests due to religious and/or political opposition to birth control (Brazil), whereas the fall was much *more* rapid, and came earlier, in China following the government-introduced 'one-child' policy (Case Study 13).

Figure 13.11

The demographic transition model

4 The timescale of the model, especially in several south-east Asian countries such as Hong Kong and Malaysia, is being squashed as they develop at a much faster rate than did the early industrialised countries.

The model can be used
- to show how the population growth of a country changes over a period of time (the UK, in Figure 13.11);
- to compare rates of growth between different countries at a given point in time.

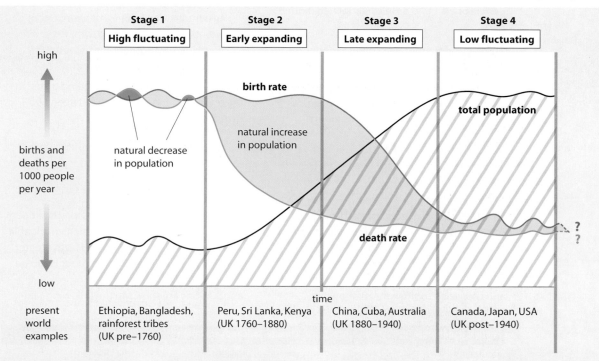

Stage 1	Stage 2	Stage 3	Stage 4
High fluctuating	**Early expanding**	**Late expanding**	**Low fluctuating**

births and deaths per 1000 people per year

birth rate

natural decrease in population

natural increase in population

total population

death rate

time

| present world examples | Ethiopia, Bangladesh, rainforest tribes (UK pre–1760) | Peru, Sri Lanka, Kenya (UK 1760–1880) | China, Cuba, Australia (UK 1880–1940) | Canada, Japan, USA (UK post–1940) |

Stage 1: Here both birth rates and death rates fluctuate at a high level (about 35 per 1000) giving a small population growth.

Birth rates are high because:

- No birth control or family planning.
- So many children die in infancy that parents tend to produce more in the hope that several will survive.
- Many children are needed to work on the land.
- Children are regarded as a sign of virility.
- Some religious beliefs (Roman Catholics, Muslims and Hindus) encourage large families.

High death rates, especially among children, are due to:

- Disease and plague (bubonic, cholera, kwashiorkor).
- Famine, uncertain food supplies, poor diet.
- Poor hygiene: no piped, clean water and no sewage disposal.
- Little medical science: few doctors, hospitals, drugs.

Stage 2: Birth rates remain high, but death rates fall rapidly to about 20 per 1000 people giving a rapid population growth.

The fall in death rates results from:

- Improved medical care: vaccinations, hospitals, doctors, new drugs and scientific inventions.
- Improved sanitation and water supply.
- Improvements in food production, both quality and quantity.
- Improved transport to move food, doctors, etc.
- A decrease in child mortality.

Stage 3: Birth rates now fall rapidly, to perhaps 20 per 1000 people, while death rates continue to fall slightly (15 per 1000 people) to give a slowly increasing population.

The fall in birth rates may be due to:

- Family planning: contraceptives, sterilisation, abortion and government incentives.
- A lower infant mortality rate leading to less pressure to have so many children.
- Increased industrialisation and mechanisation meaning fewer labourers are needed.
- Increased desire for material possessions (cars, holidays, bigger homes) and less desire for large families.
- An increased incentive for smaller families.
- Emancipation of women, enabling them to follow their own careers rather than being solely child–bearers.

Stage 4: Both birth rates (16 per 1000) and death rates (12 per 1000) remain low, fluctuating slightly to give a steady population.

(Will there be a **Stage 5** where birth rates fall below death rates to give a declining population? Some evidence suggests that this might be occurring in several western European countries).

Models form an integral and accepted part of present-day geographical thinking and teaching. Nature is highly complex and, in an attempt to understand this complexity, geographers try to develop simplified models of it.

Chorley and Haggett described a model as:

> a simplified structuring of reality which presents supposedly significant features or relationships in a generalised form . . . as such they are valuable in obscuring incidental detail and in allowing fundamental aspects of reality to appear.

They stated that a model:

> can be a theory or a law, an hypothesis or structured idea, a rôle, a relationship, or equation, a synthesis of data, a word, a graph, or some other type of hardware arranged for experimental purposes.

A good model will stand up to being tested in the real world and should fall between two extremes:

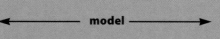

very simple and easy to work, but too generalised to be of real value ← **model** → very difficult to use, being almost as complex as reality

To achieve this balance (several though sometimes only one) critical criteria or variables are selected as a basis for the model. For example, J. H. von Thünen (page 430) chose distance from a market as his critical variable and then tried to show the relationship between this variable and the intensity of land use. If necessary, other variables may be added which, as in the case of von Thünen's navigable river and a rival market, may add both greater reality and greater complexity. Models can be used in all fields of geography. Some applications are shown in the following table.

Physical (landforms)	Climate, soils and vegetation	Human and economic
beach profile	atmospheric circulation	cities (Burgess)
slope form	heat budget	land use (von Thünen)
corrie development	seres	industrial location (Weber)
geomorphological systems (Framework 1), drainage basin, glacier budget	catena	settlement size and distribution (Christaller)
	depression	gravity models
	biome	demographic transition
	soil profiles (podosols)	economic growth (Rostow)

Throughout this book, models and theories are presented; their advantages and limitations are examined; and their applications to real-world situations are demonstrated, together with their usefulness in explaining that situation.

Q

1 What do you understand by the following terms: birth rate; death rate; natural increase; annual growth rate?

2 Use Figures 13.11 and 13.12 to answer the following questions.
   a Explain why Britain had a high birth rate and a high death rate between 1700 and 1760 (Stage **A**).
   b Why did Britain have a rapidly falling death rate between 1760 and 1880 (Stage **B**)? (Historians should be able to give dates of social legislation and health improvements.)
   c 'Between 1880 and 1940 (Stage **C**), Britain's birth rate declined rapidly while the death rate fell relatively slowly.' Suggest an explanation for this trend.
   d Explain why there has been a slightly fluctuating population growth with both birth and death rates remaining low since 1940 (Stage **D**).
   e Attempt to predict Britain's natural increase for the year 2000.
   f How closely do you consider that Britain's population growth can be compared with the demographic transition model?

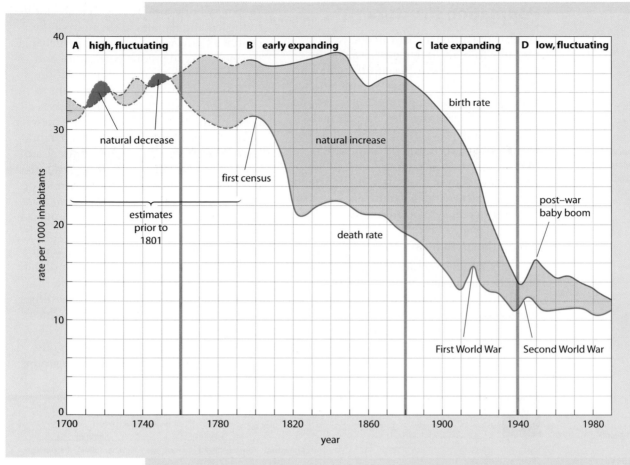

Figure 13.12

Changes in Britain's population, 1700–1990 (*after* J. H. Lowry)

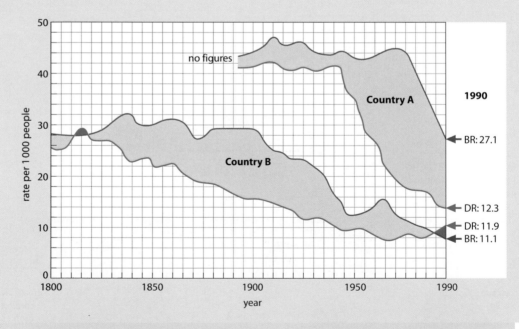

Figure 13.13

A comparison of the population cycles for Sweden and India, 1800–1990

3 Figure 13.13 shows the population cycles of Sweden and India.

a Is Country **A** Sweden or India?

b Describe and give reasons for the differences between the two graphs.

c For each country, give the approximate year in which the natural increase was at its greatest.

d What is likely to happen to the total population of each country in the next few years?

Population structure

The rate of natural increase or decrease, resulting from the difference between the birth and death rates of a country, represents only one aspect of the study of population structure. A second important aspect is population. This is important because the make-up of the population by its age and sex, together with its **life expectancy**, has implications for the future growth, economic development and social policy of a country. Differences in language, race, religion, family size, etc. can all affect a country's socio-economic welfare.

Life expectancy is the number of years that the average person born in a given area may expect to live.

Population pyramids

The population structure of a country is best illustrated by a **population** or **age–sex pyramid**. The technique normally divides the population into 5-year age groups (e.g. 0–4, 5–9, 10–14) on the vertical scale, and into males and females on the horizontal scale. The number in each age group is given as a percentage of the total population and is shown by horizontal bars with males located to the left and females to the right of the central axis. As well as showing past changes, the pyramid can predict both short- and long-term future changes in population.

Whereas the demographic transition model shows only the natural increase or decrease resulting from the balance between births and deaths, the population pyramid shows the effects of migration, the age and sex of migrants (Figure 13.40) and the effects of large-scale wars and major epidemics of disease. Figure 13.14 is a partly completed pyramid for the UK. You should complete it and then notice the following: a narrow pyramid indicating approximately equal numbers in each age group; a low birth rate (meaning fewer school places will be needed); and a low death rate (suggesting a need for more old people's homes) which together indicate a steady growth, or even a static population. Although males outnumber females in the younger age groups, it is females who live the longer.

Figure 13.14

Constructing a population pyramid

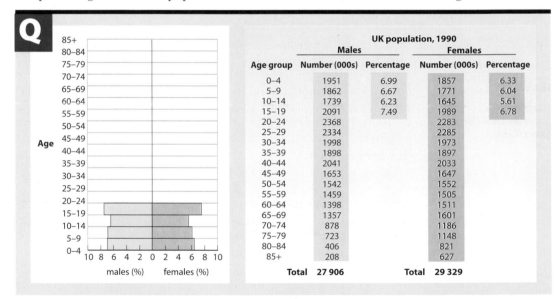

	UK population, 1990			
	Males		**Females**	
Age group	**Number (000s)**	**Percentage**	**Number (000s)**	**Percentage**
0–4	1951	6.99	1857	6.33
5–9	1862	6.67	1771	6.04
10–14	1739	6.23	1645	5.61
15–19	2091	7.49	1989	6.78
20–24	2368		2283	
25–29	2334		2285	
30–34	1998		1973	
35–39	1898		1897	
40–44	2041		2033	
45–49	1653		1647	
50–54	1542		1552	
55–59	1459		1505	
60–64	1398		1511	
65–69	1357		1601	
70–74	878		1186	
75–79	723		1148	
80–84	406		821	
85+	208		627	
Total	**27 906**		**Total**	**29 329**

Figure 13.15

Population pyramids characteristic of each stage of the demographic transition model

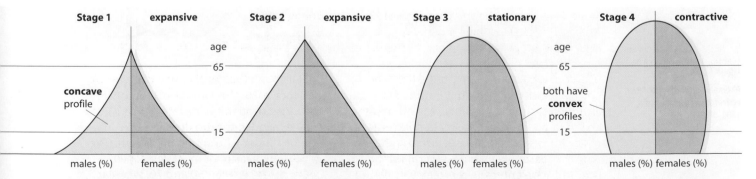

Stage 1 — expansive — concave profile — high birth rate; rapid fall in each upward age group due to high death rates; short life expectancy

Stage 2 — expansive — still a high birth rate; fall in death rate as more living in middle age; slightly longer life expectancy

Stage 3 — stationary — declining birth rate; low death rate; more people living to an older age

Stage 4 — contractive — both have convex profiles — low birth rate; low death rate; higher dependency ratio; longer life expectancy

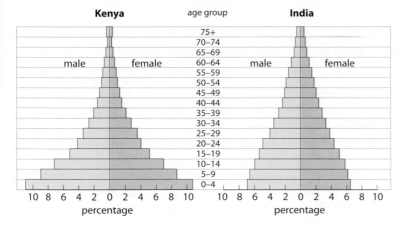

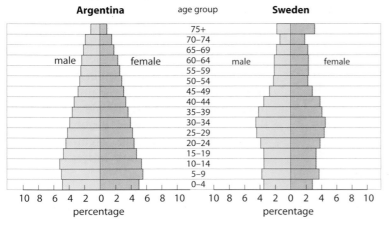

Figure 13.16

Population pyramids for selected countries, 1990 data

Infant mortality rate is the average number of children out of every 100 born alive that die under the age of one.

A model has also been produced to try to show the characteristics of four basic types of pyramid (Figure 13.15). As with most models, many countries show a transitional shape which does not fit precisely into any pattern. Figure 13.16 shows the pyramids for *selected* countries — chosen because they *do* conform closely to the model!

Stage 1 Kenya's pyramid has a concave shape, showing that the birth rate is very high. Over half the inhabitants (51 per cent) are under 15 years old; there is a rapid fall upwards in each age group showing a high death rate (including **infant mortality**) and a low life expectancy with only 2 per cent who can expect to live beyond 65.

Stage 2 India appears to have reached Stage 2 (shown by the more uniform sides). All pyramids in this stage have a broad base indicating a high birth rate but, as the infant mortality and death rates decline, more people reach middle age and the life expectancy is slightly longer. The result is that although the actual numbers of children may be the same, they form a smaller percentage of the total population (as shown by the narrower base). The large youthful population will soon enter the reproductive period and become economically active. India has 39 per cent under 15; and 3 per cent over 65.

Stage 3 Argentina has probably just reached this stage as its birth rate is declining — as shown by the almost equal numbers in the lower age groups. As the death rate is much lower, more people are able to live to an older age, and the actual growth rate becomes stable. Argentina has 26 per cent under 15; and 8 per cent over 65.

Stage 4 Sweden has a smaller proportion of its population in the pre-reproductive age groups (22 per cent under 15) and a larger proportion in the post-reproductive groups (16 per cent over 65), indicating low birth, infant mortality and death rates and a long life expectancy. As the numbers entering the reproductive age groups decline there will be, in time, a fall in the total population.

Dependency ratios

The population of a country can be divided into two categories according to their contribution to economic productivity. Those aged 15–65 years are known as the **economically active** or **working population**; those under 15 and over 65 are known as the **non-economically active population**. (Perhaps in Britain the division should be made at 16, the school-leaving age; in developing countries, however, the cut-off point is much lower as many children have to earn money from a very young age.)

The dependency ratio can be expressed as:

$$\frac{\text{children (0--14)} \text{ and elderly (65 and over)}}{\text{those of working age}} \times 100$$

e.g. UK 1971 (figures in millions):

$$\frac{13\ 387\ +\ 7\ 307}{31\ 616} \times 100 = 65.45$$

So, for every 100 people of working age there were 65.45 people dependent upon them.

By 1990 the dependency ratio had changed to:

$$\frac{10\ 799 + 8\ 683}{37\ 281} \times 100 = 52.26$$

So, although the number of elderly people had increased, this was more than offset by the larger drop in the number of children. (The dependency ratio does not take into account those who are unemployed). The dependency ratio for most developed countries is between 50 and 70, whereas for less economically developed countries it is often over 100.

Trends in population growth

1 Global trends

In Mother Earth's 46 years (Places 1, page 9), it was only "in the middle of last week, in Africa, that some manlike apes turned into ape-like men" and the world's human population slowly began to grow. In the absence of any census, this population is estimated to have been about 500 million by 1650. It was only after the Industrial Revolution in western Europe that numbers "began to multiply prodigiously". The most rapid increase has taken place over the last 70 years (Figure 13.17).

The United Nations Fund for Population Activities (UNFPA) designated 11 July 1987 as the date of the arrival of the five billionth (5000 millionth) human being on earth. Of course that 'celebration' was fictitious as nobody knows exactly how many people are living on the earth at a given moment since, in many areas, census figures are either inaccurate or non-existent. However, although that figure is approximate, it is certain that the world's population is still growing by about 150 people every minute and 96 million each year.

Recent evidence has shown that **fertility rates** in economically developing countries have, at last, begun to fall. A 1992 World Bank report claimed that the annual growth rate of the world's population, which had been 2.1 per cent per annum between 1965 and 1970, had fallen to 1.7 per cent — mainly due to China's one child per family policy (Case Study 13). This will mean that by the end of the century the world's population may be only 6300 million instead of the 7600 million it would have reached had the growth rate of 1950–80 continued. The variants given in most population projections (low, medium and high) provide an indication of the breadth of possible outcomes in terms of population change. Figures 13.17 and 13.18 both use medium projections.

2 Regional trends

What these figures fail to show is the marked variation between different areas in the world and especially between the economically developed and economically developing continents. At present, the average population growth rate for developed countries is 0.64 per cent per year compared with 2.07 per cent in those described as developing (Figures 13.18 and 13.19). To achieve population stability, the average family of today would, worldwide, have to consist of 2.3 children. While the average number is currently under two in western Europe, North America, the CIS and, for exceptional reasons (Case Study 13), China, it is over four in most of Asia and Latin America, and over six in Africa.

The fertility rate is the number of children per woman.

Figure 13.18

Global and regional trends in population growth

In global terms the major trend has been a decline in the rate of population growth from a peak of 2.1% p.a. in 1965-70 to approximately 1.7% in 1992 (World Bank). There continue to be still more persons (over 90 millions p.a.) being added in total to the population than ever before in human history. Whilst the distinction between low (or no) population growth in 'developed' countries and high population growth in 'less developed' countries continues, a major feature of the last four decades has been the wide divergence of the demographic trajectories of the 'less developed' countries. Whilst the detailed pattern is complex, in general terms there has been rapid fertility decline in East Asia, moderate decline in parts of Latin America and to a lesser extent South Asia and very limited decline in Sub-Saharan Africa. However, there have been recent encouraging indications that a number of African countries are moving towards fertility decline. Even those countries which have recently undergone rapid fertility decline continue to grow due to population momentum. Although there is always the possibility of stalling in the process of fertility decline, the main question is not whether, but when and at what total, global population will stabilise. The most authoritative estimates point towards global population rising from 5.5 billions in 1993, to 6.5 billions by 2010 possibly stabilising at 10.1 billions by the middle of the twenty-second century (World Bank 1992) even if population growth follows the rapid fertility decline scenario. Population projections beyond the near future need to be taken into consideration with especial caution.

(N. Ford, at the International Conference on Population and Development, Cairo, September 1994)

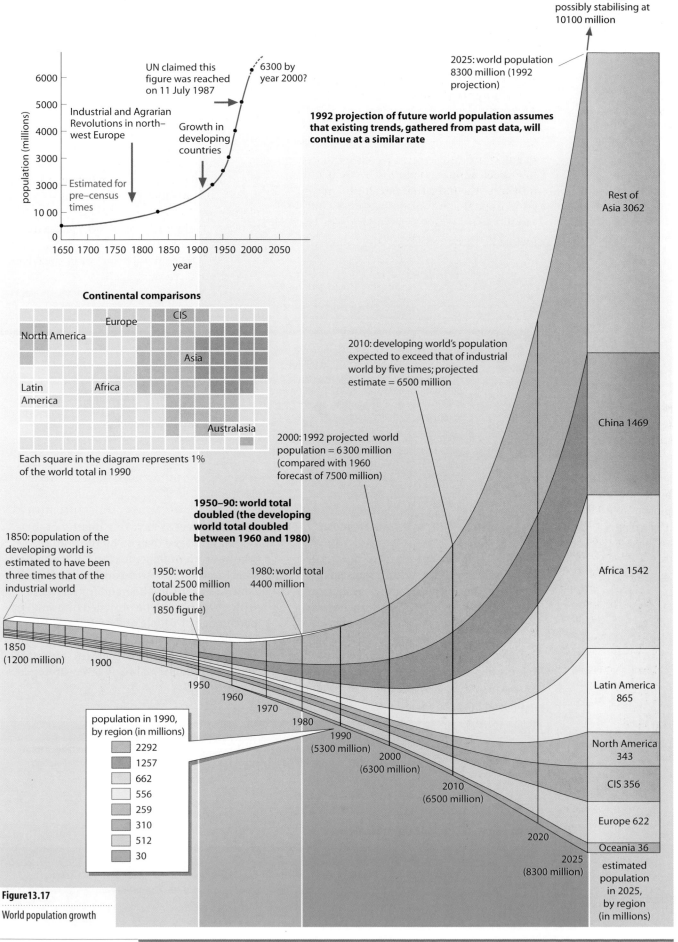

2150: world population possibly stabilising at 10100 million

2025: world population 8300 million (1992 projection)

1992 projection of future world population assumes that existing trends, gathered from past data, will continue at a similar rate

UN claimed this figure was reached on 11 July 1987

6300 by year 2000?

Industrial and Agrarian Revolutions in north–west Europe

Growth in developing countries

Estimated for pre–census times

population (millions)

6000
5000
4000
3000
3000
10 00
0

1650 1700 1750 1800 1850 1900 1950 2000 2050

year

Rest of Asia 3062

2010: developing world's population expected to exceed that of industrial world by five times; projected estimate = 6500 million

China 1469

Continental comparisons

CIS
Europe
North America
Asia
Latin America
Africa
Australasia

Each square in the diagram represents 1% of the world total in 1990

2000: 1992 projected world population = 6 300 million (compared with 1960 forecast of 7500 million)

Africa 1542

1950–90: world total doubled (the developing world total doubled between 1960 and 1980)

1850: population of the developing world is estimated to have been three times that of the industrial world

1950: world total 2500 million (double the 1850 figure)

1980: world total 4400 million

Latin America 865

1850 (1200 million)

1900

1950

1960

1970

1980

1990 (5300 million)

2000 (6300 million)

2010 (6500 million)

2020

2025 (8300 million)

North America 343

CIS 356

Europe 622

Oceania 36

estimated population in 2025, by region (in millions)

population in 1990, by region (in millions)

| 2292 |
| 1257 |
| 662 |
| 556 |
| 259 |
| 310 |
| 512 |
| 30 |

Figure 13.17

World population growth

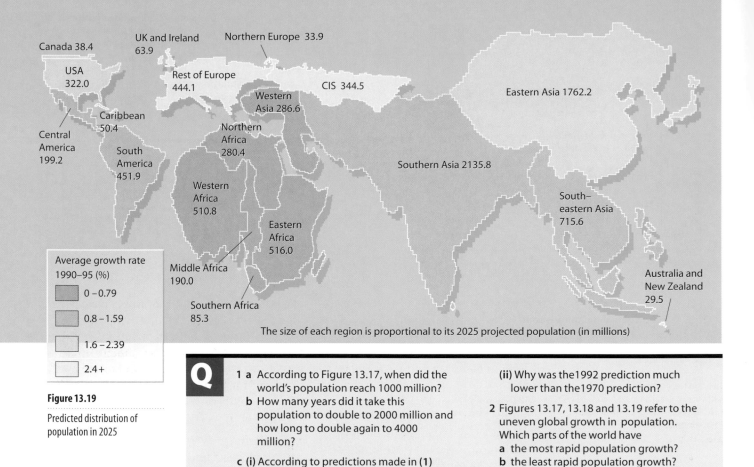

Canada 38.4

UK and Ireland 63.9

Northern Europe 33.9

USA 322.0

Rest of Europe 444.1

CIS 344.5

Eastern Asia 1762.2

Caribbean 50.4

Western Asia 286.6

Central America 199.2

Northern Africa 280.4

South America 451.9

Southern Asia 2135.8

Western Africa 510.8

South-eastern Asia 715.6

Eastern Africa 516.0

Middle Africa 190.0

Southern Africa 85.3

Australia and New Zealand 29.5

The size of each region is proportional to its 2025 projected population (in millions)

Average growth rate 1990–95 (%)

- 0 – 0.79
- 0.8 – 1.59
- 1.6 – 2.39
- 2.4 +

Figure 13.19

Predicted distribution of population in 2025

Q

1 a According to Figure 13.17, when did the world's population reach 1000 million?
  b How many years did it take this population to double to 2000 million and how long to double again to 4000 million?
  c (i) According to predictions made in (1) 1970 and (2) 1992, what was the world's population estimated to be by the year 2000?

(ii) Why was the1992 prediction much lower than the1970 prediction?

2 Figures 13.17, 13.18 and 13.19 refer to the uneven global growth in population. Which parts of the world have
  a the most rapid population growth?
  b the least rapid population growth?

3 Declining death rates and increased life expectancy

Due to improvements in medical facilities, hygiene and vaccines, death rates have declined and life expectancy has increased considerably. Figure 13.20 shows how, as a result of the falling death rate, the world's population is ageing (Britain has the highest number of over-60s in the EU) and how life expectancy is increasing. The Japanese, who can expect to live longer than any other nation, expect to have one-quarter of their population aged over 65 by the year 2025 (the equivalent figure for the UK is one-fifth). This trend is likely to mean, in 'developed' countries, a greater demand for services which will have to be provided by a smaller percentage of people of working age. In 'developing countries' it means, initially, a rapid increase in population with an associated strain on their already often stretched resources.

Figure 13.20

Consequences of the falling death rate

a the world's ageing population

age	1970 total (millions)	percentage of total world population	2000 (estimate) total (millions)	percentage of total world population
over 60	291	7.8	620	9.8
over 85	26	0.8	58	1.9

b life expectancy in Japan and the UK (years)

	1960 male	female	1992 male	female	2025 (estimate) male	female
UK	68	74	73	79	76	82
Japan	67	71	76	82	82	85

Figure 13.21

Percentage of population aged 65 and over

The replacement rate occurs when there are just sufficient children born to balance the number of people who die.

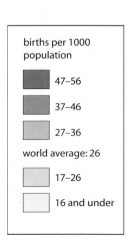

percentage aged 65 and over

over 16

8 – 15

under 8

0   5000 km

4 Zero growth rate

In western Europe, population decline is emerging as a serious problem. Except for Ireland, birth rates have dropped to a level that is too low to maintain a stable population. Throughout history, the **replacement rate** has always been exceeded — hence the continuous growth in world population. Recently, several of the more industrialised countries including Sweden, Switzerland and Japan, have been producing insufficient numbers of children to maintain overall numbers (is this the beginning of the possible Stage 5 in the demographic transition model? — Figure 13.11). The West Germans, before reunification, were concerned that their population could decrease by several million by 2035 A.D. which would reduce their competitive advantage in science and technology, the supply of unskilled labour and the security of future pensions.

(The problems of labour shortages have been alleviated, in the short term at least, by the influx of former East Germans after 1989.) On a continental basis, it is estimated that zero growth will be the norm in Europe by 2010, in North America by 2030, in China by 2070, in south-east Asia and Latin America by 2090, and in Africa by 2100.

5 Declining fertility rates

Although world fertility rates are falling, and in some places rapidly (Figure 13.18), many countries still have high birth rates (Figure 13.22). High birth rates have been considered characteristic of 'underdevelopment'. Although there appears to be a close correlation between falling birth rates and an increased level of development as measured by per capita GNP (Spearman rank; Framework 14, page 572), in 1986 UNFPA claimed that "a high birth rate was the consequence, not the cause, of poverty".

Figure 13.22

World birth rates, 1991

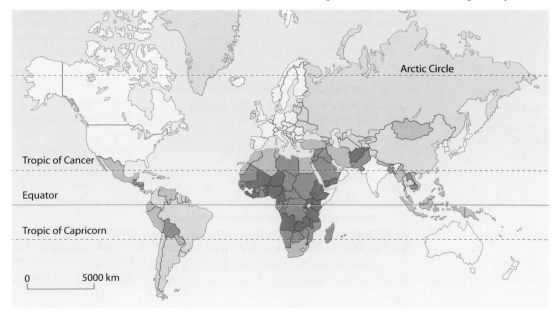

births per 1000 population

47–56

37–46

27–36

world average: 26

17–26

16 and under

0   5000 km

It is now generally recognised that the three key factors influencing fertility decline are improvements in family planning programmes, in health care and in women's education and status (arguably in that order, although they are all interrelated). The major world movement is now towards 'children by choice rather than chance', a goal which can only be achieved by giving women reproductive options.

In 1994, the Population Council, based in New York, claimed that:

'...The most direct way to bring about significant fertility declines is by implementing comprehensive and high-quality family planning programmes. Although past efforts have been substantial, there are still many countries where services are poor or even non-existent. Recent surveys in developing countries reveal that many couples who wish to delay or stop childbearing are not using contraception. About 100 million women — one in six married women in the developing countries (outside China) — have an unmet need for contraception. This unmet need is highest in countries of sub-Saharan Africa — averaging near 25 per cent — but even in Asia and Latin America, where services are more accessible, unmet need levels of around 15 per cent are typical. As a consequence many women bear more children than they want. Approximately one in four births in the developing world (again excluding China) is unwanted, and there are approximately 25 million abortions each year, many of them unsafe. Both unintended pregnancies and unsafe abortions can be prevented if women are given greater control over their sexual and reproductive lives.'

The causes of this unmet need for contraception include lack of knowledge of contraception methods and/or sources of supply; limited access to and low quality of family planning services; lack of education, especially among women; cost of contraception commodities; disapproval of husbands and family members; and opposition by religious groups. Improvements in health care include safer abortions and a reduction in infant mortality — the latter meaning that fewer children need to be born as more of them survive. Improved education raises the status of women and postpones the age of marriage. A survey of younger mothers in Brazil showed that those with secondary education had, on average, 2.5 children whereas those without such education had 6.5. Several governments, especially in south-east Asian countries, have attempted in recent years to 'encourage' couples to have fewer children. Places 31 summarises the degree of succes and the problems created by the Singaporean government's attempt to reach zero birth rate while, at the same time, encouraging higher birth rates among selected groups. Some governments, notably China (Case Study 13) have sought to strengthen the effectiveness of their population policies by a range of 'beyond family planning measures'. There is an international consensus which is generally critical of such strategies where they involve coercion.

Places 31 Singapore

The government introduced a massive family planning scheme in the late 1960s. The main objectives of this were:

- To establish family planning clinics and to provide contraceptives at minimal charge.

- To advertise through the media the need for and advantages of smaller families — a 'stop at two' policy.

- To legislate so that under certain circumstances both abortions and sterilisation could be allowed.

- To introduce social and economic incentives such as paid maternity leave, income tax relief, housing priority, cheaper health treatment and free education which would cease as the size of the family grew.

By the early 1990s, this policy had been so successful that the country had an insufficient supply of labour to fill the growing number of job vacancies and fewer young people to support an increasingly ageing population — hence the changed family planning slogan of 'stop at three if you can afford three'. The government were worried that the middle class élite were having the fewest children. A Social Development Unit was set up to encourage graduates to meet on 'blind dates', hoping that the result would be marriage. Female graduates are encouraged to have three or more children through financial benefits and large tax exemptions. Even so, the average female graduate has only one child so that she can pursue a career and live in a more expensive house/area. Low-income non-graduates only receive housing benefits if they stop at two children.

Migration: change in space and time

Migration is a movement and in human terms usually refers to a permanent change of home. It can also, however, be applied more widely to include temporary changes involving seasonal and daily movements. It includes movements both between countries and within a country. Migration affects the distribution of people over a given area as well as the total population of a region and the population structure of a country or city. The various types of migration are not easy to classify, but one method is given in Figure 13.23.

Internal and external (international) migration

Internal migration refers to population movement within a country, whereas external migration involves a movement across national boundaries and between countries. External migration, unlike internal, affects the total population of a country. The **migration balance** is the difference between the number of **emigrants** (people who leave the country) and **immigrants** (newcomers arriving in the country). Countries with a **net migration loss** lose more through emigration than they gain by immigration and, depending upon the balance between birth and death rates, may have a declining population. Countries with a **net migration gain** receive more by immigration than they lose through emigration and so are likely to have an overall population increase (assuming birth and death rates are evenly balanced).

Figure 13.23

Types of migration

Permanent	External (international):	between countries
	1 voluntary	West Indians to Britain
	2 forced (refugees)	African slaves to America, Kurds, Rwandans
	Internal:	**within a country**
	1 rural depopulation	most developing countries
	2 urban depopulation	British conurbations
	3 regional	from north-west to south-east of Britain
Semi-permanent	for several years	migrant workers in France and (former West) Germany
Seasonal	for several months or several weeks	Mexican harvesters in California, holiday-makers, university students
Daily	commuters	south-east England

Voluntary and forced migration

Voluntary migration occurs when migrants move from choice — e.g. because they are looking for an improved quality of life or personal freedom. Such movements are usually influenced by **'push and pull' factors** (page 342). Push factors are those which cause people to leave because of pressures which make them dissatisfied with their present home, while pull factors are those perceived qualities which attract people to a new settlement. When people have virtually no choice but to move from an area due to natural disasters or to economic, religious or social impositions (Figure 13.25), migration is said to be forced.

Times and frequency

Migration patterns include people who may move only once in a lifetime, people who move annually or seasonally, and people who move daily to work or school. Figure 13.23 shows the considerable variations in timescale over which migration processes can operate.

Distance

People may move locally within a city or a country or they may move between countries and continents: migration takes place at a range of spatial scales.

Migration laws and a migration model

In 1885, E. G. Ravenstein put forward seven 'laws of migration' based on his studies of migration within the UK. These laws stated that:

1 Most migrants travel short distances and their numbers decrease as distance increases (distance decay, page 379).
2 Migration occurs in waves and the vacuum left as one group of people moves out will later be filled by a counter-current of people moving in.
3 The process of dispersion (emigration) is the inverse of absorption (immigration).
4 Most migrations show a two-way movement as people move in and out: net migration flows are the balance between the two movements.
5 The longer the journey, the more likely it is that the migrant will end up in a major centre of industry or commerce.
6 Urban dwellers are less likely to move than their rural counterparts.
7 Females migrate more than males within their country of birth, but males are more likely to move further afield.

More recent global migration studies have largely accepted Ravenstein's 'laws', but have demonstrated some additional trends:

8 Most migrants follow a step movement which entails several small movements from the village level to a major city, rather than one traumatic jump.

9 People are leaving rural areas in ever-increasing numbers.

10 People move mainly for economic reasons, e.g. jobs and the opportunity to earn more money. (In many contexts in the early 1990s, it appears to be the escape from civil war.)

11 Most migrants fall into the 20–34 age range.

12 With the exception of short journeys in developed countries, males are the more mobile. (In many societies, females are still expected to remain at home.)

13 There are increasing numbers of migrants who are unable to find accommodation in the place to which they move; this forces them to live on the streets, in shanties (Places 43, page 408) and in refugee camps (page 344).

The examples in Figure 13.25 help to explain the migration model shown in Figure 13.24.

Internal migration in developed countries

Certain patterns of internal migration are more characteristic of developed countries than economically less developed countries. Three examples have been chosen to illustrate this: rural–urban movement; regional movement; and movements within and out of large urban areas.

Figure 13.24

A migration model *(after* Jones and Hornsby)

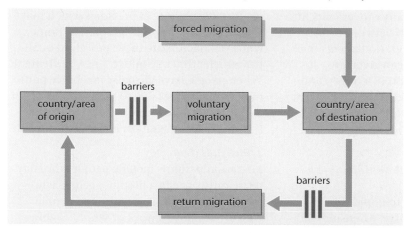

Figure 13.26

Rural and urban dwellers in the USA as a percentage of the total population, 1870–1990

Figure 13.25

Causes and examples of migration

year	urban (%)	rural (%)
1870	24	76
1900	40	60
1920	51	49
1940	56	44
1960	70	30
1980	76	24
1990	77	23

Forced migration

Religious: Jews; Pilgrim Fathers to New England

Wars: Muslims and Hindus in India and Pakistan; Rwanda

Political persecution: Ugandan Asians

Slaves or forced labour: Africans to south-east USA

Lack of food and famine: Ethiopians into the Sudan

Natural disasters: floods, earthquakes, volcanic eruptions (Mt Pinatubo, Case Study 1)

Overpopulation: Chinese in south-east Asia

Redevelopment: British inner-city slum clearance, *ujamaa* in Tanzania, (Places 33)

Resettlement: Native Americans (USA) and Amerindians (Brazil) into reservations

Environmental: Chernobyl (Ukraine), Bhopal (India)

Voluntary migration

Jobs: Bantus into South Africa, Turks into former West Germany (Places 34), Mexicans into California

Higher salaries: British doctors to the USA

Tax avoidance: British pop/rock and film stars to the USA

Opening up of new areas: American Prairies; Israelis into Negev Desert, Brasilia (Places 33)

Territorial expansion: Roman and Ottoman Empires

Trade and economic expansion: former British colonies

Retirement to a warmer climate: Americans to Florida

Social amenities and services: better schools, hospitals, entertainment

Prevention of voluntary movement

Government restrictions: immigration quotas, Berlin Wall, work permits

Lack of money: unable to afford transport to and housing in new areas

Lack of skills and education

Lack of awareness of opportunities

Illness

Threat of family division and heavy family responsibilities

Reasons for return

Racial tension in new area

Earned sufficient money to return

To be reunited with family

Foreign culture proved unacceptable

Causes of initial migration removed (political or religious persecution)

Barriers to return

Insufficient money to afford transport

Standard of living lower in original area

Racial, religious or political problems in original area

Rural–urban movement

Although rural depopulation is now a worldwide phenomenon, it has been taking place for much longer in the more developed, industrialised countries. Figure 13.26 and Places 32 describe and explain the changing balance between rural and urban dwellers in the USA since 1870.

Places 32 Rural–urban movement in the USA

The 1870s was the decade after the American Civil War and the abolition of slavery. Many black farmworkers, most of whom were sharecroppers (page 425), found it impossible either to find vacant land or to afford to buy any that might have become available. Consequently, many began to move to the cities to seek work.

During the 1930s, drought and soil erosion in the Dust Bowl of the Midwest caused many farmers to give up their land. At about the same time, the economic depression lowered prices for farm produce. Farmhands, who had always been poorly paid yet who worked harder and much longer hours compared with their counterparts in industry, were often on yearly or shorter-term contracts and so were paid off by farmers whose profits were falling. A further decline in the farm population occurred in the early 1940s when America's export markets were cut off during World War II.

During the 1950s and 1960s, the mechanisation of farming and the availability of cheap oil for fuel saw a big reduction in the number of farmworkers. The consolidation of farms into larger units and the introduction of contract farming, e.g. farmers with combine harvesters who travelled northwards in 'teams' as the cereal ripened, meant employment for even fewer people living in rural areas. Since the 1970s, there has been an increase in the number of 'suitcase farmers' — farmers who live in the city and travel out to their farms periodically. Added to this may be the limited number of schools, hospitals, and places of entertainment for members of the rural community, together with fewer job opportunities for females. Improvements in private transport have led to the decline of small service centres. Recent waves of immigrants, mainly Hispanics from Mexico, Puerto Rico and Cuba, and Asians from China, the Philippines, Japan and Korea, have been attracted to urban centres in general and to inner cities in particular.

In spite of their professed love of the 'big outdoors', most Americans seem to prefer living in urban areas — although the percentage of the white population (European origin) living in cities has declined slightly from 72.4 in 1970 to 70.8 in 1990.

Q Why has there been a reversal in the percentage of people living in rural and urban areas in the USA between 1870 and 1980? Why has this trend continued since 1980, but at a much slower rate?

Regional movement in Britain

For over a century there has been a drift of people from the north and west of Britain to the south-east of England. The early 19th century was the period of the Industrial Revolution when large numbers of people moved into large urban settlements on the coalfields of northern England, central Scotland and South Wales, and to work in the textile, steel, heavy engineering and shipbuilding industries. However, since the 1920s there has been more than a steady drift of population away from the north of Britain to the south. Some of the major reasons for this movement are listed here.

- A decline in the farming workforce and rural population, for reasons similar to those quoted in Places 32 on rural–urban movements.
- The exhaustion of supplies of raw materials (coal and iron ore).
- The decline of the basic heavy industries such as steel, textiles and shipbuilding. Many industrial towns had relied not only on one form of industry but, in some cases, on one individual firm. With no alternative employment, those wishing to work had to move south.
- Higher birth rates in the industrial cities meant more potential job-seekers.

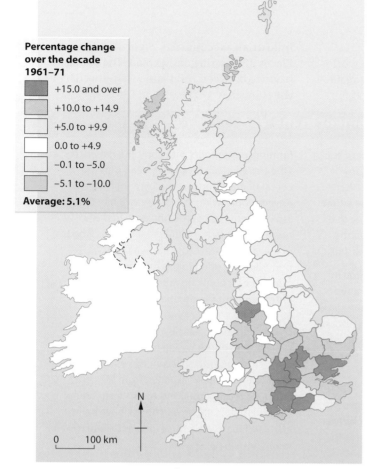

Percentage change over the decade 1961–71

- +15.0 and over
- +10.0 to +14.9
- +5.0 to +9.9
- 0.0 to +4.9
- −0.1 to −5.0
- −5.1 to −10.0

Average: 5.1%

N

0   100 km

Figure 13.27

Population changes in the UK, 1961–71 (boundaries adjusted to the 1974 changes)

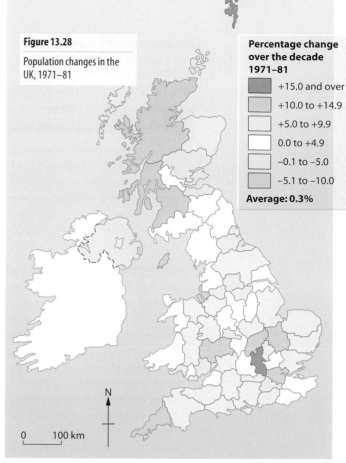

Figure 13.28

Population changes in the UK, 1971–81

Percentage change over the decade 1971–81

- +15.0 and over
- +10.0 to +14.9
- +5.0 to +9.9
- 0.0 to +4.9
- −0.1 to −5.0
- −5.1 to −10.0

Average: 0.3%

N

0   100 km

■ New post-war industries, which included car manufacturing, electrical engineering, food processing and, since the 1980s, micro-electronics and hi-tech industries, have tended to be market-orientated. They are said to be footloose, in the sense that they have a free choice of location — unlike the older industries which had to locate near to sources of raw materials and/or energy supplies.

■ The growth of service industries has been mainly in the south-east. This has resulted from the many office firms wanting a prestigious London address, the growth of government offices, the demand for hospitals, schools and shops in a region where one in five British people live and tourism taking advantage of Britain's warmer south coast.

■ Joining the EU meant increased job opportunities in the south and east, while traditional industrial areas and ports such as Glasgow, Liverpool and Bristol, which had links with the Americas, have declined.

■ Salaries were higher in the south.

■ With so much older housing, derelict land and waste tips, the quality of life is often perceived as being lower in the north, despite the beauty of its natural scenery and slower pace of life.

■ There are more social, sporting and cultural amenities in the south.

■ Communications were easier to construct in the flatter south. Motorways, railways, international airports, cross-Channel ports and the Channel Tunnel were built and/or improved as this region had the greatest wealth and population size.

Q Using Figures 13.27, 13.28 and 13.29

1 Describe and give reasons for the changes in the UK's population between 1961 and 1981.

2 Describe the changes in population between 1981 and 1991.

3 Compare the changes in population between the two periods, and suggest reasons for the differences.

Population

Movements within urban areas

Since the 1930s there has been, in Britain, a movement away from the inner cities to the suburbs — a movement accelerated first by improved public transport provision and then by the increase in private car ownership. The former included more bus services, the extension of the London Underground and the construction of the Tyne and Wear Metro. Some of the many reasons for this outward movement are summarised in Figure 13.30. The result, in human terms, has been a polarisation of groups of people within society and the accentuation of inequalities (wealth and skills) between them. (Beware, however, of the dangers of stereotyping when discussing these inequalities; Framework 11, page 404).

- The inner cities tended to be left with a higher proportion of low-income families, handicapped people, the elderly, single-parent families, people with few skills and limited qualifications, first-time home-buyers, the unemployed, recent immigrants and ethnic minorities.
- The suburbs tended to attract people moving towards middle age, married with a growing family, possessing higher skills and qualifications, earning higher salaries, in secure jobs and capable of buying their own homes and car.

Recently there has been, in part at least, a reversal of this movement and parts of some inner cities have become regenerated and 'fashionable'. This re-urbanisation is partly due to energy conservation, partly to changes in housing markets, partly to planning initiatives such as refurbished waterfronts (London, Places 41, page 402; Baltimore) and partly to new employment growth (leisure, financial services).

Figure 13.29

Population changes in the UK, 1981–91

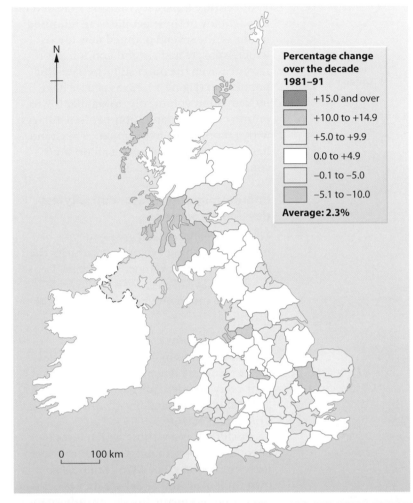

Percentage change over the decade 1981–91

- +15.0 and over
- +10.0 to +14.9
- +5.0 to +9.9
- 0.0 to +4.9
- −0.1 to −5.0
- −5.1 to −10.0

Average: 2.3%

0   100 km

	Inner city	Suburbs
Housing	Poor quality; lacking basic amenities; high density; overcrowding	Modern; high quality; with amenities; low density
Traffic	Congestion; noise and air pollution; narrow, unplanned streets; parking problems	Less congestion and pollution; wider, well-planned road system; close to motorways and ring roads
Industry	Decline in older secondary industries; cramped sites with poor access on expensive land	Growth of modern industrial estates; footloose and service industries; hypermarkets and regional shopping centres; new office blocks and hotels on spacious sites
Jobs	High unemployment; lesser-skilled jobs in traditional industries	Lower unemployment; cleaner environment; often more skilled jobs in newer high-tech industries
Open space	Limited parks and gardens	Individual gardens; more, larger parks; nearer countryside
Environment	Noise and air pollution from traffic and industry; derelict land and buildings; higher crime rate; vandalism	Cleaner; less noise and air pollution; lower crime rate; less vandalism
Social factors	Fewer, older services, e.g. schools and hospitals; ethnic and racial problems	Newer and more services; fewer ethnic and racial problems
Planning and investment	Often wholesale redevelopment/clearance; limited planning and investment	Planned, controlled development; public and private investment
Family status/wealth	Low incomes; often elderly and young; large family or none	Improved wealth and family/professional status

Figure 13.30

Some causes of migration from the inner cities to the suburbs

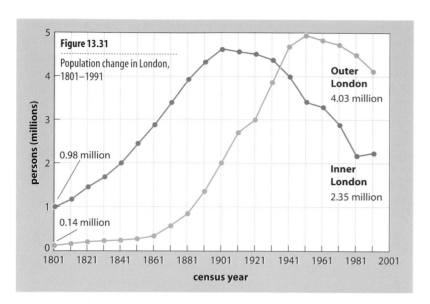

Figure 13.31

Population change in London, 1801–1991

persons (millions)

0.98 million

0.14 million

Outer London 4.03 million

Inner London 2.35 million

census year

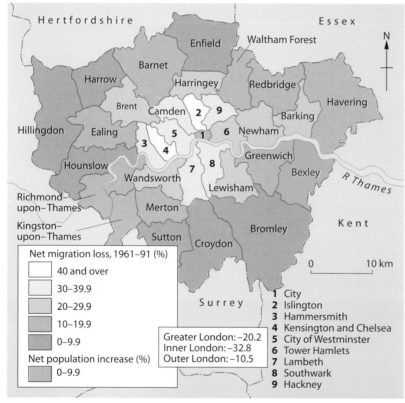

Net migration loss, 1961–91 (%)

- 40 and over
- 30–39.9
- 20–29.9
- 10–19.9
- 0–9.9

Net population increase (%)

- 0–9.9

Greater London: –20.2
Inner London: –32.8
Outer London: –10.5

1 City
2 Islington
3 Hammersmith
4 Kensington and Chelsea
5 City of Westminster
6 Tower Hamlets
7 Lambeth
8 Southwark
9 Hackney

0   10 km

Figure 13.32

Population movement in Greater London, 1961–91

Figure 13.33

Population change in UK conurbations (percentage change per decade)

Conurbation	1961–71	1971–81	1981–91
Greater London	-6.76	-10.15	-4.75
Inner London	-13.20	-17.65	-5.88
Outer London	-1.76	-5.00	-4.08
Greater Manchester	+0.33	-4.91	-5.39
Merseyside	-3.60	-8.66	-9.01
South Yorkshire	+1.47	-1.57	-4.09
Tyne and Wear	-2.58	-5.65	-4.92
West Midlands	+17.77	-5.27	-5.55
West Yorkshire	+3.10	-1.46	-2.59
Glasgow City	-13.84	-22.06	-14.51

Movements away from conurbations

Since the mid 1950s, large numbers of people have moved out of London altogether (Figure 13.31). Initially these were people who were virtually forced to move as large areas of 19th-century inner-city housing were demolished. Many of these people were rehoused in one of the several planned new towns which were created around London. More recently, even the outer suburbs have lost population (Figure 13.32) as people have moved, often voluntarily, to smaller towns, or into commuter and suburbanised villages, with a more rural environment. This trend has become characteristic of all Britain's conurbations (Figure 13.33).

Internal migration in economically less developed countries

There is usually a much greater degree of migration within developing countries than there is in more developed countries. Two examples have been chosen to illustrate this: rural–urban movement and political resettling.

Rural–urban movement

Many large cities in developing countries are growing at a rate of more than 20 per cent every decade. This growth is partly accounted for by rural 'push' and partly by urban 'pull' factors.

Push factors are those which force or encourage people to move — in this case, to leave the countryside. Many families do not own their own land or, where they do, it may have been repeatedly divided by inheritance laws until the plots have become too small to support a family. Food shortages develop if the agricultural output is too low to support the population of an area, or if crops fail. Crop failure may be the result of overcropping and overgrazing (Case Study 7 and page 462), or natural disasters such as drought (the Sahel countries), floods (Bangladesh), hurricanes (the Caribbean) and earth movements (in Andean countries). Elsewhere, farmers are encouraged to produce cash crops for export to help their country's national economy instead of growing sufficient food crops for themselves. Mechanisation reduces the number of farmers needed, while high rates of natural increase may lead to overpopulation (page 354). Some people may move because of a lack of services (schools, hospitals, water supply) or be forced to move by

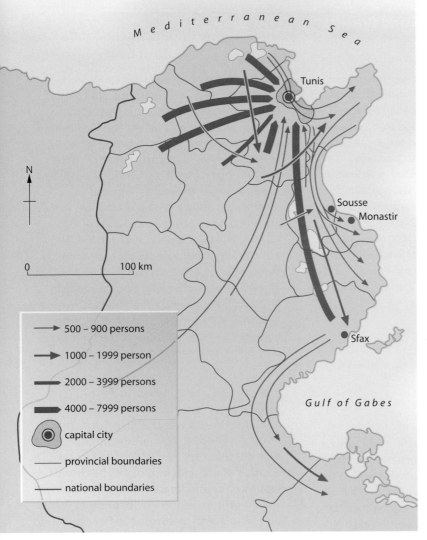

Figure 13.34

Migration patterns in Tunisia

Map legend:

→ 500 – 900 persons

→ 1000 – 1999 person

→ 2000 – 3999 persons

→ 4000 – 7999 persons

◉ capital city

— provincial boundaries

— national boundaries

their rural village first to small towns, then to larger cities and finally to a major city.

Figure 13.34 shows migration patterns in Tunisia. There are several points to notice:

- There is a greater movement of rural than of urban dwellers.
- Most migrants move to Tunis, the capital city.
- Most migrants tends to travel short distances: relatively few make long journeys (distance decay factor, page 379).
- Most move from rural, inland, desert areas to urban, coastal areas.
- A few move from Tunis to coastal towns, such as the holiday resorts of Sousse and Monastir.
- Very few migrants return to rural districts.
- There is evidence of a twofold movement into and out of Tunis and Sfax.

Political resettling

National governments may, for political reasons, direct, control or enforce movement as a result of decisions which they believe to be in the country's (or their own) best interests. Some governments have actively encouraged the development of new community settlements, especially in areas which were, at the time, sparsely populated — e.g. the creation of *kibbutzim* in Israel and of *ujamaa* in Tanzania (Places 33). Other governments have founded new capital cities in an attempt to develop new growth regions — e.g. Brasilia (Places 33), Dodoma (Tanzania) and Abuja (Nigeria) — while others have built settlements to try to strengthen their claims to an area — e.g. Israeli settlements in the West Bank.

In Brazil and the USA, minority groups of indigenous people, the Amerindians and Native Americans respectively, have been forced off their tribal lands and onto reservations. In South Africa, under *apartheid*, the black population were forced to live either in shanty settlements in urban townships or on homelands in rural areas, which lacked resources (Places 35). In the last few years, an increasing number of people have been forced to move due to so-called 'ethnic cleansing' policies enforced by several governments, as in the former Yugoslavia.

governments or the activities of multi-national companies.

Pull factors are those which encourage people to move — in this case, to the cities. People in many rural communities may have a perception of the city which in reality does not exist. People migrate to cities hoping for better housing, better job prospects, more reliable sources of food, and better services in health and education. While it is usually true that in most countries more money is spent on the urban areas — where the people who allocate the money themselves live — the present rate of urban growth far exceeds the amount of money available to provide accommodation for all the new arrivals. Recent studies seem to confirm that many migrants make a stepped movement from

 Q How closely do the patterns shown in Figure 13.34 correspond to those suggested by Ravenstein (page 337)?

Tanzania

When Tanzania became independent from British colonial rule in 1961, over 90 per cent of its population lived in dispersed farmsteads. In 1967, the government introduced a scheme to encourage people to move to *ujamaa* or family villages. The hope was that these new communities, where work and wealth were shared, would lead to self-reliance and cooperation within the local population. Between 1970 and 1976, the number of villages increased from just under 2000 to over 6200. However, although most of those who had to move had to migrate only short distances, there was considerable resentment and resistance to the plan.

Brasilia

For decades, concern had been expressed in Brazil at the relative richness and rapid population growth of the south-east of the country, around Rio de Janeiro and São Paulo, in comparison with the inland states (Figure 13.5). In 1952, Congress agreed to move the capital 1200 km inland from Rio, where most politicians preferred to live, to Brasilia — then an uninhabited site in a savanna landscape — in an attempt to open up the central areas of the country. Building began in 1957 and Brasilia was officially inaugurated as the capital in 1960. The population of the city had already exceeded 1 million by 1986, the figure initially forecast for the year AD 2000. Despite this growth, Brasilia remains a city of offices, with many of its 'weekly' inhabitants commuting back to the older centres for weekends.

External migration

Refugees

The United Nations' definition of a refugee is "a person who cannot return to his or her own country because of a well-founded fear of persecution for reasons of race, religion, nationality, political association or social grouping". Refugees, who are increasing rapidly in numbers due to internal strife (civil wars) and environmental disasters (famine), move to other countries hoping to find help and asylum. Refugees do not include 'displaced persons' who are people who have been forced to move within their own country (e.g. the former Yugoslavia).

Before World War II, the majority of refugees tended to become assimilated in their new host country but, in the last 50 years, the number of permanent refugees has risen rapidly to approximately 14 million in 1989 and 23 million in 1995 (with another 25 million 'displaced'), according to UN estimates. However, as most refugees are illegal immigrants, the UN admits that these figures could be very inaccurate. The first, apparently insoluble, refugee problem was the setting up of Palestinian Arab camps following the creation of the state of Israel in 1948. The refugee problem has intensified during 30 years of conflict in parts of south-east Asia such as Vietnam, and more recently following food shortages and political unrest in much of sub-Saharan Africa, including Ethiopia, Mozambique and Rwanda (Figure 13.35).

Half of the world's refugees are children of school age; most adult refugees are female and four-fifths are in the developing countries which have fewest resources to deal with the problem. Refugees usually live in extreme poverty and lack food, shelter, clothing, education and medical care. They rarely have citizenship, and few (if any) civil, legal or basic human rights. There is little prospect of returning home and the long periods spent in camps mean that they often lose their sense of identity and purpose.

Figure 13.35

Part of the displacement camp at Nyaconga, Rwanda

Migrant workers

As economic development has taken place at different rates in different countries, supplies of and demand for labour are uneven, and due to improvements in transport, there has been an increase in the number of people who move from one country to another in search of work. Such cross-border movements in search of work can operate at different timescales. For example,

- **Permanent** For a century and a half, the UK has received Irish workers and, since the 1950s, West Indians and citizens from the Indian sub-continent. Most of these migrants have made Britain their permanent home.
- **Semi-permanent** After World War II, several European countries experienced a severe labour shortage. In order to help rebuild their economies, countries such as France and the then West Germany, accepted cheap, semi-skilled labour from North Africa and Portugal (into France) and from Turkey, Yugoslavia and the Middle East (into Germany, Places 34).
- **Short-term and seasonal** The South African economy depends largely upon migrant black labour from adjacent nation states. Under *apartheid*, these workers were not allowed to remain in South Africa for more than six months at a time. In North America, large numbers of Mexicans find seasonal employment picking fruit and vegetables in California.
- **Daily** With the introduction of free movement for EU nationals within the EU, an increasing number of workers commute daily into adjacent countries.

Places 34   Turks into (West) Germany

A 1945–89

Sakaltutan is a village of 900 inhabitants in central Turkey. Until recently it was a poor, isolated settlement dependent upon agriculture. With a high birth rate and limited resources, the village had become overpopulated (page 354). There were too many males to work on the land (women were not expected to work outside the home) and the demand for craftsmen was limited.

When an all-weather road was built this encouraged the sale of surplus produce in the nearby large towns and led to an increase in mechanisation and a further decline in the need for agricultural labourers. The village school was expanded from one teacher to seven and farmers were able to obtain advice on how to increase output. The result was a growth in the aspirations of the villagers. Outward migration followed, either to the city of Adana, or to the capital Ankara, or overseas to West Germany.

Pforzheim is an industrial town near Stuttgart in Germany. Like other west European towns it had to be rebuilt after 1945 at a time when there were more job vacancies than workers. The extra labour needed was obtained from the poorer parts of southern Europe and the Middle East. Many of these 'guest workers' or *gastarbeiters* initially went into agriculture as most were originally farmers, but they soon turned to the relatively better-paid jobs in factories and the construction industry. These jobs were not taken by the local Germans because they were dirty, unskilled, poorly paid and often demanded long and unsociable hours.

At one stage there were 15 families from Sakaltutan living in Pforzheim. The first arrivals were males in their twenties, all of whom had had some education and were skilled at crafts, making them acceptable in the local car factory or in construction. In an attempt to earn as much money as possible, they often took accommodation in poorly equipped company hostels, missed meals and used public transport to reach work.

Although by Turkish standards the migrants were earning high salaries, they found the West German cost of living much higher and what was intended to be a short working stay became more permanent. In time, the families of *gastarbeiters* arrived and by the mid 1980s over 3 per cent of the West German workforce was Turkish (Figure 15.38).

This movement of labour had advantages and disadvantages to Sakaltutan in Turkey, the losing village/country, and to Pforzheim in West Germany, the receiving town/country. These have been summarised in Figure 13.36.

	Advantages	Disadvantages
Sakaltutan	Reduces pressure on jobs and local resources	People of working age leave
	Birth rate may be lowered as people of childbearing age leave	People with skills and education are most likely to leave
	Money may be sent back to the village: estimates suggest that 50 per cent of Sakaltutan's income is from overseas	It is mainly males who migrate and this divides families
	Migrants may develop new skills overseas which they then bring back to village	An elderly population is left, resulting in a higher death rate
		In the long term, creates dependency upon money sent back to home villages
Pforzheim	Labour shortage is solved, especially in dirty, poorly paid, unskilled jobs	Resentment towards Turks when Germany's rate of unemployment rises
	Cheaper labour for less desirable jobs	Turks form an ethnic group which does not mix: they live in poor housing, and retain their own culture
	Cultural advantages of discovering new foods, music, pastimes, etc.	Turks feel discriminated against with racial tension and police harassment
	Migrants tend to be the more economically active and skilled	Migrants may be a drain on local services
		Migrants are mainly male, which can lead to social problems

Figure 13.36

In 1973 the West German government imposed a ban on the recruitment of foreign workers, although Turks still arrived to make family reunions. When in 1980 grants were offered to Turks wishing to return home, very few took advantage of the offer. Of those who remained, fewer than 1 per cent had taken out German citizenship as that meant giving up Turkish nationality.

Figure 13.37

A Turkish village

Figure 13.38

Turkish workers in Germany

B Since 1989

During the early 1990s, when there were still 1.8 million Turks living in Germany, tension between the two communities began to increase — culminating in several arson attacks on Turkish property and the loss of several Turkish lives. The demolition of the Berlin Wall and the reunification of Germany in 1989 brought with it several problems. As large numbers of migrants arrived from the former East Germany and other former communist-bloc countries, the demand for jobs and houses in the former West Germany increased. These new arrivals took the menial, poorer-paid jobs previously held by the Turks. The cost of housing (rents) increased, as did levels of unemployment and inflation; few people anticipated the economic problems of unification. The Turks, being an ethnic group living within their own community, were blamed by the Germans for the decline in the country's economy — even though it was the Turks who suffered first and the most.

Turks from villages like Salkatutan are now turning to Saudi Arabia and Libya as places in which to work and to live.

Figure 13.39

Directions of some major international migrations since 1970

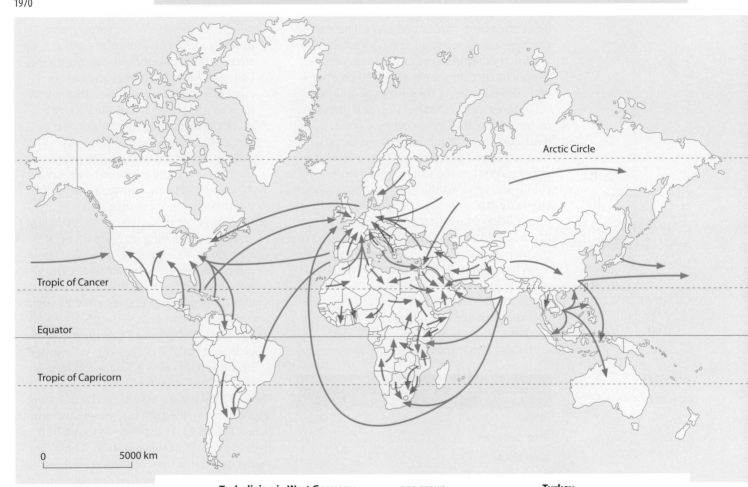

Figure 13.40

Population structure of Turks in West Germany and Turkey, 1986

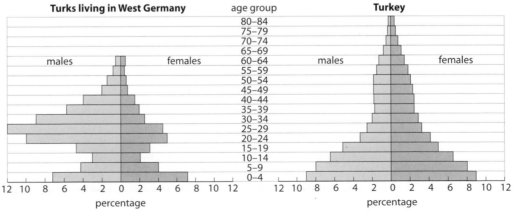

Multicultural societies

This is often a sensitive and emotive issue. Attempts to explain terms are not intended to cause insult or resentment.

The latest scientific research suggests that humans evolved in central Africa about 200 000 years ago and began, 100 000 years later, to migrate to other parts of the world. This common origin, identified by the study of genes, shows that humans are genetically homogeneous to a degree unparalleled in the animal kingdom.

Previous scientific opinion suggested that the many peoples of the modern world had descended from three main races. These were the Negroid, Mongoloid and Caucasoid. The dictionary definition of *race* is "a group of people having their own inherited character-istics distinguishing them from people of other races", e.g. colour of skin and physical features. In reality, often because of inter-marriage, the distinction between races is now so blurred that the word 'race' has little significant scientific value. Today, while colour still remains the most obvious visible characteristic, groups of people differ from one another in religion, language, nationa-lity and culture. These differences have led to the identification of many ethnic groups.

What criteria do members of various ethnic groups prefer to use when identifying themselves?

■ **Colour of skin** Whereas people of 'European' stock have long accepted being called 'white', it is only in more recent years that, in Britain, people from Africa and the Caribbean have preferred to be known collectively as 'black'. The 1971 UK census divided immigrants born in Commonwealth countries into the Old (white) and New (black) Commonwealth. (It made no allowance for children born in the UK of New Commonwealth parentage.)

■ **Place of birth (nationality)** *The Annual Abstract of Statistics* for the UK lists immigrants under the heading 'country of last residence' — thus avoiding a reference to colour. Most groups of people, in the USA for example, have been identified by their place of birth, or that of their ancestors, and are known as Chinese, Puerto Rican, etc. There is currently a major movement in the USA (and to a lesser extent in the UK) by blacks, also wishing to be identified by place of origin, to be referred to as African-Americans. Will black people in the UK eventually prefer to be known as African-Caribbean, African-British, or another term not yet invented?

■ **Language** At present, the largest group of migrants moving into the USA is the Hispanics — i.e. Spanish speakers. These migrants, mainly from Mexico, Central America and the West Indies, have been identified and grouped together by their common language.

■ **Religion** Other ethnic groups prefer to be linked with, and are easily recognised by, their religion — e.g. Jews, Sikhs, Hindus and Muslims.

The 1991 UK census asked, for the first time, respondents to identify themselves by ethnic group. Figure 13.41 lists these groups, and gives the results of this question. These results have been used elsewhere (Robinson, 1994) to describe the spatial distributions of different ethnic groups in England and Wales.

The migrations of different ethnic groups have led to the creation of multicultural societies in many parts of the world. In most countries there is at least one minority group. While such a group may be able to live in peace and harmony with the majority group, unfortunately it is more likely that there will be prejudice and discrimination leading to tensions and conflict. Four multi-cultural countries with differing levels of integration and ethnic tension are: South

Figure 13.41

Ethnic groups in Britain, 1991

Ethnic groups	Percentage in each group
White	94.5
Black-Caribbean	0.9
Black-Africa	0.4
Black-other (please describe)	0.3
Indian	1.5
Pakistani	0.9
Bangladeshi	0.3
Chinese	0.3
Any other group (please describe)	
a) Asian	0.4
b) Other	0.5
Total non-white	5.5

If the person is descended from more than one ethnic or racial group, please tick the group to which he/she considers he/she belongs.

Population

Africa (Places 35), the USA and Brazil (Places 36), and Singapore (Places 37). Remember, though, that when we look at these countries from a distance we can rarely appreciate the feelings generated by, or the successes/failures of, different state or government policies.

Places 35 South Africa

At the time of South Africa's first all-party elections in 1994, the country's total population was estimated to be 40.3 million. This total consisted of 1 m Asians, 31 m Blacks, 3.8 m Coloureds and 4.6 m Whites. The first inhabitants of the area were the San (Bushman) people and the Khoi-Khoin people. However, the majority of South Africa's black people are descended from Bantu-speakers who migrated into the area many centuries ago. The white population (most of whom are now known as Afrikaner) are descended from Dutch, German, French, British and Belgian immigrants since 1652. The Coloured people are descended from mixed relations among the European settlers, indigenous peoples and people from Madagascar, India, Indonesia and Malaysia. The Asian population arrived after 1860 and came mainly from India.

A policy of segregation between black and white originated in the first Dutch settlement, the Cape, in 1652. This practice became customary, and was established legally as apartheid by the first National Party government in 1948 when some Afrikaners in the Party united to protect their language, culture and heritage from a perceived threat by the black majority and to assert their economic and political independence from British colonial domination.

Statutory apartheid regulated the lives of all groups, but particularly of Blacks, Coloureds and Indians. The Population Registration Act categorised the nation into White, Black, Indian, Malay and Coloured citizens. Further Acts made mixed marriages illegal; prescribed segregation in restaurants; transport; schools; places of entertainment and political parties. The Group Areas Act stipulated where and with whom people could live; and the Black Authorities Act established black homelands.

The outcome of all this legislation was the unequal division of rights and resources. This included the disproportionate division of land; the unequal distribution of funding for education; and the general denial of constitutional rights for the majority of South Africans.

Legalised racial discrimination was abolished in the early 1990s. After long negotiations, the first all-party elections, held in 1994, established a multi-party Government of National Unity. The five-year interim constitution has ended the existence of the homelands and includes a section on human rights; a Reconstruction and Development Programme will address housing problems, equality in education and health care, and the encouragement of private home ownership; a Land Court is to address land claims from dispossessed people.

Figure 13.42

Segregated residential areas in two South African cities

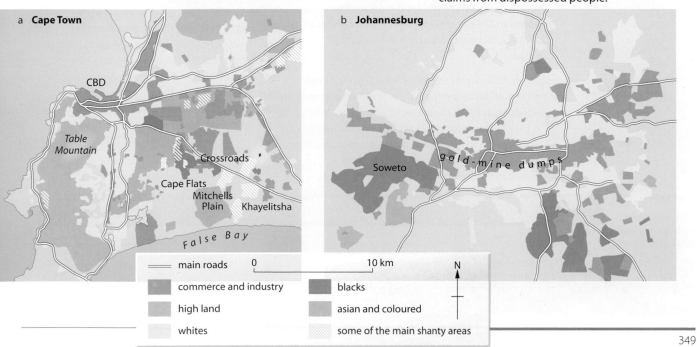

349

Figure 13.43

A white residential area in Cape Town, South Africa

Figure 13.44

Housing in the Soweto township, Johannesburg, South Africa

Figure 13.45

Shanty settlement, Khayelitsha, Cape Town, South Africa

However, the legacy of apartheid will take many years to eradicate. For example, apartheid led to:

Housing

The Group Areas Act (1950) ensured that White, Coloured and Asian communities lived in different parts of the city (Figure 13.42) with the Whites having the best residential areas (Figure 13.43). Buffer zones at least 100 m wide, often along main roads or railway lines, were created to try to prevent contact between the three groups. Blacks were treated differently. Those who had lived in the city since birth, or had worked for the same employer for 10 years, were moved to newly created townships on the urban fringes. The remainder were forced away from the cities to live on one of ten designated reserves or homelands, where the environmental advantages were minimal (drought, poor soils and a lack of raw materials). The homelands took up 13 per cent of South Africa's land; held 72 per cent of its total population; and produced 3 per cent of the country's wealth. Most Blacks living in the homelands were employed on 1-year contracts to prevent them gaining urban residential rights.

Life in the townships was no less difficult. These were built far away from White residential areas, which meant that those Blacks who found jobs in the cities had long and expensive journeys to work. Many of the original shanty towns have been bulldozed and replaced by rows of identical, single-storey houses (Figure 13.44).

These have four rooms and a backyard toilet, but only 20 per cent have electricity. Corrugated-iron roofs make the buildings hot in summer and cold in winter. The settlements lack infrastructure and services and, due to rapid population growth (high birth rates and in-migration), are surrounded by vast shanty settlements (Figure 13.45). Two of the better-known townships are Soweto in Johannesburg (1.5 million inhabitants) and Crossroads in Cape Town.

Employment

Under apartheid, Blacks were severely restricted in mobility and type of job. A man had to return to his homeland in order to apply for a job. If successful, he was given a contract to work in 'White' South Africa for 11 months after which he had to return to his homeland if he wished to renew the contract. This system prevented Blacks from becoming permanent residents in the city.

Services, employment and quality of life

Figure 13.46 illustrates some of the many differences between ethnic groups under apartheid. Schooling, for example, was free and compulsory for all Whites and Asians, but not for Coloureds or Blacks. The present government intends to introduce compulsory education for all, and will also promote non-formal education and basic adult education.

Figure 13.46

Inequality between ethnic groups in South Africa

	Infant mortality	Pupil–teacher ratio	Nurses per 1000 people (WHO minimum = 2)	Average weekly income (in rands)	Managerial jobs (%)	Labourers (%)
African	80	420:1	1.2	300	1.6	88.1
Coloured	55	26:1	1.8	650	1.8	11.1
Asian	20	23:1	1.5	1200	2.0	0.7
White	12	18:1	6.0	2000	94.6	0.1

The United States of America

Americans have long prided themselves that their country is a 'melting pot' in which peoples of all ethnic groups could be assimilated into one nation. Yet problems do exist. The indigenous Indian populations have been granted reservations, usually in areas lacking resources, although they are not forced to live on them. It is, however, only on reservations occupied by tribes such as the Navajo and Hopi that traditional Amerindian culture still flourishes. Indians who have drifted to towns often become the focus of social problems.

The African-Americans (blacks), released from slavery after the Civil War, could not find land on which to farm (Places 32). They too moved to large urban centres where they have congregated in inner-city 'ghettos' (Chicago, Figure 15.7; and Case Study 15b). Despite some improvements following the Civil Rights movement of the 1960s, this ethnic group remains socially and economically deprived in comparison with most whites.

Despite the US claim that it has an 'open-door' policy, strong restrictive laws have frequently been imposed as a barrier to immigration (Figure 13.24). This policy has been noticeable against Chinese in the 1920s, Japanese during World War II, Mexicans since the 1980s and, most recently, illegal migrants from Central America and Haiti. Many immigrant groups still identify themselves with their 'home' country and its culture, living and marrying within their own ethnic groups (Puerto Ricans in New York) and congregating to form ethnic areas — Los Angeles, for example, has its Chinatown, Japantown, Koreatown and Filipinotown.

Brazil

Most of the inhabitants of Brazil, having almost every colour of skin conceivable, regard themselves as Brazilians, and the country rightly claims that it has little racial discrimination or prejudice. The Census Department does, however, recognise the following divisions based upon colour:

1 Whites (*Branco*): anyone with a *café au lait* or lighter-coloured skin (the USA and South Africa would regard *café au lait* as black or Coloured). This group includes many of the European migrants who came from Portugal (the original colonists), Italy, Germany and Spain.

2 Mulatto (*Pardo*): darker skins but with a discernible trace of European ancestry. They are the result of mixed marriages or 'liaisons' between the early Portuguese male settlers and either female Indians or African slaves. There is pride rather than prejudice in coming from two racial backgrounds.

3 Blacks (*Preto*): those of pure African descent.

4 Orientals (*Amarelo*): those who have emigrated more recently from south and east Asia.

5 Amerindians: a continually declining, yet still distinctive, indigenous group.

All these groups mix freely, especially at football matches, in carnivals and on the beach. Yet despite the lack of racial tension there tends to be a correlation between colour and social status and jobs. Walking into a hotel on arrival in Rio, it is apparent that the baggage-carriers are black, hotel porters a slightly lighter colour and the receptionists and cashiers *café au lait*. In the army, officers are usually white and the ranks black or mulatto. Similarly, the lighter the colour of skin, the more likely it is that a person will become a doctor, bank manager, solicitor or airline pilot. Yet the author has met two Indian tourist guides, one at Manaus in the Amazon rainforest and the other at Iguaçu on the border with Paraguay, who both stated that while blacks and Amerindians had the opportunity of reaching the top in Brazil they preferred to avoid the stresses of modern life.

The three main races of Singapore have separate religions, yet each is completely tolerant of the others and most people even celebrate all three 'New Years' (Figure 13.47). Although in 1994 there was still a Chinatown (Figure 15.43), restricted to ten streets, Arab Street (four streets) and Little India (six streets), the Singapore government has pulled down most of the old houses in these areas. Ethnic concentrations have been broken up and 87 per cent of Singaporeans now live in modern high-rise flats in new towns (Places 45, page 413). Posters promote racial harmony (Figure 13.48), and all races, religions and income groups live together in what appears to be a most successful attempt to create a national unity — a unity best seen on National Day.

Figure 13.47

Ethnic and religious groups in Singapore

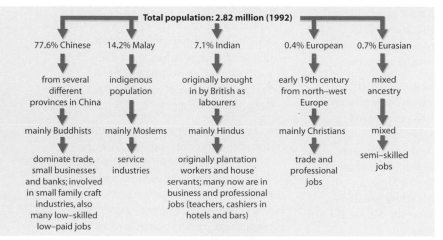

Total population: 2.82 million (1992)

77.6% Chinese	14.2% Malay	7.1% Indian	0.4% European	0.7% Eurasian
from several different provinces in China	indigenous population	originally brought in by British as labourers	early 19th century from north–west Europe	mixed ancestry
mainly Buddhists	mainly Moslems	mainly Hindus	mainly Christians	mixed
dominate trade, small businesses and banks; involved in small family craft industries, also many low–skilled low–paid jobs	service industries	originally plantation workers and house servants; many now are in business and professional jobs (teachers, cashiers in hotels and bars)	trade and professional jobs	semi–skilled jobs

Figure 13.48

Racial harmony poster, Singapore

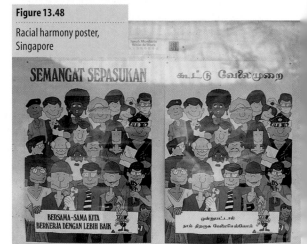

Daily migration: commuting

A commuter is a person who lives in one community and works in another. There are two types of commuting:

1 **Rural–urban**, where the commuter lives in a small town or village and travels to work in a larger town or city. There is rarely much movement in the reverse direction. The **commuter village** is sometimes also referred to as a **dormitory village** or a **suburbanised village**.

2 **Intra-urban**, where people who live in the suburbs travel into the city centre for work. This category now includes inhabitants of inner-city areas who have to make the reverse journey to edge-of-city industrial estates and regional shopping centres.

A **commuter hinterland**, or **urban field**, is the area surrounding a large town or city where the workforce lives. Patterns of commuting are likely to develop where

- Hinterlands are large, communications are fast and reliable (the London Underground), public transport is highly developed and private car ownership is high (south-east England).
- Modern housing is a long way from either the older inner-city industrial areas or from the CBD (as in the New Towns in central Scotland).
- There is a nearby city or conurbation with plenty of jobs, especially in service industries (London).
- There is no rival urban centre within easy reach (Plymouth).
- Salaries are high so that commuters can afford travelling costs.
- People feel that their need to live in a cleaner environment outweighs the disadvantages of time and cost of travel to work (people living in the Peak District and working in Sheffield or Manchester).
- Housing costs are high so that younger people are forced to look for cheaper housing further away from their work (as in south-east England).
- Flexible working hours allowing people to travel during non-rush-hour times.
- The more elderly members of the workforce buy homes in the country or near to the coast and commute until they retire (the Sussex coast).
- There have been severe job losses which force people to look for work in other areas/towns (some of the inhabitants of Cleveland work in south-east England).

What problems arise from commuting?

Most problems occur in or near to city centres. There is often traffic congestion as cars, buses and delivery vehicles all focus on the CBD during peak travel times — i.e. the morning and evening rush hours. Many older cities still have narrow, unplanned roads in the centre and/or in the area surrounding the CBD. The problems of where to park the car and the cost of leaving it for a day cause frustration and expense. The volume of traffic increases the risk of accidents and heightens levels of noise and air pollution. Road improvement schemes often mean the demolition of shops and houses and a further increase in traffic.

In large cities such as London, New York and Tokyo, many commuters may take up to 2 hours travelling each way, which leaves them tired and with little free time in the evenings. The problem is not confined to cities in developed countries. In developing countries, most jobs tend to be in the largest or capital city. As these settlements are expanding rapidly, anyone living on the outskirts may have many kilometres to travel on very congested roads and where public transport systems are inadequate.

Job ratios

This is a method of quantifying the relationship between where people work and where they live. The ratio is expressed by the following formula:

$$\frac{\text{People working in the area (residents and commuters)}}{\text{Total number of employed people living in the area}} \times \frac{100}{1}$$

If the ratio is over 100, it means that there are more commuters travelling into the area than people who already live and work there; if the ratio is less than 100, it means that there are fewer jobs available than there are people living in the area, so people have to travel to find work. For example, Castle Carrock is a small village in Cumbria. Of those employed, 23 work in the village, 21 in the nearby town of Brampton, 46 in Carlisle and 5 beyond Carlisle and Brampton. However, 3 people who work in the village live outside it. The job ratio is therefore:

$$\frac{23 + 3}{95} \times \frac{100}{1} = 27.37$$

This confirms that a high proportion of working residents in the village are commuters. If this ratio is applied to other settlements around a major city, planners may be able to predict traffic movements and possible problems.

Figure 13.49

Job ratio figures for Greater London, 1981

Job ratios

	50–74
	75–99
	100–124
	125–149
	150 and over

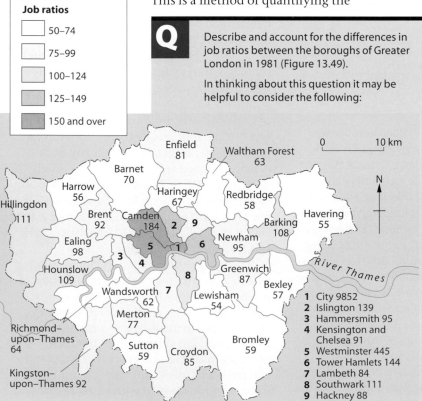

Q Describe and account for the differences in job ratios between the boroughs of Greater London in 1981 (Figure 13.49).

In thinking about this question it may be helpful to consider the following:

a Name the two boroughs with the highest job ratios and try to explain these ratios by referring to the presence of tertiary industry (offices, shops, tourism, administration and entertainment) and the lack of housing (with the exception of luxury flats in the Barbican) due to high land values.

b The three inner-city boroughs north of the river adjacent to the City were the location for the traditional secondary industries of London. Name the boroughs and give specific examples of probable industries.

c Explain the high density of housing in the four inner-city boroughs along the south bank of the Thames and the relative lack of jobs in these areas.

d Give three examples of outer suburbs with low job ratios. Give reasons for these, with reference to the types of residential area and the fact that industry tends to be found along main lines of communication.

e Name the two outer western suburbs which have high job ratios, and suggest why they are anomalous. (Communications may be an important clue.)

f How might job ratios for some of the surrounding dormitory and new towns (name two of each) assist in showing the scale of commuting from those settlements?

Optimum, over and under population

Optimum population

In theory, the **optimum population** of an area is the number of people which, when working with all the available resources, will produce the highest per capita economic return — i.e. the highest standard of living and quality of life. If the size of the population increases or decreases from the optimum, the output per capita and standard of living will fall. This concept is of a dynamic situation changing with time as technology improves, as population totals and structure change (age and sex ratios) and as new raw materials are discovered to replace old ones which are exhausted or whose values change over a period of time.

The **standard of living** of an individual or population is determined by the interaction between physical and human resources and can be expressed in the following formula:

$$\text{Standard of living} = \frac{\text{Natural resources} \left\{ \begin{array}{l} \text{minerals,} \\ \text{energy,} \\ \text{soils etc.} \end{array} \right\} \times \text{Technology}}{\text{Population}}$$

Overpopulation

Overpopulation occurs when there are too many people relative to the resources and technology locally available to maintain an 'adequate' standard of living. Bangladesh, Ethiopia and parts of China, Brazil and India are often said to be overpopulated as they have insufficient food, minerals and energy resources to sustain their populations. They suffer from localised natural disasters such as drought and famine; and are characterised by low incomes, poverty, poor living conditions and often a high level of emigration. In the case of Bangladesh (Places 38), where the population density increased from 282 people per sq km in 1950, to 704 in 1985, and to 888 in 1990, it is easier to appreciate the problem of 'too many people' than in the case of the north-east of Brazil where the density is less than 2 persons per sq km (Places 30, page 323).

Underpopulation

Underpopulation occurs when there are far more resources in an area, e.g. of food, energy and minerals, than can be used by the number of people living there. Canada, with a mid 1980s total population of 25 million, could theoretically double its population by AD 2000 and still maintain its standard of living (Places 38). Countries like Canada and Australia can export their surplus food, energy and mineral resources, have high incomes, good living conditions, and high levels of technology and immigration. It is probable that standards of living would rise, through increased production and exploitation of resources, if population were to increase.

However, when making comparisons on a global scale, there does not seem to be any direct correlation between population density and over/under population. North-east Brazil is 'overpopulated' with 2 people per sq km; whereas California, despite water problems and pollution, is 'underpopulated' with over 500 persons per sq km. Similarly, population density is not necessarily related to gross domestic product (GDP) per capita. The Netherlands and Germany both have a high GDP per capita and a high population density; Canada and Australia have a high GDP per capita and a low population density; Bangladesh and Puerto Rico have a low GDP per capita and a high population density; the Sudan and Bolivia have a low GDP per capita and low population density.

The balance of population and resources within a country may be uneven — for example, a country may have a population which is too great for one resource such as energy, yet too small to use fully a second, such as food supply. The relationships between population and resources are highly complex and the terms 'over' and 'under' population must therefore be used with extreme care.

Is Bangladesh overpopulated?

Bangladesh has a high population density of 888 per sq km (Figure 13.50). It has a high birth rate (41 per 1000) and a declining death rate (from 28 per 1000 in 1970 to 14 in 1990). This has resulted in a high and accelerating rate of natural increase, from 1.6 per cent in 1950 to 2.74 per cent in 1990 (birth and death rates, page 326). Over 20 per cent of the population is under 9 years of age. The GNP of US$220 per capita in 1992 is extremely low.

As most of the country is a flat delta, it is prone to frequent and severe natural disasters. Flooding of the Ganges and Brahmaputra rivers is due mainly to the monsoon rains and partly to deforestation in the Himalayas. Flood damage is also caused by tropical cyclones in the Bay of Bengal (Places 15, page 132 and 24, page 221). Most (71 per cent) of the inhabitants are farmers and live in rural communities (86 per cent). There is a shortage of industry, services and raw materials (Bangladesh has no energy or mineral resources of note), and the transport network is poorly developed. The low level of literacy has led to limited internal innovation and a lack of capital has meant that the country can ill afford to buy technical skills from overseas.

Figure 13.50

High population densities in Bangladesh

Figure 13.51

Low population densities in Canada

Is Canada underpopulated?

Canada's population density is low — 3 people per sq km (Figure 13.51). It has low birth and death rates of 13 and 8 per 1000 respectively, giving a rate of natural increase of only 1.0 per cent a year. Less than 15 per cent of the population is aged under 9 years. The GNP of US$1920 per capita in 1992 is extremely high. Natural disasters are rare. Relatively few (4 per cent) of the inhabitants are farmers or live in rural areas (24 per cent). Canada has developed industries, services, energy supplies, mineral resources (of which there are large reserves) and an effective transport network. The high levels of literacy and national wealth have enabled the country to develop its own technology and to import modern innovations.

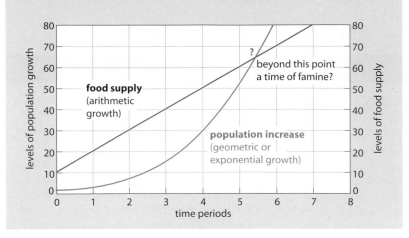

Figure 13.52

Relationships between population growth and food supply (*after* Malthus)

The carrying capacity is the largest population of humans/animals/plants that a particular area/environment/ecosystem can carry or support.

Theories relating to world population and food supply

Malthus

Thomas Malthus was a British demographer who believed that there was a finite optimum population size in relation to food supply and that an increase in population beyond that point would lead to a decline in living standards and to 'war, famine and disease'. He published his views in 1798 and although, fortunately, many of his pessimistic predictions have not come to pass, they form an interesting theory and provide a possible warning for the future. His theory was based on two principles.

1 Human population, if unchecked, grows at a **geometric** or **exponential rate**, i.e. 1 → 2 → 4 → 8 → 16 → 32, etc.

2 Food supply, at best, only increases at an **arithmetic rate**, i.e. 1 → 2 → 3 → 4 → 5 → 6, etc. Malthus considered that this must be so because yields from a given field could not go on increasing forever and the amount of land available is finite.

Malthus demonstrated that any rise in population, however small, would mean that eventually population would exceed increases in food supply. This is shown in Figure 13.52, where the exponential curve intersects the arithmetic curve. Malthus therefore suggested that after five years, the ratio of population to food supply would increase to 16:5, and after six years to 32:6. He suggested that once a ceiling had been reached, further growth in population would be curbed by negative (preventive) or by positive checks.

Preventive (or **negative**) **checks** were methods of limiting population growth and included abstinence from, or a postponement of, marriage which would lower the fertility rate. Malthus noted a correlation between wheat prices and marriage rates (remember that this was the late 18th century): as food became more expensive, fewer people got married.

Positive checks were ways in which the population would be reduced in size by such events as a famine, disease and war, all of which would increase the mortality rate and reduce life expectancy.

The carrying capacity of the environment

The concept of a population ceiling, first suggested by Malthus, is of a saturation level where the population equals the **carrying capacity** of the local environment. Three models portray what might happen as a population, growing exponentially, approaches the carrying capacity of the land (Figure 13.53).

1 The rate of increase may be unchanged until the ceiling is reached, at which point the increase drops to zero. This highly unlikely situation is unsupported by evidence from either human or animal populations.

2 Here, more realistically, the population increase begins to taper off as the carrying capacity is approached, and then to level off when the ceiling is reached. It is claimed that populations which are large in size, have long lives and low fertility rates conform to this 'S' curve pattern.

3 In this instance, the rapid rise in population overshoots the carrying capacity, resulting in a sudden check — e.g. famine and reduced birth rates — which causes a dramatic fall in the total population. After this, the population recovers and fluctuates around, eventually settling down at, the carrying capacity. This 'J' curve appears more applicable to populations which are small in number, have short lives and high fertility levels.

1 How do you think that Figure 13.54 helps to explain Malthus's theory?

2 Africa, with the most rapid population increase in the world, has experienced several almost continuous famines since 1973. Have these famines been inevitable, according to Malthus's theory; have they been induced by human mismanagement of resources; or have they been the result of natural disasters beyond human control and prediction?

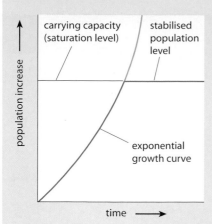

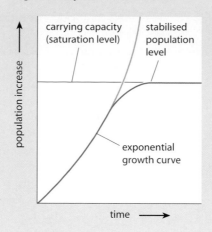

b gradual adjustment: the 'S' curve

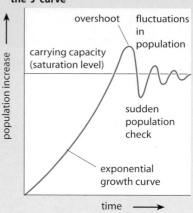

c fluctuating, gradual adjustment: the 'J' curve

Figure 13.53

Three models illustrating the relationship between an exponentially growing population and an environment with a limited carrying capacity

The links between rapid population growth and economic development

These were summarised by Ford at the International Conference on Population and Development, held in Cairo in 1994.

"It is convenient to view the main lines of thinking about links between population growth and economic development in terms of the three broad perspectives of 'Transition theory', 'Orthodoxy' (or 'Neo-Malthusianism') and 'Revisionism'. Transition theory provided the first comprehensive explanation of fertility change, which viewed transition from high to low birth and death rates, as being achieved through industrialisation removing the social and structural 'props' which supported high fertility. This view and its policy implications were expressed by some countries at the 1974 World Population Conference (Bucharest) in terms of the slogan 'development is the

Figure 13.54

Population growth and population checks

best contraceptive'. Neo-Malthusianism (1986), which arose in recognition of the dramatic acceleration in population growth following declines in mortality rates in developing countries, questioned transition theory postulated on the grounds that rapid population growth could itself impede economic development by exacerbating social and economic problems. Transition theory was further undermined by the empirical findings that fertility decline could often take place prior to appreciable economic development, partly in response to a range of social forces. By the time of the 1984 World Population Conference (Mexico City) an emphasis upon taking positive steps to reduce population growth (largely through family planning programmes) had become the consensus view among both international agencies and the governments of many less-developed countries. The general consensus view articulated the need for population policies in integration with other development strategies. The 'Revisionist' view (Boserup) argued that economic growth could be stimulated by the search for innovation to increase production in the face of population pressure."

Boserup was a Danish economist who, in 1965, put forward an alternative theory to that of Malthus. She asserted that in a pre-industrial society an increase in population was likely to stimulate change in agrarian technology and result in increased food production — i.e. 'necessity is the mother of invention'. She suggested that as population increased, farming became more intensive due to innovation and the introduction of new methods and technology.

Case Study 13

Population control in China

In 1990, 23 per cent of the world's population lived in China. Over 90 per cent of these belonged to the dominant Han people; the remainder comprise 56 small minority groups.

Figure 13.55 shows the demographic cycle for China since the formation of the People's Republic in 1950; and Figure 13.56 shows China's age structure, using estimates for mid 1993. On both diagrams, labels **A–E** show changes in state attitudes and policies towards population growth.

Figure 13.55

Demographic cycle of China, since 1950

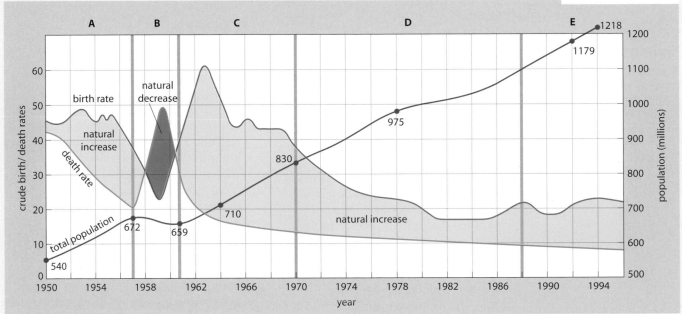

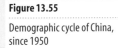

Figure 13.55

Demographic cycle of China, since 1950

A The high birth rate of the 1950s was a response to the state philosophy that 'a large population gives a strong nation' and people were encouraged to have as many children as possible. At the same time, death rates were falling, mainly due to improved food supplies and medical care.

B The period between 1959 and 1961 coincided with the 'Great Leap Forward'. It was a time when industrial production had to be increased at all costs and little attention was paid to farming (Places 49, page 428). The result was a catastrophic famine in which an estimated 20 million people died; infant mortality rates rose and birth rates fell.

C During the 1960s, attempts to control population growth were thwarted by the Cultural Revolution. Every three years, China's population increased by 55 million — i.e. by the same amount as the total UK population.

D State family planning programmes were introduced in the 1970s. By 1975 the average family size had fallen to three children, but this was still regarded by the state as being too many. The state began an advertising campaign for Wan-xi-shao — i.e 'later, longer, fewer' (later marriages, longer gaps between children, and fewer children). Even if the family size had been reduced to two, it would still have meant that China's population would, due to the millions of couples in the reproductive age group, have doubled within 50 years. By 1979 the government, wishing to stop population growth altogether, began to 'encourage' a 'one child per family' policy (Figure 13.57) so that total population in the year 2000 would be 1200 million. Inducements to have only one child included free education, priority housing, pension and family benefits. Not only were these

lost after a second child was born, but fines of up to 15 per cent of the family's annual income were imposed. In addition, the marriage-able age was set at 22 for men and 20 for women, with couples having to apply to the state for permission to be married, and again to have a child.

Although the birth rate fell from 40 per 1000 in 1968 to 17 per 1000 by 1980, the government's policy was often resisted, especially in rural areas. Abortions became compulsory for a second pregnancy (an estimated 10 million a year) and there were frequent reports of coerced sterilisations and of female infanticide where the first-born was a girl (girls were considered less useful for working in the fields) in the hope of a later child being a boy (Figure 13.58). The state became alarmed at the emergence of a generation of 'little Emperors' — spoilt single children who were greedy, bad-tempered and lazy.

E In 1987 the government began to relax its rigid policy in response to intermittent outrage about cases of coercion and brutality in implementing population goals. The latest attitude (mid 1994) still enforces the minimum age for marriage and restricts families living in urban areas to one child. However, a second child is allowed in rural areas if the firstborn is a girl and providing there is a 4-year gap between births. There are two exceptions. The first allows an extra child if one already born is proved to be handicapped (mentally or physically). The second permits minority groups to have as many children as they wish.

China's family size had fallen from 5.8 to 2.4 in 20 years; the figure is 1.7 in urban areas (better education, stronger state control), compared with 2.7 in rural areas. Even so, the 1200 million goal for 2000 has already been exceeded. Although there has been some relaxation of control, and perhaps some change in attitudes, a report from Hong Kong (September 1994) claimed that many babies born above the legal number, and first-born babies if girls, were still being left in public places to die.

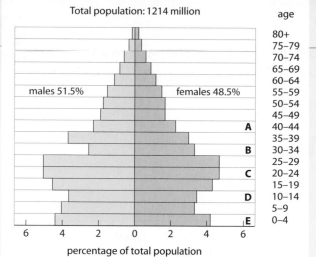

Total population: 1214 million

males 51.5%   females 48.5%

percentage of total population

Figure 13.56

Age structure of Chinese population, mid 1993 estimate

Figure 13.57

Poster promoting China's 'one-child' policy

Figure 13.58

Problems of the 'one-child' policy

Discriminatory abortion is creating a worrying imbalance in China's sex ratio, many couples terminating a pregnancy if they suspect the baby will be a girl. Recognising that ultrasound techniques were being abused for this purpose, authorities have made it illegal to tell parents the sex of their child before birth, but as Chinese doctors are highly underpaid and are used to accepting gifts then . . .

The state has now banned both forced and late abortions. Termination, even when a pregnancy is 'outside the plan' is no longer compulsory and officials can only fine families who ignore family planning advice. However there are still reports from rural areas of over-zealous task forces performing forced abortions. Even in the larger cities, abortions as late as seven months are common. To reduce these pressures, the authorities do try to encourage, often forcibly in rural areas, women who have had their quota to undergo tubal ligation.

South China Morning Post, 1994

References

Annual Abstract of Statistics 1993 (1993) HMSO.

Barke, M. and O'Hare, G. (1984) *The Third World*. Oliver & Boyd.

Bradford, M. G. and Kent, W. A. (1977) *Human Geography*. Oxford University Press.

Briggs, K. (1982) *Human Geography*. Hodder & Stoughton.

Hornby, W. F. and Jones, M. (1980) *An Introduction to Population Geography*. Cambridge University Press.

Key Data 1993/94 (1994) HMSO.

McBride, P. (1980) *Human Geography*. Blackie.

Migration (1983) Open University, BBC Television (plus booklet).

North–South: A Programme for Survival. Report of the Brandt Commission. (1980) Pan Books.

Robinson, V. (1994) The geography of ethnic minorities, *Geography Review*, 7(4), p. 10.

The New Geographical Digest (1992) George Philip & Sons.

Waugh, D. (1994) *The Wider World*. Thomas Nelson.

Chapter 14

Settlement

> *"The largest single step in the ascent of man is the change from nomad to village agriculture."*
> J. Bronowski, *The Ascent of Man*, 1973

Early settlement

About 8000 BC, at the end of the last ice age, the world's population consisted of small bands of hunters and collectors living mainly in subtropical lands and at a subsistence level (page 438). These groups of people, who were usually migratory, could only support themselves if the whole community was involved in the search for food. At this time two major technological changes, known as the 'Neolithic revolution', turned the migratory hunter-collector into a sedentary farmer. The first was the domestication of animals (sheep, goats and cattle) and the second the cultivation of cereals (wheat, rice and maize). Slow improvements in early farming gradually led to food surpluses and enabled an increasing proportion of the community to specialise in non-farming tasks.

The evolution in farming appears to have taken place independently, but at about the same time, in three river basins: the Tigris–Euphrates (in Mesopotamia), the Nile

and the Indus (Figure 14.1). These areas had similar natural advantages:

- hills surrounding the basins provided pasture for domestic animals;
- flat, flood plains next to large rivers;
- rich, fertile silt deposited by the rivers during times of flood;
- a relatively dry but not too dry climate which maintained soil fertility (i.e. limited leaching) and enabled mud from the rivers to be used to build houses (climatically, these areas were more moist than they are today);
- a warm subtropical climate; and
- a permanent water supply from the rivers for domestic use and, as farming developed, for irrigation.

By 1500 BC, larger towns and urban centres had developed with an increasingly wider range of functions. Administrators were needed to organise the collection of crops and the distribution of food supplies; traders exchanged surplus goods with other urban centres; and early engineers introduced irrigation systems. Craftsmen were required to make farming equipment and household articles — the oldest-known pottery and woven textiles were found at Catal Huyuk in

Figure 14.1

Civilisations and cities before 1500 BC

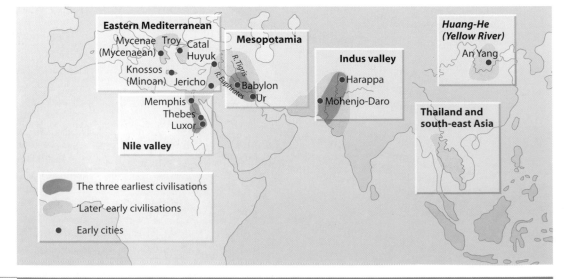

present-day Turkey — and copper and bronze were being worked by 3000 BC. As towns continued to grow, it became necessary to have a legal system and an army for defence.

Although there is divergence of opinion over the exact dates, Figure 14.2 gives a chronological sequence of early settlements.

Figure 14.2

A chronology of early settlement

Approximate date BC		Near East { Tigris–Euphrates / Nile / Indus	Rest of world
9000		Hunters and collectors	
8000	8500	First domesticated animals and cereals	Northern Europe recovering from the last ice age
	8300	Jericho: first walled city	
7000			
6000	6250	Catal Huyuk: first pottery and woven textiles; became largest city in world	
5000	5500	Growth of villages in Mesopotamia Growth of many villages in Nile and Indus valleys	
	5000	Early methods of irrigation	Rice cultivation in south-east Asia
4000		Bronze casting	
3000	3500	Invention of the wheel and plough in Mesopotamia and the sail in Egypt	First Chinese city
	3000	Cities in Mesopotamia	First crops grown in central Africa; bronze worked in Thailand
2000	2600	Pyramids	
	2000	Minoan civilisation in Crete	Metal-working in the Andes
1000	1600	Mycenaean civilisation in Greece	

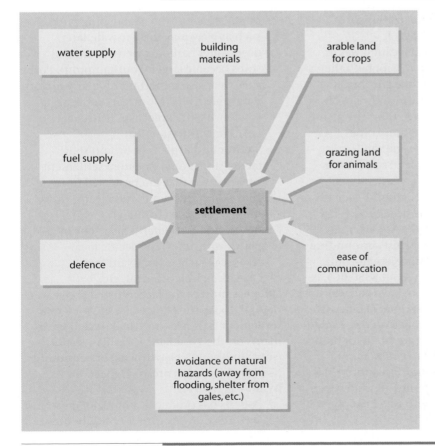

Figure 14.3

Settlement location factors

Site and situation of early settlements

Site describes the characteristics of the actual point at which a settlement is located and was of major importance in the initial establishment and growth of a village or town. **Situation** describes the location of a place relative to its surroundings (neighbouring settlements, rivers and uplands). Situation, along with human and political factors, determined whether or not a particular settlement remained small or grew into a larger town or city (Case Study 14).

Early settlements developed in a rural economy which aimed at self-sufficiency, largely because transport systems were limited. While the most significant factors in determining the site of a village include those shown in Figure 14.3 and described below, remember that several factors would usually operate together when a choice in the location of a settlement was being made.

Among the most important factors are:

- **Water supply** A nearby, guaranteed supply was essential as water is needed daily throughout the year and is heavy to carry any distance. In earlier times, rivers were sufficiently clean to give a safe, permanent supply. In lowland Britain, many early villages were located along the spring line at the foot of a chalk or limestone escarpment (Figure 8.9). In regions where rainfall is limited or unreliable, people settled where the water table was near to the surface (a desert oasis, Figure 14.7a) enabling shallow wells to be dug. Such settlement sites are known as **wet-point** or **water-seeking sites**.

- **Flood avoidance** Elsewhere, the problem may have been too much water. In the English Fenlands, or on coastal marshes, villages were built on mounds which formed natural islands (Ely). Other settlements were built on river terraces (page 76) which were above the flood level and, in some cases, avoided those diseases associated with stagnant water. Such sites are known as **dry-point** or **water-avoiding sites**.

- **Building materials** Materials were heavy and bulky to move and, as transport was poorly developed, it was important to build settlements close to a supply of stone, wood and/or clay.

- **Food supply** The ideal location was in an area which was suitable for both the rearing of animals and the growing of crops — such as the scarps and vales of south-east England (page 183). The quality and quantity of farm produce often depended upon climate and soil fertility.

- **Relief** Flat, low-lying land such as the North German Plain was easier to build on than steeper, higher ground such as the Alps. However, the need for defence sometimes overruled this consideration.

- **Defence** Protection against surrounding tribes was often essential. Jericho, built over 10 000 years ago (about 8350 BC), is the oldest city known to have had walls. In Britain, the two best types of defensive site were those surrounded on three sides by water (Durham, Figure 14.7b) or built upon high ground with commanding views over the surrounding countryside (Edinburgh). Hill-top sites may, however, have had problems with water supply (Figure 14.7c).

- **Fuel supply** Even tropical areas need fuel for cooking purposes as well as for warmth during colder nights. In most early settlements, firewood was the main source — and still is in many of the least economically developed areas, such as the Sahel (page 495).

- **Nodal points** Sites where several valleys meet were often occupied by settlements which became **route centres** (Carlisle and Paris). **Confluence towns** are found where two rivers join (Khartoum at the junction of the White Nile and the Blue Nile, St Louis at the junction of the Mississippi and the Missouri (Figure 3.62)). Settlements on sites which command routes through the hills or mountains are known as **gap towns** (Dorking and Carcassonne).

- **Bridging-points** Settlements have tended to grow where routes had to cross rivers, initially where the river was shallow enough to be forded (Oxford) and later where the site was suitable for a bridge to be built. Of great significance for trade and transport was the lowest bridging-point before a river entered the sea (Newcastle upon Tyne).

- **Harbours** Sheltered sea inlets and river estuaries provided suitable sites for the establishment of coastal fishing ports, such as Newquay in Cornwall; later, deep-water harbours were required as ships became larger (Southampton). Port sites were also important on many major navigable rivers (Montreal on the St Lawrence) and large lakes (the Great Lakes in North America).

- **Shelter and aspect** In Britain south-facing slopes offer favoured settlement sites because they are protected from cold, northerly winds and receive maximum insolation (Torquay).

- **Resources** Settlements also grew in places with access to specific local resources such as salt (Nantwich, Cheshire), iron ore, coal, etc.

Whereas most of the factors listed above were natural, today the choice of a site for a new settlement is usually **political** (Israeli settlements on the West Bank, Brasilia), **social** (some of Britain's New Towns) or **economic** (Carajas for its iron ore and Iguaçu for its hydro-electricity).

Roberts has produced a model (Figure 14.4) which draws together not only site and

Settlement

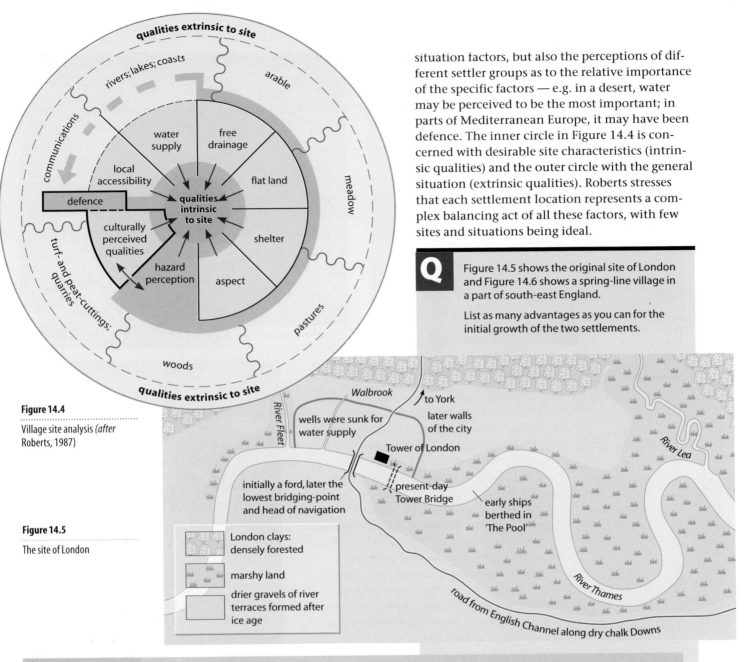

Figure 14.4

Village site analysis (*after* Roberts, 1987)

Figure 14.5

The site of London

situation factors, but also the perceptions of different settler groups as to the relative importance of the specific factors — e.g. in a desert, water may be perceived to be the most important; in parts of Mediterranean Europe, it may have been defence. The inner circle in Figure 14.4 is concerned with desirable site characteristics (intrinsic qualities) and the outer circle with the general situation (extrinsic qualities). Roberts stresses that each settlement location represents a complex balancing act of all these factors, with few sites and situations being ideal.

Q Figure 14.5 shows the original site of London and Figure 14.6 shows a spring-line village in a part of south-east England.

List as many advantages as you can for the initial growth of the two settlements.

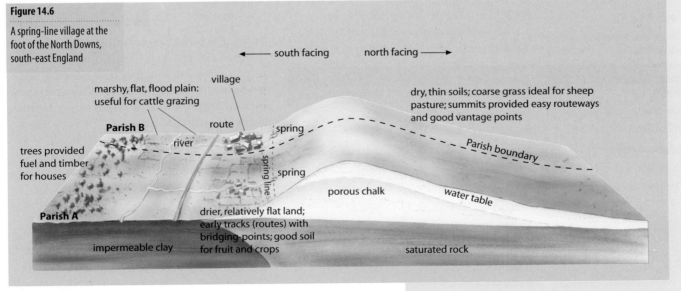

Figure 14.6

A spring-line village at the foot of the North Downs, south-east England

a An oasis: Morocco

b Durham in meander loop

Figure 14.7

Settlement sites

Functions of settlements

As early settlements grew in size, each one tended to develop a specific function or functions. The **function** of a town relates to its economic and social development and refers to its main activities. There are problems in defining and determining a town's main function and often, due to a lack of data such as employment and/or income figures, subjective decisions have to be made. As settlements are very diverse, it helps to try to group together those with a similar function (Framework 5, page 151). Over the years, numerous attempts have been made to classify settlements based on function (by Houston, Money, Harris and Nelson), but these tended to refer to places in industrialised countries and are no longer applicable to post-industrial societies. Further problems arose when the growth of some settlements was based upon an activity which no longer

c A hilltop defensive site:
  Andalucia, Spain

exists (the former coalmining villages of north-east England and South Wales), or where the original function has changed over **time** (a Cornish fishing village may now be a holiday resort). Functions may also differ between continents — i.e. there is also a difference over **space**. Finally it should be realised that today, and especially in the more developed countries, towns and cities are multifunctional — even if one or two functions tend to be predominant. Bearing in mind that the value to geographers of classifying settlements based on function has declined, Figure 14.8 has been included, as much as anything, as a checklist of possible functions.

Figure 14.8

Classification of settlement based on function

Rural	Urban	
Global	**Developed countries**	**Developing countries**
Market and agricultural	Mining	Administration
Route centre/transport	Manufacturing/industrial	Marketing/agricultural
Small service town	Route centre/transport	Route centre/port
Defensive	Retail/wholesale	Mining
Dormitory/overspill/satellite	Religious/cultural	Commercial
	Trade/commerce/financial	Religious
	Administration	Residential
	Resort/recreation	
	Residential	
	New towns	

Differences between urban and rural settlement

Figure 14.9 shows the commonly accepted types of settlement, but hides the divergence of opinion as to how and where to draw the borders between each type. Several methods have been suggested in trying to define the difference between a village, or rural settlement, and a town, or urban settlement.

- **Population size** There is a wide discrepancy of views over the minimum size of population required to enable a settlement to be termed a town, e.g. in Denmark it is considered to be 250 people, in Ireland 500, in France 2000, in the USA 2500, in Spain 10 000 and in Japan 30 000. In India, where many villages are larger than British towns, a figure of less than 25 per cent engaged in agriculture is taken to be the dividing point.
- **Economic** Rural settlements have traditionally been defined as places where most of the workforce are farmers or are engaged in other primary activities (mining and forestry). In contrast, most of the workforce in urban areas are employed in secondary and service industries. However, many rural areas have now become commuter/dormitory settlements for people working in adjacent urban areas or, even more recently, a location for smaller, footloose industries, such as high-tech industries.
- **Services** The provision of services, such as schools, hospitals, shops, public transport and banks, is usually limited, at times absent, in rural areas (Figure 14.18).
- **Land use** In rural areas, settlements are widely spaced with open land between adjacent villages. Within each village there may be individual farms as well as residential areas and possibly small-scale industry. In urban areas, settlements are often packed closely together and within towns there is a greater mixture of land use with residential, industrial, services and open-space provision.
- **Social** Rural settlements, especially those in more remote areas, tend to have more inhabitants in the over 65 age group, whereas the highest proportion in urban areas lies within the economically active age group (page 331) or those under secondary school age.

It has becoming increasingly more difficult to differentiate between villages and towns, especially where urban areas have spread outwards into the **rural fringe**. It is, therefore, more realistic to talk about a transition zone from 'strongly rural' to 'strongly urban'. Cloke (1979) devised an **index of rurality** based upon 16 variables taken largely from census data for England and Wales. These variables included people aged over 65; proportion employed in primary, secondary and tertiary sectors; population density; population mobility (those moving home in the previous 5 years); proportion commuting; and distance from a large town (Figure 14.17). Cloke then identified four categories:

- extreme rural (parts of south-west England, central Wales, East Anglia and the northern Pennines);
- intermediate rural;
- intermediate non-rural; and
- extreme non-rural (mainly suburbanised villages (page 369) around London in Surrey, Hertfordshire and Essex).

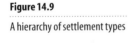

Figure 14.9

A hierarchy of settlement types

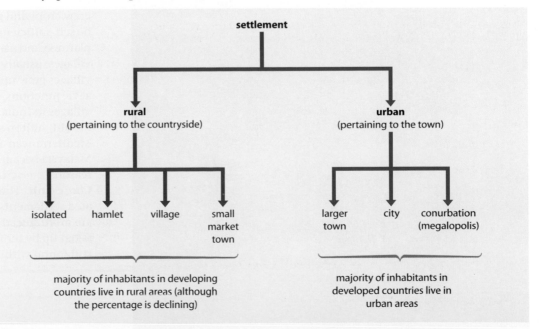

Figure 14.10

An isolated settlement in the Amazon rainforest

Figure 14.11

Dispersed settlement in North Yorkshire

Rural settlement

Morphology

Geographers have become increasingly interested in the **morphology**, i.e. the pattern or shape, of settlements. Although village shapes vary spatially in Britain and across the world it has been, again traditionally, possible to identify seven types. (Remember that, as in other classifications, some geographers may identify more or fewer categories).

1 **Isolated** This refers to an individual building, usually found in an area of extreme physical difficulty where the natural resources are insufficient to maintain more than a few inhabitants, e.g. the Amazon rainforests where tribes live in a communal home called a *maloca* (Figure 14.10). Isolated houses may also be found in planned pioneer areas such as on the Canadian Prairies where the land was divided into small squares each with its own farm buildings.

2 **Dispersed** Settlement is described as dispersed when there is a scatter of individual farms and houses across an area; there are either no nucleations present, or they are so small that they consist only of two or three buildings forming a hamlet (Figure 14.11). Each farm or hamlet may be separated from the next by 2 or 3 km of open space or farmland. In the Scottish Highlands and Islands, some communities consist of crofts spaced out alongside a road or raised beach. Hamlets are common in rural areas of northern Britain, on the North German Plain (where their name *urweiler* means 'primaeval hamlet') and in sub-Saharan Africa.

3 **Nucleated** Nucleated settlement is common in many rural parts of the world where buildings have been grouped closely together for economic, social or defensive purposes (Figure 14.12). In Britain, where recent evidence suggests that nucleation only took place after the year 1000, villages were surrounded by their farmland, where the inhabitants grew crops and grazed animals in order to be self-sufficient; this led to an unplanned and variable spacing of villages, usually 3–5 km apart. Some villages grew up around crossroads and at 'T' junctions, as is the case of many villages in India. Many border villages in Britain, hilltop settlements around the Mediterranean Sea, and kampongs in Malaysia became nucleated for defensive reasons.

4 **Loose-knit** These are similar to nucleated settlements except that the buildings are more spread out, possibly due to space taken up by individual farms which are still found within the village itself.

Figure 14.12

Nucleated settlement in Sumatra, Indonesia

5 **Linear**, or **ribbon** Where the buildings are strung out along a main line of communication or along a confined river valley (Figure 14.13), the settlement is described as linear. **Street villages**, planned linear villages, were common in medieval England. Unplanned linear settlements also developed on long, narrow, flood-avoidance sites — e.g. along the raised beaches of western Scotland and on river terraces, as in London. Later, unplanned linear settlement grew up along the floors of the narrow coalmining valleys of South Wales and on main roads leading out of Britain's urban areas following the increase in private car ownership and the development of public transport. In the Netherlands, Malaysia and Thailand, houses have been built along canals and waterways.

6 **Ring and 'green' villages** Ring villages are found in many parts of sub-Saharan Africa and the Amazon rainforest (Figure 14.14). Houses were built around a central area which was left open for tribal meetings and communal life. In Kenya, the Maasai built their houses around an

Figure 14.13

Linear (ribbon) settlement: Coombe Martin, Devon

area into which their cattle were driven for protection during most nights. In England, many villages have been built around a central green.

7 **Planned** Although many early settlements were planned (Pompeii, York), the apparently random shape of many British villages appears to suggest that they were not. More recently, villages surrounding large urban areas in, for example, Britain and the Netherlands, have expanded and become suburbanised, having small and often crescent-shaped estates. Figure 14.15 shows a planned settlement.

If you study maps of village plans, it is very likely that you will find many settlements with a mixture of the above shapes — e.g. a village may have a nucleated centre, a planned estate on its edges and a linear pattern extending along the road leading to the nearest large town.

Roberts (1987) suggested a different basis for classification (Figure 14.16). Even so, he concedes there are difficulties in trying to fit a particular village into a specific category, as when determining if a strip of grass is large enough to be called a green, and concludes that many villages are **composite** (or **polyfocal**), incorporating several plans and phases of development.

Figure 14.14

Ring village: Kraito, in the Amazon rainforest

Figure 14.15

A planned settlement: the New Town of Milton Keynes

Basic shape	Plan and morphology		Village green
Linear (in a row)	regular	———	with
			without
	irregular	— — —	with
			without
Agglomerated (more nucleated)	regular grid	⊞	with
			without
	regular radial	✳	with
			without
	irregular grid	⁻'⁻'⁻	with
		-|-|-	without
	irregular agglomerated	●	with
			without

Figure 14.16

Village forms (after Roberts, 1987)

Dispersed and nucleated rural settlement

Whether settlement is dispersed or nucleated depends upon local physical conditions; economic factors such as the time and distance between places; and social factors which include who owns the land and how the people of the area live and work on it.

Causes of dispersion

The more extreme the physical conditions and possible hardship of an area, the more probable it is that the settlement will be dispersed. Similarly, dispersed settlement develops in areas where natural resources are limited and insufficient to support many people. This lack of resources could include a limited water supply (the Carboniferous limestone outcrops of the Pennines); forested areas (the Canadian Shield and the Amazon basin); and marginal farmland (the Scottish Highlands and the Sahel countries), where pastoral farming is limited by the quality and quantity of available grass. Areas with physical difficulties are also less likely to have good transport networks.

Forms of land tenure can also result in dispersed dwellings, especially in those parts of the world where inheritance laws have meant that the farm is successively divided between several sons. Similar patterns, though with larger farm units, can be found in pioneer areas such as the Canadian Prairies and the Dutch polders.

The 'agrarian revolution' in Britain in the 18th century ended the open-field system, in which strips were owned individually but the crops and animals were controlled by the community. It was replaced by enclosing several fields which were owned by a farmer who became responsible for all the decisions affecting that farm; new farmhouses were sometimes built outside the village.

Two other changes at about the same time increased the incidence of dispersed settlement. The first was the growth of large estates belonging to wealthy landowners. The second was the extension of farming in hilly areas, in the 18th and again in the 19th century, to produce the extra food needed to feed the rapidly growing urban areas. Much moorland in the Pennines was walled; while fenland areas, previously of limited value, were drained and farmed. Areas of downland were also put under the plough. Increased mechanisation reduced labour needs, resulting in overpopulation and, eventually, out-migration.

Finally, settlement was more likely to develop a dispersed pattern where there was less risk of war or civil unrest as there was then less need for people to group together for protection.

Causes of nucleation

The majority of humans have always preferred to live together in groups, as witnessed by the cities of ancient Mesopotamia and Egypt, and the present-day conurbations and cities with more than 1 million inhabitants (page 385). Two major reasons for people to group together have been either a limited or an excess water supply. Settlements have grown up around springs, as at the foot of chalk escarpments in southern England (Figure 14.6), and at waterholes and oases in the desert (Figure 14.7a). Settlements have also been built on mounds in marshy fenland regions and on river terraces above the level of flooding (Figures 14.5 and 3.49).

A further cause was the need to group together for defence and protection. Examples of defensive settlements include living in walled cities on relatively flat plains (Jericho and York); behind stockades (African kraals); in hill-top villages in southern Italy and Greece (Figure 14.7c); or in meander loops, taking advantage of a natural water barrier (Durham, Figure 14.7b).

In Anglo-Saxon England, when many villages had their origin, the feudal open-field system of farming encouraged nucleation: the local lord could better supervise his serfs if they were clustered around him; while the serf, living in the village, was probably equidistant from his fragmented strips of farmland (Case Study 14). Today, the more intensive the nature of farming, the more nucleated the settlement tends to be. People like to be as near as possible to services so that the larger and more nucleated the village, the more likely it is to have a wide range of services such as a primary school, shops and a public house (Figure 14.18).

Transport and routeways have always had a major influence on the clustering of dwellings. Buildings tend to be grouped together at crossroads and 'T' junctions; controlling a gap through hills; at bridging-points and along main roads, waterways and railways. Compact settlement patterns are also found in areas with an important local resource (a Durham coalmining or a North Wales slate quarry village), or where

1 Briefly describe the differences between a dispersed and a nucleated settlement pattern.

2 Illustrating your answer with specific examples from any part of the world, explain why some rural settlement patterns are dispersed whereas others are nucleated.

there was an abundance of building materials. More recently, many governments have encouraged new, nucleated settlements in an attempt to achieve large-scale self-sufficiency. Examples may be found as far afield as the Soviet collective farm, the Chinese commune, the Tanzanian *ujamaa* (Places 33, page 344) and the Israeli *kibbutz*.

Changes in rural settlement in Britain

Within the British Isles, there are areas, especially those nearer to urban centres, where the rural population is increasing and others, usually in more remote locations, where the rural population is decreasing (rural depopulation). These population changes affect the size, morphology and functions of villages. Figure 14.17 shows that there is some relationship between the type and rate of change in a rural settlement and its distance from, and accessibility to, a large urban area.

Accessibility to urban centres
As public and private transport improved during the inter-war period (1919–39), British cities expanded into the surrounding countryside at a rapid and uncontrolled rate. In an attempt to prevent this urban sprawl, a **green belt** was created around London following the 1947 Town and Country Planning Act. The concept of a green belt, later applied to most of Britain's conurbations, was to restrict the erection of houses and other buildings and to preserve and conserve areas of countryside for farming and recreational purposes.

Beyond the green belt, **new towns** and **overspill towns** were built, initially to accommodate new arrivals seeking work in the nearby city and, later, those forced to leave it due to various redevelopment schemes. These new settlements, designed to become self-supporting both economically and socially, developed urban characteristics and functions. New towns, overspill and green belts were part of a wider land-use planning process which aimed to manage urban growth.

Meanwhile, despite the 1968 Town and Country Planning Act, uncontrolled growth also continued in many small villages beyond the green belt. Referred to during the inter-war period as **dormitory** or **commuter villages** (page 352), these settlements have increasingly adopted some of the characteristics of nearby urban areas and have been termed **suburbanised villages**. Figure 14.18 lists some of the changes which occur as a village becomes increasingly suburbanised.

Figure 14.17

Rural settlements and distance from large urban areas

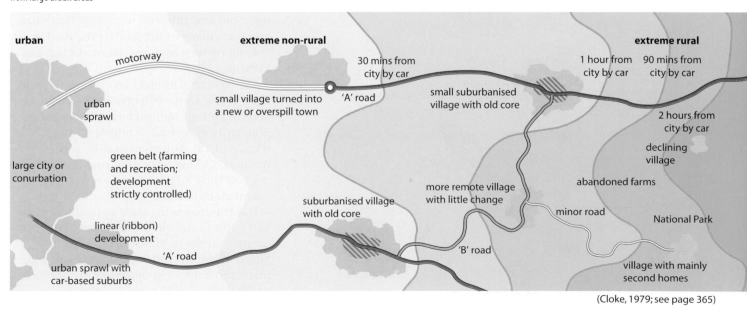

(Cloke, 1979; see page 365)

Characteristic	Extreme non-rural (increasingly suburbanised)	Original village	Extreme rural (increasingly depopulated)
Housing	Many new detached houses, semi-detached houses and bungalows; renovated barns and cottages; expensive estates	Detached, stone-built houses/cottages with slate/thatch roofs; some farms, many over 200 years old; barns	Poor housing lacking basic amenities; old stone houses, some derelict, some converted into holiday/second homes
Population structure	Young/middle-aged married couples with children; very few born in village; professional/executive groups; some wealthy retired people	An ageing population; most born in village; labouring/manual groups	Mainly elderly/retired; born and lived all life locally; labouring/manual groups; younger people have moved away
Employment	New light industry (hi-tech and food processing); good salaries; many commuters (well-paid); tourist shops	Farming and other primary activities (forestry, mining); low-paid local jobs	Low-paid; unemployment; farming jobs (declining if in marginal areas) and other primary activities; some tourist-related jobs
Transport	Good bus service (unless reduced by private car); most families have one or two cars; improved roads	Bus service (limited); some cars; narrow/winding roads	No public transport; poor roads
Services	More shops; enlarged school; modern public houses/restaurants; garage	Village shop; small junior school; public house; village hall	Shop and school closed; perhaps a public house
Community/Social	Local community swamped; division between locals and newcomers; may be deserted during day (commuters absent)	Close-knit community (many are related)	a small community; more isolated
Environment	Increase in noise and pollution, especially from traffic; loss of farmland/open space	Quiet, relatively pollution-free	Quiet; increase in conserved areas (National Parks/forestry)

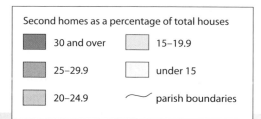

Second homes as a percentage of total houses

- 30 and over
- 25–29.9
- 20–24.9
- 15–19.9
- under 15
- parish boundaries

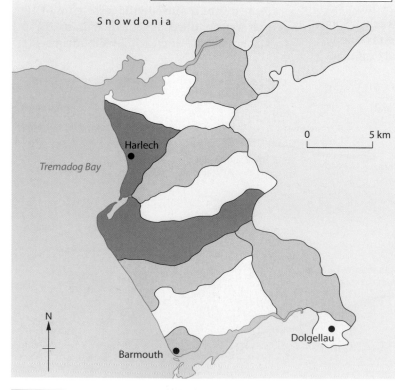

Less accessible settlements

These villages are further in distance from, or have poorer transport links to, the nearest city — i.e. they are beyond commuting range. This makes the journey longer in time, more expensive and less convenient. Though these villages may be relatively stable in size, their social and economic make-up is changing. Many in the younger age groups move out, pushed by a shortage of jobs and social life. They are replaced by retired people seeking quietness and a pleasant environment but who often do not realise that rural areas lack many of the services required by the elderly such as shops, buses, doctors and libraries.

Villages in National Parks and other areas of attractive scenery in upland or coastal areas are being changed by the increased popularity of **second** or **holiday homes** (Figure 14.19). The more wealthy urban dwellers, seeking relaxation away from the stress of their local working and living environment, buy vacant properties in villages. While this may bring trade to the local shop and improve the quality of some buildings, it means that the local inhabitants can no longer afford the inflated house prices, and many properties may stand empty for much of the year.

Remote areas

These areas suffer from a population loss which leaves houses empty and villages decreasing in size (Figure 14.18). Resultant problems include a lack of job opportunities, fewer services and poor transport facilities. Employment is often limited to the shrinking primary industries which are low-paid and lack future prospects. The cost of providing services to remote areas is high and there is often insufficient demand to keep the local shop or village school open. With fewer inhabitants to use public transport, bus services may decline or stop altogether, forcing people to move to more accessible areas.

Measuring settlement patterns

Nearest neighbour analysis

Settlements often appear on maps as dots. Dot distributions are commonly used in geography, yet their patterns are often difficult to describe. Sometimes patterns are obvious, such as when settlements are extremely nucleated or dispersed (Figure 14.20). As, in reality, the pattern is likely to lie between these two extremes, then any description will be subjective. One way in which a pattern can be measured objectively is by using nearest neighbour analysis.

This technique was devised by a botanist who wished to describe patterns of plant distributions. It can be used to identify a tendency towards nucleation (clustering) or dispersion for settlements, shops, industry, etc., as well as plants. Nearest neighbour

analysis gives a precision which enables one region to be compared with another and allows changes in distribution to be compared over a period of time. It is, however, only a technique and therefore does *not* offer any explanation of patterns.

The formula used in nearest neighbour analysis produces a figure (expressed as *Rn*) which measures the extent to which a particular pattern is clustered (nucleated), random, or regular (uniform) (Figure 14.20).

- **Clustering** occurs when all the dots are very close to the same point. An example of this in Britain is on coalfields where mining villages tended to coalesce. In an extreme case, *Rn* would be 0.
- **Random** distributions occur where there is no pattern at all. *Rn* then equals 1.0. The usual pattern for settlement is one which is predominantly random with a tendency either towards clustering or regularity.
- **Regular** patterns are perfectly uniform. If ever found in reality, they would have an *Rn* value of 2.15 which would mean that each dot (settlement) was equidistant from all its neighbours. The closest example of this in Britain is the distribution of market towns in East Anglia.

Using nearest neighbour analysis
Figure 14.21 shows settlements in part of north-east Warwickshire and south-west Leicestershire, an area of the English Midlands where it might be expected that there would be evidence of regularity in the distribution.

Figure 14.20

Nearest neighbour values *(Rn)*

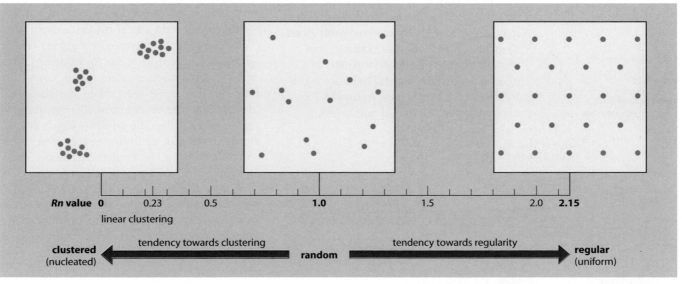

Figure 14.21

Nearest neighbour analysis: a worked example for part of north-east Warwickshire and south-west Leicestershire

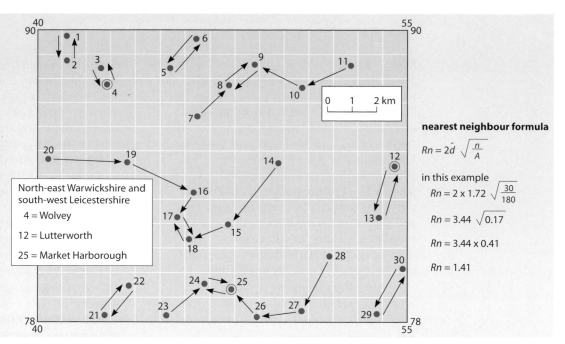

North-east Warwickshire and south-west Leicestershire

4 = Wolvey

12 = Lutterworth

25 = Market Harborough

nearest neighbour formula

$$Rn = 2\bar{d}\sqrt{\frac{n}{A}}$$

in this example

$$Rn = 2 \times 1.72\sqrt{\frac{30}{180}}$$

$$Rn = 3.44\sqrt{0.17}$$

$$Rn = 3.44 \times 0.41$$

$$Rn = 1.41$$

Settlement number	Nearest neighbour	Distance (km)
1	2	1.0
2	1	1.0
3	4	0.6
4	3	0.6
5	6	1.6
6	5	1.6
7	8	1.8
8	9	1.3
9	8	1.3
10	9	2.1
11	10	2.2
12	13	2.2
13	12	2.2
14	15	3.3
15	18	1.7
16	17	1.3
17	18	1.0
18	17	1.0
19	16	3.0
20	19	3.2
21	22	1.6
22	21	1.6
23	24	2.1
24	25	1.1
25	24	1.1
26	25	1.5
27	26	1.8
28	27	2.5
29	30	2.2
30	29	2.2
		Σ51.7

1 The settlements in the study area were located. (The minimum number recommended for a nearest neighbour analysis is 30.) Each settlement was given a number.

2 The nearest neighbour formula was applied. This formula is

$$Rn = 2\bar{d}\sqrt{\frac{n}{A}}$$

where:

Rn = the description of the distribution;

$\bar{d}$ = the mean distance between the nearest neighbours (km);

n = the number of points (settlements) in the study area; and

A = the area under study (km²).

3 To find $\bar{d}$, measure the straight-line distance between each settlement and its nearest neighbour, e.g. settlement 1 to 2, settlement 2 to 1, settlement 3 to 4 and so on. One point may have more than one nearest neighbour (settlement 8) and two points may be each others' nearest neighbour (settlements 1 and 2). In this example, the mean distance between all the pairs of nearest neighbours was 1.72 km — i.e. the total distance between each pair (51.7 km) divided by the number of points (30).

4 Find the total area of the map — i.e. 15 km x 12 km = 180 km².

5 Calculate the nearest neighbour statistic, Rn, by substituting the formula. This has already been done in Figure 14.21 and gives an Rn value of 1.41.

6 Using this Rn value, refer back to Figure 14.20 to determine how clustered or regular is the pattern. A value of 1.41 shows that there is a fairly strong tend-ency towards a regular pattern of settlement.

7 However, there is a possibility that this pattern has occurred by chance. Referring to Figure 14.22, it is apparent that the values of Rn must lie outside the shaded area before a distribution of clustering or regularity can be accepted as significant. Values lying in the shaded area at the 95 per cent probability level show a random distribution. (*Note*: with fewer than 30 settlements, it becomes increasingly difficult to say with any confidence that the distribution is clustered or regular.) The graph confirms that our Rn value of 1.41 has a significant element of regularity.

How can the nearest neighbour statistic be used to compare two or more distributions? Figure 14.22 shows the Rn value for three areas in England, including that for our worked example, the English Midlands. The Rn statistic of 1.57 for part of East Anglia shows that the area has a more pronounced pattern of regularity than the Midlands. An Rn value of 0.61 for part of the Durham coalfield indicates that it has a significant tendency towards a clustered distribution.

Figure 14.22

Interpretation of the Rn statistic: significant values

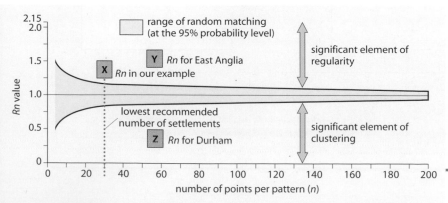

Limitations and problems

As noted earlier, nearest neighbour analysis is a useful statistical technique but it has to be used with care. In particular, the following should be taken considered:

1 The size of the area chosen is critical. Comparisons will be valid only if each area chosen is of the same size.

2 The area chosen should not be too large, as this lowers the *Rn* value (i.e. it exaggerates the degree of clustering), or too small, as this increases the *Rn* value (i.e. it exaggerates the level of regularity).

3 Distortion is likely to occur in valleys, where nearest neighbours may be separated by a river, or where spring-line settlements are found in a linear pattern as at the foot of a scarp slope (Figure 14.6).

4 Which settlement sizes are to be included? Are hamlets acceptable, or is the village to be the smallest size? If so, when is a hamlet large enough to be called a village?

5 There may be difficulty in determining the centre of a settlement for measurement purposes, especially if it has a linear or a loose-knit morphology.

6 The boundary of an area is significant. If the area is a small island or lies on an outcrop of a particular rock, there is little problem; but if, as in Figure 14.21, the area is part of a larger region, the boundaries must have been chosen arbitrarily (in this instance by predetermined grid lines). In such a case, it is likely that the nearest neighbour of some of the points (e.g. number 20) will be off the map. There is disagreement as to whether those points nearest to the boundary of the map should be included, but perhaps of more importance is the need to be consistent in approach and to be aware of the problems and limitations.

Despite these problems, nearest neighbour analysis forms a useful basis for further investigation into why any clustering or regularity of settlement has taken place.

The rank–size rule

This is an attempt to find a numerical relationship between the population size of settlements within an area such as a country or county. The rule states that **the size of settlements is inversely proportional to their rank.** Settlements are ranked in descending order of population size with the largest city placed first. The assumption is that the second-ranked city will have a population one-half that of the first-ranked, the third-ranked city a population one-third of the first-ranked, the fourth-ranked one-quarter of the largest city, and so on.

The rank–size rule is expressed by the formula:

$$Pn = Pl \div n \text{ (or } R)$$

where:

Pn	= the population of the city;
Pl	= the population of the largest (primate) city;
n (or *R*)	= the rank–size of the city.

For example, if the largest city has a population of 1 000 000, then:

the second-largest city will be 1 000 000 ÷ 2, i.e. 500 000;

the third-largest city will be 1 000 000 ÷ 3, i.e. 333 333; and

the fourth-largest city will be 1 000 000 ÷ 4, i.e. 250 000.

If such a perfect negative relationship actually occurred (Framework 14, page 572), it would produce a steeply downward-sloping, smooth, concave curve on an arithmetic graph (Figure 14.23a). However, it is more usual to plot the rank–size distribution on a logarithmic scale, in which case the perfect negative relationship would appear as a straight line sloping downwards at an angle of 45° (Figure 14.23b).

Figure 14.23

The rank—size rule

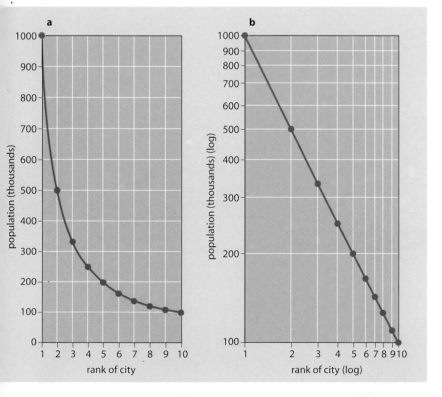

Variations from the rank–size rule

In reality, it is rare to find a close correlation between the city size of a country and the rank–size rule. There are, however, two major variations from the rank–size rule.

Primate distribution (urban primacy) is found where the largest city, often the capital, completely dominates a country or region (in terms of population size, economic development, wealth, services and cultural activities). In such a case, the primate city will have a population size many times greater than that of the second-largest city (Buenos Aires in Figure 14.24). Montevideo in Uruguay is seventeen times larger than the second-largest city, and Lima in Peru is eleven times larger.

Binary distribution occurs where there are two very large cities of almost equal size within the same country: one may be the capital and the other the chief port or major industrial centre. Examples of binary distribution include Madrid and Barcelona in Spain and Quito and Guayaquil in Equador.

It has been suggested (though there are many exceptions) that the rank–size rule is more likely to operate if the country is developed; has been urbanised for a long time; is large in size; and has a complex and stable economic and political organisation. In contrast, primate distribution is more likely to be found (also with exceptions, including France and Austria) in countries which are small in size; less developed; former colonies of European countries; only recently urbanised; and which have experienced recent changes in political organisation and/or boundaries.

Two schools of thought exist concerning the causes of variation in urban primacy. One suggests that as a city begins to dominate a country it attracts people, trade, industry and services at an increasingly rapid rate and at the expense of rival cities (arguably this is more applicable to economically less developed countries). The other claims that as a country becomes more urbanised and industrialised, the growth of several cities tends to be stimulated, thus reducing the importance of the primate city (arguably more applicable to economically more developed countries where some of the largest cities are now experiencing urban depopulation, page 342).

Q Figure 14.24 shows the rank—size graph for Brazil. The larger table shows the actual populations for the ten largest cities in four other countries.

Figure 14.24

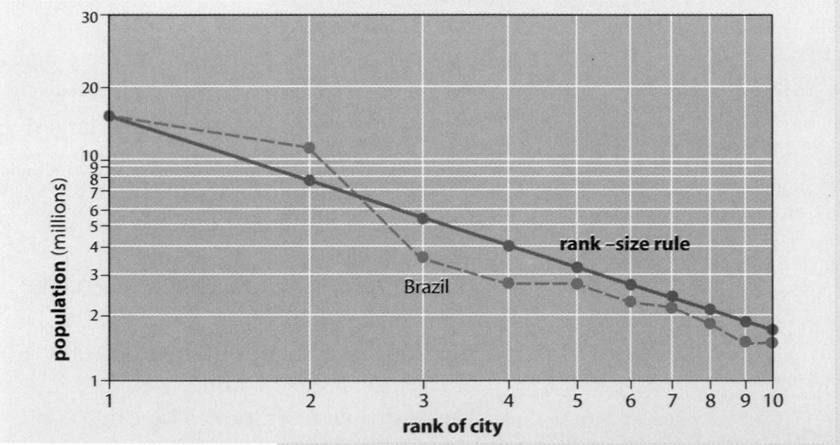

Brazil 1990 (thousands)			
Rank	City	Actual	Estimated
1	São Paulo	16832	
2	Rio de Janeiro	11141	8416
3	Belo Horizonte	3446	5611
4	Recife	2945	4208
5	Porto Alegre	2924	3366
6	Salvador	2362	2805
7	Fortaleza	2169	2405
8	Curitiba	1926	2104
9	Brasilia	1597	1870
10	Nova Iguaçu	1325	1683

1 For each country, work out the estimated population based on the rank—size rule, and then plot it on a logarithmic graph.

2 Describe the resultant relationship between actual city size and the estimated rank size. Which, if any, of the five countries appears to be an anomaly?

Rank	USA 1988	Actual	Italy 1987	Actual	Argentina 1988	Actual	Japan (1990)	Actual
1	New York	18120	Rome	2817	Buenos Aires	11126	Tokyo	11936
2	Los Angeles	13770	Milan	1464	Cordoba	1134	Yokohama	3220
3	Chicago	8181	Naples	1203	Rosario	1071	Osaka	2624
4	San Francisco	6042	Turin	1012	Mendoza	707	Nagoya	2155
5	Philadelphia	5963	Palermo	731	La Plata	630	Sapporo	1672
6	Detroit	4620	Genoa	715	Tucuman	603	Kobe	1477
7	Dallas	3776	Bologna	422	Mar del Plata	504	Kyoto	1461
8	Boston	3736	Florence	417	San Juan	346	Fukuoka	1237
9	Washington	3734	Catania	370	Santa Fe	330	Kawasaki	1174
10	Houston	3642	Bari	357	Salta	328	Hiroshima	1086

Figure 14.25

Size, spacing and functions of settlements

Central place	Population	Distance apart (km)	Sphere of influence (km²)	Functions (services)
Hamlet	10–20	2	—	probably none
Village	1 000	7	45	church, post office, shop, junior school
Small town	20 000	21	415	shops, churches, senior school, bank, doctor
Large town	100 000	35	1 200	shopping centre, small hospital, banks, senior schools
City	500 000	100	12 000	shopping complex, cathedral, large hospital, football team, large bus and rail station, cinemas, theatre
Conurbation	1 million	200	35 000	shopping complexes, several CBDs
Capital or primate city	several million	—	whole country	government offices, all other functions

Notes: the distances and service areas have been taken from Christaller's work in southern Germany (1933) with, in some cases, a rounding-off of figures for simplicity. The population figures and functions are more applicable to the UK and the present time. Populations, distances and service areas vary between and within countries and should be taken as comparative and approximate rather than absolute. All places in the hierarchy have all the services of the settlements below them.

Central place theory

A **central place** is a settlement which provides goods and services. It may vary in size from a small village to a conurbation or primate city (Figures 14.25 and 14.26) and forms a link in a hierarchy. The area around each settlement which comes under its economic, social and political influence is referred to as its **sphere of influence, urban field** or **hinterland**. The extent of the sphere of influence will depend upon the spacing, size and functions of the surrounding central places.

Functional hierarchies

Four generalisations may be made regarding the spacing, size and functions of settlements.

1 The larger the settlements are in size, the fewer in number they will be — i.e. there are many small villages, but relatively few large cities.
2 The larger the settlements grow in size, the greater the distance between them — i.e. villages are usually found close together, while cities are spaced much further apart.
3 As a settlement increases in size, the range and number of its functions will increase (Figure 14.26).
4 As a settlement increases in size, the number of higher-order services will also increase — i.e. a greater degree of specialisation occurs in the services (Figure 14.25).

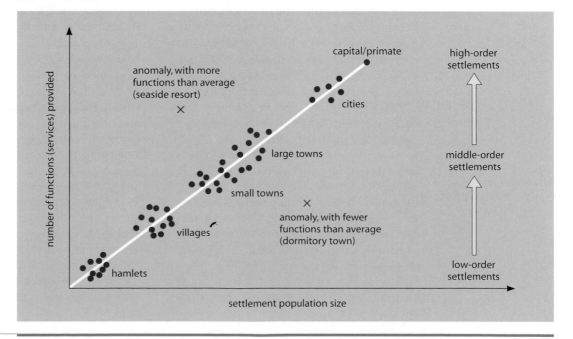

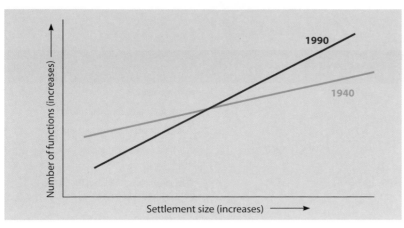

Figure 14.27

Relationship between the number of functions and settlement size in the UK, 1940 and 1990

Central place functions are activities, mainly within the tertiary sector, that market goods and services from central places for the benefit of local customers and clients drawn from a wider hinterland

The range and threshold of central place functions

The **range** of a good or service is the maximum distance that people are prepared to travel to obtain it. It is dependent upon the value of the good, the length of the journey, and the frequency that the service is needed. People are not prepared to travel as far to buy a newspaper (a low-order item), which they need daily, as they are to buy furniture (a high-order item), which they might purchase only once every several years. Low-order functions, such as corner shops and primary schools, need to be spaced closely together as people are less willing and less able to travel far to use them. High-order functions, such as regional shopping centres and hospitals, are likely to be widely spaced as people are more prepared to travel considerable distances to them (page 532).

The **threshold** of a good or service is the minimum number of people required to support it. It is assumed, incorrectly in practice, that people will always use the service located nearest to them (the nearest superstore). As a rule, the more specialised the service, the greater the number of people needed to make it profitable or viable. It has been suggested that, in the UK, about 300 people are necessary for a village shop, 500 for a primary school, 2500 for a doctor, 10 000 for a senior school or a small chemist's shop, 25 000 for a shoe shop, 50 000 for a small department store, 60 000 for large supermarket, 100 000 for a large department store, and over 1 million for a university. Services locate where they can maximise the number of people in their catchment area and maximise the distance from their nearest rival. Threshold analysis was used by planners of British New Towns who equated, for example, 20 000 people with a cinema, 10 000 people with a swimming pool and 100 000 people with a theatre.

Changes in population size and number of functions

Figure 14.27 shows that over the last 50 years in the UK there has been a decrease in the number of services available in small settlements and an increase in the number of functions provided by large settlements. This may be due to many factors — for example:

- Small villages are no longer able to support their former functions (village shop) as the greater wealth and mobility (car ownership) of some rural populations enable them to travel further to larger centres where they can obtain, in a single visit, both high- and low-order goods.
- Domestic changes (deep freezers, convenience foods) mean that rural householders need no longer make use of daily, low-order services previously available in their village.
- As larger settlements attract an increasingly larger threshold population, they can increase the variety and number of functions and, by reducing costs (supermarkets), are likely to attract even more customers.
- In areas experiencing rural depopulation, villages may no longer have a population large enough to maintain existing services.

Christaller's model of central places

Walter Christaller was a German who, in 1933, published a book in which he attempted to demonstrate a sense of order in the spacing and function of settlements. He suggested that there was a pattern in the distribution and location of settlements of different sizes and also in the ways in which they provided services to the inhabitants living within their sphere of influence. Regardless of the level of service provided, he termed each settlement a **central place**. Although Christaller's **central place theory** was based upon investigations in southern Germany, and it was not translated into English until 1966, his work has contributed a great deal to the search for order in the study of settlements.

The two principles underlying Christaller's theory were the **range** and the **threshold** of goods and services. He made a set of assumptions which were similar to those of two earlier German economists, von Thünen (agricultural land use model, page 430) and Weber (industrial location theory, page 507).

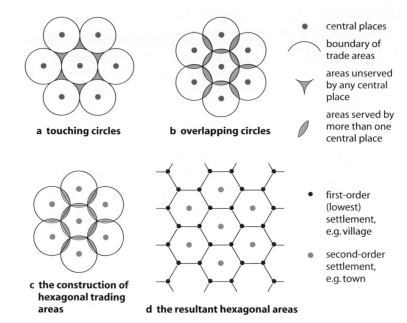

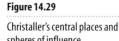

a touching circles

b overlapping circles

- central places
- boundary of trade areas
- areas unserved by any central place
- areas served by more than one central place

c the construction of hexagonal trading areas

d the resultant hexagonal areas

- first-order (lowest) settlement, e.g. village
- second-order settlement, e.g. town

Figure 14.28

Constructing spheres of influence around settlements (*after* Christaller)

These assumptions were that:

- There was unbounded flat land so that transport was equally easy and cheap in all directions. Transport costs were proportional to distance from the central place and there was only one form of transport.
- Population was evenly distributed across the plain.
- Resources were evenly distributed across the plain.
- Goods and services were always obtained from the nearest central place so as to minimise distance travelled.
- All customers had the same purchasing power (income) and made similar demands for goods.
- Some central places offered only low-order goods, for which people were not prepared to travel far, and so had a small sphere of influence. Other central places

offered higher-order goods, for which people would travel further, and so they had much larger spheres of influence. The higher-order central places provided both higher- and lower-order goods.

- No excess profit would be made by any one central place, and each would locate as far away as possible from a rival to maximise profits.

The ideal shape for the sphere of influence of a central place is circular, as then the distances from it to all points on the boundary are equal. If the circles touch at their circumferences, they leave gaps which are unserved by any central place (Figure 14.28a); if the circles are drawn so that there are no gaps, they necessarily overlap (Figure 14.28b) — which also violates the basic assumptions of the model. To overcome this problem, the overlapping circles are modified to become touching hexagons (Figure 14.28c). A hexagon is almost as efficient as a circle in terms of accessibility from all points of the plain and is considerably more efficient than a square or triangle (Figure 14.28d). A hexagonal pattern also produces the ideal shape for superimposing the trading areas of central places with different levels of function — the village, town and city of Christaller's hierarchy. Figure 14.29 shows a large trade area for a third-order central place, a smaller trade area for the six second-order central places, and even smaller trade areas for the 24 first-order central places.

By arranging the hexagons in different ways, Christaller was able to produce three different patterns of service or trading areas. He called these $k = 3$, $k = 4$ and $k = 7$, where k is the number of places dependent upon the next-highest-order central place.

The following should be noted at this point.

- Where $k = 3$, the trade area of the third-order (i.e. highest) central place is three times the area of the second-order central place, which in turn is three times larger than the trade area of the first-order (lowest) central place.
- Where $k = 4$, the trade area of the third-order central place is four times the area of the second-order central place, which is four times larger than the trade area of the first-order central place.
- Where $k = 7$, the trade area of each order is seven times greater than the order beneath it.

Figure 14.29

Christaller's central places and spheres of influence

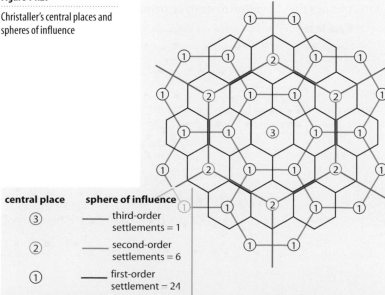

central place	sphere of influence
③	third-order settlements = 1
②	second-order settlements = 6
①	first-order settlement = 24

Figure 14.30

Christaller's *k* = 3

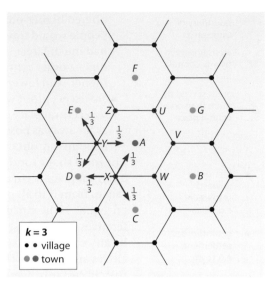

k = 3
• • village
● ● town

k = 3

The arrangement of the hexagons in this case is the same as given in Figure 14.29 and the explanation of how *k* = 3 is reached is shown in Figure 14.30, where:

A is the central place or third-order settlement;

B, C, D, E, F and *G* are 6 second-order settlements surrounding *A*; and

U, V, W, X, Y and *Z* are some of the 24 first-order settlements which lie between *A* and the second-order settlements.

It is assumed that one-third of the inhabitants of *Y* will go to *A* to shop, one-third to *D* and one-third to *E*. Similarly, one-third of people living at *X* will shop at *A*, one-third at *D* and one-third at *C*. This means that *A* will take one-third of the customers from each of *U,V,W, X, Y* and *Z* (6 x 1/3 = 2) plus all of its own customers (1). In total, *A* therefore serves the equivalent of three central places (2 + 1).

Christaller based the *k* = 3 pattern on a **marketing principle** which maximises the number of central places and thus brings the supply of higher-order goods and services as close as possible to all the dependent settlements and therefore to the inhabitants of the trade area.

k = 4

In this case, the size of the hexagon is slightly larger and it has been re-orientated (Figure 14.31). The first-order settlements, again labelled *U, V, W, X, Y* and *Z*, are now located at the mid-points of the sides of the hexagon instead of at the apexes as in *k* = 3. Customers from *Y* now have a choice of only two markets, *A* and *N*, and it is assumed that half of those customers will go to *A* and half to *N*. Similarly, half of the customers from *X* will go to *A* and the other half to *M*. *A* will therefore take half of the customers from each of the six settlements at *U, V, W, X, Y* and *Z* (6 x ½ = 3) plus all of its own customers (1) to serve the equivalent of four central places (3 + 1). This pattern is based on a **traffic principle**, whereby travel between two centres is made as easy and as cheap as possible. The central places are located so that the maximum number may lie on routes between the larger settlements.

k = 7

Here the pattern shows the same high-order central place, *A*, but all the lower-order settlements, *U,V,W,X, Y* and *Z*, lie within the hexagon or trade area (Figure 14.32). In this case, all of the customers from the six smaller settlements will go to *A* (6 x 1 = 6), together with all of the inhabitants of *A* (1). This means that *A* serves seven central places (6 + 1). As this system makes it efficient to organise or control several places, and as the loyalties of the inhabitants of the lower-order settlements to a higher one are not divided, it is referred to as the **administrative principle**.

Figure 14.31

Christaller's *k* = 4

Figure 14.32

Christaller's *k* = 7

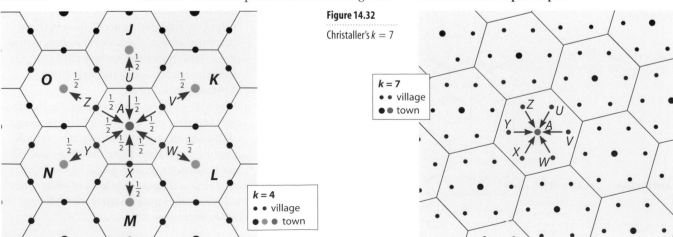

k = 4
• • village
● ● ● town

k = 7
• • village
● ● town

Why, with the possible exception of the reclaimed Dutch polders, can no perfect example of Christaller's model be found in the real world? The answer lies mainly in the basic assumptions the model.

- Large areas of flat land rarely exist and the presence of relief barriers or routes along valleys means that transport is channelled in certain directions. There is more than one form of transport; costs are not proportional to distance; and both systems and types of transport have changed since Christaller's day.
- People and wealth are not evenly distributed.
- People do not always go to the nearest central place — for example, they may choose to travel much further to a new edge-of-city hypermarket.
- People do not all have the same purchasing power, or needs.
- Governments often have control over the location of industry and of new towns.
- Perfect competition is unreal and some firms make greater profits than others.
- Christaller saw each central place as having a particular function whereas, in reality, places may have several functions which can change over time.
- The model does not seem to fit industrial areas, although there is some correlation with flat farming areas in East Anglia, the Netherlands and the Canadian Prairies.

Christaller has, however, provided us with an objective model with which we can test the real world. His theories have helped geographers and planners to locate new services such as retail outlets and roads.

Newton's law of gravity
Any two bodies attract one another with a force that is proportional to the product of their masses and inversely proportional to the square of the distance between them.

Reilly's law
Two centres attract trade from intermediate places in direct proportion to the size of the centres and in inverse proportion to the square of the distances from the two centres to the intermediate place.

Interaction or gravity models

These models, derived from Newton's law of gravity, seek to predict the degree of interaction between two places. When used geographically, the words 'bodies' and 'masses' are replaced by 'towns' and 'population' respectively.

The interaction model in geography therefore is based upon the idea that as the size of one or both of the towns increases, there will also be an increase in movement between them. The further apart the two towns are, however, the less will be the movement between them. This phenomenon is known as **distance decay**.

This model can be used to estimate
1 traffic flows (page 380);

2 migration between two areas;
3 the number of people likely to use one central place, e.g. a shopping area, in preference to a rival central place.

It can also be used to determine the sphere of influence of each central place by estimating where the **breaking point** between two settlements will be — i.e. the point at which customers find it preferable, because of distance, time and expense considerations, to travel to one centre rather than the other.

Reilly's law of retail gravitation (1931)
Reilly's interaction breaking-point is a method used to draw boundary lines showing the limits of the trading areas of two adjacent towns or shopping centres. Unlike Christaller, Reilly suggested that there were no fixed trade areas, that these areas could vary in size and shape, and that they could overlap.

This can be expressed by the formula:

$$Db = \frac{Dab}{1 + \sqrt{\dfrac{Pa}{Pb}}}$$

or similarly

$$djk = \frac{dij}{1 + \sqrt{\dfrac{Pi}{Pj}}}$$

where:

Db (or djk) = the breaking-point between towns A and B;
Dab (or dij) = the distance (or time) between towns A and B;
Pa (or Pi) = the population of town A (the larger town); and
Pb (or Pj) = the population of town B (the smaller town).

Taking as an example Grimsby–Cleethorpes which has a population of 131 000 and Lincoln, 71 km away, with a population of 75 000, the formula can be written as:

$$Db = \frac{71}{1 + \sqrt{\dfrac{131\,000}{75\,000}}}$$

which means that

$$Db = \frac{71}{1 + 1.32}$$
$$\therefore Db = 30.58$$

Thus the breaking-point is 30.58 km from Lincoln (town B) and 40.42 km from Grimsby–Cleethorpes (town A). This is shown in Figure 14.33.

Figure 14.33

Reilly's breaking-point
between settlements of
different sizes

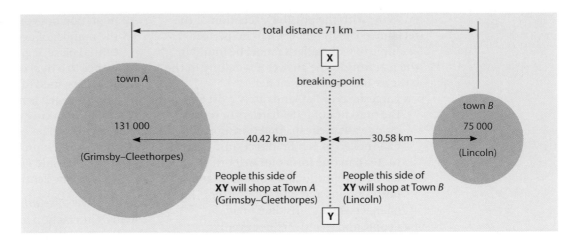

Q Scunthorpe has a population of 71 000 and is 59 km from Grimsby–Cleethorpes and 56 km from Lincoln. What is the breaking-point between Scunthorpe and each of the other two settlements?

Limitations of Reilly's model

As with other models, Reilly's model is based on assumptions which are not always applicable to the real world. In this case, the assumptions are that:

- the larger the town, the stronger its attraction; and that
- people shop in a logical way, seeking the centre which is nearest to them in terms of time and distance.

These assumptions may not always be true. For example,

- there may be traffic congestion on the way to the larger town and, once there, car parking may be more difficult and expensive;
- the smaller town may have fewer but better quality shops;
- the smaller centre may be cleaner, more modern, safer and less congested; and
- the smaller town may advertise its services more effectively.

A variation on Reilly's law of retail gravitation

Like central place theory, Reilly's law seems to fit rural areas better than closely packed, densely populated urban areas. One of several variations on Reilly's law of retail gravitation is based upon the drawing power of shopping centres (i.e. the number and type of shops in each) rather than distance between the two towns. (Other variations include retail floorspace and retail sales.)

The version based on the drawing power of shopping centres has the formula:

$$Db = \frac{Dab}{1 + \sqrt{\dfrac{Sa}{Sb}}}$$

where:

Sa = the number of shops in town A; and
Sb = the number of shops in town B.

Referring to our original example, suppose Grimsby–Cleethorpes has 800 shops and Lincoln has 300 shops. The formula could then be written:

$$Db = \frac{71}{1 + \sqrt{\dfrac{800}{300}}}$$

$$\therefore Db = 27$$

This means that out of every 71 shoppers, 44 would go to Grimsby–Cleethorpes and 27 to Lincoln.

Measuring settlement patterns: conclusion

Nearest neighbour analysis, the rank–size rule, Christaller's central place theory and the interaction models are all difficult to observe in the real world. Their value lies in the fact that they form hypotheses against which reality can be tested — provided you do not seek to *make* reality fit them! (Framework 7, page 285). Also, they offer objective methods of measuring differences between real-world places. When theory and reality diverge, the geographer can search for an explanation for the differences. An important shared characteristic of these approaches is that they aim to find order in spatial distributions.

Case Study 14

The evolution of settlement in Britain

When Britain's first census was taken in 1801, almost 80 per cent of the population still lived in hamlets and villages. (The corresponding figure in the 1991 census was 7 per cent.) Most people have their own mental image of a 'traditional' hamlet, village, or market town. However, in reality, the development of rural settlement has been so dynamic and complex that, due to differences in site, form (morphology) and function, there is no such thing as a typical rural settlement (Figure 14.34) — nor is there a 'typical' urban settlement.

Iron Age settlements

Palaeolithic Man left behind flint tools, but few marks on the landscape. The first people to alter their natural surroundings were those of the Neolithic period, the Bronze Age and the Iron Age (Figure 11.18). They began, despite limited technology, to clear woodlands and to leave a legacy of stone circles, tumuli, barrows, hillforts (Figure 14.35) and settlement sites. The hillfort built on the volcanic sill at Drumadoon (Figure 1.33) had a fine panorama of an enemy approaching from the sea while the steep

Figure 14.34

Villages in the British landscape

There is a tendency to think of country life as stable, conservative and unchanging but this is far from the truth. Settlements, like the people who live in them, are mortal. There is, however, no recognisable expected life-span, and a village can survive for twenty or two thousand years depending on its ability to adapt to changing economic and social conditions. In addition to extant village communities there are in Britain thousands of former occupation sites which have been abandoned.

Rural settlement in the past reflected the ever-changing relationship between man and his environment. Human society is never completely static and the settlements which serve it can never remain absolutely still for very long; and before a well-balance form of settlement becomes generally established, new forces will be at work altering that form. The forces which created our hamlets and villages have involved factors as varied as the pace of technological change, the nature of local authority, inheritance customs, the presence of arable or pasture, and the availability of building materials. Village history tells a story of fluctuating expansion, decline and movement. Sometimes reflecting national factors such as pestilence, economic changes and social development and sometimes purely local events, such as the silting up of a river estuary or the bankruptcy of a local entrepreneur. Such factors have combined to give each village a unique history and plan.

(T. Rowley, 1978)

cliffs prevented any frontal attack. Hillforts may, however, have only been settled during times of attack as few had a guaranteed water supply. Not all Iron Age settlements were hillforts; some forts were located in lowland areas, while other settlements may have had a religious or market function as opposed to a military one.

Figure 14.35

Maiden Castle hillfort, Dorset, England

Romano-British settlements

While the Romans preferred to live in well planned towns or in large rural villas, it is clear that at the same time many nucleated villages existed in lowland Britain, many of which showed evidence of Roman influence by having well planned streets. One characteristic feature of Romano-British villages was the presence of small-scale industrial activity — usually pottery production and iron-working.

Anglo-Saxon settlements

Although many English village and town names have Anglo-Saxon origins, it does not prove that they existed during those times. Most Anglo-Saxon settlements were sited in clearings in the natural forest, on 'islands' in marshy areas or near to the coast. Archaeological evidence suggests that most settlements were likely to have consisted of several farms grouped together to form self-contained hamlets. The houses, or rather huts, were rectangular in shape and built from local materials -— wood for the frame from the forest, mud and wattle (interlaced twigs and branches) for the walls from the river and forest, and thatch for the roof from local reeds or straw left over after the harvest. The huts, which were shared with the animals in winter, may have been protected by a stone or wooden wind-break. It was only by late Anglo-Saxon times that larger nucleated villages, with their open fields worked in strips by a heavy plough drawn by oxen, became more commonplace.

Medieval settlements

By medieval times, each village was dominated by a large farm, or manor, house in which the lord of the manor lived. The village would have contained several peasant cottages, built with materials similar to those of Anglo-Saxon homes, a church, a house for the priest, a blacksmith's forge and a mill. Surrounding the village were (usually) three large open fields — open because they had neither hedges nor fences as boundaries. Each field was divided into numerous, long, narrow strips, shared between the peasants. Two of the fields were likely to be growing cereals such as wheat, barley and rye (mainly for bread), while the third was left fallow (allowed to rest). The crops were rotated so that each field was left fallow every third year — the three-field system of crop rotation. When the fields were ploughed, a ridge was formed about 0.3 m above an adjacent furrow. Over many years of ploughing, the ridges built up so that they can still be recognised in our present-day landscape (Figure 14.36).

In the scarp-and-vale areas of south-east England (page 183), the villages were often close together along the spring lines. The parish boundaries were laid out between each village and parallel to each other, so that each individual parish had a long, narrow strip of land extending across the clay vale and over the chalk escarpment (Figure 14.6). This allowed each parish to be self-contained by having a permanent water supply together with land suitable for both rearing animals and growing crops. Although individual parishes no longer need be self-supporting, the old boundaries still remain.

Figure 14.36

Ridge and furrow, south-east Leicestershire

By medieval times, a network of small market towns had also developed (see Christaller's $k = 3$; page 378). Many of these towns incorporate Roman (Chester) or Anglo-Saxon (Much Wenlock) features, indicating their earlier origin as settlements. Figure 14.37 shows an 1810 map of the City of Carlisle on which many recurring elements of the medieval town can still be clearly seen.

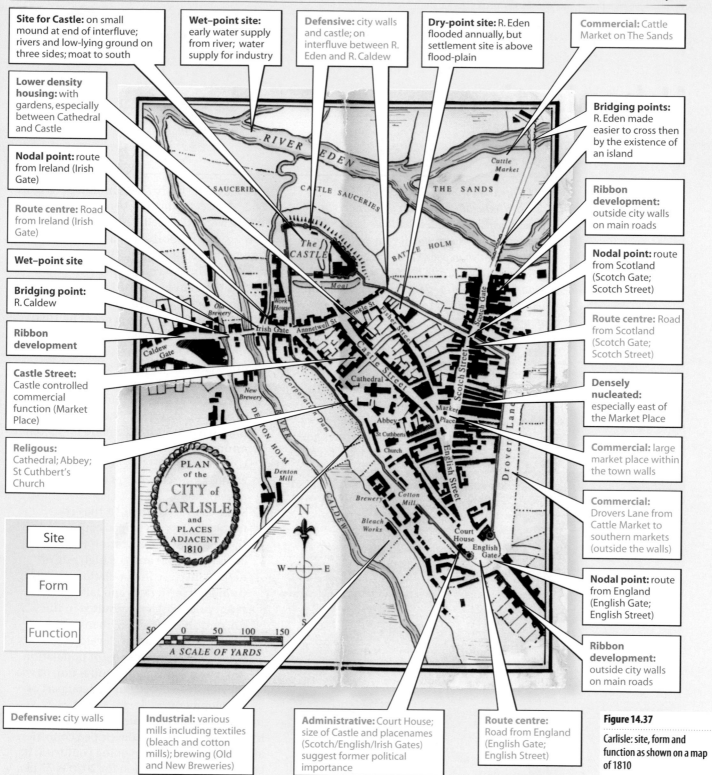

Site for Castle: on small mound at end of interfluve; rivers and low-lying ground on three sides; moat to south

Wet–point site: early water supply from river; water supply for industry

Defensive: city walls and castle; on interfluve between R. Eden and R. Caldew

Dry-point site: R. Eden flooded annually, but settlement site is above flood-plain

Commercial: Cattle Market on The Sands

Lower density housing: with gardens, especially between Cathedral and Castle

Nodal point: route from Ireland (Irish Gate)

Route centre: Road from Ireland (Irish Gate)

Wet–point site

Bridging point: R. Caldew

Ribbon development

Castle Street: Castle controlled commercial function (Market Place)

Religous: Cathedral; Abbey; St Cuthbert's Church

Bridging points: R. Eden made easier to cross then by the existence of an island

Ribbon development: outside city walls on main roads

Nodal point: route from Scotland (Scotch Gate; Scotch Street)

Route centre: Road from Scotland (Scotch Gate; Scotch Street)

Densely nucleated: especially east of the Market Place

Commercial: large market place within the town walls

Commercial: Drovers Lane from Cattle Market to southern markets (outside the walls)

Nodal point: route from England (English Gate; English Street)

Ribbon development: outside city walls on main roads

Site

Form

Function

PLAN of the CITY of CARLISLE and PLACES ADJACENT 1810

A SCALE OF YARDS

Defensive: city walls

Industrial: various mills including textiles (bleach and cotton mills); brewing (Old and New Breweries)

Administrative: Court House; size of Castle and placenames (Scotch/English/Irish Gates) suggest former political importance

Route centre: Road from England (English Gate; English Street)

Figure 14.37

Carlisle: site, form and function as shown on a map of 1810

References

Bradford, M. G. and Kent, W. A. (1977) *Human Geography: Theories and Their Applications*. Oxford University Press.

Briggs, K. (1982) *Human Geography: Concepts and Applications*. Hodder & Stoughton.

Hornby, W. F. and Jones, M. (1991) *Settlement Geography*. Cambridge University Press.

Philips Geographical Digest 1992-93 (1993) Heinemann-Philips.

Roberts, B. (1987) *The Making of the English Village*. Longman.

Rowley, T. (1978) *Villages in the English Landscape*. J. M. Dent & Son.

Wilson, J. (1984) *Statistics in Geography for 'A' level Students*. Schofield & Sims.

Chapter 15

Urbanisation

"The invasion from the countryside . . . is overwhelming the ability of city planners and governments to provide affordable land, water, sanitation, transport, building materials and food for the urban poor. Cities such as Bangkok, Bogota, Bombay, Cairo, Delhi, Lagos and Manila each have over one million people living in illegally developed squatter settlements or shanty towns."

L. Timberlake, *Only One Earth*, 1987

Figure 15.1

The proportion of world population living in urban areas

Urbanisation is defined as the process whereby an increasing proportion of the world's, a nation's or a region's population lives in urban areas. There is not, however, any global agreement as to what constitutes an 'urban area'. Urbanisation began at least as far back as the fourth millennium BC (Figure 14.2). However, the number of people living in urban areas formed, until fairly recently, only a small proportion of a country's or a region's total population. One estimate claims that in 1800 only 3 per cent of the world's population were urban dwellers — a figure which had risen, according to the latest UN estimate, to 45 per cent by 1990 and could reach, before 2025, 60 per cent (Figure 15.1). Rapid urbanisation has occurred twice in time and space.

1 During the 19th century, in what are now referred to as the economically more developed countries, where industrialisation led to a huge demand for labour in mining and manufacturing centres. Urbanisation was, in these parts of the world, a consequence of economic development.

2 Since the 1950s when, in the economically less developed countries, the twin processes of migration from rural areas (page 342) and the high rate of natural increase in population (resulting from high birth rates and falling death rates, page 326) have resulted in the uncontrolled growth of many cities. Urbanisation is, in the developing countries, a consequence of population movement and growth and is not, as was previously believed, an integral part of development.

In 1990, the UN estimated that 73 per cent lived in urban areas in developed countries, and 37 per cent in developing countries. Its corresponding prediction for 2020 is 77 per cent and 53 per cent (Figures 15.1 and 15.2). Figure 15.3 shows national levels of urbanisation in 1990.

Simultaneous with urbanisation has been the growth of very large cities. Whereas there were only two cities with a population exceeding 1 million in 1900 (London and Paris), there were 70 by 1950 and 295 in 1990

	Urban population (percentage)			
Area	**1950**	**1970**	**1990**	**2020 (estimate)**
World	29.2	37.1	45.2	57.4
More developed regions	53.8	66.6	73.0	77.2
Less developed regions	17.0	25.4	37.1	53.1
Europe and CIS	56.3	66.7	73.4	76.7
North America	63.9	73.8	74.3	78.9
Oceania	61.3	70.8	71.3	75.1
Latin America	41.0	57.4	75.1	83.0
Asia (excluding CIS)	16.4	24.1	28.2	49.3
Africa	15.7	22.5	33.9	52.2

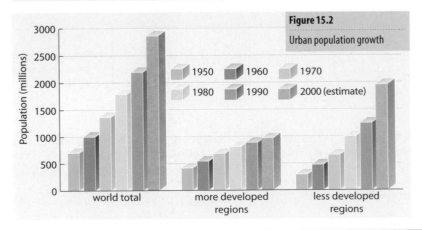

Figure 15.2

Urban population growth

1950 1960 1970
1980 1990 2000 (estimate)

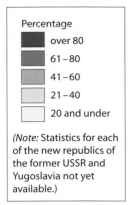

Percentage

- over 80
- 61–80
- 41–60
- 21–40
- 20 and under

(Note: Statistics for each of the new republics of the former USSR and Yugoslavia not yet available.)

Figure 15.3

National levels of urbanisation, 1990

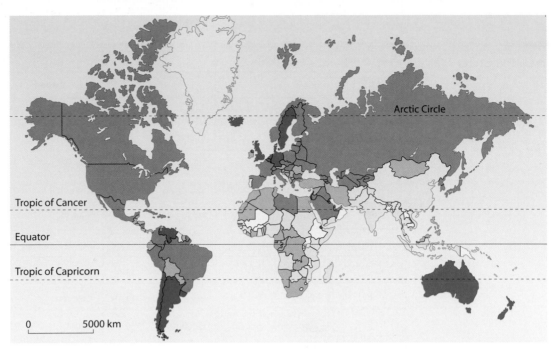

- over 5 million
- over 3 million
- over 1 million

Figure 15.4

Distribution of world cities with populations over 1 million, 1990

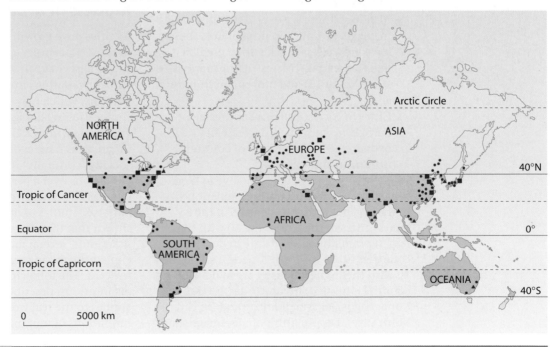

(according to Philip's *Geographical Digest*). The number of cities exceeding 5 million (31) and 10 million (8) is also increasing.

Also changing is the spatial distribution of these 'million cities' (Figure 15.4). Prior to 1940, the majority of cities with over 1 million inhabitants were found in the temperate latitudes of the northern hemisphere — i.e. in developed regions. Since 1950, there has been a dramatic increase in the proportion of 'million cities' found within the tropics and in other developing regions. Whereas in 1950 71 per cent of these cities lay north of 40°N, by 1990 this figure had fallen to 26 per cent and the mean latitude of these largest cities had changed from 39°20'N to 31°43'N. Figure 15.5 gives the rank order of the world's largest cities. By 2000, according to estimates, the two largest cities will be in Latin America and nine of the top ten will be in Latin America and Asia. Figure 15.5 also gives two different 'estimates' for 2000 (Framework 12, page 411). The uncertainty in giving present figures and predicting future ones can result from the use of different criteria to define the size of an urban area; the unreliability of the census data; differences in the interpretations of estimated natural increase changes (made annually between each 10-year census); and difficulties in obtaining accurate migration figures.

Figure 15.5

The world's largest cities

Rank order	1970		1985		Year 2000			
Figures in millions					**Estimate 1**		**Estimate 2**	
1	New York	16.5	Tokyo	23.0	Mexico City	31.0	Mexico City	26.0
2	Tokyo	13.4	Mexico City	18.7	São Paulo	25.8	São Paulo	24.1
3	London	10.5	New York	18.2	Tokyo	23.7	Tokyo	17.0
4	Shanghai	10.0	São Paulo	16.8	Shanghai	23.7	Calcutta	16.6
5	Mexico City	8.6	Shanghai	13.3	New York	23.4	Bombay	16.0
6	Los Angeles	8.4	Los Angeles	12.8	Rio de Janeiro	19.0	New York	15.5
7	Buenos Aires	8.4	Buenos Aires	11.6	Bombay	16.8	Shanghai	13.8
8	Paris	8.4	Rio de Janeiro	11.1	Calcutta	16.4	Rio do Janeiro	13.5
9	São Paulo	7.1	Calcutta	9.2	Seoul	13.7	Seoul	13.54
10	Moscow	7.1	Bombay	8.2	Delhi	11.8	Delhi	13.3

Latin America	Europe	North America	Asia

There have been several noticeable trends in the growth of the largest cities since the mid 1980s.

- The fastest-growing cities are in Latin America and in the Japan–Korea region.
- Most of the fastest-growing cities are located in less developed countries, with cities in the least developed countries often having the greatest increase.
- The rate of growth in most of the developing regions was less rapid than had previously been predicted.
- Although there are over 70 Chinese cities in excess of 1 million inhabitants, these have grown less quickly than predicted due to state policies restricting family size (Case Study 13) and movement from rural areas.
- Cities in North America and western Europe are showing a decrease in overall size — the process of **counter-urbanisation** (page 342).

What affects most people who live in large urban areas is not the actual population size of the city but rather its density. Of the 85 largest cities in the world, the 10 with the lowest population density are in developed countries (9 are in North America) while the 10 with the highest density are in developing countries (headed, in rank order, by Hong Kong, Lagos, Jakarta, Bombay and Ho Chi Minh City). Despite this popular image that the largest cities in developing countries are growing so rapidly, it should be remembered that over one-third of the inhabitants, especially in India and China, still live in smaller towns of 20 000–100 000 inhabitants. **Over-urbanisation** occurs when migrants are driven from rural areas to large cities where slow economic growth does not allow the provision of sufficient jobs or shelter (page 409).

Models of urban structure

As cities have grown in area and population in the 20th century, geographers and sociologists have tried to identify and to explain variations in spatial patterns. Spatial patterns, which may show differences and similarities in land use and/or social groupings within a city, reflect how various urban areas have evolved economically and socially (culturally) in response to changing conditions over a period of time. While each city has its own distinctive pattern, or patterns, studies of other urban areas have shown that they too exhibit similar patterns. As a result, several models describing and explaining urban structure have been put forward. The following section lists the basic assumptions, describes the theory behind, applies to the real world, and gives the limitations of four urban models.

Urbanisation

1 Burgess, 1924

Burgess attempted to identify areas within Chicago based upon the outward expansion of the city and the socio-economic groupings of its inhabitants (Places 39).

Basic assumptions
Although the main aim of his model was to describe residential structures of a city, geographers have subsequently presumed that Burgess made certain assumptions. These included:

■ The city was built on flat land which therefore gave equal advantages in all directions, i.e. morphological features such as river valleys were removed.
■ Transport systems were of limited significance being equally easy, rapid and cheap in every direction.
■ Land values were highest in the centre of the city and declined rapidly outwards to give a zoning of urban functions and land use.
■ The oldest buildings were in, or close to, the city centre. Buildings became progressively newer towards the city boundary.
■ Cities contained a variety of well-defined socio-economic and ethnic areas.

■ The poorer classes had to live near to the city centre and places of work as they could not afford transport or expensive housing.
■ There were no concentrations of heavy industry.

Burgess's concentric zones
The resultant model (Figure 15.6) shows five concentric zones:

1 The **central business district** (CBD) contains the major shops and offices; it is the centre for commerce and entertainment, and the focus for transport routes.
2 The **transition** or **twilight zone** is where the oldest housing is either deteriorating into slum property or being 'invaded' by light industry. The inhabitants tend to be of poorer social groups and first-generation immigrants.
3 Areas of **low-class housing** are occupied by those who have 'escaped' from zone 2, or by second-generation immigrants who work in nearby factories. They are compelled to live near to their place of work to reduce travelling costs and rent. In modern Britain, these zones are equated with the inner cities.
4 **Medium-class housing** of higher quality which, in present-day Britain, would include inter-war private semi-detached houses and council estates.
5 **High-class housing** occupied by people who can afford the expensive properties and the high cost of commuting. This zone also includes the commuter (suburbanised) villages beyond the city boundary — although there were very few of these when Burgess produced his model in 1924.

Limitations
Urban models, like all models (Framework 9, page 328), have limitations and are therefore open to criticism. Despite the advantage of simplicity, and his own admission that it was specific to one place (Chicago, Places 39) and one period of time (the 1920s), the Burgess model has been criticised (Figure 15.16) — in some instances on grounds which did not exist when it was put forward.

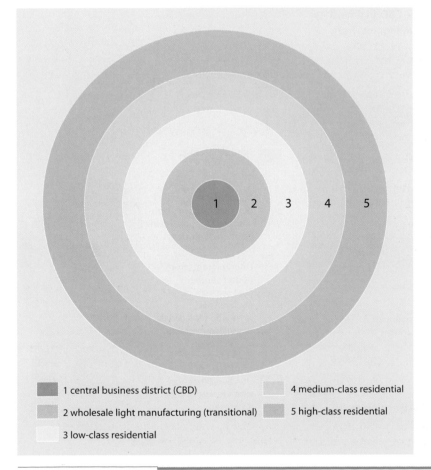

1 central business district (CBD)
2 wholesale light manufacturing (transitional)
3 low-class residential
4 medium-class residential
5 high-class residential

Figure 15.6

The Burgess concentric model

Burgess, in producing his model, was influenced by the emerging science of plant ecology at the University of Chicago. He made analogies with such ecological processes as the **invasion** of an area by competing groups, **competition** between the invaders and the natural groups, and the eventual **dominance** of the area by the invaders which allowed them to **succeed** the natural groups.

Relating this to urban geography, Burgess suggested that people living in the inner zone were **invaded** by newcomers and, in face of this **competition** by immigrants who became **dominant** there, **succeeded** to the next outer zone — a process also referred to as **centrifugal movement.** The energy to maintain this dynamic system came from a continual supply of immigrants to the centre, and existing groups being forced (or choosing) to move towards the periphery.

Chicago lies on the shores of Lake Michigan with its CBD, known as the 'Loop', facing the lake. Surrounding the CBD, the city's housing developed a distinctive pattern (Figure 15.7). The initial migrants, from north-western Europe, settled around the CBD. In time, they were replaced by newer immigrants from southern Europe (especially Italy) and by Jews who were, in turn, replaced by blacks from the American south (Figure 15.8). This led to the creation of a series of income, social and ethnic zones radiating outwards from the centre. These zones showed

1 That wealth, as seen by the quality of housing, increased towards the outskirts of the city. People with the highest incomes lived in the newest property (on the north-west fringe) while those with the lowest incomes occupied the poorest housing next to the CBD.

2 That people in their early twenties or over 60 tended to live close to the CBD, while middle-aged people and families with young children tended to live nearer to the city boundary.

3 That areas of ethnic segregation existed, with the early white immigrants, whose wealth had tended to increase in relation to the length of time they had lived in the city, living towards the outskirts and non-white groups living nearer to the city centre — e.g. in China Town and the black belt.

Figure 15.7

Urban areas of Chicago (*after* Burgess)

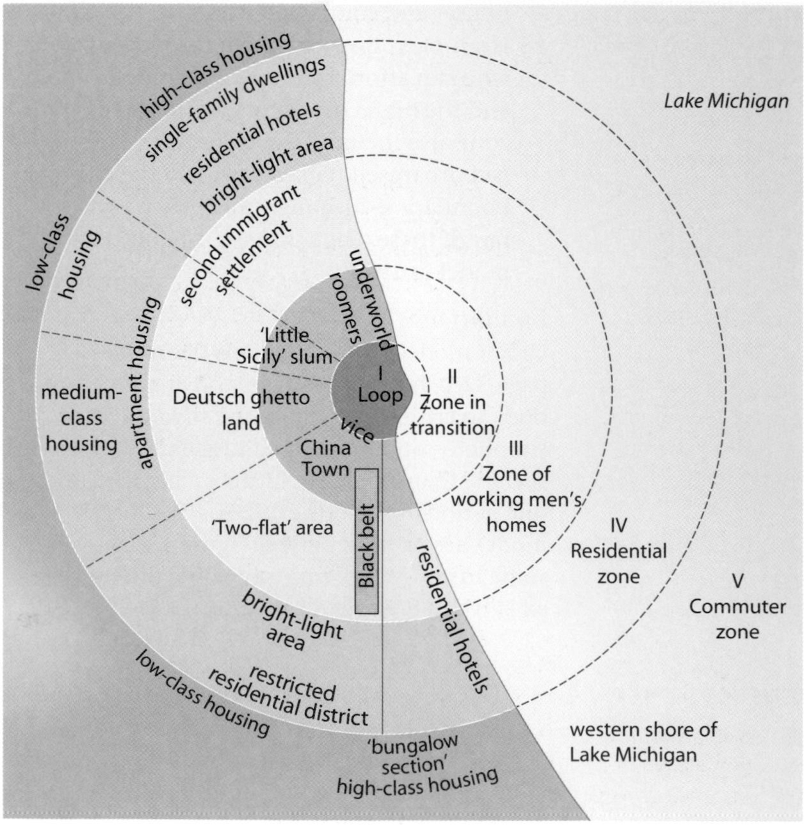

Figure 15.8

Centrifugal movement in Chicago

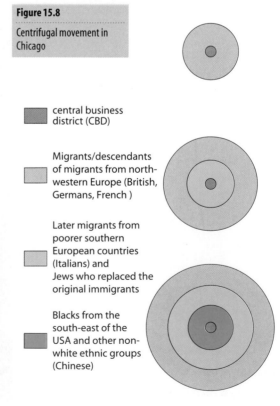

- central business district (CBD)

- Migrants/descendants of migrants from north-western Europe (British, Germans, French)

- Later migrants from poorer southern European countries (Italians) and Jews who replaced the original immigrants

- Blacks from the south-east of the USA and other non-white ethnic groups (Chinese)

high-class housing
single-family dwellings
residential hotels
bright-light area
second immigrant settlement
low-class housing
apartment housing
'Little Sicily' slum
underworld roomers
Deutsch ghetto land
medium-class housing
China Town
vice
Black belt
'Two-flat' area
bright-light area
restricted residential district
low-class residential housing
'bungalow section' high-class housing
I Loop
II Zone in transition
III Zone of working men's homes
IV Residential zone
V Commuter zone
residential hotels
Lake Michigan
western shore of Lake Michigan

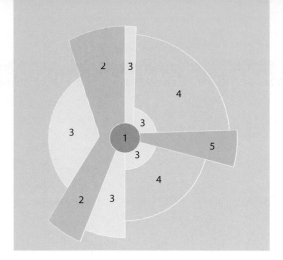

1	CBD (central business district)
2	wholesale light manufacturing (transitional)
3	low-class residential
4	medium-class residential
5	high-class residential

Figure 15.9

The Hoyt sector model

Figure 15.10

Urban areas of Calgary, 1961 (*after* Hoyt)

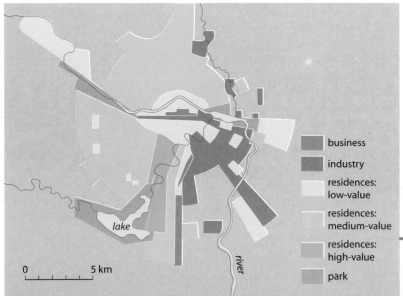

	business
	industry
	residences: low-value
	residences: medium-value
	residences: high-value
	park

lake

river

0   5 km

2 Hoyt, 1939

Hoyt's model was based on the mapping of eight housing variables for 142 cities in the USA. He tried to account for changes in, and the distribution of, residential patterns.

Basic assumptions

Hoyt made the same implicit assumptions as had Burgess, with the addition of three new factors:

- Wealthy people, who could afford the highest rates, chose the best sites, i.e. competition based on 'ability to pay' resolved land use conflicts.
- Wealthy residents could afford private cars or public transport and so lived further from industry and nearer to main roads.
- Similar land uses attracted other similar land uses, concentrating a function in a particular area and repelling others. This process led to a 'sector' development.

Hoyt's sector model

Hoyt suggested that areas of highest rent tended to be alongside main lines of communication and that the city grew in a series of wedges (Figure 15.9). He also claimed that once an area had developed a distinctive land use, or function, it tended to retain that land use as the city extended outwards, e.g. if an

area north of the CBD was one of low-class housing in the 19th century, then the northern suburbs of the late 20th century would also be likely to consist of low-class estates. Calgary, in Canada, is the standard example of a city exhibiting the characteristics of Hoyt's model (Figure 15.10).

Limitations

Many criticisms are similar to those made of Burgess's model and have been summarised in Figure 15.16. It should be remembered that this model was put forward before the rapid growth of the car-based suburb, the swallowing-up of small villages by urban growth, the redevelopment of inner-city areas and the relocation of shopping, industry and office accommodation on edge-of-city sites.

3 Mann, 1965

Mann tried to apply the Burgess and Hoyt models to three industrial towns in England: Huddersfield, Nottingham and Sheffield. His compromise model (Figure 15.11) combined the ideas of Burgess's concentric zones and Hoyt's sectors. Mann assumed that because the prevailing winds blow from the southwest, the high-class housing would be in the south-western part of the city and industry, with its smoke (this was before Clean Air Acts), would be located to the north-east of the CBD. His conclusions can be summarised as follows.

- The twilight zone was not concentric to the CBD but lay to one side of the city which allowed, elsewhere, more wealthy residential areas.
- Heavy industry was found in sectors along main lines of communication.
- Low-class housing should be called the 'zone of older housing' (age-based classification, rather than social).
- Higher-class or, in Hoyt's terms, 'modern' housing was usually found away from industry and smoke.
- Local government (political) played a role in slum clearance and gentrification. This led to large council estates which took the working class/low incomes to the city edge (opposite of the Burgess model).

Robson (1975) applied Mann's model to a north-eastern industrial town, Sunderland (Figure 15.12), and to Belfast. Mann's model does show, despite its small sample, that a variety of approaches are possible to the study of urban structures.

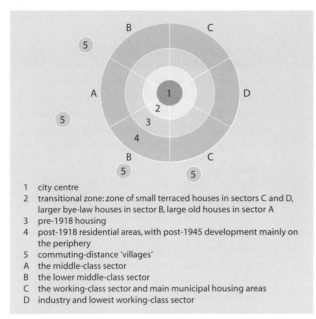

1 city centre
2 transitional zone: zone of small terraced houses in sectors C and D, larger bye-law houses in sector B, large old houses in sector A
3 pre-1918 housing
4 post-1918 residential areas, with post-1945 development mainly on the periphery
5 commuting-distance 'villages'
A the middle-class sector
B the lower middle-class sector
C the working-class sector and main municipal housing areas
D industry and lowest working-class sector

Figure 15.11

Mann's model of urban structure

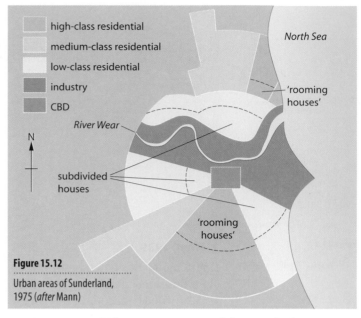

Figure 15.12

Urban areas of Sunderland, 1975 (*after* Mann)

4 Ullman and Harris, 1945

Ullman and Harris set out to produce a more realistic model than those of Burgess and Hoyt but consequently ended with one that was more complex (Figure 15.13) — and more complex models may become descriptive rather than predictive if they match reality too closely in a specific example (Framework 9, page 328).

Basic assumptions

■ Modern cities have a more complex structure than that suggested by Burgess and Hoyt.

■ Cities do not grow from one CBD, but from several independent nuclei.

■ Each nucleus acts as a growth point, and probably has a function different from other nuclei within that city. (In London, the City is financial; Westminster is government and administration; the West End is retailing and entertainment; and Dockland was industrial.)

■ In time, there will be an outward growth from each nucleus until they merge as one large urban centre (Barnet and Croydon now form part of Greater London; Figure 13.6).

■ If the city becomes too large and congested, some functions may be dispersed to new nuclei. (In Greater London, edge-of-city retailing takes place at Brent Cross and new industry has developed close to Heathrow Airport/M25/M4.)

Multiple nuclei developed as a response to the need for maximum accessibility to a centre, to keep certain types of land use apart, for differences in land values and, more recently, to decentralise (Places 40).

Urban structure models: conclusions

The four models described were put forward to try to explain differences in structure within cities in the developed world. It must be remembered that

a Each model will have its limitations (Figure 15.16).

b If you make a study of your local town or city, you must avoid the temptation of saying that it *fits* one of the models — at best it will show characteristics of one or possibly two. Each city is unique and will have its own structure — a pattern not necessarily derived according to any existing model (Framework 9, page 328).

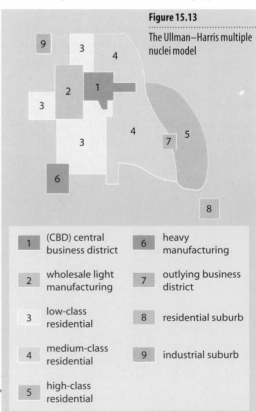

Figure 15.13

The Ullman–Harris multiple nuclei model

1	(CBD) central business district	6	heavy manufacturing
2	wholesale light manufacturing	7	outlying business district
3	low-class residential	8	residential suburb
4	medium-class residential	9	industrial suburb
5	high-class residential		

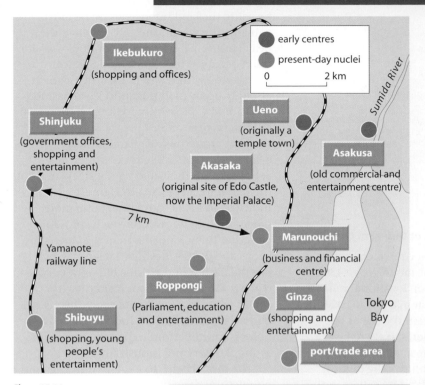

early centres

present-day nuclei

0   2 km

Ikebukuro
(shopping and offices)

Sumida River

Shinjuku
(government offices, shopping and entertainment)

Ueno
(originally a temple town)

Asakusa
(old commercial and entertainment centre)

Akasaka
(original site of Edo Castle, now the Imperial Palace)

7 km

Yamanote railway line

Marunouchi
(business and financial centre)

Roppongi
(Parliament, education and entertainment)

Ginza
(shopping and entertainment)

Tokyo Bay

Shibuyu
(shopping, young people's entertainment)

port/trade area

Figure 15.14

Multiple nuclei in Tokyo

Tokyo began to grow in the late 16th century around the castle of the Edo Shogunate (near the present Imperial Palace, Figure 15.14). Later religious, cultural and financial districts developed to the north-east. Over the centuries, the mainly wooden-built city was destroyed several times, including during the 1923 Kanto earthquake (140 000 deaths) and by US aircraft in 1945. The modern city has no single CBD but, rather, has several nuclei each with its own specialist land use and functions — government offices, shopping, finance, entertainment, education and transport (Figure 15.15). Most of these nuclei are linked by one of Tokyo's many railways which forms a circle with a diameter of 7 km.

Figure 15.15

Tokyo
a the Shinjuko district
b the Ginza district

Figure 15.16

Limitations/criticisms of the four urban models

	Burgess	Hoyt	Mann	Ullman–Harris
1	zones, in reality, are never as clear-cut as shown on each model			
2	each zone usually contains more than one type of land use/housing			
3	no consideration of characteristics of cities outside USA and north-west Europe			
	based on 1 USA city	based on 142 USA cities	based on 3 English cities (in north and Midlands)	based on cities in economically more developed world
4	redevelopment schemes and modern edge-of-city developments are not included (most of the models pre-date these developments)			
5	based mainly on housing: other types of land use neglected		industry not always to north-east of British cities	
6	cities not always built upon flat plains			
7	tended to ignore transport			

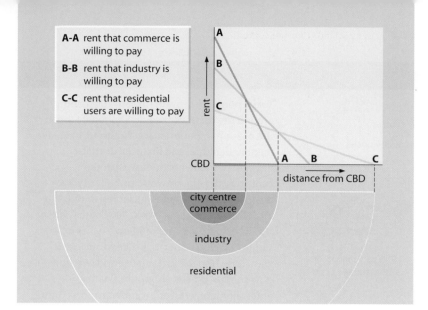

Figure 15.17

Bid–rent curves

The land value model or bid–rent theory

This model is the urban equivalent of von Thünen's rural land use model (page 430) in that both are based upon locational rent. The main assumption is that in a free market the highest bidder will obtain the use of the land. The highest bidder is likely to be the one who can obtain the maximum profit from that site and so can pay the highest rent. Competition for land is keenest in the city centre. Figure 15.17 shows the locational rent that three different land users are prepared to pay for land at various distances from the city centre.

The most expensive or 'prime' sites in most cities are in the CBD, mainly because of its accessibility and the shortage of space there. Shops, especially department stores, conduct their business using a relatively small amount of groundspace and due to their high rate of sales and turnover they can

bid a high price for the land (for which they try to compensate by building upwards and by using the land intensively). The most valuable site within the CBD is called the peak land value intersection or PLVI — a site often occupied by a Marks and Spencer store! Competing with retailers are offices which also rely upon good transport systems and, traditionally, proximity to other commercial buildings (page 532).

Away from the CBD, land rapidly becomes less attractive for commercial activities — as indicated by the steep angle of the bid–rent curve (**AA**) in Figure 15.17. Industry, partly because it takes up more space and uses it less intensively, bids for land that is less valuable than that prized by shops and offices. Residential land, which has the flattest of the three bid–rent curves (**CC**), is found further out from the city centre where the land values have decreased due to less competition. Individual house-holders cannot afford to pay the same rents as shopkeepers and industrialists.

The model helps to explain housing (and population) density. People who cannot afford to commute have to live near to the CBD where, due to higher land values, they can only obtain small plots which results in high housing densities. People who can afford to commute are able live nearer the city boundary where, due to lower land values, they can buy much larger plots of land which creates areas of low housing densities. Figure 15.18 shows the predicted land use pattern when land values decrease rapidly and at a constant rate from the city centre. The resultant pattern is similar to that suggested by Burgess (Figure 15.6).

One basis of this model is 'the more accessible the site, the higher its land value'. Rents will therefore be greater along main routes leading out of the city and along outer ring roads. Where two of these routes cross, there may be a secondary or subsidiary land value peak (Figure 15.19). Here the land use is likely to be a small suburban shopping parade or a small industrial estate. The 'retail revolution' of the 1980s (page 533), which led to the development of large edge-of-city shopping complexes (MetroCentre in Gateshead, Places 71, page 534; and Brent Cross in north London), has altered this pattern. Similarly, large industrial estates and science parks (Places 66, page 516) have been located near to motorway interchanges.

Figure 15.18

Urban land use patterns based on land values

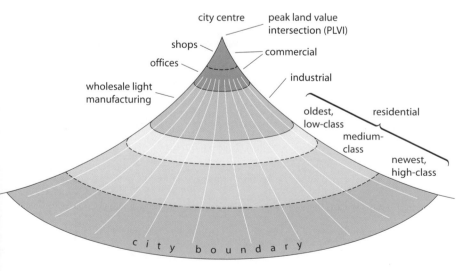

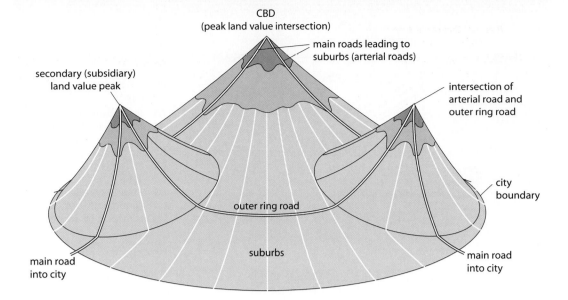

Functional zones within a city

Different parts of a city usually have their own specific functions (Figure 15.14). These functions may depend upon:

- the age of the area: buildings usually get older towards the city centre except that most CBDs and many old inner-city areas have been redeveloped and modernised;
- land values: these increase rapidly from the city boundary in towards the CBD (Figure 15.17);
- accessibility: some functions are more dependent upon transport than others.

While each urban area will have its own unique pattern of functional zones and land use, most British cities exhibit similar characteristics. These characteristics have been summarised and simplified in Figure 15.21 where:

Zone A = the CBD (shops and offices).

Zone B = Old inner city (including, before redevelopment, 19th-century/low-cost/low-class housing, industry and warehousing).

Zone C = Inter-war (medium-class housing).

Zone D = Suburbs (modern/high-cost/high-class housing, open space, new industrial estates/science and business parks, shopping complexes and office blocks).

Figure 15.20
....

Land use in a British city: an idealised transect from the CBD to the city boundary

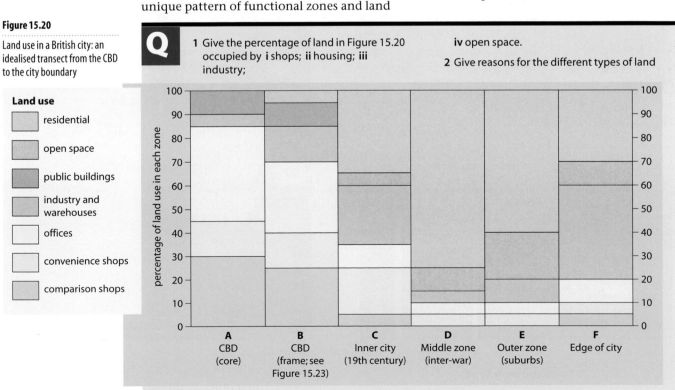

Q

1 Give the percentage of land in Figure 15.20 occupied by **i** shops; **ii** housing; **iii** industry;   **iv** open space.

2 Give reasons for the different types of land

Land use

- residential
- open space
- public buildings
- industry and warehouses
- offices
- convenience shops
- comparison shops

A CBD

Figure 15.21

Functional zones in a
British city

A1 Indoor shopping mall (St Enoch's Centre, Glasgow)
A2 High-rise office development (The City of London)

B1 An inner-city corner shop (Leicester)
B2 Nineteenth-century terraced housing (Newcastle upon Tyne)
B3 Inner-city redevelopment (Sheffield)
B4 Nineteenth-century industry and transport (Manchester)

B Inner city

C Inter-war areas

D Edge of city

1   Shopping types

2   Residential styles

C1 A suburban shopping parade
C2 Inter-war semi-detached private housing (Leeds)
C3 Inter-war council housing estate
C4 Public open space

D1 Edge-of-city shopping complex (Lakeside shopping centre, Dartford)
D2 A modern private housing estate (Wirral)
D3 Post-war edge-of-city council housing estate (North Paddington, London)
D4 Business/science park (Guildford)

2 Residential styles

3 Other land uses

The central business district (CBD)

The CBD is regarded as the centre for retailing, office location and service activities (banking and finance). It contains the principal commercial streets and main public buildings and forms the **core** of a city's business and commercial activities. Some large cities, such as London and Tokyo (Figure 15.14), may have more than one CBD and may have experienced a decentralisation of offices (page 536) and a relocation of retailing (page 534).

The delimitation of the CBD

This is a practical exercise which might be carried out in your local town. Several methods have been suggested, based on the pioneer work of Murphy and Vance in North America, as no single approach appears to be completely satisfactory.

What are the main characteristics of the CBD?

- It contains the major retailing outlets. The principal department stores and specialist shops with the highest turnover and requiring largest threshold populations compete for the prime sites.
- It contains a high proportion of the city's main offices (Figure 15.21A2).
- It contains the tallest buildings in the city (more typical in North America) mainly due to the high rents which result from the competition for land (Figure 15.17).
- It has the greatest number and concentration of pedestrians.
- It has the greatest volume and concentration of traffic. The city centre grew at the meeting point of the major lines of communication into the city and therefore had the greatest accessibility. Much of present-day planning aims to limit the number of vehicles entering this zone.
- It has the highest land values in the city, including the PLVI (Figure 15.18).
- It is constantly undergoing change. New shopping centres, taller office blocks and new traffic schemes seem to be announced daily. Some of the grandiose schemes of the early 1960s are now viewed as out of date and unattractive (Birmingham's Bull Ring; London's Paternoster Square at St Paul's). Many have since been demolished and rebuilt. Recent studies have shown the CBD of many cities to be advancing in some directions (**zone of assimilation**) and retreating in

others (**zone of discard**). The zone of assimilation is usually towards the higher-status residential districts whereas the zone of discard tends to be nearer the industrial and poorer-quality residential areas (Figure 15.23). There is also a significant trend for retailing to be static, or even declining (due to competition from out-of-town developments), while offices, banks and insurance are increasing in terms of space taken and income generated.

Mapping the characteristics of the CBD

Taking the seven characteristics listed above, the following methods might be used in fieldwork study.

- **Land use mapping of shops**
1 Plot the location of all the shops. Where the ratio of shops to other properties is more than 1:3, count that area as being within the CBD. This straightforward exercise is based upon the (UK) Census of Information definition that over 33 per cent of buildings in the CBD are connected with retailing.
2 An alternative method is to include within the CBD all shops which are within 100 m (or any distance which you judge to be suitable) of adjacent shops. This may produce a central 'core' and several smaller groupings.
3 A third possibility might be to take the mean frontage (in metres) of, for example, the middle five buildings or shop units in a block. Shop frontages are likely to be greatest near to the PLVI where most department stores are located.

- **Land use mapping of offices**
The first method described above could be repeated using offices instead of shops, and a ratio of 1:10. This recognises that, at ground floor level, offices are less numerous than shops. Banks and building societies should be included in this count.

- **Height of buildings**
Another simple technique is to plot the height of individual buildings, or the mean of a group of buildings in the centre of a block. Most cities tend to have a sharp decline in building height at the edge of the CBD.

- **Number of pedestrians**
This needs to be a group activity — the more groups the better! Each group counts the number of pedestrians passing a given point at a given time (e.g. 1100–1105 hours). The

greater the number of sites (ideally chosen by using random numbers, Framework 4, page 145), the greater the accuracy of the survey. Define a pedestrian as someone of school age and over, walking into, out of or past a shop on your side of the street. These criteria may be altered as long as they are applied by all the groups.

- **Accessibility to traffic** This is similar to the previous survey except that here vehicles are counted. Make sure all class groups have the same definition of a vehicle, e.g. do you include a bicycle and/or pram?
- **Land values** These might be expected to decline outwards at a uniform rate in which case there would be no 'natural breaks' by which to delimit the area. Providing rateable values can be obtained (try the rates office) and there is the time to process them (or a sample of the total), this is often a good indicator of the CBD. To overcome problems of variations in the floor space of different buildings (a

Marks and Spencer store will cover more space than a shoe shop), you will need to work out the rateable index (*RI*).

$$RI = \frac{\text{gross rateable value of the property}}{\text{ground floor area of the property}}$$

It may be useful to take the PLVI point and call this 100 per cent, and then convert the rateable index for all other properties as a percentage of the PLVI. It has been suggested than a figure of 5 per cent delimits the CBD for a North American urban area and 20 per cent for a British city.

- **Changing land use and functions** This is a mapwork exercise using an old map of the central area (shopping maps are produced by GOAD plans) and superimposing onto it present-day land uses. Look for evidence of zones of assimilation and discard.
- **Central business index** This is probably the best method as it involves a combination of land use characteristics, building height and land values. The problem is in obtaining the necessary data, i.e.
1 the total floor area of all central or CBD functions;
2 the total ground floor area (central and non-central functions);
3 the total floor area (upstairs floor area as well as the ground floor).

You may laboriously work this out from a large-scale plan, or choose to compromise by taking the mean of a sample of buildings in each block. From these data, two indices can be derived:

The **central business height index**, or **CBHI**, which is expressed as:

$$CBHI = \frac{\text{total floor area of all CBD functions}}{\text{total ground floor area}}$$

The **central business intensity index**, or **CBII**, which is expressed as:

$$CBII = \frac{\text{total floor area of all CBD functions}}{\text{total floor area}} \times \frac{100}{1}$$

To be considered part of the CBD, the CBHI of a plot should be over 1.0 and the CBII over 50 per cent (Figure 15.22). Some studies identify the central core of the CBD as having a CBHI over 2.0 and a CBII over 70 per cent.

Figure 15.22

The central business index

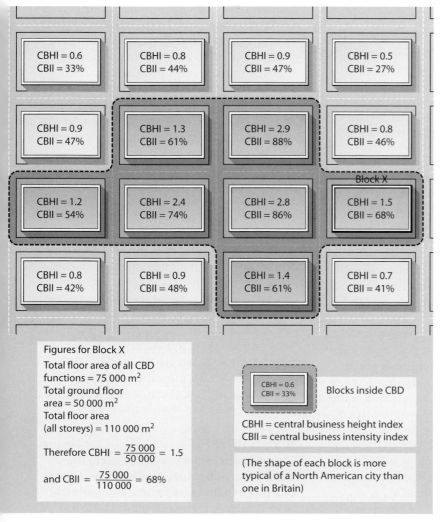

Figures for Block X
Total floor area of all CBD functions = 75 000 m²
Total ground floor area = 50 000 m²
Total floor area (all storeys) = 110 000 m²

Therefore $CBHI = \frac{75\,000}{50\,000} = 1.5$

and $CBII = \frac{75\,000}{110\,000} = 68\%$

CBHI = 0.6
CBII = 33%   Blocks inside CBD

CBHI = central business height index
CBII = central business intensity index

(The shape of each block is more typical of a North American city than one in Britain)

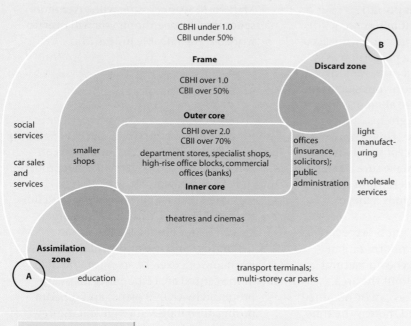

CBHI under 1.0
CBII under 50%

Frame

CBHI over 1.0
CBII over 50%

Outer core

CBHI over 2.0
CBII over 70%

department stores, specialist shops,
high-rise office blocks, commercial
offices (banks)

Inner core

theatres and cinemas

Discard zone

offices
(insurance,
solicitors);
public
administration

light
manufact-
uring

wholesale
services

social
services

car sales
and
services

smaller
shops

**Assimilation
zone**

A

education

transport terminals;
multi-storey car parks

B

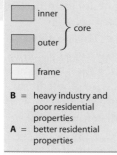

inner ⎱ core
outer ⎰

frame

B = heavy industry and
poor residential
properties
A = better residential
properties

Figure 15.23

The core and frame concept for
the CBD

Plotting the data

Careful consideration should be given as to
which cartographic technique is best applied
to each set of collected data. You may wish to
use one or several of the following: land use
maps, isolines, choropleths, flow graphs,
histograms, bar graphs, scattergraphs and
transects. Alternatively, you may be able to
devise a technique of your own. You may
save time and produce results which are
easier to compare by using tracing overlays
and/or a computer.

Delimiting the CBD: conclusions

If you have carried out your own survey,
your report might include comments on the
following questions:

1 What problems did you encounter in
   collecting and refining the data?
2 Which of the methods used in collecting
   the data appeared to give the most, and
   the least, accurate delimitation of the
   CBD?
3 In your town, was there an obvious CBD;
   did you find an inner and an outer core
   (Figure 15.23)? Was there evidence of
   zones of assimilation and discard? Were
   there any specific functional zones other
   than shops and offices? Was the area of
   the CBD similar to your mental map
   (your preconceived picture) of its limits?
4 What refinements would you make to the
   techniques used if you had to repeat this
   task in another urban area?

Industrial zones

Industry within urban areas has changed its
location over time. In the early 19th century,
it was usually sited within city centres, e.g.
textile firms, slaughter houses and food
processing. However, as the Industrial
Revolution saw the growth in size and
number of factories, and later when shops
began to compete for space in the city centre,
industry moved centrifugally outwards into
what today is the inner city (Places 39).
Inner-city areas could provide the large
quantity of unskilled labour needed for textile mills, steelworks and heavy engineering.
The land was cheaper and had not yet been
built upon. Factories were also located next
to main lines of communications: originally,
rivers and canals, then railways and finally
roads (Figure 15.21, B4). Firms including
bakeries, dairies, printing (newspapers) and
furniture, which have strong links with the
city centre, are still found here.

Between the 1950s and the 1980s this
zone increasingly suffered from industrial
decline as older, traditional industries closed
down and others moved to edge-of-city sites.
In Britain, recent changes in government
policy have led to attempts to regenerate
industry in these areas through initiatives
such as Enterprise Zones, derelict land grants
and urban development corporations (page
401). Even so, the replacement industries are
often on a small scale and compete for space
with warehouses and DIY shops.

Most modern industry is 'light' and clean
in comparison to that of the last century and
has moved to greenfield sites near to the city
boundary (Figure 15.21, D4). Trading estates
and modern business and science parks are
located on large areas of relatively cheap land
where firms have built new premises and use
up-to-date equipment (Places 66, page 516).
The internal road systems are purpose-built for
cars and lorries and are linked to nearby motorway interchanges and other main roads. The
wide range of skills and increasing demand for
female labour are satisfied by local housing
estates. Most of the industries are 'footloose'
and include hi-tech, electronics, assembling,
food processing and distributive firms.

Residential zones

The Industrial Revolution also led to the
rapid growth in urban population and the
outward expansion of towns. Long, straight
rows of terraced houses (Figure 15.21, B2)

were constructed as close as possible to the nearby factories where most of the occupants worked. The closeness was essential as neither private nor public transport was yet available. Houses and factories competed for space. As a result, houses were small, sometimes with only one room upstairs and one downstairs or they were built 'back to back'. The absence of gardens and public open space added to the high housing density.

By the 1950s, many of these inner-city areas, the low-class/low-income houses of the urban models, had become slums. Wholesale clearances saw large areas flattened by bulldozers and redeveloped with high-rise blocks of flats (Figure 15.21, B3). Within 20 years, the previously unforeseen social problems of these flats led to a change in policy. Under urban renewal, older housing was improved, rather than replaced, by adding bathrooms, kitchens, hot water and indoor toilets. More recently, some inner-city areas, especially those with

a waterfront location, have undergone a renaissance which has made them fashionable and expensive (Places 41, page 402).

The outward growth of the city continued both during the inter-war period when, aided by the development of private and public transport, large estates of semi-detached houses were built (the medium-class houses of the urban models, Figure 15.21, C3) and after the 1950s. Many of the present edge-of-city estates consist of low-density, private housing. Due to low land values (Figure 15.18), the houses are large in size, and have gardens and access to open space (Figure 15.21, D2). Other estates were created by local councils in an attempt to rehouse those people forced to move during the inner-city clearances. These estates, a mixture of high- and low-rise buildings (Figure 15.21, D3), have a high density and, like some older inner-city areas, are now experiencing extreme social and economic problems (page 403).

Framework 10   Values and attitudes

In the past, some teachers, perhaps because of their religious or political opinions, felt it necessary to pass on their own values to students, while the majority tended to avoid considering the role of values in geography in an attempt to remain neutral and non-doctrinaire. Today it is being accepted, perhaps among some with reservations, that in order to understand the character of places and the behaviour of people in relation to their environments, it is necessary to consider the motivations, values and emotions of those people involved.

The present author has tried, rightly or wrongly, to maintain a 'neutral' stance. Some would claim that what has been included in this book has been influenced by the author's own values and attitudes, e.g. a belief in the fundamental role of physical geography in an understanding of environmental problems, a preference for living in a semi-rural area rather than an inner city. Criticism could be levelled for using personal experiences as exemplars as **Places** and **Case Studies**. What the author has tried to do is to present readers with information in the hope that they may become more aware of their own values in relation to the behaviour of others and to enable them to discuss, with

fewer prejudices and preconceptions, the foundations of their own values.

This may be illustrated with reference to the following section on inner cities which is structured as follows:

1 **The problem of inner cities** Will these issues be seen differently by the inhabitant of an inner-city area and a person living in a rural environment?

2 **The image of an inner-city area** Will a description of inner-city problems give a negative picture of the quality of life in those environments and in doing so perpetuate the problems, or could it help in the understanding and tackling of them?

3 **Possible solutions to the inner-city problem** Would solutions proposed by inner-city residents be similar to those suggested and implemented by the government or the local authority?

4 **What successes have government schemes had?** Your answer to this may depend upon your own political views. Conservatives point out many achievements since 1980; Labour, the Liberal Democrats and other opposition parties claim that little has been done. Who, if either, is correct?

Issues in Britain's inner cities

The widest definition of an inner city is 'an area found in older cities, surrounding the CBD, where the prevailing economic, social and environmental conditions pose severe problems'. While much activity in inner cities may be positive, such as London's Notting Hill Carnival and the lively multi-cultural society of Brixton, and despite the increasing number which are experiencing a renaissance (Places 41), the more likely perception of such areas by non-residents is usually negative. To them, inner cities have an image of poverty, dirt, crime, overcrowding, unemployment, poor housing conditions and racial tension (Framework 11, page 404). While a description of these problems *may* present a negative picture (Framework 10, page 399) it is necessary to identify them if people are to begin to understand the difficulties and to offer workable solutions.

■ **Economic problems** Inner cities have long suffered from a lack of investment — especially after 1945 when much money was channelled into the 'new towns'. Traditional industries declined, while those remaining shed much of their labour force due to improved technology and falling demand (shipbuilding and textiles). A subsequent decline and shift in trade affected ports such as Liverpool and London. Relatively few new industries have sought an inner-city location partly due to the environment and partly because the former labour force often lacked the relevant skills needed in the new service and hi-tech industries.

■ **Social inequalities** Burgess, in his urban model, accepted that there were well-defined socio-economic and ethnic areas within a city (Figure 15.6). The segregation of different groups of people in British cities may result from differences in wealth, class, colour, religion, education and the quality of the environment. Much inner-city housing remains high density (although this should not automatically mean poor quality) and is pre-1914 in age. Relatively few occupants can afford to buy their homes and the houses may be in a poor state of repair. The small-area census statistics (1991) reveal the following characteristics.

1 A lack of basic amenities with over 1 million dwellings still lacking either a bathroom, WC or hot water.

2 Overcrowding as a result of either large families living in small houses or, where slum clearances have taken place, large numbers living in poorly built high-rise flats. (The census figures do not include the increasing number of homeless.)

3 Higher death and infant mortality rates, a lower life expectancy and a greater incidence of illness than in other residential environments.

4 A predominance of low-income, semi-skilled and manual workers.

5 A higher incidence of single-parent families, elderly people, children in care and free school meals than other districts.

6 Concentrations of ethnic minorities which may cause tension. The Scarman Report, following the 1981 riots, concluded that racial discrimination was a major issue.

■ **Environmental factors** Inner-city areas may suffer from noise and air pollution caused by heavy traffic and the few remaining factories; visual pollution in the form of derelict factories and houses, waste land, rubbish, vandalism and graffiti; and water pollution in rivers and old canals. Although some open space has been created by the clearance of slum property, there is likely to be a general absence of trees and grass.

Indicators of welfare and deprivation
Deprivation is defined by the Department of the Environment, as when 'an individual's well-being falls below a level generally regarded as a reasonable minimum for Britain today'. Estimates suggest that 5–6 million people live under such conditions (Figure 15.24). Holterman (1975) used 18 chosen indicators from the 1971 census figures. He noted that while Inner London ranked highest for sub-standard housing, Clydeside scored as the worst area for nearly all of the other indi-cators (which included aspects of education, health, employment, housing and delinquency). Smith (1987) used five indicators to highlight economic and social inequalities between whites and blacks in Atlanta, USA. These were wealth (mean family income); house prices (median prices of owner-occupied houses); overcrowding (houses with over 1 person per room); education (median of completed school years); and health (infant mortality rates)(Case Study 15b).

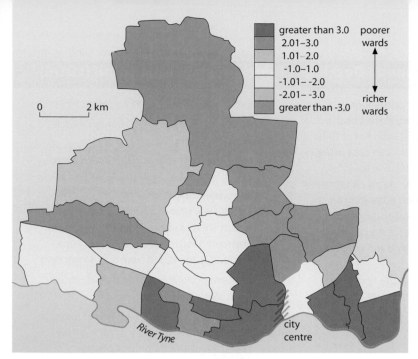

Figure 15.24

Deprivation index by wards, Newcastle upon Tyne, 1991

Legend:
- greater than 3.0 — poorer wards
- 2.01–3.0
- 1.01–2.0
- -1.0–1.0
- -1.01–-2.0
- -2.01–-3.0
- greater than -3.0 — richer wards

Government policies for the inner cities

When launching the 'Action for Cities' policy in March 1988, Mrs Thatcher said: "In partnership with the people and the private sector, we intend to step up the pace of renewal and regeneration to make our inner cities much better places in which to live, work and invest". The inner-city programme specified six main aims:

1 to enhance job prospects and the ability of residents to compete for them;
2 to bring land and buildings back into use;
3 to improve housing conditions;
4 to encourage private sector investment;
5 to encourage self-help and improve social fabric; and
6 to improve environmental quality.

It planned to achieve these aims through six main programmes (a seventh was added in 1991).

1 **The Urban Programme** This gives 75 per cent grants to 57 of the most needy local authorities who are trying to tackle underlying economic, social and environmental problems. Within the first year, 1989/90, over 9000 projects were funded.

2 **Derelict land grants** Reclamation schemes have included turning the redundant Edge Hill railway sidings (Liverpool) into Wavertree Technology Park; converting derelict mills, railway land and old housing at Listerhills (Bradford) into a £40 million warehouse and distribution centre; and the creation of National Garden Festivals (Liverpool 1984, Stoke-on-Trent 1986, Glasgow 1988, Gateshead 1990 and Ebbw Vale

1992). This has perhaps been the most successful scheme.

3 **Land registers** These list unused and underused land.

4 **Enterprise Zones (EZs)** EZs were created to try to stimulate economic activity by lifting certain tax burdens, e.g. exemption from paying rates for 10 years after their designation, 100 per cent grants for machinery and buildings, and relaxing or speeding up applications for planning permission (Figure 19.5).

5 **Grants for urban development** The City Grant bridges the gap between costs and value on completion, enabling investors to make a reasonable capital return.

6 **Urban Development Corporations (UDCs)** These have spearheaded the government's attempts to regenerate areas which contain substantial amounts of derelict, unused or underused land and buildings. They aim to encourage maximum private-sector investment and development. UDCs have powers to acquire, reclaim and service land and to restore buildings to effective use. They promote new industrial and housing developments and support community facilities. The first two, the London Docklands (LDDC, Places 41) and the Merseyside Development Corporations (MDC) were set up in 1981. Four more were created in 1986 (Trafford Park in Greater Manchester; on Teesside; in the West Midlands; and in Tyne and Wear), with three more designated in 1987 (Bristol, Leeds and Central Manchester) and one in 1988 (the Lower Don Valley, Sheffield).

7 **City Challenge** This was, in effect, to be a competition between 15 selected inner cities to see who could suggest the best 'high quality schemes aimed at revamping squalid housing and industrial estates and rebuilding derelict sites' and who could raise the most money from private enterprise (no more was to be made available by the government). The best schemes were put forward by 11 'Pacemaker' authorities: Barnsley/Doncaster (Dearne Valley), Bradford, Lewisham (Deptford), Liverpool, Manchester (Hulme), Middlesbrough, Newcastle upon Tyne, Nottingham, Tower Hamlets (Brick Lane), Wirral and Wolverhampton.

The object of an urban development corporation was, according to the Local Government Planning and Land Act of 1980, "to secure the regeneration of its area, by bringing land and buildings into effective use, encouraging the development of existing and new industry and commerce, creating an attractive environment and ensuring that housing and social facilities are available to encourage people to live and work in the area".

Success for the Government?

The LDDC published, in 1991, a booklet entitled 'A decade of success'. Among its many successes, it claimed

Physical regeneration

- 600 hectares of derelict land reclaimed and £155m spent on land acquisition;
- 90 km of new roads, the opening of the Dockland Light Railway (10 minutes to Central London), the London City Airport, and new bus routes and river bus;
- £142m expenditure on public utility services: gas, electricity, water, drainage.

Economic regeneration

- £1120m public-sector and £8420m private-sector investment;
- 2.5m sq m of commercial and industrial redevelopment completed or under construction with first stage of Canary Wharf opened in summer 1991;
- the number of businesses had doubled from 1100 in 1981 to 2300 in 1990, e.g. several newspaper corporations, the Stock Exchange, Limehouse ITV studios;
- 41 000 more jobs (17 000 newly created, 24 000 relocations);
- total employment nearly doubled from 27 000 in 1981 to 53 000 in 1990.

Social regeneration

- 15 200 new homes completed, 5300 homes refurbished or improved, £72m spent on improving local housing, private home ownership increased from 5 to 44 per cent;
- resident population more than doubled from 39 429 to 61 500;
- new shopping facilities at Surrey Quays, Hays Galleria and Tobacco Dock; a new Post-16 and a Technology College.

Environmental regeneration

- 100 000 trees planted, 130 hectares of green space and 17 conservation areas;
- 90 km of waterfront.

Failure for local people?

The scheme came too late to prevent many previous inhabitants from leaving the area due to slum clearances and the lack of job opportunities. Many of the new homes are extremely expensive ('Yuppy-land') which puts them beyond the price range of local people and, by attracting the better-off, has led to a sharp rise in the cost of living and a destruction of the close-knit 'Eastenders' community. Many of the new jobs are hi-tech and skilled, reducing opportunities for the relatively low-skilled local population. Residents complain at the lack of hospitals and services for the elderly.

Figure 15.25

Regeneration in the Isle of Dogs, London

Figure 15.26

Canary Wharf, Isle of Dogs, London

1991—95

The early 1990s, a time of international recession, badly affected the LDDC. The property crash, the failure of Olympia and York (which left large parts of Canary Wharf unoccupied), a growing criticism of the transport system, a fall in investment and a rise in unemployment suggested to many people that the LDDC achievements had come to an end. However, by 1995, indications of an upturn in the economy suggest that more firms and people might move into the area now that office rents and house prices have fallen and the Jubilee Line has been given its go-ahead. Only time will tell (perhaps by the time you come to read this section) if the initial scheme will be totally successful, partly successful, or become a huge white elephant.

Issues in Britain's edge-of-city council estates

Whereas much government policy and funding have been focused upon the inner cities, there is increasing evidence that poverty, unemployment and social stress may be even higher on edge-of-city council estates. These estates (Figure 15.21, D3), built on greenfield sites during the 1950s and 1960s, were created to house people forced to move by inner-city redevelopment schemes. Today:

- The physical fabric of the buildings, many originally built using cheap materials and methods, is deteriorating rapidly. Local councils, without the financial help given to the inner cities, are trying to upgrade selected estates as and when they can.
- Many estates include high-rise buildings which, as in the inner cities, have created feelings of isolation and stress-related illnesses. Flats and maisonettes have not proved popular under the Right-to-Buy schemes and have been too expensive for most of the occupants to consider buying.
- The low level of car ownership and high bus fares have increased the feeling of isolation from jobs, shops and entertainment.
- Levels of unemployment often exceed 30 per cent. There are also many low-income families; many elderly living on small pensions; and up to two-thirds of house-holds may be receiving housing benefit.
- The environmental quality of the estates is poor, often with a lack of open space.
- The estates tend to have high levels of marital problems, drug-taking, petty crime and vandalism and low levels of academic attainment and aspiration.

Figure 15.27

Social, economic and ethnic segregation in a city in the developed world

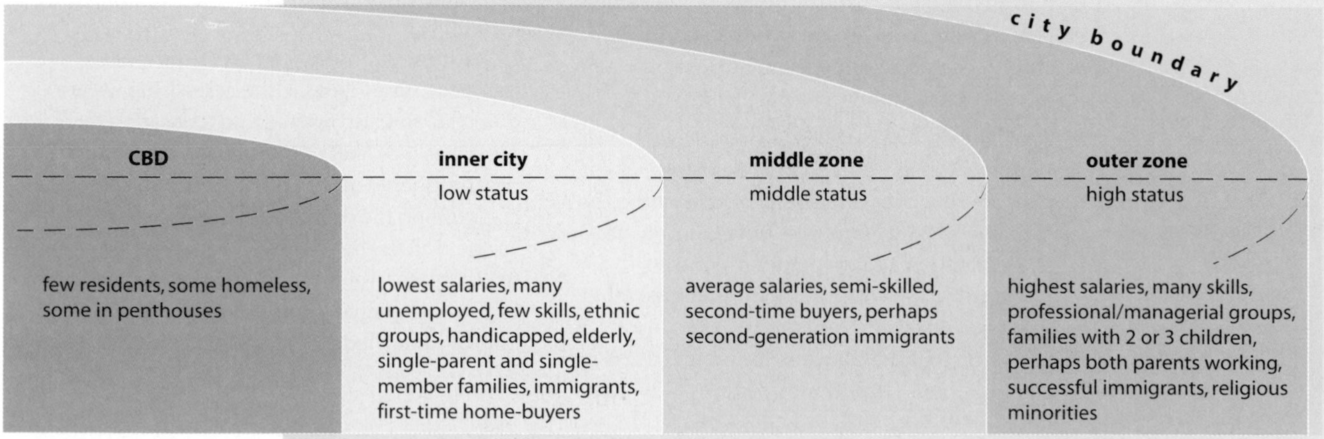

Q Figure 15.27 is a model showing the social, economic and ethnic groups living in a British city.

1 a Why do so few people live in the CBD?
  b Give reasons for the differences in the groups living in the inner city and those in the outer zone.
  c Under what conditions might people move from (i) the inner city to the outer zone; and (ii) the outer zone to the inner zone?

2 a What schemes have the British government introduced to try to deal with inner-city problems?
  b How successful do you think these schemes have been?

CBD	inner city — low status	middle zone — middle status	outer zone — high status
few residents, some homeless, some in penthouses	lowest salaries, many unemployed, few skills, ethnic groups, handicapped, elderly, single-parent and single-member families, immigrants, first-time home-buyers	average salaries, semi-skilled, second-time buyers, perhaps second-generation immigrants	highest salaries, many skills, professional/managerial groups, families with 2 or 3 children, perhaps both parents working, successful immigrants, religious minorities

city boundary

One of several dangers which may result from putting forward geographical models and from making generalisations is that of creating stereotypes. For example,

a Urban models suggest that all housing in inner-city areas is low-class/low-income (Burgess) and that all elderly and single-parent families live in this zone while families with two or three children only live in the outer zone (Figure 15.27).

b Different groups of people tend to develop their own customs and ways of life. By putting such characteristics together, we make mental pictures and develop preconceptions of different groups of people — i.e. we create stereotypes.

The following unsupported, emotive statements may not only be grossly inaccurate, they may also be considered, by many, to be offensive.

■ The Germans, on holiday, are always first to the swimming pool and dining room.

■ All Italians drive cars dangerously.

■ All Chinese and Japanese are small.

■ *Favelas* are shanty settlements whose inhabitants have no chance of improving their living conditions.

■ The inhabitants of a *favela* can only survive by a life of crime (see below).

■ The Amazon Amerindian way of life remains undeveloped as the people are lazy and unintelligent (see below).

The following accounts are based on the author's experience in Brazil in 1986.

Example One
"According to books which I had read in Britain and advice given to me by guides in São Paulo, *favelas* were to be avoided at all costs. Any stranger entering one was sure to lose his watch, jewellery and money and was likely to be a victim of physical violence.

With this in mind, I set off in a taxi to take photographs of several *favelas*. On reaching the first *favela*, to my horror the driver turned into the settlement and we bumped along an unmade track. He kept stopping and indicating that I should take photographs. Expecting at each stop that the car would be attacked and my camera stolen, I hastily took pictures — which

turned out to be over-exposed because, not daring to open windows, I took them through the windscreen and looking into the sun!

Suddenly the taxi spluttered and stopped. In one movement, I had hidden my camera and was outside trying to push the car. I raised my eyes to find three well-built males helping me to push the car. Which one would hit me first? I smiled and they smiled. I pointed to each one in turn and called him after one of Brazil's football players and then referred to myself as 'Lineker'. Huge smiles, big pats on the back and comments like '*Ingleesh amigo*' were only halted by the car re-starting. As we drove away, I began to question my original stereotyped view of a *favela* inhabitant."

Example Two
"I was surprised to find, on landing at Manaus airport in the middle of the Amazon rainforest, that our courier was an Amerindian. He dashed around quickly getting our party organised and our luggage collected. (He certainly did not seem to be slow or lazy.) He later admitted, and proved, that he could to speak in seven languages. (Hardly the sign of someone unintelligent — how many can you speak?). I asked him why so few Amerindians appeared to have good jobs and why he kept talking about returning to the jungle. His reply was simple: 'to avoid hassle'. He considered that the Indian lifestyle was preferable to the Western one with its quest for material possessions. Had he returned to the jungle, he would have rejoined his family and become a shifting cultivator living in harmony with the environment (Places 52, page 441). Is that traditional way of life really less demanding of intelligence than that imposed by invading timber and beefburger multinationals engaged in the destruction of large tracts of rainforest?"

From these examples, we can see how easy it is to accept stereotypes without realising we are doing so, and also how seeing a situation for ourselves may lead us to question our original picture. Should geographers take a role in overcoming the problems of stereotyped images (on the basis of which planning decisions, for example, may be made) by helping to provide relatively unbiased information to improve knowledge and understanding?

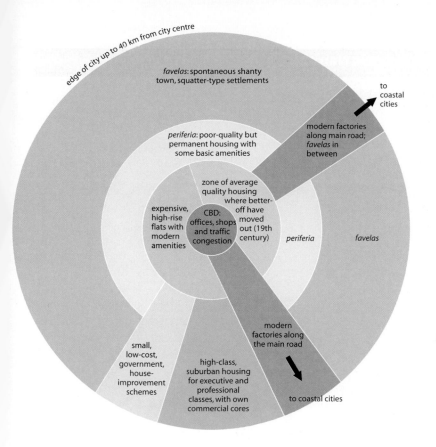

favelas: spontaneous shanty town, squatter-type settlements

periferia: poor-quality but permanent housing with some basic amenities

modern factories along main road; *favelas* in between

to coastal cities

zone of average quality housing where better-off have moved out (19th century)

expensive, high-rise flats with modern amenities

CBD: offices, shops and traffic congestion

periferia

favelas

small, low-cost, government, house-improvement schemes

high-class, suburban housing for executive and professional classes, with own commercial cores

modern factories along the main road

to coastal cities

Figure 15.28

Model showing land use and residential areas in Brazilian cities (excluding Brasilia): the zoning of housing, with the more affluent living near to the CBD and the poorest further from the centre, is typical of cities in developing countries (*after* Waugh, 1983)

The simplistic term 'developing city' has been used to indicate a city in a less economically developed country

Figure 15.29

The slums of Howrah, Calcutta

Cities in developing countries

Cities in economically less developed countries, which have grown rapidly in the last few decades (page 385), have developed different structures from those of older settlements in developed countries. Despite some observed similarities between most developing cities, few attempts have been made to produce models to explain them. Clarke has proposed a model for West African cities, McGhee for south-east Asia and the present author (based on two television programmes on São Paulo and Belo Horizonte together with some limited fieldwork) for Brazil (Figure 15.28).

Functional zones in developing cities

The CBD is similar to those of 'Western' cities except that congestion and competition for space are even greater (São Paulo, Cairo and Nairobi, Places 42).

Inner zone In pre-industrial and/or colonial times, the wealthy landowners, merchants and administrators built large and luxurious homes around the CBD. While the condition of some of these houses may have deteriorated with time, the well-off have continued to live in this inner zone — often in high-security, modern, high-rise apartments, sometimes in well-guarded, detached houses.

Middle zone This is similar to that in a developed city in that it provides the 'in between' housing, except that here it is of much poorer quality. In many cases, it consists of self-constructed homes to which the authorities *may* have added some of the basic infrastructure amenities such as running water, sewerage and electricity (the *periferia* in Figure 15.28 and the 'site and service' schemes on page 411).

Outer zone Unlike that in the developed city, the location of the 'lower-class zone' is reversed as the quality of housing decreases rapidly with distance from the city centre. This is where migrants from the rural areas live, usually in shanty towns (the *favelas* of Brazil, Places 43) which lack basic amenities. Where groups of better-off inhabitants have moved to the suburbs, possibly to avoid the congestion and pollution of central areas, they live together in well-guarded communities with their own commercial cores.

Industry This has either been planned within the inner zone or has grown spontaneously along main lines of communication leading out of the city.

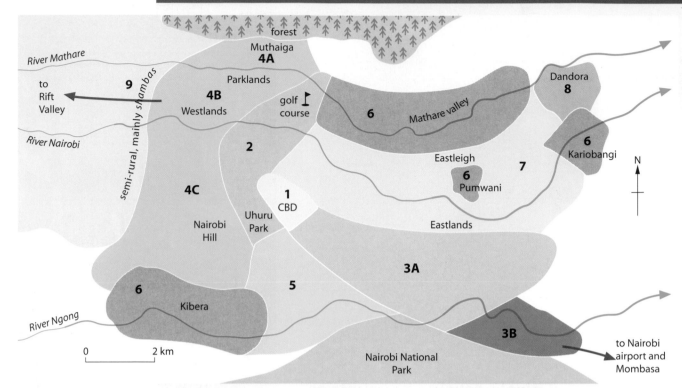

Figure 15.30

Functional zones and residential areas in Nairobi

Figure 15.31

The CBD, Nairobi (zone 1)

In 1899 a railway, being built between Mombasa, on the coast, and Lake Victoria, reached a small river which the Maasai called *enairobi* (meaning cool). The land which surrounded the river was swampy, malarial and uninhabited. Despite these seemingly unfavourable conditions, a railway station was built and, less than a century later, the settlement at Nairobi had grown to over 1.5 million people. The present-day functional zones (Figure 15.30) show the early legacy of Nairobi as a colonial settlement and the more recent characteristics associated with a rapidly growing city in an economically developing country.

1 CBD This is the centre for administration; it includes the Parliament Buildings, the prestigious Kenyatta International Conference Centre, commerce and shopping (Figure 15.31). Also located here are large hotels and, in the north, the University and the National Theatre.

2 Open space Immediately to the west and north of Nairobi's CBD, unlike in developed cities, are several large areas of open space. These include Uhuru (Freedom) Park and several other parks, sports grounds and a golf course. Other areas of open space, notably the Nairobi National Park to the south and the Karura Forest to the north, lie outside the city boundary.

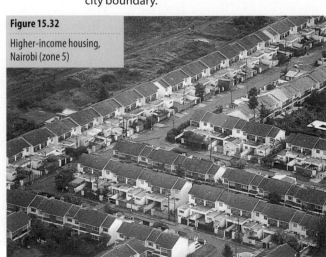

Figure 15.32

Higher-income housing, Nairobi (zone 5)

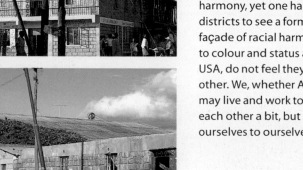

Figure 15.33

Shanty settlement, Mathare Valley, Nairobi (zone 6)

Figure 15.34

Inside the shanty settlement, Kibera, Nairobi (zone 6)

3 Industrial zone Early industry, much of which is formal, grew up in a sector which borders the railway linking Nairobi with the port of Mombasa (**3A** in Figure 15.30). The main industries, most of which are formal (Figure 19.32), include engineering, chemicals, clothing and food processing. A modern industrial area (**3B**) extends alongside the airport road and contains many well-known multinational firms.

4 High-income residential Wealthy European colonists and, later, immigrant Asians lived on ridges of highland to the north and west of the CBD where they built large houses above the malarial swamps. Today, Europeans tend to concentrate in Muthaiga (**4A**) and the Asians and more wealthy Africans in Parklands and

Westlands (**4B**). Westlands, with its shops and restaurants, forms a small secondary core while several large hotels are located on Nairobi Hill (**4C**). Many of the largest private properties have their own security guards.

5 Middle-income residential The southern sector was originally built for Asians who worked in the adjacent industrial zone. The estates, which were planned, are now mainly occupied by those Africans who have found full-time employment.

6 Shanty settlements As in other developing cities, shanty settlements have grown up away from the CBD on land which had previously been considered unusable — in Nairobi, this was on the narrow, swampy, flood plains of the Rivers Mathare and Ngong. The two largest settlements are those which extend for several kilometres along the Mathare valley (Figure 15.33) and in Kibera (Figure 15.34). Estimates suggest that over 100 000 people, almost exclusively African, live in each area. They find work in informal industries (page 524).

7 Low-income residential These areas include flats, 3–5 storeys in height and council-built (Figure 15.35), and former shanty settlements to which the council has added a water supply, sewage and electricity.

8 Self-help housing Under this scheme (page 411), the council provided basic amenities and, at a cheap price, building materials. In Dandora (Figure 15.36), which has over 120 000 residents, relatively wealthy people bought plots of land and built up to six houses around a central courtyard. The council then installed a tap and a toilet in each courtyard and added electricity and roads to the estate. The 'owner' is able to sell or rent the houses which are not needed by his/her family.

In 1993, an article in Nairobi's daily newspaper *The Nation* stated that "Kenya has been hailed as Africa's leading example of multi-racial harmony, yet one has only to tour its residential districts to see a form of 'apartheid'. Despite a façade of racial harmony, people live according to colour and status and, unlike in the UK or USA, do not feel they have to mix with each other. We, whether African, Asian or European, may live and work together and may even like each other a bit, but we like our space and keep ourselves to ourselves."

Figure 15.35

Low-income, council-built housing, Nairobi (zone 7)

Figure 15.36

Dandora 'site and services scheme', Nairobi (zone 8)

Calcutta's *bustees*

Although over 100 000 people live and sleep on Calcutta's streets, one in three inhabitants of the city lives in a *bustee* (Figure 15.29). These dwellings are built from wattle, with tiled roofs and mud floors — materials which are not particularly effective in combating the heavy monsoon rains. The houses, packed closely together, are separated by narrow alleys. Inside, there is often only one room, no bigger than an average British bathroom. In this room the family, often up to eight in number, live, eat and sleep. Yet, despite this overcrowding, the interiors of the dwellings are clean and tidy. The houses are owned by landlords who readily evict those *bustee* families who cannot pay the rent.

Rio de Janeiro's *favelas*

A *favela* is a wildflower that grew on the steep *morros*, or hillsides, which surround and are found within Rio de Janeiro. Today, these same *morros* are covered in *favelas* or shanty settlements (Figure 15.37). The *favelados*, the inhabitants, are squatters who have no legal right to the land they live on. They live in houses constructed from any materials available — wood, corrugated iron, and even cardboard.

Some houses may have two rooms, one for living in and the other for sleeping. There is no running water, sewerage or electricity, and very few local jobs, schools, health facilities or forms of public transport. The land upon which the *favelas* are built is too steep for normal houses. The most favoured sites are at the foot of the hills near to the main roads and water supply, although these may receive sewage running in open drains downhill from more recently built homes above them. Often there is only one water pump for hundreds of people and those living at the top of the hill (with fine views over the tourist beaches of Copacabana and Ipanema!) need to carry water in cans several times a day. When it rains, mudslides and flash floods occur on the unstable slopes (page 48). These can carry away the flimsy houses (over 200 people were killed in February 1988).

Several attempts have been made to clean up the *favelas* or to remove them altogether. In some cases, new homes have been built for the *favelados*, but over 40 km away, in areas lacking jobs and transport. In other examples, the evicted inhabitants have simply moved back again as they had nowhere else to go. In 1992 almost 1 million people lived in *favelas* — about 10 per cent of the total population of Rio.

Figure 15.37

Favelas in Rio de Janeiro, Brazil

Problems resulting from rapid growth

The 'pull' and rapid growth of cities in the developing world has led to serious problems in providing housing, basic services and jobs; problems accentuated by a much wider gulf between the minority rich and the majority poor than exists in the developed world. (Remember that developing cities do have positive as well as negative features.)

Housing

Despite some promising initiatives, most authorities have been unable to provide adequate shelter and services for the rapidly growing urban population and so the majority of the poor have to fend for themselves and to survive by their own efforts. Estimates suggest that one-third of the urban dwellers in developing countries either cannot afford or cannot find accommodation which meets basic health and safety standards. Consequently, they are faced with three alternatives: to sleep on pavements or in public places; to rent a single room if they have some resources; or to build themselves a shelter, possibly with the help of a local craftsman, on land which they do not own and on which they have no permission to build (Figure 15.33 and Places 43).

A growing number have adopted the third option and have set up home in illegal 'shanty towns' where they face the constant threat of eviction or the demolition of their homes. Estimates in 1991 suggested (no accurate figures are available so nobody knows how many people actually do live in squatter settlements) that 30 per cent of Rio de Janeiro's total population live in shanty towns. The equivalent figure is 25 per cent in São Paulo, 45 per cent in Mexico City, 40 per cent in Bombay and 60 per cent in Calcutta.

In time, some squatter settlements may develop into residential areas of 'adequate' standards (the *periferia* in Figure 15.28 and Dandora in Places 42). Rather than trying to build new housing, city councils find it cheaper and easier to add water supplies, sewerage systems, electricity and public services (refuse disposal, street lighting) to existing shanties, and to allow occupants to obtain legal tenure of the land (page 411).

Services

Only small areas within many developing cities have running water and mains sewerage. Rubbish, dumped in the streets, is rarely collected. When heavy rains fall, especially in the monsoon countries, the drains are inadequate to carry the surplus water away. The lack of electricity hinders industrial growth as well as affecting the material standard of living in homes. There is a shortage of schools and teachers, and of hospitals, doctors and nurses. Police, fire and ambulance services are unreliable. Shops may sell only essentials, and food may be exposed to heat and infection-carrying flies.

Pollution and health

Drinking water is often contaminated with sewage which may give rise to outbreaks of cholera, typhoid and dysentery. The uncollected rubbish is an ideal breeding-ground for disease. Many children have worms and suffer from malnutrition as their diet lacks fresh vegetables, protein, calories and vitamins. Local industry is rarely subjected to pollution controls and so discharges waste products into the air which may cause respiratory diseases, and/or into water supplies. The constant struggle for survival often causes stress-related illnesses. It is not surprising that in these rapidly growing urban areas infant mortality is high and life expectancy is low.

Unemployment and underemployment

New arrivals to a city far outnumber the jobs available and so high unemployment rates result. As manufacturing industry is limited, full-time occupations are concentrated in service industries such as the police, the army, cleaning, security guards and the civil service. The majority of people who do work are in the informal sector — i.e. they have to find their own form of employment (page 524). Informal jobs may include street trading (selling food or drinks), food processing, services (shoe-cleaning) and local crafts (making furniture and clothes, often out of waste products). Most of these people are underemployed and live at a subsistence level.

Transport

Few developing cities can afford elaborate public transport systems and what they have tend to be outdated and overused. The wealthy have cars, but the road network is often inadequate to cope with the large volume of traffic. Local traditional forms of transport (rickshaws, bullock carts, donkeys) compete in some cities with other road users (Places 75, page 558).

Figure 15.38

Apartments in the medieval centre of Cairo, Egypt

1990 figures put Cairo's population somewhere between 12 and 16 million (Framework 12). In comparison with cities in other developing countries, Cairo has relatively few squatter settlements. Instead, most newcomers disappear into the medieval centre of the old city to live in overcrowded, two-roomed apartments within tall blocks of flats (Figure 15.38) or in roof-top slums (the desert climate allows roofs to be flat). There are few public services and washing hangs everywhere. Many

migrants live in the 'City of the Dead', a huge Muslim cemetery (Figure 15.39). People actually live within the tombs, which are often cleaner than the city apartments, but are over a kilometre from the nearest water supply. New flats are being built remarkably quickly but these are in the suburbs and their rents are unaffordable and the cost of the journey to work in the city centre is too high for most people.

Cairo's narrow, unplanned streets were not built for the volume of motorised traffic which is noted for its noise and pollution. Pollution also comes from the breakdown of sewerage systems which were built in the early 20th century, or their total absence, while the many small factories emit their wastes into the air and streets. Although small workshops appear to be everywhere, over 35 per cent of the population are without full-time jobs. Workshops are set up in backyards, within houses and on rooftops. Donkey carts take rubbish to dumps on the edge of the city, providing jobs for refuse collectors and people who sort out the bottles, plastic and paper on the dump ready for recycling in local industry. The rubbish dumps, limited clean water, poor sewerage system, industrial pollution and high-density housing all contribute to create potential health hazards.

The Cairo authorities are trying to overcome these problems by extending the old sewerage system and also building a new one (making this, possibly, the world's biggest public-health engineering scheme); organising refuse collection; constructing an underground metro system; building many new roads (including an outer ring road) and erecting numerous high-rise apartment blocks (whose safety was questioned after several collapsed during the 1993 earthquake). Most of these projects have become possible since the production of electricity from the High Dam at Aswan in southern Egypt. Even so these schemes have failed to keep pace with population growth and so five satellite or new towns are being built in the surrounding desert. Perhaps the solution to Cairo's problems may be both to reduce migration from the countryside and to accelerate the already falling birth rate.

Figure 15.39

The 'City of the Dead', Cairo, Egypt

Accurate and reliable statistics are often difficult to obtain, even for developed countries. Some of the least reliable are for population and those presented in this book, e.g. birth rates and urban populations, should be used with considerable caution. It is suggested that Britain's 1991 census could have a margin of error, even with supposedly high levels of refinement, of ±1 per cent. (It is believed that many people failed to complete the relevant forms and so are a 'lost' statistic.) The size of this error increases considerably in developing countries. For example . . .

What is Cairo's population?

The following sources all produced estimates for Cairo's 1986 population (unless otherwise stated):

a The *New Geographical Digest* (George Philip) puts Cairo at 6 325 000 and El Giza (the suburb surrounding the Pyramids and considered by many as part of Cairo) at 1 858 000, making a combined total of 8 183 000.

b The United States Bureau of Census: 8 595 000.

c Reader's Digest (UN figures): 9 770 000.

d *Geofile* (No. 88, 1987): 10 000 000 — 12 000 000.

e A Granada Television programme for schools on Cairo: 15 000 000.

f Official Cairo Tourist guide (1989): 15 000 000.

The first problem in obtaining accurate figures is that Cairo's most recent census was in 1976 and since then the size has had to be estimated. Different organisations use different criteria. A second problem is in determining whether the statistic is for a specified city area or if it includes developments which have sprawled beyond Cairo's 'official' limits ,such as El Giza. It would be highly unlikely that every inhabitant was consulted for the census as many migrants to the city 'lose' themselves in the high-density housing of the 'old town' and an estimated 500 000 live in and around the tombs of the 'City of the Dead' (Figure 15.39).

Recent estimates are based upon birth and death rates, but these too are most unreliable. Egypt had no consistent method of registering births until 1986 when a new law was passed in an attempt to ensure that all births took place in hospital and so would be recorded. However, methods of recording natural increase remain suspect and there is no effective means of counting migrants from rural areas.

Government housing

Upgrading and self-help schemes
A policy of wholesale demolition of squatter settlements, as was attempted in Rio de Janeiro (Places 43) and South Africa (Places 35, page 340), is often a mistaken one. Squatters have shown that they are capable of constructing cheap accommodation for themselves, but that they cannot provide the essential basic services. In Latin America, and less successfully in Africa and south-east Asia, governments have, albeit reluctantly, at times accepted that shanties are permanent and that it is cheaper and easier to improve them by adding basic amenities than it is to build new houses.

The concept of **'site and services'**, funded by the World Bank and several voluntary organisations, encourages local people to become involved in self-help projects. This approach seems to be most appropriate in the poorer countries whose governments cannot afford large rehousing schemes. One such scheme, in Dandora in Nairobi (Figure 15.36) was briefly described in Places 42.

A similar scheme in Lusaka (Zambia) encourages about 25 individuals to group together. They are given a standpipe and 8 hectares of land. If the group digs ditches and foundations then, with the money saved, the authorities will lay water and drainage pipes and construct the houses. Moreover, if local craftsmen are prepared to build the shells of the houses, the group will be supplied with low-priced building materials and the extra money saved by the authorities may be used to add electricity and to tarmac the roads. In some cases, a small clinic and school may be added.

Several schemes in São Paulo's *periferia* (Figure 15.40) have enabled running water,

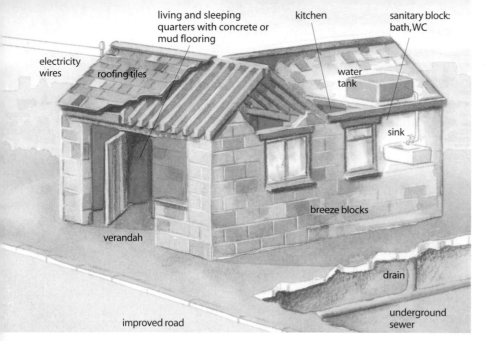

living and sleeping quarters with concrete or mud flooring

kitchen

sanitary block: bath, WC

electricity wires

roofing tiles

water tank

sink

breeze blocks

verandah

drain

improved road

underground sewer

Figure 15.40

A 'sites and services' scheme, São Paulo, Brazil

Figure 15.41

A self-help scheme in Sri Lanka linked to Intermediate Technology

main drains and electricity to be added to houses, with street lighting and improved roads if there was any surplus money. The result over a lengthy period of time has been an upgrading of living conditions, although the people are still poor, and the introduction of some shops and small-scale industry.

Such schemes can create a community spirit, can improve the skills of local people and can result in cheap-to-erect accommodation. Yet their success often depends upon the motivation and skills of the local people and the use of appropriate and cheap building materials under expert guidance.

Intermediate Technology (IT) and 'materials for shelter'

IT helps people in Africa, Asia and South America to develop and use technologies and methods which give them more control over their lives and which contribute to the long-term development of their communities. Several of IT's projects involve investigating, developing and promoting a range of building materials suitable and affordable for self-help schemes (Figures 15.41 and 15.42). An IT-sponsored scheme in India prolongs the lives of thatch roofs by coating them with a waterproof compound of copper sulphate and cashew nut resin. In Kenya, the Maasai are under increasing internal pressure to give up their semi-nomadic way of life and settle in permanent houses. IT has responded to this situation by working closely with the Maasai in helping to modify their traditional houses by adding a concrete mix to the cow-dung roof (which always seemed to leak), inserting a small chimney to remove smoke (all cooking is done inside the house), improving lighting (previously each house had only one minute opening as a 'window'), and using chicken wire as a framework for the walls (Places 70, page 527). It also provides, in several parts of the world, technical assistance in the mining, quarrying and processing of local raw materials which can be used for building.

Relocation housing and new towns

Some of the more wealthy developing countries, such as Venezuela with its oil revenue and Hong Kong and Singapore with their income from trade and finance, have made considerable efforts to provide new homes to replace squatter settlements. In most cases, high-rise blocks of flats have been built on sites as close as close as possible to the CBD or in new towns beyond the city boundary (Places 45).

Figure 15.42

Production of low-cost roofing tiles in Kenya

Faced with a large and rapidly increasing number of slum dwellers, and an overcrowded, unplanned, central area, the Singapore government set up, in 1960, the Housing and Development Board (HDB). The HDB cleared old property near to the CBD, especially in the Chinese, Arab and Indian ethnic areas, and created purpose-built estates (with 10 000 — 30 000 people) within a series of new towns (each of up to 250 000 people). By 1994 there were 14 new towns, all within 12 km of the CBD.

In both cases, the HDB constructed housing units of 1–3 rooms in closely packed high-rise flats (Figure 15.43). The flats were initially for low-income families and rents were kept to a minimum. However, one-quarter of every wage-earner's salary is automatically deducted and individually credited by the government into a central pension fund (CPF). Western-style welfare benefits are regarded as an anti-work ethic, but Singaporeans can use their CPF capital to buy their own apartment or flat. Since 1974 the HDB have built many 4-room and 5-room units for the average and higher income groups who have then been expected to buy their own property. In 1994, 87 per cent of Singaporeans lived in government-built housing and 80 per cent owned their own homes.

The large estates are functional in design and were developed on the neighbourhood concept of British new towns. Each estate contains much greenery and is well provided with amenities such as shops, schools, banks, medical and community centres. Where several estates are in close proximity, better services are provided such as department stores and

Figure 15.43

Early high-rise flats on the edge of China Town, Singapore

entertainment facilities. All the new towns have been linked to, and are within half an hour of, the city centre by the MRT (mass rapid transport railway). Each estate has its own light industries producing, usually, clothing, food products and high-technology goods. The estates, like everywhere else in Singapore, are models of cleanliness with lawns trimmed and even the oldest apartments being constantly repainted. They are free of litter and graffiti: the minimum fine for each is $500 Singapore (about £130). The lifts are clean and almost always work. As in many other countries with high-rise flats, there was the initial problem of the lifts being used as toilets. The HDB put sensors in the floors which were activated by salt in the urine. This locked the doors automatically and set off an alarm. The offenders were **very** heavily fined and the problem was quickly solved!

By 1994 the HDB had built over 700 000 flats. Its aim now is to provide every householder with a minimum flat size of 3 rooms. This is being achieved by pulling down and replacing some of the earliest apartment blocks, merging adjacent flats to make them bigger, and building even more estates. To ensure all Singaporeans have a home, the HDB has been buying 3-roomed flats on the open market and selling them at discount prices to low-income families. The newest estates (Figure 15.44) have been architect-designed and are set in considerable areas of open space.

Figure 15.44

A new estate in Bishan, Singapore

Case Study 15

Los Angeles a. Physical hazards

Figure 15.45

Los Angeles: hazard city

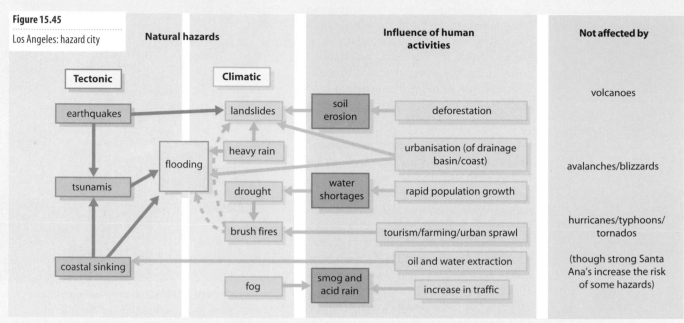

For several generations, southern California was seen as America's promised land. Now it seems that the 'sunshine state' is cursed by natural disasters such as earthquake, fire, drought and flood (by rivers and tsunamis) — disasters which, in part, are created or exacerbated by the life-style and economic activities of its inhabitants (Figure 15.45). Los Angeles, with a population in excess of 13 million, has become known as 'hazard city'.

Earthquakes

Not only does the San Andreas Fault, marking the conservative boundary between the Pacific and North American Plates, cross southern California (Places 4, page 20), but Los Angeles itself has been built over a myriad of transform faults (Figure 15.46). Although the most violent earthquakes are predicted to occur at any point along the San Andreas Fault between Los Angeles and San Francisco, earth

movements frequently occur along most of the lesser-known faults. The most recent earthquake to affect Los Angeles occurred in February 1994. It registered 6.7 on the Richter scale, lasted for 30 seconds, and was followed by aftershocks lasting several days. The quake killed 60 people, injured several thousand, caused buildings and sections of freeways to collapse, ignited fires following a gas explosion, and left 500 000 homes without power and 200 000 without water (Figure 15.47).

Figure 15.46

Major faults in the Los Angeles basin

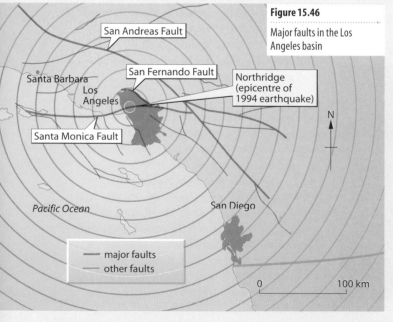

Figure 15.47

The effects of the earthquake in February 1994

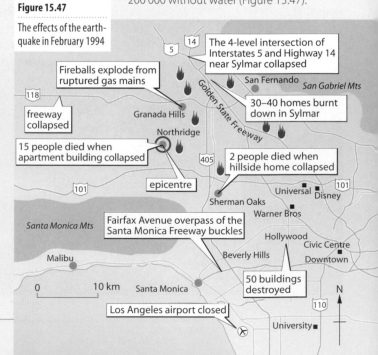

Tsunamis

Tsunamis are large tidal waves triggered off by submarine earthquakes which can travel across oceans at great speed. The 1964 Alaskan earthquake caused considerable damage in several Californian coastal regions. Although Los Angeles has escaped so far, it is considered to be a tsunami hazard prone area.

Sinking coastline

The threat of coastal flooding has increased due to crustal subsidence. Although this may, in part, be due to tectonic processes, the main cause has been the extraction of oil and, to a much lesser extent, subterranean water. Parts of Long Beach have sunk by up to 10 m since 1926. Although this sinking has now been checked, parts of the harbour area lie below sea-level and are protected from flooding by a large sea-wall.

Landslides and mudflows

Landslides and mudflows occur almost annually during the winter rainfall season within the city boundary of Los Angeles. They have increased in number and frequency due to effects of urbanisation such as the removal of vegetation from, and the cutting of roads through, steep hillsides and by channelling rivers (Figure 3.8). Figure 15.48 describes the effects of one mudflow (February 1994). Landslides are frequent along coastal cliffs.

Heavy rain

Winter storms bring rain and strong winds. These are especially severe during an El Nino event (Case Study 9). Although most rivers in the Los Angeles basin are short in length and seasonal, they can transport large volumes of water during times of flood. Deforestation and brush fires on the steep, surrounding hillsides and rapid urbanisation (page 56) have increased surface runoff. Large dams have been built to try to hold back flood water but even so the flood risk remains. In February 1992 (during an El Nino event) eight people died and dozens of cars and caravans were poured out to sea when, following two days of torrential rain, flood waters poured through a caravan park to the south of Malibu. Heavy rain also triggers off landslides and mudflows.

A fierce winter storm has brought more misery and destruction to areas of southern California already devastated by brushfires and earthquakes. Torrents of mud have trapped residents and washed away cars. More than a dozen expensive beachfront homes were swamped by landslides in the star colony of Malibu, 30 km west of Los Angeles, as rain-soaked hillsides, stripped of their vegetation by last year's fires, suddenly gave way. Some people became trapped in their homes and had to be rescued by city council workers who picked them up in the scoops of bulldozers and earthmovers. Hundreds of people were forced to evacuate their homes.

Exclusive oceanside homes of stars and entertainment industry executives in the Malibu colony faced double jeopardy – mudslides from the mountains and 2.5 m waves crashing in from the Pacific. Police reported walls of mud sweeping across the Pacific Coast Highway, through the front doors of homes and carrying furniture out of ocean-facing doors and onto the beach. Parts of the highway were buried in over 1 m of mud and water, trapping some drivers in their cars. More than a dozen parked cars were swept away. People who lost power in last month's earthquake found themselves in the dark again as the storms knocked out electricity to more than 3000 homes in Malibu.

Figure 15.48

River of mud, Los Angeles

Drought

The long, dry summers associated with the Mediterranean climate may be ideal for tourists but, as the population of Los Angeles continues to grow, they put tremendous pressure on the limited water resources. Much of the city's water comes, via the Colorado aqueduct, from the River Colorado 400 km to the east. So much water is now extracted from the river that, in very dry years, it almost dries up before reaching the sea (Case Study 3b, page 85).

Brush fires

Much of the Los Angeles basin is covered in drought-resistant (xerophytic) chapparal, or brush vegetation (page 300). By the autumn, after six months without rain, this vegetation becomes tinder dry. Santa Ana's are hot, dry winds which owe their high temperatures to adiabatic heating as they descend from the mountains. Their heat and extreme low humidity cause discomfort to humans and increase the dryness in vegetation. A careless spark or an electrical storm can prove sufficient to set off serious fires. In September 1970 a fire, 56 km in width, swept down from the Santa Monica Mountains (Figure 15.47) into Malibu. Some 72 000 hectares of brush and 295 houses were destroyed together with the loss of three human lives. In November 1993 the homes of many American film and recording stars were destroyed in another severe brush fire (Figure 15.49).

Fog and smog

Advection fog (page 205) occurs when cool air from the cold offshore Californian current drifts inland where it meets warm air. Fog can form most afternoons between May and October as the strength of the sea-breeze increases (page 223). This event can cause temperature inversions to occur (page 200) where warm air becomes trapped under cold air. When many pollutants from Los Angeles' traffic, power stations and industry are released into the air, the result is smog (Figure 9.26) and, when they return to earth, acid rain. Smog in Los Angeles can be a major health problem (Figure 15.50); indeed, during the 1990 Olympics there was great concern for athletes competing in longer-distance races such as the marathon.

Figure 15.49

Brush fire near Malibu, California

Figure 15.50

Smog over Los Angeles, California

Case Study 15

b.Urban deprivation

The statistics in Figure 15.51 come from the 1991 census data for Watts, a district which lies 2 km south of the wealthy downtown (CBD) of Los Angeles (Figure 15.52). Does this information match up with the stereotyped image (Framework 11, page 404) of Los Angeles which has so often been projected on our TV screens? The vision of film stars with huge ranch-style homes and a luxurious standard of living is perhaps more familiar.

Average income	$516 per month ($6200 per year)
Average rent	$625 per month
Age of houses	50+ years (in spite of earthquakes)
Persons/hectare	100+
12.7%	unemployment rate
10%	population moved in from outside USA since 1986
37%	population are Hispanic
52%	population are Afro-American
28%	population live below the poverty level

Figure 15.51

Watts district, Los Angeles: some 1991 statistics

Figure 15.52

Los Angeles

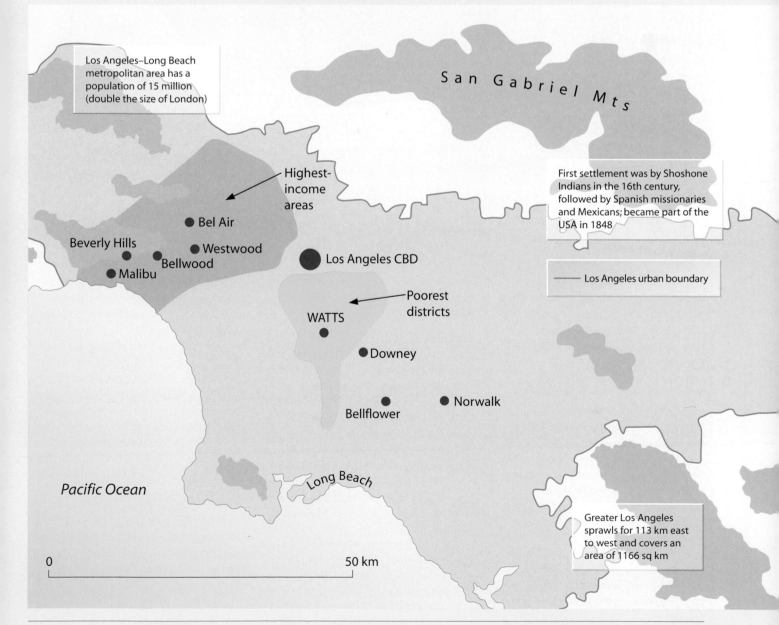

Los Angeles–Long Beach metropolitan area has a population of 15 million (double the size of London)

San Gabriel Mts

Highest-income areas

First settlement was by Shoshone Indians in the 16th century, followed by Spanish missionaries and Mexicans; became part of the USA in 1848

Bel Air

Beverly Hills

Westwood

Bellwood

Malibu

Los Angeles CBD

——— Los Angeles urban boundary

Poorest districts

WATTS

Downey

Bellflower

Norwalk

Pacific Ocean

Long Beach

Greater Los Angeles sprawls for 113 km east to west and covers an area of 1166 sq km

0   50 km

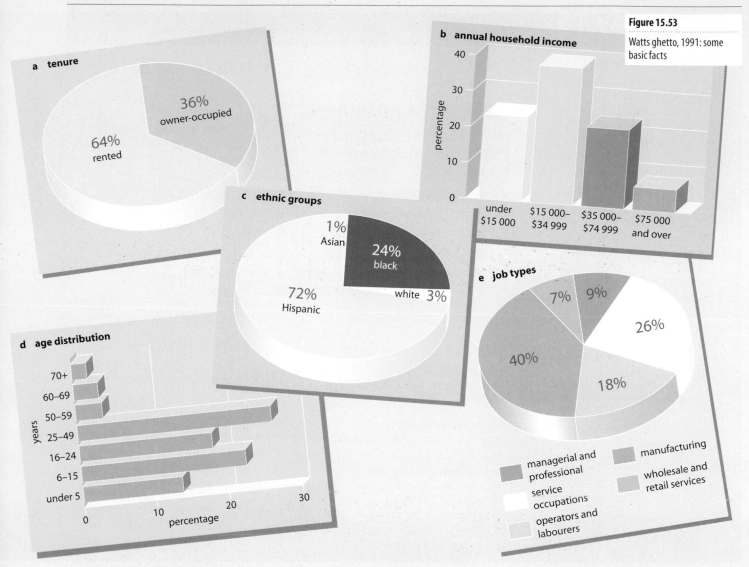

Figure 15.53

Watts ghetto, 1991: some basic facts

a tenure

36% owner-occupied

64% rented

b annual household income

percentage

40
30
20
10
0

under $15 000 | $15 000–$34 999 | $35 000–$74 999 | $75 000 and over

c ethnic groups

1% Asian

24% black

72% Hispanic

white 3%

d age distribution

years

70+
60–69
50–59
25–49
16–24
6–15
under 5

0   10   20   30
percentage

e job types

7%   9%
26%
40%
18%

managerial and professional

service occupations

operators and labourers

manufacturing

wholesale and retail services

Los Angeles has grown very rapidly in the last 70 years as a major centre for the film industry and for aeronautical and space engineering. Southern California became one of the fastest-growing regions in the USA following the decline of the traditional industrial areas in the north-east. Migrants from all parts of the USA moved in to share in its growing prosperity. Downtown Los Angeles has many huge skyscrapers, typical of the centres of cities which are the headquarters of oil companies, engineering corporations and property developers, as well as TV and film companies. Prosperity has brought wealth to residential areas such as Bel Air and Malibu, to the north-west of the city centre (Figure 15.52) — but not to other districts, including Watts.

Watts became known as a deprived area in the 1960s when riots occurred as a consequence of discrimination over housing, education and employment. The situation has altered relatively little since then.

A profile of Watts

One characteristic of life in California is the high rent demanded for housing. Two-thirds of the families in Watts (Figure 15.53a) rent their homes and most are paying more than 35 per cent of their income in rent. Homes are crowded and often have only one bedroom, although most have kitchens and bathrooms. The low average household income (Figure 15.53b) means that all family members aged over 16 need to work. This may not be easy, especially if skills are demanded. Only one-third of the population of Watts were high-school graduates in 1991, in contrast to almost 95 per cent in wealthier districts.

It is to an area such as Watts that migrants, legal and illegal, mainly Hispanics from Central America are drawn (Figure 15.53c). Most are young (Figure 15.53cw), have little money and few qualifications. They are attracted by California's wealthy image (stereotype), but the reality when they arrive is often different. Unless they have a Green Card from the Department of Immigration they cannot work legally and they cannot claim welfare. This means that they are forced to take very low-paid unskilled jobs (Figure 15.53e), often in the informal sector (page 524), and they have to move around to avoid detection. Unless they have means of their own, they will have difficulty in obtaining housing. Language difficulties make it hard for some to obtain permanent jobs. Spanish is spoken by two-thirds of the population of Watts.

Figure 15.54

Poor quality environment in Los Angeles

Figure 15.55

High quality environment in Los Angeles

Low educational standards, lack of vocational qualifications and poor health and housing conditions are characteristic of the Afro-American (black) communities and of minority groups such as the Hispanic, Vietnamese and Korean immigrants. There have been difficulties in assimilation; competition for jobs is high and the unemployment rate is three times as high as that in Bel Air or Beverly Hills. In addition there are fewer health facilities; 22 per cent of the over-16s in Watts are registered disabled or have long-term illnesses.

Why are people forced to remain in these conditions?

Recession in Southern California has caused the closure of factories and the loss of jobs in construction and associated industries. Unskilled workers are laid off; therefore less money enters the local economy and the tax revenues of the city and State governments are reduced. It appears that the poorer areas suffer first if cutbacks are made in local services. The black and Hispanic communities believe that they are deprived because of their race. They demand justice for all. One cause of friction

has been alleged heavy-handed policing within Los Angeles, with blacks and Hispanics being arrested on slender evidence. There are few legitimate opportunities and it is easy for young people to drift into crime. Gang warfare is a fact of life in the ghettos.

The local authorities, on the other hand, say that they are attempting to help deprived communities, but that they must also do more to help themselves. A big issue at present is a suggestion that welfare payments be withdrawn after two years of unemployment. This is intended to force the jobless back into work.

It is a long way from Watts to Bel Air: from deprivation to affluence (Figure 15.56). Deprivation may be seen as the gap between those who can and those who cannot obtain for themselves the standards and conditions necessary for living in their particular society.

Figure 15.56

Watts and Bel Air, 1991

		Watts	Bel Air
Average income (US$)		6700	$84000
Unemployment (%)		12.7	4.1
Ethnic composition:	white (%)	3.5	90.8
	black (%)	25.0	2.0
	Hispanic (%)	69.75	3.6
	Asian (%)	1.75	3.6

References

Barke, M. and O'Hare, G. (1984) *The Third World*. Oliver & Boyd.

Bradford, M. G. and Kent, W. A. (1982) *Human Geography: Theory and Applications*. Hodder & Stoughton.

Brazil: City of Newcomers (1980) BBC Television.

Brazil: Skyscrapers and Slums (1980) BBC Television.

Decade of Achievement 1981-1991 (1991) London Docklands.

Hornby, W. F. and Jones, M. (1991) *Settlement Geography*. Cambridge University Press.

North-South: A Programme for Survival. Report of the Brandt Commission (1980) Pan.

People and Places: Cairo (1987) Granada Television.

Philip's Geographical Digest 1992–93 (1993) Heinemann-Philips.

Prosser, R. (1992) *Human Systems and the Environment*. Thomas Nelson.

Salterthwaite, D. (1988) *The Global Housing Crisis*. CPC.

Small-area Statistics (1993) HMSO.

Urban deprivation, *Geographical Magazine*, July 1993.

Waugh, D. (1994) *The Wider World*. Thomas Nelson.

Chapter 16

Farming and food supply

"But of all the occupations by which gain is secured, none is better than agriculture, none more profitable, none more delightful, none more becoming to a free man."

Cicero, *De Officiis*, 1.51

"Behold, there shall come seven years of great plenty throughout all the land of Egypt: and there shall arise after them seven years of famine; and all the plenty shall be forgotten in the land of Egypt; and the famine shall consume the land ..."

The Bible, Genesis 41: 29, 30

"He who slaughters his cows today shall thirst for milk tomorrow."

Muslim proverb

Figure 16.1

The optima and limits model
(after McCarty and Lindberg)

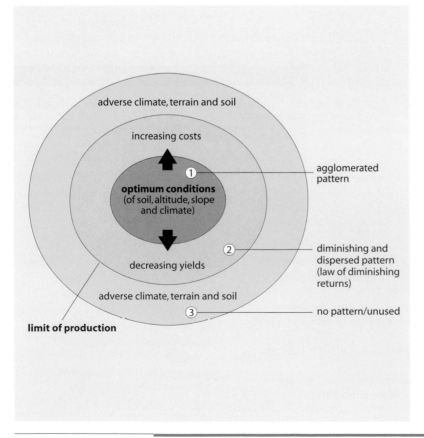

The location of different types of agriculture at all scales depends upon the interaction of physical, cultural and economic factors. Where individual farmers in a market economy (capitalist system) or the state in a centrally planned economy have a knowledge, or understanding, of these three influences, then decisions may be made. How these decisions are reached involves a fourth factor: the behavioural element.

Environmental factors affecting farming

Although there has been a movement away from the view that agriculture is controlled solely by physical conditions, it must be accepted that environmental factors do exert a major influence in determining the type of farming practised in any particular area.

In 1966 McCarty and Lindberg produced their optima and limits model, an adaptation of which appears in Figure 16.1. They suggested that there was an optimum or ideal location for each specific type of farming based on climate, soils, slopes and altitude. As distance increases from this optimum, conditions become less than ideal — i.e. too wet or dry; too steep or high; too hot or cold; or a less suitable soil. Consequently, the profitability of producing the crop or rearing animals is reduced, and the **law of diminishing returns** operates when either the output decreases or the cost of maintaining high yields becomes prohibitive. Eventually a point is reached where physical conditions are too extreme to permit production on an economically viable scale, and later at even a subsistence level (page 437). McCarty and Lindberg applied their model to the cotton belt of the USA (Figure 16.2), but it can equally be adapted to account for the growth of spring wheat on the Canadian Prairies (Figure 16.3). Can you suggest other regions where the model might be successfully tested?

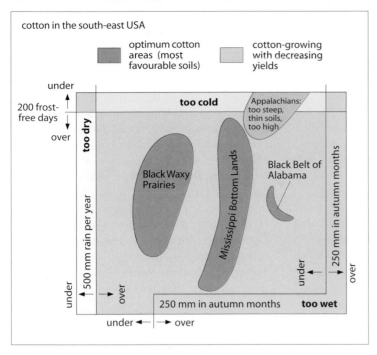

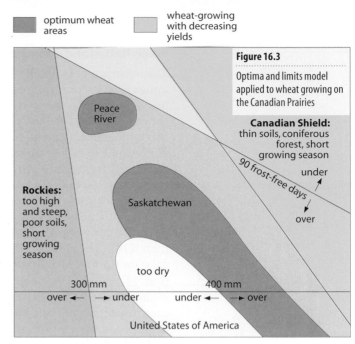

Figure 16.2

Optima and limits model applied to the former cotton belt in the south-eastern USA

Figure 16.3

Optima and limits model applied to wheat growing on the Canadian Prairies

Temperature

This is critical for plant growth because each plant or crop type requires a minimum growing temperature and a minimum growing season. In temperate latitudes, the critical temperature is 6°C. Below this figure, members of the grass family, which include most cereals, cannot grow — an exception is rye, a hardy cereal, which may be grown in more northerly latitudes.

In Britain, wheat, barley and grass begin to grow only when the average temperature rises above 6°C — which coincides with the beginning of the growing season. The growing season is defined as the number of days between the last severe frost of spring and the first of autumn. It is therefore synonymous with the number of frost-free days which are required for plant growth. Figures 16.2 and 16.3 show that cotton needs a minimum of 200, and spring wheat 90. Barley can be grown further north in Britain than wheat, and oats further north than barley because wheat requires the longest growing season of the three and oats the shortest. Frost is more likely to occur in hollows and valleys. It has beneficial effects as it breaks up the soil and kills pests in winter, but it may also damage plants and destroy fruit blossom in spring.

Within the tropics there is a continuous growing season, provided moisture is available. As well as decreasing with distance from the Equator, both temperatures and the length of the growing season decrease with height above sea-level. This produces a succession of natural vegetation types according to altitude, although many have been modified for farming purposes (Figure 16.4).

Figure 16.4

The effect of altitude on farming and vegetation

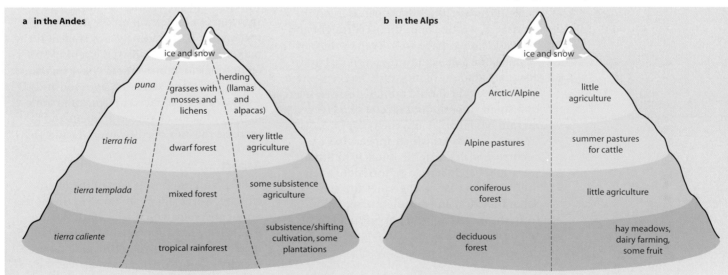

Precipitation and water supply

The mean annual rainfall for an area determines whether its farming is likely to be based upon tree crops, grass or cereals, or irrigation. The relevance and effectiveness of this annual total depends on temperatures and the rate of evapotranspiration. Few crops can grow in temperate latitudes where there is less than 250 mm a year or in the tropics where the equivalent figure is 500 mm. However, the seasonal distribution of rainfall is usually more significant for agriculture than is the annual total. Wheat is able to grow on the Canadian Prairies (Places 56, page 447) because the summer rainfall maximum means that water is available during the growing season. The Mediterranean lands of southern Europe have relatively high annual totals, yet the growth of grasses is restricted by the summer drought. Some crops require high rainfall totals during their ripening period (maize in the American corn belt), whereas for others a dry period before and during harvesting is vital (coffee).

The type of precipitation is also important (page 55). Long, steady periods of rain allow the water to infiltrate into the soil, making moisture available for plant use. Short, heavy downpours can lead to surface runoff and soil erosion and so are less effective for plants. Hail, falling during heavy convectional storms in summer in places such as the Canadian Prairies, can destroy crops. Snow, in comparison, can be beneficial as its insulates the ground from extreme cold in winter and provides moisture on melting in spring. In Britain we tend to take rain for granted, forgetting that in many parts of the world amounts and occurrence are very unreliable (Figure 9.29). India depends upon the monsoon — if this fails, there is drought and a risk of famine (page 462). Even in the best of years, the Sahel countries receive a barely adequate amount of moisture. The ecosystem is so fragile that should rainfall decrease even by a small amount (and several years recently no rain has fallen at all), then crops fail disastrously — an event which appears to be occurring with greater frequency. In Britain, we would barely notice a shortfall of a few millimetres a year: in the Sahel and sub-Saharan Africa, an equivalent fluctuation from the mean can ruin harvests and cause the deaths of many animals (Figure 16.61).

Wind

Strong winds increase evapotranspiration rates which allows the soil to dry out and to become vulnerable to erosion. Several localised winds have harmful effects on farming: the *mistral* brings cold air to the south of France; the *khamsin* is a dry, dust-laden wind found in Egypt; Santa Ana's can cause brush fires in California (Case Study 15a): hurricanes, typhoons and tornados can all destroy crops by their sheer strength. Other winds are beneficial to agriculture: the föhn and chinook (page 224) melt snows in the Alps and on the Prairies respectively, so increasing the length of the growing season.

Altitude

The growth of various crops is controlled by the decrease in temperature with height. In Britain few grasses, including those grown for hay, can give commercial yields at heights exceeding 300 m, whereas in the Himalayas, in a warmer latitude, wheat can ripen at 3000 m. As height increases, so too does exposure to wind and the amounts of cloud, snow and rain while the length of the growing season decreases. Soils take longer to develop as there are fewer mixing agents; humus takes longer to break down and leaching is more likely to occur. Those high-altitude areas where soils have developed are prone to erosion (Case Study 10, page 258).

Angle of slope (gradient)

Slope (see catena, page 256) affects the depth of soil, its moisture content and its pH (acidity, page 248), and therefore the type of crop which can be grown on it. It influences erosion and is a limitation on the use of machinery. Until recently, a 5° slope was the maximum for mechanised ploughing but technological improvements have increased this to 11°. Many steep slopes in south-east Asia have been terraced to overcome some of the problems of a steep gradient and to increase the area of cultivation (Figure 16.31).

Aspect

Aspect is an important part of the micro-climate. **Adret** slopes are those in the northern hemisphere which face south (Figure 9.11). They have appreciably higher temperatures and drier soils than the **ubac** slopes which face north. The adret receives the maximum incoming radiation and sunshine, whereas the ubac may be permanently in the shade. Crops and trees both grow to higher altitudes on the adret slopes.

Soils (edaphic factors)

Farming depends upon the depth, stoniness, water-retention capacity, aeration, texture, structure, pH, leaching and mineral content of the soil (Chapter 10). Three examples help to show the extent of the soil's influence on farming:

1 Clay soils tend to be heavy, acidic, poorly drained, cold and ideally should be left under permanent grass.
2 Sandy soils tend to be lighter, less acidic, perhaps too well-drained, warmer and more suited to vegetables and fruit.
3 Lime soils (chalk) are light in texture, alkaline, dry and give high cereal yields.

Although soils can be improved, e.g. by adding lime to clay and clay to sands, and by applying fertiliser, there is a limit to the increase in their productivity — i.e. the law of diminishing returns operates.

Places 46 Effects of precipitation and water supply on farming in Northern Kenya

The Rendille tribe live on a flat, rocky plain in northern Kenya where the only obvious vegetation is a few small trees and thorn bushes. Their traditional way of life has been to herd sheep, goats and camels, moving about constantly in search of water. (See Places 51, page 440.)

"On the government map of Kenya, the realities of the Rendille's land are summarised in a few words 'Koroli Desert', it says, and just above this is the warning 'Liable to Flood'. There are two rainy seasons here: the long rains in April and May and the short rains in November. But the word 'season' suggests that the rains are much more predictable and steady than they are in reality. Add together rainfall from the long and the short rains and you arrive at only 150 mm on the Rendille's central plains in an average

year. But the word average means nothing here, because 'normal variation' from that average can bring only 35 mm of rain one year and 450 mm the next. Variation from place to place is even more erratic than variation from year to year. Rains can be heavy when they do come, and water often rushes off the baked ground in flash floods; thus the apparent contradiction of a flood-prone desert.

It may suddenly rain in a valley for the first time in ten years; and it may not rain there for another decade. Therefore, the Rendille do not so much follow the rains as chase them, rushing to get their animals on to new grasses, which are more easily digested and converted into milk than are the drier, older shoots."

(L. Timberlake, *Only One Earth*, p. 92)

Figure 16.5

Relationships between land use, altitude and slope in south-east Arran

Q 1 Account for the relationship between the height of the land and selected types of land use in south-east Arran as shown in Figure 16.5a.

2 Account for the relationship between the angle of slope and selected types of land use in south-east Arran as shown in Figure 16.5b.

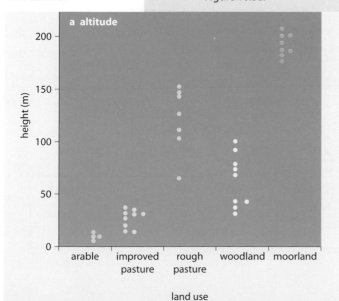

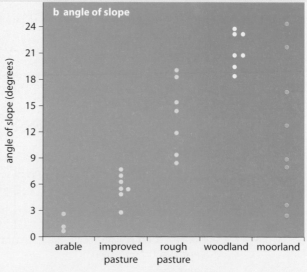

Global warming

Despite uncertainty as to the exact effects of global warming, scientists agree that the greenhouse effect will not only lead to an increase in temperature but also to changes in rainfall patterns. The global increase in temperature will allow many parts of the world to grow crops which at present are too cold for them: wheat will grow in more northerly latitudes in Canada and Russia, while maize, vines, oranges and peaches may flourish in southern England (Case Study 9, page 234). Of greater significance will be the changes in precipitation with some places becoming wetter and more stormy (Australia and south-east Asia) while others are likely to become drier (the wheat-growing areas of the American Prairies and Russian Steppes).

Places 47 Physical controls on farming in the former Soviet Union

Although the former Soviet Union is the largest country in the world, physical controls of climate, relief and soils have restricted farming to relatively small parts of the country. Of the land area of 22.27 million sq km, only 27 per cent was farmed in 1989 (10 per cent arable and 17 per cent pastoral), mainly in the deciduous forest belt, where the land had been cleared, and on the Steppes. The remaining 73 per cent (non-farmed) consisted of forest (42 per cent), tundra, desert and semi-desert (Figure 16.6).

After World War II, farmers were offered incentives to exceed their production targets. This task was most difficult for those farmers who were 'encouraged' to develop the 'virgin lands' (Figure 16.6), in such states as Kazakhstan, by ploughing up the natural grassland in order to grow wheat and other cereals. Unfortunately, the unreliable rainfall, with totals often less than 500 mm a year, did not guarantee reliable crop yields. Later, to help cereal production, irrigation schemes were begun. These have since been extended into semi-desert areas where cotton is now grown. This necessitated the Soviets constructing large-scale transfer schemes by which water from rivers in the wetter parts of the country was diverted to areas suffering a deficiency.

Future water-transfer schemes are even more ambitious and may never reach fruition as they involve diverting water from the northward-flowing Pechora, Ob and Yenisei rivers towards the south. Apart from the cost, environmentalists fear that this could result in the saline Arctic Ocean receiving less cold river water and then being warmed up sufficiently to cause the pack ice to melt and sea-levels to rise.

Figure 16.6

Physical controls on farming in the former Soviet Union

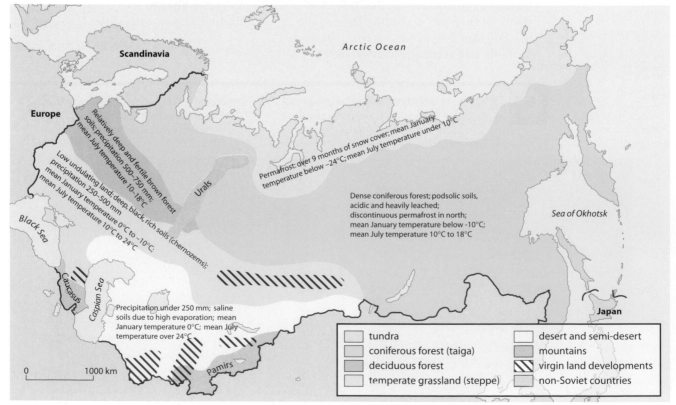

Cultural (human) factors affecting farming

Land tenure

Farmers may be owner-occupiers, tenants, landless labourers or state employees on the land which they farm. The **latifundia** system is still common to most Latin American countries. The land is organised into large, centrally managed estates worked by peasants who are semi-serfs. Even in the mid 1980s it was estimated that in Brazil 70 per cent of the land belonged to 3 per cent of the landowners. Land is worked by the landless labourers among the peasantry who sell their labour, when conditions permit, for substandard wages on the large estates or commercial plantations.

Other peasant farmers in Latin America have some land of their own held under insecure tenure arrangements. This land may be owned by the farmer, but it is more likely to have been rented from a local landowner or pawned to a moneylender. This latter type of tenancy takes two forms: cash-tenancy and share-cropping. **Cash-tenancy** is when farmers have to give as much as 80 per cent of their income or a fixed pre-arranged rent to the landowner. If the farmer has a short-term lease, he tends to overcrop the area and cannot afford to use fertiliser or to maintain farm buildings. If the lease is long-term, the farmer may try to invest but this often leads to serious debt. **Share-cropping** is when the farmer has to give part of his crop direct to the landowner. As this fraction is usually a large one, the farmer works hard with little incentive and remains poor. This system operated in the cotton belt of the USA following the abolition of slavery and still persists in places. Both forms of tenancy, together with that of latifundia, resemble feudal systems found in earlier times in western Europe. The plantation is a variant form of the large estate system in that it is usually operated commercially, producing crops for the world market rather than for local use as in latifundia. On some plantations (oil palm in Malaysia), the labourers are landless but are given a fixed wage; on others (sugar in Fiji), they are smallholders as well as receiving a payment.

In economically more developed, capitalist countries, many farmers are owner-occupiers — i.e. they own, or have a mortgage on, the farm where they live and work. Such a system should, in theory, provide maximum incentives for the farmer to become more efficient and to improve his land and buildings.

In sharp contrast to the neo-feudalist (latifundia, cash-tenancy and share-cropping) and capitalist systems of land tenure is the socialist system. In the former USSR, and certainly before the *perestroika* of the late 1980s and the subsequent break-up of the Soviet Union, the individual farmer and the company-run estates were replaced by the *kolkhoz* (collective farm) and *sovkhoz* (state farm) system of organisation and management (Places 48). Another form of socialist tenure is the **commune** system which operated during the early years of communism in China (Places 49).

Places 48   Land tenure in the former Soviet Union (USSR)

The state ownership of land was fundamental to communist ideology. In the former USSR, the belief was that large farms were preferable to smallholdings and that cooperative enterprise was better than that undertaken by individuals. During the 1920s, the USSR, and especially the rapidly growing industrialised towns, experienced severe food shortages. In response, Stalin introduced the **kolkhoz** system by which farmers, who worked small farms, were forced to join with their neighbours to form large **collective** units. The land was then leased from the government and managed by an elected chairperson and a committee representing up to 400 families. The farm produce was collected and sold to the state at a fixed price. Income from this was then equally divided amongst all the families. Once all the families had fulfilled their collective tasks, they could grow, on minute plots, and sell their own produce such as vegetables, milk and eggs.

The *kolkhoz* were mainly confined to the Ukraine, the western part of the Steppes which contained the country's best farmland (Figure 16.6). Despite considerable resistance from the Russian peasants, the number of individually owned farms fell from 256 million in 1928 to 0.2 million in 1940. The peasants, resentful at losing their land, found little incentive in a 'shared income economy' and often failed to meet the production targets set by the socialist state. The *kolkhoz* were overpopulated in relation to their inputs and resources and after 1940 were increasingly replaced by the **sovkhoz** system.

A sovkhoz was an even larger state-owned farm (Figure 16.7) where the workers were paid a weekly wage in the same way as labourers in a factory. The Soviet government favoured this system as it was able to control and set production targets, determine which type of

crop was to be grown or animal reared on each unit, and could control the sale and marketing of produce. The state provided the necessary capital resources, such as machinery and fertiliser (Figure 16.8). The sovkhoz workers, who lived in blocks of flats (Figure 16.9), were divided into 'brigades', each of which specialised in a particular job — mechanics, tractor drivers, accountants. Each 'farm' was self-contained, with its own shops, schools, libraries and places of recreation.

Figure 16.7

Changes in land tenure in the USSR, 1940–75

	Percentage of all farmland	Number in 1940	Number in 1975	Size of community (families)	Average size (hectares)
Kolkhoz	38	130 000	27 000	400	500
Sovkhoz	61	4 200	24 000	4000–10 000	1 500
Privately owned	1	2 000	0?	1	1

Figure 16.8

A state grain centre in the Kustanay region of Kazakhstan

Figure 16.9

Housing on a *sovkhoz* in Byelorussia

For most of this century, the individual farmer as a decision-maker has not existed in the USSR, although this situation began to change in the late 1980s with, initially, the introduction of *perestroika* and, later, the break-up of the USSR.

We have already seen on several occasions that geography is dynamic — i.e. its subject matter is constantly changing. Farming in the former USSR may prove to be a classic example. The

first edition of this book was written at a time when the then president of the USSR, Mikhail Gorbachev, was attempting to revitalise the Soviet economy by 'injecting a dose of capitalism'. Gorbachev has gone (and has been forgotten?), but have his policies on reducing state controls gone with him? You will be in a better position to describe any changes and to assess their significance at the time when you read this section.

Inheritance laws and the fragmentation of holdings

In several countries, inheritance laws have meant that on the death of a farmer the land is divided equally between all his sons (rarely between daughters). Also, dowry customs may include the giving of land with a daughter on marriage. Such traditions have led to the sub-division of farms into numerous scattered and small fields. In Britain, fragmentation of land parcels may also result from the legacy of the open-field system (page 382) or, more recently, from farmers buying up individual fields as they come onto the market. Fragmentation results in much time being wasted in moving from one distant field to another, and may cause problems of access. It may, however, be of benefit as it can enable a wider range of crops to be grown.

Farm size

Inheritance laws, as described above, tend to reduce the size of individual farms so that, often, they can operate only at subsistence level or below. In most of the EU and North America, the trend is for farm sizes to increase in order to make them, as the activity becomes more an **agribusiness**, more efficient and profitable. Capital-intensive farms use much machinery, fertiliser, etc. and have a wide choice in types of production.

In south-east Asia and parts of Latin America and Africa, the rapid expansion of population is having the reverse effect. Farms, already inadequate in size, are being further divided and fragmented, making them too small for mechanisation (even if the farmers could afford machines). They are increasingly limited in the types of production possible, and output in certain areas, such as sub-Saharan Africa, is falling. Although farms of only 1 ha can support families in parts of south-east Asia where intensive rice production occurs and several croppings a year are possible, the average plot size in many parts of Taiwan, Nepal and South Korea has fallen to under 0.5 ha (about the size of a football pitch). In comparison, farms of several hundred hectares are needed to support a single family in those parts of the world where farming is marginal (upland sheep farming in Britain, cattle ranching in northern Australia). Bearing in mind the dangers of making generalisations (Framework 8, page 322), Figure 16.10 gives some of the spatial variations, and reasons for these variations, between large and small farms. Differences in farm size also affect other types of land use and the landscape.

Economic factors affecting farming

However favourable the physical environment may be, it is of limited value until human resources are added to it. Economic man, the term introduced by von Thünen, applies resources to maximise profits. Yet these resources are often available only in developed countries or where farming is carried out on a commercial scale.

Transport

This includes the types of transport available, the time taken and the cost of moving raw materials to the farm and produce to the market. For perishable commodities, like milk and fresh fruit, the need for speedy transport to the market demands an efficient transport network, while for bulky goods, like potatoes, transport costs must be lower for output to be profitable. In both cases, the items should ideally be grown as near to their market as possible.

Markets

The role of markets is closely linked with transport (perishable and bulky goods). Market demand depends upon the size and affluence of the market population, its religious and cultural beliefs (fish consumed in Catholic countries, abstinence from pork by Jews), its preferred diet, changes in taste and fashion over time (vegetarianism) and health scares (BSE).

Figure 16.10

Reasons for spatial variations in farm size

Large farms are often...	extensive on more marginal land	commercial in the EU and North America	animal grazing (sheep, cattle ranching); plantations; and temperate cereals (wheat)	further from large cities	areas of low population density and/or under-populated	increasing in size and efficiency due to amalgamation and mechanisation
Small farms are often...	intensive on flat, fertile land	subsistence in Asia, Latin America and Africa	tropical crops (rice); and market gardening	nearer large cities	areas of high population density and/or overpopulated	decreasing in size and efficiency due to fragmention and hand labour

Capital

Most economically developed countries, with their supporting banking systems, private investment and government subsidies, have large reserves of readily available finance, which over time have been used to build up **capital-intensive** types of farming (Figure 16.24) such as dairying, market gardening and mechanised cereal growing. Capital is often obtained at relatively low interest rates but remains subject to the law of diminishing returns. In other words, the increase in input ceases to give a corresponding increase in output, whether that output is measured in fertiliser, capital investment in machinery, or hours of work expended.

Farmers in developing countries, often lacking support from financial institutions and having limited capital resources of their own, have to resort to **labour-intensive** methods of farming (Figure 16.24). A farmer wishing to borrow money may have to pay exorbitant interest rates and may easily become caught up in a spiral of debt. The purchase of a tractor or harvester can prove a liability rather than a safe investment in areas of uncertain environmental, economic and political conditions.

Technology

Technological developments such as new strains of seed, cross-breeding of animals, improved machinery and irrigation may extend the area of optimal conditions and the limits of production (Green Revolution, page 464). Lacking in capital and expertise, developing countries are rarely able to take advantage of these advances and so the gap between them and the economically developed world continues to increase.

Governments

We have already seen that in centrally planned economies it is the state, not the individual, which makes the major farming decisions (Places 48 and 49). In the UK, farmers have been helped by government subsidies. Initially, organisations such as the Milk and Egg Marketing Boards ensured that British farmers got a guaranteed price for their products. Today, most decisions affecting British farmers are made by the EU. Sometimes EU policy benefits British farmers (support grants to hill farmers) and sometimes it reduces their income (reduction in milk quotas). Certainly countries in the EU have improved yields, evident by their food surpluses (page 448) and have adapted farming types to suit demand. Governments are also responsible for agricultural training schemes and for giving advice on new methods.

Places 49 Farming in China

Pre-1949

Before the establishment of the People's Republic in 1949, farming in China was typical of south-east Asia — i.e. it was mostly intensive subsistence (page 442). Farms were extremely small and fragmented with the many tenants having to pay up to half of their limited produce to rich, often absentee landlords. Cultivation was manual or using oxen. Despite long hours of intensive work, the output per worker was very low. The need for food meant that most farmland was arable, with livestock restricted to those kept for working purposes or which could live on farm waste (chickens and pigs). The type of crops grown, and the number of croppings per year, changed from north to south as both rainfall totals and the length of the growing season increase (Figure 16.11).

People's communes, 1958

After taking power in 1949, the communists confiscated land from the large landowners and divided it amongst the peasants. However, most plots proved too small to support individual farmers. After several interim experiments, the government created the 'people's communes'. The communes, which were meant to become self-sufficient units, were organised into a three-tier hierarchy with communist officials directing all aspects of life and work.

Production teams (foot of hierarchy), consisting of some 50 families (300 people), were responsible for, on average, 20 ha. Each team was responsible for its own finances and, after the payment of taxes for welfare services, it could distribute any surplus income.

Brigade teams (centre of hierarchy) were formed from 10 production teams and consisted of around 3000 people and covered 200 ha. Although responsible for overall planning, they left the farming details to the production team.

The commune (head of hierarchy) was composed of five brigade teams — i.e. 15 000 people and 100 ha. Members of the commune elected a people's council, who elected a sub-committee to ensure that production targets,

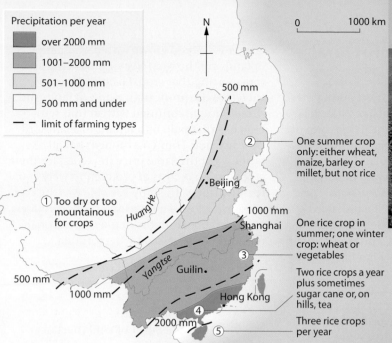

Figure 16.11

Farming in China: the relationship between precipitation and farming type

Precipitation per year
- over 2000 mm
- 1001–2000 mm
- 501–1000 mm
- 500 mm and under
- – – limit of farming types

① Too dry or too mountainous for crops

② One summer crop only: either wheat, maize, barley or millet, but not rice

③ One rice crop in summer; one winter crop: wheat or vegetables

④ Two rice crops a year plus sometimes sugar cane or, on hills, tea

⑤ Three rice crops per year

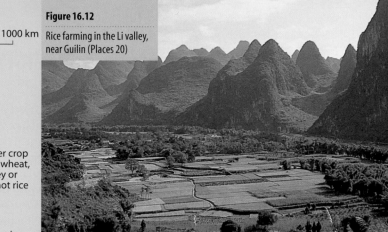

Figure 16.12

Rice farming in the Li valley, near Guilin (Places 20)

Figure 16.13

Typical farmhouse near Guilin, China

set by the Central Planning Committee (the government) in a series of Five Year Plans, were met. The committee was also responsible for providing an adequate food supply to make the unit self-supporting (crops, livestock, fruit and fish), for providing small-scale industry (mainly food processing and making farm implements), organising housing and services (hospital, schools) and for flood control and irrigation systems. Most communes had a research centre which trained workers to use new forms of machinery, fertiliser and strains of seed correctly (Green Revolution, page 464). By pooling their resources, farmers were able to increase yields per hectare.

Responsibility system, 1979

The introduction in 1979 of this more flexible approach, which encouraged farming families to become more 'responsible', preceded the abolition of the commune system in 1982. Under it, individual farmers were given land in their own village or district. They then had to take out contracts with the government, initially for 3 years but now extended to 15, to deliver a fixed amount of produce. To help meet their quota, individual farmers were given tools and seed. Once farmers had fulfilled their quotas, they could sell the remainder of their produce on the open market for their own profit. The immediate effect, due to farmers working much harder, was an increase in yields by an average of 6 per cent per year throughout the 1980s. Rural markets thrived and some farmers have become quite wealthy. Profits were used to buy better seed and machinery and to create village industry. Although most farmers have improved their standard of living, admittedly from an extremely low base, those living near to large cities (large nearby market) and in the south of the country (climatic advantages and proximity to the open market of Hong Kong) have benefitted the most.

Farming in the Li valley, Guilin

Rice is still the staple crop (Figure 16.12) with, due to the favourable climate, two crops being grown each year (May to early August and late August to November). During the colder winter, a third crop is grown — usually a vegetable for either the family (beans) or the pigs (spinach). Farmers are increasingly beginning to diversify into specialised farming by growing fruit (oranges, watermelons, peaches and water-chestnut) for the rapidly growing nearby tourist/industrial city of Guilin (specialist cash crops in other parts of southern China include oil-seed rape, cotton and sugar cane). Most farmers still use water buffalo for ploughing and to provide fertiliser and, for their own consumption, have a fish farm (grass carp) and keep ducks, chickens and pigs.

Von Thünen's model of rural land use

Heinrich von Thünen, who lived during the early 19th century, owned a large estate near to the town of Rostock (on the Baltic coast of present-day Germany). He became interested in how and why agricultural land use varied with distance from a market and published his ideas in a book entitled *The Isolated State* (1826). To simplify his ideas, he produced a model in which he recognised that the patterns of land use around a market resulted from competition with other land uses. Like other models, von Thünen's makes several simplifying assumptions. These include:

- The existence of an isolated state, cut off from the rest of the world (transport was poorly developed in the early 19th century).
- In this state, one large urban market (or central place) was dominant. All farmers received the same price for a particular product at any one time.
- The state occupied a broad, flat, featureless plain which was uniform in soil fertility and climate and over which transport was equally easy in all directions.
- There was only form of transport available. (In 1826 this was the horse and cart.)
- The cost of transport was directly proportional to distance.
- The farmers acted as 'economic men' wishing to maximise their profits and all having equal knowledge of the needs of the market.

In his model, von Thünen tried to show that with increasing distance from the market:

a the intensity of production decreased; and

b the type of land use varied.

Both concepts were based upon **locational rent** (*LR*) which von Thünen referred to as **economic rent**. Locational rent is the difference between the revenue received by a farmer for a crop grown on a particular piece of land and the total cost of producing and transporting that crop. Locational rent is therefore the profit from a unit of land, and should not be confused with **actual rent** which is that paid by a tenant to a landlord.

Since von Thünen assumed that all farmers got the same price (revenue) for their crops and that costs of production were equal for all farmers, the only variable was the cost of transport which increased proportionately with distance from the market. Locational rent can be expressed by the formula:

$$LR = Y(m - c - td)$$

where:

- LR = locational rent;
- Y = yield per unit of land (hectares);
- m = market price per unit of commodity;
- c = production cost per unit of land (ha);
- t = transport cost per unit of commodity;
- d = distance from the market.

Since Y, m, c and t are constants, it is possible to work out by how much the LR for a commodity decreases as the distance from the market increases. Figure 16.14 shows that LR (profit) will be at its maximum at **M** (the market), where there are no transport costs. LR decreases from **M** to **X** with diminishing returns, until at **X** (the **margin of cultivation**) the farmer ceases production because revenue and costs are the same — i.e. there is no profit.

Details of von Thünen's theory

Von Thünen tried to account for the location of several crops in relation to the market. He suggested that

a bulky crops, such as potatoes, should be grown close to the market as their extra weight would increase transport costs;

b perishable goods, such as vegetables and dairy produce, should also be produced as near as possible to the market (he wrote before refrigeration had been introduced); and

c intensive crops should be grown nearer to the market than extensive crops (Figure 16.15).

Consequently, bulky, perishable and intensive crops (or commodities) will have steep R lines (Figures 16.14, 16.15 and 16.16).

Figure 16.14

The relationship between locational (economic) rent and distance from the market

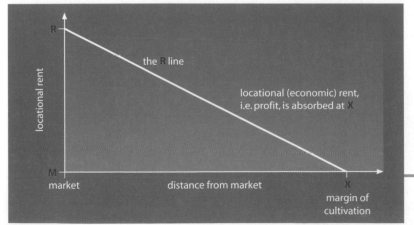

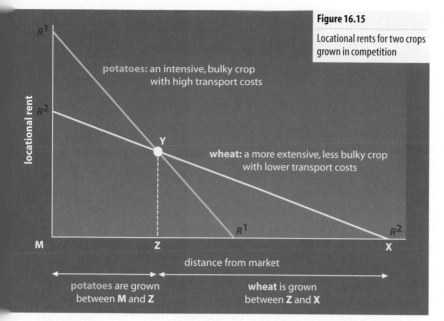

R^1
R^2

locational rent

potatoes: an intensive, bulky crop
with high transport costs

Y

wheat: a more extensive, less bulky crop
with lower transport costs

R^1 R^2

M   Z   X

distance from market

potatoes are grown
between **M** and **Z**

wheat is grown
between **Z** and **X**

Figure 16.15

Locational rents for two crops
grown in competition

Figure 16.16

Locational rents for three
commodities in competition

Figure 16.15 shows the result of two crops,
potatoes and wheat, grown in competition.
The two *R* lines, showing the locational rent

Farm product	Market price per unit of commodity	Production costs per unit of land (hectares)	Transport costs per unit of commodity	Profit if grown at market
Potatoes	100	30	10	70
Wheat	65	20	3	45
Wool	45	15	1	30

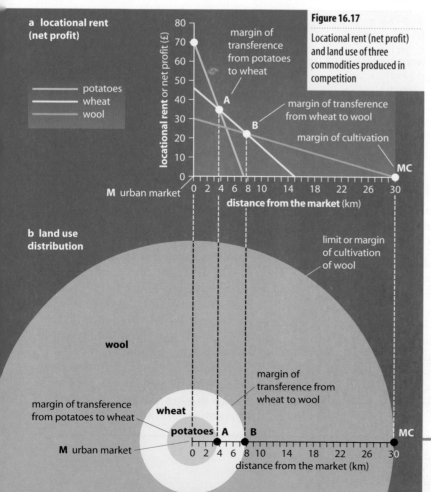

**a locational rent
(net profit)**

— potatoes
— wheat
— wool

locational rent or net profit (£)

margin of
transference
from potatoes
to wheat

A

B

margin of transference
from wheat to wool

margin of cultivation

MC

M urban market

distance from the market (km)

**b land use
distribution**

limit or margin
of cultivation
of wool

wool

margin of
transference from
wheat to wool

margin of transference
from potatoes to wheat

wheat

potatoes A B

MC

M urban market

distance from the market (km)

Figure 16.17

Locational rent (net profit)
and land use of three
commodities produced in
competition

or profit for each crop, intersect at **Y**. If a
perpendicular is drawn from **Y** to **Z**, locational rent can be translated into land use.
Potatoes, an intensive, bulky crop, are grown
near to the market (between **M** and **Z**) as their
transport costs are high. Wheat, a more
extensively farmed and less bulky crop, is
grown further away (between **Z** and **X**)
because it incurs lower transport costs.

What happens if three crops are grown in
competition? This is the combination of von
Thünen's two concepts: variation of intensity
and type of land use with distance from
market. Let us suppose that wool is produced in addition to potatoes and wheat
(Figure 16.16).

Potatoes give the greatest profit if grown
at the market, and wool the least. However,
as potatoes cost £10 to transport every kilometre, after 7 km their profit will have been
absorbed in these costs (£70 profit – £70
transport = £0). This has been plotted in
Figure 16.17a which is a **net profit graph**.
Wheat costs £3/km to transport and so can
be moved 15 km before it becomes unprofitable (£45 profit – £45 transport = £0).
Wool, costing only £1/km to transport, can
be taken 30 km before it, too, becomes
unprofitable. Figure 16.17 also shows that —
although potatoes can be grown profitably
for up to 7 km from the market — at point **A**,
only 3.5 km from the market, wheat farming
becomes equally profitable and that, beyond
that point, wheat farming is more lucrative.
Similarly, wheat can be grown up to 15 km
from the market, but beyond 7.5 km it is less
profitable than, and is therefore replaced by,
wool. The point at which one type of land
use is replaced by another is called the
margin of transference.

The types of land use can now be plotted
spatially. Figure 16.17b shows three concentric circles, with the market as the
common central point. As on the graph,
potatoes will be grown within 3.5 km of the
market. This is because competition for land,
and consequently land values, are greatest
here so only the most intensive farming is
likely to make a profit. The plan also shows
that wheat is grown between 3.5 and 7.5 km
from the market, while between 7.5 and 30
km, where the land is cheaper, farming is
extensive and wool becomes the main product. Von Thünen's land use model is therefore based on a series of concentric circles
around a central market.

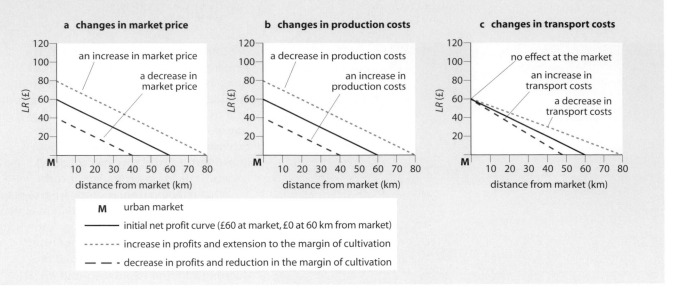

a **changes in market price**

b **changes in production costs**

c **changes in transport costs**

M urban market

——— initial net profit curve (£60 at market, £0 at 60 km from market)

------- increase in profits and extension to the margin of cultivation

— — · decrease in profits and reduction in the margin of cultivation

Figure 16.18

Some causes of variation in locational rent

The formula for locational rent (page 430) assumed that market prices (*m*), production costs (*c*) and transport costs (*t*) were all constant. What would happen to a crop's area of production if each of these in turn were to alter?

If the market price falls or the cost of production increases, there is a decrease in both the profit and the margin of cultivation of that crop (Figure 16.18a and b). Conversely, if the market price rises or the costs of

production decrease, profits would rise, leading to an extension in the margin of cultivation. Changes in transport costs will not affect any farm at the market (Figure 16.18c) but an increase in transport costs reduces profits for distant farms, causing a decrease in the margin of cultivation. Conversely, a fall in transport costs makes those distant farms more profitable and enables them to extend their margin of cultivation.

Q

1 a Draw a graph to show net profit curves for market gardening, dairying and wheat using the following data:
locational rent (profit) for market gardening is £160 at the market and £0 at 40 km;
locational rent for dairying is £120 at the market and £0 at 60 km;
locational rent for wheat is £80 at the market and £0 at 80 km.

b Give the two marginal distances for market gardening and dairying.

2 Label on your graph:
a the margin of transference from market gardening to dairying;
b the margin of transference from dairying to wheat;
c the margin of cultivation for wheat.

3 Draw a diagram, using three concentric circles, to show the location and extent of the three types of land use surrounding the central market.

4 a Explain how the graph and diagram illustrate the locational rent for market gardening, dairying and wheat.
b Explain why land use changes at the two margins of transference.

Von Thünen's land use model

Von Thünen combined his conclusions on how the intensity of production decreased and the type of land use varied with distance from the market to create his model (Figure 16.19a). He suggested six types of land use which were located by concentric circles.

1 Market gardening (horticulture) and dairying were practised nearest to the city, due to the perishability of the produce. Cattle were kept indoors for most of the year and provided manure for the fields.

2 Wood was a bulky product much in demand as a source of fuel and as a building material within the town (there was no electricity when von Thünen was writing). It was also expensive to transport.

3 An area with a 6-year crop rotation based on the intensive cultivation of crops (rye, potatoes, clover, rye, barley and vetch) and with no fallow period.

4 Cereal farming was less intensive as the 7-year rotation system relied increasingly on animal grazing (pasture, rye, pasture, barley, pasture, oats and fallow).

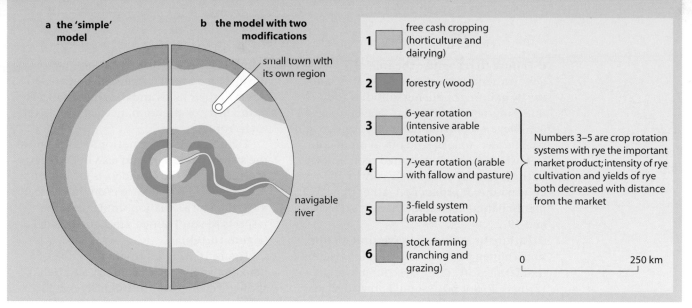

a the 'simple' model

b the model with two modifications

small town with its own region

navigable river

1 free cash cropping (horticulture and dairying)

2 forestry (wood)

3 6-year rotation (intensive arable rotation)

4 7-year rotation (arable with fallow and pasture)

5 3-field system (arable rotation)

6 stock farming (ranching and grazing)

Numbers 3–5 are crop rotation systems with rye the important market product; intensity of rye cultivation and yields of rye both decreased with distance from the market

0   250 km

Figure 16.19

The von Thünen land use model

5 Extensive farming based on a 3-field crop rotation (rye, pasture and fallow). Products were less bulky and perishable to transport and could bear the high transport costs.

6 Ranching with some rye for on-farm consumption. This zone extended to the margins of cultivation, beyond which was wasteland.

Modifications to the model

Later, von Thünen added two modifications in an attempt to make the model more realistic (Figure 16.19b). This immediately distorted the land use pattern and made it more complex. The inclusion of a navigable river allowed an alternative, cheaper and faster form of transport than his original horse and cart. The result was a linear, rather than a circular, pattern and an extension of the margin of cultivation. The addition of a secondary urban market involved the creation of a small trading area which would compete, in a minor way, with the main city.

Later still, von Thünen relaxed other assumptions. He accepted that climate and soils affected production costs and yields (though he never moved from his concept of the featureless plain) and that, as farmers do not always make rational decisions, it was necessary to introduce individual behavioural elements.

Why is it difficult to apply von Thünen's ideas to the modern world?

As no model is perfect, all are open to criticism. Criticisms of von Thünen's model can be grouped under four headings:

a Oversimplification There are very few places with flat, featureless plains and where such landscapes do occur they are likely to

contain several markets rather than one. As large areas with homogeneous climate and soils rarely exist, certain locations will be more favourable than others. Similarly, the 'isolated state' is rarely found in the modern world — Albania may be nearest to this situation — and there is much competition for markets both within and between countries. Von Thünen accepted that while his model simplified real-world situations, the addition of two variables immediately made it more complex (Figure 16.19b).

b Outdatedness As the model was produced 170 years ago, critics claim it is out-dated and of limited value in modern farming economics. Certainly since 1826 there have been significant advances in technology, changing uses of resources, pressures created by population growth, and the emergence of differing economic policies. The invention of motorised vehicles, trains and aeroplanes has revolutionised transport, often increasing accessibility in one particular direction and making the movement of goods quicker and relatively cheaper. Milk tankers and refrigerated lorries allow perishable goods to be produced further from the market (London uses fresh milk from Devon) and stored for longer (the EU's food mountains). The use of wood as a fuel in developed countries has been replaced by gas and electricity and so trees need not be grown so near to the market, while supplies of timber in developing countries are being rapidly consumed and not always replaced. Improved farming techniques using fertilisers and irrigation have improved yields and extended the margins of cultivation. Elsewhere, farmland has been taken over by urban growth or used by competitors who obtain higher economic rents.

c Failure to recognise the role of government Governments can alter land use by granting/reducing subsidies and imposing/removing quotas. The EU (page 455) have, recently, reduced milk quotas and paid farmers to take land out of production (set-aside). Centrally planned economies, as in the former USSR, directly control the types and amount of production rather than manipulating market mechanisms (Places 48 and 49).

d Failure to include behavioural factors Von Thünen has been criticised for assuming that farmers are 'rational economic men'. Farmers do not possess full knowledge, may not always make rational or consistent decisions, may prefer to enjoy increased leisure time rather than seeking to maximise profits and may be reluctant to adopt new methods. Farmers, as human beings, may have different levels of ability, ambition, capital and experience and none can predict changes in the weather, government policies or demand for their product.

How relevant is von Thünen's theory to the modern world?
Chapter 14 concluded with the observation that although theories are difficult to observe in the real world they *are* useful because reality can be measured and compared against them. This section and the questions which follow look at examples at different scales (local and international) which show similarities with von Thünen's model (see also Figure 16.37).

Figure 16.20 takes, at a **local** level, a relatively remote, present-day hill village in the Mediterranean lands of Europe. Many villages in southern Italy, Spain and Greece have hilltop sites (in contrast to von Thünen's featureless plain) where, usually, transport links are poor, affluence is limited and the village provides the main, perhaps the only, market (Figure 14.7c). As the distance from the village increases, the amount of farmland used, and the yields

from it, decrease. Two critical local factors are the distance which farmers are prepared to travel to their fields and the amount of time, or intensity of attention, needed to cultivate each crop.

Figure 16.21, at the **national** level, shows the spatial pattern of land uses in Uruguay. The capital city, Montivideo, is located on the coast, and Fray Bentos is on the navigable Rio Uruguay: a situation similar in some respects to von Thünen's modified model (Figure 16.19b).

Figure 16.22 is a simplified version of a map, drawn by van Valkenburg and Held, showing changes in the intensity of agriculture in Europe — the international level. They took eight crops: barley, hay, maize, oats, potatoes, rye, sugar beet and wheat (notice that only wheat is typically Mediterranean) and for each worked out the average yield per acre for each country. They also gave, as an index of 100, the average yield for Europe as a whole. The map shows a core, or optimum zone (Figure 16.1), beyond which are a series of zones of decreasing intensity. Areas with the lowest yields lie on the periphery of Europe where, as in the optima and limits model, the climate is less suitable (too cold in northern Scandinavia, summer drought in the Mediterranean); the terrain is inhospitable (the Alps and the Pyrenees); and the soils are infertile and eroded (the Apennines).

Conclusions
Von Thünen's land use model still has some modern relevance, particularly at the local level, provided its limitations are understood and accepted. His concept of locational rent, which is useful in studying urban as well as rural land use (page 392) is still applicable today. Concentric circles of land use may not exist around every urban centre, yet many areas, perhaps particularly at the local scale, do have patterns which show a similarity to the von Thünen land use model, and can be partially explained by it.

---

1 Figure 16.20 shows some of the farming characteristics in an area surrounding a hilltop village in southern Europe.
  a Describe and attempt to explain the general pattern of land use.
  b To what extent and for what reasons does this general pattern of land use

  i conform to; and
  ii differ from the principles of the von Thünen model (i.e. how closely are changes in the intensity of production and type of land use related to distance from the market)?

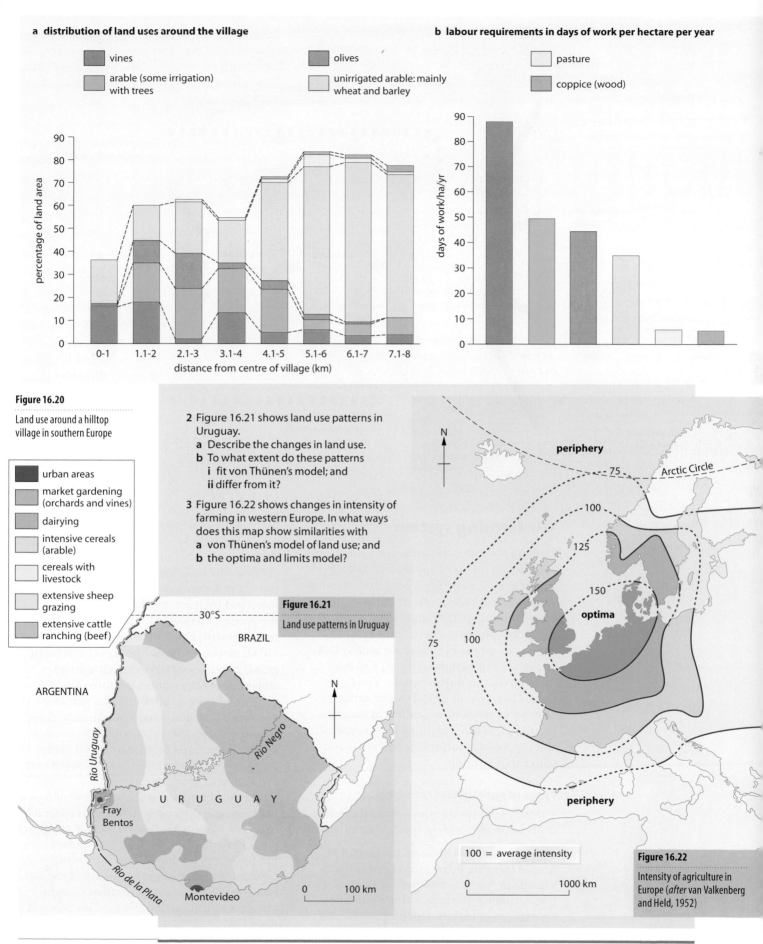

a distribution of land uses around the village

- vines
- arable (some irrigation) with trees
- olives
- unirrigated arable: mainly wheat and barley

b labour requirements in days of work per hectare per year

- pasture
- coppice (wood)

distance from centre of village (km)

percentage of land area

days of work/ha/yr

Figure 16.20

Land use around a hilltop village in southern Europe

2 Figure 16.21 shows land use patterns in Uruguay.
  a Describe the changes in land use.
  b To what extent do these patterns
   i fit von Thünen's model; and
   ii differ from it?

3 Figure 16.22 shows changes in intensity of farming in western Europe. In what ways does this map show similarities with
  a von Thünen's model of land use; and
  b the optima and limits model?

- urban areas
- market gardening (orchards and vines)
- dairying
- intensive cereals (arable)
- cereals with livestock
- extensive sheep grazing
- extensive cattle ranching (beef)

Figure 16.21

Land use patterns in Uruguay

ARGENTINA

BRAZIL

30°S

URUGUAY

Rio Uruguay

Rio Negro

Fray Bentos

Rio de la Plata

Montevideo

0   100 km

periphery

Arctic Circle

75

100

125

150

optima

75

100

periphery

100 = average intensity

0   1000 km

Figure 16.22

Intensity of agriculture in Europe (*after* van Valkenberg and Held, 1952)

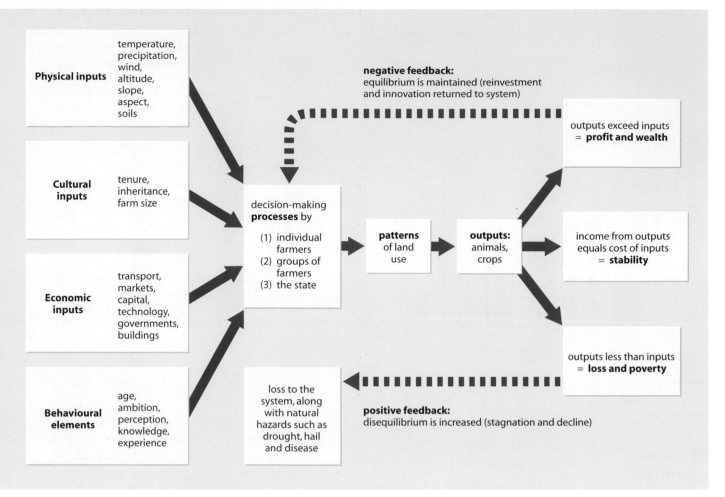

Figure 16.23

The farming system

The farming system

Farming is another example of a system, and one which you may have studied already (Framework 1, page 39). The system diagram (Figure 16.23) shows how physical, cultural, economic and behavioural factors form the inputs. In areas where farming is less developed, physical factors are usually more important but as human inputs increase, these physical controls become less significant. This system model can be applied to all types of farming, regardless of scale or location. It is the variations in inputs which are responsible for the different types and patterns of farming.

Types of agricultural economy

The simplest classifications show the contrasts between different types of farming.

1 Arable, pastoral and mixed farming

Arable farming is the growing of crops, usually on flatter land where soils are of a higher quality. It was the development of new strains of cereals which led to the first permanent settlements in the Tigris–Euphrates, Nile and Indus valleys (Figure 14.1). Much later, in the mid 19th century, the building of the railways across the Prairies, Pampas and parts of Australia led to a rapid increase in the global area 'under the plough' (page 446). Today, there are few areas left with a potential for arable farming. This fact, coupled with the rapid increase in global population, has led to continued concern over the world's ability to feed its present and future inhabitants, a fear first voiced by Malthus (page 356). Already, there has been a decrease in the amount of arable land in some parts of the world, especially those parts of Africa affected by drought and soil erosion (Figure 16.59).

Pastoral farming is the raising of animals, usually on land which is less favourable to arable farming (i.e. colder, wetter, steeper and higher land). However, if the grazed area has too many animals on it, its carrying capacity is exceeded or the quality of the soil and grass is not maintained, then erosion and desertification may result (Case Study 7, page 175).

Mixed farming is the growing of crops and the rearing of animals together. It is practised on a commercial scale in developed countries, where it reduces the financial risks of relying upon a single crop or animal (monoculture), and at a subsistence level in developing countries, where it reduces the risks of food shortage.

2 Subsistence and commercial farming

Subsistence farming is the provision of food by farmers only for their own family or local community — there is no surplus (Places 53, page 442). The main priority of subsistence farmers is self-survival which they try to achieve, whenever possible, by growing/rearing a wide range of crops/animals. The fact that subsistence farmers are rarely able to improve their output is due to a lack of capital, land and technology — and not to a lack of effort or ability. They are the most vulnerable to food shortages.

Commercial farming takes place on a large, profit-making scale. Commercial farmers, or the companies for whom they work, seek to maximise yields per hectare. This is often achieved, especially within the tropics, by growing a single crop or rearing one type of animal (Places 54, page 444). Cash-cropping operates successfully where transport is well developed, domestic markets are large and expanding, and there are opportunities for international trade (Places 55, page 445, and 56, page 447).

3 Shifting and sedentary farming

Many of the earliest farmers moved to new land every few years, due to a reduction in yields and also reduced success in hunting and gathering supplementary foods. **Shifting cultivation** is now limited to a few places where there are low population densities and a limited demand for food; where soils are poor and become exhausted after three or four years of cultivation (Places 52, page 441); or where there is a seasonal movement of animals in search of pasture (Places 51, page 440). However, farming over most of the world is now **sedentary**, i.e. farmers remain in one place to look after their crops or to rear their animals.

4 Extensive and intensive cultivation

These terms have already been used in describing von Thünen's model (Figure 16.15). **Extensive farming** is carried out on a large scale, whereas **intensive farming** is usually relatively small-scale. Farming is extensive or intensive depending upon the relationship between three factors of production: labour, capital and land (Figure 16.24). Extensive farming occurs when:

- Amounts of labour and capital are small in relation to the area being farmed. In the Amazon Basin (Places 52), for example, the yields per hectare and the output per farmer are both low (Figure 16.24a).
- The amount of labour is still limited but the input of capital may be high. In the Canadian Prairies (Places 56), for example, the yields per hectare are often low but the output per farmer is high (Figure 16.24b).

Intensive farming occurs when:

- The amount of labour is high, even if the input of capital is low in relation to the area farmed. In the Ganges valley (Places 53), for example, the yields per hectare may be high although the output per farmer is often low (Figure 16.24c).
- The amount of capital is high, but the input of labour is low. In the Netherlands (Places 57, page 448), for example, both the yields per hectare and the output per farmer are high (Figure 16.24d).

Figure 16.24

Extensive and intensive farming *(after* Briggs)

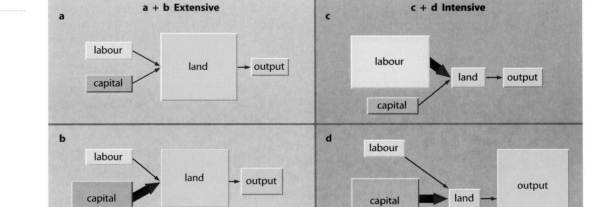

437

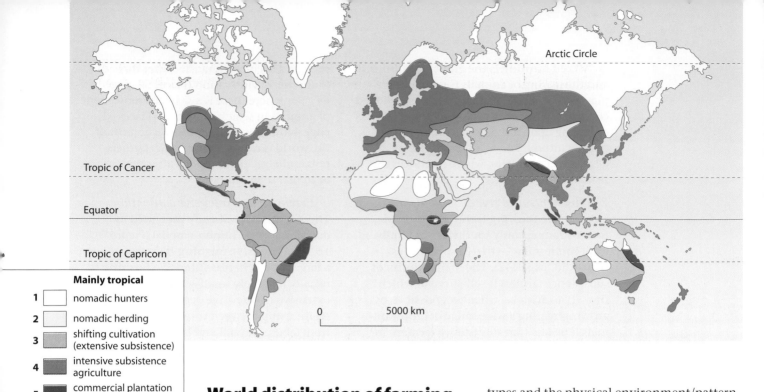

Mainly tropical

1	nomadic hunters
2	nomadic herding
3	shifting cultivation (extensive subsistence)
4	intensive subsistence agriculture
5	commercial plantation agriculture

Mainly temperate

6	livestock ranching (commercial pastoral)
7	cereal cultivation (commercial grain)
8	intensive commercial (mixed)
9	Mediterranean agriculture
10	irrigation
11	unsuitable for agriculture

Figure 16.25

Location of the world's major farming types

Arctic Circle

Tropic of Cancer

Equator

Tropic of Capricorn

0   5000 km

World distribution of farming types

There is no widely accepted consensus as to how the major types of world farming should be classified or recognised (Framework 5, page 151). There is disagreement over the basis used in attempting a classification (intensity, land use, tropical or temperate, level of human input, the degree of commercialisation); the actual number and nomenclature of farming types; and the exact distribution and location of the major types.

You should be aware that:

1 Boundaries between farming types, as drawn on a map, are usually very arbitrary.
2 One type of farming merges gradually with a neighbouring type: there are few rigid boundaries.
3 Several types of farming may occur within each broad area — as in West Africa, where sedentary cultivators live alongside nomadic herdsmen.
4 A specialised crop may be grown locally — e.g. a plantation crop in an area otherwise used by subsistence farmers.
5 Types of farming alter over a period of time with changes in economies, rainfall, soil characteristics, behavioural patterns and politics.

Figure 16.25 suggests one classification and shows the generalised location and distribution of farming types based upon the four variables described in the previous section. On a continental scale, this map demonstrates a close relationship between farming types and the physical environment/pattern of biomes (page 284). It disguises, however, the important human-economic factors which operate at a more local level.

The following section describes the main characteristics of each of these categories of farming together with the conditions favouring their development. A specific example is used in each case (which should be supplemented by wider reading) together with an account of recent changes or problems within that agricultural economy.

1 Hunters and gatherers

Some classifications ignore this group on the grounds that it is considered to be a relict way of life, with the original lifestyle now largely, or totally, destroyed by contact with the outside world. Others feel that it does not constitute a 'true' farming type, as no crops or domesticated animals are involved. It is included here as, before the advent of sedentary farming, all early societies had to rely upon hunting birds and animals, catching fish, and collecting berries, nuts and fruit in order to survive … which is surely why we rely upon farming today. There are now very few hunter-gatherer societies remaining — the Bushmen of the Kalahari, the Pygmies of central Africa, several Amerindian tribes in the Brazilian rainforest, and the Australian Aborigines (Places 50). All have a varied diet resulting from their intimate knowledge of the environment, but each needs an extensive area from which to obtain their basic needs.

When Captain Cook landed in Australia in 1770 there were about 300 000 Aborigines living on the island. Of Melanesian descent, there is evidence that they had lived there for over 50 000 years. Their rock paintings are older than those at Lascaux in France and pre-date the Egyptian pyramids by 10 000 years. The majority lived close to the northern coast or inland near to rivers (many of which were seasonal) or Ayers Rock which was sacred to them (Figure 7.6). They had strong religious beliefs and family ties.

Figure 16.26

An Australian aborigine

The Aborigines' diet, which some suggest contained over 350 different items of food, included emus and their eggs; kangaroos and wallabies; grubs and ants; acacia seeds, roots and grasses. Those living nearer rivers caught fish and crocodiles, while others on the coast ate turtles, crabs and mussels. Their main hunting weapons were boomerangs, long spears, shields, stone axes, fishing nets and fish hooks. As the food supply was limited, the population remained stable in size.

By 1930, only 70 000 Aborigines remained and their lifestyle was severely threatened. Recently, they have been given land rights to one-third of the Northern Territory (including Ayers Rock) and one-fifth of South Australia. Their demands for more land and improved civil rights were vocally expressed during the bicentennial celebrations of 1988. Today, relatively few Aborigines survive in their natural environment as economic and population pressures (numbers have now increased to 160 000) have forced many to leave the reserves and migrate to urban centres. There, Aborigines tend to experience discrimination and poverty, an infant mortality rate twice the national average and a life expectancy 20 years less than the national average. They have poor housing and education and, with few rights, many fall victim to social problems. Indeed, Aborigines face all the problems of an immigrant ethnic minority when in fact they were the indigenous inhabitants.

2 Nomadic herding

In areas where the climate is too extreme to support permanent settled agriculture, farmers become **nomadic pastoralists**. They live in inhospitable environments where vegetation is sparse and the climate is arid or cold (Places 51). The movement of most present-day nomads is determined by the seasonal nature of rainfall and the need to find new sources of grass for their animals (the Bedouin and Tuareg in the Sahara, the Fulani in western Africa and the Maasai in Kenya). The Lapps of northern Finland have to move when their pastures become snow-covered in winter, while the Fulani may migrate to avoid the tse-tse fly.

There are two forms of nomadism. **Total nomadism** is where the nomad has no permanent home, while **semi-nomads** may live seasonally in a village. There is no ownership of land and the nomads may travel extensive distances, even across national frontiers, in search of fresh pasture. There may be no clear migratory pattern, but migration routes increase in size under adverse conditions, e.g. during droughts in the Sahel. The animals are the source of life. Depending upon the area, they may provide milk, meat and blood as food for the tribe; wool and skins for family shelter and clothing; dung for fuel; mounts and pack animals for transport; and products for barter. Just as sedentary farmers will not sell their land unless in dire economic difficulty, similarly pastoralists will not part with their animals, retaining them to regenerate the herd when conditions improve.

Figure 16.27

Rendille herders at a shallow, hand-dug well

Figure 16.28

Rendille camels and goats at a water hole

Rainfall is too low and unreliable in northern Kenya to support settled agriculture (Places 46, page 423). Over the years, the Rendille have learned how to survive in an extreme environment (Figure 16.27). All they need are their animals (camels, goats and a few cattle): all their animals need is water and grass. The tribe are constantly on the lookout for rain which usually comes in the form of heavy, localised downpours. Once the rain has been observed or reported, the tribe pack their limited possessions onto camels (a job organised by the women) and head off, perhaps on a journey of several days, to an area of new grass growth. In the past, this movement prevented overgrazing as grazed areas were given time to recover. Camels, and to a much lesser extent goats, can survive lengthy periods without water by storing it within their bodies or by absorbing it from edible plants — food supply is as important as water. Humans, who can go longer than animals without food but much less long without water, rely upon the camels for milk and blood, and the goats for milk and occasional meat. Indeed, the main diet of blood and milk avoids the necessity of cooking and the need to find firewood.

But the Rendille way of life is changing. Land is becoming overpopulated and resources overstretched as the numbers of people and animals increase and as water supplies and vegetation become scarcer. Consequently, as the droughts of recent years continue, pastoralists are forced to move to small towns, such as Korr. Here there is a school, health centre, better housing, jobs, a food supply and a permanent supply of water from a deep well. The deep well waters hundreds of animals, many of which are brought considerable distances each day. However, the increase in animal numbers has resulted in overgrazing and the increase in townspeople has led to the clearance of all nearby trees for firewood. This has resulted in an increase in soil erosion, creating a desert area extending 150 km around the town. Although attempts are being made to dig more wells to disperse the population, travelling shops now take provisions to the pastoralists, and the tribespeople have been shown how to sell their animals at fairer prices, many Rendille are still moving to Korr to live. There the children, having been educated, remain, looking for jobs, with the result that there are fewer pastoralists left to herd the animals (Figure 16.28).

3   Shifting cultivation (extensive subsistence agriculture)

Subsistence farming was the traditional type of agriculture in most tropical countries before the arrival of Europeans and remains so in many of the less economically developed countries and in more isolated regions. The inputs to this system are extremely limited. Only relatively few labourers are needed (although they may have to work intensively), technology is limited (possibly to axes) and capital is not involved. Over a period of years, extensive areas of land may be used as the tribes have to move on to new sites. Outputs are also very low with, often, only sufficient being grown for the immediate needs of the family, tribe or local community.

The most extensive form of subsistence farming is shifting cultivation which is still practised in the tropical rainforests (the *milpa* of Latin America and *ladang* of south-east Asia) and, occasionally, in the wooded savannas (the *chitimene* of central Africa). The areas covered are becoming smaller, due to forest clearances, and are mainly limited to less accessible places within the Amazon Basin (Places 52), Central America, Zaire and parts of Indonesia. Shifting cultivation, where it still exists, is the most energy-efficient of all farming systems as well as operating in close harmony with its environment.

Figure 16.29

Shifting cultivator clearing the rainforest

Figure 16.30

Crops grown in *chagras* (fields) around the *maloca* (communal house)

With the help of stone axes and machetes, the Amerindians clear a small area of about 1 hectare in the forest (Figures 16.29 and 16.30). Sometimes the largest trees are left standing to protect young crops from the sun's heat and the heavy rain; so also are those which provide food, such as the banana and kola nut. The buttress roots of the giant trees are left (Figure 12.5). After being allowed to dry, the felled trees

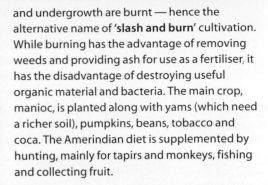

and undergrowth are burnt — hence the alternative name of '**slash and burn**' cultivation. While burning has the advantage of removing weeds and providing ash for use as a fertiliser, it has the disadvantage of destroying useful organic material and bacteria. The main crop, manioc, is planted along with yams (which need a richer soil), pumpkins, beans, tobacco and coca. The Amerindian diet is supplemented by hunting, mainly for tapirs and monkeys, fishing and collecting fruit.

The productivity of the rainforest depends upon the rapid and unbroken recycling of nutrients (Figure 12.8). Once the forest has been cleared, the nutrient cycle is broken. The heavy, afternoon, convectional rainstorms hit the unprotected earth causing erosion and leaching. With the source of humus removed, and in the absence of fertiliser and animal manure, the ferralitic soils (page 294) rapidly lose their fertility. Within 4 or 5 years, the decline in crop yields and the re-infestation of the area by weeds force the tribe to shift to another part of the forest. Although shifting cultivation appears to be a wasteful use of land, it has no long-term adverse effect upon the environment as, in most places, humus can built up sufficiently to allow the land to be re-used within 25 years if necessary.

The traditional Amerindian way of life is being threatened by the destruction of the rainforest. As land is being cleared for highways, cattle ranches, commercial timber, hydro-electric schemes, reservoirs and mineral exploitation, the Amerindians are pushed further into the forest or forced to live on reservations. Of the estimated 6 million Amerindians living in the Amazon rainforest when the European colonists arrived, only about 4 per cent remain today as a result of illnesses caught from and death inflicted by the invaders. Recent government policy of encouraging the in-migration of landless farmers from other parts of the country, together with the development of extensive commercial cattle ranching, has meant that sedentary farming is rapidly replacing shifting cultivation. After just a few years, as should have been foreseen, large tracts of some cattle ranches and many individual farms have already been abandoned as their soils have become infertile and eroded.

4 Intensive subsistence farming

This involves the maximum use of the land with neither fallow nor any wasted space. Yields, especially in south-east Asia, are high enough to support a high population density — up to 2000 per sq km in parts of Java and Bangladesh. The highest-yielding crop is rice which is grown chiefly on river flood plains (the Ganges) and in river deltas (the Mekong and Irrawaddy). In both cases, the peak river flow, which follows the monsoon rains, is trapped behind bunds, or walls (Places 53). Where flat land is limited, rice is grown on terraces cut into steep hillsides, especially those where soils have formed from weathered volcanic rock as in Indonesia and the Philippines (Figure 16.31). Upland rice, or dry padi, is easier to grow but, as it gives lower yields, it can support fewer people. Rice requires a growing season of only 100 days which means that the constant high temperatures of south-east Asia enable two, and sometimes even three, croppings a year (Figure 16.11).

The high population density, rapid population growth and large family size in many south-east Asian countries mean that, despite the high yields, there is little surplus rice for sale. The farms, due to population pressure and inheritance laws (page 427), are often as small as 1 hectare. Many farmers are tenants and have to pay a proportion of their crops to a landlord. Labour is intensive and it has been estimated that it takes 2000 hours per year to farm each 1 hectare plot. Most tasks, due to a lack of capital, have to be done by hand or with the help of water buffalo. The buffalo are often overworked and their manure is frequently used as a fuel rather than being returned to the land as fertiliser. Poor transport systems hinder the marketing of any surplus crops after a good harvest and can delay food relief during the times of food shortage which may result from the extremes of the monsoon climate: drought and flood.

Figure 16.31

Rice cultivation on terraced hillsides, Bali

Places 53 The Ganges valley: intensive subsistence cultivation of rice

Rice, with its high nutritional value, can form up to 90 per cent of the total diet in some parts of the flat Ganges valley in northern India and western Bangladesh. Padi, or wet rice, needs a rich soil and is grown in silt which is deposited annually by the river during the time of the monsoon floods. The monsoon climate (page 222) has an all-year growing season but, although 'winters' are warm enough for an extra crop of rice to be grown, water supply is often a problem. During the rainy season from July to October, the *kharif* crops of rice, millet and maize are grown. Rice is planted as soon as the monsoon rains have flooded the padi fields and is harvested in October when the rains have stopped and the land has dried out. During the dry season from November to April, the *rabi*

Figure 16.32

Rice cultivation on the flood plain of the River Ganges

crops of wheat, barley and peas are grown and harvested. Where water is available for longer periods, a second rice crop may be grown.

Rice growing is labour intensive with much manual effort needed to construct the bunds (embankments); to build irrigation channels; to prepare the fields; and to plant, weed and harvest the crop. The bunds between the fields are stabilised by tree crops. The tall coconut palm is not only a source of food, drink and sugar, but also acts as a cover crop protecting the smaller banana and other trees which have been planted on the bunds. The flooded padi fields may be stocked with fish which add protein to the human diet and fertiliser to the soil.

In 1964, many Indian farmers and their families were short of food, lacked a balanced diet and had an extremely low standard of living. The government, with limited resources, made a conscious decision to try to improve farm technology and crop yields by implementing Western-type farming techniques and introducing new hybrid varieties of rice and wheat — the so-called Green Revolution. Although yields have increased and food shortages have been lessened, the 'green revolution' is not considered to be, in this part of the world, a social, environmental or political success (pages 464–5).

5 Tropical commercial (plantation) agriculture

Figure 16.33

A rubber plantation in Malaysia

Plantations were developed in tropical areas, usually where rainfall was sufficient for trees to be the natural vegetation, by European and North American merchants in the 18th and 19th centuries. Large areas of forest were cleared and a single bush or tree crop was planted in rows (Figure 16.33) — hence the term monoculture (page 260). This so-called **cash crop** was grown for export and was not used or consumed locally (Places 54).

Plantations needed a high capital input to clear, drain and irrigate the land; to build estate roads, schools, hospitals and houses; and to bridge the several years before the crop could be harvested. Although plantations were often located in areas of low population density, they needed much manual labour. The owners and managers were invariably white. Black and Asian workers, obtained locally or brought in as slaves or indentured labour from other countries, were engaged as they were prepared, or forced, to work for minimum wages. They were also capable of working in the hot, humid climate. Today, many plantations, producing most of the world's rubber, coffee, tea, cocoa, palm oil, bananas, sugar cane and tobacco, are owned and operated by large multinational companies (Figure 16.34).

Advantages	Disadvantages
Higher standards of living for the local workforce	Exploitation of local workforce, minimal wages
Capital for machines, fertiliser and transport provided initially by colonial power, now the multinational corporations	Cash crops grown instead of food crops: local population have to import foodstuffs
Use of fertilisers and pesticides improves output	Most produce is sent overseas to the parent country
Increases local employment	Most profit returns to Europe and North America
Housing, schools, health service and transport provided, also often electricity and a water supply	Dangers of relying on monoculture: fluctuations in world prices and demand
	Overuse of land has led, in places, to soil exhaustion and erosion

Figure 16.34

The advantages and disadvantages of plantation farming

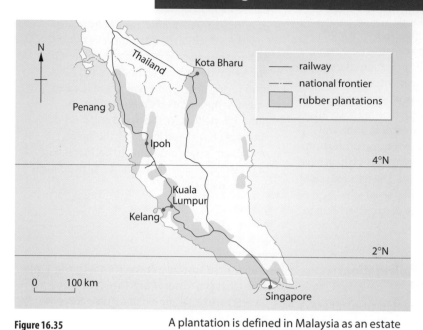

Figure 16.35

Location of rubber plantations in Malaysia

Figure 16.36

Tapping a rubber tree

A plantation is defined in Malaysia as an estate exceeding 40 ha in size. Many extend over several thousand hectares. Rubber is indigenous to the Amazon basin, but some seeds were smuggled out of Brazil in 1877, brought to Kew Gardens in London to germinate and then sent out to what is now Malaysia. The trees thrive in a hot, wet climate, growing best on the gentle lower slopes of the mountains forming the spine of the Malay peninsula (Figure 16.35). Rubber tends not to be grown on the coasts where the land is swampy, but near to the relatively few railway lines and the main ports. The 'cheap' labour needed to clear the forest, work in the nurseries, plant new trees and tap the mature trees was provided by the poorer Malays and immigrants from India.

Young trees, having begun their lives in a nursery, are planted out in rows. The time it takes for them to mature sufficiently to be tapped has been reduced from 7 to 5 years. Rubber is made from latex, a white liquid obtained from within the trunk. Each day the tapper makes a small incision, less than 0.5 m long, into the bark of mature trees. The latex runs downwards to be collected in a cup (Figure 16.36). The ideal time to tap the tree is at 5 a.m. when the internal pressure is at its greatest; the worst times are after 11 a.m. when the heat prevents the liquid from flowing, and during rain which dilutes the latex. An experienced tapper can cover 2 ha (500–600 trees) in a morning and is paid per litre of latex collected, not per tree tapped. Each tree gives on average 200 ml (7 fluid ounces) daily. Therefore, 600 trees should give 120 litres for which the tapper will received $20 Malaysian (just under £5).

Many tappers, both male and female, try to find a second job in the afternoon to increase their income. Each plantation is likely to have its own processing factory where the latex is turned into sheets ready for transport, often along good roads, to the nearest port. Rubber trees produce less latex after 25 years and so are felled, chopped up, and allowed to decay into a mulch which acts as a fertiliser into which new trees are planted.

The Malaysian government has now taken over all the large estates, formerly run by such multinationals as Dunlop and Guthries, having seen them as a relict of colonialism. Publicly owned companies now only account for 30 per cent of the land devoted to rubber: the remainder is in the form of smallholdings. In the early 1970s, the Federal Land Development Authority (FELDA) was set up. Under FELDA, the forest is cleared and land divided into 5-ha plots which are allocated to farmers. For the first 4 years, the government does all the work, supervising planting and caring for the young trees. The farmer is then put in charge, but is still provided with free fertiliser and pesticide as the trees are too young to provide any income. Once the crop is ready, it is bought and marketed by the government. Most farmers live in *kampongs* (small villages) and grow subsistence crops around their homes. Some keep Brahman cattle which are allowed to graze between the rubber trees, keeping down the undergrowth. As their income increases, farmers can buy their smallholding. Some smaller units have amalgamated as cooperatives to help share costs.

Since World War II, the world demand for rubber has steadily declined, mainly due to competition from synthetic rubber. Since 1988, however, this trend has shown signs of reversal due to the increased demand for protection against AIDS and the increase in family planning. In 1994 Malaysia claimed to have manufactured 60 per cent of the world's condoms!

6 Extensive commercial pastoralism (livestock ranching)

Livestock ranching returns the lowest net profit per hectare of any commercial type of farming. It is practised in more remote areas where other forms of land use are limited and where there are extensive areas of cheaper land with sufficient grass to support large numbers of animals. It is found mainly in areas with a low population density and aims to give the maximum output from minimum inputs — i.e. there is a relatively small capital investment in comparison to the size of the farm or ranch, but output per farmworker is high. This type of farming includes commercial sheep farming (in central Australia, Canterbury Plains in New Zealand, Patagonia, upland Britain) and commercial cattle ranching (Places 55), mainly for beef (in the Pampas, American Midwest, northern Australia and, more recently, Amazonia and Central America). It corresponds, therefore, to the outer land use zone of von Thünen's model (Figure 16.19) and does not include commercial dairying which, being more intensive, is found nearer to the urban market (Places 57, page 448).

The raising of beef cattle is causing considerable environmental concern. It is a cause of deforestation (uses 40 per cent of the cleared forest in Amazonia), desertification and soil erosion (overgrazing) and global warming (release of methane). It also takes more water and feed to produce one pound of beef than the equivalent amount of any other food or animal product.

Places 55 Beef cattle on the South American Pampas: extensive commercial pastoralism

The Pampas covers Uruguay and northern Argentina. The area receives 500–1200 mm of rainfall a year; enough to support a temperate grassland vegetation. During the warmer summer months the water supply has to be supplemented from underground sources, while in the cooler, drier winter much of the grass dies down. Temperatures are never too hot to dry up the grass in summer, nor low enough to prevent its growth in winter. The relief is flat and soils are often deep and rich having been deposited by rivers which cross the plain (Figure 16.37). The grasses help to maintain fertility by providing humus when they die back.

Many ranches, or *estancias,* exceed 100 sq km and keep over 20 000 head of cattle. Most are owned by businessmen or large companies based in the larger cities and are run by a manager with the help of cowboys or gauchos.

Figure 16.37

Land use on the South American Pampas: one of the few areas to show a zonation similar to that suggested by von Thünen

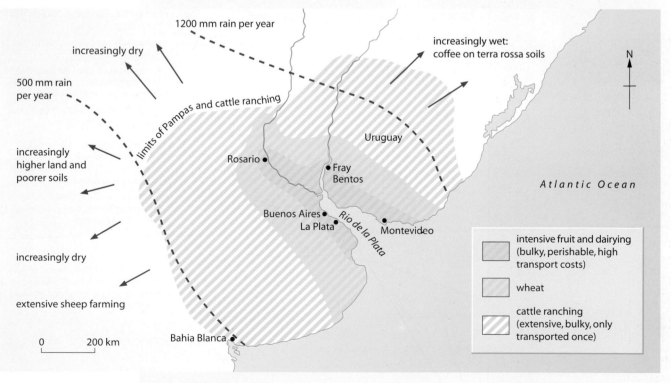

Several economic improvements have been added to the natural physical advantages. Alfalfa, a leguminous, moisture-retaining crop, is grown to feed the cattle when the natural grasses die down. Barbed wire, for field boundaries, was essential where rainfall was insufficient for the growth of hedges. Pedigree bulls were brought from Europe to improve the local breeds and later British Hereford cattle were crossed with Asian Brahman bulls to give a beef cow capable of living in warm and drier conditions. Initially, due to distances from world markets, cattle were reared for their hides. It was only after the construction of a railway network, linking places on the Pampas to the chief ports, that canned products such as corned beef became important. Later still, the introduction of refrigerated wagons and ships meant that frozen beef could be exported to the more industrialised countries.

7 Extensive commercial grain farming

As shown on the map of the Pampas (Figure 16.37) and in the von Thünen model (Figure 16.19), cereals utilise the land use zone closer to the urban market than commercial ranching. Grain is grown commercially on the American Prairies (Places 56), the Russian Steppes (Figure 16.6) and parts of Australia, Argentina and north-west Europe (Figure 16.25). In most of those areas, productivity per hectare is low but per farmworker it is high.

It was the introduction and cultivation of new strains of cereals which led to the first permanent settlements (Figure 14.1) and, later, it was a reliance upon these cereals to provide a staple diet which allowed steady population growth in Europe, Russia and south-east Asia (Figure 16.38). A demand for increased cereal production came, in the mid 19th century, from those countries experiencing rapid industrialisation and urban growth. This demand was met following the building of railways in Argentina, Australia and across North America (Figure 16.38). More recent demands have, so far, been met by the 'green revolution' in south-east Asia (page 464) and increases in irrigation and mechanisation.

Figure 16.38

Changes in the world's arable areas, 1870–1990

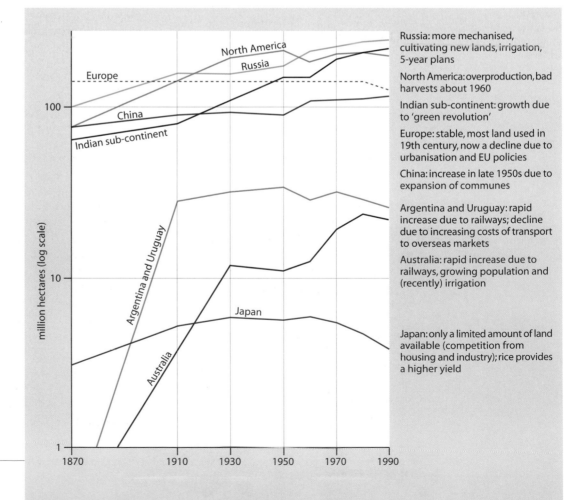

The Prairies have already been referred to in the optima and limits model (Figure 16.3). Although this area has many favourable physical characteristics, it also has disadvantages (Case Study 12b, page 314). Wheat, the major crop, ripens well during the long, sunny, summer days, while the winter frosts help to break up the soil. However, the growing season is short and in the north falls below the minimum requirement of 90 days. Precipitation is low, about 500 mm, but though most of this falls during the growing season there is a danger of hail ruining the crop, and droughts occur periodically. The winter snows may come as blizzards but they do insulate the ground from severe cold and provide moisture on melting in spring. The chinook wind (page 224) melts the snow in spring and helps to extend the growing season, but tornados in summer can damage the crop. The relief is gently undulating, which aids machinery and transport. The grassland vegetation has decayed over the centuries to give a black (chernozem) or very dark brown (prairie) soil (pages 303–304). However, if the natural vegetation is totally removed, the soil becomes vulnerable to erosion by wind and convectional rainstorms.

Figure 16.39

Extensive commercial cereal farming on the Canadian Prairies

When European settlers first arrived, they drove out the local Indians, who had survived by hunting bison, and introduced cattle. The world price for cereals increased in the 1860s and demand from the industrialised countries in western Europe rose. The trans-American railways were built in response to the increased demand (and profits to be made) and vast areas of land were ploughed up and given over to wheat. The flat terrain enabled straight, fast lines of communication to be built (essential as most of the crop had to be exported) and the land was divided into sections measuring 1 square mile (1.6 sq km). In the wetter east, each farm was allocated a quarter or a half section; while in the drier west, farmers received at least one full section.

The input of capital has always been high in the Prairies as farming is highly mechanised (Figure 16.39). Mechanisation has reduced the need for labour although a migrant force, with combine harvesters, now travels northwards in late summer as the cereals ripen. Seed varieties have been improved, and have been made disease-resistant, drought-resistant and faster-growing. Fertilisers and pesticides are used to increase yields and the harvested wheat is stored in huge elevators while awaiting transport via the adjacent railway.

In the last two decades, wheat has become less of a specialist crop and the area upon which it is grown has decreased. Many farms have diversified to produce sugar beet, flax, dairy produce and beef (Case Study 12b).

8 Intensive commercial (mixed) agriculture

This corresponds with von Thünen's inner zone where dairying, market gardening (horticulture) and fruit all compete for land closest to the market. All three have high transport costs, are perishable, bulky, and are in daily demand by the urban population. Similarly, all three require frequent attention, particularly dairy cows which need milking twice daily, and market gardening. Although this type of land use is most common in the eastern USA and north-west Europe (Places 57), it can also be found around every large city in the world. Intensive commercial farming needs considerable amounts of capital to invest in high technology and numerous workers: it is labour intensive. The average farm size used to be under 10 ha but recently this has been found to be uneconomic and amalgamations have been encouraged by the American government and the EU in order to maximise profits. This type of farming gives the highest output per hectare and the highest productivity per farmworker.

The EU and food surpluses

As farming in the EU continued to become more efficient, output has increased. Farmers were paid subsidies, or a guaranteed minimum price, for their produce. The result was the overproduction of certain commodities for the European and domestic market, but at a price beyond the reach of developing countries. Attempts to slim down the 'mountains' and 'lakes' have met with some success (Figure 16.40). The EU has spent most money on dairy products. In 1986 it cost almost £1 million a day to store butter and when some of it was sold to the then USSR at 7p/lb there were political repercussions. This action, together with some limited sales at reduced prices to the poor and the elderly within the Community, used up large amounts of the EU budget. Although the decision in 1986 to reduce milk output by 9.5 per cent by 1989 may have appeared sensible, it should be remembered that this has forced many dairy farmers to sell up and to slaughter some 5 million cattle. In 1988, attention was turned to cereals. It was agreed that £35/ha should be paid to arable farmers if they 'set-aside' their land (i.e. took it out of production) instead of growing crops on it (page 455). The beef mountain and wine lake were tackled after 1989.

Figure 16.40

EU food surpluses

Commodity	January 1986	January 1992
	(figures in thousand of tonnes unless otherwise stated)	
Butter	1400	300
Skimmed milk powder	800	0
Beef	500	800
Cereals	15000	7000
Wine/alcohol	4000 (hectolitres)	2500 (hectolitres)

Places 57 The Western Netherlands: intensive commercial farming

Most of the western Netherlands, stretching from Rotterdam to beyond Amsterdam, lies 2–6 m below sea-level. Reclaimed several centuries ago from the sea, peat lakes or areas regularly flooded by rivers, this land is referred to as the **old polders**. Today, they form a flat area drained by canals which run above the general level of the land. Excess water from the fields is pumped (originally by windmills) by diesel and electric pumps into the canals. With 440 persons per sq km in 1990 (compared to only 360 in 1975), the Netherlands has the highest population density in Europe. Consequently, with farm land at a premium, the cost of reclamation so high and the proximity of a large domestic urban market, intensive demands are made on the use of the land (Figure 16.41).

There are three major types of farming on the old polders.

■ Dairying is most intensive to the north of Amsterdam, in the 'Green Heart' and in the south-west of Friesland. It is favoured by mild winters, which allow grass to grow for most of the year; the evenly distributed rainfall, which provides lush grass; the flat land; and the proximity of the *Randstad* conurbation. Most of the cattle are Friesians. Some of the milk is used fresh but most is turned into cheeses (the well-known Gouda and Edam) and butter. Most farms have installed computer systems to control animal feeding.

■ The land between The Hague and Rotterdam (Figure 16.41) is a mass of glasshouses where **horticulture** is practised on individual holdings averaging only 1 ha. The cost of production is exceptionally high. Oil and natural

Farming and food supply

gas-fired central heating maintain high temperatures and sprinklers provide water. Heating, moisture and ventilation are all controlled by computerised systems. Machinery is used for weeding and removing dead flowers, and the soils are heavily fertilised and manured. Sometimes plants are grown through a black plastic mulch (heat-absorbing) which has the effect of advancing their growth and thus extending the cropping season to meet market demand for fresh produce. Several crops a year can be grown in the glasshouses, i.e. cut flowers in spring, tomatoes and cucumbers in summer, and lettuce in autumn and winter.

■ The sandier soils between Leiden and Haarlem are used to grow **bulbs**. Tulips, hyacinths and daffodils, protected from the prevailing winds by the coastal sand dunes, are grown on farms averaging 8 ha. The flowers form a tourist attraction, especially in spring (Figure 16.42) and bulbs are exported all over Europe from nearby Schipol Airport.

Figure 16.41

Agricultural land use in the western Netherlands

Legend:
- sand dunes
- arable on the new polders
- arable
- mainly pasture
- horticulture
- woodland and heath
- fresh water
- Randstad conurbation
- 'Green Heart'
- major dikes and dams

0   50 km

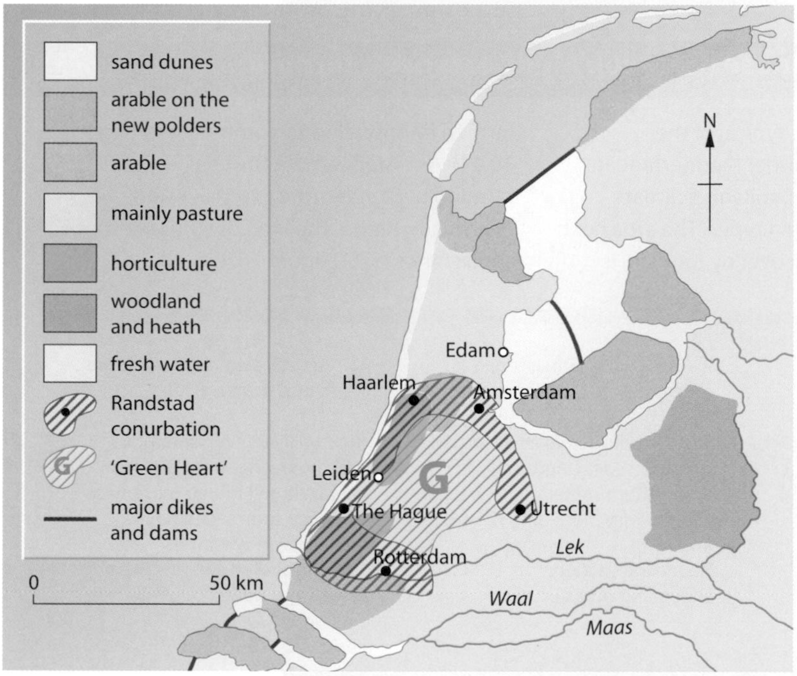

Figure 16.42

Intensive farming on reclaimed polder land in the western Netherlands

9 Mediterranean agriculture

A distinctive type of farming has developed in areas surrounding the Mediterranean Sea. Winters are mild and wet allowing the growth of cereals and the production of early spring vegetables or *primeurs*. Summers are hot, enabling fruit to ripen, but tend to be too dry for the growth of cereals and grass. As rainfall amounts decrease and the length of the dry season increases from west to east and from north to south, irrigation becomes more important. River valleys and their deltas (the Po, Rhône and Guadalquivir) provide rich alluvium, but many parts of the Mediterranean are mountainous with steep slopes and thin rendzina soils (page 253). Due to earlier deforestation, many of these slopes have suffered from soil erosion. Frosts are rare at lower levels though the cold mistral and bora winds may damage crops (Figure 12.22).

Farming tends to be labour intensive but with limited capital. There are still many absentee landlords (latifundia, page 425) and outputs per hectare and per farmworker are usually low. Most farms tend to be small in size. Land use frequently shows (Figure 16.20) that crops which need most attention are grown nearest to the farmhouse or village, and that land use is more closely linked to the physical environment than controlled by human inputs (Places 58). Many village gardens and surrounding fields are devoted to citrus fruits, such as oranges, lemons and grapefruit, as these have thick waxy skins to protect the seeds and to reduce moisture loss. These fruits are also grown commercially where water supply is more reliable, e.g. oranges in Spain around Seville and on *huertas* (irrigated farms) near Valencia, lemons in Sicily and grapefruit in Israel.

Vines, another labour intensive crop, and olives, the 'yardstick' of the Mediterranean climate, are both adapted to the physical conditions. They can tolerate thin, poor, dry soils and hot, dry summers by having long roots and protective bark. Wheat may be grown in the wetter winter period in fields further from the village as it needs less attention, while sheep and goats are reared on the scrub and poorer quality grass of the steeper hillsides. Grass becomes too dry in summer to support cattle and so milk and beef are scarce in the local diet.

Apart from central Chile, other areas experiencing a Mediterranean climate have developed a more commercialised type of farming based upon irrigation and mechanisation. Central California supports agribusiness based on a large, affluent, domestic market which is, in terms of scale, organisation and productivity, the ultimate in the capitalist system. Southern Australia produces dried fruit to overcome the problem of distance from world markets. All Mediterranean areas have now become important wine producers.

Places 58 Peloponnese (Greece): Mediterranean farming

Figure 16.43 is a transect, typical of the Peloponnese and many other Mediterranean areas, showing how relief, soils and climate affect land use and farming types. The area next to the coast, unless taken over by tourism, is farmed intensively and commercially (Figure 16.44). As distance from the coast increases, farming becomes more extensive and eventually, before the limit of cultivation, at a subsistence level (Figure 16.45).

Figure 16.43

Land use and farming types in the Peloponnese (not to scale)

A Intensive commercial farming

Coastal plain: flat with deep, often alluvial, soil (washed down from hills by seasonal rivers)
Hot, dry summers; mild, wet winters with no frost
Some mechanisation; irrigation needed in summer
Citrus fruits (oranges, clementines and mandarins); peaches and some figs

B Extensive commercial farming

Undulating land with small hills: soils quite deep and relatively fertile (terra rossa)
Larger farms (villages on hills originally for defence, now above best farmland)
Similar climate, with a slight risk of frost in winter
Some mechanisation, but donkeys still used
Olives and vines with, occasionally, tobacco

C Extensive, subsistence farming

Steeper hillsides covered in scrub: thin, poor soils (rendzina)
Warm, dry summers; cool, wet winters with increasing risk of frost
No mechanisation
Sheep and goats

D Virtually no farming/some rough grazing

Steep hillsides, mountainous, with poor, discontinuous scrub: very little soil (much erosion)
Cool, dry summers; cold, wet and windy winters with a risk of snow
Sheep and goats (mainly in summer)

Figure 16.45

Farming near Mycenae, Greece: orange and olive groves in foreground; rough grazing on hillsides beyond

Figure 16.44

Intensive fruit, mainly oranges, next to the coast, settlement on higher, less fertile land in mid distance. Deforested hills in background, Greece

10 Irrigation

Irrigation is the provision of a supply of water from a river, lake or underground source to enable an area of land to be cultivated. It may be needed where

1 Rainfall is limited and where evapotranspiration exceeds precipitation — i.e. in semi-arid and arid lands such as the Peruvian desert (Places 18, page 164) and the Nile valley (Places 59).
2 There is a seasonal water shortage due to drought, as in southern California with its Mediterranean climate (Case Study 15a, page 414).
3 Amounts of rainfall are unreliable, as in the Sahel countries (Figure 9.29).
4 Farming is intensive, either subsistence or commercial, despite high annual rainfall totals, e.g. the rice-growing areas of south-east Asia.

In economically more developed countries, large dams may be built from which pipelines and canals may transport water many kilometres to a dense network of field channels (Case Study 3b, page 85). The flow of water is likely to be computer-controlled. Unfortunately, it is the economically less developed countries, lacking in capital and technology, which suffer most severely from water deficiencies. Unless they can obtain funds from overseas, most of their schemes are extremely labour intensive as they have to be constructed and operated by hand.

Place 59 The Nile valley in Egypt: irrigation

From the times of the Pharaohs until very recently, water for irrigation was obtained from the River Nile by two methods. First, each autumn, the annual floodwater was allowed to cover the land, where it remained trapped behind small bunds until it had deposited its silt. Secondly, during the rest of the year when river levels were low, water could be lifted 1–2 m by a **shaduf**, **saquia** (sakia) **wheel** or **Archimedes screw**. However, the Egyptians had long wished to control the Nile so that its level would remain relatively constant throughout the year. Although barrages of increasing size had been built during the early 20th century, it was the rapid increase in population (which doubled from 25 to 50 million between 1960 and 1987) and the accompanying demand for food that led to the building of the Aswan High Dam (opened 1971) and several new schemes to irrigate the desert near Cairo (late 1980s).

The main purpose of the High Dam was to hold back the annual floodwaters generated by the summer rains in the Ethiopian Highlands. Some water is released throughout the year, allowing an extra crop to be grown, while any surplus is saved as an insurance against a failure of the rains. The river regime below Aswan is now more constant allowing trade and cruise ships to travel on it at all times. Two and sometimes three crops can now be grown annually in the lower Nile valley (Figure 16.46). Yields have increased and extra income is gained from cash crops of cotton, maize, sugar cane, potatoes and citrus fruits. The dam incorporates a hydroelectric power station which provides Egypt with almost a third of its energy needs for domestic and industrial purposes. Lake Nasser is important for fishing and tourism.

Figure 16.46

Landsat photo of the Nile delta: the River Nile is shown black, the irrigated crops in magenta red, and Cairo and its south-western extension to the pyramids at Giza is pale blue. Numerous dry valleys can be seen in the desert (Chapter 7)

Figure 16.47

Boom irrigation in Saudi Arabia: a giant sprinkler rotates around a central pivot

Following the construction of the Aswan High Dam (Figure 16.48), Egypt has modernised its methods of irrigation. Electricity is now used to power pumps which, by raising water to higher levels, allow a strip of land up to 12 km wide on both sides of the Nile to be irrigated. **Drip irrigation** utilises plastic pipes in which small holes have been made; these are laid over the ground and water drips onto the plants in a much less wasteful manner as less evaporates or drains away. Between the Nile and the Suez Canal, **boom irrigation** has been introduced (Figure 16.47 shows this method in use in the Libyan Sahara), creating fields several hectares in diameter.

However, the Dam has created several problems. Environmentally, the cessation of the Nile flood has also meant ending the annual deposition of fertile silt on the fields, which means fertiliser has now to be added; without the silt, the delta has begun to retreat — having lost its supply of sediment. The number of

bilharzia snails has increased as a consequence of the greater number of irrigation channels. Economically and socially, development has encouraged farmers to grow cash crops instead of providing a better diet for themselves and costs have increased due to the need to buy fertiliser. Clay is no longer available for making bricks in the traditional manner. The drought during the 1980s in Ethiopia meant a dramatic decrease in flood water brought down by the Blue Nile and River Atbara (Figure 16.48). By 1988 Lake Nasser was only 40 per cent full; ships were having difficulty travelling up the shallow Nile; and there was talk of having to close the turbines which generated the hydro-electricity (since then lake levels have risen). Large areas of land reclaimed after 1971 have become saline and have been allowed to return to desert. The UN Food and Agriculture Organisation (FAO) claims that between 1974 and 1984 there was an actual *decrease* of 12 per cent in land under irrigation and that salinisation was affecting 30–40 per cent of the remaining irrigated land.

Figure 16.48

The Nile: sources and uses of water

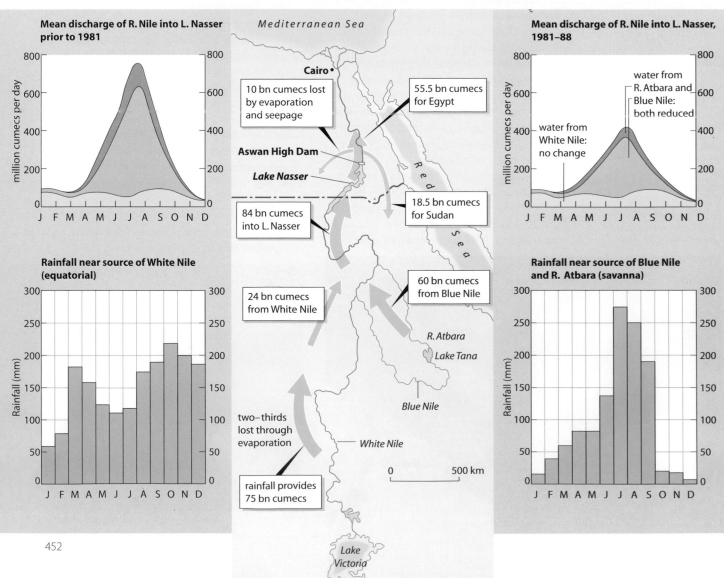

Q

1 For one (or several) of the ten types of farming described above:
  a describe its characteristics;
  b describe the conditions favouring its development; and
  c describe any present-day problems.

2 With reference to specific examples, discuss the extent to which the characteristics of farming systems may be influenced by each of the following:
  a *either* the amount and seasonal distribution of precipitation *or* drought;
  b relief features and soil types;
  c availability of capital;
  d mean temperatures and seasonal variations in temperature;
  e transport and marketing costs;
  f population density;
  g slope gradient; and
  h land tenure.

Farming types and economic development

Throughout the section on types of farming, several fundamental assumptions have been made. These included the generalisations that: 'the poorest countries are those who, because they have the lowest inputs of capital and technology, have the lowest outputs'; and 'the wealthier countries are those who can afford the highest inputs giving them the maximum yield, or profit, per person'. Is it really possible to make a simple correlation between wealth (the standard of living) and the type of agriculture?

Figure 16.49 shows 15 countries selected (*not* chosen randomly) as representing the main types of farming and used as examples in the previous section. Using the five variables A–E, it is possible to postulate four hypotheses:

1 The less developed a country (i.e. the lower its GDP per capita), the greater the percentage of its population involved in agriculture.

2 The less developed a country, the greater the percentage of its GDP is made up from agriculture.

3 The less developed a country, the less fertiliser it will use.

4 The less developed a country, the less mechanised will be its farming (fewer tractors per head of population, for example).

As with all data, there are considerations which you should remember when drawing conclusions from Figure 16.49.

■ The countries were selected with some bias in order to cover all the main types of farming.

■ GDP is not the only indicator of wealth or development (pages 570–1).

■ GDP figures are not necessarily accurate and may be derived from different criteria (page 569).

■ There may be several different types of farming in each country.

Q

Figure 16.49 shows the types of farming, GDP and agricultural production figures for 15 selected countries.

Country	Major farming type	A	B	C	D	E
1 Ethiopia	Nomadic herding	135	77	45	1	10882
2 Bangladesh	Intensive subsistence	158	82	48	26	20581
3 China	Centrally planned	218	56	45	47	1247
4 India	Intensive subsistence	233	60	29	37	1480
5 Kenya	Nomadic herding/subsistence	279	76	27	14	3006
6 Egypt	Irrigation	711	49	19	361	1117
7 Uruguay	Extensive commercial ranching	1331	11	12	3	188
8 Malaysia	Commercial plantation	1566	45	20	111	1900
9 CIS	Centrally planned	2588	14	20	38	102
10 Greece	Mediterranean	2966	34	16	69	142
11 Spain	Mediterranean	3853	14	6	47	65
12 Argentina	Extensive commercial ranching	4124	12	13	1	120
13 UK	Intensive commercial	6514	2	2	140	96
14 Netherlands	Intensive commercial	7716	4	4	340	77
15 Canada	Extensive commercial grain	13034	4	3	32	38

A Gross domestic product (GDP) per capita in US$, in rank order
B Percentage of population engaged in agriculture
C Percentage of GDP derived from agriculture
D Kg of fertiliser used per hectare of population
E Number of people in country for each tractor

Figure 16.49

1 Draw four scattergraphs to see if there appears to be any correlation between the GDP of the 15 countries and the four other variables (see Figure 22.6).

2 The countries have already been ranked in order of their GDP. Rank the other four variables and for each use the Spearman rank correlation test (Framework 14, page 572) to calculate the correlation coefficient (Rn) between them and the GDP.

3 Using your results to questions 1 and 2:
   a Which variable shows the closest correlation with GDP per capita?
   b How well do you consider that your results support each of the four hypotheses given above?
   c Describe and try to account for any outstanding anomalies. For example, why do Argentina and Canada use so little fertiliser? Why does Egypt use so much fertiliser? Why does Malaysia have so few tractors?
   d With how much confidence can you say that 'Farming in developing countries is traditional and non-mechanised with a small output per worker despite the large numbers it employs. As countries become more developed the numbers in agriculture decline, the methods become more scientific and mechanised and there is a high output per worker and per unit area'?

4 Figure 16.50 is an alternative method of showing links between various types of farming and levels of economic development. It was a group effort by a number of sixth-formers. Criticise their model and try to suggest other ways by which farming systems and levels of development may be compared.

Figure 16.50

Farming types and levels of development

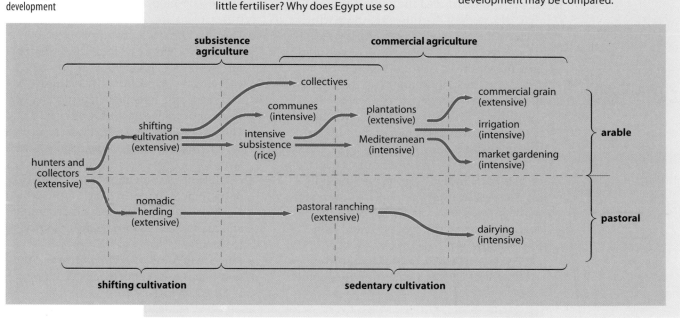

Recent changes in the Common Agricultural Policy

When the UK joined the European Community in 1973, it was given five years to implement the Common Agricultural Policy (CAP). As a result, most major decisions affecting British farming are now made in Brussels and not by the British government or by individual farmers. The five basic aims of the CAP were to

1 increase agricultural productivity and to improve self-sufficiency;
2 maintain jobs on the land, preferably on family farms;
3 improve the standard of living (income) of farmers and farmworkers;
4 stabilise markets; and
5 keep consumer food prices stable and reasonable.

Most of these aims have been achieved, except in the more marginal areas of upland and southern Europe (Figure 16.22), but at the expense of disrupting world trade, having insufficient regard for the environment, and creating large food surpluses in certain commodities (Figure 16.40). The CAP has upset economically developing countries by imposing trade tariffs, to restrict them dumping their cheaper foodstuffs, and non-EU economically developed countries, by granting generous subsidies to EU farmers. Since 1992 the CAP has undergone a series of reforms in order to solve some of these problems and has introduced policies aimed at

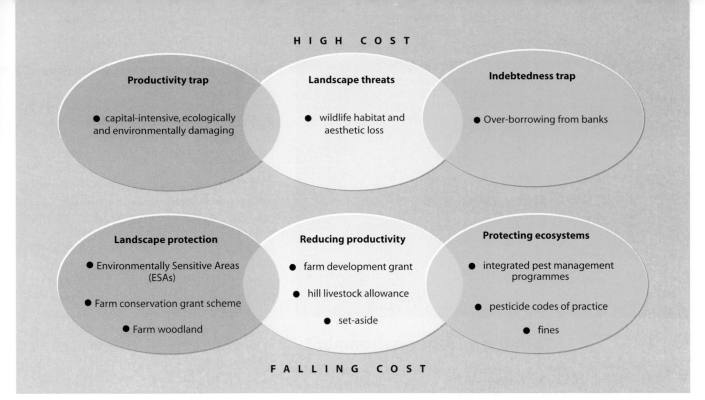

Productivity trap

- capital-intensive, ecologically and environmentally damaging

Landscape threats

- wildlife habitat and aesthetic loss

Indebtedness trap

- Over-borrowing from banks

Landscape protection

- Environmentally Sensitive Areas (ESAs)
- Farm conservation grant scheme
- Farm woodland

Reducing productivity

- farm development grant
- hill livestock allowance
- set-aside

Protecting ecosystems

- integrated pest management programmes
- pesticide codes of practice
- fines

FALLING COST

Figure 16.51

The CAP in the 1980s and 1990s (Geofile, September 1994)

encouraging the de-intensification of farming and the protection of the environment (Figure 16.51).

- **Subsidies** guaranteed farmers a minimum price and an assured market for their produce. Farmers tended, therefore, to overproduce (hence the EUs 'mountains and lakes'), and the payment of subsidies became a drain on EU finances. (In 1985 the CAP consumed 70 per cent of the EU budget for just 3 per cent of its GDP.) There is now, beginning in 1993, a 3-year progressive reduction of 29 per cent on subsidies for cereals and 18 per cent on beef.
- **Quotas** were introduced in 1984 to reduce milk output. Although output had decreased by 20 per cent by 1992, UK dairy farmers were still producing 20 per cent more milk than was being consumed — hence the announcement of a further 2 per cent cut, spread over the following 2 years. Quotas have also been imposed to maximise the amount of beef to be produced and the number of sheep reared.
- **Set-aside** was initially introduced on a voluntary basis in 1988, to try to reduce overproduction in arable crops. Farmers who took over 20 per cent of their cultivated land out of production (pasture and fallow land were not included) received £200 per hectare (£253/ha in 1993) provided that the land was then left fallow; was turned into woodland (the Farm Woodland Management scheme paid up

to £505/ha for conifers and up to £1375/ha for broadleaved woodland); or was put to other non-agricultural use (golf courses, caravan sites and nature trails). Despite these inducements, cereals were still being overproduced in 1992 and so the EU decreed that all farmers with farms exceeding 20 hectares must withdraw 15 per cent of their land from use if they wished to continue to receive subsidies. In 1994, an area larger than Nottingham was set aside.

- **Environmentally friendly farming** was to be encouraged (not made compulsory) in all EU countries (see next section).
- The **European Economic Area** will see the amalgamation of the EU and EFTA (Figure 21.24) in which the 12 present members of the EU and the 3 new members (Sweden, Finland and Austria) will probably be joined by Switzerland and Iceland, creating an internal market of over 350 million people. The GATT agreement (General Agreement on Tariffs and Trade), signed in December 1993 after 7 years of debate, should, by reducing subsidies and trade tariffs, help world trade by making internal EU markets more accessible to non-EU countries.

It is hoped that these policies will reduce the CAP budget; bring production more into line with EU consumption; target farming communities in greatest financial difficulty (i.e. those in marginal lands); encourage farm diversification; and improve the quality of produce.

Figure 16.52

A rural landscape with trees and hedges, Dorset

Figure 16.52

A rural landscape with trees and hedges, Dorset

Figure 16.53

How eutrophication can upset the ecosystem

Farming and the environment

Numerous pressure groups are claiming that the traditional British countryside is being spoilt, yet the countryside of today is not 'traditional' — it has always been changing. The primeval forests, regarded as Britain's climatic climax vegetation (page 264), were largely cleared, initially for sheep farming and later for the cultivation of cereals. Although there is evidence that hedges were used as field boundaries by the Anglo-Saxons, it was much later that land was 'enclosed' by planting hedges and building dry stone walls (page 368). It is this 18th- and 19th-century landscape which has become looked upon, incorrectly, as the traditional or natural environment (Figure 16.52). However, the rate of change has never been faster than in recent years. Estimates by the Nature Conservancy Council suggest that, between 1949 and 1990, 40 per cent of the remaining ancient broadleaved woodlands, 25 per cent of hedgerows, 30 per cent of heaths, 60 per cent of wetlands and 30 per cent of moors have 'disappeared'. While most accusing fingers point to the intensification of agriculture, together with afforestation and building programmes, as the major causes, it should be remembered that farmland too is under threat from rival land users (Figure 17.1).

Farming as a threat to the environment

a The use of chemicals

Fertiliser, slurry and pesticides all contribute to the pollution of the environmental system. Fertiliser, in the form of mineral compounds which contain elements essential for plant growth, is widely used to produce a healthy crop and increase yields. If too much nitrogenous fertiliser or animal waste (manure) is added to the soil, some remains unabsorbed by the plants and may be leached to contaminate underground water supplies and rivers. Where chemical fertiliser accumulates in lakes and rivers, the water becomes enriched with nutrients (eutrophication) and the ecosystem is upset (Figure 16.53). In parts of north-west Europe, levels of nitrates in groundwater are above EU safety limits.

In Britain, the Water Authorities claim that slurry (farmyard effluent) is now the major pollutant of, and killer of life in, rivers. After several decades in which the quality of river water had improved, the last few years have seen levels of pollution again increasing, especially in farming areas. Even so, the Water Authorities themselves asked permission in 1989 to *increase* the discharge of sewage into rivers.

Pesticides and herbicides are applied to crops to control pests, diseases and weeds. Estimates suggest that, without pesticides, cereal yields would be reduced by 25 per cent after one year and 45 per cent after three. The Friends of the Earth claim that pesticides are injurious to health and, although there have been no human fatalities reported in Britain in the last 15 years, there are many incidents in developing countries resulting from a lack of instruction, fewer safety regulations and faulty equipment. A UN report (1994) claims that 25 million agricultural workers in developing countries (3 per cent of the total workforce) experience pesticide poisoning each year. Pesticides are blamed for the rapid decrease in Britain's bee and butterfly populations, and an up to 80 per cent reduction in 800 species of fauna in the Paris basin.

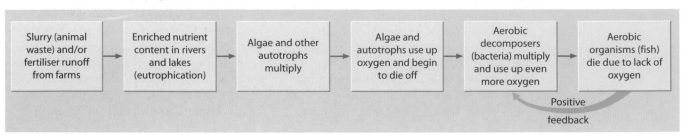

| Slurry (animal waste) and/or fertiliser runoff from farms | → | Enriched nutrient content in rivers and lakes (eutrophication) | → | Algae and other autotrophs multiply | → | Algae and autotrophs use up oxygen and begin to die off | → | Aerobic decomposers (bacteria) multiply and use up even more oxygen | → | Aerobic organisms (fish) die due to lack of oxygen |

Positive feedback

Pesticides can dissipate in the air as vapour, in water as runoff, or in soil by leaching to the groundwater.

b The loss of natural habitats

The most emotive outcries against farmers have been at their clearances of hedges, ponds and wetlands. These clearances mean a loss of habitat for wildlife and a destruction of ecosystems, some of which may have taken centuries to develop and, being fragile, may never recover or be replaced. As stated earlier, over 25 per cent of British hedgerows were removed between 1949 and 1990 — in Norfolk, the figure was over 40 per cent. Figure 16.54 lists some of the arguments for and against the removal of hedgerows and the drainage of ponds/wetlands. Figures 16.52 and 16.55 show the contrast between a landscape with trees and hedges, and one where they have been removed.

Farming can increase soil erosion. The rate of erosion is determined by climate, topography, soil type and vegetation cover (Case Study 10, page 258), but it is accelerated by poor farming practices (overcropping and overgrazing) and deforestation. (The conditions favouring wind erosion are given on page 166 and the processes of wind transportation are shown in Figure 7.8). In Britain, wind erosion tends to be restricted to parts of East Anglia and the Fens where the natural vegetation cover, including hedges, has been removed and where soils are light or peaty. Water erosion is most likely to occur after periods of prolonged and heavy rainfall, on soils with less than 35 per cent clay content, in large and steeply-sloping fields and where deep ploughing has exposed the soil.

Arable farming, especially when ploughing is done in the autumn, removes the protective vegetation cover. The intensification of farming, and overcropping, in areas of highly erodible soils in the USA have led to a decrease in yields and an estimated loss of one-third of the country's topsoil — much of it from the Dust Bowl during the 1930s. Deforestation in tropical rainforests, mountainous and semi-arid areas — Brazil, Nepal and the Sahel, respectively — also accelerates soil erosion.

Figure 16.54

The case for and against hedgerows and ponds in a farming environment

For	Against
■ **Hedgerows**	
Form part of the attractive, traditional British landscape	Are not traditional and were initially planted by farmers
Form a habitat for wildlife: birds, insects and plants (Large Blue butterfly is extinct, 10 other species are endangered)	Harbour pests and weeds
Act as windbreaks (and snowbreaks)	Costly and time-consuming to maintain
Roots bind soil together, reducing erosion by water and wind	Take up space which could be used for crops
	Limit size of field machinery (combine harvesters need an 8-m turning circle)
■ **Ponds**	
Form a habitat for wildlife: birds, fish and plants	Take up land that could be used more profitably
Add to the attractiveness of the natural environment	Stagnant water may harbour disease
i.e. Concern is environmental	**i.e. Concern is economic**

Figure 16.55

An agricultural landscape without trees or hedges, Wiltshire

Figure 16.56

Salinisation in California

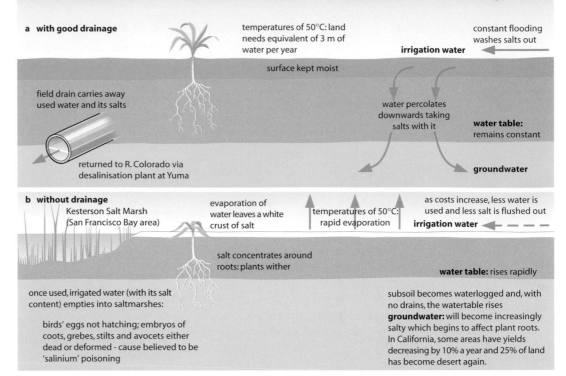

Irrigation (Places 59) also needs the surplus water to be drained away. Without this careful, and often expensive, management, the soil can become increasingly saline and waterlogged (Figure 16.56). As the water table rises it brings, through capillary action (page 252), dissolved salts into the topsoil. These affect the roots of crops, which are intolerant of salt, so that over a period of time they die. Where water is brought to the surface and then evaporates, a crust of salt is left on the surface and the area may revert to desert. To date, only rough estimates have been made of the amount of irrigated land now affected by salinisation, but figures suggest that it may be as high as 40 per cent in Pakistan and Egypt, and 30 per cent in California.

Attempts by farming to improve the environment

a Environmental improvement schemes

The EU and British government have introduced several schemes by which financial incentives have been offered to farmers who try to improve their environment, e.g. the 'set-aside', woodland management and Environmentally Sensitive Area (ESA) schemes.

Many parts of the British environment are benefitting from 'set-aside'. Soils which are left under either permanent or rotational fallow have a protective vegetation cover and are allowed the opportunity to improve their humus content. The non-agricultural use of land can include the conversion of farmland into nature trails and reserves, and the

restoration of ponds and wetlands. The woodland management scheme has increased the number of trees and small woods. In 1985, a 3-year experimental scheme was set up jointly by the Ministry of Agriculture and the Countryside Commission on the Halvergate Marshes in the Norfolk Broads. The two parties represented the conflicting interests of farmers and conservationists. At the same time, the Countryside Commission and the Nature Conservancy Council (NCC) looked at 46 'search' areas where it was considered that farmed landscapes may be under threat from changing farming practices (Figure 16.57). Many of these have now been designated as 'Environmentally Sensitive Areas' (ESAs), because of the historical importance and the habitat value of their landscapes. Farmers in the 31 ESAs designated by 1992 are eligible for two levels of payment: a lower level, paid on condition that they maintain the present landscape; and a higher level, if they make environmental improvements such as replanting hedges and restoring ponds.

b Organic farming

October 1994 was designated 'Organic harvest month'. Since the mid 1980s, a small, but increasing, number of British farmers have turned to organic farming. Organic farming is, compared with conventional farming, self-sustaining in that it produces more energy than it consumes and it does not destroy itself by misusing soil and water resources. It rules out the use of artificial (chemical) fertilisers, herbicides and

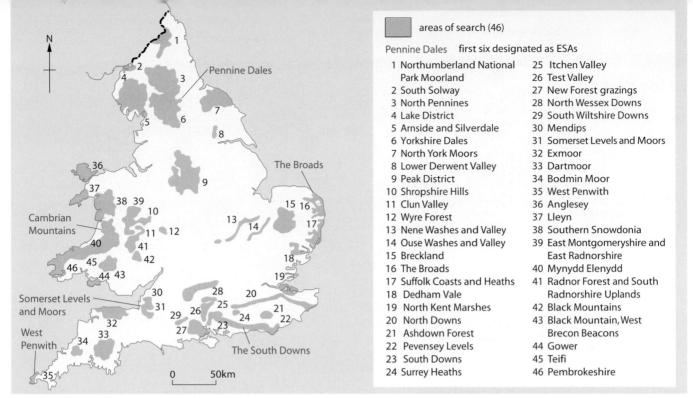

Figure 16.57

'Environmentally Sensitive Areas' (ESAs) in England and Wales

areas of search (46)

Pennine Dales   first six designated as ESAs

1 Northumberland National Park Moorland
2 South Solway
3 North Pennines
4 Lake District
5 Arnside and Silverdale
6 Yorkshire Dales
7 North York Moors
8 Lower Derwent Valley
9 Peak District
10 Shropshire Hills
11 Clun Valley
12 Wyre Forest
13 Nene Washes and Valley
14 Ouse Washes and Valley
15 Breckland
16 The Broads
17 Suffolk Coasts and Heaths
18 Dedham Vale
19 North Kent Marshes
20 North Downs
21 Ashdown Forest
22 Pevensey Levels
23 South Downs
24 Surrey Heaths
25 Itchen Valley
26 Test Valley
27 New Forest grazings
28 North Wessex Downs
29 South Wiltshire Downs
30 Mendips
31 Somerset Levels and Moors
32 Exmoor
33 Dartmoor
34 Bodmin Moor
35 West Penwith
36 Anglesey
37 Lleyn
38 Southern Snowdonia
39 East Montgomeryshire and East Radnorshire
40 Mynydd Elenydd
41 Radnor Forest and South Radnorshire Uplands
42 Black Mountains
43 Black Mountain, West Brecon Beacons
44 Gower
45 Teifi
46 Pembrokeshire

pesticides, favouring instead only animal and green manures (compost) and mineral fertilisers (rock salt, fish and bone meal). These natural fertilisers put organic matter back into the soil enabling it to retain more moisture during dry periods and allowing better drainage and aeration during wetter spells. Organic farming involves the intensive use of both land and labour. It is a mixed farming system which involves crop rotations and the use of fallow land. It is less likely to cause soil erosion or exhaustion as the soils contain, in comparison with non-organic farms, more organic material (humus), earthworms and bacteria. It is also less likely to harm the environment as there will be no nitrate runoff (no eutrophication in rivers) and less loss of wildlife (no pesticides to kill butterflies and bees)(Places 60).

Organic farming has its problems. If it replaces a conventional farming system, yields can drop considerably in the first two years, when artificial fertiliser is no longer used, although they soon rise again as the quality of the soil improves. Also, during the transition period, farmers cannot market any goods as 'organic': they must wait until they meet the Soil Association's standards before receiving its label guaranteeing the authenticity of their produce. Weeds can increase, without herbicides, and may have to be controlled by hand labour or by being covered with either mulch or polythene. This means that, although organic farming is helpful to the environment and, arguably, less harmful to human health, its produce is more expensive to buy.

Places 60 An organic farm in southern England

Colin Hutchin farms near Taunton in Somerset. His 130-hectare, mixed farm is hilly and well-wooded, with thick hedges and hay meadows rich in herbs. He keeps a herd of white Charolais cows, as well as other beef breeds and a flock of sheep. He therefore has no need to buy fertiliser. He devotes 40 hectares of the farm to growing vegetables which he is then able to sell to local wholesalers. The fertility of his land is maintained by adding lime, slag and animal manure. When parts of the farm are left fallow, to allow them to rest and be replenished, large

areas are given over to clover. Colin has a rotation pattern of two straw crops followed by a root crop and then either grass and clover (clover returns nitrogen to the soil) or peas or beans (also leguminous crops). He believes that this rotation not only maintains a natural soil fertility, but also reduces pests and diseases by changing the crop before they can take hold. He also relies on ladybirds to do the job of pesticides as they keep down the number of harmful blackfly and aphids.

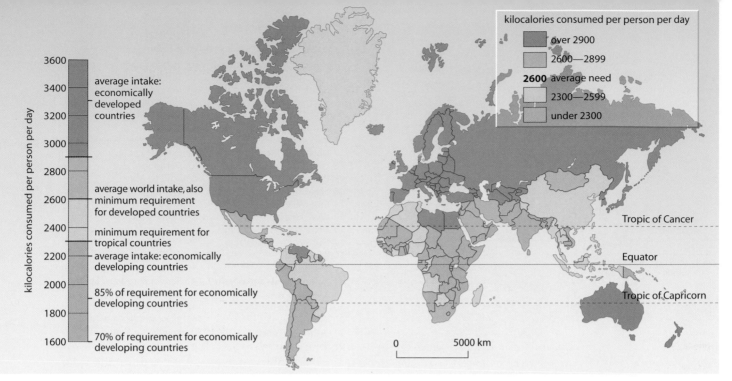

kilocalories consumed per person per day
- over 2900
- 2600—2899
- **2600** average need
- 2300—2599
- under 2300

kilocalories consumed per person per day (vertical axis, 1600–3600)

- 3600
- average intake: economically developed countries
- average world intake, also minimum requirement for developed countries (2600)
- minimum requirement for tropical countries (2300)
- average intake: economically developing countries
- 85% of requirement for economically developing countries
- 70% of requirement for economically developing countries

Tropic of Cancer
Equator
Tropic of Capricorn

0   5000 km

Figure 16.58

World food supply in the late 1980s: average kilocalorie consumption per person/day, by country

Food Supplies

Diet and health

It is 200 years since Malthus expressed his fears that world population would outstrip food supply (page 356). Today, despite assurances from various international bodies such as the Food and Agriculture Organisation (FAO) that there is still sufficient food for everyone, it is estimated that three-quarters of the world's population is inadequately fed, and that the majority of these live in less economically developed countries. The problem is, therefore, the unevenness in the distribution of food supplies: massive surpluses exist in North America and the EU; and there are shortages in many developing countries.

This uneven distribution is reflected in Figure 16.58 which shows variations in kilocalorie intake throughout the world. Dieticians calculate that the average adult in temperate latitudes requires 2600 kilocalories a day, compared with 2300 kilocalories for someone living within the tropics. The FAO reports that the actual average intake for the economically more developed world is 3300 kilocalories, but only 2200 kilocalories in less developed countries. However, the quantity of food consumed is not always as important as the quality and balance of the diet. A good diet should contain different types of food to build and maintain the body, and to provide energy to allow the body to work. A balanced diet should contain

- **proteins**, such as meat, eggs and milk, to build and renew body tissues;

- **carbohydrates**, which include cereals, sugar, fats, meat and potatoes, to provide energy; and
- **vitamins** and **minerals**, as found in dairy produce, fruit, fish and vegetables, which prevent many diseases.

Malnutrition and undernutrition, often caused by poverty, affect many people including even a surprisingly high number in developed countries. Malnutrition may not be a primary cause of death, but by reducing the ability of the body to function properly, it reduces the capacity to work and means that people, and especially children, become less resistant to disease and more likely to fall ill. Nutritional diseases, which include rickets (vitamin D deficiency), beri-beri (vitamin B1 deficiency and common in rice-dependent China), kwashiorkor (protein deficiency) and marasmus (shortage of protein and calories), can reduce resistance to intestinal parasitic diseases, malaria and typhoid. In contrast, people in developed countries are at risk from over-eating and from an unbalanced diet which often contains too many animal fats which can cause heart disease.

Trends in food supply

Between the early 1950s and 1986 world food output increased more rapidly than did world population. The increase in output was more rapid in the developing countries, albeit from a much lower base, than in developed countries. The main exceptions to this generalisation were several African countries where food output per person actually fell

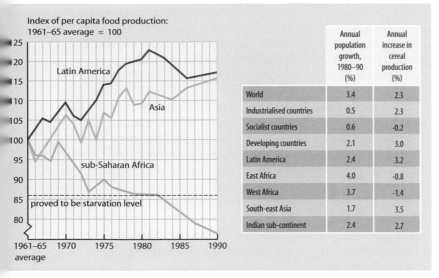

Index of per capita food production:
1961–65 average = 100

	Annual population growth, 1980–90 (%)	Annual increase in cereal production (%)
World	3.4	2.3
Industrialised countries	0.5	2.3
Socialist countries	0.6	-0.2
Developing countries	2.1	3.0
Latin America	2.4	3.2
East Africa	4.0	-0.8
West Africa	3.7	-1.4
South-east Asia	1.7	3.5
Indian sub-continent	2.4	2.7

Figure 16.59

Population growth and cereal production

(Figure 16.59). In 1986 the FAO announced that although there was sufficient food to feed everyone in the world 1 kg of food per day, this could not be achieved due to its uneven distribution and its rapidly rising cost. Disparities on a continental scale meant that while there was 5 kg per person for North America, 3.5 kg for Oceania and 2 kg for western Europe, there was only 1 kg for Latin America and south-east Asia and less than 0.5 kg for Africa.

Between 1986 and 1988, mainly due to below-average global rainfall totals, it was estimated that the world's food reserves fell from 459 million tonnes (101 days' supply) to 240 million tonnes (54 days' supply). Recent FAO reports (1993) suggest that, despite continuing regional increases in food production, over 1 000 million (1 bn) people now live at, or below, starvation level (Figure 16.60). These figures are likely to continue to rise due to such factors as:

- the increasing number of mouths to be fed;
- the attempts by the EU to reduce over-production and its stockpiles of food by lowering farm subsidies and taking arable land out of production;
- increasing world political instability, especially in several African countries (causing major refugee problems) and following the break-up of the former USSR;
- the world recession and consequent reduction in world trade; and
- global warming, which is blamed for increasing long-term drought in sub-Saharan Africa and short-term droughts in the major cereal producing areas of the USA and the CIS.

Food shortages in many parts of sub-Saharan Africa

The population of sub-Saharan Africa, with its high birth rates and falling death rates, is growing faster than anywhere else in the world. With 71 per cent of the labour force in agriculture and 77 per cent of the population living in rural areas, the income, nutrition and health of most Africans is closely tied to farming. In an area where, due to limited capital and technology, the use of new seeds, fertilisers, pesticides, machinery and irrigation is the lowest in the world, agriculture is almost wholly reliant upon an environment which is not naturally favourable. The soils in many areas have fertility constraints, low water-holding capacity and are vulnerable to erosion. High evapotranspiration rates harm crops, as do the unreliable rains which may cause flooding one year and then fail for several years (Figure 16.61). While the periods of water budget deficiency (drought) are getting longer and more frequent, experts argue as to whether this is part of a natural climatic cycle, a reduction of moisture in the air as a consequence of deforestation, or the effects of global warming.

Figure 16.60

Children awaiting food aid: Somalia

Traditional farming methods, such as shifting cultivation and nomadic pastoralism, use extensive areas of land for a limited period and then abandon them for several years before, probably, returning to them at a later date. With increases in population, fallow periods have been reduced and land has been overgrazed or overcropped. This, together with the destruction of forests for fuelwood, has allowed accelerated erosion and desertification. Efforts to increase food production have been impeded by a lack of money for fertilisers, seeds and tools. Even when overseas financial aid has been given, it has often had to be channelled towards unsuitable projects such as promoting monoculture, increasing cattle herds on marginal land and ploughing soils which would be better left with a protective vegetation cover. Financial aid schemes from overseas may create problems as the recipient country is likely to fall into debt, while the donor expects cash crops to be grown for export rather than food crops for local consumption. The diet often lacks protein and, during times of shortage, people cannot afford to buy food. Animals are likely to be attacked by the tse-tse fly, crops in the field by the locust, and crops in storage by rats and fungi. To add to these difficulties, several countries are torn by civil war and administrative corruption both of which interrupt farming and the distribution of relief supplies.

Figure 16.61

Famine in drought-gripped Africa, 1992

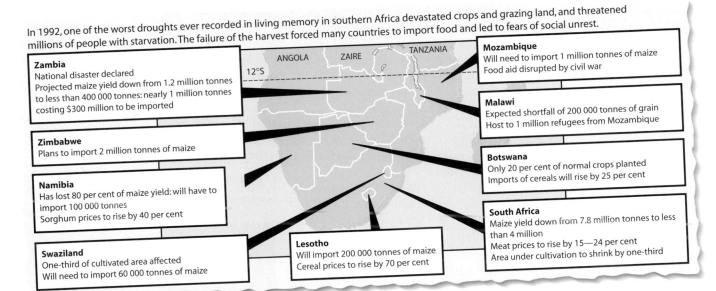

In 1992, one of the worst droughts ever recorded in living memory in southern Africa devastated crops and grazing land, and threatened millions of people with starvation. The failure of the harvest forced many countries to import food and led to fears of social unrest.

Zambia
National disaster declared
Projected maize yield down from 1.2 million tonnes to less than 400 000 tonnes: nearly 1 million tonnes costing $300 million to be imported

Zimbabwe
Plans to import 2 million tonnes of maize

Namibia
Has lost 80 per cent of maize yield: will have to import 100 000 tonnes
Sorghum prices to rise by 40 per cent

Swaziland
One-third of cultivated area affected
Will need to import 60 000 tonnes of maize

Lesotho
Will import 200 000 tonnes of maize
Cereal prices to rise by 70 per cent

Mozambique
Will need to import 1 million tonnes of maize
Food aid disrupted by civil war

Malawi
Expected shortfall of 200 000 tonnes of grain
Host to 1 million refugees from Mozambique

Botswana
Only 20 per cent of normal crops planted
Imports of cereals will rise by 25 per cent

South Africa
Maize yield down from 7.8 million tonnes to less than 4 million
Meat prices to rise by 15—24 per cent
Area under cultivation to shrink by one-third

ANGOLA   ZAIRE   TANZANIA
12°S

Famine

The notion that famine means a total food shortage (as implied in the introductory Biblical quote to this chapter) has been challenged as recent studies suggest that famine only affects certain groups in society (the poorest, least skilled and unemployed). Even during the worst times of famine, some food still appears in local markets — but at a price beyond the reach of most people.

Amartya Sen, an economist, says that "the idea that the causation of famines can be neatly split into 'natural' and 'man-made' ones would appear to be a bit of a non-starter". Indeed, it is now more widely accepted that most famines result from a combination of natural events and human mismanagement (Figure 16.62). Sen talked about the 'entitlement' of workers. He considered that workers have 'endowments', which may be goods, resources, skills, food and the ability to work, which are used to obtain items which they want or need. 'Entitlement failure' occurs when the endowment is insufficient to avoid starvation — i.e. certain groups in the community cannot afford to buy food, rather than there being no food available. Famine is therefore a decline in access to food (due to cost of transport and/or food supplies) rather than in a decline in the available food supply. For example, the 1974 flood in Bangladesh (Figure 16.62) did not totally destroy the rice crop but led to fears that, as the harvest would be poor, rice prices would rise. This, combined with rising economic inflation, resulted in a decrease in rural jobs and wages and a collapse in the entitlement of rural workers.

Figure 16.62

Famine and food supply in
Bangladesh

Q Figure 16.62 shows some of the causes of
famine which affected Bangladesh in the
mid 1980s.

1 In what ways is famine in that country
beyond human control?

2 How is the problem aggravated by social
and economic factors?

3 What, if anything, can be done to overcome
the problems of famine in the short term
and the long term?

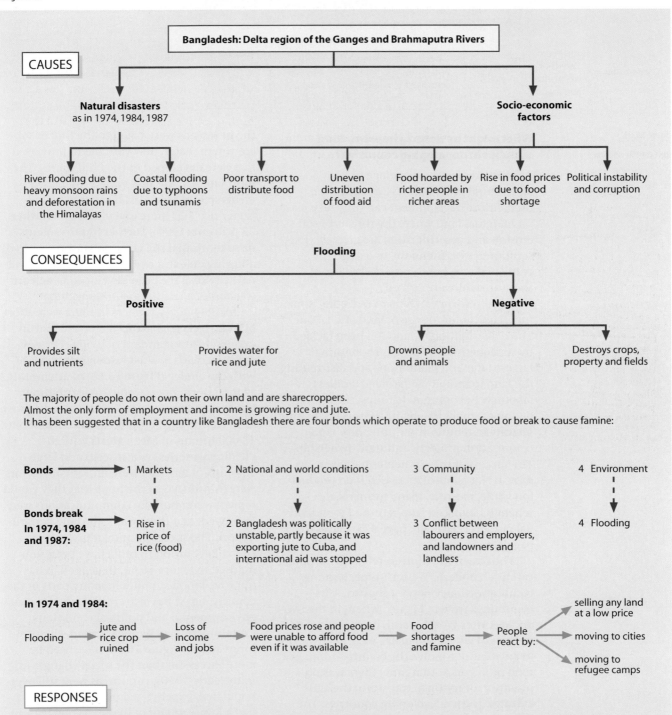

Bangladesh: Delta region of the Ganges and Brahmaputra Rivers

CAUSES

Natural disasters
as in 1974, 1984, 1987

**Socio-economic
factors**

River flooding due to
heavy monsoon rains
and deforestation in
the Himalayas

Coastal flooding
due to typhoons
and tsunamis

Poor transport to
distribute food

Uneven
distribution
of food aid

Food hoarded by
richer people in
richer areas

Rise in food prices
due to food
shortage

Political instability
and corruption

CONSEQUENCES

Flooding

Positive

Negative

Provides silt
and nutrients

Provides water for
rice and jute

Drowns people
and animals

Destroys crops,
property and fields

The majority of people do not own their own land and are sharecroppers.
Almost the only form of employment and income is growing rice and jute.
It has been suggested that in a country like Bangladesh there are four bonds which operate to produce food or break to cause famine:

Bonds → 1 Markets

2 National and world conditions

3 Community

4 Environment

Bonds break
In 1974, 1984
and 1987: → 1 Rise in
price of
rice (food)

2 Bangladesh was politically
unstable, partly because it was
exporting jute to Cuba, and
international aid was stopped

3 Conflict between
labourers and employers,
and landowners and
landless

4 Flooding

In 1974 and 1984:

Flooding → jute and
rice crop
ruined → Loss of
income
and jobs → Food prices rose and people
were unable to afford food
even if it was available → Food
shortages
and famine → People
react by: → selling any land
at a low price

→ moving to cities

→ moving to
refugee camps

RESPONSES

1 **Nijera Kori** (self-help schemes) prices for agricultural work linked to food prices. In theory, if food prices rose then so would agricultural
wages, *but* there is no minimum wage; people worked before a price was fixed; if there was a flood, then there was no work
and no wage.

2 **Government help** The state tried to improve transport, stockpile food in high-risk areas and improve food output — *but* do the real
poor get the aid or is it kept by the distributors?

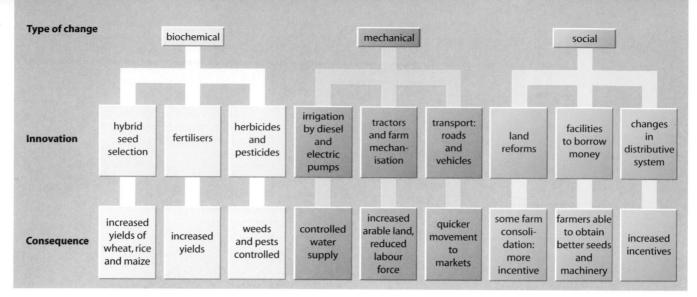

Type of change	biochemical			mechanical			social		
Innovation	hybrid seed selection	fertilisers	herbicides and pesticides	irrigation by diesel and electric pumps	tractors and farm mechan-isation	transport: roads and vehicles	land reforms	facilities to borrow money	changes in distributive system
Consequence	increased yields of wheat, rice and maize	increased yields	weeds and pests controlled	controlled water supply	increased arable land, reduced labour force	quicker movement to markets	some farm consoli-dation: more incentive	farmers able to obtain better seeds and machinery	increased incentives

Figure 16.63

The Green Revolution

What might be done to improve food supplies in developing countries?

As most areas with an average or high agricultural potential have already been used, future development can only take place on marginal land where the threats of soil erosion and desertification are greatest. The solution is not, therefore, to extend the cultivated area but to make better use of those areas already farmed.

Land reform can help to overcome some inefficiencies in the use of land and labour. The redistribution of land has been tackled by such methods as the expropriation of large estates and plantations and distributing the land to individual farmers, landless labourers or communal groups; the consolidation of small, fragmented farms; increasing security of tenure; attempting new land colonisation projects; and state ownership. The success of these schemes has been mixed. Not all have increased food production while, recently, many former state schemes have seen land returned to individual farmers (Places 48, page 425; Places 49, page 428).

The **Green Revolution** refers to the application of modern, Western-type, farming techniques to developing countries. Its beginnings were in Mexico when, in the two decades after World War II, new varieties or hybrids of wheat and maize were developed in an attempt to solve the country's domestic food problem. At that time, there was no intention of trying to transform the agriculture of other developing countries. The new strains of wheat produced dwarf plants capable of withstanding strong winds, heavy rain and diseases (especially the 'rusts' which had attacked large areas). Yields of wheat and maize tripled and doubled respectively, and the new seeds were taken to the Indian sub-continent. Later, new varieties of improved rice were developed in the Philippines. The most famous, the IR-8 variety, increased yields sixfold at its first harvest. Another 'super rice' has increased yields by a further 25 per cent (1994). Further improvements have shortened the growing season required, allowing an extra rice crop to be grown, and new strains have been developed which are tolerant of a less than optimum climate.

In 1964 many farmers in India were short of food, lacked a balanced diet and had an extremely low standard of living. The government, with limited resources, was faced with the choice (Figure 16.63) of attempting a land reform programme (redistributing land to landless farmers) or trying to improve farm technology. It opted for the latter. Some 18 000 tonnes of Mexican HYV (high-yielding varieties) wheat seeds and large amounts of fertiliser were imported. Tractors were introduced in the hope that they would replace water buffalo; communications were improved; and there was some land consolidation. The successes and failures of the Green Revolution in India have been summarised in Figure 16.64. In general, it has improved food supplies in many parts of the country, but it has also created adverse social, environmental and political conditions. Certainly the biochemical and mechanical innovations (Figure 16.63) have been far more successful than the social changes as, in the early 1990s, rural areas were still over-populated, experienced out-migration and had a low standard of living. As one television commentator said: "The poorer families can probably now afford two meals a day instead of one, while the rich can educate their children".

Successes	Failures
Wheat and rice yields have doubled	HYV seeds need heavy application of fertilisers and pesticides which have increased costs, encouraged weed growth and polluted water supplies
Often an extra crop per year	Extra irrigation is not always possible and can cause salinisation
Rice, wheat and maize have varied the diet	HYV not suited to waterlogged soils
Dwarf plants can withstand heavy rain and wind	Farmers unable to afford tractors, seed and fertiliser have become relatively poorer
Farmers able to afford tractors, seed and fertiliser now have a higher standard of living	Farmers with less than 1 ha of land have usually become poorer
Farmers with more than 1 ha of land have usually become more wealthy	Farmers who have to borrow money to buy seed and fertiliser are likely to get into debt
The need for fertiliser has created new industries and local jobs	Still only a few tractors, partly due to cost and shortage of fuel
Some road improvements	Mechanisation has increased rural unemployment
Area under irrigation has increased	Some HYV crops are less palatable to eat
Some land consolidation	

Conclusions	
A production and economic success which has lessened but not eliminated the threat of food shortages	Social, environmental and political failure: bigger gap between rich and poor

Figure 16.64

An appraisal of the Green Revolution in the Indian sub-continent

Appropriate technology (page 499) is needed to replace the many, often well-intentioned schemes which involved importing capital and technology from the more developed countries. Appropriate technology, often funded by non-governmental organisations such as the British based Intermediate Technology (Places 70) seeks to develop small-scale, sustainable projects which are appropriate to the local climate and environment, and the wealth, skills and needs of local people. This means

- Not large dams and irrigation schemes, but more wells so that people do not migrate to the few existing ones, drip irrigation as this wastes less water, stone lines (Figures 16.65 and 10.41) and check dams (Figure 10.42). Stones are laid, following the contours, even on gentle slopes in Burkina Faso while small dams built of loess are constructed across gulleys in northern China. In both cases, surface runoff is trapped giving water time to infiltrate into the soil and allowing silt to be deposited behind the barriers. These simple methods, taking up only 5 per cent of farmland, have increased crop yields by over 50 per cent.
- Not chemical fertiliser, but cheaper organic fertiliser from local animals (who also can provide meat and milk in the diet). Unfortunately, in many parts of Africa dung is needed as fuel instead of being returned to the fields.
- Not tractors, but simple, reliable, agricultural tools made, and maintained, locally.
- Not cash crops (often monoculture) on large estates, but smallholdings where both cash crops (income) and subsistence crops (food supply) can be grown. Mixed farming and crop rotation are less likely to cause soil erosion and exhaustion. Intercropping can protect crops and increase yields (smaller plants protected by tree crops).

Figure 16.65

Stone lines in Burkina Faso

Farming a. Farming in Western Normandy

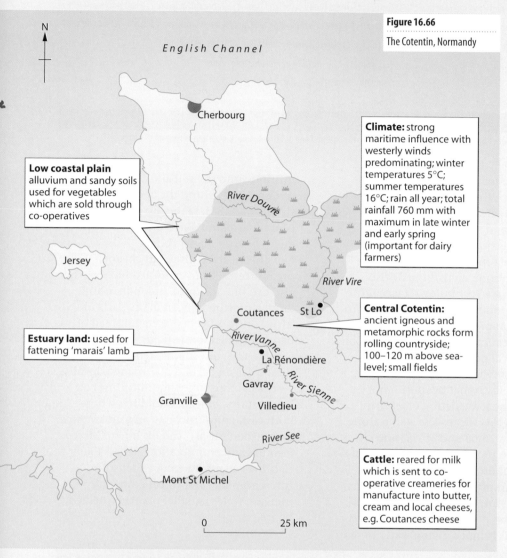

Figure 16.66

The Cotentin, Normandy

Low coastal plain alluvium and sandy soils used for vegetables which are sold through co-operatives

Climate: strong maritime influence with westerly winds predominating; winter temperatures 5°C; summer temperatures 16°C; rain all year; total rainfall 760 mm with maximum in late winter and early spring (important for dairy farmers)

Estuary land: used for fattening 'marais' lamb

Central Cotentin: ancient igneous and metamorphic rocks form rolling countryside; 100–120 m above sea-level; small fields

Cattle: reared for milk which is sent to co-operative creameries for manufacture into butter, cream and local cheeses, e.g. Coutances cheese

0   25 km

The Cotentin lies between the Vire estuary and Mont St Michel Bay (Figure16.66). It is mainly an agricultural region, although tourism is gaining in importance. The maritime climate, with rain (760 mm per year) occurring at all seasons and reaching a maximum in the late winter and spring months, is important for the farming. The maximum occurs just as temperatures are rising and the grass is starting to grow. This has been the basis of the successful dairy farming industry. Cattle are reared for their milk from which Normandy butter is made in addition to many local cheeses and cream. Most farms also produce fodder for their cattle, either in the form of silage in the late spring or as crops of corn in the late summer.

La Rénondière is a typical Cotentin dairy farm (Figure 16.67). It lies at 71 m above sea-level in a small valley whose stream flows into the River Vanne 0.75 km to the north. The land slopes very gently; fields are small and bounded by dense hedges; and most of the farm can be ploughed except for a small area in the valley bottom which becomes very wet. The Normandy-style farmhouse of grey stone covered in creeper, with white shutters, faces south. It is sheltered from the westerly winds, as are most of the buildings grouped around it.

The farm is 44 hectares in area. This is large for Normandy, where the average size is between 15 and 24 ha. Cattle are kept on 4 ha close to the farm; the rest of the land is used for producing fodder for the animals. The present herd consists of 46 cattle — mainly Friesian, with some traditional Normandy cows. The black and white Friesians have high milk yields, but the Normandy cows have better quality milk with high cream content. They are kept outdoors all year round with some protection in the winter. The cattle in milk are brought to the dairy twice a day and they produce on average 116 litres per cow per day (Figure 16.68). The small milking parlour is similar to many in the region. It holds 8 cows at a time, and is simpler than large dairies in the English Midlands or on dairy farms close to Paris. The milk is kept under refrigeration on the farm until collected by the creamery lorry each day in summer, but every two days at other times of the year (Figure 16.69).

The cows are artificially inseminated and produce one calf a year. Bull calves are sold in Gavray market for veal, and female calves are sold or used to replenish the herd. Cattle are carefully checked for yield and as this drops off they are replaced. Several of the present herd are over 10 years old.

The present farmer has been on the farm for 13 years, but it was farmed earlier by his parents and grandparents. All the work is done by the farmer, his wife (she is in charge of the dairy) and his father. Neighbours help during silage making. There is a strong tradition of dairy farming in the region.

On the western side of the Cotentin, there is a low-lying plain approximately 15–60 m above sea-level. It contains areas of sandy soils which are important for producing vegetables, including carrots, leeks, sweet corn, lettuce and tomatoes. These vegetables are marketed through co-operatives in the larger towns of the region, as well as in Paris and the UK.

The lowlands along the estuary of the Sienne and the Vanne are used as grazing land for the 'marais lamb'; large flocks of

Figure 16.67

A typical Normandy farmhouse

Figure 16.68

A small milking parlour

Figure 16.69

A cooperative creamery in Normandy

sheep are fattened on the marshes, providing yet another income for the farmers of the region.

In addition to their regular enterprises, many Normandy farmers breed and train trotting ponies — making regular visits to the long open sandy beaches to train them at low tide. As in Britain, bed and breakfast accommodation during the short tourist season from June to the end of August provides an additional source of income.

A major issue facing farmers in this part of France is the steady loss of people from the land. Many small farmers are going out of business, leaving houses empty. As in other peripheral regions of Europe, young people are moving to the cities. There is evidence that one or two wealthier large farmers are buying up vacant land. Some of the villages contain summer homes, owned by Parisians, with a number of British residents both in holiday and permanent homes. Prices for some houses without land have been low, encouraging overseas buyers. Villages still contain their bakery and shop, often with a butcher, but children are being forced to travel increasing distances to school. These features of rural life are common to many remoter areas within the European Union.

The impact of EU regulations can be seen. Milk quotas in line with EU rulings have been set by the government (page 455). They are generally higher than in the UK, perhaps due to the political strength of the farmers, and are an established part of the farm economy. However, they are generally unpopular with local farmers. This attitude was clearly conveyed by Mme Leboisellier at La Rénondière who threw up her hands in a typical Gallic gesture when interviewed on the subject. Prices for dairy cattle in northern France at present are favourable to the farmer.

Subsidies for lamb encourage the producer to maintain flocks. Demand for lamb is high, as is shown by the high prices in the supermarkets. Sheep farmers in the region took action in 1993 against imports of cheaper British lamb. They feared that their way of life was being threatened.

The issue of 'set-aside' has become important, with farmers leaving more of their land uncropped as they gain additional subsidies from the EU (page 455).

b. Farming in Gwent

Great Llwygy is an intensive sheep farm with the farming year (Figure 16.70) organised around the flock of 340 breeding ewes and 120 ewe lambs. A pedigree flock of Blue-faced Leicesters produces breeding rams for sale. The main flock of cross-bred Welsh Mule sheep is further crossed to produce a good-quality lamb for meat. There are also 20–25 beef cattle which are kept to sell as strong store animals in March in Abergavenny market.

The availability of good-quality feed is basic to the farming system. The ewes must be in good condition when put to the rams in October and November. This dictates the number and quality of the following year's lamb crop. A very dry summer, as in 1992, can reduce lamb numbers by 17–20 per cent. A high percentage of twin and triplet lambs is required. Sheep remain out on grass in the autumn as late as possible, but additional silage may be fed to them if the weather is poor. Housing the flock in winter not only protects them before and during the lambing season, but allows the land to recover from heavy use. The impact of the animals on the wet ground can severely damage future grazing.

An important part of the feed is silage produced on the farm. Fertiliser is added to the grass in spring and early summer when the temperature rises and the grass begins to grow. Some 15 tonnes were used in 1994, but this will have to be increased if the number of sheep is raised. Fields that will produce silage are shut off early to protect the grass. This means better silage if produced by mid May; the quality drops off rapidly after that time. Contractors are employed to cut and bale the silage in large black plastic bags, for storage until required. Hay is made by the farmer using his own baler and turner, when there is maximum daylight towards the end of June. Unreliable weather makes it difficult to predict dates for a contractor. It needs a week of warm dry weather.

High technology is used on the farm in January, when the ewe flock is scanned to check the number of ewes bearing triplets, twins or single lambs. These are separated out into groups and those with multiple lambs are fed additional concentrates, as silage does not provide sufficient energy for these ewes. This is important as it does not waste expensive extra supplement on ewes who do not require it.

The main output of the farm is lamb for meat; only a small proportion of the farm income is derived from the wool clip. Prices have increased slightly and it is now economic to hire contractors to do the shearing at the end of May. Approximately 2 kg wool is produced by each sheep. Lambs are sold in Abergavenny market every two weeks from the middle of July to late September when prices have fallen. Store lambs (small lambs for fattening) are sold in Hereford market. Some older stores are kept for sale around Christmas time. Ewes are kept for 5–6 years as long as they produce good lambs. Their value depreciates by £10 a year.

The flock has to be watched carefully throughout the year and the lambs are wormed at 3-weekly intervals from May to August. Chemical sprays are used to prevent flies attacking the lambs in May, July and September. The main sheep dip is carried out in October. The farm is small in comparison with many extensive high moorland farms (Figure 16.71), but this enables Joe Binns to work it himself with the minimum of help. One additional worker is employed during the lambing season, and contractors are used where it is uneconomic to keep expensive machinery on the farm.

Costs of additional feed, fertiliser, veterinary care, fuel, scanning, shearing and silage-making have to be set against the main output of lamb production. Enterprises which are not cost effective,

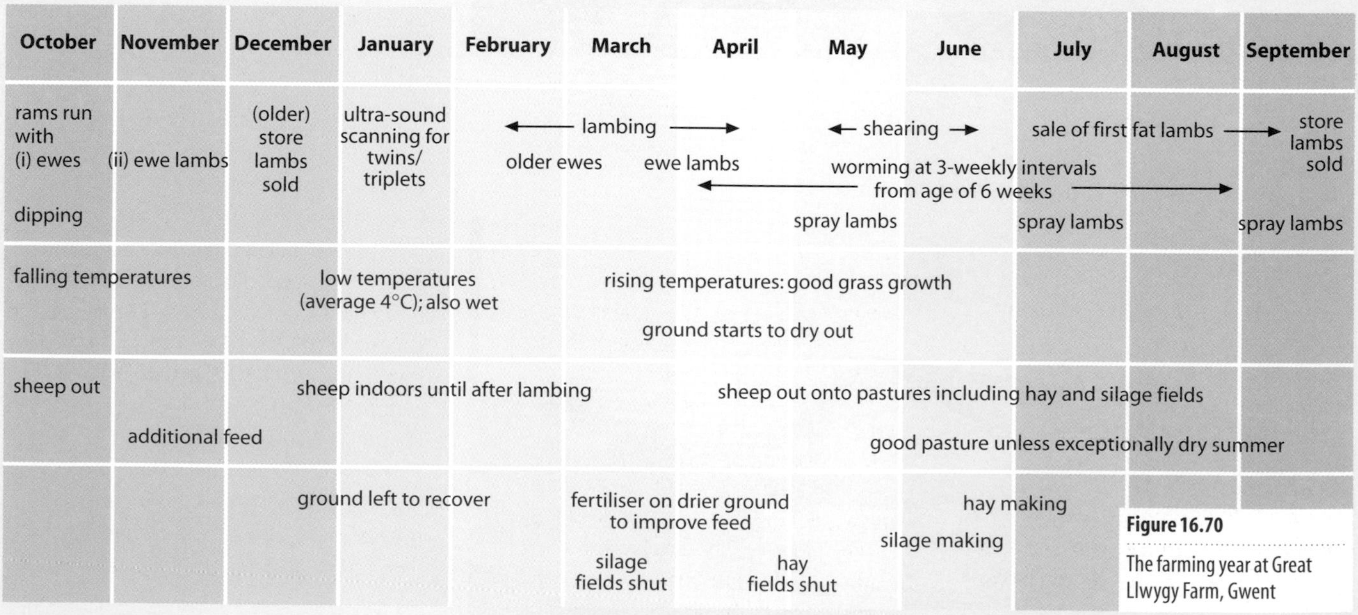

October	November	December	January	February	March	April	May	June	July	August	September
rams run with (i) ewes	(ii) ewe lambs	(older) store lambs sold	ultra-sound scanning for twins/ triplets	← lambing → older ewes		ewe lambs	← shearing →		sale of first fat lambs →		store lambs sold
dipping							worming at 3-weekly intervals ← from age of 6 weeks →				
							spray lambs		spray lambs		spray lambs
falling temperatures		low temperatures (average 4°C); also wet			rising temperatures: good grass growth						
						ground starts to dry out					
sheep out		sheep indoors until after lambing				sheep out onto pastures including hay and silage fields					
	additional feed							good pasture unless exceptionally dry summer			
		ground left to recover			fertiliser on drier ground to improve feed			hay making			
							silage making				
					silage fields shut	hay fields shut					

Figure 16.70

The farming year at Great Llwygy Farm, Gwent

Figure 16.71

Great Llwygy Farm, Gwent: the photograph, taken in early autumn, shows almost the full extent of the farm

such as pig rearing for sale to local butchers, have been abandoned as uneconomic.

The European Union (EU) now provides a £23 subsidy for all ewes on a farm. The farm is also classified as an upland Less Favoured area, giving an additional subsidy. These help to keep upland marginal farms such as Great Llwygy in business. There has been no effective increase in lamb prices (outputs) to the farmer since 1983, although costs of transport to market, fertiliser and feedstuffs (inputs) have all increased. Like many farmers in eastern Wales, Joe Binns has been forced to diversify, modernising and letting a cottage to visitors. He also provides an educational visit for students.

There are other constraints upon further development. It would be useful to increase the size of the flock, but that would mean additional feed, grazing, silage and possibly extra labour. Two fields, approximately 10 hectares, are rented. The cattle graze on half and silage is made on the rest, providing almost all the requirements for these animals. All the possible improved pasture has been developed. At the moment, there are few opportunities to rent or buy additional land nearby.

The National Park authorities exercise control over developments on farms within the Park boundaries. In practice, this has been minimal where the traditional environment is not threatened. Joe Binns, like

many farmers, has planted trees and makes provision for conservation through his farming methods.

As on the farm featured in Case Study 16A, there is serious concern about the number of EU regulations, with bureaucracy, and with the type of subsidies being provided. The present ewe subsidy is paid regardless of whether ewes produce lambs. This could encourage some farmers to keep flocks artificially high. Older farmers are leaving the land and there are few young people taking their place. The working life of a hill farmer is not easy and is rarely economically rewarding, but can give great personal satisfaction.

References

Barke, M. and O'Hare, G. (1984) *The Third World*. Oliver & Boyd.

Bradford, M. G. and Kent, W. A. (1977) *Human Geography*. Oxford University Press.

Briggs, K. (1982) *Human Geography*. Hodder & Stoughton.

Changing Agricultural Policies: Europe in the 1990s (1994) Geofile No. 250, Mary Glasgow Publications.

Environmentally Sensitive Areas in the UK (1987) Geofile No. 85, Mary Glasgow Publications.

IDG Bulletin (1993) IDG, Rijksuniversiteit, Utrecht.

Macmillan, B. (1991) Famine: the unnatural disaster, *Geography Review*, 5(1), pp. 18-21.

McBride, P. (1980) *Human Geography*. Nelson-Blackie.

Philip's Geographical Digest, 1992-93 (1993) Heinemann/Philip.

Prosser, R. (1992) *Human Systems and the Environment*. Thomas Nelson.

Timberlake, L. (1987) *Only One Earth*. Earthscan/BBC Books.

Third World Studies: Pastoralism in the Sahel (1983) The Open University.

Third World Studies: The Green Revolution in India (1985) The Open University.

Waugh, D. (1994) *The Wider World*. Thomas Nelson.

Chapter 17

Rural land use

"Nor rural sights alone, but rural sounds, exhilarate the spirit"

William Cowper

The term **rural** refers to those less densely populated parts of a country which are recognised by their visual 'countryside' components. Areas defined by this perception will depend upon whether attention is directed to economic criteria (a high dependence upon agriculture for income), social and demographic factors (the 'rural way of life' and low population density) or spatial criteria (remoteness from urban centres). Usually it is impossible to give a single, clear definition of rural areas as, in reality, they often merge into urban centres (the rural–urban fringe) and differ between countries.

Although generalisations may lead to oversimplifications (Framework 8, page 322), it is useful to identify three main types of rural area.

1 Where there is relatively little demand for land, certain rural activities can be carried out on an extensive scale, e.g. arable farming in the Canadian Prairies and forestry on the Canadian Shield.

2 In many areas, especially in economically developing countries, there is considerable pressure upon the land which results in its intensive use. Where human competition for land use becomes too great to sustain everyone, the area is said to be overpopulated (page 354). This often leads to rural depopulation, e.g. the movement to urban centres in Latin American countries (page 342).

3 In many economically developed countries, competition for land is greater in urban than in rural areas. The resultant high land values and declining quality of life are leading to a repopulation of the countryside (urban depopulation), e.g. migration out of New York and London (page 342).

Figure 17.1 shows some of the major competitors for land in a rural area. In many parts of the world, farming takes up the majority of the land and, especially in developing countries, employs most of the population.

Figure 17.1

Competition for rural land use and the need for management of rural resources

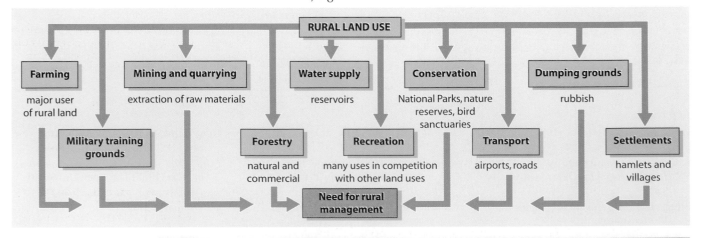

Q Using the above descriptions, which of the following do you perceive or consider to be 'rural': a Lake District valley, a parish in south-east England, a drought-stricken area in Ethiopia, part of the northern coniferous forest (Canadian Shield), the Ganges valley, an oasis in the Sahara, and a village in the Bolivian Andes? Explain your answers.

Forestry

In Britain

Neolithic farmers began the clearance of Britain's primeval forests about 3000 years ago. Aided by the development of axes, some clearances may have been on a scale not dissimilar from that in parts of the tropical rainforests of today. In 1919, with less than 4 per cent of the UK covered in trees, the Forestry Commission was set up to begin a controlled replanting scheme. Since then the policy has been to look towards an economic profit over the long term and to try to protect the environment. The area under trees is increasing, as in other countries in western Europe and North America, and had reached 10 per cent in the early 1990s.

Commercial forestry is easier to operate in Britain, Scandinavia and Canada than it is in developing countries as they are nearer to the major world markets. Softwoods are in greater demand than tropical hardwoods and, regardless of whether grown naturally or following replanting, as there are few species within a large area, selection and felling is made easy. Afforestation in north-ern Britain begins by growing young trees in nurseries. These, on reaching a height of 50 cm, are planted in rows on land which has been cleared and drained. The young trees are regularly sprayed with fertiliser and pesticide. The lower and dead branches are removed, while the poorer-quality trees are felled. This carefully controlled process of thinning gives more light and room for the healthier trees to grow — although overthinning can increase the risk of wind damage. The forester fells trees with modern lightweight chainsaws and uses machinery which can remove branches and bark, and cut the trunks into standard lengths ready for transport, by lorry, out of the forest. The land is then cleared ready for another 'crop' of young trees to be planted.

The softwoods of northern Britain take 40–60 years to mature so afforestation is an investment for the future. Most early plantations (before the 1980s) were not attractive either for human recreation or as a habitat for wildlife. Since then, a strong environmental lobby has ensured that modern plantations are carefully landscaped (Figure 17.2).

Figure 17.2

Managing an upland British forest (Kielder) in the 1990s

summits left clear for heather moorlands which provide a habitat for grouse and golden eagles

mature woodland forms a habitat for tawny owls and provides food for short-eared owls

trees planted at different times as differences in height are scenically more attractive

a variety of species and a lower density of trees replanted: helps to encourage more bird life which feeds on insects and so reduces the need for pesticide spraying

only small areas cleared at one time to reduce 'scars'

grassland provides a habitat for short-eared owls and food for tawny owls

winding forest road

land beside roads/tracks cleared to a width of 100 m and left as grass or planted with attractive deciduous trees

ponds created

cleared forest: branches left to rot; it takes 10 years for the nutrients to be returned to the soil

land next to river left clear for migrating animals such as deer

In developing countries

Commercial forestry is a relatively new venture in the tropics. It is usually controlled by multinationals based overseas who look for an immediate economic profit and have little thought for the long-term future or the environment. The forested areas in Latin America, tropical Africa and south-east Asia are rapidly decreasing. Figures giving the rate of forest clearances are diverse and unreliable but the UN suggests that for every 29 ha deforested in Africa, only 1 ha is replanted (the ratio is 11:1 in tropical Latin America). The forest is totally cleared by chainsaw, bulldozer and fire: there is no selection of trees to be felled. The secondary succession (page 294) is of poorer quality trees as little restocking is undertaken. Where afforestation of hardwoods does take place, there is often insufficient money for fertiliser and pesticide. The hope for the future may lie in **agroforestry**, where trees and food crops are grown alongside each other. Forest soils, normally rated unsuitable for crops, can be improved by growing leguminous tree species. Commercial forestry is more difficult to operate in developing countries as they are distant from world markets, the demand for hardwood is less than for softwood and although there are several hundred species in a small area only a few are of economic value.

The threat of the destruction of the rainforests has become a major global concern. Some of the consequences of deforestation are described in Places 61 and Case Study 11 (page 286).

Places 61   Forestry in Ethiopia, Amazonia and Malaysia

Ethiopia

Early in the 20th century, 40 per cent of Ethiopia was forested. Today the figure is under 2 per cent. In 1901, a traveller described part of Ethiopia as being "most fertile and in the heights of commercial prosperity with the whole of the valleys and lower slopes of the mountains one vast grain field. The neighbouring mountains are still well-wooded. The numerous springs and small rivers give ample water for domestic and irrigation purposes, and the water meadows produce an inexhaustible supply of good grass the whole year." In 1985, the same area was described as "a vast barren plain with eddies of spiralling dust that was once topsoil. The mountains were bare of vegetation and the river courses dry." As the trees and bushes were cleared, less rainfall was intercepted and surface runoff increased, resulting in less water for the soil, animals and plants.

Amazonia

The clearance of the rainforests means a loss of habitat to many Indian tribes, birds, insects, reptiles and animals. Over half of our drugs, including one from a species of periwinkle which is used to treat leukaemia in children, come from this region. It is possible that we are clearing away a possible cure for AIDS and other as yet incurable diseases. (Despite the rainforests being the world's richest repository of medical plants, only 2 per cent have so far been studied for potential health properties).

Without tree cover, the fragile soils are rapidly leached of their minerals, making them useless for crops and vulnerable to erosion. The Amazon forest supplies one-third of the world's oxygen and stores one-quarter of the world's fresh water — both would be lost if the region was totally deforested. The burning of the forest not only reduces the amount of oxygen given off, but increases the release of carbon dioxide (a contributory cause of global warming). It has also been suggested that the decrease in evapotranspiration, and subsequently rainfall, caused by deforestation could also have serious climatic repercussions — could the Amazon basin become another Ethiopia?

Malaysia: a model for the future?

Malaysia has several thousand species of tree, mainly hardwoods, with timber and logs being the country's third-largest export. However, the government has imposed strict controls, and the Forestry Department "manages the nation's forest areas to ensure sufficient supply of wood and other forest produce and manages and implements forest activities that would help to sustain and increase the productivity of the forest" (Malaysia Official Year Book, 1993). The Department insists that trees reach a specific height, age and girth before they can be felled (Figures 17.3 and 17.4). Logging companies are given contracts only on agreement that they will replant the same number of trees as they remove. Many newly planted hardwoods are

Rural land use

Figure 17.3

Logging operations in Malaysia

Figure 17.4

Timber lorries in the Malaysian rainforest

ready for harvesting within 20–25 years due to the rapid local growing conditions. Further experiments are being made with acacias and rattan, both of which grow even faster. Consequently, two-thirds of Malaysia is still forested and as most of the remaining third is under tree crops such as rubber, oil palm and coca, stocks are being successfully maintained. Even so, Malaysia's rapid industrialisation (Case Study 19, page 528) is causing increased deforestation, especially around the capital of Kuala Lumpur.

Mining and quarrying

Quarrying and mining have been, even since the Neolithic (when flint was excavated from chalk pits), Bronze and Iron Ages, an integral part of civilisation. It was often through the extraction and processing of minerals that many of today's 'developed' countries first became industrialised while to some 'developing' countries the export of their mineral wealth provides the only hope of raising their standard of living. The modern world depends upon 80 major minerals of which 18 are in relatively short supply, including lead, sulphur, tin, tungsten and zinc. Minerals are a finite, non-renewable **resource** which means that, although no essential mineral is expected to run out in the immediate future, their **reserves** are continually in decline. Although many items in our daily lives originated as minerals extracted from the ground, no mineral can be quarried or mined without some cost to local communities and the environment. Extractive industries provide local jobs and create national wealth, but they also cause inconvenience, landscape scars, waste tips, loss of natural habitats and various forms and levels of pollution.

The most convenient methods of mining are **open-cast** and **quarrying**. In open-cast mining, all the vegetation and topsoil are removed, thus destroying wildlife habitats and preventing other types of economic activity such as farming. Sand and gravel are extracted from depressions which, although shallow, often reach down to the water table, as in the Lea valley in north-east London. Coal and iron ore are often obtained from deeper depressions using drag-line excavators which are capable of removing 1500 tonnes per hour (Figure 17.5). Often, the worst scars (eyesores) result from quarrying into hillsides to extract 'hard rocks' such as limestone (Figure 8.15) and slate (Figure 17.6 and Places 62). There is usually greater economic and political pressure for open-cast coalmining than to quarry any other resource: it is the cheapest method of obtaining a strategic energy resource, but none generates greater social and environmental opposition. The increased demand for aggregates for road building and cement manufacture has led to the go-ahead being given for superquarries to be developed in the Scottish Highlands. One prospective quarry on Harris (1994) may see the eventual removal of a mountain and the creation of a new loch.

Mining involves the construction of either horizontal **adit mines**, where the mineral is exposed on valley sides, or vertical **shaft mines**, where seams or veins are deeper.

Resources are the total amount of a mineral in the earth's crust. The quantity and quality are determined by geology.

Reserves are the amount of a mineral which can be economically recovered.

Figure 17.5

Opencast mining for coal, West Virginia, USA

Figure 17.6

The Dionorwic slate quarries, Gwynedd, North Wales

Deep mining still affects local communities and the environment either by the piling up of rock waste to form tips — of coal in South Wales valleys (Aberfan, Figure 2.25) and china clay in Cornwall, for example — or by causing surface subsidence — as in some Cheshire saltworkings. Waste can also be carried into rivers where it can cause flooding by blocking channels and, when it contains poisonous substances, can kill fish and plants and contaminate drinking water supplies. This was highlighted in early 1992 when flood waters from Cornwall's last working tin mine, Wheal Jane, flowed into rivers and to the coast carrying with them arsenic and cadmium.

Places 62   Slate in north Wales and tin in Malaysia

Slate: Blaenau Ffestiniog, north Wales

The Oakley slate quarries were first worked in 1818. By the 1840s, the most easily obtained slate had been won and mining began. The introduction of steam power and the building of the Ffestiniog railway led to the export of 52 million slates from Porthmadog in 1873. At the quarry's peak productivity, 2000 men and boys were employed on seven different levels. Each level was steeply inclined into the hillsides and was worked to a depth of 500 m. Apart from farming, the slate mines were the sole providers of employment, and Blaenau Ffestiniog's population peaked at 12 000. Working in candlelight in damp and dusty conditions for up to 12 hours a day, and with rock falls common (pressure release, page 35) the life expectancy of miners was short. By the turn of the century, the manufacture of clay roof tiles heralded the beginning of the industry's decline and in 1971 the mine at Blaenau closed. A decade later, renamed Gloddfa Ganol, the underground galleries were re-opened to tourists, some of whom arrived via the narrow-gauge Ffestiniog railway.

As the mines closed, many people became either unemployed or were forced to move to seek work — the present population of Blaenau is under 500. Even so, as a tourist attraction, the slate mines are still the largest single employer. As in many British mining towns, the older houses are rows of cottages coated grey with the dust of earlier economic activity. Dust still blows into the cottages, but the inhabitants no longer die from silicosis. Above the town rise the large and unsightly spoil heaps (Figure 17.7), though they appear more stable than the coal tips which affected Aberfan (Case Study 2). On average, 10 tonnes of waste were created for every tonne of usable slate. Discarded machinery and unused buildings add to the scene of dereliction, though some are being restored as tourist attractions. Fortunately, there are few signs of the subsidence which is common to other mining areas.

Figure 17.7

Spoil heaps above Blaenau Ffestiniog, Gwynedd, North Wales

Figure 17.8

Disused tin mine, Malaysia

Figure 17.9

Sunway Lagoon Theme Park, near Kuala Lumpur, Malaysia

Tin: Malaysia

Malaysia is the world's major producer of tin (1994), but the mineral is no longer one of the country's major exports. Early tin mining was typical of the colonial trade: British settlers brought in the capital, machinery and technology; supervised the mining; and exported the tin for refining. Malaysia itself received few advantages. Most tin was obtained by open-cast methods together with the use of hydraulic jets.

Since independence, the mines have been nationalised and operated under the Malaysian government. However, the total output, and the number of working mines and tin-workers have all fallen rapidly due to the depletion of reserves and the low world market price for tin. Many of the old mines now form large tracts of land either covered in spoil (Figure 17.8) or left flooded. Disused buildings, overhead 'railways' and machinery often still remain.

One of the largest abandoned mines lies 15 km south of Kuala Lumpur in an area of rapidly growing housing and industry. It has been converted into a theme park complete with the world's longest aerial ropeway as well as water slides, musical fountains, a pirate ship and various water sports (Figure 17.9).

Recreation and tourism

Recreation has become an integral part of everyday life and tourism a major source of employment and income in many rural areas of Britain, other EU countries and North America (Case Study 20). In these parts of the world, the rapid growth in the tourist industry in the last three or four decades has been linked to the fact that the majority of people work shorter hours; have more disposable wealth; retire earlier; receive longer, paid holidays; and have better mobility. The opportunities to relax in rural and coastal areas and to travel overseas have been created through package holidays, cheaper air fares, advertising, improved roads and the increase in private transport. Other reasons include the need for a break by those people in employment who work under considerable pressures, the increased desire (perpetuated through the media) to travel to more distant places and to sample different types of lifestyle and activity, the fashion to take more than one holiday a year, a growth in mini-breaks and the development of national, country and theme parks.

These changes have put tremendous pressures upon many areas of countryside. Careful planning and management are now essential if visitors to rural areas are to maximise their leisure time spent there without

a coming into conflict with people who live and work in the area, other land users such as the Forestry Commission and Water Authorities (Figure 17.1) and other visitors wishing to participate in different recreational activities; and

b ruining the area and amenities which they want to enjoy.

The natural resources of rural areas which attract visitors are increasingly having to be protected from overuse and misuse, and managed for specific leisure activities.

It is possible to identify three levels of recreation in rural areas:

1 High-intensity areas where recreation is the major consideration (the Lea valley in north-east London and various theme parks, such as Alton Towers).
2 Average-intensity areas where a balance must be kept between tourism and other land users, and between recreation and conservation (the Peak District National Park, the New Forest and certain reservoirs open to public use). See Figure 17.10.
3 Low-intensity areas, usually of high scenic value, where conservation of land and wildlife is given top priority (upland parts of Snowdonia, the Kielder Forest and the Slimbridge Wildfowl Trust).

Is it possible to plan and manage the countryside to provide access to places of natural beauty and wildlife and to encourage energetic, and possibly noisy, pursuits while at the same time protecting that environment and without imposing undue restrictions? Robin Arvill wrote as long ago as 1967:

"To do this requires sound knowledge of those features contributing to the quality of an area — the special landscapes, sites for intensive recreation, countryside treasures, nature reserves and sanctuaries, buildings of artistic, architectural or historic interest, and caves, archaeological, geological and other attractions. In sum, a 'heritage inventory' is essential to the effective planning and development of an area. And where the heritage is deficient in desirable features, then the plan should provide for the creation — as far as possible — of these values. This could embrace open spaces, public parks and gardens, woods, water areas, individual buildings of distinction and town areas of character. In general, in the development of towns and resorts, incompatible uses should be separated, traffic segregated from pedestrians, buildings zoned to create centres of quality for community activities, and much more insight shown in reconciling development with the natural constraints of land, air, water and wildlife."

Figure 17.10

Some possible causes of conflict in a National Park

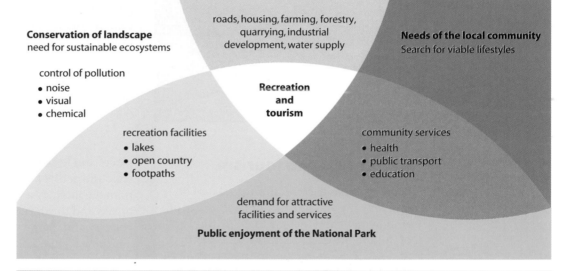

Conservation of landscape
need for sustainable ecosystems

control of pollution
• noise
• visual
• chemical

recreation facilities
• lakes
• open country
• footpaths

roads, housing, farming, forestry, quarrying, industrial development, water supply

Recreation and tourism

Needs of the local community
Search for viable lifestyles

community services
• health
• public transport
• education

demand for attractive facilities and services

Public enjoyment of the National Park

Q

Several of the questions that follow should give you scope for individual research and the expression of your own values (Framework 10, page 329). You may wish to consult Figure 17.10.

1 What types of problems may tourism create for:
   a farmers;
   b forestry workers;
   c Water Authorities; and
   d ornithologists?

2 Describe the possible impact of tourism on each of the following environments:
   a a mountainous National Park;
   b a lake or reservoir;
   c a coastal area;
   d a wetland; and
   e a small village.

3 What are the economic, social and environmental advantages and disadvantages of increased tourism in:
   a a British National Park;
   b an Alpine ski resort; and
   c a developing country?

4 Choose either a small area of countryside near to where you live or, if you live in a large urban area, a well-known National Park. Carefully describe how you would plan the area both to maximise and to protect its resources.

Having completed any sampling exercise (Framework 4, page 144), it is important to remember that patterns exhibited may not necessarily reflect the parent population. In other words, the results may have been obtained purely by chance. Having determined the mean of the sample size, it is possible to calculate the difference between it and the mean of the parent population by assuming that the parent population will conform to the normal distribution curve of Figure 6.36. However, while the sample mean must be liable to some error as it was based on a sample, it is possible to estimate this error by using a formula which calculates the **standard error of the mean** (*SE*).

$$SE\bar{x} = \frac{\sigma \ (standard\ deviation\ of\ parent\ population)}{\sqrt{n}\ (the\ square\ root\ of\ number\ of\ samples)}$$

where: $\bar{x}$ = mean of the parent population.

We can then state the reliability of the relationship between the sample mean and the parent mean within the three confidence levels of 68, 95 and 99 per cent (Framework 4). Unfortunately, when sampling, the standard deviation of the parent population is not available and so to get the standard error we have to use the standard deviation of the sample — i.e. using *s* rather than σ. Although this introduces a margin of error, it will be small if *n* is large (*n* should be at least 30).

For example: a sample of 50 pebbles was taken from a spit off the coast of eastern England. The mean pebble diameter was found to be 2.7 cm and the standard deviation 0.4 cm. What would be the mean diameter of the total population (all the pebbles) at that point on the spit?

$$SE = \frac{0.4}{\sqrt{50}} = \frac{0.4}{7.07} = 0.06 \text{ (to two decimal places)}$$

This means we can say:

1 with 68 per cent confidence, that the mean diameter will lie between 2.7 cm ± 0.06 cm, i.e. 2.64 to 2.76 cm;

2 with 95 per cent confidence, that the mean diameter will lie between 2.7 cm ± 2 x *SE* (2 x 0.06 = 0.12 cm), i.e. 2.58 to 2.82 cm;

3 with 99 per cent confidence, that the mean diameter will lie between 2.7 cm ± 3 x *SE* (3 x 0.06 = 0.18 cm), i.e. 2.52 to 2.88 cm.

If we wanted to be more accurate, or to reduce the range of error, then we would need to take a larger sample. Had we taken 100 values in the above example we would have had

$$SE = \frac{0.4}{\sqrt{100}} = \frac{0.4}{10.00} = 0.04 \text{ (to two decimal places)}$$

which means we can now say with 68 per cent confidence that the mean diameter size will lie between 2.7 cm ± 0.04 cm (i.e. 2.66 to 2.74 cm). Of course, this also means there is a 32 per cent chance that the mean of the parent population is **not** within these values. This is why most statistical techniques in geography require answers at the 95 per cent confidence level.

This standard error formula is applicable only when sampling actual values (**interval** or **measured data**). If we wish to make a count to discover the frequency of occurrence where the data are **binomial** (i.e. they could be placed into one of two categories), we have to use the **bionomial standard error**. For example, we may wish to determine how much of an area of sand dune is covered in vegetation and how much is *not* covered in vegetation. When using binary data, the sample population estimates are given as percentages, not actual quantities — i.e. *x* per cent of points on the sand dune *were* covered by vegetation; *x* per cent of points on the sand dune were *not* covered by vegetation.

The formula for calculating standard error using binomial data is

$$SE = \sqrt{\frac{p \times q}{n}}$$

where: p = the percentage of occurrence of points in one category;

q = the percentage of points not in that category; and

n = the number of points in the sample.

A random sample of 50 points was taken over an area of sand dunes similar to those found at Morfa Harlech (Figure 6.31). Of the 50 points, 32 lay on vegetation and 18 on non-vegetation (sand) which, expressed as a percentage, was

64 per cent and 36 per cent respectively. How confident can we be about the accuracy of the sample?

$$SE = \sqrt{\frac{64 \times 36}{50}} = \sqrt{46.08} = 6.79$$

As the sample found 64 per cent of the sand dunes to be covered in vegetation and knowing the standard error to be ± 6.79, we can say:

1 with 68 per cent confidence, that the vegetated area will lie between 64 per cent ± 6.79, i.e. between 57.21 and 70.79 per cent;

2 with 95 per cent confidence, that it will lie between 64 per cent ± 2 x SE (2 x 6.79 = 13.58), i.e. between 50.42 and 77.58 per cent will be vegetated; and

3 with 99 per cent confidence, that it will lie between 64 per cent ± 3 x SE (3 x 6.79 = 20.37), i.e. between 43.63 and 84.37 per cent.

Minimum sample size

It seems obvious that the larger the size of the sample, the greater is the probability that it accurately reflects the distribution of the parent population. It is equally obvious that the larger the sample, the more costly and time consuming it is likely to be to obtain. There is, however, a method to determine the minimum sample size needed to get a satisfactory degree of accuracy for a specific task — e.g. to find the mean diameter of pebbles on a spit, or the amount of vegetation cover on sand dunes. This is achieved by reversing the two standard error calculations.

For measured data: Imagine you wish to know the mean diameter of pebbles at a given point on a spit to within ± 0.1 cm at the 99 per cent confidence level.

The 99 per cent confidence level is 3 x SE.

$$3\,SE = \frac{3s}{\sqrt{n}}$$

i.e. $3s = 0.1\sqrt{n}$

i.e. $\frac{3s}{0.1} = \sqrt{n}$

We determined earlier that s (standard deviation of the sample) for the pebble size was 0.4, and so by substitution we get:

$$\frac{1.2}{0.1} = \sqrt{n}$$

i.e. $12 = \sqrt{n}$

$$n = 12^2$$

$$n = 144$$

We would need, therefore, to measure the diameter of 144 pebbles to get an estimate of the parent population at the 99 per cent confidence level.

For binomial data: How many sample values are needed to estimate the area of sand dunes which is vegetated, with an accuracy which would be within 5 per cent of the actual area (i.e. at the 95 per cent confidence level)?

$$n = \frac{p \times q}{(SE)^2}$$

Again by substitution we get:

$$n = \frac{64 \times 36}{(5)^2}$$

$$n = 92.16$$

We would therefore have to take a sample of 93 values to achieve results within 5 per cent of the parent population.

Study Figure 17.11 which shows the land use within a 40 sq km sector of west Norfolk. The area is divided into three main soil types in which a cut-off channel separates the zone of drained fen peats to the west from the zones of free-draining brown earths and of podsolic soils to the east. The soil zones are defined on the map.

1 Undertake a systematic sample of the five land use categories defined in the table for the zone of drained fen peats and for the zone of podsolic soils in order to estimate the percentage of land occupied by each use. Use all the grid intersections within the respective areas as sampling points, including those on the margins of the map. Record your tallies in the spaces provided in the table and convert the tallies into percentages.

Note: the four grid intersections falling within the cut-off channel and its enclosing banks are not to be considered.

Rural land use

2 For the zone of the drained fen peats, calculate the standard error of the estimate for the arable land and for the market gardening land.

3 For the zone of podsolic soils, calculate the standard error of the estimate for the grassland and for the woodland.

4 Comment upon the usefulness of the standard error calculations undertaken in answers to 2 and 3.

Figure 17.11

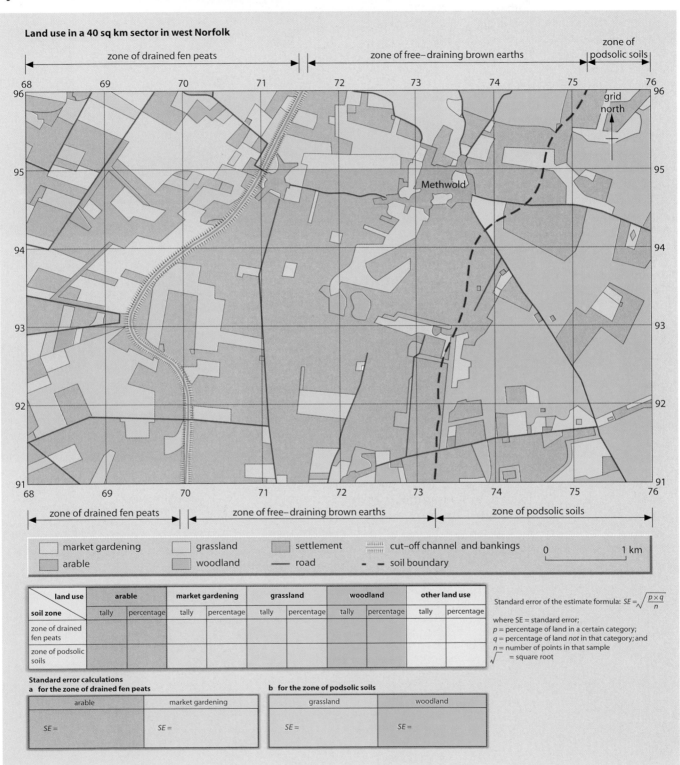

Land use in a 40 sq km sector in west Norfolk

land use	arable		market gardening		grassland		woodland		other land use	
soil zone	tally	percentage	tally	percentage	tally	percentage	tally	percentage	tally	percentage
zone of drained fen peats										
zone of podsolic soils										

Standard error of the estimate formula: $SE = \sqrt{\dfrac{p \times q}{n}}$

where SE = standard error;
p = percentage of land in a certain category;
q = percentage of land *not* in that category; and
n = number of points in that sample
$\sqrt{}$ = square root

Standard error calculations
a for the zone of drained fen peats

arable	market gardening
SE =	SE =

b for the zone of podsolic soils

grassland	woodland
SE =	SE =

Case Study 17

Designing an integrated conservation and development project: the coastal ecosystems of Tanzania

This Case Study draws together several topics discussed within the book. It discusses an island (coasts) with three interrelated ecosystems (biogeography) offering alternative land use possibilities (rural land use), where the number of people (population) is increasing and wishing to improve their standard of living (world development), thus putting pressure upon natural resources (rural land use). If economic development is to be maintained, raw materials are to be conserved for future generations, and conflicts between different types of land uses are to be avoided, then careful management of the island is essential.

Figure 17.12

The Tanzanian coastline

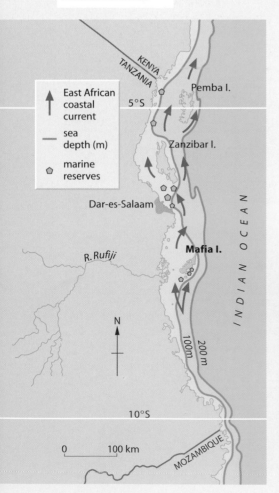

Tanzania's coastal ecosystems

Tanzania's 900-km coastline is narrow and, for about two-thirds of its length, consists of fringing and patch reefs which are situated close to the land (Figure 17.12). This system is broken by freshwater outlets such as the River Rufiji (East Africa's largest river) which culminates in a 1022 sq km delta and extensive mangrove stand. The coral reef and mangrove swamp are separated by a lagoon. The coral reef, lagoon and mangrove swamp form three ecosystems (Figure 17.13). The relationship between these three ecosystems is shown in Figure 17.14.

Figure 17.13

Examples of three tropical coastline ecosystems

a Mangroves
b Seagrass
c Coral

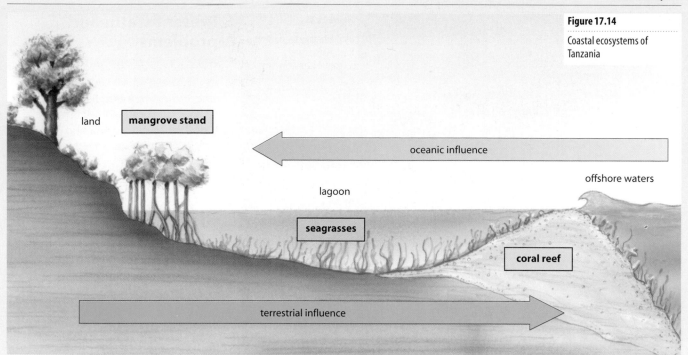

Figure 17.14

Coastal ecosystems of Tanzania

Mangrove

Mangrove is a generic term; related by lifestyle, these terrestrial trees live at the seaward end of often highly sediment-loaded, freshwater outlets in the tropics and sub-tropics. With characteristic interlacing stilt or aerial roots that trap sediments (Figure 17.13a), mangroves are able to colonise and thus stabilise the shore and create new land. Mangrove swamps support much life in the nutrient-rich water and relative safety of the canopy. Mangrove wood is strong, resistant to insect attack and can be used as building poles, or in shorter lengths in traditional fish traps, or for firewood.

Seagrasses

Seagrasses (true grasses) and algae (marine plants) occur in lagoons situated in the lee of barrier or fringing reefs where the water is sufficiently clear and shallow to allow photosynthetic light to reach the sediments (Figure 17.13b). These lagoons are a very productive environment — although few organisms use seagrasses as a direct food source, the nutrients released once the plants die form an important part of the food chain (page 275). Many commercially important species of fish obtain their food either directly from algae or from the life supported by this community. Shallow lagoons are also important as areas where the commercial cultivation of seaweeds can take place.

Corals

Corals (Figure 17.13c) are organisms that are similar to sea anemones; each lives within a stony cell which, collectively, form colonies. In association with algae and other organisms, they build coral reefs in environments providing a firm base, warm water (above 20°C) and photosythetic light. Corals ambush animal and particulate food and flourish best in nutrient-poor water where faster-growing algae can be naturally grazed by snails, fish and other animals. Coral reefs have been described, due to their richness of species and biological productivity, as the marine equivalent of the terrestrial rainforests.

Fragility and importance

The degree of interdependence, interrelation and exchange between these habitats is very high; the health of one being dependent on the condition of the others:

- Destruction of corals leads to loss of a naturally repairing system of breakwaters which protect the inshore ecosystems from erosion.
- The overexploitation of mangroves can lead to loss of erosion controls and higher offshore nutrient and sediment loads.
- Removal of seagrass and algae areas releases sediments and nutrients to

choke coral reefs. Mangrove stands are exposed to increased wave action.

Many organisms, particularly fishes, migrate between these habitats, seeking shelter in the coral reefs during the day and feeding over the seagrass areas at night. The more sheltered areas are also very important spawning and nursery grounds.

Main aims of coastal management

In the context of these three ecosystems, the objectives of sustainable integrated coastal management can be summarised as follows:

- Physical protection of the coastal areas from forces leading to erosion and habitat loss.
- Harnessed productivity from the coastal areas — vital for the support of local populations, and for generating income from commercial stocks.
- Recreation and tourist development — Tanzania has a national Economic Recovery Programme, part of which envisages the expansion of foreign currency earnings. Tourism is an important part of this goal, and the coral reefs in particular are a considerable tourist attraction.

Figure 17.15

Dynamite damage on Kitutia Reef, Mafia Island

Figure 17.16

Burning coral to release lime, Tanzania

Present status and problems

The resource-rich coastal zone of Tanzania has long been settled by communities which have exploited its resources without damage to the natural environment. Population levels were low, and within the carrying capacity of the environment (page 356).

In recent times, however, development and technological change have led to an increase both in population totals and in their rate of consumption of coastal resources, resulting in scarcities and conflicts. In turn, these have had negative social effects as economic decline has been fuelled by the widespread loss of, or drastic alteration to, the quality and diversity of the coastal resources.

Some of the many economic activities threatening the fragile coastal ecosystems are:

- **Discharge of raw sewage and effluent:** a particular problem around and down-current of industrial and urban areas.
- **Dynamite fishing:** the illegal use of dynamite to kill and stun fish. Recognised since the 1960s, this is a particular problem around Dar-es-Salaam. It destroys the physical structure of the reef and kills virtually all organisms within 15 m of the blast centre (Figure 17.15).
- **Coral mining:** the removal of live coral for the tourist and curio trade, and of fossilised coral rock for building purposes (Places 29, page 280). The material is selected according to growth form, collected by boat at low water and left to dry. For lime, the rock is burned over a fire made from locally collected wood (Figure 17.16). When mixed in various consistencies with sand, water and ground coral rock, it forms products ranging from a stable, concrete-like mixture to a fine plaster.
- **Hydrocarbon extraction:** limited at present to the Songo Songo gas field. To further national economic development and reduce foreign dependence, new fuel sources are needed. Much of the coastal geology is thought to be suitable, although test

Figure 17.17

The fishery industry, Tanzania

drilling has not so far yielded other commercially exploitable reserves.

- **Seaweed farming:** important as a means of diversifying income generation, but suffers from the problems associated with cash crops. If employed on a large scale, it could lead to biodiversity loss through the creation of monoculture communities; also, little is known about the effects on the environment of introduced varieties of seaweed.

- **Salt production:** by evaporation. Hyper-saline seawater is boiled, usually using mangrove wood for fuel because it is readily available. The process is crude and can quickly denude large areas of mature trees.

- **Fisheries:** at all scales, from subsistence to commercial, taking fin-fish, octopus, sea-cucumber, crayfish and edible shellfish. This is a complex set of activities, involving many tens of exploited species, a variety of gears and boats using several types of fishing ground, and varying with the season and the state of the tide (Figure 17.17).

- **Tourism:** a rapidly growing industry and one that the government is keen to promote. Coastal tourism includes game fishing, 'sea-faris', diving and snorkelling, as well as typical beach activities. Tourists, per capita, are major consumers of resources (potable water, fuel and foods); their presence can damage the natural environment (insensitive hotel development, poor diving etiquette, the curio trade) and lead to cultural conflicts (dress code and beach access) (Figure 17.18).

Controlling and managing activities such as these is important, but not not easy. Two examples — fishing and tourism — will illustrate this.

1 Fishing The management of fisheries, especially in and around marine parks, is vital given its status as a primary industry, major source of affordable protein and employer of many people. However, it is extremely difficult to contain this diverse and unpredictable industry within sustainable levels. A favoured solution to the complexity is to take simple measures which are either activity-dependent or time-dependent or, more usually, a mixture of the two.

- **Activity-dependent measures:** since environmental degradation reduces productivity, it is most important to ban gears that are known to damage the environment, particularly the spawning and nursery grounds, and which can catch endangered species (turtles).

- **Time-dependent measures:** many fish migrate and knowledge of these patterns can allow breeding stocks to be avoided. It is most important to preserve the juveniles and spawning adults. This is usually achieved with 'closed seasons'.

2 Tourism In order to give serious consideration to the use of the coast and its wildlife for tourist purposes, especially in the context of marine parks, decision-makers must assess what constitutes an acceptable financial return from such resources. As well as economic considerations, other key concerns include the technical and ecological/biological soundness of any exploitation. Are the species recognised as endangered? What is the resource's 'tourist carrying capacity'? Are the proposed activities culturally acceptable to local people? Will all parts of the community benefit equally from the resources generated by tourism?

Initial management responses

"The survival of our wildlife is a matter of grave concern to all of us in Africa. These wild creatures and the wild places they inhabit are not only sources of wonder and inspiration, but are an integral part of our national resources and of our future livelihood and well being" (Julius Nyerere, first President of Tanzania).

In response to the actual and perceived pressures, eight areas of natural richness were designated as Marine Reserves by the Ministry of Natural Resources in 1975. Within these areas, no extractive uses were permitted: only tourism and scientific activities were allowed. Outside these relatively tiny areas, existing fisheries regulations applied.

The Marine Reserves were called 'paper parks' because they had so little functional value. When planned, inadequate consideration was given to their objectives, to the desires and requirements of local people, and to the possible integration of international aims. Thus the enforcement agency, the Department of Fisheries, was beset by a poor mandate under which to work and a severe lack of financial and other resources. Corruption was widespread; fines were small; and the courts were reluctant to confiscate fishing equipment, often the only source of income generation for a family.

The new management approach

The aims of the new approach to coastal management are to satisfy economic, social and environmental objectives in order to ensure:

- the maximum sustainable economic benefit from the long-term yield of natural resources;
- the maintenance of the conditions and productivity of the natural environment; and
- the allocation of resources between competing uses and users.

These aims are often seen as contradictory and the major problem is how to cope with the diverse requirements of the many different user groups who utilise what are finite resources in a wide variety of different ways.

The original Marine Reserves aimed at protecting small areas through a rigid management strategy, but they were unable to cope with the multifaceted threats to the natural environment. To achieve the management aims listed above, requires an area large enough to justify investment in an integrated conservation and development project (ICDP). Each ecosystem has its own management requirements for conservation purposes and thus the large, well-defined area is zoned according to its uses and users. A management plan is devised for each zone, detailing the activities which are permitted within it, and this plan is integrated with the overall goals for the entire project area.

The Mafia Island Marine Park

The condition of much of Tanzania's coastal ecosystems is critical. To develop a comprehensive new system of management requires a suitable location to develop a pilot strategy, solving real management issues, which once seen to work can be extended to the rest of the coastline.

Mafia Island, which forms part of the coral reef, lies 60 km south of Dar-es-Salaam and opposite the Rufiji delta (Figure 17.12). In many ways, it is an ideal location for the establishment of Tanzania's first National Marine Park. For example:

Figure 17.18

Tourism and the trade in curios, Tanzania

Figure 17.19

Part of the Mafia Island Marine Park

- **Hydrographical location:** This is especially important in a national and regional (i.e. east African) context. Mafia Island and the Kitutia Marine Reserve (Figure 17.20) in particular have potential 'seed-bank' properties with respect to many of the over-exploited areas further north. The predominantly north-flowing East African coastal current is thought to play an important role in spreading marine plankton and larvae.

Developing a management model

To achieve an understanding of the nature and condition of the resources in a proposed management area — and of those using the resources — the following considerations should first be explored:

- **Political factors:** What is the political scale and structure of the area (international, national, state/provincial)? Is it stable? Can it provide finance or secure funding? Who will pay? Who will advise (UN agency, non-governmental organisation)? Are there powerful interest groups that may not, by nature, be objective?
- **Physical factors:** What physical features does the area contain (reefs, shoals, islands, sandbars, mangroves, estuaries)? Are there existing records of these (maps, charts, aerial photographs, satellite imagery)?
- **Biological factors:** What biological communities exist (composition, distribution, dominants, diversity, commercially important species)? In what condition are they? Are there records of change with time, indications of significant damage/overuse/overexploitation? Are there feeding or breeding sites for fragile or endangered species? Are there species of cultural and/or commercial importance?
- **Socio-economic factors:** What are the current uses of the area? Who uses it? Are they traditional local users/commercial interests/government interests? How are the resources exploited? Are these practices sustainable or destructive in the long term?

- **Administration:** The area comprises a single administrative district. This simplifies the exchange of information and means that many decisions can be made on-site.
- **Close-knit community:** The 30 000 population of the district is organised through a Muslim system of elders and exhibits a high degree of solidarity. It must be remembered that to protect resources, the welfare of the resource users is essential. Their local knowledge and continuous presence are vital in the five main areas of information-gathering, consultation, decision-making, initiating action and evaluation. Much impetus for the designation of Mafia Island as a Marine National Park has come from the local population who have been encouraged to adopt a participatory approach.
- **Unspoiled natural resources:** The marine resources around the Island are among the richest in east Africa. So far, they have escaped many of the

disastrous fishing methods which have contributed to widespread habitat loss in areas nearer to large population centres. Mafia Island's resources are not only marine, but also include coastal forest remnants, and populations of bats, dikker, honey bees, fish eagles and even hippopotamuses; all are important intrinsically and for the tourist market.
- **Geographical isolation:** The over-exploitation of coastal resources elsewhere in Tanzania is an increasing threat to Mafia Island, and the local population's traditional virtual monopoly of local fishing rights. The widespread scarcity of fish has increased its commodity value. As inshore areas become depleted, fishermen from Dar-es-Salaam and elsewhere gradually extend their search to more distant, previously uneconomic, fishing grounds such as Mafia Island. This process leads to competition for resources and dispute.

None of these considerations should alone determine the location of an ICDP. Given the scarcity of funding, project areas should always be selected with great care.

Once an area has been chosen, four stages lead to a practical plan for its establishment as a multi-user management area:

I The definition of management goals

Management goals for any ICDP would normally include conservation, sustainable resource use and economic development.

II The establishment of an administrative authority

This process provides the park with its first human representation. This group will expand and evolve as the needs of the park change. Initially, it will probably include representatives responsible for research, legislation and financial accounting; political and local representation; and advisors from the funding agency.

III The formulation of a management strategy and objectives

At this stage, a detailed assessment is made of the physical and human characteristics of the area, after which it is subdivided into small zones, each with its own flexible management plan. Zone plans can range from areas of total human exclusion to those in which the harvesting of resources is permitted. Examples of zone regimes are:

- **Core areas:** access is restricted. Core areas may include biological zones, where habitats and genetic diversity are protected; species zones, where particular animal and/or plant species are protected; and scientific research zones, where non-destructive scientific research is allowed in relatively undisturbed areas.
- **Buffer areas:** areas between core areas and the outer boundaries, where harmonious relationships between the natural environment and people are promoted through active, adaptive management. Buffer areas include:
  - **Sustained yield harvest zones:** where exploitation is allowed at a sustainable level, using minimally or non-destructive methods.
  - **Fisheries management zones:** where particular fish stocks are monitored and catch levels controlled.
  - **Water quality control zones:** where potential or existing forms of pollution are closely monitored and controlled.
  - **Tourist management zones:** where recreational uses of an area are allowed which are consistent with its carrying capacity.
  - **Education zones:** where instructional activities are allowed for the benefit of students, the general public and tourists, sometimes in combination with other zones.
  - **Cultural zones:** where cultural activities or monuments are preserved.

From this time onwards, it is important for the administrative authority to assess its performance, in terms of costs and benefits to the area and its users. This can only be accomplished through the objective monitoring of all aspects of the project.

IV The development of legislation

Legislation is vital to achieve the broad management objectives of the park. If enacted through parliament and with precedence over previous often-conflictory bylaws and legal notices, it will give the project a high political profile and add to its status, nationally and perhaps internationally. It is important that the legislation should not be divided along the traditional land/sea split, since the two are part of the same physical and human systems (Framework 1, page 39).

Mafia Island: the future

The efforts to stop widespread coral mining on Mafia Island have recently been boosted — e.g. contractors extending Mafia's airstrip have switched from using coral rock to the more expensive concrete. The additional cost to this infrastructure project, primarily aimed at encouraging tourism, will be met by the intended users. This is a small but encouraging example of the kind of integrated approach which is needed.

On Mafia Island, the development of resource use, the protection of the ecosystem and the development of tourism are all seen as vitally important goals. The area deemed most suitable for tourism almost totally overlaps the best-conserved ecosystems (Figure 17.21). The possible conflicts of interests resulting from this, provide a starting-point for discussions involving all interested parties. Ultimately, definition of the exact extent of the zones and their permitted uses and users will depend on the relative priority given to the different objectives defined by the ICDP.

Figure 17.20

Muslim elders, Mafia Island

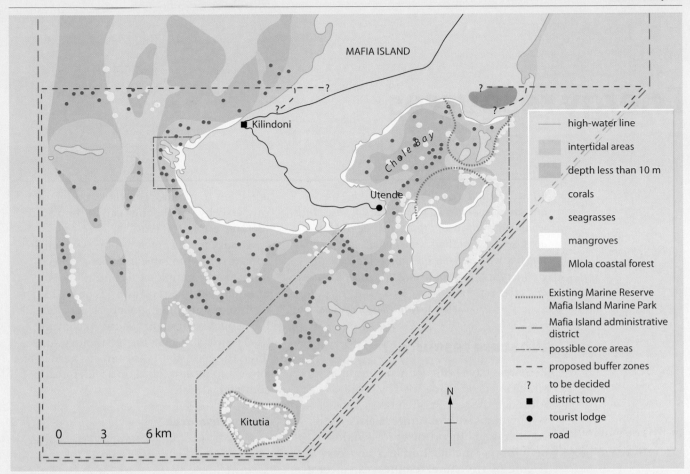

Figure 17.21

The proposed Mafia Island Marine Park

The Mafia Island Marine Park project, even though the legislation is yet to be enacted, is seen as one of the few examples of effective community management. In this project, identifying and resolving the many present and envisaged conflicts have to be given high priority. Conservationists and other concerned people are embarking on a long-term process — one in which a modicum of faith is very much required.

The ICDP approach: the future

The logic of sustainable exploitation of the earth's resources seems simple and obvious; yet it is a philosophy that has eluded industrial and high-tech societies which rely on innovation to improve yields.

These societies have regarded the sea as a free-for-all hunting ground. Other societies still have the opportunity to pursue a different course and to avoid dependency on the West by enhancing their rich traditions of sustainable exploitation of the environment.

The science of protecting ecosystems and their productivity and exploitation levels is far from exact. The example of the Mafia Island ICDP illustrates a common approach, but its solutions should never be thought of as being final — it is simply an improvement on what was done before.

It is vital to develop the science of management, since conflicts over the allocation of resources can only become more severe and more common. Accurate scientific data are also required, and the understanding and cooperation of communities and other user groups. Only recently has data collection been expanded to include modern thinking on the economics of natural resources and the environment, the value of indigenous knowledge, gender issues, and so on.

A change of philosophy is also needed. Large, traditional, 'top–down' initiatives developed to feed political ambition have to be seen for what they are. It is only through people wanting change in their own backyard that change will really occur. Politicians should grasp the 'bottom–up' approach and champion the wishes of those they represent. Their goal, as well as that of other decision-makers, is to help the people to guarantee their own future.

References

Arvill, R. (1967) *Man and the Environment*. Pan Books.

Pickering, K. and Owen, L. (1994) *Global Environmental Issues*. Routledge.

Science in Geography, Books 1–4 (1974) Oxford University Press.

Wilson, J. (1984) *Statistics in 'A'-level Geography*. Schofield & Sims.

Chapter 18

Energy resources

> *"If each Indian were to start consuming the amount of commercial energy a Briton does, that would mean the world finding the equivalent of an extra 3190 million tonnes of oil each year. Imagine what consuming that would do to the greenhouse effect, not to mention its effect on oil and other reserves."*
>
> **Mark Tully,** *No Full Stops in India*, 1991

What are resources?

Resources have been defined as features which are needed and used by people. Although the term is often taken to be synonymous with **natural resources**, geographers and others often broaden this definition to include **human resources** (Figure 18.1). Natural resources can include raw materials, climate and soils. Human resources may be sub-divided into people and capital. A further distinction can be made between **non-renewable resources**, which are finite as their exploitation can lead to the exhaustion of supplies (oil), and **renewable resources**, which, being a 'flow' of nature, can be used over and over again (solar energy). As in any classification, there are 'grey' areas. For example, forests and soils are, if left to nature, renewable; but, if used carelessly by people, they can be destroyed (deforestation, soil erosion).

Reserves are known resources which are considered exploitable under current economic and technological conditions. For example, North Sea oil and gas needed a new technology and high global prices before they could be brought ashore; in contrast, tidal power still lacks the technology, and often the accessibility to markets, needed to allow it to be developed on a widespread, commercial scale.

Energy resources

The sun is the primary source of the earth's energy. Without energy, nothing can live and no work can be done. Coal, oil and natural gas, which account for an estimated 87.8 per cent of the global energy used in 1990 (Figure 18.2), are forms of stored solar energy produced, over thousands of years, by photosynthesis in green plants. As these three types of energy, which are referred to as **fossil fuels**, take long periods of time to form and to be replenished, they are classified as non-renewable. As will be seen later, these fuels have been relatively easy to develop and

Figure 18.1

A classification of resources

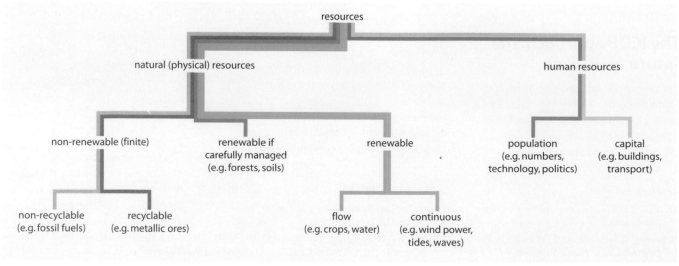

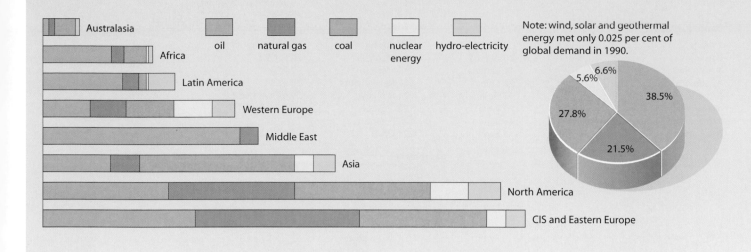

Note: wind, solar and geothermal energy met only 0.025 per cent of global demand in 1990.

Figure 18.2

World energy production, 1990

cheap to use, but they have become major polluters of the environment. Nuclear energy is a fourth non-renewable source but, as it uses uranium, it is not a fossil fuel.

Renewable sources of energy are mainly forces of nature which can be used continually, are sustainable and cause minimal environmental pollution. They include running water, waves, tides, wind, the sun, geothermal, biogas and biomass. At present, with the exception of running water (hydro-electricity) and the wind, there are economic and technical problems in converting their potential into forms which can be used.

World energy producers and consumers

It has been estimated that the world annually consumes an amount of fossil fuel that took nature about 1 million years to produce, and that the rate of consumption increased four-fold between 1950 and 1990 (Figure 18.3). This consumption of energy is not evenly distributed over the globe (Figure 18.2). At present, the 77 per cent of people living in the 'developing' countries consume less than 30 per cent of the total energy supply. Although recently the consumption of energy in 'developed' countries has begun to slow down, due partly to industrial decline and environmental concerns, it has been increasing more rapidly in 'developing' countries with their rapid population growth and aspirations to raise their standard of living. This led to a conflict of interest between groups of countries at the Rio Earth Summit conference. The 'industrialised' countries, with only 23 per cent of the world's population yet consuming almost 70 per cent of the total energy, wished to see resources conserved and, belatedly, the environment protected. The 'developing' countries, who blame the industrialised countries for most of the world's pollution and depletion of resources, considered that it was now their turn to use energy resources, often regardless of the environment, in order to develop economically and to improve their way of life.

The world's reliance upon fossil fuels (Figure 18.2) is likely to continue well into the next century. However, while the economically recoverable reserves of coal remain high, the similar life expectancies of oil and natural gas are much shorter (coal: about 200–400 years; oil: about 50 years; natural gas: about 120 years). The distribution of recoverable fossil fuels is spread very unevenly across the globe, with the former USSR being well endowed with coal and natural gas; North America and parts of Asia with coal; and the Middle East with oil and natural gas (Figure 18.4). As these producers are not always major consumers, there is a considerable world movement of, and trade in, fossil fuels (Figure 18.5).

Figure 18.3

World energy use, 1950–90

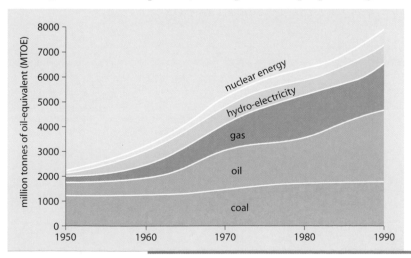

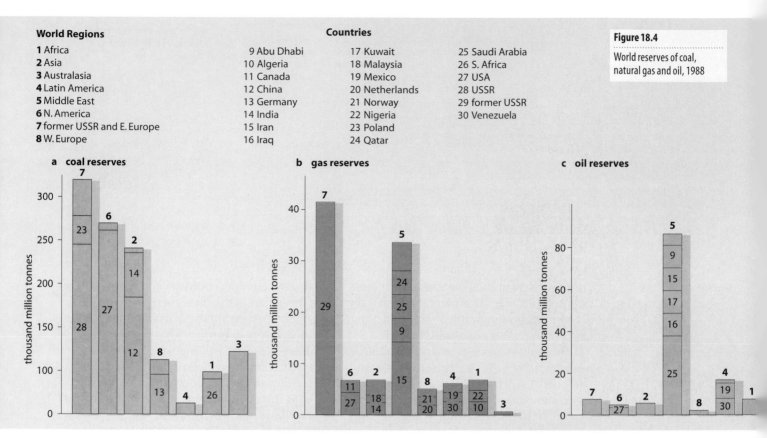

World Regions

1 Africa
2 Asia
3 Australasia
4 Latin America
5 Middle East
6 N. America
7 former USSR and E. Europe
8 W. Europe

Countries

9 Abu Dhabi	17 Kuwait	25 Saudi Arabia
10 Algeria	18 Malaysia	26 S. Africa
11 Canada	19 Mexico	27 USA
12 China	20 Netherlands	28 USSR
13 Germany	21 Norway	29 former USSR
14 India	22 Nigeria	30 Venezuela
15 Iran	23 Poland	
16 Iraq	24 Qatar	

Figure 18.4

World reserves of coal, natural gas and oil, 1988

a coal reserves

b gas reserves

c oil reserves

UK energy consumption

Figure 18.5

Location and movement of the world's fossil fuels

- oilfields
- oil movements
- natural gas fields
- gas movements
- coal production
- coal movements

Britain has always been fortunate in having its own energy sources. In the Middle Ages, fast-flowing rivers were used to turn water-wheels while, in the early 19th century, the use of steam, from coal, enabled Britain to become the world's first industrialised country. Just when the most accessible and cheapest supplies of coal began to run short, natural gas (1965) and oil (1970) were discovered in the North Sea and improvements in technology enabled the controversial production of nuclear power. Looking ahead to a time when the UK's fossil fuels do become less available (300 years for coal and 40 years for oil and natural gas), Britain's seas and weather have the potential to provide renewable sources of energy using the wind, waves and tides.

The total energy consumption in the UK in 1992 was equivalent to 207.7 million tonnes of oil, of which 90.1 per cent was from fossil fuels (Figure 18.6). In that same year, the amount of energy obtained from non-fossil fuels (nuclear, hydro-electric and wind power) amounted to 9.9 per cent. Figure 18.7 shows how the sources of Britain's energy have changed since 1950 and are predicted to change by the year 2020.

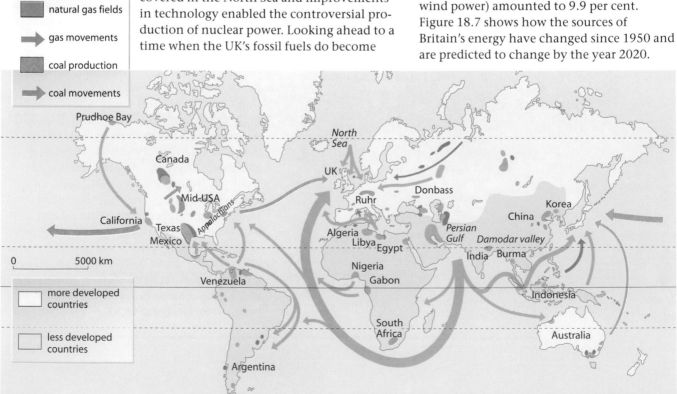

- more developed countries
- less developed countries

0   5000 km

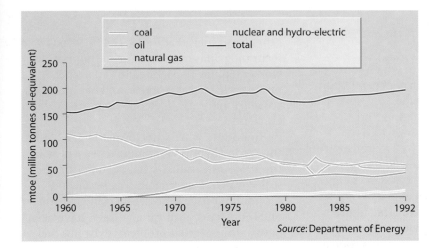

Figure 18.6

UK primary energy consumption, 1960–92

Energy consumption by final user has also changed over time. Between 1972 and 1992 there was a decline in the amount of energy consumed by industry from 42 per cent of the total to 25 per cent, but an increase in the amount used by transport from 22 per cent to 31 per cent. Domestic demand has fallen slightly due to energy efficiency schemes (double-glazing and roof and cavity wall insulation).

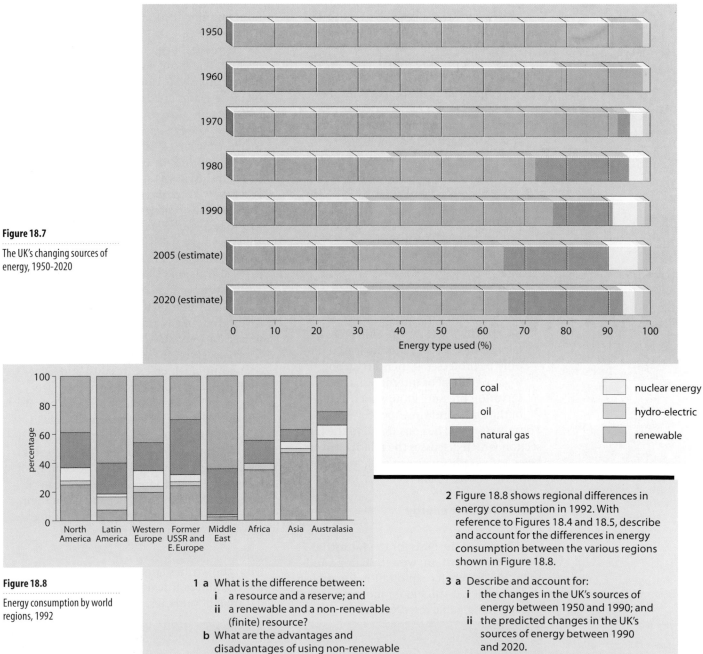

Figure 18.7

The UK's changing sources of energy, 1950-2020

Figure 18.8

Energy consumption by world regions, 1992

1 a What is the difference between:
   i   a resource and a reserve; and
   ii a renewable and a non-renewable (finite) resource?
  b What are the advantages and disadvantages of using non-renewable resources?

2 Figure 18.8 shows regional differences in energy consumption in 1992. With reference to Figures 18.4 and 18.5, describe and account for the differences in energy consumption between the various regions shown in Figure 18.8.

3 a Describe and account for:
   i   the changes in the UK's sources of energy between 1950 and 1990; and
   ii the predicted changes in the UK's sources of energy between 1990 and 2020.
  b Why has the UK relied so heavily upon the use of fossil fuels?

Sources of energy

Decisions by countries as to which source, or sources, of energy to use may depend upon several factors. These include:

- Availability, quality, lifetime and sustainability of the resource.
- Cost of harnessing, as well as transporting (importing or within the country), the source of energy: some types of energy, such as oil, may be too expensive for less wealthy countries; while others, such as tides, may as yet be uneconomical to use.
- Technology needed to harness a source of energy: this may, like costs, be beyond some of the less developed countries (nuclear energy), or may yet have to be developed (wave energy).
- Demands of the final user: in less developed countries, energy may be needed mainly for domestic purposes; in more developed countries, it is needed for transport, agriculture and industry.
- Size, as well as the affluence, of the local market.
- Accessibility of the local market to the source.
- Political decisions: for example, which type of energy to utilise or to develop (nuclear), or whether to deny its sale to rival countries (embargoes).
- Competition from other forms of available energy.
- Environment: this may be adversely affected by the use of specific types of energy, such as coal and nuclear; it may only be protected if there are strongly organised local or international conservation pressure groups such as Friends of the Earth and Greenpeace.

These factors will be considered in the next section which discusses the relative advantages and disadvantages of each available or potential source of energy.

Non-renewable energy

Coal

Coal provided the basis for the Industrial Revolution in Britain, western Europe and the USA. Despite its exploitation for almost two centuries, it still has far more economically recoverable reserves than any of the other fossil fuels (Figure 18.4). Improved technology has increased the output per worker (Figure 17.5), has allowed deeper mining with fewer workers, and has made conversion for use as electricity more efficient.

In Britain, where pit closures began before the Second World War and reached a peak in the 1960s, the industry has been drastically scaled down (Figure 18.9). The social and economic consequences, especially in former one-industry coal-mining villages, have been devastating. Similar problems exist in other old mining communities in Belgium, the Ruhr and the Appalachians (USA) (Figure 18.5).

There have been many reasons for this decline. The most easily accessible deposits have been used up and many of the remaining seams are dangerous, due to faulting, and uneconomic to work. Costs have risen due to expensive machinery and increased wages. The demand for coal has fallen for industrial use (the decline of such heavy industries as steel), domestic use (oil- and gas-fired central heating) and power stations (now preferring gas). British coal has had to face increased competition from cheaper imports (USA and Australia), alternative methods of generating electricity (gas-fired) and cleaner forms of energy. Political decisions have seen subsidies paid to the nuclear power industry and a greater investment in gas rather than coal-fired power stations (the 'dash for gas' policy). Green pressures have also led to a decline in coal mining, which creates dust and leaves spoil tips, and using coal to produce electricity, as this releases sulphur dioxide and carbon dioxide which are blamed for acid rain and the greenhouse effect. In places such as China and India, with large reserves and a rapidly growing demand for energy, it is essential to introduce 'clean coal technologies', but these will depend upon their development, and possibly their funding, by the industrialised countries.

Oil

Oil is the world's largest business, with commercial and political influence transcending national boundaries. Indeed, several of the

	1947		1994	
	working collieries	employees	working collieries	employees
Northumberland and Durham	188	148 000	0	under 2 000 (open-cast)
UK	958	718 400	19	10 800

Figure 18.9

Decline of coal mining in the UK

Figure 18.10

Milford Haven oil refineries

largest multinational enterprises are oil companies. Oil, like other fossil fuels, is not even in its distribution (Figure 18.5), and is often found in areas which are either distant from world markets or have a hostile environment — such as the Arctic (Alaska), tropical rainforests (Nigeria and Indonesia), deserts (Algeria and the Middle East) and under stormy sea (North Sea). This means that oil exploration and exploitation is expensive, as is the cost of its transport by pipeline or tanker to world markets. Oil, with its fluctuating prices, has been a major drain on the financial reserves of many developed countries and has been beyond the reach of most developing countries. Countries where oil is at present exploited can only expect a short 'economic boom' as, apart from several states in the Middle East where production may continue for 100 years, most world reserves are predicted to become exhausted within 30 years.

New technology has had to be developed to tap less accessible reserves. Before oil could be recovered from under the North Sea, large concrete platforms, capable of withstanding severe winter storms, had to be designed and constructed. Each platform, supported by four towers, had to be large enough to accommodate a drilling rig, process plant, power plant, helicopter landing-pad and living and sleeping quarters for its crew. The towers may either be used to store oil or may be filled with ballast to provide extra anchorage and stability. Two 90cm trunk pipelines were laid, by a specially designed pipe-laying barge, over an uneven sea-bed to Sullom Voe in the Shetlands.

On a global scale, oil production and distribution are affected by political and military decisions. OPEC (Figure 21.24) is a major influence in fixing oil prices and determining production — although even it is helpless in the face of international conflicts

such as Suez (1956), the Iran–Iraq War (early 1980s) and the Gulf War (1991). Closer to home, recent British and EU fuel policies have favoured the gas and nuclear industries at the expense of oil. Oil is used in power stations, by industry, for central heating and by transport. Although it is considered less harmful to the environment than coal, it still poses many threats. Oil tankers can run aground during bad weather (Braer, 1993) releasing their contents which pollute beaches and kill wildlife (Exxon Valdez, 1989), while explosions can cause the loss of human life (Alpha Piper rig, 1988). To try to reduce the dangers of possible spillages and explosions, oil refineries have often been built on low-value land adjacent to deep, sheltered tidal estuaries, well away from large centres of population (Figures 18.10 and 18.12). During 1991 there were, in England and Wales alone, 5288 reported cases of inland water pollution by oil and petroleum products. Road transport, using petrol, is a major contributor to the greenhouse effect.

Natural gas

Natural gas has become the fastest-growing energy resource, providing an alternative to coal and oil. Latest estimates suggest that global reserves will last another 120 years. Gas is often found in close proximity to oil-fields (Figure 18.5) and therefore experiences similar problems in terms of production and transport costs and requirements for new technology. At present it is considered to be the cheapest and cleanest of the fossil fuels, which is why the British government has favoured new power stations being gas-fired and older ones being converted to gas (mainly at the expense of coal). As gas causes less atmospheric pollution, it also helps the British government to achieve its promised 60 per cent reduction of greenhouse-gas emissions from power stations by the year 2000.

Nuclear energy

During the early 1950s, nuclear energy, with its slogan 'atoms for peace', was seen by many to be a sustainable, inexpensive and clean energy resource and by others as a dangerous military aid and a threat to the environment.

Figure 18.11

Major users of nuclear energy, 1991

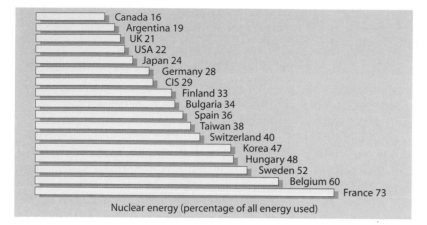

Nuclear energy uses, as its raw material, uranium. Compared with fossil fuels, uranium is only needed in relatively small amounts (50 tonnes of uranium a year, compared with 500 tonnes of coal per hour for coal-fired power stations). Uranium has a much longer lifetime than coal, oil or gas and can be moved more easily and cheaply. However, the development of nuclear energy, with its new technology, specially designed power stations and essential safety measures, has been very expensive. As a result, it has only been adopted by wealthy countries, and even then only by those (Figure 18.11) lacking fossil fuels (Japan) or with significant energy deficits (UK, USA and France). As a source of electricity, it is fed into the National Grid; but, as a source of energy, it cannot be used by transport or for heating. The decision to develop nuclear power has been, universally, a political decision. In Britain, for example, the first nuclear programme was initiated in 1955 by a government which felt that the scheme would be uneconomic. Later, two further programmes were announced, respectively, by Labour in 1964 and the Conservatives in 1979. Since the 1950s, most money spent on developing energy in the UK has been spent on nuclear power whereas, according to conservation groups, it should have been spent on renewable, non-polluting resources. Nuclear power is a major issue in Japan, where the government plans to build more power stations. Japan is desperately short of its own sources

Figure 18.12

Location of Britain's major power stations

of power, but experienced first-hand the worst side of nuclear power when two atomic bombs were dropped on Hiroshima and Nagasaki in 1945.

Although nuclear power stations produce fewer greenhouse gases than thermal (coal-, oil- and gas-fired) power stations, they do present potential risks in three main areas: routine emissions of radio-activity, waste disposal and radio-active contamination accidents. Routine emissions have been linked with clusters of increased leukaemia, especially in children, around several power stations (notably Sellafield and Dounreay). Radio-active waste has to be stored safely, either deep underground or at Sellafield. Every radio-active substance has a 'half-life' — i.e. the time it takes for half of its radio-activity to die away. Iodine, with a half-life of 8 days, becomes 'safe' relatively quickly. In contrast, plutonium 239, produced by nuclear reactors, has a half-life of 250 000 years and may still be dangerous after 500 000 years. The two worst radio-active accidents resulted from the melt-down of reactor cores at Three Mile Island in the USA (1979) and at Chernobyl in the Ukraine (1986). It was mainly for economic and safety reasons that British nuclear power stations (Figure 18.12) were built on coasts and estuaries where there is water for cooling and cheap, easily reclaimable land well away from major centres of population.

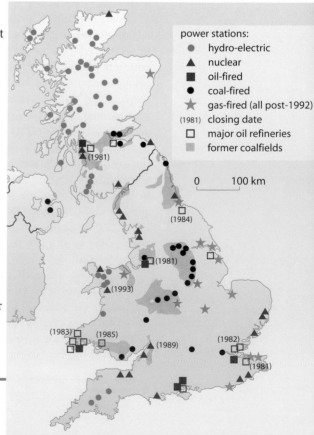

Energy resources

Figure 18.13

Collecting fuelwood

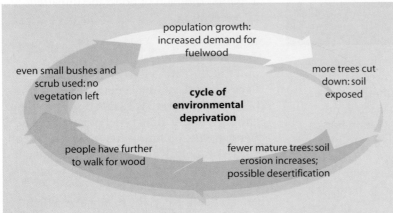

population growth: increased demand for fuelwood

more trees cut down: soil exposed

cycle of environmental deprivation

fewer mature trees: soil erosion increases; possible desertification

people have further to walk for wood

even small bushes and scrub used: no vegetation left

Figure 18.14

The cycle of environmental deprivation

Fuelwood

Trees are a sustainable resource provided that those cut down are replaced, or that sufficient time is allowed for natural regeneration. The major source of energy in many of the least developed countries, especially those in Africa, is fuelwood. In these parts of the world, fuelwood — a fossil fuel — is needed daily for cooking and heating. As nearby supplies are used up, collecting fuelwood becomes an increasingly time-consuming task; in extreme cases, it may take all day (Figure 18.13 and page 500). Many of these countries have a rapid population growth, which adds greater pressure to their often meagre resources, and lack the capital and technology to develop or buy alternative resources. In places where the demand for fuelwood outstrips the supply, and where there is neither the money to replant nor the time for regeneration, the risk of desertification and irreversible damage to the environment increases (Figure 18.14 and Case Study 7, page 175).

Renewable energy

With the depletion of oil and gas reserves by the second quarter of next century, renewable energy resources are likely to become more attractive. They will become more cost-competitive, offer greater energy diversity, and therefore security, and promise a cleaner environment.

Hydro-electricity

Hydro-electricity is the most widely used renewable energy resource. Its availability depends upon an assured supply of fast-flowing water which may be obtained from rainfall spread evenly throughout the year or by building dams and storing water in large reservoirs. The initial investment costs and levels of technology needed to build new dams and power stations, to install turbines and to erect pylons and cables for the transport of the electricity to often-distant markets, are high. However, once a scheme is operative, the 'natural, continual, renewable' flow of water makes its electricity cheaper than that produced by fossil fuels.

Although the production of hydro-electricity is perceived as 'clean', it can still have very damaging effects upon the environment. The creation of reservoirs can mean large areas of vegetation being cleared (Tucurui in Amazonia), wildlife habitats (Kariba in Zimbabwe) and agricultural land (Volta in Ghana) being lost, and people being forced to move home (Aswan in Egypt). Where new reservoirs drown vegetation, the resultant lake is likely to become acidic and anaerobic. Dams can be a flood risk if they collapse or overflow (Places 6, page 43) and have been linked to increased earthquake activity (Colorado basin, Case Study 3b, page 85; and India). Despite these negative aspects, many countries rely upon large, sometimes prestigious, schemes or, increasingly in less developed countries, on smaller projects using more appropriate levels of technology (Case Study 18).

Wind

Wind is an underused energy resource, especially in places where winds are strong, steady and reliable, and where the landscape is either high or, as on coasts facing prevailing winds, exposed. In 1994, California produced 80 per cent of the world's wind-generated power (Figure 18.15), although this was only 1 per cent of the USA's annual energy need. At that same time, Britain had

Figure 18.15

Wind farms at Altamont Pass, California

in operation 21 wind farms, the collective name for groups of wind-powered turbines, although plans had been submitted for a further 1200 (Figure 18.16). Each wind-driven turbine stands over 30 m in height (soon to be to over 50 m) and, with blades exceeding 35 m in diameter, is erected up to 200 m from adjacent turbines. Wind farms are capital-intensive but cheap to run. It is

Figure 18.16

Windpower in the UK, 1994

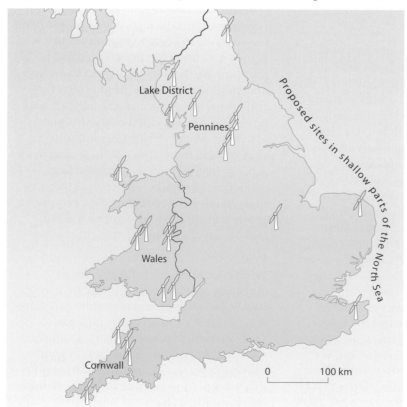

estimated that the relative cost of producing electricity from gas is 3p a kilowatt hour, from wind and coal 4p, and from nuclear power 8p. However, it will take 7000 wind turbines to produce the same amount as one nuclear power station, and up to 100 000 wind farms to produce the same amount of energy as at present obtained from all Britain's nuclear power stations.

Wind power is pollution-free and the land between the turbines can still be farmed. There is no waste and it does not contribute to acid rain or to global warming. However, environmentalists are now less united in their enthusiasm and support for wind power as many actual and proposed wind farms are in areas of scenic attraction which often also contain important wildlife habitats. Local residents complain of noise and impaired radio reception.

Solar energy

The sun, as stated earlier, is the primary source of the earth's energy. Estimates suggest that the annual energy received from the sun (insolation), is 15 000 times greater than the current global energy supply. Solar energy is safe, pollution-free, efficient and of limitless supply. Unfortunately, it is expensive to construct solar 'stations' although many individual homes have had solar panels added, especially in climates which are warmer and sunnier than in Britain. It is hoped, globally, that future improvements in technology will result in reduced production costs. This would enable many developing countries, especially those lying within the tropics, to rely increasingly upon solar energy. In Britain, the solar energy option is less favourable partly due to the greater amount of cloud cover and partly to the long hours of darkness in winter when demand for energy is at its highest.

Tidal energy

Tidal energy, of all the renewable resources, is the most reliable and predictable but, to date, only places with the maximum tidal range offer potential sites. Incoming tides in two adopted schemes, the Rance estuary in north-west France and the Bay of Fundy in Canada, turn turbines, the blades of which are reversed to harness the receding tide. Although the technology is now proven, economic and environmental costs are high. In Britain, the Severn barrage has been discussed for several decades. If built, it would

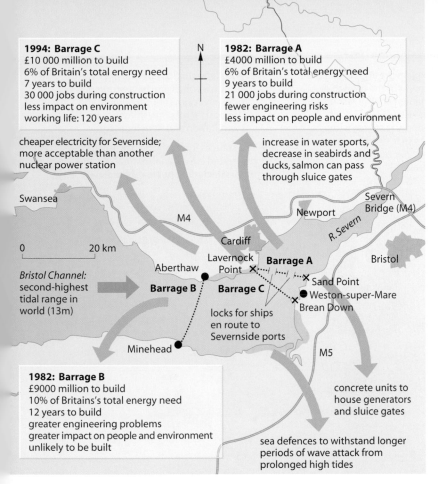

1994: Barrage C
£10 000 million to build
6% of Britain's total energy need
7 years to build
30 000 jobs during construction
less impact on environment
working life: 120 years

cheaper electricity for Severnside; more acceptable than another nuclear power station

Swansea

N

1982: Barrage A
£4000 million to build
6% of Britain's total energy need
9 years to build
21 000 jobs during construction
fewer engineering risks
less impact on people and environment

increase in water sports, decrease in seabirds and ducks, salmon can pass through sluice gates

Severn Bridge (M4)

M4   Newport

0   20 km

Cardiff

Bristol Channel:
second-highest tidal range in world (13m)

Aberthaw

Lavernock Point ✕

Barrage A

Bristol

Barrage B   **Barrage C**

✕ Sand Point
● Weston-super-Mare
✕ Brean Down

locks for ships en route to Severnside ports

Minehead

M5

1982: Barrage B
£9000 million to build
10% of Britains's total energy need
12 years to build
greater engineering problems
greater impact on people and environment
unlikely to be built

concrete units to house generators and sluice gates

sea defences to withstand longer periods of wave attack from prolonged high tides

Figure 18.17

Proposals for tidal barrages across the Severn estuary

provide the energy of five nuclear power stations but would cost the equivalent of six to construct (Figure 18.17). Other estuaries with large tidal ranges include the Mersey, Morecambe Bay and the Solway Firth. Tidal barrages do not contribute to global warming or to acid rain, but are not favoured by environmentalists because they result in the permanent flooding of marshland and wetland habitats, favoured as feeding grounds by migratory birds, and have an adverse affect on spawning fish.

Figure 18.18

A geothermal electricity generating station

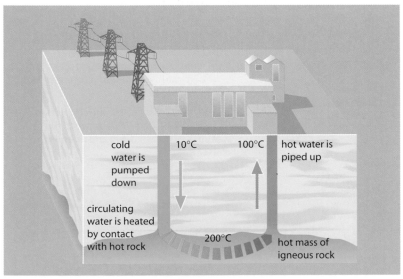

cold water is pumped down

10°C   100°C   hot water is piped up

circulating water is heated by contact with hot rock

200°C   hot mass of igneous rock

Wave power

Waves are created by the transfer of energy from winds which blow over them (page 124). In western Europe, winter storm waves from the Atlantic Ocean transfer large amounts of energy towards the coast where it may, in the future, be used to generate electricity. Although, at present, two small 'shore-line' commercial wave power stations are in operation (in Norway and on the Scottish island of Islay), 'offshore' devices have potential energy levels several times greater — if they can be designed to withstand the power of the waves.

Geothermal energy

Several countries, especially those located in active volcanic areas, obtain energy from heated rocks and molten magma at depth under the earth's surface — for example, Iceland, New Zealand, Kenya and several countries in central America. Cold water (Figure 18.18) is pumped downwards, is heated naturally and is then returned to the surface as steam which can generate electricity. Geothermal energy does pose environmental problems as carbon dioxide and hydrogen sulphide emissions may be high, the water supply can become saline, and earth movements can damage power stations.

Biomass

Biomass refers to energy obtained from organic matter — i.e. crops, plants and animal waste. Brazil uses ethanol, a type of alcohol obtained from sugar cane, instead of petrol as a cheaper car fuel. The EU is at present conducting similar experiments using oil-seed rape. In poorer parts of Africa, animal dung is allowed to ferment to produce methane gas. Although the methane provides a vital domestic fuel, it is a greenhouse gas and the dung can no longer be used as a fertiliser. In more developed countries, it is now possible to convert the methane gas released at landfill 'waste' sites into energy. The present production of biomass energy in the USA is equivalent to 10 nuclear power stations.

Ocean thermal energy conversion

In tropical seas, there is a considerable temperature difference between surface water and that found at depth. The American government is funding an experimental and highly ambitious scheme off the coast of India aimed at converting this heat difference into energy.

Q

1 a Describe and explain the impacts on the natural environment of the growth or decline of one named energy resource in a country you have studied.

b For any single energy resource with which you are familiar, examine the problems of its development and use in a variety of countries.

c With reference to any one energy resource which you have studied, show how environmental, economic and political factors may influence its exploitation and development.

2 The sources from which energy can be generated are oil, coal, natural gas and others (including nuclear, hydro-electricity, solar and wind). An index of diversification can be calculated to measure the degree to which a country depends on one or more energy source. The index varies between 0% for no diversification (only one fuel is used) and 100% for total diversification (when the four sources are used in equal amounts). With reference to Figure 18.19, describe the world pattern of diversification in 1985.

3 Electricity may be generated in different ways and by using a variety of fuels.

a By reference to specific countries, suggest reasons for the relative importance of fossil fuels, hydro-electric power and nuclear power in the generation of electricity within individual countries.

b Describe in detail other schemes which are being developed to generate electricity, and comment on their feasibility.

4 Figure 18.20 shows a possible correlation (Framework 14, page 572) between the wealth of a country (GNP) and its energy consumption.

a What is the GNP of the UK, Brazil and India?

b What is the energy consumption per person in the USA, Brazil and Tanzania?

c What does the graph show of the relationship between energy consumption per person and wealth?

d Why is Saudi Arabia an anomaly?

Figure 18.19

Indices of diversification, 1985

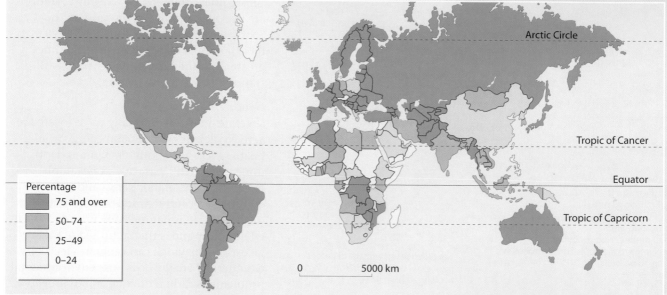

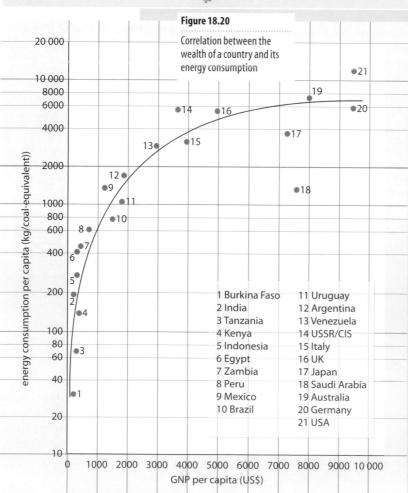

Figure 18.20

Correlation between the wealth of a country and its energy consumption

1 Burkina Faso
2 India
3 Tanzania
4 Kenya
5 Indonesia
6 Egypt
7 Zambia
8 Peru
9 Mexico
10 Brazil
11 Uruguay
12 Argentina
13 Venezuela
14 USSR/CIS
15 Italy
16 UK
17 Japan
18 Saudi Arabia
19 Australia
20 Germany
21 USA

Case Study 18

Appropriate technology: micro-hydro in Nepal

"An Appropriate Technology is exactly what it says — a technology appropriate or suitable to the situation in which it is used (page 526). If that situation is a highly industrialised urban centre the appropriate technology may well be 'high tech'. If however the situation is a remote Nepalese village 'appropriateness' will be measured in the following terms:

- Is it culturally acceptable?
- Is it what people really want?
- Is it affordable?
- Is it cheaper or better than alternatives?
- Can it be made and repaired with local material, by local people?
- Does it create new jobs or protect existing ones?
- Is it environmentally sound?

For many decades 'Aid' meant sending out the same large-scale, expensive, labour-saving technologies that we use: huge hydro-electric schemes, coal-fired power stations, diesel-powered generators. In some cases, for example towns and industrial areas, these have been appropriate. But such schemes do not reach the poorer communities in the rural areas. What was needed was some way of using local resources appropriately, and best of all some way of using renewable resources to decrease the need for reliance on outside help. Wind, solar and biogas energy are possibilities, but another resource widely available and already in use for thousands of years is water. Water is attracting much attention in the search for renewable sources of energy. However, despite continuing public outrage at the devastating impact of large hydro-electric schemes on people's livelihoods and the environment (page 495), vast sums of money continue to be pumped into big dams and other inappropriate power generation plans. On the other hand, the intermediate approach, through small-scale hydro, has no negative impact on the environment, offers positive benefits to the local community, and uses local resources and skills."

(Intermediate Technology, 1990)

Intermediate Technology and micro-hydro in Nepal

"The small Himalayan kingdom of Nepal ranks as one of the ten poorest countries in the world. Around 90 per cent of its 19 million people earn their living from farming, often at a subsistence level. The Himalayas offer Nepal one vast resource — the thousands of streams which pour down from the mountains all year round. Nepali people have harnessed the power in these rivers for centuries, albeit on a small scale (Figure 18.21).

About, 20 years ago, two local engineering workshops began to build small, steel, hydro power schemes for remote villages. These turbines have the advantage of producing more power than the traditional mills, as well as being able to run a range of agricultural processing machines (Figure 18.22). IT first became involved in Nepal's micro-hydro sector in the late 1970s when the local manufacturers asked for help in using their micro-hydro schemes to generate electricity.

In the mid 1980s, IT ran two training courses on micro-hydro power aimed at improving the technical ability of the nine new water turbine manufacturers that had been established in Nepal. These courses were very successful and prompted an agreement between IT and the Agricultural Development Bank (the agency which funds micro-hydro power in Nepal) to collaborate on the development of small water turbines for rural areas. This work not only improved and extended the range and number of micro-hydro schemes in Nepal, but also established IT as a leader in the field. In 1990 IT was included in a government task force investigating the whole area of rural electrification; and in 1992 IT was asked by the government to help establish an independent agency to promote all types of appropriate energy in rural areas of the country."

(Intermediate Technology, 1990)

Figure 18.21
Cross-section of a traditional Nepali water mill

1 chute delivering the water to the paddles of the wheel
2 grain hopper (basket)
3 device to keep the grain moving
4 metal piece to lock top of shaft in upper millstone
5 grinding stones
6 metal shaft
7 thick wooden hub
8 wooden horizontal wheel, with obliquely set paddles attached to hub
9 metal pin and bottom piece
10 lifting device to adjust gap between millstones

Figure 18.22
Cross-section of a modern Nepali water turbine

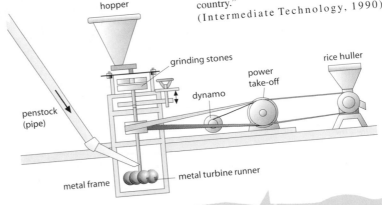

"Water power, harnessed using water wheels or *ghattas*, has been used for centuries for grinding corn. The micro-hydro system the village now has improved the efficiency of milling, so that what would take a woman four hours to grind by *ghatta* can be done in fifteen minutes. The power can also be used for dehusking rice and extracting oil from sunflower seeds (Figure 18.23).

The mechanical power produced by Ghandruk's micro-hydro system is also converted to electric power, which is distributed to every house in the village. Apart from the obvious benefit of lighting, many households are starting to use electric cookers or *bijuli dekchis*, which work like slow cookers (Figure 18.23).

Women are turning to *bijuli dekchis* because they reduce smoke levels in the kitchen, they save time by reducing the amount of firewood the family needs to collect, and they are more convenient and cook faster than traditional stoves. In a country ravaged by deforestation — villagers spend up to 12 hours on a round trip to collect wood — fuel saving is becoming more and more important.

Micro-hydro schemes like the one in Ghandruk work because the community has 'ownership' of the scheme by participating in its planning, installation and management; because the machinery needed can be made and maintained by local manufacturers using local materials available in the country; and because production and consumption are linked within a community.

The lives of villagers all over Nepal are literally being lit up by micro-hydro schemes, and the country could serve as a model for decentralised, sustainable energy production. Already, 700 mechanical and 100 electrical schemes have been installed.

Much of the impetus for the development of hydro in Nepal initially stemmed from the absence of fossil fuel reserves to exploit. However, if the Government can resist the temptations of big dam schemes and the dollars being thrown at them by the big, international donor agencies, it could have the last laugh watching the rest of the world scrabble for the last of fossil fuel reserves."

(Intermediate Technology, 1992)

References

Department of the Environment (1992) *The UK Environment*. HMSO.

Intermediate Technology (various dates) *Micro-hydro in Nepal*. IT.

Philip's Geographical Digest, 1992–93 (1993) Heinemann/Philip.

Pickering, K. and Owen, L. (1994) *Global Environmental Issues*. London: Routledge.

Tully, M. (1991) *No Full Stops in India*. Penguin Books.

Crop processing

A woman grinds corn for the family meal

An elderly Nepalese woman and young girl hull rice with a traditional foot-powered *dhiki*.

Cooking

Cooking on an open fire.

Living

Light comes from a single kerosene lamp.

Power for Life

Figure 18.23

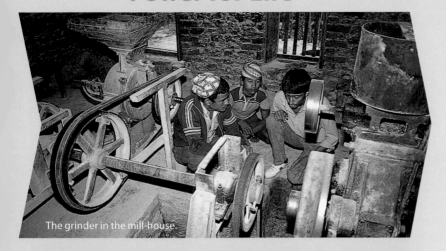

The grinder in the mill-house.

Grinding enough corn to feed a family for just three days takes fifteen hours when it is done by hand.

By taking corn to the grinder in the mill-house — usually a popular meeting place for villagers — three days' worth of corn can be ground in just 15 minutes.

Rice being hulled mechanically in a 3kW mill.

For thousands of women, the supply of power releases them from the many labour-intensive and time-consuming tasks they previously had to carry out by hand.

Villagers can now hull their rice mechanically with this 3-kw mill, driven by a micro-hydro turbine. Time is saved and quality and productivity increased.

Low-wattage electric cookers.

Cooking on an open fire burns up a great deal of wood (which is becoming increasingly scarce) and gives off a lot of thick smoke. As a result, the villagers not only have to walk long distances to collect their fuel, but many women and children suffer from serious lung disorders.

IT is helping to develop two low-wattage electric cookers which have been specifically designed to make use of 'off-peak' electricity. The *bijuli dekchi* heats water during off-peak times for use in cooking later on, while the heat storage cooker stores the energy available during off-peak periods and releases it at mealtimes for cooking. Both save fuelwood and help to reduce deforestation.

A Nepalese family enjoying the benefits of electric light.

Kerosene lamps are costly to run, and those who can afford them have to collect fuel in cans from towns which are usually several days' walk away.

With electric light, children and adults can improve their education by learning to read and write in the evenings. Electric light is also cheaper, cleaner and brighter than kerosene.

Chapter 19

Manufacturing industries

"We need methods and equipment which are cheap enough so that they are accessible to virtually everyone; suitable for small scale production; and compatible with man's need for creativity. Out of these three characteristics is born non-violence and a relationship of man to nature which guarantees permanence. If one of these three is neglected, things are bound to go wrong."

E. F. Schumacher, *Small is Beautiful,* **1974**

What is meant by industry? In its widest and more traditional sense, the word industry is used to cover all forms of economic activity: **primary** (farming, fishing, mining and forestry); **secondary** (manufacturing and construction); **tertiary** (back-up services such as administration, retailing and transport); and **quaternary** (high technology and information services). In this chapter, the use of the term 'industry' has been confined to its narrowest definition — i.e. manufacturing. Manufacturing industry includes the processing of raw materials (iron ore, timber) and of semi-processed materials (steel, pulp).

It needs be pointed out, however, that while this definition may be convenient, it does create several major problems. At present, only some 21 per cent of the UK's working population are employed in manufacturing, a trend which is repeated across most of the developed market economies. This shift from an industrial to a post-industrial society is shown in Figure 19.1. In reality, it is also unrealistic to draw boundaries between 'manufacturing' and 'services' (Chapter 20). Not only are the two integrated in reality through linkages (page 518 and Figure 19.2), buyer–supplier relations, etc., but many people who are officially classified as working in the manufacturing sector have occupations which are service based (salespeople, administrators) within 'manufacturing' sector firms. It can be argued, with much justification, that it is conceptually (and empirically) unrealistic to sever manufacturing from services. This distinction becomes particularly problematic when discussing, for example, high-tech developments along the M4 (Places 66, page 516) as, by their nature, many firms are 'information-intensive' and knowledge based rather than production or materials based; or when describing the differences between the 'formal' and 'informal' sectors in less economically developed, less industrialised countries (page 523).

Figure 19.1

Towards a post-industrial economy: employment structure in the UK, 1841-1991

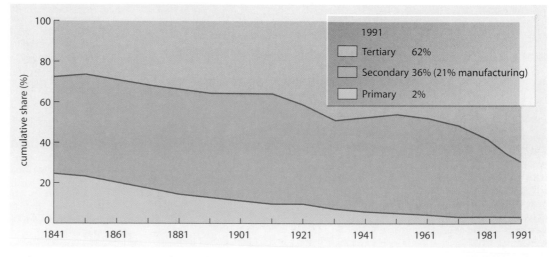

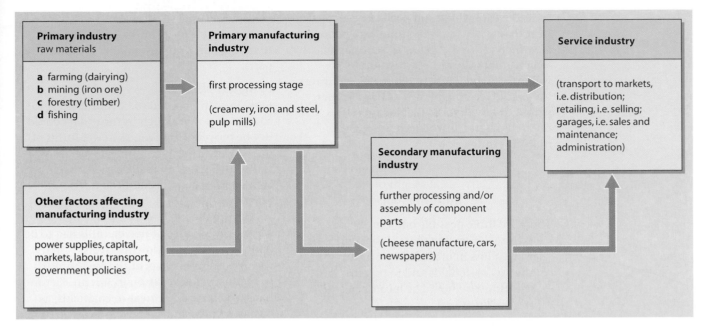

| Primary industry
raw materials

a farming (dairying)
b mining (iron ore)
c forestry (timber)
d fishing | Primary manufacturing industry

first processing stage

(creamery, iron and steel, pulp mills) | | Service industry

(transport to markets, i.e. distribution; retailing, i.e. selling; garages, i.e. sales and maintenance; administration) |

Other factors affecting manufacturing industry

power supplies, capital, markets, labour, transport, government policies

Secondary manufacturing industry

further processing and/or assembly of component parts

(cheese manufacture, cars, newspapers)

Figure 19.2

Linkages between various sections and types of industry

The location of industry

The processes which contribute to determine the location and distribution of industry are more complex and dynamic than those affecting agriculture. This means that the making of generalisations becomes less easy and the dangers of stereotyping increase. Some reasons for this complexity include:

- Some locations were chosen before the Industrial Revolution and many more during it. Initial factors favouring a location may no longer apply today. For example, the original raw materials may now be exhausted (iron ore and coal in west Cumbria), or replaced by new innovations (cotton replaced by synthetic fibres) and sources of energy (water power replaced by electricity).
- New locational factors which were not applicable last century include cheaper and more efficient transport systems, the movement of energy in the form of electricity, new techniques and automation.
- Some industries have developed from older industries and are linked to these former patterns of production even when the modern product is different (the Mazda Car Corporation began as a cork-making and then a machine-tools firm).
- Before the 20th century, industry was usually financed and organised by individual **entrepreneurs** but decisions concerning modern industry are often made by the state or **multinational companies** (MNCS or **transnationals**, page 523). Increasingly such decisions are being

made far away from the location of a particular factory.

- Many factories now produce a single component and therefore are a part of a much larger organisation which they supply.
- The sites of many early factories were chosen by individual preference or by chance — i.e. the founder of a firm just happened to live at, or to like, a particular location (Unilever at Port Sunlight and Rowntree at York). Other industries have found their present locations as a result of 'trial and error'.

Factors affecting the location of manufacturing industry

Raw materials

Industry in 19th-century Britain was often located close to raw materials (ironworks near iron ore), sources of power (coalfields) or ports (to process imports), mainly due to the immobility of the raw materials which were heavy and costly to move when transport was expensive and inefficient. In contrast, today's industries are rarely tied to the location of raw materials and so are described as **footloose**. There is now a greater efficiency in the use of raw materials; power is more mobile; transport of raw materials, finished products and the workforce is more efficient and relatively cheaper; components for many modern, and especially hi-tech, industries are relatively small in size and light in weight; and some firms may simply rely on assembling component parts made elsewhere.

An entrepreneur initiates and organises, usually for profit, an enterprise or business; this includes risk-taking, deciding what goods will be produced or services provided, the scale of production, and marketing.

Industries which still need to be located near to raw materials are those using materials which are heavy, bulky or perishable; which are low in value in relation to their weight; or which lose weight or bulk during the manufacturing process. **Alfred Weber**, whose theory of industrial location is referred to later, introduced the term **material index** or **MI**.

$$MI = \frac{\text{total weight of raw materials}}{\text{total weight of finished product}}$$

There are three possible outcomes.

1 If the MI is greater than 1, there must be a weight loss in manufacture. In this case, the raw material is said to be **gross** and the industry should be located near to that raw material, e.g. iron and steel:

$$MI = \frac{6 \text{ tonnes raw material}}{1 \text{ tonne finished steel}} = 6.0$$

2 If the MI is less than 1, there must be a gain in weight during manufacture. This time the industry should be located near to the market, e.g. brewing:

$$M1 = \frac{1 \text{ tonne raw material}}{5 \text{ tonnes beer}} = 0.2$$

3 Where the MI is exactly 1, the raw material must be **pure** as it does not lose or gain weight during manufacture. This type of industry could therefore be located at the raw material, the market or any intermediate point.

Industries which lose weight during manufacture include food processing (butter has only one-fifth the weight of milk, refined sugar is only one-eighth the weight of the cane), smelting of ores (copper ore is less than 1 per cent pure copper, iron ore has a 30–60 per cent iron content; Places 64, page 513) and forestry (paper has much less mass than trees; Places 63, page 513). Industries which gain weight in manufacture include those adding water (brewing and cement), and those assembling component parts (cars; Places 65, page 515; and electrical goods;

Places 66, page 516). In these cases, the end product is more bulky and expensive to move than its many smaller constituent parts.

Power supplies

Early industry tended to be located near to sources of power which in those days could not be moved. However, as newer forms of power were introduced and the means of transporting it were made easier and cheaper, this locational factor has become less important (Figure 19.3).

During the medieval period, when water was a prime source of power, mills had to be built alongside fast-flowing rivers. When steam power took over at the beginning of the Industrial Revolution in Britain, factories had to be built on or near to coalfields, as coal was bulky and expensive to move. When canals and railways were constructed to move coal, new industries were located along transport routes. By the mid 20th century, oil (relatively cheap before the 1973 Middle East War) was being increasingly used as it could be transported easily by tanker or pipeline. This began to free industry from the coalfields and to offer it a wider choice of location (except for such oil-based industries as petrochemicals). Today, oil, coal, natural gas, nuclear and hydro-electric power can all be used to produce electricity to feed the National Grid. Electricity, in addition to its cleanliness and flexibility has the advantage that it can be transferred economically over considerable distances either to the long-established industrial areas, where activity is maintained by **geographical inertia**, or to new areas of growth. (In 1900, electricity could be transmitted economically only 59 km; today, the distance is over 1500 km.)

Transport

Transport costs were once a major consideration when locating an industry. Weber based his industrial location theory on the premise that transport costs were directly related to distance (compare von Thünen's assumptions, page 430). Since then, new forms of transport have been introduced, including lorries (for door-to-door delivery), railways (preferable for bulky goods) and air (where speed is essential). Meanwhile, transport networks have improved, with the building of motorways, and methods of handling goods have become more efficient through containerisation. For the average British

Figure 19.3

Power supply and the location of iron and steel works

Period	Source of power	Examples of location
early iron industry	charcoal	wooded areas (the Weald, the Forest of Dean)
later iron industry	waterwheels	fast-flowing rivers (R. Don, Sheffield)
early steel industry	coal	coalfields (South Wales, north-east England)
present-day steel industry	electricity	coastal (Port Talbot, Newport)

Manufacturing industries

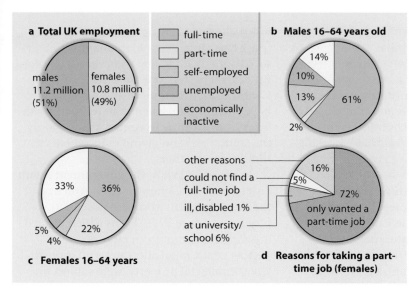

Figure 19.4

UK employment data, 1993

firm, transport costs are now only 2–3 per cent of their total expenditure. Consequently, raw materials can be transported further and finished goods sold in more distant markets without any considerable increase in costs. Firms previously tied to coastal sites relying on overseas materials or markets, or earlier nodal points, now have a freer choice of location — although many still favour growth-point axes such as the M4 corridor (Places 66, page 516).

There are two types of transport costs from an economic point of view. Terminal costs refer to the time and equipment needed to handle and store goods (cranes and warehouses) and the costs of providing the transport system (building railway tracks, docks and motorways). Long-haul costs refer to the cost of actually moving the goods (lorries, petrol and salaries). These costs and the advantages and disadvantages of different forms of transport are summarised in Figure 21.1.

Markets

Today, the pull of a large market is more important than the location of raw materials and power supplies; indeed, it has been suggested that flexibility and rapid response to changing market signals are perhaps the most important determinants of location.

Industries will locate near to markets if:

1 The product becomes more bulky with manufacture or there are many linkage industries involved (the assembling of motor vehicles).

2 The product becomes more perishable after processing (bread is more perishable than flour); if it is sensitive to changing fashion (clothes); or if it has a short life-span (daily newspapers).

3 The market is very large (north-eastern states of the USA, or south-east England).

4 The market is wealthy.

5 Prestige is important (publishing).

Labour supply

Labour varies spatially in its cost, availability and quality.

In the 19th century, a huge force of semi-skilled, mainly male, workers operated in large-scale 'heavy' industries doing manual jobs in steelworks, shipyards and textile mills. Today, there are fewer semi-skilled and more highly skilled workers operating in small-scale 'light' industries which increasingly rely upon machines, computers and robots. The cost of labour, especially in EU countries can be high — accounting for 10–40 per cent of total production costs. Three consequences of this have been the introduction of mechanisation to reduce human inputs, the exploitation of female labour and the use of 'cheap' labour in developing countries.

Traditionally, labour has been relatively immobile. Although there was a drift to the towns during the Industrial Revolution, since the First World War British people have expected respective governments to bring jobs to them rather than they themselves having to move for jobs. Certain industries need specialist training (cutlery, furniture-making and electronics) and so are difficult to develop in a new area with an untrained labourforce. If you wished, for example, to open a cutlery business you would probably choose Sheffield in preference to Exeter or Aberdeen. However, as industries become more mechanised and transport improvements allow greater mobility, firms can locate, and their workforce can travel, more freely. A problem in the mid 1990s in Britain is that, despite a high level of unemployment, there is a shortage of highly skilled people, partly due to a lack of training.

Similarly, the roles of women and trade unions have both changed. An increasing number of women are now either seeking career jobs or are prepared to work part-time (Figure 19.4) or flexi-time, even if they often have to accept lower salaries than males. The role of trade unions has declined significantly as their membership numbers have fallen with the decline of the large 'heavy' industries.

Capital

Capital may be in three forms.

1 **Working capital** (money) which is acquired from a firm's profits, shareholders or financial institutions such as banks. Money is mobile and can be used within and exchanged between countries. Location is rarely constrained by working capital unless money is to be borrowed from the government who might direct industry to certain areas (see below). In Britain, capital is more readily available in the City of London where most of the financial institutions are based.

2 **Physical** or **fixed capital** refers to buildings and equipment. This form of capital is not mobile — i.e. it was invested for a specific use.

3 **Social capital** is linked to the workforce's out-of-work needs rather than to the factory or office itself. Houses, hospitals, schools, shops and recreational amenities are social capital which may attract a firm, particularly its management, to an area.

Government policies

These should aim to even out differences in employment, income levels and investment within a country. In 1934, regional assistance was granted in Britain to Clydeside, west Cumberland, North-East England and South Wales (compare these areas with the 1993 Assisted Areas in Figure 19.5). This was followed by the setting up of trading/industrial estates in areas of high unemployment such as the Team Valley (Gateshead), Trafford Park (Manchester) and Treforest (north of Cardiff). Since then successive governments have tried various schemes to try to dissuade firms from locating in certain areas (mainly the South East) and giving inducements if they locate in others.

In the early 1980s, the government set up Enterprise Zones where unemployment and the state of the environment posed serious problems (Figure 19.5). Firms which located in EZs were exempt from rates, received 100 per cent capital allowances on industrial and commercial property, were subject to simplified planning procedures (so long as health, safety and pollution control standards were guaranteed), and were exempt from industrial training levies. Further policies were introduced later in the 1980s to try to help badly hit docklands (London and Liverpool) and other inner-city areas (page 401; and Places 41, page 402). Governments have also made direct attempts to attract foreign firms (Nissan and Toyota) and to encourage private–public partnership.

Land

In the 19th century, extensive areas of flat land were needed for the large factory units. Today, although modern industry is usually smaller in terms of land area occupied, it prefers cheaper land, less congested and cramped sites and improved accessibility as found on green-field sites on the edges of cities and in smaller towns. Recent British government policy has aimed to attract industry to derelict and underused sites, especially in the inner cities (page 401) in order to utilise the existing infrastructure and to reduce pressure on green-field sites.

Environment

The 1980s saw an increasing demand by both managers and workforce to live and work in a more pleasant environment. This has led to firms seeking locations in smaller towns within easy reach of open countryside and away from the commuting problems and high land values of south-east England. There is still, however, a common perception (stereotype) that 'the north' is all Coronation Streets and 'the south' all quiet rural villages. Consequently, potential employers have to work hard to 'sell' any social or environmental advantages which their area possesses.

Figure 19.5

Assisted areas (Latest Designation, 1993) and Enterprise Zones (1995)

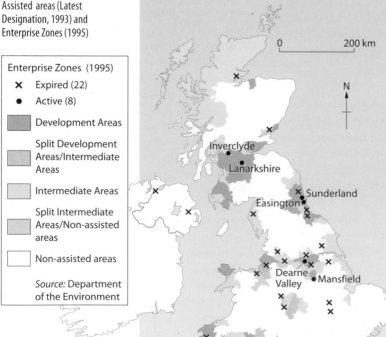

Enterprise Zones (1995)
- ✗ Expired (22)
- ● Active (8)
- Development Areas
- Split Development Areas/Intermediate Areas
- Intermediate Areas
- Split Intermediate Areas/Non-assisted areas
- Non-assisted areas

Source: Department of the Environment

0   200 km

N

Inverclyde
Lanarkshire
Sunderland
Easington
Dearne Valley
Mansfield
Swansea
North-west Kent

Theories of industrial location

We have already seen that "models form an integral and accepted part of present-day geographical thinking" (Framework 9, page 328). Models, as in many branches of geography, have been formulated in an attempt to try to explain, in a generalised, simplified way, some of the complexities affecting industrial location. Two of the more commonly quoted models are those based upon:

1 the industrialist who seeks the lowest-cost location (LCL) — (Weber, 1909); and
2 the industrialist who seeks the area which will give the highest profit — (Smith, 1971).

Before looking more closely at these two models, it must be remembered that, as models, they form theoretical frameworks which may be difficult to observe in the real world, but against which reality can be tested. Mainly due to constant changes in the world economy, the validity of both 'least cost' and 'profit maximisation' have come under question. What is important, however, is that you should

a recognise that there is no single body of theory, i.e. model, to explain the real world but, rather, there exists a variety of approaches; and
b begin to appreciate the complexity of industrial organisation in the real world and not to assume that conceptual models actually work.

Weber's model of industrial location

Alfred Weber was a German 'spatial economist' who, in 1909, devised a model to try to explain and predict the location of industry. Like von Thünen before him and Christaller later, Weber tried to find a sense of order from apparent chaos and made assumptions to simplify the real world in order to produce his model.

These assumptions were that:
■ There was an isolated state with flat relief, a uniform transport system in all directions, a uniform climate, and a uniform cultural, political and economic system.
■ Most of the raw materials were not evenly distributed across the plain (this differs from von Thünen). Those which were evenly distributed (water, clay) he called **ubiquitous materials**. As these did not have to be transported, firms using them could locate as near to the market as was possible. Those raw materials which were not evenly distributed he called **localised materials**. He divided these into two types; gross and pure (page 504).
■ The size and location of markets were fixed.
■ Transport costs were a function of the mass (weight) of the raw material and the distance it had to be moved. This was expressed in tonnes per kilometre (t/km).
■ Labour was found in several fixed locations on the plain. At each point it was paid the same rates, had equivalent skills, was immobile and in large supply. Similarly, entrepreneurs had equal knowledge, related to their industry, and motivation.
■ Perfect competition existed over the plain (i.e. markets and raw materials were unlimited) which meant that no single manufacturer could influence prices (i.e. there was no monopoly). As revenue would therefore be similar across the plain, the best site would be the one with the minimal production costs (i.e. the least-cost location or LCL).

Possible least-cost locations

Weber produced two types of locational diagram. A straight line was sufficient to show examples where only one of the raw materials was localised (it could be pure or gross). However, when two localised raw materials were involved, he introduced the idea of the **locational triangle**. Figure 19.6 summarises the nine possible variations based on the type of raw material involved.

1 One gross localised raw material. As there is weight loss during manufacture (the material index for a gross raw material is more than 1) then it is cheaper to locate the factory at the source of the raw material – there is no point in paying transport costs if some of the material will be left as waste after production (Figure 19.7a).
2 a One ubiquitous raw material or b one pure localised raw material gaining weight on manufacture (MI less than 1). If the raw material is found all over the plain (ubiquitous) then transport is unnecessary as it is already found at the market. If a pure material gains mass on manufacture then it is cheaper to move it rather than the finished product and so again the LCL will be at the market (Figure 19.7b).

3 One pure localised raw material. If this neither gains nor loses weight during manufacture (MI = 1), the LCL can be either at the market, at the location of the raw material, or at any intermediate point (Figure 19.7c).

4 Two ubiquitous (gross or pure) raw materials. As these are found everywhere, they do not have to be transported and so the LCL is at the market.

5 Two raw materials: one ubiquitous and one pure and localised. The LCL is at the market because the ubiquitous material is already there and so only the pure localised material has to be transported (Figure 19.8a). It will be cheaper to move one raw material than the more cumbersome final product.

6 Two raw materials: one ubiquitous and the other gross and localised. The ubiquitous material is available at every location. As the gross material loses weight, the LCL could theoretically be at any intermediate point between its source and the market. However, if the mass of the product is greater than that of the localised raw material, the LCL is at the market; if it is less, the LCL is at location of the raw material; and if it is the same,

the LCL is at the mid-point (Figure 19.8b).

7 Two raw materials: both localised and pure. In the unlikely event of the two raw materials lying to the same side of and in line with the market, the LCL will be at the market. If the materials do not conform with this arrangement but form a triangle with the market (Figure 19.9), the LCL is at an intermediate point near to the market. This is because the weight and therefore the transport costs of the raw material are the same as, or less than, those of the product.

8 Two localised raw materials: one pure and one gross. In this case, the industry will locate at an intermediate point (Figure 19.10a). The greater the loss of weight during production, the nearer the LCL will be to the source of the gross material.

9 Two raw materials: both localised and gross. If both raw materials have an equal loss of weight, the LCL will be equidistant between these two sources but closer to them than to the market (Figure 19.10b1). However, if one raw material loses more mass than the other, the industry is more likely to be located closer to it (Figure 19.10b2).

Figure 19.6

Least-cost locations dependent upon types of raw material

Type(s) of raw material (RM) MI = material index	LCL at raw material	LCL at any intermediate point	LCL at market
1 one gross localised RM MI >1	■		
2 one RM gaining weight or one ubiquitous RM MI <1			■
3 one pure localised RM MI = 1		■	
4 two ubiquitous RMs (pure or gross)			■
5 two RMs (one ubiquitous and one pure)			■
6 two RMs (one ubiquitous, one gross)		(could be any site, according to amount of weight loss)	
7 two RMs (both pure)			■
8 two RMs (one pure, one gross)	■ (if big weight loss)	■ (if a small weight loss)	
9 two RMs (both gross)	■ (at RM with greatest weight loss)	■ (equal weight loss)	

Figure 19.7

Least-cost locations with one raw material

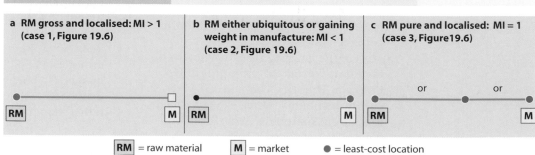

a RM gross and localised: MI > 1 (case 1, Figure 19.6)

b RM either ubiquitous or gaining weight in manufacture: MI < 1 (case 2, Figure 19.6)

c RM pure and localised: MI = 1 (case 3, Figure19.6)

RM = raw material   M = market   ● = least-cost location

Figure 19.8

Least-cost locations with two raw materials one of which is ubiquitous

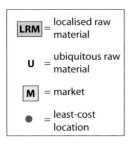

LRM = localised raw material

U = ubiquitous raw material

M = market

● = least-cost location

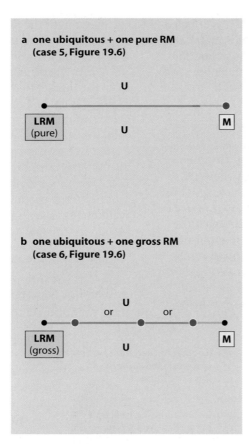

a one ubiquitous + one pure RM (case 5, Figure 19.6)

U

LRM (pure)   U   M

b one ubiquitous + one gross RM (case 6, Figure 19.6)

U
or   or
LRM (gross)   U   M

Weber's industrial triangle: the concept is illustrated by three pieces of string, tied at one end by a knot and having a weight to represent the weights of each of the raw materials and of the final product.

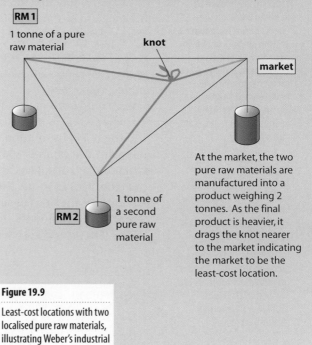

RM 1

1 tonne of a pure raw material

knot

market

RM 2   1 tonne of a second pure raw material

At the market, the two pure raw materials are manufactured into a product weighing 2 tonnes. As the final product is heavier, it drags the knot nearer to the market indicating the market to be the least-cost location.

Figure 19.9

Least-cost locations with two localised pure raw materials, illustrating Weber's industrial triangle (case 7, Figure 19.6)

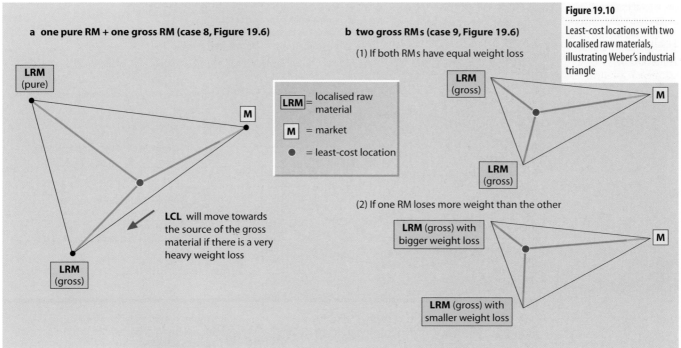

a one pure RM + one gross RM (case 8, Figure 19.6)

LRM (pure)

M

LCL will move towards the source of the gross material if there is a very heavy weight loss

LRM (gross)

LRM = localised raw material

M = market

● = least-cost location

b two gross RMs (case 9, Figure 19.6)

(1) If both RMs have equal weight loss

LRM (gross)

M

LRM (gross)

(2) If one RM loses more weight than the other

LRM (gross) with bigger weight loss

M

LRM (gross) with smaller weight loss

Figure 19.10

Least-cost locations with two localised raw materials, illustrating Weber's industrial triangle

Q

1 What is a gross material? What is its material index (MI)?

2 What is a pure material? What is its material index?

3 Which of the following materials are gross and which are pure: iron ore; sand and gravel; oil; television components; sugar beet?

Weber claimed that four factors affected production costs: the cost of raw materials and the cost of transporting them and the finished product, together with labour costs and agglomeration/deglomeration economies (pages 509, 511).

Spatial distribution of transport costs
As transport costs lay at the heart of his model, Weber had to devise a technique which could both measure and map the spatial differences in these costs in order to find the LCL. His solution was to produce a map with two types of contour-type lines which he called **isotims** and **isodapanes**.

Figure 19.11a shows the costs of transporting 1 tonne of a raw material (R) as concentric circles. In this example, it will cost 5 t/km (tonne/kilometres) to transport

the material to the market. Figure 19.11b shows, also by concentric circles, the cost of transporting 1 tonne of the finished product (P). The total cost of moving the product from the market to the source of the raw material is again 5 t/km. By superimposing these two maps it is possible to show the total transport costs (Figure 19.11c).

If a factory were to be built at **X** (Figure 19.11c), its transport costs would be 7 t/km (i.e. 2 t/km for moving the raw material plus 5 t/km for the product). A factory built at **Y** would have lower transport costs of 6 t/km (4t/km for the raw material plus 2t/km for the product). However, the LCL in this case may be at the source of the raw material, the market or any intermediate point in a straight line between the two because all these points lie on the 5 t/km isodapane.

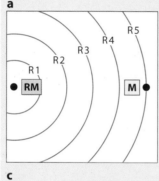

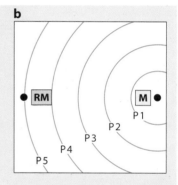

a **isotims showing transport costs of a raw material, pure and localised (tonne km)**

b **isotims showing transport costs of finished product (tonne km)**

c **isodapanes showing total transport costs (RM + finished product) (tonne km)**

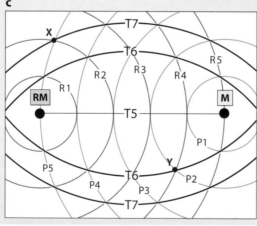

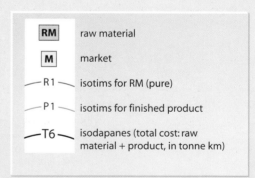

RM	raw material
M	market
R 1	isotims for RM (pure)
P 1	isotims for finished product
T6	isodapanes (total cost: raw material + product, in tonne km)

Figure 19.11

Isotims and isodapanes

1 The statements below describe the raw materials used by four different industrial plants, **A, B, C, D.**
Plant **A** uses two gross, ubiquitous, raw materials.
Plant **B** uses two pure, localised, raw materials.
Plant **C** uses one gross, localised, raw material.
Plant **D** uses one pure, localised, raw material and one gross, ubiquitous, raw material.

a State the material index for each of the four plants (i.e. is it >1, 1, or <1?).
b Describe the location of each of the four plants (i.e. has it a market location, a raw material location or an intermediate location?).

2 Show with the aid of appropriate diagrams, how the LCL in Weber's locational triangle may be:
a near to the market;
b near to the raw material; and
c at an intermediate point.

Figure 19.12

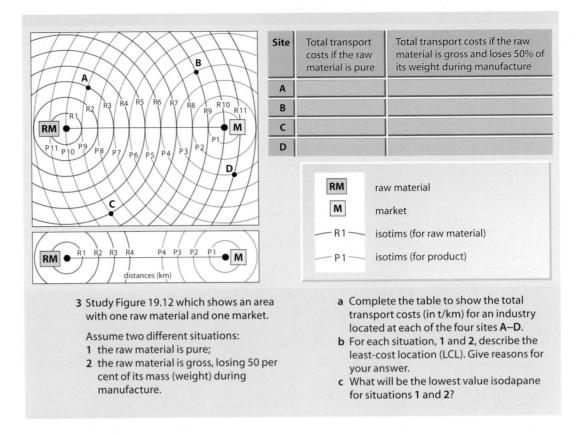

Site	Total transport costs if the raw material is pure	Total transport costs if the raw material is gross and loses 50% of its weight during manufacture
A		
B		
C		
D		

RM	raw material
M	market
R1	isotims (for raw material)
P1	isotims (for product)

3 Study Figure 19.12 which shows an area with one raw material and one market.

Assume two different situations:
1 the raw material is pure;
2 the raw material is gross, losing 50 per cent of its mass (weight) during manufacture.

a Complete the table to show the total transport costs (in t/km) for an industry located at each of the four sites **A–D**.
b For each situation, **1** and **2**, describe the least-cost location (LCL). Give reasons for your answer.
c What will be the lowest value isodapane for situations **1** and **2**?

The effects of labour costs and agglomeration economies

It has been stated that Weber considered that four factors affected production costs: we have seen the effects of the costs of raw materials and transport — let us now look at labour costs and agglomeration economies.

- **Labour costs** Weber considered the question of whether any savings made by moving to an area of cheaper or more efficient labour would offset the increase in transport costs incurred by moving away from the LCL. He plotted isodapanes showing the increase in transport costs resulting from such a move. He then introduced the idea of the **critical isodapane** as being the point at which savings made by reduced labour costs equalled the losses brought about by extra transport costs. If the cheaper labour lay within the area of the critical isodapane, it would be profitable to move away from the LCL in order to use this labour.

- **Agglomeration economies** Figure 19.13 shows the critical isodapanes for three firms. It would become profitable for all the firms to locate within the central area formed by the overlapping of all three critical isodapanes. It may be slightly more profitable for firms **A** and **B**, but less profitable for firm **C**, to locate within the purple area. However, it would not be additionally profitable for any firm to move if none of the isodapanes overlapped.

Figure 19.13

Critical isodapanes and agglomeration economies

Critical isodapanes for firms A, B and C

Firm B

Firm A

Firm C

Firms A and B might agglomerate here, but it would not be worthwhile for firm C (beyond its critical isodapane)

Intersection of 3 critical isodapanes means it would be worth the 3 firms agglomerating in this area

Criticisms of Weber's model

The point has already been made with previous examples and on page 507 that no model is perfect and all have their critics. Criticisms of Weber's industrial location model include:

- It no longer relates to modern conditions — such as the present extent of government intervention (grants, aid to Enterprise Zones), improvements in and reduced costs of transport, technological advances in processing raw materials, the development of new types of industry other than those directly involved in the processing of raw materials, the increased mobility of labour and the increased complexity of industrial organisation (multinationals instead of single-product firms).
- Each country evolves its own industrial patterns and may be in different stages of economic development (page 569).
- There are basic misconceptions in his original assumptions. For example there are changes over time and space in demand and price; there are variations in transport systems; perfect competition is unreal as markets vary in size and change over a period of time; and decisions made by industrialists (who do not all have the same knowledge) may not always be rational (von Thünen's 'economic man', page 430).
- Weber's material index was a crude measure and applicable only to primary processing or to industries with a very high or very low index.

Bradford and Kent (1977) claim that Weber's work, apart from its overemphasis on transport costs and leaving some gaps in our understanding of the system as a whole, has not yet been superseded. Rather, "others have added important principles which, taken together with those of Weber, help to explain a much more complex industrial world".

Smith's area of maximum profit

An alternative approach to that of Weber was put forward by **David Smith** in 1971. He suggested that, as profits could be made anywhere where total revenue exceeds total costs, there would be a wider area where production was still profitable (Figure 19.14) — although there would be a point of **maximum profit**. To introduce his **space–cost curves**, Smith also used isodapanes. By taking a cross-section he was able to identify, spatially, margins of profitability. He reasoned that firms rarely located at the ideal, or LCL, site (which had probably already been taken), but somewhere between the two profit margins. In other words, firms choose a **sub-optimal location** because they have imperfect knowledge about production and market demand; they have imperfect decision-makers who do not always act rationally; and they may be tempted or encouraged to locate in areas of high unemployment.

Industrial location: changing patterns

Four different types of industry have been selected as exemplars to try to demonstrate how the importance of different factors affecting the location of industry have changed through time. Their choice may reinforce the generalisation, by no means true in every case, that the more important locational factors in the 19th century were physical, while in modern industry they tend to be human and economic. The four industries are:

1 A primary manufacturing industry where, due to weight loss, the presence of raw materials and sources of energy is more important than the market and other economic factors (Places 63).
2 A secondary manufacturing industry initially tied to raw materials and sources of energy but in which economic and political factors have become increasingly more important (Places 64).
3 A secondary manufacturing industry where the nearness of a market and labour supply is more important than the presence of raw materials and sources of energy (Places 65).
4 Modern secondary (quaternary) manufacturing industries where human and economic factors are the most important (Places 66).

Figure 19.14

Space–cost curve to show the area of maximum profit (*after* Smith, 1971)

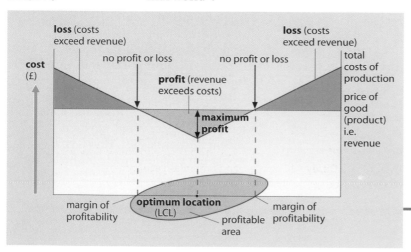

Places 63 Wood pulp and paper in Sweden

There are three stages in this industry: the felling of trees, the processing of wood pulp (primary processing), and the manufacture of paper (secondary processing). In Sweden, most pulp and paper mills (Figure 19.15) are located at river mouths on the Gulf of Bothnia (Figure 19.16). Timber is a gross raw material which loses much of its weight during processing; it is bulky to transport; and it requires much water to turn it

Figure 19.15

Pulp mill on the Gulf of Bothnia

into pulp. Towns such as Sundsvall and Kramfors are ideally situated (Figure 19.16): the natural coniferous forests provide the timber; the fast-flowing Rivers Ljungan, Indals and Angerman which initially provided cheap water transport for the logs are a source of the necessary and cheap hydro-electricity; and the Gulf of Bothnia provides an easy export route. Paper has a higher value than pulp and it is convenient and cheaper to have integrated mills.

Weber's agglomeration economies seem to operate with the clustering of so many mills. Smith's concept of an area of profitability rather than an individual least-cost location also seems applicable as the forest is extensive and there are several processing centres.

Figure 19.16

Location of wood pulp and paper factories in central Sweden

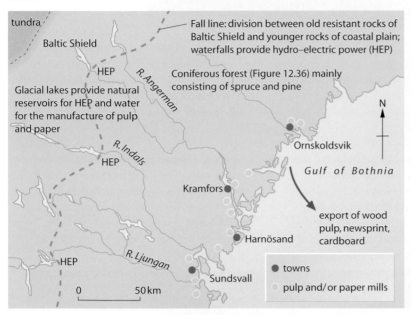

tundra

Baltic Shield

HEP

Glacial lakes provide natural reservoirs for HEP and water for the manufacture of pulp and paper

R. Angerman

Fall line: division between old resistant rocks of Baltic Shield and younger rocks of coastal plain; waterfalls provide hydro–electric power (HEP)

Coniferous forest (Figure 12.36) mainly consisting of spruce and pine

N

R. Indals

HEP

Ornskoldsvik

Gulf of Bothnia

Kramfors

export of wood pulp, newsprint, cardboard

Harnösand

R. Ljungan

HEP

Sundsvall

● towns

○ pulp and/or paper mills

0   50 km

Places 64 Iron and steel in the UK

Although the early iron and later steel industries were tied to raw materials, modern integrated iron and steelworks have adopted new locations as the sources of both ore and energy have changed.

- **Before AD 1600** Iron-making was originally sited where there were surface outcrops of iron ore and abundant wood for use as charcoal (the Weald, the Forest of Dean, Figure 19.17a). Locations were at the source of these two raw materials as they had a high material index, were bulky and expensive to transport, had a limited market and could not be moved far owing to the poor transport system.

- **Before AD 1700** Local ores in the Sheffield area were turned into iron by using fast-flowing rivers to turn waterwheels as water provided a cheaper source of energy.

- **After AD 1700** In 1709, Abraham Derby discovered that coke could be used to

smelt iron ore efficiently. At this time, it took 8 tonnes of coal and 4 t of ore to produce 1 t of iron, and so new furnaces were located on coalfields. One of the first areas to develop was South Wales where bands of iron ore (blackband ores) were found between seams of coal. The advantages possessed by South Wales at that time are shown in Figure 19.18a. Later, the industry extended into other British coalfields. When local ores became exhausted, the industry continued in the same locations because of geographical inertia, a pool of local skilled labour, a local market using iron as a raw material, improved techniques reducing the amount of coal needed (2 t per 1 t of final product), improved and cheaper transport systems (rail and canal) which brought distant mined iron ore; and the beginnings of agglomeration economies.

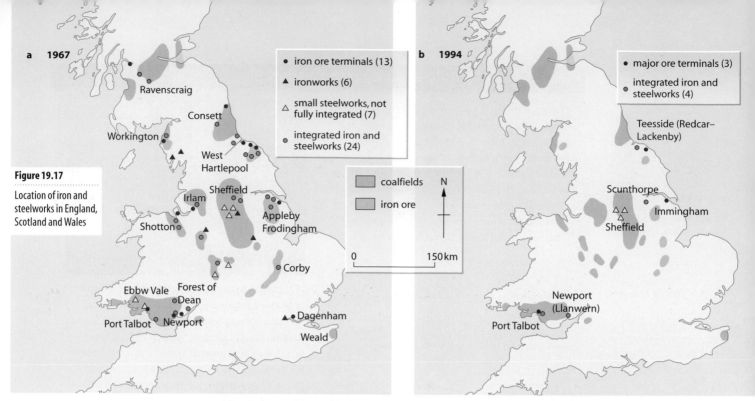

Figure 19.17

Location of iron and steelworks in England, Scotland and Wales

Maps:
- **a** 1967
  - Ravenscraig
  - Workington
  - Consett
  - West Hartlepool
  - Irlam
  - Sheffield
  - Shotton
  - Appleby Frodingham
  - Corby
  - Ebbw Vale
  - Forest of Dean
  - Port Talbot
  - Newport
  - Dagenham
  - Weald

  Legend:
  - • iron ore terminals (13)
  - ▲ ironworks (6)
  - △ small steelworks, not fully integrated (7)
  - • integrated iron and steelworks (24)

- **b** 1994
  - Teesside (Redcar–Lackenby)
  - Scunthorpe
  - Immingham
  - Sheffield
  - Newport (Llanwern)
  - Port Talbot

  Legend:
  - • major ore terminals (3)
  - ▪ integrated iron and steelworks (4)

Shared legend:
- coalfields
- iron ore
- N
- 0 150 km

Figure 19.18

Growth, decline and changing location of iron and steelworks in South Wales

■ **After 1850** Until the 1880s, the low ore and high phosphorous content of deposits found in the Jurassic limestone, extending from the Cleveland Hills to Oxfordshire, had not been touched. After 1879, the **Gilchrist–Thomas process** allowed this ore to be smelted economically. As iron ore now had a higher material index than coal it was more expensive to move. As a result, new steelworks were opened on Teesside, near to the Cleveland Hills deposits, and at Scunthorpe and Corby, on the ore fields. However, the major markets remained on the coalfields.

Period of time			**a** Location of early 19th-century iron foundries in South Wales (e.g. Ebbw Vale)	**b** Disadvantages of these early locations by 1960 (e.g. Ebbw Vale)	**c** Location of integrated steelworks of the early 1990s (Port Talbot and Llanwern, Newport)
Physical	Raw materials	Coal	mined locally in valleys	older mines closing	little now needed; imported
		Iron ore	found within the coal measures	had to be imported: long way from coast	imported from N Africa and N America
		Limestone	found locally	found locally	found locally
		Water	for power and effluent: local rivers	insufficient for cooling	for cooling: coastal sites
	Energy/fuel		charcoal for early smelting, later rivers to drive machinery; then coal	electricity from national grid	electricity from national grid using coal, oil, natural gas and nuclear power
	Natural routes		materials local; export routes via the valleys	poor; restricted by narrow valleys	coastal sites
	Site and land		narrow valley floor locations	cramped sites; little flat land	large areas of flat, low potential farmland
Human and economic	Labour		large quantities of semi-skilled labour	still large numbers of semi-skilled workers	still relatively large numbers but with a higher level of skill; fewer due to high-tech/mechanisation
	Capital		local entrepreneurs	no investment	government and EU incentives
	Markets		local	difficult to reach Midlands and ports	tin plate industry (Llanelli) and the Midland car industry
	Transport		little needed; some canals; low costs	poor; old-fashioned; isolated	M4; purpose-built ports
	Geographical inertia		not applicable	not strong enough	tradition of high-quality goods
	Economies of scale		not applicable	worked against the inland sites	two large steelworks more economical than numerous small iron foundries
	Government policy		not applicable	Ebbw Vale kept open by government help	having the capital, governments can determine locations and closures and provide heavy investment
	Technology		small scale: mainly manual	out-of-date	high technology: computers, lasers, etc.

■ **After 1950** With iron ore still the major raw material (less than 1 t of coal was now needed to produce 1 t of steel), but with deposits in the UK largely exhausted, Britain became increasingly reliant upon imported ores. This meant that new **integrated steelworks** were located on coastal sites (Figure 19.17b). By 1980, the only two remaining inland sites, at Ravenscraig and Scunthorpe, had been linked to new, nearby ore terminals. (Ravenscraig was forced to close in the early 1990s.) The present day advantages of the two coastal South Wales steelworks are shown in Figure 19.18c.

Until the 1950s, the iron and steel industry satisfied much of Weber's theory. After that, unforeseen by him 40 years earlier, three new elements became increasingly important in the location of new steelworks: government intervention, improved technology and reduced transport costs. Since the nationalisation of the industry, the government has made the major decisions concerning the location of modern integrated plants and which are to be kept open and which are to close. Improved technology has meant less reliance upon raw materials and labour, while reduced transport costs have allowed raw materials to be imported. The introduction of the oxygen furnace in the 1950s has meant that relatively little energy (coal) is now needed, while the reduction in the workforce has cut production costs but increased unemployment — an economic gain but a social loss.

Places 65   Car assembly in Japan

Japan's production of 9.9 million cars in 1992, which was 24.6 per cent of the world's total, made it the world leader ahead of the USA (6.0 m) and Germany (4.4 m). This has been achieved despite a lack of basic raw materials.

Japan has very limited energy resources for, although it produces hydro-electricity and nuclear energy, it has to import virtually all its coal, oil and natural gas requirements. Similarly, most of the iron ore and coking coal needed to manufacture steel also has to be imported. The result has been the location of the major steelworks on tidal sites found around the country's many deep and sheltered natural harbours. As only 17 per cent of the country is flat enough for economic development (for homes, industry and agriculture), most of the population also has to live in coastal areas and around the harbours. The five major conurbations, linked by modern communications, provide both the workforce and the large, affluent, local markets needed for such steel-based products as cars (Figure 19.20). Within these conurbations, especially Keihin, Chukyo and Setouchi, are numerous firms engaged in making car component parts. This agglomeration of firms limits transport costs and conforms with Weber's concept that industries gaining weight through processing (car assembly) are best located at the market. As many of the smaller, older and original firms have amalgamated into large-scale companies, the extra space required for their factories has had to come from land reclaimed from the sea (Figure 19.19). These new locations, despite the high costs of reclamation, make excellent sites from which to export finished cars to all parts of the world.

Figure 19.19

Mazda's new Hofu car plant, built upon land reclaimed from the sea

Figure 19.20

Major industrial areas in Japan

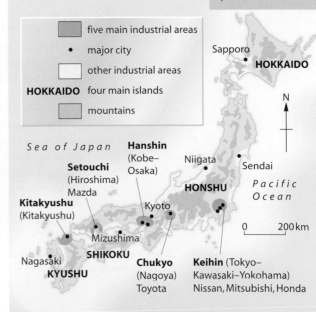

The large local labour force contains both skilled and semi-skilled workers. It is well-educated and industrious, with workers who are very loyal to their firms. The car industry, which has received considerable government financial assistance, has an organisation which centres around teamworking, worker involvement, total-quality management and 'just-in-time' production. (This is when various component parts arrive just as they are needed on the assembly line, thus avoiding the need to store or to overproduce.) The Japanese car industry has a high level of automation and uses the most modern technology: it produces three times the number of cars per worker as does western Europe. The assembled cars are reliable and universally acceptable in design which means, together with the shift from mass production to lean production, that the Japanese have gained strong footholds in world markets. To expand further into these markets, the Japanese have either built overseas assembly plants or have amalgamated with local companies so that more cars can be produced close to the large urban markets within western Europe and the USA.

Places 66   Hi-tech industries along the M4 and M11 corridors

The term **high technology** refers to industries which have developed within the last 25 years and whose processing techniques often involve micro-electronics, but may include medical instruments, biotechnology and pharmaceuticals. These industries, which collectively fit into the **quaternary** sector (page 502), usually demand high inputs of information, expertise and research and development (R&D). They are also said to be **footloose** in that, not being tied to raw materials, they have a free choice of location. However, they do tend to occur in clusters in particular areas, forming what Weber would have called 'agglomerated economies', and they locate, according to Smith's terminology, in areas of maximum profit such as the M4 and M11 corridors in England (also Silicon Glen in Scotland, Silicon Valley in California and the Côte d'Azur in France). By locating close together, hi-tech firms can exchange ideas and information and share basic amenities such as connecting roads.

The major concentrations of hi-tech industries in Britain are along the M4 westwards from London to Reading, Newbury ('Video Valley'), Bristol (Aztec West) and into South Wales, and the M11 northwards to Cambridge (Figure 19.21). Transport is convenient due to the proximity of several motorways and mainline railways, together with the four main London airports. Transport costs are, in any case, relatively insignificant as the raw materials (silicon chips) are lightweight and the final products (computers) are high in value and small in bulk. Even so, it has been argued that the main reasons for hi-tech development in this part of Britain were the presence of government-sponsored research establishments at Harwell and Aldermaston and of government aerospace contractors in the Bristol area.

Most firms which have located here claim that the major factor affecting their decision was the availability of two types of labour:

- highly skilled research scientists and engineers, the majority of whom are university graduates or qualified technicians. These specialists, whose abilities are in short supply, can often dictate areas where they want to live and work — i.e. areas of high environmental, social and cultural quality. The proximity of several universities (Figure 19.21) provides a pool of skilled labour and facilities for R&D.

- females who (according to stereotyping) can perform delicate work and often prefer part-time/flexible jobs (Figure 19.4). Female labour tends to be plentiful as an increasing number of career-minded women are among those who have recently moved out of London into surrounding new and overspill towns.

Science Parks are often joint ventures between universities and local authorities. They are usually located adjacent to universities on edge-of-town green-field sites where, because the land is of lower value, there is plenty of space for car parking, landscaping (ornamental gardens and lakes) and possible future expansion. The Cambridge Science Park (Figures 19.22 and 19.23) has been developed in conjunction with Trinity College, Cambridge.

Opened in 1972, the success of early firms soon attracted more (agglomeration economies), so that by 1992 there were amost 100 companies employing over 2500 people. Existing companies can be divided into those making scientific instruments (38 per cent), electronics (30 per cent) and drugs and pharmaceuticals (22 per cent).

Figure 19.21

The M4 and M11 Corridors

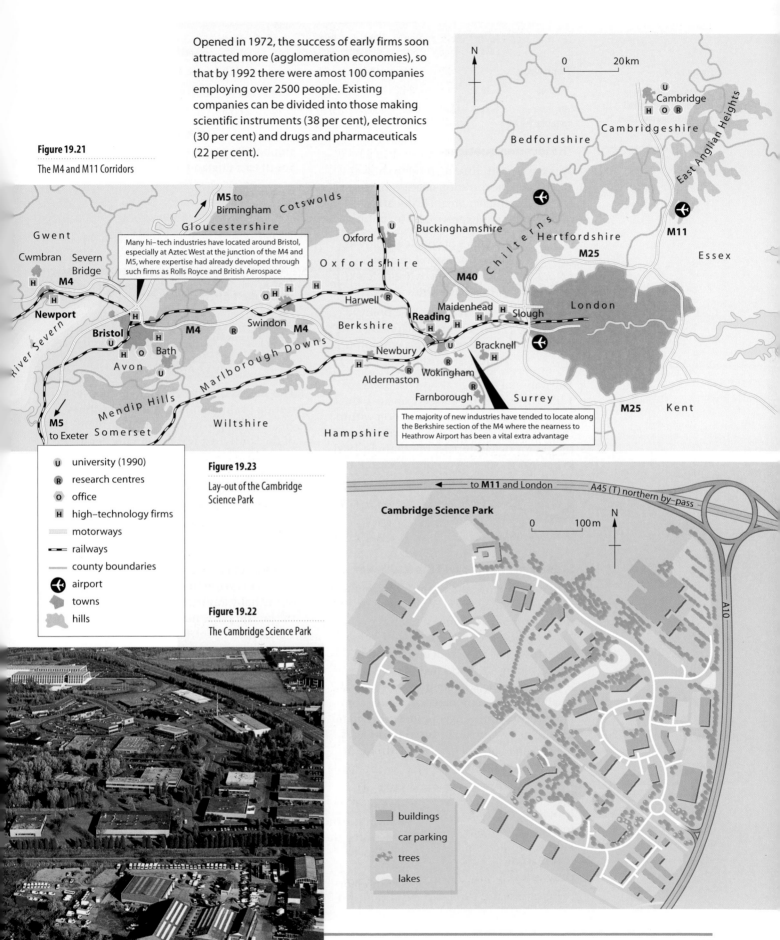

Many hi–tech industries have located around Bristol, especially at Aztec West at the junction of the M4 and M5, where expertise had already developed through such firms as Rolls Royce and British Aerospace

The majority of new industries have tended to locate along the Berkshire section of the M4 where the nearness to Heathrow Airport has been a vital extra advantage

U	university (1990)
R	research centres
O	office
H	high–technology firms
≡	motorways
⊷	railways
—	county boundaries
✈	airport
	towns
	hills

Figure 19.23

Lay-out of the Cambridge Science Park

Figure 19.22

The Cambridge Science Park

Cambridge Science Park

to **M11** and London   A45 (T) northern by-pass

0   100m

	buildings
	car parking
	trees
	lakes

Industry	Region										Great Britain
	South East	East Anglia	South West	West Midlands	East Midlands	Yorkshire and Humberside	North West	North	Wales	Scotland	
Shipbuilding	39.5	3.5	19.5	0	1.5	7.3	10.0	48.3	1.6	43.0	174.2
Textiles	19.6	2.6	8.5	16.6	85.4	65.0	64.2	11.0	8.5	41.7	323.1
All manufacturing	1683.5	186.0	395.7	800.7	533.4	578.9	799.8	339.4	238.2	502.0	6057.6

Figure 19.24

UK employment in ship-building, textiles and all manufacturing industries, by region, 1981

The location quotient

This is a quantitative method to show the degree of concentration of a specific industry within a particular area. The **location quotient** (LQ) is expressed by the formula:

$$LQ = \frac{\left(\dfrac{\text{number of people employed in industry } A \text{ in area } X}{\text{number of people employed in all industries in area } X} \right)}{\left(\dfrac{\text{number of people employed in industry } A \text{ nationally}}{\text{number of people employed in all industries nationally}} \right)}$$

The national average always has an LQ of 1.00. The higher the index of an area, the greater the degree of concentration within that area. Figure 19.24 gives employment figures for two industries in Britain in 1981. What was the degree of concentration of shipbuilding in the North of England and in South East England for that year?

North of England:

$$LQ = \frac{\left(\dfrac{48.3}{339.4} \right)}{\left(\dfrac{174.2}{6057.6} \right)} = \frac{0.141}{0.029} = 4.90$$

South East England:

$$LQ = \frac{\left(\dfrac{39.5}{1683.5} \right)}{\left(\dfrac{174.2}{6057.6} \right)} = \frac{0.023}{0.029} = 0.79$$

This means that the North had a concentration of shipbuilding almost five times greater than the national average, while that of the South East was marginally below the national average.

Q 1 What are the location quotients for the textile industry in:

a the North West of England; and
b Wales?

Industrial linkages and the multiplier

When Weber introduced the term 'agglomeration economies', he acknowledged that many firms made financial savings by locating close to, and linking with, other industries. The success of one firm may attract a range of associated or similar type industries (cutlery in Sheffield), or several small firms may combine to produce component parts for a larger product (car manufacture in Coventry). **Industrial linkages** may be divided into **backward linkages** and **forward linkages**:

backward linkages
to firms providing raw materials or component parts

← FACTORY →

forward linkages
to firms further processing the product or using it as a component part

A more detailed classification of industrial linkages is given in Figure 19.25. The more industrially advanced a region or country, the greater is the number of its linkages. Developing countries have few linkages partly because of their limited number of industries and partly because few industries go beyond the first stage in processing — the simple chain in Figure 19.25a. Industrial linkages may result in:

- energy savings;
- reduced transport costs;
- waste products from one industry forming a raw material for another;
- energy given off by one process being used elsewhere;
- economies of scale where several firms buy in bulk or share distribution costs;
- improved communications, services and financial investment;
- higher levels of skill and further research; and

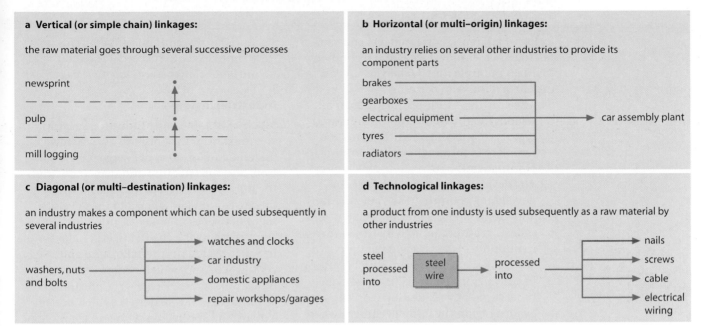

a Vertical (or simple chain) linkages:

the raw material goes through several successive processes

newsprint

pulp

mill logging

b Horizontal (or multi–origin) linkages:

an industry relies on several other industries to provide its component parts

brakes
gearboxes
electrical equipment → car assembly plant
tyres
radiators

c Diagonal (or multi–destination) linkages:

an industry makes a component which can be used subsequently in several industries

washers, nuts and bolts →
watches and clocks
car industry
domestic appliances
repair workshops/garages

d Technological linkages:

a product from one industy is used subsequently as a raw material by other industries

steel processed into → steel wire → processed into →
nails
screws
cable
electrical wiring

Figure 19.25

Types of industrial linkage

■ a stronger political bargaining position for government aid (the securing of EU funding now depends upon having a network of linked organisations).

Cole (1994) stresses the increasingly critical importance of local linkages in ensuring competitive success. In the fashion industry in Nottingham's Lace Market, for example, 85 per cent of all firms are linked to others in the Lace Market (suppliers, manufacturers, retailers, etc.).

The multiplier effect and Myrdal's model of cumulative causation

If a large firm, or a specialised type of industry, is successful in an area, it may generate a **multiplier effect**. Its success will attract other forms of economic development creating jobs, services and wealth —

a case of 'success breeds success'. This circular and cumulative process was used by **Gunnar Myrdal**, a Swedish economist writing in the mid 1950s, to explain why inequalities were likely to develop between regions and countries. Figure 19.26 is a simplified version of his model.

Myrdal suggested that a new or expanding industry in an area would create more jobs and so increase the spending power of the local population. If, for example, a firm employed a further 200 workers and each worker came from a family of four, there would be 800 people demanding housing, schools, shops and hospitals. This would create more jobs in the service and construction industries as well as attracting more firms linked to the original industry. As **growth poles**, or points, develop there will be an

Figure 19.26

A simplified version of Myrdal's model to show the development of an industrial region

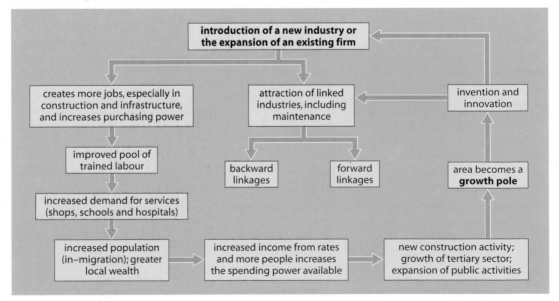

influx of migrants, entrepreneurs and capital, together with new ideas and technology. Myrdal's multiplier model may be used to explain a number of patterns.

1 The growth of 19th-century industrial regions (South Wales and the Ruhr) and districts (cutlery in Sheffield, guns and jewellery in Birmingham and clothing in Nottingham).

2 The development of growth poles in developing countries (São Paulo in Brazil and the Damodar Valley in India), where increased economic activity led, in turn, to multiplier effects, agglomeration economies and an upward spiral resulting in core regions (Places 67). At the same time, cumulative causation worked against regions near the **periphery** where Myrdal's **backwash effects** included a lack of investment and job opportunities.

3 The creation of modern government regional policies which encourage the siting of new, large, key industries in either peripheral, less developed (Trombetas and Carajas in Amazonia) or high unemployment (Nissan and Toyota in England) areas in the hope of

stimulating economic growth. This policy is more likely to succeed if the industries are labour intensive (Places 77, page 577).

Industrial regions

Much of Britain's early industrial success stemmed from the presence of basic raw materials and sources of energy for the early iron, and the later iron and steel, industries; the mass production of materials using the processed iron and steel; and the development of overseas markets. During the 19th century it was the coalfields, especially those in South Wales, northern England and central Scotland, which became the core industrial regions. However, as the initial advantages of raw materials (which became exhausted), specialised skills and technology (no longer needed as the traditional heavy industries declined) and the ability to export manufactured goods (in the face of growing overseas competition) were lost, these early industrial regions have become more peripheral. Recent attempts to revive their economic fortunes have met with varying success (Places 67).

Places 67   South Wales

Pre-1920: industrial growth creating a core region

The growth of industry in South Wales was based upon readily obtainable supplies of raw materials (Figure 19.18a). Coking coal and blackband iron ore were frequently found together, exposed as horizontal seams outcropping on steep valley sides. Their proximity to each other meant that the area around Merthyr Tydfil and Ebbw Vale (Figure 19.27) was ideally suited for industrial

development (Weber's least-cost location for two gross raw materials; Figure 19.10b). Added to this was the presence of limestone only a few kilometres to the north, and the expertise of the local population in iron-making where water wheels, driven by fast-flowing rivers, had earlier been used to power the blast furnace bellows. By the time the more accessible coal had been used up, mining techniques had improved sufficiently to allow shafts to be sunk vertically into the valley floors. When local supplies of

Figure 19.27

Early industrial development in South Wales

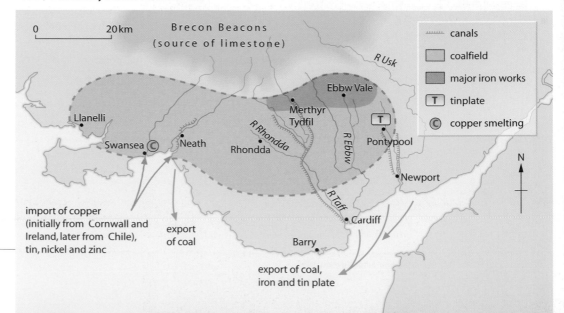

520

iron ore became exhausted, there were ports nearby through which substitute ore could be imported.

"… Thus began the spread of the well-known industrial landscape of the Valleys. Pits crammed themselves into the narrow valley bottoms, vying for space with canals, housing and, later, railways and roads. Housing began to trail up the valley sides, line upon line of terraces pressed against the steep slopes (Figure 19.28). The opening-up of the underground coal seams resulted in massive immigration, much of it from rural areas. Working conditions, living conditions and wages were deplorable while health and safety standards underground were poor. Housing was overcrowded as the provision of homes, financed by the local entrepreneur ironmasters, lagged far behind the supply of jobs."

The rapid increase in coalmining and iron-working partly resulted from the growth of large overseas markets as both products were mainly exported. Transport to the Welsh ports first involved simply allowing trucks to run downhill under gravity. Later, canals and then railways were used to move the bulky materials. While Barry, Cardiff and Newport developed as exporting ports, Swansea and Neath grew from smelting the imported ores of copper, nickel and zinc.

The inter-war and immediate post-war years: depression and industrial decline

Just as the existence of raw materials and overseas markets had led to the growth of local industry, so did their loss hasten its decline. Iron ore had long since been exhausted and it increasingly became the turn of coal, even though there were still over 500 collieries employing 260 000 miners in 1925. The steelworks which had replaced the iron foundries had been built upon the same inland, cramped sites; as they became less competitive mainly due to rising transport costs, so they became increasingly dependent upon government support. Overseas markets were lost as rival industrial regions with lower costs and more up-to-date technology were developed overseas. The difficulties of an economy reliant on a narrow industrial base, dependent upon an increasingly out-of-date infrastructure, and unable to compete with overseas competition, led to major economic, social and environmental problems.

Towards the present: industrial diversification in a peripheral area

Steel-making and non-ferrous metal smelting have been maintained, partly due to geographical inertia, despite a significant fall in output and workers. The centre of gravity for steel-making has moved to the coast and so the two South Wales integrated works (half of Britain's remaining four) are located at Port Talbot and Llanwern (Newport). Tin plate, using local steel, is produced at Trostre near Llanelli (the Felindre works near Swansea closed in 1989), while the Mond nickel works near Swansea is the world's largest (Figure 19.29).

The major factor to have affected industry in the region in the last 40 years has been government intervention (or lack of it, depending upon your political views). The Special Areas Act of 1934 saw the first government assistance which set up industrial estates at Treforest, Merthyr Tydfil and Rhondda (Figure 19.29), while Cwmbran became one of Britain's first New Towns (1949). Much of the former coalfield remains a Development Area (Figure 19.5). The last NCB colliery closed in 1994, although the Tower Colliery re-opened privately in 1995. There are still, despite environmental concerns, several areas of open-cast mining (page 473). Two local areas of exceptionally high unemployment, Swansea and Milford Haven Waterway, were designated two of Britain's 27 Enterprise Zones (page 506). The Swansea EZ includes five parks — the Enterprise (commerce and light industry), Leisure (recreation facilities), Riverside (heritage and environmental schemes), City (retailing)

Figure 19.28

Industry, communications and terraced housing strung along the valley floor and lower valley sides: Llanillth, Ebbw Vale, South Wales

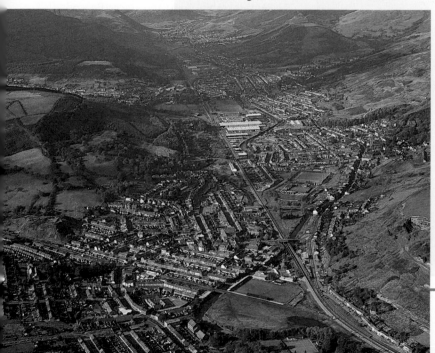

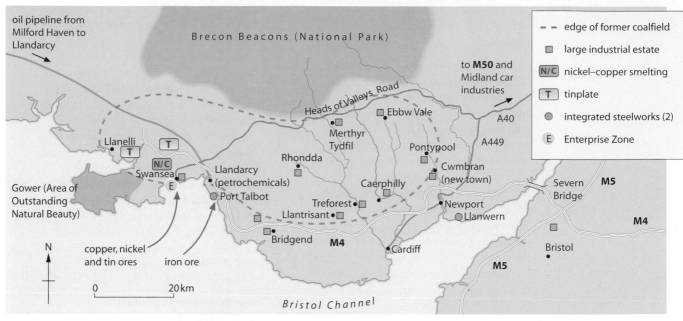

oil pipeline from
Milford Haven to
Llandarcy

Brecon Beacons (National Park)

to **M50** and
Midland car
industries

Legend:
- − − edge of former coalfield
- ▪ large industrial estate
- N/C nickel–copper smelting
- T tinplate
- ● integrated steelworks (2)
- E Enterprise Zone

Gower (Area of
Outstanding
Natural Beauty)

Heads of Valleys Road

Ebbw Vale

Merthyr Tydfil

Rhondda

Pontypool

Cwmbran (new town)

A40

A449

Severn Bridge

M5

Llanelli

T

T

N/C

Swansea

E

Llandarcy (petrochemicals)

Port Talbot

Caerphilly

Treforest

Llantrisant

Newport

Llanwern

M4

Bristol

M5

N

copper, nickel and tin ores

iron ore

Bridgend

M4

Cardiff

0   20 km

Bristol Channel

Figure 19.29

Recent industrial development in South Wales

Figure 19.30

A modern industrial estate in South Wales

and Maritime (housing and cultural) Parks. The Ford Motor Company took advantage of government incentives to build two plants in the region one of which, at Bridgend, has been expanded. It was government policy which built an integrated steelworks at Ebbw Vale in 1938, and which closed it in 1979. The futures of Port Talbot (highly modernised and hopefully secure) and Newport are also in government hands. A policy to decentralise some government departments has seen vehicle licensing moved to Swansea and the Royal Mint to Llantrisant (Figure 19.29). Improvements in communications include the M4, the Heads of Valleys Road, the InterCity rail link and Cardiff airport — some of which were financed by EU funds.

Most new industrial development has taken place on the edge of the coalfield where communications are better and access to markets is easier. The region has a large pool of skilled and semi-skilled labour, although people need retraining for the new-style industries which are mainly hi-tech and services. This pool of labour, together with the efforts of the Welsh Development Agency (WDA), has attracted much overseas investment — mainly from Japan, but also from the USA and Germany. At present, there are over 40, mainly hi-tech, Japanese firms in South Wales employing 12 000 people (Figure 19.30). These industries are subject to decisions made outside Wales and employ only a small percentage of those previously engaged in the mining and steel industries.

The Pembrokeshire Coast and Brecon Beacons National Parks and the Gower Peninsula provide many environmentally attractive areas favourable for tourism. Money has also been spent on landscaping old industrial areas which had been scarred either by metal-smelting industries (lower Swansea Valley) or by slag (Ebbw Vale) and colliery waste tips (Aberfan). The Ebbw Vale Garden Festival (1992), sited on part of the former steelworks, was part of a larger scheme aimed at creating new jobs, improving housing, renovating old properties and improving the local environment. Other schemes, often funded by the WDA, include tourist and cultural facilities such as the Welsh Industrial and Maritime Museum in Cardiff's newly created Marina area.

SONY

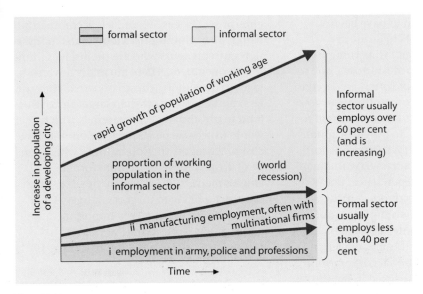

legend: formal sector ▬▬ | informal sector ☐

Figure 19.31

Growth in the informal sector

Figure 19.32

Differences between the 'formal' and 'informal' sectors

	Formal	Informal
Description	Employee of a large firm	Self-employed
	Often a multinational	Small-scale/family enterprise
	Much capital involved	Little capital involved
	Capital-intensive with relatively few workers; mechanised	Labour-intensive with the use of very few tools
	Expensive raw materials	Using cheap or recycled waste materials
	A guaranteed standard in the final product	Often a low standard in quality of goods
	Regular hours (often long) and wages (often low)	Irregular hours and uncertain wages
	Fixed prices	Prices rarely fixed and so negotiable (bartering)
	Jobs done in factories	Jobs often done in the home (cottage industry) or on the streets
	Government and multinational help	No government assistance
	Legal	Often outside the law (illegal)
	Usually males	Often children and females
Type of job	Manufacturing: both local and multinational companies	Distributive (street peddlers and small stalls)
	Government-created jobs such as the police, army and civil service	Services (shoe cleaners, selling clothes and fruit)
		Small-scale industry (food processing, dress and furniture repairers)
Advantages	Uses some skilled and many low-skilled workers	Employs many thousands of low-skilled workers
	Provides permanent jobs and regular wages	Jobs may provide some training and skills which might lead to better jobs in the future
	Produces goods for the more wealthy (food, cars) within their own country so that profits may remain within the country	Any profit will be used within the city: the products will be for local use by the lower-paid people
	Waste materials provide raw materials for the informal sector	Uses local and waste materials

Industry in economically less developed countries

In cities in economically less developed countries, the number of people seeking work far outweighs the number of jobs available. As these cities continue to grow, either through natural increase or in-migration, the job situation gets continually worse. In the early 1990s, the UN estimated that there were over 500 million people unemployed in developing countries and only 20 million, and at a time of recession, in economically more developed countries. The UN also estimates that in developing countries, on average, only about 40 per cent of those people with jobs work in the **formal sector** (Figure 19.31). These jobs, which are permanent and relatively well paid, include those offered by the state (police, army and civil service) or by overseas-run **multinational (transnational) corporations**. The remaining 60 per cent, a figure which the UN claims is rising, have to seek work in the **informal sector**. The main differences between the formal and informal sectors are listed in Figure 19.32.

Multinational (transnational) corporations

A multinational, or transnational, company is one which operates in many different countries regardless of national boundaries. The headquarters is usually in an economically more developed country with branch factories spread across the world but, increasingly, in economically less developed countries. Multinationals are believed to employ over 30 million people worldwide, and indirectly to influence an even greater number. The largest 100 multinationals, led by oil companies and car manufacturers, controlled nearly half of the world's manufacturing in 1990. Several of the largest have a turnover higher than all of Africa's total GNP.

Multinationals, with their capital and technology, have the 'power' to choose what they consider to be the ideal locations for their factories. This choice will be made at two levels: the most suitable country and the most suitable place within that country. The choice of a country usually depends upon political factors. Most governments, regardless of the level of economic development within their country, are prepared to offer financial inducements to attract multinationals which they see as providers of jobs and a means of increasing exports. (Sony,

Figure 19.30, were reputed to have been offered better inducements to locate at Bridgend than in Barcelona.) Governments of economically less developed countries, due to a greater economic need, are often prepared to impose fewer restrictions on multinationals because they often have to rely upon them to develop natural resources, to provide capital and technology (machinery, skills, transport), to create jobs and to gain access to world markets. Despite political independence, many poorer countries remain economically dependent (neo-colonialism) upon the large multinationals (together with international banks and foreign aid). Some of the advantages and disadvantages of multinational corporations to developing countries are listed in Figure 19.33.

Multinationals, having selected a country, then have to decide where to locate within that country. If the country is economically developed, the location is likely to be where financial inducements are greatest, land values are low, transport is well developed, and levels of skill and unemployment are high (Japanese companies in South Wales, page 522; a Korean company, Samsung, in north-east England). If the country is economically less developed, the location is more likely to be in the primate city (page 374), especially if that city is also the capital or the chief port. A capital city location, with an international airport, allows quick access to the companies' overseas headquarters; and a port location enables easier export of manufactured goods. Should several multinational companies locate in the same area, the multiplier effect is likely to result in the development of a core region (Places 67).

The informal sector

A large and growing number of people with work in developing countries have found or created their own jobs in the informal sector (Figure 19.32).

"This sector covers a wide variety of activities meeting local demands for a wide range of goods and services. It contains sole proprietors, cottage industries, self-employed artisans and even moonlighters. They are manufacturers, traders, transporters, builders, tailors, shoemakers, mechanics, electricians (Places 68), plumbers, flower-sellers and many other activities. The (Kenyan) government have recognised the importance of these small-scale 'jua kali'

enterprises (Places 69) and a few commercial banks are beginning to extend loans to these new entrepreneurs who are themselves forming co-operatives. There are many advantages in developing these concerns. They use less capital per worker than larger firms; they tend to use and re-cycle materials which would otherwise be waste; they provide low-cost practical on-the-job training which can be of great value later in more formal employment, and as they are flexible, they can react quickly to market changes. Their enterprising spirit is a very important national human resource."

(Central Bank of Kenya, 1991)

Figure 19.33

Advantages and disadvantages of multinational (transnational) corporations

Advantages to a country
Creates jobs and uses local labour
Local workforce receive a guaranteed income
Improves the level of education and technical skill of local people
Brings inward investment and foreign currency into the country
Companies provide expensive machinery and introduce modern technology
The resultant increase in GNP/personal income can lead to increased demand for consumer goods, services and new industries (Figure 19.26)
Leads to the development of mineral and energy resources
Prestige value (e.g. Volta project)
Widens the country's economic base
Improvements, often financed locally by the company, in services (schools, hospitals and transport)

Disadvantages to a country
Numbers employed often small in comparison to amount of investment
Local labour force often poorly paid and expected to work long hours
Relatively few skilled jobs go to local people
Most of profits go overseas (outflow of wealth)
Mechanisation reduces size of workforce
GNP grows less quickly than that of the country having the company's headquarters, widening the gap between the 'rich' and the 'poor' countries
Raw materials are often exported rather than processed locally; manufactured goods are usually for export rather than for the local market
Money may be better spent improving social conditions (housing, diet and sanitation)
Decisions are made outside the country and the company can pull out at any time
Insufficient attention paid to the health and safety of workforce or to the protection of the environment

Places 68   Don Mariano: an electrician in Guatemala

Don Mariano, a refugee from Nicaragua, lives with his family in a shanty settlement in San Jose, Guatemala. His hut, which was built with boards and roofed with corrugated iron, also contains his 'workshop'. It is a separate room with a worktable and an electric outlet. In this workshop he repairs all kinds of electrical appliances, heating plates, electric heaters, radios and television sets. His equipment is totally inadequate — two screwdrivers and one pair of pliers. For a Voltmeter he uses an old lamp — he can tell the voltage from its brightness. He does not own a blowtorch. Don Mariano's primary problem is not a lack of skill but a lack of equipment. If he had the money he could buy a blowtorch and a voltmeter. If he had those he could do more repairs and earn more money. If …

(*adapted from* Timberlake, *Only One Earth*, 1987)

Places 69   *'Jua Kali'* workshops in Nairobi, Kenya

'Jua Kali' means 'under the hot sun'. Although there are many smaller *Jua Kali* in Nairobi, the largest is near to the bus station where, it is estimated, over 1000 workers create jobs for themselves (Figure 19.34). The plot of land upon which the metal workshops have been built measures about 300 metres by 100 metres.

Figure 19.34

'Jua Kali' workshops

The first workshops were spontaneous and built illegally as their owners did not seek permission to use the land which did not belong to them. As more workshops were set up and the site developed, the government was faced with the option of either bulldozing the temporary buildings, as governments had done to shanty settlements in other developing countries, or encouraging and supporting local initiative. Realising that the informal workshops created jobs in a city where work was hard to find, the government opted to help. The Prime Minister himself became personally involved by organising the erection of huge metal sheds which protected the workers from 'the hot sun' and occasional heavy rain.

Groups of people are employed touring the city collecting scrap. The scrap is melted down, in charcoal stoves, and then hammered into various shapes including metal boxes and drums, stoves and other cooking utensils, locks and water barrels, lamps and poultry water troughs (Figure 19.35). Most of the workers are under 25 and have had at least some primary education. The technology they use is appropriate and sustainable, suited to their skills and the availability of raw materials and capital. Most of the products are sold locally and at affordable prices.

It is estimated that there are approximately 600 000 people engaged in 350 000 small-scale *'Jua Kali'* enterprise units in Kenya. This figure needs to be compared with the 180 000 recorded as employed in large-scale manufacturing and the 2.2 million total in all areas of the non-agricultural economy. *'Jua Kali'* form, therefore, a most significant part of the total employment picture.

Figure 19.35

'Jua Kali' end products

Intermediate (appropriate) technology

Dr E. F. Schumacher developed the concept of **intermediate technology** as an alternative course for development for poor people in the 1960s. He founded the Intermediate Technology Development Group (ITDG) in 1966 and published his ideas in his book, *Small is Beautiful* (1973). Schumacher himself wrote:

> "If you want to go places, start from where you are.
>
> If you are poor, start with something cheap.
>
> If you are uneducated, start with something relatively simple.
>
> If you live in a poor environment, and poverty makes markets small, start with something small.
>
> If you are unemployed, start using labour power, because any productive use of it is better than letting it lie idle.
>
> In other words, we must learn to recognise boundaries of poverty.
>
> A project that does not fit, educationally and organisationally, into the environment, will be an economic failure and a cause for disruption."

In 1988 the ITDG stated that:

> "Essentially, this alternative course for development is based on a local, small-scale rather than national, large-scale approach. It is based on millions of low-cost workplaces where people live — in the rural areas — using technologies which can be made and controlled by the people who use them and which enable those people to be more productive and earn money."

These ideas challenged the conventional views of the time on aid. Schumacher said "The best aid to give is intellectual aid, a gift of useful knowledge … The gift of material goods makes people dependent, but the gift of knowledge makes them free — provided it is the right kind of knowledge, of course."

To illustrate this he quoted an old proverb:

> "Give a man a fish and you feed him for a day; teach him how to fish and he can feed himself for life."

The first part of this might be seen as the traditional view of aid where 'giving' leads to dependency. The second part, 'teaching', is a move in the direction of self-sufficiency and self-respect. Schumacher added a further dimension to the proverb by saying "teach him to make his own fishing tackle and you have helped him to become not only self-supporting but also self-reliant and independent".

In most developing countries, not only are hi-tech industries too expensive to develop, they are also usually inappropriate to the needs of local people and the environment in which they live. Examples of intermediate, or **appropriate technology** as it is now known (Places 70) include:

- Labour-intensive projects; since, with so many people already being either unemployed or underemployed, it is of little value to replace workers by machines.
- Projects encouraging technology which is sustainable and the use of tools and techniques designed to take advantage of local resources of knowledge and skills.
- The development of local, low-cost schemes using technologies which local people can afford, manage and control rather than expensive, imported techniques.
- Developing projects which are in harmony with the environment.

 Q

1 What are the main differences between the formal and informal sectors of employment?

2 a How do sites chosen by multinational (transnational) corporations in developed countries differ from those chosen in developing countries?
   b Why do governments of many developing countries wish to attract multinational companies?
   c What problems may be created within a developing country following the opening of a new multinational project?

3 Why is the informal sector so important to developing countries such as Kenya?

4 a Why do ITDG consider that aid is not the solution for poor people?
   b Explain what is meant by 'giving leads to dependency, teaching to self-sufficiency and self-respect'.
   c Why have the schemes described in Places 70 been successful?

Intermediate Technology (IT) is a British charitable organisation which works with people in developing countries, especially those living in rural areas, by helping them to acquire the tools and techniques needed if they are to work themselves out of poverty. IT helps people to meet their basic needs of food, clothes, housing, energy and employment. IT uses, and adds to, local knowledge by providing technical advice, training, equipment and financial support so that people can become, in Schumacher's words, "more self-sufficient and independent". At present, IT is working in response to requests from local groups, often women, in India, Nepal, Bangladesh, Sri Lanka, Kenya, Sudan, Malawi, Zimbabwe and Peru. The following extracts are from IT publications concerning their operations in Kenya.

1 The most abundant building source, soil, has great advantages as a building material because it can be easily compressed, it absorbs heat well and transmits it slowly. Although it is vulnerable to erosion, soil can be 'stabilised' using lime, natural fibres or a small amount of cement. In Kenya soil blocks stabilised with cement are replacing the more expensive concrete blocks or industrially produced bricks.

2 The Maasai are slowly being forced to live in permanent homes, but few can afford a new house. IT is helping them to find affordable ways to upgrade their existing houses (Figure 19.36). Women have told us that their primary concern is a leaky roof; they dread being woken up by their husbands and ordered to climb up in the middle of the night to smear more cow-dung over a crack in the roof. One answer is a thin layer of cement reinforced with chicken wire laid over the old flat mud roof. Plastering is a job the women can do themselves. By incorporating a gutter and a water-jar women are spared another tiresome chore — collecting water from a spring or river possibly several kilometres away. By adding a small window and a chimney cowl, the inside of the house can be made lighter, less smoky and more healthy.

3 Kenyan women are almost entirely dependent upon wood as fuel for cooking. In the cities women must buy wood or charcoal (transported from rural areas) to use on their traditional stoves, while rural women spend long hours gathering the fuel they need. At the same time, Kenya's resources of wood are diminishing (page 495). An improved cooking stove (*jiko*) has been developed which drastically reduces the use of charcoal for urban households (Figure 19.37). It is based on traditional design and is made by local metal workers, from scrap metal, and incorporates a ceramic lining, produced by local potters. This development has improved the lives of many people:

■ More than 40 000 households are now making sizeable savings in their fuel costs. In many homes the new stove pays for itself in about a month. Moreover, the improved stove reduces smoke and fumes in the kitchen and contributes to improved women's health.

■ Large numbers of small-scale metal workers and potters are now engaged in manufacturing the *jiko* and have found new support from the government which makes their business more stable.

Widespread use of the stove also has a potential impact on the environment, as any saving in urban fuel consumption helps relieve the pressure on rural wood resources.

Figure 19.36

Maasai House Improvement Project

Figure 19.37

Improved cooking stoves (*jiko*)

Case Study 19

Recent industrial development in China and Malaysia

Pacific Asia is the most dynamic and productive region in the developing world. The economic growth of the region, fuelled by industrialisation, began in Japan, continued with the emergence of the 'four little tigers' of Hong Kong, Singapore, South Korea and Taiwan, and has, more recently, expanded to include China, Malaysia, the Philippines and Thailand. The rapid economic growth of these Pacific Rim countries, with their 900 million consumers and economies growing at a rate several times faster than the world average, is changing the structure of world business and industry.

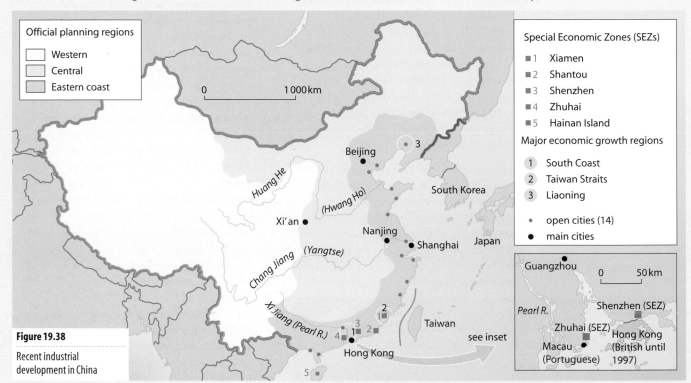

Figure 19.38

Recent industrial development in China

China

When the People's Republic of China was declared in 1949, the country's economy was still feudal with virtually no agricultural or industrial development. There was a very low standard of living; most of the land and wealth was in the hands of a small minority; and life expectancy was merely 32 years.

1949–76: the commune system State control was overseen by Mao Zedong. The government dictated where each factory had to be located, how many workers would be employed there, who would have jobs and what products would be made. The state controlled who was responsible for each stage in the manufacturing process, how much was to be produced, how much the wages would be, where the product would be sold and at what price. There were no individual

entrepreneurs or decision-makers. The First Five-year Plan (1953) concentrated on developing labour-intensive heavy industry and discouraged the production of consumer goods. China's huge coal reserves became the basis for an iron and steel industry from which ships, textile machinery, tractors and locomotives were produced. In 1958, Mao launched the 'Great Leap Forward' in an attempt to mobilise China's people and resources. In order to meet production targets, factory workers were organised, like their counterparts in agriculture, into self-sufficient communes. Mao's later attempts to create a classless society (the 'Cultural Revolution' of 1966–76) led to the virtual destruction of the country's economy.

1979 to the present: the responsibility system In 1979 the government, under

Deng Xiaoping, began replacing the regimentation of farm production through communes with the responsibility system. This system, which incorporated capitalist ideas, turned China's peasants into tenant farmers, and was later applied to industry with the result that total production and its quality improved. Encouragement was given to individual entrepreneurs to set up their own firms, especially in consumer goods. State-owned companies, still accounting for 70 per cent of the total firms, were allowed, once their production targets had been met, to sell surplus stock and to share the profits between their workers. Competition between firms was encouraged and an 'open-door' policy allowed overseas firms to invest and settle in the country (Pepsi-Cola at Shenzhen and Volkswagen in Shanghai).

Since 1979, five **special economic zones (SEZs)** and 14 **open cities** have been created (Figure 19.38). These offer tax concessions and low labour and land costs to overseas firms — especially those involved in hi-tech industries. It is not coincidental that the five SEZs are in the southeast of the country, near to Hong Kong which handles over 40 per cent of China's trade; and the open cities, although more widespread, all have coastal locations. These coastal areas have received most internal investment as well as having 'imported' capital, technology and entrepreneurial skills from surrounding countries. The result has been the emergence and dominance of three regional economies (Figure 19.38):

1 **South China,** which includes three of the SEZs and has strong links with Hong Kong (Figure 19.39).

2 **Taiwan Straits,** centred in the province of Fujian and with its unofficial links (for political reasons) with Taiwan.

3 **Liaoning Province,** with its strong links with Japan and South Korea, and their many multinationals.

China's economic achievements over the last 15 years have been stunning. Economic growth has averaged nearly 9 per cent per annum for more than a decade; trade is doubling every 5 years (albeit from a small base); and inflation has been at a manageable 6 per cent. Even so, economic development has been concentrated along the east coast; it has taken place with little concern for the environment (Figure 19.40); most people in employment receive low wages (Figure 19.41); and the country still presents "a snarl of paradox and contradiction" (Figure 19.42).

Figure 19.40

Industrial Polluters in China to be Prosecuted

BEIJING — In a drive to halt deteriorating air quality and polluted land and rivers, a government official warned on Sunday of criminal penalties against those who violate environmental laws and regulations.

He said, in a speech marking World Environment Day, that the discharge of pollutants into the environment harmed the basic interests of all people and their offspring, and he compared the crime to smuggling narcotics or marketing fake medicine.

An official newspaper said on Saturday that China's rapid economic boom had poisoned the air with acid rain and industrial gases, resulting in deteriorating air quality and people's health.

In an unusually candid report, China admitted its incidence of lung cancer and bronchial pneumonia had soared, respiratory disease was the No 1 rural killer and a massive amount of farm land was polluted beyond use.
(South China Morning Post, May 1994)

Shenzhen: a Special Economic Zone

Shenzhen is the most dramatic example of China's 'open door' policy, an area that is more Westernised than anywhere else in the country. In 1980, when the Special Economic Zone was established, Shenzhen was a rural community of 30 000 people, supplying farm produce to Hong Kong. Since then it has become a major industrial centre, with a population estimated at three million. Workers flocked there from other parts of China, drawn by the liberal lifestyle and wages five times the national average.

Shenzhen's rapid growth is due to the closeness of Hong Kong, which lies just across the border to the south. Hong Kong is short of land and labour, so Shenzhen offers an ideal opportunity for its industries to expand. Rents are cheaper, as are labour costs — only half the level in

Hong Kong — though productivity and quality tend to be lower too. Most firms use Shenzhen simply for assembling processed materials or components, producing items such as clothing, hardware, food and drink, plastics and a wide range of electrical goods. So far, Shenzhen has failed to attract much high-tech industry or research and development from abroad.

Two-thirds of Shenzhen's industrial output comes from foreign-owned enterprises or joint-venture schemes. The vast majority of investments come from Hong Kong, which has also funded many projects to develop housing, hotels and tourist facilities in the zone. Each week, 100 000 Hong Kong citizens cross the border to buy cheap goods and for entertainment; some are even buying flats in Shenzhen. But Shenzhen's phenomenal growth has outstripped the provision of basic services. Land, water and electricity are all in short supply. Shenzhen plans to spend £5 billion on its infrastructure up to the turn of the century: already a new power station has been built and Shenzhen's own airport at Hwang Tian was opened in 1991.

(Geographical Magazine, September 1994)

Figure 19.39

"We were taken to a show-place factory, manufacturing jewellery, presumably because it was modern, well organised, non-polluting and it encouraged us to spend money. The company, Myers, employed 1200 factory workers and 200 clerical assistants. Most of the factory workers were girls and boys, in seemingly equal numbers, aged between 18 and 25. The factory work, which was highly skilled as it included the polishing of stones and the assembling of jewellery, put a tremendous strain upon the eyes. The working day was from 0800 to 1200 hours and 1330 to 1730 hours, and there was a 5-day week. The workers received 300 yen (about £28) a month — but any stone lost or broken had to be paid for out of their salary. As the workers became older (mid 20s) they were either promoted into clerical jobs or found work elsewhere. In the showroom over twenty assistants were hoping (expecting!) that we would buy some of the merchandise".

(Judith and David Waugh, May 1994)

Figure 19.41

Working conditions in a Guilin factory

"Even for experienced hands, China presents a snarl of paradox and contradiction: a great economic success while the majority of its people are still very poor; an ageing and feeble 1930s-style Stalinist gerontocracy presiding over the world's most dynamic and fastest growing economy; a highly centralised state tugged apart by regional economies and vulnerable to changes in leadership. It is praised for its promising future, damned for its human rights record. The temptation for a corporation to take a substantial position in China is nearly irresistible, yet at present there are only a few instances of multinational corporate success and profitability."

(James Abegglen, *Sea Change*, 1994)

Figure 19.42

Modern China

Malaysia

The Malaysian government has a 25-year vision that, by the year 2020, their country will have become a fully developed nation. Under their Prime Minister, Dr Mahathir, they seem well on course to achieve this target.

Malaysia's economic growth rate has, since 1990, averaged 8 per cent per annum (Figure 19.43). While the government feels confident that this rate can be maintained, it is concerned that, in 1994, the population is only 19 million (Figure 19.43). Malaysia is underpopulated (page 354 and Places 38, page 355) — i.e. it has more resources than it can use effectively. It has a surplus (Figure 19.44) of energy sources (oil and natural gas), tin (Places 62, page 475), timber and agricultural products. (It is the world's leading producer of palm oil and third major producer of rubber.) However, Malaysia's present population is insufficient to attract many kinds of new industry or to build up industries that depend upon a large domestic market. Although employment rates are low (having fallen from 18 per cent in 1984 to under 4 per cent in 1994), the country is faced with an acute shortage of labour, especially for less skilled jobs. The Malaysian government is trying to overcome the labour shortage by

1 Reversing its earlier family planning programme and now encouraging people to have larger families (compare Places 31, page 336). The current development plan is dependent upon the population having reached 70 million by the end of next century.

2 Attracting cheap, unskilled labour from overseas, especially Filipinos for the construction industry and Indonesians and Bangladeshis for factory work (compare Places 34, page 343).

Malaysia's industrial strategy emphasises the development of high-value goods for both the domestic market and export (Proton cars, Figure 19.46), together with both resource-based (rubber) and high-technology industries (Figure 19.45). It was announced in June 1994 that "Malaysia is now the top choice as an investment area in the Asian region, ahead of Japan, Taiwan, Hong Kong and other tiger economies. This is primarily because of its pivotal position as a gateway to Asean (Figure 21.24), a springboard to East Asia, affordable land and liberal investment rules" (*New Straits Times*).

At present, industry is confined to special designated areas such as in the new town of Shah Alam. This policy is good environmentally, as only certain tracts of primary forest or farmland are taken over, but has the social disadvantage of concentrating jobs within a few limited areas. As firms new to Shah Alam are not expected to pay taxes for the first 10 years, many of the world's better-known multinational corporations have located there. The town, with its wide dual carriageways, is linked to three-lane motorways which lead to the adjacent capital (Kuala Lumpur), chief port (Klang) and the international airport (Subang). When a new international airport is opened in time for the 1998 Commonwealth Games, Subang airport will be used solely for freight. The largest single firm at Shah Alam is Proton who assemble cars in association with the Japanese firm Mitsubishi. The first models, Saga (1987) and Iswara, had all their parts made, and staff trained, in Japan. The present model, the Wira, now has its spare parts made in Shah Alam. Two announcements made in 1994 were that Wira cars were also to be assembled in China and that future Proton models would have all their parts made within Malaysia. Although many hi-tech firms, mainly assembling imported components and therefore having few backward or forward linkages (page 518), have also located at Shah Alam, the two Malaysian centres for electrical and electronic products are Penang ('silicon' island), which specialises in disk drives, and Malacca.

Figure 19.43

Population and economic growth rates of Malaysia, 1991–94

	1991	1992	1993	1994 (estimate)
Population (millions)	18.1	18.6	19.0	19.6
Economic growth rate (%)	8.7	7.8	8.0	8.2–10.0
GNP per capita (US$)	6796	7554	8318	—
GNP growth rate (%)	9.2	11.2	10.1	12.5

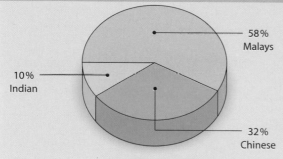

58% Malays

10% Indian

32% Chinese

Population growth rates

Malays:	2.0%	(fewer use contraceptives)
Chinese:	1.7%	(later marriages; greater use of contraceptives)
Indians:	2.1%	

19.44

Malaysia's exports, 1993

RM 120.225 million (estimate)

- Textiles, clothing and footwear 4.7%
- Electrical and electronic products 41.7%
- Other commodities 6.7%
- Rubber 1.8%
- Palm oil 4.9%
- Timber 6.3%
- Other manufactured goods 24.6%
- Tin 0.5%
- Crude oil 6.5%
- Natural gas 2.3%

Figure 19.45

New hi-tech industry, Penang

Figure 19.46

Car assembly (Proton)

References

Abegglen, J. (1994) *Sea Change*. The Free Press.

Barke, M. and O'Hare, G. (1984) *The Third World*. Oliver & Boyd.

Bradford, M. G. and Kent, W. A. (1977) *Human Geography: Theories and Their Applications*. Oxford University Press.

Briggs, K. (1982) *Human Geography*. Hodder & Stoughton.

Central Bank of Kenya (1991) *Kenya: Land of Opportunity*. Central Bank of Kenya.

Drakakis-Smith, D. (1992) *Pacific Asia*. Routledge.

IT (1987) *Intermediate Technology*. IT Publications (ITDG).

Malaysia Official Year Book 1993 (1993) Ministry of Information.

McBride, P. (1980) *Human Geography*. Nelson-Blackie.

Prosser, R. (1992) *Human Systems and the Environment*. Thomas Nelson.

Raw, M. (1993) *Manufacturing Industry: The Impact of Change*. Collins.

Schumacher, E. F. (1974) *Small is Beautiful*. Abacus.

Timberlake, L. (1987) *Only One Earth*. Earthscan/BBC.

Waugh, D. (1987) *The Wider World*. Thomas Nelson.

Chapter 20

Service industries

"L'Angleterre est une nation de boutiquiers."

Napoleon I

"We're all going on a summer holiday."

Cliff Richard

What is meant by 'services'?

It is possible to identify three different uses of the term 'services'.

1 Service **industries**, which make up the service sector, consist of firms or enterprises whose final output is non-material, irrespective of the types of occupation within that firm or enterprise — tourism, for example, would include people making holiday souvenirs.

2 Service **occupations** occur in all sectors of the economy, not just in service industries, and include, for example, accountants, secretaries, sales staff, cleaners and maintenance workers employed in primary industries and with manufacturing firms (page 502).

3 Service **products** can be produced by both service and manufacturing firms, e.g. an instruction guide to go with a video machine.

All societies, past and present, have contained a service sector. However, there does appear to be a strong positive correlation between this sector and levels of economic development: as a country develops economically, the greater the number of its inhabitants employed in the service sector (Framework 14, page 572). In recent years, the proportion of people working in this sector in Britain, has shown a considerable increase (Figure 19.1).

This chapter focuses briefly upon three important service industries: **retailing**, **finance** (offices), and **leisure and tourism**; while a further two: **transport** and **trade**, make up the following chapter.

Retailing

Traditional shopping patterns

Traditionally, as neatly summarised by Prosser, "Retailing in British cities has been based upon a well established hierarchy, from the CBD or 'High Street' at the top, through major district centres, local suburban centres, to neighbourhood parades and the local corner shop. Using numbers of outlets, floor space, type and range of goods, for example, as measures of size or 'mass', Christaller's Central Place (page 376 and Figure 14.30) and Gravity Models (page 379) have been applied to the hierarchal structure, relating mass to spatial distribution of shopping centres and their spheres of influence." Within this hierarchy were two main types of shop:

1 Those selling **convenience** or **low-order goods** which are bought frequently, usually daily, and are not sufficiently high in value to attract customers from further than the immediate catchment area, e.g. newsagents and small chain stores.

2 Those selling **comparison** or **high-order goods** which are purchased less frequently but which need a much higher threshold population, e.g. goods found in department stores and specialist shops.

The preferred location of these two types of shop was usually determined by the frequency of visit, their accessibility and the cost of land and, therefore, rent (page 392).

Convenience shops are commonly located in housing estates, both in the inner city and the suburbs, and in neighbourhood units so as to be within easy reach of their customers — often within walking distance. With a lower turnover of goods than retail units in the CBD, they may have to charge higher prices but their rent and rates are lower. Ideally, they are located along suburban arterial roads or at a crossroads for easier access and, possibly, to encourage

Source: Institute of Grocery Distribution.

impulse buying by motorists driving into the CBD (Figure 15.19). Convenience shops are located in inner cities where the corner shop (Figure 15.21B1) caters for a population which cannot afford high transport costs; in suburban shopping parades (Figure 15.21C1) where the inhabitants live a long way from the central shopping area; and along side-streets in the CBD where they take advantage of lower rents to provide daily essentials for those who work in the city centre.

Comparison shops need a large threshold population (page 376) and therefore have to attract people from the whole urban area and beyond. As they bid for a central location, they must have a high turnover in order to pay the high rents. This central area has traditionally afforded the greatest accessibility for shoppers with public transport competing with the private motorist. Large department stores and specialist shops usually locate within the CBD (Figure 15.21A1), although comparison shops may also locate in the more affluent suburbs.

The retailing revolution

There have been so many significant changes in retailing since the early 1970s that, collectively, they have been referred to as a 'revolution'. These changes include the shopping behaviour of customers, the organisation of the retail industry, the character of the shopping environment, the planning policies of local and central government and, most important of all, the location and introduction of new retail outlets (superstores, hypermarkets, regional shopping centres and retail parks). The 1970s saw the emergence of superstores, often within existing shopping centres, and hypermarkets, usually on new edge-of-city sites (Figure 20.1). The early 1980s saw the development of non-food retail parks which included furniture, DIY, carpet stores and garden centres (Figure 20.2). The late 1980s saw the growth of large regional shopping centres (Places 71). Since 1980, the numbers employed in the retail industry have doubled and now exceed 2 million.

Town centres

Congestion and other traffic problems have meant that urban access is now less easy than in the past. Many city centres have had to undergo constant change either to try to attract more customers (Birmingham's Bull Ring and Newcastle's Eldon Square) or, as is more usual, to restrict losses to the new out-of-town centres. Most city centres now contain under-cover malls, where shoppers can compare styles and prices in the warmth and dry, and pedestrianised areas, which are either traffic-free or have access limited to delivery vehicles and public transport. Many local councils are encouraging environmental improvements other than the reduction in pollution caused by traffic. Nottingham, for example, has recently spent money refurbishing St Peter's Square, making the area cleaner and more attractive by planting shrubs and other plants, giving planning permission for open-air cafes and employing staff whose early morning job is to ensure the area is clean and undamaged. These measures, together with a national trend in which land uses and functions in the 'High Street' are becoming more varied, are an attempt to win back lost shoppers and to revitalise the CBD (Figure 20.3).

Out-of-town shopping centres

Between 1980 and 1990 over 80 per cent of new retail floorspace was built on out-of-town greenfield sites. Shopping outlets locating here took advantage of economies of scale, lower rents and a pleasant, planned shopping environment. Hypermarkets and superstores were built on cheap land at, or beyond, the city margins, allowing them to buy space for their immediate use, for possible future expansion and for the necessary large car parking areas. Their aim was to capture a high threshold population, especially one which is wealthy and mobile. The ideal location is often adjacent to a main road leading directly to a motorway interchange as this facilitates access for both customers and delivery drivers. The early outlets were mainly food stores, and included the present 'big five' of Sainsbury, Tesco, Asda, Argyll and Gateway. They all offer a full range of goods, convenience and comparison, under one roof so that shoppers can complete their purchases in the warmth, at times suitable to the working family, and in a cleaner, less congested environment.

Later developments have included retail parks and regional shopping centres. Retail parks, which have also been attracted to redeveloped inner-city areas, concentrate mainly upon the sale of non-food, non-convenience items (e.g. B & Q, MFI, Do-it-all, Comet, Toys 'Я' Us). Regional shopping centres not only sell 'everything' under one roof, but often include restaurants, children's play areas and cinemas. The four largest, Gateshead MetroCentre (Places 71), Sheffield Meadowhall, Dudley Merry Hill and Thurrock Lakeside, all cover over 100 000 sq m (1.2 million sq feet). These centres are highly controversial as not only do they take a large amount of business away from the local city centre, they also attract, literally, coach-loads of shoppers from places over 150 km away. Research suggests that, in 1994, 30 per cent of retail sales were from out-of-town locations, compared with 5 per cent in 1980. In February 1994, with a fifth regional centre, at Dartford Bluewater, awaiting a final go-ahead — and to the annoyance of property developers — the government announced (as has the French) its intention to 'discourage' further out-of-town developments where "they are likely to make an unacceptable impact on the vitality of traditional town centres". It is unlikely that planning permission will be given for any future large-scale scheme.

The latest threats to the 'High Street' and to what many consider to be an already overcrowded food-store market, are Aldi (German) and Netto (Danish) who are now offering cut-price, no-frills shopping, mainly in less affluent urban areas.

Places 71 Gateshead's MetroCentre

Sir John Hall, whose brainchild the MetroCentre is, was a pioneer among modern entrepreneurs in recognising that shopping could become part of a family activity linked to the wider leisure experience as people's mobility, disposable income and leisure time (for those with jobs) increased. The MetroCentre was the prototype for a new concept in retailing in Britain — the regional shopping centre. Its success was assured after Marks & Spencer opted to make the MetroCentre its first out-of-town location.

Location
The site, before development, was marshland. This meant, together with its edge-of-Gateshead location, that a large area of land was both available and relatively cheap to buy.

Access
The site is adjacent to the western by-pass which now forms part of the main north–south trunk road, the A1, which avoids central Newcastle and Gateshead. It has 10 000 free car parking spaces with special facilities for the disabled motorist, 100 buses per hour, and 69 trains daily. The centre has had both a bus and a railway station purpose built (Figure 20.4).

Figure 20.4

Aerial view of the MetroCentre site, Gateshead: to the right is the A1 (Western by-pass) and to the top left the Newcastle–Carlisle railway, with station, and the River Tyne.

Figure 20.5

The internal environment of the MetroCentre

Figure 20.7

Traffic and visitors to the MetroCentre

Figure 20.6

Planned layout of the MetroCentre

Local and central government planning

The MetroCentre took advantage of tax concessions by locating in a government Enterprise Zone (page 506). The local council were highly cooperative, seeing the scheme as a major source of revenue and of employment — there are over 6000 employees.

Shopping environment

There are now over 300 retail units set in a pleasant shopping environment (Figure 20.5) with wide tree-lined malls, air conditioning, 1 sq km of glazed roof to let in natural light (supplemented by modern lighting in 'old world' lamps), numerous seats for relaxing, window boxes, escalators, and lifts for the disabled. A street market atmosphere has been created by traders selling from barrows and there are over 40 eating places offering all kinds of cuisine from Mexican to Chinese.

Additional amenities

Leisure is a vital part of the scheme. There is a 'children's village' and crèche, a 10-screen cinema, a 'space city' for computer and space enthusiasts, a covered 'fantasy-land' with all the attractions of the fair without the worries of the British climate, and theme areas such as 'Antiques Village' and 'The Roman Forum'. The complex also has a 150-bedroom hotel and leisure centre, an office block, an exhibition centre and a petrol station (Figure 20.6).

Visitors

MetroCentre's sphere of influence extends from Carlisle to Scandinavia, from York to Scotland. Since its opening in 1987, the centre has shown a steady increase in both volume of traffic — a record 46 000 on 21 December 1992 — and number of visitors — over 26 million in 1994 (Figure 20.7).

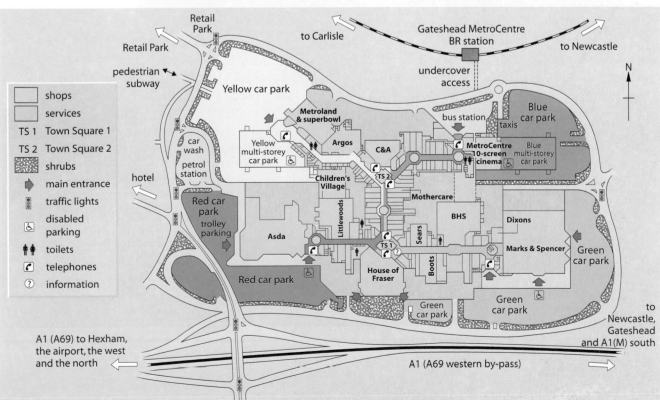

Financial institutions and offices

Financial institutions employ large numbers of people, especially in world centres such as New York, Tokyo, Hong Kong and London. These institutions, which include banking, insurance and accountancy, operate within offices. Traditionally offices have sought city centre locations regardless of the country's level of development (compare Tokyo, Figure 15.15; London, Figure 15.21A2; Hong Kong, Figure 20.8; and Nairobi, Figure 15.31). Company head offices and major institutions such as the stock exchange locate in the capital city. As offices use land intensively, they compete with shops for prime sites within city centres (Figure 15.18). Increasingly, due to high land values, they have had to locate in ever-taller office blocks. Elsewhere in city centres, offices may locate above shops in the main street or on ground floor sites in side-streets running off the main shopping thoroughfares. Banks can afford prime corner sites, while building societies and estate agents vie for high visibility locations. A city centre office location may have been desirable for prestige reasons, for ease of access for clients and staff, for proximity to other **functional links** (banks, insurance and entertainment) and sources of data and information, and for face-to-face contact.

The recent trend, in London for example, has been for offices to seek sites away from the CBD. Initially firms moved their offices to the New Towns which were created around London after the Second World War. Later the decentralisation of government offices saw a movement to more distant regions (DHSS to Newcastle, car licenses to Swansea, Giro Bank to Bootle). More recent movements from 'the City' have been to environmentally more attractive sites with either an out-of-town or a small rural town location, or to the Isle of Dogs Enterprise Zone (Places 41, page 402). These firms may have chosen, or been forced, to move due to the high rents, congestion and pollution of the city centre. They seek sites, often for the same reasons as new retailing outlets, where land values are lower; there is more space for car parking and possible future expansion; the working environment is more pleasant; and there is a greater proximity to motorways and, for overseas clients, airports. New technology has allowed the easier transfer of data and has reduced the need for face-to-face contact while increased computerisation has often led to a reduced workforce (banking) but one which is more highly skilled. Many new office locations are on purpose-built business, office or science parks (Figure 19.22; Places 66, page 516).

Figure 20.8

A wide-angled view of office development on Hong Kong island

Service industries

Leisure and tourism

Leisure and tourism is the world's second-largest industry in terms of money generated and is expected to be the EU's major source of employment by the turn of the century. Leisure can be defined as 'time at one's own disposal' and involves activities which we do and enjoy voluntarily. Whereas leisure can be enjoyed at home, tourism involves visiting places. Although we may make unplanned visits to places ourselves, tourism refers to organised touring undertaken on a commercial basis. Even so, it is often difficult to differentiate leisure and tourism from other forms of employment and sources of income. For example, how can hotels in central London differentiate between income from residents on business and those on holiday, a fish and chip shop in Blackpool between visitors and local residents, and farmers on a Greek or West Indian island selling food which ends up in a local hotel? Leisure and tourism, therefore, includes a wide range of products and services. These may be classified, according to the GNVQ syllabus on 'Leisure and Tourism', as in Figure 20.9.

Figure 20.9

GNVQ syllabus for 'Leisure and Tourism'

	Examples include:
Accommodation, catering and hospitality	hotels, restaurants, catering operations within leisure and tourist facilities
Entertainment and education	schools, clubs, venues, adult education, leisure classes, museums
Travel and tourism	travel agents, coach, taxi, aircraft and ferry operators, guides, tourist attractions
Sports and recreation	sports centres, swimming pools, gymnasiums, activity centres, coaching centres, mountain guides

In economically more developed countries, there has been a so-called explosion in the demand for both recreation and tourism. This explosion resulted initially from a growth in personal incomes, increased leisure time, paid holidays becoming part of the contractual working agreement, a shorter working week, earlier retirement, increased mobility (private cars, charter and long-haul flights), improved transport (motorways and international airports), package holidays and unreliable weather at home. It has continued where large groups in the community have become more affluent, where there is a need to escape the pressures of everyday life, due to TV travel programmes and advertising, due to changing fashions with emphasis on 'health' and 'exotic places', due to a desire to experience other cultures, and as a result of competition between multinational companies within the leisure industry.

Leisure and tourism therefore generates a major source of both employment and income. The British Tourist Authority claimed that, in 1992, the industry earned the UK over £25 billion, employed 1.5 million people, received 18.5 million overseas visitors (an 8 per cent increase over the previous year) and earned the country £7.9 billion in foreign exchange (Figure 20.10). The rising demand for recreation and holidays has already spread to the newly industrialised countries (Hong Kong, Singapore) and is an emerging trend amongst an affluent élite and an emerging middle class in many economically less developed countries. Tourism is seen by several developing countries as offering an opportunity, perhaps the only opportunity, to earn foreign currency, create local employment, increase local income and improve domestic services.

Figure 20.10

Tourism in the UK

Tourism is worth more than £25 bn a year and employs 1.5 million people in the UK. It earned £7.9 bn in 1992 in foreign exchange earnings and £9.5 bn in 1993.

Volume and value	1991	1992
Visitors from overseas	17.1 m	18.5 m
Expenditure by visitors from overseas	£7.4 bn	£7.9 bn

Origins of overseas visitors (m)	
USA	2.75
France	2.48
Germany	2.27
Irish Republic	1.42
Netherlands	1.00

Income from overseas visitors (£m)	
USA	£1490
Germany	£636
Middle East	£537
Irish Republic	£424
France	£420
Australia/New Zealand	£406

Top towns visited by overseas visitors (staying visits, in m)	
London	8.97
Edinburgh	0.74
Glasgow	0.37
Birmingham	0.32
Oxford	0.30
Manchester	0.29
York	0.28
Cambridge	0.27
Bath	0.24
Brighton/Hove	0.24

Sources: Top Towns Survey, 1991; IPS, 1993.

a the ten most popular historic properties in England, 1993

1	Tower of London	2 332 468
2	St Paul's Cathedral	1 900 000
3	Roman Baths and Pump Room, Bath	898 142
4	State Apartments, Windsor Castle	813 059
5	Warwick Castle	751 026
6	Stonehenge, Wiltshire	668 607
7	Shakespeare's birthplace, Stratford	606 697
8	Hampton Court Palace	576 664
9	Leeds Castle, Kent	533 000
10	Beaulieu Castle, Hampshire	481 223

Note: Buckingham Palace, opened for the first time, attracted 377 000 in only 56 days.

b domestic holidays taken by UK residents, by region, 1992

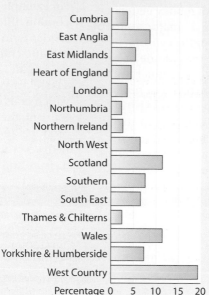

c foreign holidays taken by UK residents, 1992

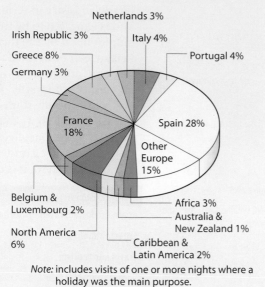

Note: includes visits of one or more nights where a holiday was the main purpose.

Figure 20.11

The uneven development of tourism

As the demands for recreation and tourism increase, so too will their impact on other socio-economic structures in society (which may be for the better or the worse), tourist environments and wildlife habitats. They will compete for space and resources with

- local people living and working in the area — farmers, quarry workers, foresters, water and river authority employees;
- other visitors wishing to pursue different recreational activities — water skiers, windsurfers, anglers and bird watchers all visiting the same lake; and
- wildlife habitats.

The development of recreation and tourist facilities creates pressure on specific places and environments in both urban and rural areas. Places with special interest or appeal that are very popular with visitors and which tend to become overcrowded at peak times are known as **honeypots**. Honeypots may include, in urban areas, sports stadia (Old Trafford), concert halls (Albert Hall) and historic buildings (Tower of London) and, in rural areas, places of attractive scenery (Lake Windermere) and historic interest (Stonehenge) (Figure 20.11). The problem of overcrowding within certain American National Parks (Yellowstone), together with congestion on access roads, has become so acute that permits are needed for entry and quotas are imposed on areas which are ecologically vulnerable (Case Study 20).

Sometimes planners encourage the development of honeypots, especially in British National Parks and African safari parks, to ensure that such sites have adequate visitor amenities (car parks, picnic areas, toilets, accommodation). It is now widely accepted than leisure amenities and tourist areas need to be carefully managed if the maximum number of people are to obtain the maximum amount of enjoyment and satisfaction. There are also signs, albeit small at present, of attempts to develop 'eco' or '**green**' **tourism** which will safeguard natural and human environments, be sustainable, and enable local people to share the economic and social benefits.

The development of tourism and the provision of leisure amenities, whether in a region, a country or between countries, is never even. This uneven development may result from a variety of physical, economic, social, technological and environmental factors. Most British tourists are attracted, if staying within Britain, to the West Country; if going overseas, to Spain (Figure 20.11). So far as developing countries are concerned, only those which are able to offer 'sun, sand, sea and scenery' (Jamaica, Sri Lanka), wildlife (Kenya), ancient civilisations (Egypt) or a widely different culture (Thailand) are likely to be able to participate in any tourist 'boom'.

The world-wide growth and extension of tourism has also created a wide range of

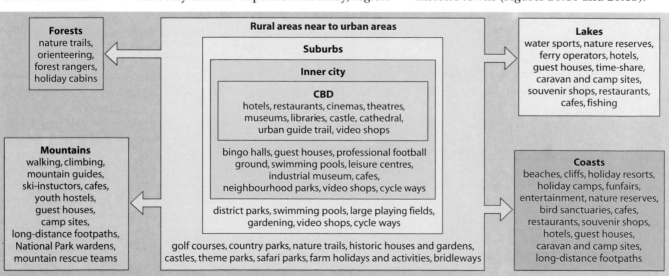

		Stage 1	Stage 2	Stage 3	Stage 4
Growth in tourism					?
		1960s	**1970s**	**1980s**	**1990s**
Tourists from UK to Spain		1960 = 0.4 million	1971 = 3.0 m	1984 =6.2 m 1988 =7.5 m	a1990 =7.0 m
State of, and changes in, tourism		very few tourists	rapid increase in tourism; government encouragement	carrying capacity reached; tourists outstrip resources e.g.water supply and sewage	decline (world recession); prices too high; cheaper upper-market hotels elsewhere
Local employment		mainly in farming and fishing	construction workers; jobs in hotels, cafes, shops; decline in farming and fishing	mainly tourism: up to 70 per cent in some places	unemployment increases as tourism declines (20 per cent); farmers use irrigation
Holiday accommodation		limited accommodation; very few hotels and apartments; some holiday cottages and campsites	large hotels built (using breeze blocks and concrete); more apartment blocks and villas	more large hotels built, also apart-ments, time-share and luxury villas	older hotels looking dirty and run down; fall in house prices; only high-class hotels allowed to be built
Infrastructure (amenities & activities)		limited access and few amenities; poor roads; limited streetlighting and electricity	some road improvements but congestion in towns; bars, discos, restaurants and shops added	E340 opened: 'the Highway of Death'; more congestion in towns; marinas and golf courses built	bars/cafes closing; Malaga by-pass and new air terminal opened
Landscape & environment		clean, unspoilt beaches; warm sea with relatively little pollution; pleasant villages; quiet with little visual pollution	farmland built upon; wildlife frightened away; beaches and sea less clean	mountains hidden behind hotels; litter on beaches; polluted seas (sewage); crime (drugs,vandalism, muggings); noise from traffic and tourists	attempts to clean up beaches and sea (EU Blue Flag beaches); new public parks and gardens opened; nature reserves

Figure 20.12

Life-cycle of a holiday area: tourists from the UK to the Costa del Sol, Spain, 1960s–90s

Figure 20.13

Types and location of various leisure and tourist facilities

environmental, social and economic prob-lems. Environmentally, forests have been cleared for holiday resorts; coral reefs and wildlife habitats are being destroyed (Case Study 18, page 499); overused areas suffer from soil erosion and various forms of pollu-tion. Socially, tourism can be perceived as destructive (loss of local cultures) and exploitive (local people performing for tourists, the role of women). Economically, tourism, especially in developing countries, can see the bulk of any profit being sent back overseas (neo-colonialism), can make a coun-try over-reliant upon one source of income, and can accentuate the gap in wealth between the 'rich' tourist and the 'poor' resident.

Yet, despite some of the obvious disad-vantages of tourism, the nightmare scenario for every tourism-dependent country, region or resort is that tourists will find somewhere else to visit and to spend their money. New resorts develop; old resorts may become run-down; fashions change; countries receive a bad press; economic recessions occur; currency rates alter. As a result, it has been suggested that resorts pass through a cycle such as that illustrated for Spain's Costa del Sol (Figure 20.12).

Most people with limited leisure time tend to seek recreational amenities near their homes. As the majority of British people live in towns and cities, many amenities are located either within or near to urban areas (Figure 20.13). People with more leisure time tend to travel further afield — to scenic rural areas, often with added amenities (coasts, mountains, National Parks), and also to historic towns (Figures 20.10 and 20.13).

Forests
nature trails, orienteering, forest rangers, holiday cabins

Rural areas near to urban areas

Suburbs

Inner city

CBD
hotels, restaurants, cinemas, theatres, museums, libraries, castle, cathedral, urban guide trail, video shops

bingo halls, guest houses, professional football ground, swimming pools, leisure centres, industrial museum, cafes, neighbourhood parks, video shops, cycle ways

district parks, swimming pools, large playing fields, gardening, video shops, cycle ways

golf courses, country parks, nature trails, historic houses and gardens, castles, theme parks, safari parks, farm holidays and activities, bridleways

Lakes
water sports, nature reserves, ferry operators, hotels, guest houses, time-share, caravan and camp sites, souvenir shops, restaurants, cafes, fishing

Mountains
walking, climbing, mountain guides, ski-instuctors, cafes, youth hostels, guest houses, camp sites, long-distance footpaths, National Park wardens, mountain rescue teams

Coasts
beaches, cliffs, holiday resorts, holiday camps, funfairs, entertainment, nature reserves, bird sanctuaries, cafes, restaurants, souvenir shops, hotels, guest houses, caravan and camp sites, long-distance footpaths

Case Study 20

Recreation in the western USA

Yellowstone was the world's first National Park and is probably the best known. It was set up by the US government in 1872 following reports sent by trappers and traders such as Jim Bridger and James Cody, and later confirmed by government expeditions which reported on the embarrassment of riches to be seen there — geysers, hot springs, canyons, lava flows, and wild life such as bears, elk and buffalo. Today there are comfortable lodges, hotels and tourist villages. Well-made roads allow visitors to see many of the park's attractions in two well designated 120 km loops, but it takes a minimum of 2 days. Even this does not do justice to the scenery and wilderness. You may get caught in lines of traffic as bison cross a road (Figure 20.14), or have to queue to park to see some of the geysers and hot springs. As for Old Faithful (Figure 20.16), the most famous of the geysers, there will probably be hundreds of people waiting to see it blow (this happens approximately every 85 minutes). People are kept away from the bears who have now retreated to the safety of the remote

north-east of the park.

The pressure on such honey-pots as Yellowstone or Yosemite and Grand Canyon is a serious concern to the

environmentalists who wish the wilderness to remain in its natural state.

Is this what National Parks were established for? How has this come about?

Figure 20.14

Yellowstone National Park, 1991: bison grazing on the verge; the vegetation is recovering from the 1988 brush fire

Figure 20.15

National Parks and Recreation Areas in the western USA

Arcs are 600 km radii from San Francisco, Los Angeles and Salt Lake City

N.P. - National Park

N.M. - National Monument

N.R.A. - National Recreation Area

Figure 20.16

Old Faithful

Figure 20.17

Arches National Park: the
Three Gossips

Figure 20.15 shows the location of the National Parks and National Recreation Areas in the south-western USA. The circles are drawn at a 600-km radius from San Francisco, Los Angeles and Salt Lake City. Over 30 m people live within 500 km of the major National Parks and Recreation Areas. Add the numbers of overseas tourists who visit the western USA each year and you can recognise that there is pressure on some of the most famous areas. Yellowstone, Yosemite and Grand Canyon (Figure 7.19) each have over 5 m visitors a year. Grand Canyon attracts so many rafting and canoeing enthusiasts that it could be 7 years before a private party is granted a permit for the incredible journey through the Canyon. The creation of the National Recreation Areas at Lake Mead and Lake Powell (Figures 3.66 and 20.16) have allowed over 3 m recorded overnight stays a year since 1985. These were remote wilderness areas until the construction of the great dams to control the flow of the River Colorado 30 years ago (Case Study3b). Now good, all-weather roads lead to these almost uninhabited areas.

The numbers visiting the five National Parks within Utah — Zion, Bryce Canyon, Arches (Figure 20.17) ,Capitol Reef and Canyonlands, with their diverse desert and river-eroded scenery, approached 5 m in 1994. Many of these were people using motor caravans, or people with cars relying on bed and breakfast accommodation or lodges close to the Parks. It is possible to visit all these areas within a 2-week stay in the south-west, but most visitors prefer to sample one or two and spend time hiking, mountain biking, or sightseeing in a smaller area.

The impact of improved communications by air and road has meant an increase in development on the edges of the Parks. Settlement within the Parks is not permitted unless linked to park management.

West Yellowstone, once a 'frontier town' and railhead for north-west Wyoming and the Northern Rockies, is now a fast-growing resort with many second homes (which have had the impact of raising local house prices), smart boutiques, and supermarkets serving the summer and winter tourist trade. Jackson,Wyoming, close to the magnificent Grand Tetons Park 100 km south of Yellowstone (Figure 20.18) is a busy tourist centre all year round as well as being the home of a number of politicians and film stars such as Harrison Ford. New developments of smart 'condos' fringe the town on the lower slopes of the hills which in winter become fashionable ski-slopes. Further south, on the roads from Los Angeles towards the River Colorado, are many small settlements which have grown up in response to increased leisure and recreation time and the desire for second homes. It is an 8-hour journey from Los Angeles to Lake Mead.

The rapid population increase since 1980 in the Sun Belt States adds further pressure on the Parks and Recreation Areas. Phoenix, Arizona, is the centre of a growing hi-tec industrial region. It is only 3 hours' drive to the South Rim of Grand Canyon. It is the focus for many 'retirees' from the north-east USA who, during the hot summer months, move north in their camper vans towards the cooler Northern Rockies and southern Canada — a modern form of transhumance?

The environmental movement wishes to reduce further development. This could affect government grazing permits (Case Study 12b) as well as forested land in Central Utah ,Wyoming and Montana. Perhaps there is a need for all future developments in the mountain and desert areas of the west to be controlled by one organisation. National Forests and National Parks are under separate control, with State authorities controlling designated wilderness regions.

Figure 20.18

Grand Tetons National Park:
Mt Moran

Native American groups on the south side of Grand Canyon have expressed concern at the encroachment on several of their traditional sites, for example Rainbow Arch (Figure 3.71).

Pressure on the most popular Parks has a number of effects. In Yellowstone, where there is more animal wildlife than in others, there have been cases where visitors have been attacked by buffalo and, when backpacking in the north-east of the Park, by bears. Elk are numerous but shy animals. They do not like to be disturbed when grazing — too many onlookers cause them to move into remote wilderness. Noise pollution from cars, vans, people and radios needs to be controlled. Hence the Parks do not allow camping at undesignated sites, unless permission has been granted. At one time, bears were a regular nuisance around camp sites looking for easy food from dustbins!

Although distances within the west are considerable, the large number of visitors to the Parks frequently means that the best viewing points are rapidly filled. Park authorities are quick to act when unauthorised tracks and parking areas are used. Alternative routes have to be provided during peak periods.

Recreation and tourism are all year round activities in the Sierra Nevadas, the Great Basin and the Rockies. Snowfall is usually high. In November 1994, for example, 1.5 m of snow had fallen in Salt Lake City, unusually early, with much more in the mountains. Ski resorts are open from Thanksgiving in November through to Easter time. Snow is light and powdery providing excellent downhill as well as cross-country ski conditions. The US national ski team has its headquarters in Park City in the Wasatch range close to Salt Lake City, which hopes soon to host the Winter Olympics. Snowmobiling and cross-country skiing are very popular throughout the region. It is possible to visit Grand Canyon in the winter, though North Rim is closed because of difficulties of access, except by four-wheel-drive vehicle or snowmobile.

One should not forget indoor recreation and the large numbers of visitors to Las Vegas which has become a wealthy city in the desert.

Glen Canyon National Recreation Area: Lake Powell

Glen Canyon (Figure 20.19) received almost 5 m visitors in the first 9 months of 1994.

There has been a strong emphasis on the importance of recreational use of land and water in the management of Lake Powell since its creation. Provision has been made for many different recreational opportunities — boating, water skiing, wind surfing, sailing, fishing, swimming and camping on the shoreline (Figure 20.20). There are wilderness trails, and back-country roads which can only be used by four-wheel-drive vehicles, but which give access to remote historical and archaeological sites. Houseboats are popular, providing both transport and accommodation.

Some of the main tourist attractions and facilities of the Glen Canyon National Recreation Area are shown in Figure 20.21. They require extremely careful management. There is greatest pressure on use between the end of May and the beginning of September. Parking lots and access roads together with adequate camp sites have to be provided, as well as fuel and marina facilities. Environmental damage is already evident on popular stretches of the shoreline, with litter from the camp sites far too common. 'Adopt-a-Canyon' has become a powerful slogan, encouraging visitors to boat out everything that they take in.

Figure 20.19

Glen Canyon National Recreation Area: Bullfrog Ferry, Lake Powell

Figure 20.20

A 7-hour drive from Salt Lake City to the south across increasingly desolate plateaus brings you to Bull Frog marina on the sparkling blue lake which winds between tall red sandstone cliffs. Here is a luxury hotel with magnificent views of the Lake, as well as a large marina for power boats as well as yachts.

Wind surfers and sailing boats give way to the ferry that takes passengers and cars to Halls Crossing and the camp and caravan sites. Stores provide for visitors needs, both on and off the water. A Park Ranger centre close by acts as a coastguard on the Lake which has a shoreline of 1800 miles and over 900 canyons. There is a smaller settlement at Hite close to an early crossing of the Colorado. The main visitor centre on the Lake is 150 km SW close to the great dam with Wahweap marina and the town of Page with its small airport. There is also a small marina and refuelling centre at Dangling Rope, halfway between the dam and Hall's Crossing.

(Morris Tours brochure, 1993, Salt Lake City)

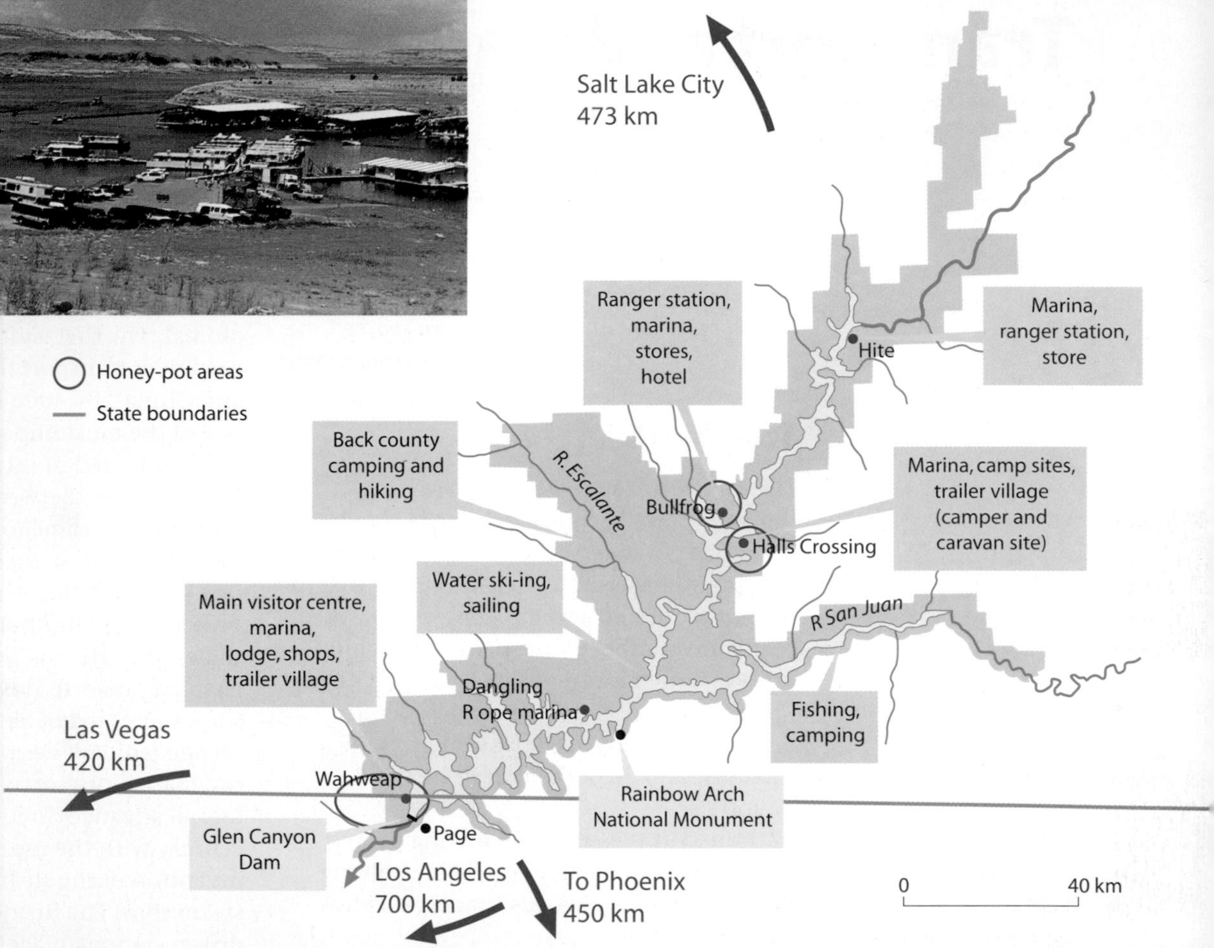

Figure 20.21

Glen Canyon National Recreation Area

Figure 20.22

Lake Powell: afternoon storm clouds gathering over Hall's Crossing

Salt Lake City
473 km

◯ Honey-pot areas

— State boundaries

Ranger station, marina, stores, hotel

Marina, ranger station, store

Hite

Back county camping and hiking

R. Escalante

Marina, camp sites, trailer village (camper and caravan site)

Bullfrog

Halls Crossing

Water ski-ing, sailing

R San Juan

Main visitor centre, marina, lodge, shops, trailer village

Dangling Rope marina

Fishing, camping

Las Vegas
420 km

Wahweap

Rainbow Arch National Monument

Glen Canyon Dam

Page

0   40 km

Los Angeles
700 km

To Phoenix
450 km

Water quality and lake use need careful monitoring. When the wind is strong and the wash of pleasure boats is reflected off the canyon walls, it is impossible to water ski and the conditions can lead to a very rough boat ride. Speed limits are enforced and there are now designated areas for some activities.

The lake level fluctuates with the amount of power and irrigation water required. This means that some surfaces are exposed after flooding. Thickets of tamarisk are invading these areas; they provide leaf litter, attract flies and produce hayfever-causing pollen. When the lake levels rise, this brings problems for water sports as the vegetation is partly submerged.

During the winter months, the main recreational use is by fishermen and there is little activity at the marinas. Day temperatures average 8°C from November to February, with plenty of sunshine. From May to September, temperatures rarely fall below 27°C and often exceed 35°C — so the plateau winds onto the water are very welcome. Local thunder storms over the lake can be a problem for water-sports enthusiasts (Figure 20.22).

This is a remote area from the viewpoint of British students, but to Americans it has become a national recreation centre in most senses of the phrase. People in the trailer parks come from all the different States, even Alaska!

References

Bradford, M. and Kent, A. (1993) *Understanding Human Geography*. Oxford University Press.

Prosser, R. (1992) *Human Systems and the Environment*. Thomas Nelson.

Waugh, D. (1994) *The Wider World*. Thomas Nelson.

Transport and trade

> *"I will build a motor car for the great multitude so low in price that no man will be unable to own one … and enjoy with his family the blessing of hours of pleasure in God's great open spaces."*
>
> **(attributed to Henry Ford, c.1900)**

Transport

The following references to the role of transport have been made at various stages in this text.

1. It may be viewed as an indicator of wealth and economic development, as measured by the number of cars per 1000 people. While the more developed countries have only 25 per cent of the world's population, they have 88 per cent of its rail traffic and 72 per cent of its cars and lorries.

2. Transport is essential in linking people, resources and activities and in enabling the exchange of goods (trade) and ideas (information).

3. Transport was considered to be a major factor in industrial location (Weber) and in determining agricultural land use (von Thünen). The relative decrease in transport costs to the average firm or farm has made this a less significant locational and land use factor since the 1950s (page 504).

4. In early economic/geographical theory, costs were thought to be proportional to distance (von Thünen's central market and Christaller's central place), especially on a flat plain where transport was equally easy and cheap in all directions. Later, costs were regarded as a function of a raw material's weight as well as the distance it had to be moved (Weber).

It is now accepted that, as transport costs comprise terminal costs plus haulage costs (Figure 21.1), cost declines with distance.

Characteristics of modern transport systems

Early tribespeople had to travel on foot or by animal. The first civilisations, after 6000 BC, used river transport (Egypt, Mesopotamia and China). By 4000 BC the wheel, arguably one of the most important inventions ever, was being used in ceremonial processions in Mesopotamia. Between then and the late 18th century came four significant technological advances: engineered roads (for military purposes, but also used for trade); the ocean-going sailing ship (for exploration and trade); the artificial pound lock (which enabled canals to traverse hills); and fixed track routes to enable heavier loads to be carried (colliery wagonways of Tyneside, originally horse-drawn). A greater technological advance came at the end of the 18th century with the use of steam power and the invention of the steam railway engine and steam ship. The first journey by a petrol-driven car took place in 1885 and the first powered flight in 1903.

A comparison of the characteristics of the major forms of presentday transport — canal, ocean shipping, rail, road, air and pipeline — is given in Figure 21.1. Each type has advantages and disadvantages over rival forms of transport. Figure 21.1 also refers to terminal and haulage costs. Terminal costs are fixed regardless of the length of time of journey and are highest for ocean transport and lowest for road transport. Haulage costs, which increase with distance but decrease with the amount of cargo handled, are lowest for water transport and highest for road and air (Figure 21.2). Figure 21.3 shows the amount of freight and passengers moved by the various forms of transport in the UK.

		Canals and rivers	Ocean transport and deep sea ports	Rail	Road	Air	Pipelines
Physical	**Weather**	Canals can freeze in winter. Drought/heavy rains make rivers unnavigable.	Storms, fog. Icebergs in North Atlantic.	Very cold (frozen points). Heavy snow (blocks line). Heavy rain can cause landslides.	Fog and ice both can cause accidents/pile-ups. Cross-winds for big lorries; snow blocks routes; sun can dazzle.	Fog, icing and snow: less since planes have had automatic pilots. Airports better if sheltered from wind and away from hills and areas of low cloud.	Not greatly affected.
	Relief	Width of channels. Need flat land or gentle gradients. Soft rock/soil for digging, problems with deltas. Rivers must be slow-flowing, have a constant discharge and have no rapids.	Harbours need to be deep, wide and sheltered. Tidal problems.	Cannot negotiate steep gradients so have to avoid hills. Estuaries can be obstacles. Flooding in valleys.	Avoids/takes detours around high land. Valleys may flood. May go around estuaries.	Large areas of flat land for runways, terminal buildings and warehousing. Firm foundations. Ideally, cheap farmland or land needing reclamation. Seas and mountains not a barrier.	Difficult to lay, then relief is not a problem.
Economic	**Speed/time**	Slowest form of transport. Long detours and possible delays at locks.	Slow form of transport, yet most economical.	Fast over medium-length distances.	Fast over short distances and on motorways. Urban delays.	Fastest over long distances, not over short ones due to delays getting to and passing through airports.	Very fast as continuous flow.
	Running or haulage costs (wages and fuel): increase with distance	Often family barge. Limited fuel use means the cheapest form of transport over lengthy journeys.	Expense (oil used as fuel) increases with distance.	Relatively cheap over medium-length journeys. Fuel and wages quite high.	Cheapest over shorter distances. Haulage costs rapidly increase with distance.	Very expensive, yet speed makes it competitive over very long distances.	Cheapest as no labour is involved (provided diameter is large).
	Terminal costs (loading and unloading costs and dues): no change with distance	Canals expensive to build and to maintain, unless natural waterways used.	Ports expensive. Harbour dues/taxes. Expensive to build specialised ships. Less since containers. Cheapest over long distances.	Building and maintenance of track/stations/signalling/rolling stock are very expensive.	Expensive building and maintenance costs, especially motorways. Car tax instead of dues, but roads built by community taxation therefore lower overheads.	Very expensive to build and maintain airports. High airport dues. Planes expensive to purchase and maintain.	Very expensive to build.
	Number of routes	Relatively few. Inflexible.	Relatively few ports, inflexible due to increased specialisation of ships. Links to hinterland.	Decreasing, as limited to linking main cities and power stations. Not very flexible. Recent increase in urban rail and new high-speed 'InterCity' routes.	Many and at different grades. Great flexibility, most in urban and industrial areas.	Often only a few: internal and international airports/routes. Not very flexible because of safety.	Limited to key routes. Inflexible and one-way flows.
	Goods and/or passengers carried	Heavy, bulky, non-perishable, low-value goods. Present-day tourists.	Heavy, bulky, non-perishable low-value goods. Cruise passengers.	InterCity passengers. Heavy, bulky (chemicals, coal) and rapid (mail) goods. Can carry several hundred passengers. Dependable and safe.	Many passengers. Perishable, smaller loads by lorry. Relatively few people carried by one bus or car.	Mainly passengers. Freight is light (mail), perishable (fruit) or high-value (watches).	Bulk liquid (oil, gas, slurry, liquid cement, water).
	Congestion	Very little.	Very little.	Little track congestion other than commuter trains.	Congestion in urban areas and at holiday times.	Usually little, other than at peak holiday times.	None.
	Convenience and comfort	Neither very convenient, unless for leisure/relaxation, nor very comfortable.	Not very convenient. Cruise liners very comfortable.	Not very, other than InterCity. Needs transport to and from station. Comfortable for passengers.	Door-to-door (except for some city centre destinations): most convenient and flexible. Safety is questionable; strain for drivers, but independent and private.	Country to country. Jet lag if more than three time zones crossed. Relaxing over shorter journeys; cramped, dehydrating and tiring over longer journeys.	Raw material or port to industry.
Environmental	**Environmental problems**	Some oil discharged, but relatively few problems.	Tankers discharging oil. Much land needed for ports, hard-standing and warehousing.	Noise and visual pollution limited to narrow belts. Noise decreases with welded rails, increases with high-speed trains. Electric trains cause less pollution.	A major cause of noise and air pollution. Effect on ozone layer, acid rain, and global warming (greenhouse effect). Uses up land, especially farmland. Structural damage caused by vibrations.	High noise levels. Some air pollution. Uses up much land for airports.	Few once 'buried'. Eyesore on surface.

Figure 21.1

Comparable characteristics of transport systems

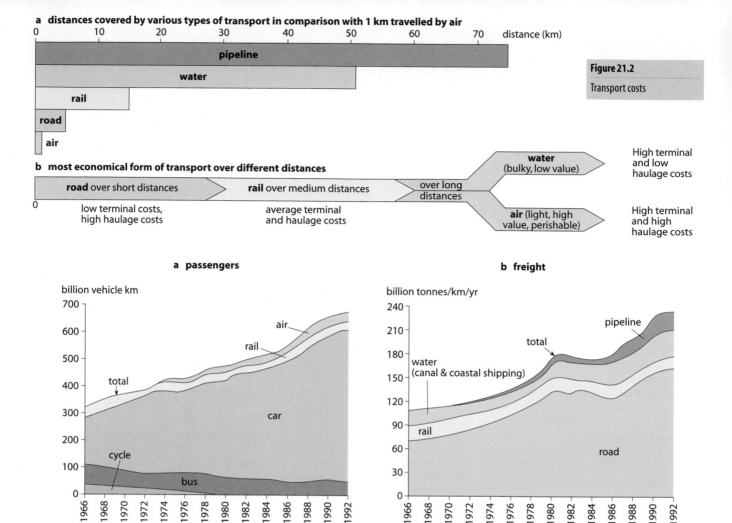

a distances covered by various types of transport in comparison with 1 km travelled by air

distance (km)

pipeline
water
rail
road
air

b most economical form of transport over different distances

road over short distances | **rail** over medium distances | over long distances

water (bulky, low value) — High terminal and low haulage costs

air (light, high value, perishable) — High terminal and high haulage costs

low terminal costs, high haulage costs | average terminal and haulage costs

a passengers

billion vehicle km

total
air
rail
car
cycle
bus

year

b freight

billion tonnes/km/yr

pipeline
total
water (canal & coastal shipping)
rail
road

year

Source: DTp

Figure 21.3

Passenger and freight traffic in the UK, 1966–92

Q

1 Assess the relative importance of physical, economic and environmental factors in the development of the following types of transport: **a** water; **b** railways; **c** road; **d** air.

2 Figure 21.4 shows the relationship between distance and costs of transporting goods.
  a Why do transport costs increase more slowly as distance increases?
  b Outline three factors, other than distance, which affect transport rates.
  c How might a government alter the structure of transport costs in order to encourage the carriage of goods by rail?

cost per unit volume

lorry
barge
train

distance (km)

Figure 21.4

Types of transport

Inland waterways

The major canal-building period in western Europe was in the late 18th and early 19th centuries. The virtual monopoly by the canals was shortlived due to competition, first from rail and later from road. However, since the 1970s the inland waterways have been rejuvenated. This began with the increase in road and rail charges caused by the rise in oil prices following the 1973 Middle East War and an appreciation that, despite its slowness and inflexibility of routes, water transport remains the cheapest form for bulky, non-perishable goods. Huge 4400-tonne barges, propelled by push-barges, have been introduced by the EU to carry the equivalent of 110 railtrucks or 220 x 20-t lorries. To accommodate these 'super-barges', the EU governments have financed the widening and deepening of canals and the linking of the Mediterranean, North and Black Seas. Smaller canals, including those in Venice and Amsterdam, are benefiting from increased tourism.

The Chao Phraya River provides a major waterway as it passes through Bangkok on its way to its delta in the Gulf of Thailand. Large flat-bottomed boats carry cargo from ocean ships which cannot negotiate the shallow water. These wooden boats, resembling a modern Noah's Ark, also provide permanent homes for large Thai families. Much smaller vessels, also acting as family homes, pull up to a dozen barges each (Figure 21.5). Between these commercial ships ply passenger ferries and numerous 'long boats' (similar to long rowing boats but powered by a car's engine) which carry the daily goods of the local people. Parts of the older city are a maze of smaller *klongs* (canals) along the banks of which the homes of many of Bangkok's inhabitants are built on stilts.

Located some 80 km west of Bangkok is one of Thailand's floating markets. Here women from all over the delta region meet daily to exchange their produce (Figure 21.6). They paddle their small boats along the network of *klongs* as these provide the only form of transport between the villages. Their income is, today, being increased by sales to tourists.

Figure 21.5

Barges on the Chao Phraya River, Bangkok, Thailand

Figure 21.6

The floating market, Dannoen Saduak, Thailand

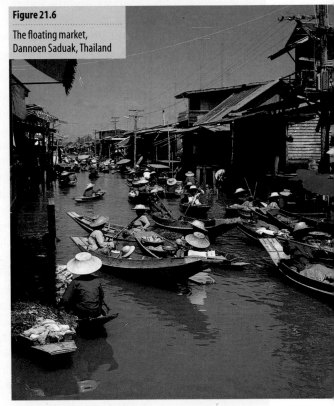

Ocean shipping

Many ports in western Europe and eastern North America developed either by trading with each other across the Atlantic or by importing raw materials from colonies and other developing countries and exporting manufactured goods in return. Recently, several major changes have occurred in British ports which have led to the growth of those in the south and east of England (with the exception of the port of London) and a corresponding decline elsewhere. The increase in ship size, especially purpose-built **bulk carriers** for oil and iron ore, has meant that wider and deeper estuaries are needed. The recession in world trade (page 561) has led to a general decrease in the number of ships and ports needed.

A ship berthed at a quayside is not earning money. Two innovations have enabled the turn-around time (the time taken to unload and load cargo) to be shortened. The first was the development of **roll-on/roll-off** (Ro-Ro) methods where lorries loaded with freight are driven on board so reducing the need for cranes. Rail Ro-Ro ferries have also been introduced (Dover–Dunkirk, Denmark to Sweden) in order to save time. The second innovation was the introduction of **containerisation** (unitised traffic) in which goods are packed into containers of a specified size at the factory, taken by train or lorry to the container port, and easily and quickly loaded by specialist equipment (Figure 21.7). Containers have to be of internationally specified standard sizes (3 m, 6 m or 12 m long; 2.5 m wide; and 2 m, 2.5 m or 3 m high). Some of the larger container ships (containers are usually stacked on deck) can hold upwards of 4000 units. Unfortunately, these technological advances have decimated the dock labour force and reduced the number of ships required.

Felixstowe was Britain's ninth-largest port in 1960, handling only 1 per cent of the country's seaborne trade. By 1992 it was the largest container port, the second-largest in terms of foreign trade by value and in Ro-Ro units, and the fourth-largest for foreign trade by tonnage. Reasons for this growth include: the presence of the deep water necessary for handling large vessels, regardless of the state of the tide; its coastal site, so that no time is lost sailing up an estuary; its position on the major trade routes with the EU and Scandinavia (its main links are with the container ports of Rotterdam and Zeebrugge); the availability of Ro-Ro facilities and five container terminals; its site on marshy, poor quality, cheap land, with space for future expansion; good road links with its hinterland; and good industrial relations (no restrictive industrial practices).

Places 73 The port of Singapore

On founding the modern port of Singapore in 1819, Sir Stamford Raffles decreed that it was open to all maritime nations. By 1992 over 600 shipping lines with links to over 800 ports had taken advantage of that decree and Singapore had become the world's busiest port in terms of shipping tonnage. At any given time, over 400 ships are likely to be in port with one arriving or weighing anchor every 10 minutes (81 000 vessels during 1992). It takes less than 1 second to handle 1 tonne of cargo, and cargo is handled during every second of the year. Warehouses are automated and computerised. Vessels vary from modern supertankers and container ships to the more traditional bumboats and junks. In 1992 the port handled over 7 million 6-m container units. To save docking time, harbour pilots now fly out to incoming vessels by helicopter.

Singapore is a **free port**, open to all countries, with seven free trade zones (six for seaborne cargo and one at Changi International Airport). In these zones, goods can be made or assembled without payment of import or export duties, and profits can be sent back to the parent company without being taxed. Many hi-tech firms assemble their products there and sell them at very competitive prices. (A visiting Japanese student was surprised to find his own country's brand names selling more cheaply in Singapore than in Japan.) Singapore is the world's largest manufacturer and exporter of disks. However, the port's largest money-earner is oil, a resource which the country does not possess. This is because the port imports crude oil from the Middle East, Indonesia and Malaysia, refines it within the free port (Figure 21.8) and then exports a range of oil products. It is the world's third-largest oil refining centre.

Figure 21.7

Containers at Singapore

Figure 21.8

The port of Singapore

Transport and trade

Rail transport

Railways became the dominant form of transport in Britain during the Industrial Revolution. Although they are still the most convenient method of moving bulky goods and large numbers of passengers (Figure 21.1), the inflexibility of their routes and terminals has left them at a considerable disadvantage relative to road transport as cities have grown and industry has changed its location. Railways were constructed when cities were much smaller and industry was sited near to the mainline station in what is now the inner city. Modern industry now tends to seek an out-of-town or small town location — sites which usually do not possess a rail terminal. The east-coast, main line route between London and Edinburgh has now been electrified, the entire route controlled from just seven modern signalling centres, and a new high-speed electric locomotive introduced. The Channel (Fixed Link) Tunnel project (Places 74, page 551), funded in Britain mainly by the private sector, will eventually allow passengers to travel between London and Paris in 2 hours 30 minutes. By then the British Rail network, if present government policies go ahead, will have been privatised and deregulated.

Rail transport has recently regained some trade in both North America and western Europe, party because of congestion and other road problems and partly because of successful modernisation attempts. France (the TGV), Germany (Transrapid), Italy and Spain have all built new high-speed routes, but the track and rollingstock are not always compatible. The introduction of freightliners has enabled containers of internationally agreed standard size to be carried by rail to specialist ports which saves time and reduces labour costs and the risk of theft and damage.

Road transport

The major advantage of road transport is its flexibility which allows it to operate from door-to-door, especially over short distances, and at the most competitive price (Figure 21.2). Yet while the building of motorways and urban roads seems to be essential if an area wishes to improve its accessibility and stimulate economic activity, the results of such construction often increase the amount of traffic and accentuate environmental problems (Figure 21.1). Older urban areas, built in a pre-automobile era, experience the modern problems of congestion, such as unplanned, narrow streets (unless redeveloped), car parking difficulties, delays caused by buses and delivery lorries, air pollution from exhausts, health problems (asthma), noise and visual pollution, roadworks (the repairing of old utility services), and old buildings being shaken by passing traffic. Even where redevelopment has occurred there is still heavy pollution; land is taken up by road widening; land values next to urban motorways are reduced; and the risk of accidents seems as great.

At present 22 million households in Britain have regular use of a car. Almost two families in three have one car and one in four have two cars (Figure 21.9). The number of cars on Britain's roads in 1990 is predicted to increase by over 50 per cent by the year 2025. The 19 million cars (22 million vehicles in total) on the roads is likely to increase by 20 per cent to 23 million by the turn of the century if the car continues to be a symbol of economic success for most British people. It is this growth in car ownership which has enabled retail outlets to relocate in out-of-town, regional shopping centres (page 534) and industry to relocate in industrial, science and business parks (pages 516–7), and which has given individual families greater freedom of movement and choice in where they live, shop, work and spend their leisure time.

Since 1975 the length of Britain's motorways has increased by just under 100 km per year and principal roads by 150 km per year. Unfortunately the creation of outer ring roads to divert traffic from city centres, and motorways to link large urban areas, have generated their own traffic (Figure 21.10). It is, for example, often quicker and easier for shoppers living on the edge of an urban area to travel by motorway to a regional shopping centre (Places 71, page 534) than it is to make the shorter journey to their local, possibly crowded, city centre. Several British cities have already implemented **rapid transit systems** by which the car is replaced by public transport. These, at present, involve the use of underground railways (the Tyne and Wear Metro, 1980), light railways (the London Docklands Light Railway, 1987; Places 41, page 402), and modern 'supertrams' (Greater Manchester Metrolink, 1992; South Yorkshire Supertram, 1995). Other schemes, including those

Figure 21.10

Traffic congestion on the M25, London

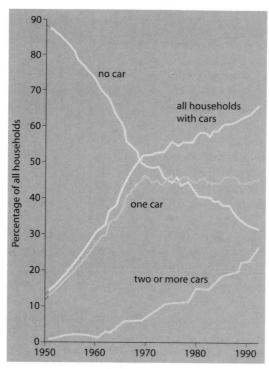

Figure 21.9

The increase in car ownership, 1950-93

submitted by Bristol, Birmingham, Croydon, Leeds and Nottingham, have already received government approval. Such schemes are only likely to survive if they make a profit, regardless of their relative success in reducing traffic problems in the urban area they serve.

Air transport

Air transport has the highest terminal charges, high haulage costs (fuel) and affects large numbers of people living near to airports. Its advantages (Figure 21.1) include speed over long distances for passengers, tourists and business people, and for freight, if it is of high value, light in weight and perishable. Air transport is also important to countries of considerable size (Brazil); where ground terrain is difficult (Sahara desert); when crossing short stretches of sea (London to Dublin, Belfast, Amsterdam and, before the Channel tunnel, Paris); where relief aid is necessary following a major human (Rwanda) or natural (earthquake) disaster or international incident (Kuwait).

Recent deregulation of airlines (USA since 1978, the EU in 1993) has led, due to increased competition, to the availability of more routes and lower fares. Although this has caused a decrease in the number of airlines, it has encouraged more people to travel by air, apart from at times of recession.

Scheduled international and domestic traffic handled by British carriers has increased steadily over the years causing, during peak holiday periods, congestion at airports and competition for air-space. By early next century, technological changes are likely to result in a fleet of larger planes capable of carrying twice the number of passengers of the 'jumbo' jet. These new 1000-seater jets are likely to be limited to key long-haul routes and between airports with long runways and a shortage of take-off slots.

Transport and the environment

As the demand for greater mobility by an ever-increasing number of people, both in Britain and across the world, continues to grow, so too does the need for improved traffic systems and more efficient forms of transport. However, improvements which appear to allow people this greater mobility invariably seem to have adverse effects upon the environment. The unanswerable question is: 'How can we accommodate the ever-growing movement of people and goods without causing serious and irreversible damage to the environment?' Places 74 considers three contemporary issues involving rail, road and air in south-east England.

1　The Channel Tunnel Rail Link

In January 1994, 7 years after the 110-km, high-speed Channel rail link between Kent and London was first mooted, the government finally confirmed the chosen route (Figure 21.11). Well, almost the chosen route, for, although Britain was already years behind its European counterparts in building the link, decisions were still deferred as to whether the route would go through, or around, Ashford; whether Ebbsfleet or possibly Stratford or Rainham would be an intermediate station; and how much the government would contribute towards the scheme (rumour suggested no more than one-third of the total cost). It is unlikely that work will begin before 1997 or will be completed by 2002. The main reasons for the delay in the past have been economic and environmental. British Rail's initial favoured option, the cheapest, cut through some of the most attractive scenery and residential areas of Kent, as well as through several marginal parliamentary constituencies. Further delay is likely due to the expected difficulties in raising private investment (the estimated cost of £1.7bn in 1989 had risen to almost £3bn by 1994).

Figure 21.11

The Channel Tunnel Rail Link (as of 11 January 1995) and its potential environmental impact

Maidstone	Ashford	Paris	Brussels
27 mins from London (55 mins for commuters at present)	35 mins from London (70 mins for commuters at present)	After tunnel opened 3 hrs 0 mins. After tunnel link (2002) 2 hrs 27 mins	2 hrs 40 mins 2 hrs 07 mins

Little benefit to people/businesses north of London; cargo wagons too large for continental track.

15-km tunnel through north London: reduces noise; no buildings to be demolished.

Possible stations

Tunnel under Thames to Thurrock Marshes.

New track needed to carry containers 3 m in height at 200 km/hour: 90 decibels; noise pollution.

Viaduct over Medway crossing: 225 km/hr, 90 decibels.

Small tunnel under North Downs: 200 km/hr.

Commuters from Kent increased by 25 per cent between 1984 and 1994: need to relieve extra congestion resulting from this increase.

Rail route by-passes Kent villages; should reduce traffic on roads and cause less air pollution.

Area of Outstanding Natural Beauty (AONB).

110 km (68 miles) from St Pancras

Six international platforms; 3 domestic platforms (not completed before 2002).

New route will be financed mainly by private sector; money will have to come from commuters and international travellers.

Likely site for intermediate station.

New track 225 km/hr: 90 decibels.

Small tunnels at Hollingbourne and Sandway.

Freight terminal: increase in jobs; proposed new enlarged international station.

Uncertainty over final route.

Village and good quality farmland lost to construction of Channel terminal.

proposed Channel Tunnel rail link	——— surface
	┄┄┄┄ tunnels

Q　Figure 21.11 shows the confirmed Channel Tunnel Rail Link route (1995). Using the information provided, or researched by yourselves, answer the following questions.

1　To what extent do you think that the rail link is of economic advantage: **a** to Britain as a whole; **b** to south-east England?

2　How successfully does the confirmed route minimise environmental problems?

3　Which do you consider to be the more important: an efficient rail link to benefit business people and tourists, or the preservation of the environment? Give reasons for your answer.

2 The M25 orbital road

The M25 was built in the 1980s to reduce the volume of traffic in central London, to relieve the overcrowded North and South Circular (outer) ring roads, to link motorways converging on the capital and to link London's four main airports, especially those at Gatwick and Heathrow (Figure 21.12). It was built with three lanes in anticipation of its being used by 80 000 vehicles/day by the turn of the century. To try to reduce objections from environmentalists, landowners and residents, the route meanders, was in parts landscaped and, between the M11 and A12, had bridges and underpasses to allow the movement of deer.

The motorway, on its busiest sections, already carries over 200 000 vehicles daily, confirming the theory that new roads attract more traffic and so increase, rather than reduce, congestion (the 'record' jam on the M25 is a 35-km, 3-lane standstill). The M25 accounts for 14 per cent of all Britain's motorway traffic and is especially congested in the section between the access roads to Gatwick and Heathrow airports. A fourth lane has already been constructed between the A3 and the M4 on land previously set aside for this purpose as planners felt that if such a proposal had been made initially it would have been defeated on environmental grounds. Further plans, released in 1994, proposed further widening of the M25, in places up to dual 7 lanes. Such a scheme could destroy 10 Sites of Special Scientific Interest (SSSIs), 47 ancient woodlands and 81 wildlife sites, as well as increasing air pollution and traffic noise. However, a change in government thinking in early 1995 suggests that these schemes are unlikely to be implemented. One earlier bottleneck, the Dartford Tunnel under the River Thames, has been relieved by the building of the Queen Elizabeth II Bridge. This bridge was financed by the private sector which will recover its investment through the imposition of tolls, after which the bridge will revert to public ownership.

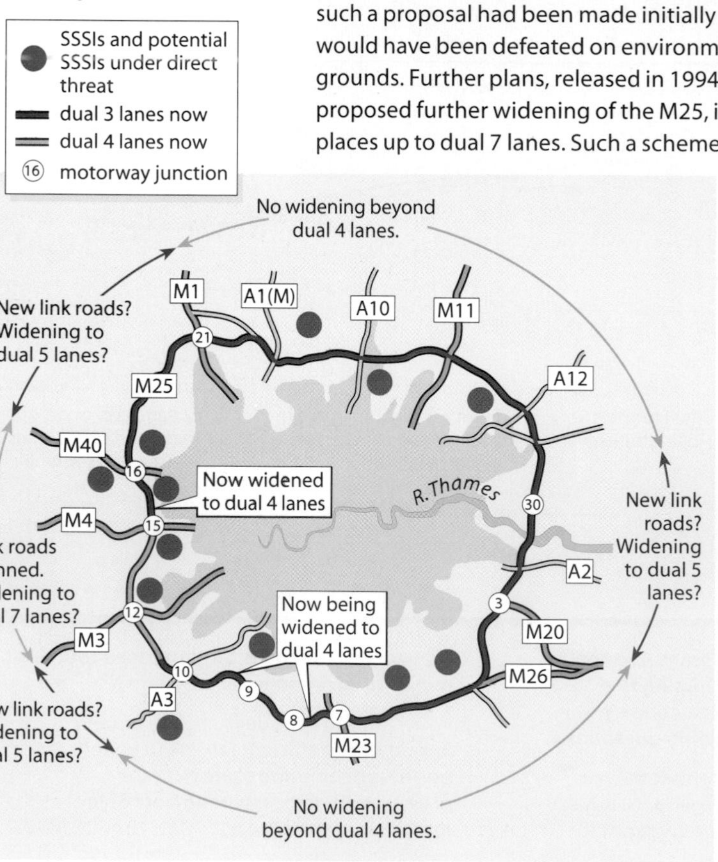

3 London's airports

Delays for passengers at airports and the safety factor, involving both the increasing use of airspace over London and the limited number of available runways, are major concerns. Despite the opening of the City Airport (Isle of Dogs) and increased flights from Stansted, both Heathrow and Gatwick will be at full capacity before the year 2000 and, although environmentalists would disagree, a new runway will have to be built 'somewhere'.

Although an all-party committee at Westminster recommended (March 1989) that a second runway should be built at Gatwick, the latest proposal (1994) by the British Airways Authority (BAA) is to build a fifth terminal at Heathrow on land at present occupied by a sewage and sludge farm (Figure 21.13). The BAA claims that the new terminal is essential if

Figure 21.12

Reshaping the M25

- ● SSSIs and potential SSSIs under direct threat
- ━ dual 3 lanes now
- ═ dual 4 lanes now
- ⑯ motorway junction

No widening beyond dual 4 lanes.

New link roads? Widening to dual 5 lanes?

Now widened to dual 4 lanes

Link roads planned. Widening to dual 7 lanes?

New link roads? Widening to dual 5 lanes?

Now being widened to dual 4 lanes

No widening beyond dual 4 lanes.

New link roads? Widening to dual 5 lanes?

M1 · A1(M) · A10 · M11 · M25 · M40 · M4 · M3 · A3 · M23 · A12 · R. Thames · A2 · M20 · M26

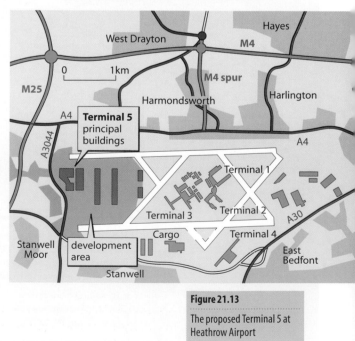

Figure 21.13

The proposed Terminal 5 at Heathrow Airport

London's airports are to handle the predicted 165 million passengers in 2016 (compared with 69 million in 1993). Terminal 5, if built, would handle 30 million passengers a year, have two stations (for the Underground and the new Heathrow Express) and a new network of roads. Before that, BAA is preparing for the costliest, longest and most bitter planning battle of the decade. Already 3800 people have submitted objections in advance of a public enquiry which is due to open in May 1995 and which is expected to last until at least December 1996. Opponents list, among their many objections, the increase of flights from 400 000 to 450 000 a year over an area which already has planes landing or taking off for 75 per cent of the time, and the increase in traffic on access roads (the dual 7-lane M25).

The Royal Commission on Transport published its report in October 1994. It set a range of targets and made recommendations as to how these might be achieved (Figure 21.14).

Figure 21.14

The Royal Commission's Report on Transport, October 1994

Targets

- Increase the proportion of journeys by public transport (measured in total passenger km) from 12 per cent in 1993 to 20 per cent in 2002 and 30 per cent in 2020.
- Raise the proportion of freight carried by rail from 6.5 per cent in 1993 to 10 per cent by 2000, and 20 per cent by 2010.
- Increase the fuel efficiency of new cars sold in Britain by 40 per cent by 2005.
- Cut the proportion of journeys in London made by cars from 50 per cent in 1993 to 35 per cent by 2020; corresponding reductions elsewhere.
- Raise urban bicycle use fourfold by 2000, from 2.5 per cent of all journeys to 10 per cent.
- Achieve full compliance by 2002 with World Health Organisation air-quality guidelines for pollution produced mainly by vehicles. Freeze emissions of carbon dioxide, the principal global-warming pollutant, from road and rail at their 1990 level by 2000, and reduce them by 20 per cent by 2020.
- Reduce road and rail noise.

Recommendations

- Increase fuel duty annually in order to double the real price of fuel by 2005, and base the annual car vehicle excise duty (road tax) on a car's efficiency.
- Reduce planned spending on motorways and other trunk roads to half its present level.
- Introduce tougher emissions checks during the annual MOT test, which should be compulsory when the vehicle is 1 year old.
- Reduce road tax for vehicles already meeting the more stringent pollution standards.
- Stop sales of unleaded super-premium petrol because of its high content of cancer-causing benzene.
- Introduce stronger protection from road building for cherished landscapes and wildlife sites.
- Enforce speed limits more strictly and, by 1996, require 90 km/hr (56 mph) speed limits on lorries.
- Give bigger subsidies to the railways and more government support to modern tram schemes.
- Modify a rail link from the Channel tunnel to Scotland to enable freight trains to carry standard cargo containers.
- Consider giving incentives to bus companies switching to less-polluting natural gas and give government support to the development of electric vehicles.

Transport routes and networks

A single route location

The shortest distance between two points on the earth's surface is, due to the world being a sphere, a **great circle route**. This is why planes flying from London to airports adjacent to the Pacific Ocean (Los Angeles, Tokyo) fly over polar regions, and was one reason why the Titanic struck an iceberg well to the north of a straight-line route from Liverpool to New York. However, although over short distances the shortest route is approximately a straight line, in practice, direct routes have rarely been constructed, even by the Romans. Haggett (1969) suggested that deviations from a straight-line route were for either positive or negative reasons.

Figure 21.15

Single route location: a positive deviation

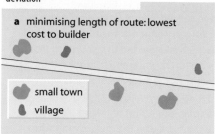

a minimising length of route: lowest cost to builder

small town

village

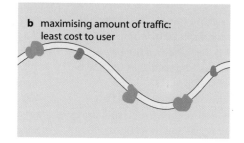

b maximising amount of traffic: least cost to user

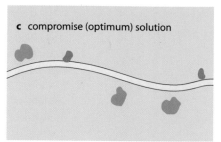

c compromise (optimum) solution

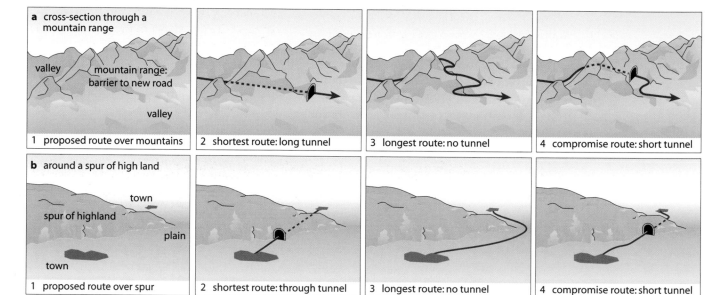

a cross-section through a mountain range

valley

mountain range: barrier to new road

valley

1 proposed route over mountains

2 shortest route: long tunnel

3 longest route: no tunnel

4 compromise route: short tunnel

b around a spur of high land

spur of highland

town

plain

town

1 proposed route over spur

2 shortest route: through tunnel

3 longest route: no tunnel

4 compromise route: short tunnel

Figure 21.16

Single route location: a negative deviation

Positive deviations are when routes are diverted to gain more traffic. Figure 21.15a shows that although (assuming there are no physical obstacles) the straight-line route is cheapest to build, having the shortest distance, it is the most expensive to the user. In comparison, the least-cost route for the user (Figure 21.15b) is likely to be the longest and most expensive to build. The selected route will probably be a compromise between these two extremes (Figure 21.15c), provided that it gains acceptance from a third party — the environmental lobby.

Negative deviations result from routes being diverted to avoid physical obstacles such as a mountain range (Figure 21.16a) or a spur (Figure 21.16b).

Builder and user costs

The least-cost network to the **builder** is likely to be the one with the shortest overall length (Figure 21.17a). These are more likely to be found in areas with a sparse population and low traffic density (Australia); in highland areas where routes are funnelled along valley floors (Lake District); or where, as in the case of motorways, the construction cost per unit length is extremely high.

The least-cost network to the **user** is the one where the traveller can move as quickly and directly as possible between any point in the network (Figure 21.17b). These are likely to occur in conurbations where high population and traffic densities demand numerous alternative routes.

Figure 21.17

Builder and user costs

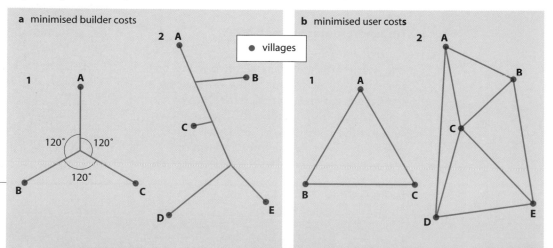

a minimised builder costs

b minimised user costs

villages

A network is defined as a group of places joined together by a number of routes to form a system.

Transport networks

Network patterns vary greatly between and within countries, so it is necessary to adopt a universally acceptable method of description to allow comparison between different systems. Network analysis is achieved by showing the network as a series of straight lines. Drawing a **topological map** preserves the pattern of routes and junctions but distorts directions and distances (like the London Underground map). The resultant map (Figure 21.18a) consists of vertices and edges. **Vertices** or **nodes** are points in the network. **Edges** or **links** form the direct lines of transport between these points.

Vertices (Figure 21.18a) may be a point of origin or destination (Swansea), a significant place *en route* (Cardiff), or a junction of two

or more routes (Bristol). This division into vertices and edges helps to describe the **accessibility** of a place and the **efficiency** of the network. Figure 21.18b gives the accessibility matrix for points on the topological map. The aim is to follow a route between two places, passing through as few vertices as possible, since they can cause congestion and delay. The **Shimbel index** (which is 14 for Bath) is the total number of edges needed to connect any one place with all the other vertices. The place(s) with the lowest Shimbel index (Newport and Bristol with 11) has/have the highest **centrality** and is/are the most accessible. If the Shimbel index for all the vertices is plotted it is possible to draw an isopleth map to show the relative accessibility of places (isopleths for 12, 15, 18 and 21 could be added to Figure 21.18a).

Figure 21.18

A topological map and accessibility matrix

a topological map

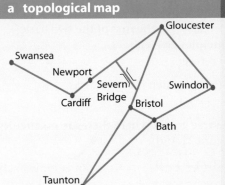

b accessibility matrix

	Bath	Bristol	Cardiff	Gloucester	Newport	Swansea	Swindon	Taunton	TOTAL
Bath	—	1	3	2	2	4	1	1	14
Bristol	1	—	2	1	1	3	2	1	11
Cardiff	3	2	—	2	1	1	3	3	15
Gloucester	2	1	2	—	1	3	1	2	12
Newport	2	1	1	1	—	2	2	2	11
Swansea	4	3	1	3	2	—	4	4	21
Swindon	1	2	3	1	2	4	—	2	15
Taunton	1	1	3	2	2	4	2	—	15

Shimbel index

Figure 21.19

Types of network

Types of network

Networks are shown by means of **planar graphs** —i.e. where all vertices and edges are in the same plane, so that while two edges

may meet at a vertex they cannot cross over each other. Figure 21.19 shows six different methods of linking, or not linking, four vertices.

a In a **null graph**, none of the vertices is linked: they may divided by mountain barriers or political frontiers.

b A **connected graph** is where every vertex is linked directly to the adjacent vertex in the network, but only indirectly to other vertices (a cross-city bus route, or the Central Line in the London Underground).

c A **circuit**, or **closed path**, follows a 'circular' route linking up all the vertices (a milkround beginning and ending at a dairy, or the Circle Line in the London Underground). It is the shortest possible distance linking all points.

d In a **tree graph**, all the vertices are linked, but there are no circuits. Each edge, leading to a terminal point, acts as a branch line (motorways radiating from London).

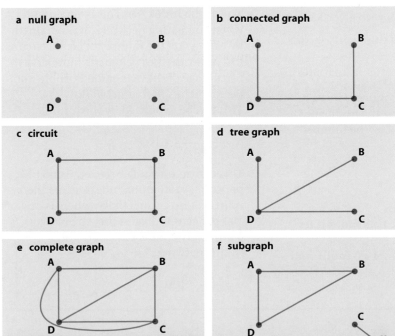

(edges cannot overlap in a planar graph)

555

e A **complete graph** is where each vertex is directly linked to every other vertex. It is more likely to be found in a developed country and equates with Christaller's $k = 3$ model, page 378.

f A **subgraph** is where part of the network is detached from the rest. This incomplete system may be associated with the pre-take-off stage of the Rostow model (Figure 22.9).

The connectivity of networks

Connectivity is a means of **measuring** the **efficiency** of a network (the more vertices which are connected with each other, the more efficient the network) and **comparing**, quantitatively, different networks. Before looking at four methods of measuring connectivity, two basic principles should be noted.

■ The **minimum** number of edges, e (minimum), needed to link all the vertices in a network is 1 less than the total number of vertices (v).
Therefore:

$$e_{(min)} = v - 1$$

e.g. in the tree and connected graphs with 4 vertices, $e = 3$.

■ The **maximum** number of edges, e (maximum), which can exist in a network without a duplication of routes (linkages) is found by subtracting 2 from the number of vertices and then multiplying by 3.
Therefore:

$$e_{(max)} = 3(v - 2)$$

e.g. in a complete graph with 4 vertices,

$$e = 3(4 - 2) = 6.$$

Measuring connectivity

■ **The beta index** (β) This is the easiest method of measuring connectivity as it simply involves dividing the number of edges (e) in the network by the number of vertices (v).

Therefore: $\beta = \dfrac{e}{v}$

Three situations can result:

1 In simple networks, the beta index is less than 1.0,
e.g. in Figure 21.19a, $0 \div 4 = 0$ (null), and in Figures 21.19b and d, $3 \div 4 = 0.75$ (tree and connected).

2 In a network with one circuit, the beta index is 1.0.
e.g. in Figure 21.19c, $4 \div 4 = 1.0$ (circuit).

3 In a complete network, the beta index is greater than 1.0,
e.g. in Figure 21.19e, $6 \div 4 = 1.5$ (complete).

As can be seen, the higher the value of the beta index, the greater the degree of connectivity and the more efficient the system.

■ **The cyclomatic number** (c) This refers to the number of circuits in a given network. The more edges there are in a network, the greater the number of circuits is likely to be. The formula is derived from the fact that a tree graph plus one edge gives a circuit. Therefore it can be assumed that the number of circuits = the number of edges (e) minus the number of edges needed to form a tree ($v - 1$) — i.e. the first of the two basic principles given above.
Therefore:

$e - (v - 1)$, which is the same as
$c = e - v + 1$.

In a tree graph, where there are no circuits (Figure 21.19d):

$c = 3 - 4 + 1 = 0.$

In a complete graph, where there are three circuits (Figure 21.19e):

$c = 6 - 4 + 1 = 3.$

A disadvantage of this method is that networks with very different forms may have the same cyclomatic number.

■ **The alpha index** (α) This is very useful when comparing networks. It is obtained by expressing the number of actual circuits in a network (the cyclomatic number) as a percentage of the maximum possible number of circuits in that network.
This can be expressed by the formula:

$$\alpha = \frac{(e - v + 1)}{2v - 5} \times 100.$$

This can give a value between 0 and 100 per cent. A low alpha index means there is little or no connectivity, whereas one of 100 per cent indicates that the network is completely connected.
In this network:

$$\alpha = \frac{2}{8 - 5} \times 100.$$

Therefore:

$\alpha = 66\%$.

■ **The gamma index** (γ) This index, considered less useful than the alpha index, compares the actual number of edges in a network with the maximum possible number of edges in that network. This is expressed as:

$$e_{(max)} = \frac{e}{3(v-2)} \times 100.$$

Using the same two diagrams as for the cyclomatic number, the maximum links in the circuit will be:

$$e_{(max)} = 3(4-2) = 6.$$

However, in the tree graph, where there are only three edges:

$$e_{(max)} = \frac{3}{3(4-2)} \times 100 = 50\%.$$

The gamma index ranges from a minimum value which varies, but is always above zero, and 100 per cent. This proves a weakness when comparing different transport networks or the same network over a period of time. It is of more value when comparing the efficiency of networks of different sizes.

■ The **detour index** (DI) When using topological maps, the edges which link the vertices are drawn as straight lines. Yet it has been shown that in reality communications rarely follow the most direct route. The efficiency of a route can be measured by determining how far it deviates from the straight-line route. This is calculated by using the formula:

$$DI = \frac{\text{Shortest possible actual route distance}}{\text{Direct (straight-line) distance}}$$

The lowest possible index is 1.00 (some texts suggest this is given as a percentage). A detour index of 2.00 means that the actual route is twice as long as the most direct route.

Desire line maps In transport studies, a **desire line** is a straight line joining two places between which people, or goods, travel. On a desire-line map, one line is drawn for each separate journey or movement. The number and closeness of desire lines indicate the frequency and intensity of journeys. We have already seen, for example, that when shopping for convenience goods journeys are usually short in distance and frequent in occurrence, whereas journeys made for comparison goods are much longer and more infrequent.

Figure 21.20

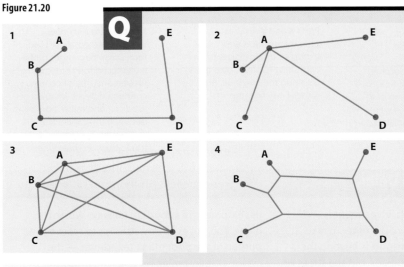

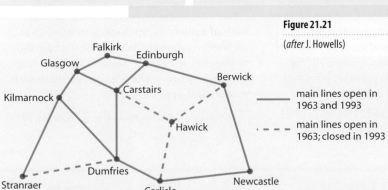

Figure 21.21

(*after* J. Howells)

main lines open in 1963 and 1993

main lines open in 1963; closed in 1993

1 Figure 21.20 shows four transport networks with five vertices or nodes. Describe briefly the main advantages and disadvantages to the user and the builder of each of these networks.

2 Figure 21.21 is a topological map showing part of a mainline rail network.
   a Using the beta index, calculate the connectivity of the network in 1963 and 1993.
   b Explain what the difference in the indices between the two dates means.
   c Which vertices were the most accessible in 1963 and which in 1993?
   d For the rail network in 1993, work out the alpha index and the cyclomatic number.

3 'Transport links often make considerable detours from the straight-line route between two places.'

   With reference to specific examples, discuss this statement in relation to each of the following: water transport; motorways; air routes; railways. (You may wish to consider physical, climatic, economic, human, political and conservation factors.)

A model showing the evolution of a transport network in a developing country

In 1963 Taaffe, Morrill and Gould put forward a sequential model to show how a transport network might evolve in a developing part of the world. Their model, based on studies in Nigeria and Ghana, has been summarised in Figure 21.22. The early stages reflected the needs of a controlling colonial power to:

1 link politically and militarily the centres of administration, which were on the coast, with inland centres;
2 exploit mineral resources;
3 develop plantations and increase agricultural production; and
4 carry freight in preference to passengers.

The model has since been applied to other developing countries, especially in east Africa, South America, Africa and Malaysia. In many cases, only Stage 2 has been reached, mainly because following independence these countries had insufficient money to improve and modernise their networks beyond the main urban areas, except perhaps, to a prestigious airport. To reach Stages 3 and 4 (Brazil and Malaysia), it appears that a country needs the development of a manufacturing industry and the emergence of an affluent sector in the population.

Figure 21.22

Model showing the evolution of a transport network in a developing part of the world (*after* Taaffe, Morrill and Gould)

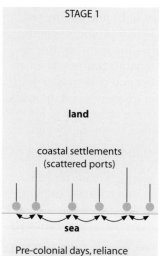

STAGE 1

Pre-colonial days, reliance upon local sea-borne trade with tracks leading inland; port development limited by shallow seas, heavy surf, few natural inlets and deltas at river mouths.

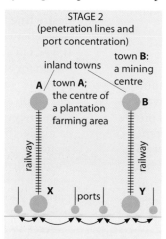

STAGE 2 (penetration lines and port concentration)

Early colonial development included routes inland, usually rail, to newly developed towns associated with the collection of a cash crop or the exploitation of a mineral: both of which were exported to the colonial power.

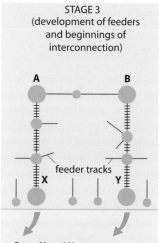

STAGE 3 (development of feeders and beginnings of interconnection)

Ports X and Y grow as primary products are exported; intermediate towns with short feeder routes grow up along the major inland routes (very few imports).

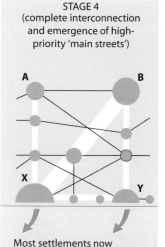

STAGE 4 (complete interconnection and emergence of high-priority 'main streets')

Most settlements now interconnected; there is an increase in passenger traffic as opposed to earlier goods movements; major centres connected by all-weather roads, air and rail; major link is between the capital **B** and major port, **X**; some feeder tracks may be abandoned.

Places 75   Traffic in Delhi

"In spite of all the controls, investment is already slanted in favour of the élite. To take one example, private transport is beyond the wildest dreams of most Indians, but the streets of Delhi are nevertheless clogged up with Japanese-designed cars and scooters which can compete in the international market. For the less affluent there are only decrepit, outdated and fuel-inefficient buses quite incapable of providing an efficient service even if the roads were cleared for them. There has been no development of suburban railways worth the name, and not even any attempt to relieve the burden of the poor man's taxi-driver — the cycle rickshaw puller. He does not enjoy the fruits of modern aerodynamics, metallurgy or engineering. His vehicle hasn't changed in the twenty-five years I have known Delhi — it's still inordinately heavy, and doesn't even have such modern aids as gears."

(Mark Tulley, *No Full Stops in India*, 1991)

Interaction (gravity) models

As described on page 379, these models try to predict the degree of interaction, or movement, between pairs of places. They are based on two premises. First, as the sizes of both towns increase, so too will the amount of movement between them. Secondly, the further apart the two towns are, the less will be the movement between them (distance decay). The model predicts that movement will be greatest where two large cities are close together.

In transport terms, the gravity model may be used to estimate two features of a system:

■ The **boundary** between two towns beyond which people find it easier (less distance, less time or less cost) to travel to one centre rather than the other (Reilly's breaking point, Figure 14.33).

Figure 21.23

Interaction (gravity) model applied to an area of eastern Scotland

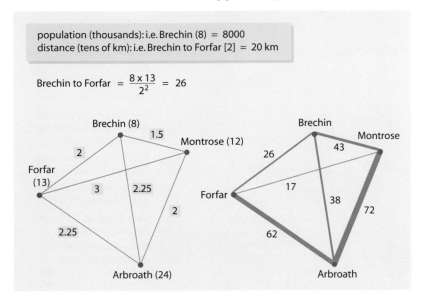

population (thousands): i.e. Brechin (8) = 8000
distance (tens of km): i.e. Brechin to Forfar [2] = 20 km

Brechin to Forfar $= \dfrac{8 \times 13}{2^2} = 26$

■ The **volume** of traffic (passengers, freight or, as initially suggested by Christaller, telephone calls) between various towns. Geographers are interested not only in the direction of flow of people and goods, but also in predicting the volume of movement. The latter can be estimated by multiplying the population of two towns and dividing the product by the square of the distance between them.

This can be expressed by the formula:

$$\text{movement A to B} = \frac{p\text{A} \times p\text{B}}{d^2}$$

where: pA is the population of Town A;
pB is the population of town B; and
d is the distance between towns A and B.

A worked example is given in Figure 21.23. It shows that on the basis of the gravity model, more buses or trains (public transport) and better roads (private transport) are needed, for example, to cope with the volume of movement between Montrose and Arbroath than between Montrose and Forfar.

International trade

The development of international trade

Trading results from the uneven distribution of raw materials over the earth's surface. Today, trade plays a major role in the economy of most countries as none has large enough supplies of the full range of minerals, fuels and foods to make it self-sufficient.

While early peoples bartered in order to exchange goods, the first civilisations found that making transactions with money proved a method by which wealth could be accumulated. By 1500 BC, the Canaanites had learnt to become 'middlemen' in the trade between Egypt, Asia Minor and Mycenae (Greece), while their descendants, the Phoenicians, traded five centuries later around and beyond the Mediterranean Sea. Until this time, trading was mainly the simple exchange of raw materials:

| Area A grew wheat | ⇄ | Area B reared animals |

Subsequently, certain communities proved to be more inventive than others and began to process some of these raw materials:

| Country A grew cotton and mined iron ore | ⇄ | Country B manufactured textiles and smelted iron |

Such exchanges led to the emergence of European colonial powers who used the raw materials from their colonies to establish their own domestic industries. This saw the beginnings of modern international trade and the widening gap between those countries producing the raw materials, with the exception of oil, and those who made a much larger profit by manufacturing goods (Figure 21.26). Trade developed further as new and improved forms of transport were introduced, such as transcontinental railways, air traffic and refrigerated ships.

Since the 1960s there has been, especially among the industrialised nations of Europe and North America, a tendency to specialise in particular aspects of manufacturing. This creates greater benefits (the **law of comparative advantage**) than trying to compete with other countries who have equal, or better, opportunities. This can be seen in the modern car industry:

Country A:		Country B:		Country C:
brakes,	→	tyres,	→	clutches, oil
distributors	←	carburettors	←	pumps and
and glass		and		pistons
		headlights		

The law of comparative advantage operated in 19th-century Britain. Britain could have produced more of its own food but, having a greater comparative advantage over other countries in manufacturing, opted to concentrate on industry where greater profits could be made.

In recent decades, there has also been a grouping together of trading partners to form common markets such as the EU, EFTA, OPEC, LAIA, ASEAN, NAFTA and, until 1990, COMECON. These trade associations enable goods to pass more easily and cheaply within an enlarged 'domestic' or local market. While Figure 21.24 gives the definitions and locations of the major trade groupings, it should be remembered that these associations are in a constant state of change as individual nations seek the grouping which will best serve their own interests.

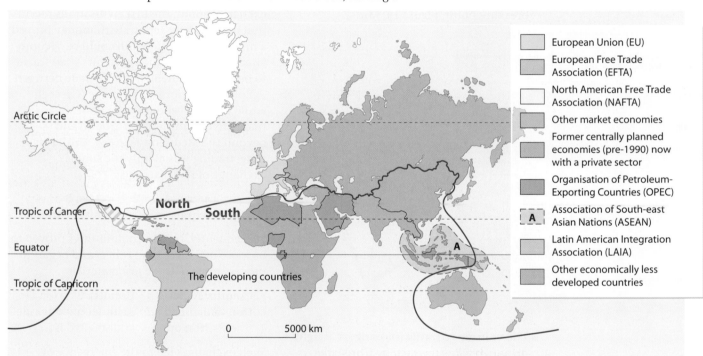

Figure 21.24

Major world trading groups, 1994

Directions of international trade

Figure 21.25 gives information concerning the directions of world trade. The data provided, the latest available, were for the period just prior to the break-up of the former USSR and the changes in eastern Europe. The following points should be noted:

- Nearly half of the world's trade was between the advanced market economies (the EU, the USA and Japan).
- The advanced market economies themselves were involved in 85 per cent of total world trade.
- There was relatively little trade between the centrally planned economies and the rest of the world.
- The advanced market economies received more imports (oil) from the **OPEC** countries than they exported to them.
- The advanced market economies had relatively little trade with the economically less developed countries although they exported more to them in value, although less in weight, than they imported.
- There was relatively little trade between the economically developing countries, many of whom have low rates of economic growth and limited economic potential.

Figure 21.25

Simplified pattern of world
trade in the mid 1980s

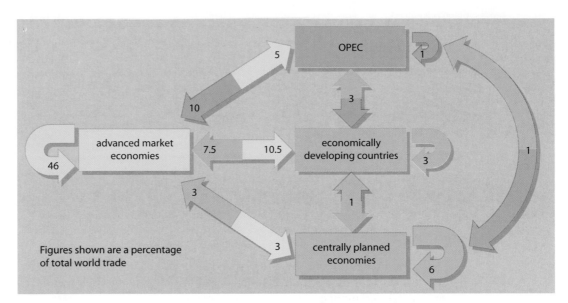

Figures shown are a percentage
of total world trade

Changes in world trade during the early 1990s

- The volume of trade has slumped due to a worldwide recession.
- The economically most developed countries (the USA, Japan and within Europe) have faced increasingly strong competition from the newly industrialising countries (NICs) of south-east Asia. So far, this competition has come mainly from the 'little tigers' of South Korea, Singapore, Hong Kong and Taiwan (page 528) but it is also predicted to come increasingly from Malaysia (Case Study 19, page 528), Thailand and China. Of the two major world trade systems, that of the **Pacific Rim** (the USA and eastern Asia) now exceeds that of the North Atlantic (the USA and Europe).
- Manufactured goods account for just over half of world trade by value although there has been a shift towards products which require higher levels of technology.
- Despite the growing number of trading groups, world trade is still dominated by the EU (44 per cent in 1992), North America (13 per cent) and Japan (9 per cent).
- Although there was an increase in trading between the economically developing countries, eight of those (e.g. Brazil and Mexico) accounted for nearly two-thirds of the 'south–south' total.
- The OPEC share showed a slight decrease as, in a world economic recession, the demand for oil falls.
- The trade deficit for many developing countries increased further as they continued to export unprocessed raw materials, often in smaller amounts. Many raw materials are vulnerable to changes in market price and world demand.
- Many developing countries have maintained historic links with their former colonial power. Many also have a high **specialisation index**. This index gives the percentage of total exports, by value, accounted for by the most valuable category (Figure 21.26) and illustrates the frequent reliance upon only one or two commodities, usually a mineral or a foodstuff, as the source of income (Gambia 88 per cent copper, Mauritius 90 per cent sugar cane).
- The trade deficit of several industrialised countries increased partly due to a fall in world demand and partly from increased competition.
- Primary products, exported from developing countries, became increasingly more vulnerable to changes in market prices and world demand. Often any comparative advantage held by countries exporting raw materials was lost as few possessed their own means of transport.
- World trade continued to become increasingly dominated by large multinational (transnational) companies.

Trade links between economically developed, economically developing and OPEC countries

Figure 21.26 gives an indication of the nature and value of exports of three contrasting economic groupings and shows the six customary categories into which the many items

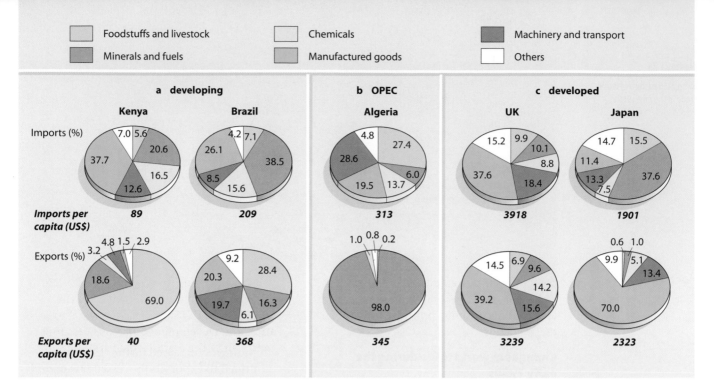

Figure 21.26

Imports and exports of selected countries, 1990

of world trade are manageably placed. The developed countries of the EU, the USA and Japan often have manufactured goods, machinery and transport equipment accounting for over half their exports. This has enabled these countries to accumulate the capital and technology needed to buy and to process requisite raw materials (fuels and minerals). In contrast, while most developing countries have some manufacturing, it is often primary processing or is operated by multinationals taking advantage of cheap labour rates (pages 523–525).

The world market in fuels (70 per cent of which is oil) is dominated by the OPEC countries. Most oil is exported to the energy-short advanced market economies in Europe (excluding the UK and Norway) and to Japan. As developing countries rarely have the capital to afford these fuels, their economic development tends to be further retarded. The pattern of mineral exports is less obvious with both developed (Australia and Canada) and developing (Jamaica, Zambia) countries being major exporters. It is again, however, the advanced market economies which are the chief importers.

Agricultural products often account for over three-quarters of a developing country's exports, yet an increasing number of African states are having to import cereals as their food production decreases (page 461). Developed countries have to import tropical goods but, on balance, export almost as much agricultural produce as they import. The USA, Canada and Australia, with their

extensive agricultural systems, are net exporters accounting for over 75 per cent of the world's wheat (page 447), while others, like the Netherlands and Denmark (page 448), use their farmland intensively to obtain high yields. Most OPEC countries have to rely on imports as their natural environments rarely favour agriculture.

Developing countries have, for many years, made demands for a fairer trading system. One request is for higher and fixed prices for primary products as this might prevent a further widening of their trade gap. A second priority is better access to markets within developed countries. At present, if there is a world recession or if developing countries begin to increase their trade too much with industrialised countries, the latter often impose quotas to limit the number of goods imported or add tariffs so that the prices of imported goods increase and they become less competitive. While the developed countries thus protect their own jobs, they cause increased depression within developing countries. The knock-on effect can be that the developing countries, with a reduced income, will have less money to spend on goods manufactured in industrialised countries and the resultant trade recession will adversely affect both parties. Other demands have included changes in the international monetary system to eliminate fluctuations in currency exchange rates; to encourage industrialised countries to share their technology; to stop developed countries 'dumping' their unwanted and

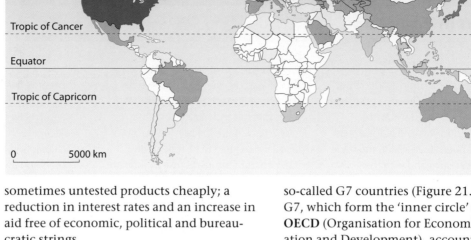

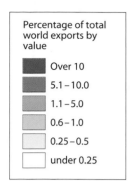

Percentage of total world exports by value

■	Over 10
■	5.1 – 10.0
■	1.1 – 5.0
■	0.6 – 1.0
■	0.25 – 0.5
□	under 0.25

0   5000 km

Figure 21.27

Share of the world's trade, 1989

Balance of payments includes the balance of trade together with any invisible earnings, i.e. income from banking and insurance, tourism, remittances from migrant workers abroad, professional advice and air/sea transport.

sometimes untested products cheaply; a reduction in interest rates and an increase in aid free of economic, political and bureaucratic strings.

GATT (the General Agreement on Trade and Tariffs) was set up in 1948 with the aim of reducing tariffs (import duties) and to provide a forum for discussing problems of international trade. The latest round of GATT negotiations, which began in Uruguay in 1986, had the ambitious target of trying to replace the various protectionist blocs and associations (Figure 21.24) with one large global free trade area. Although some 150 nations attended the meetings, in reality the representatives of most of the developing countries were often little more than spectators. Much of the arguing was between, and the main decisions were made by, the

so-called G7 countries (Figure 21.28). The G7, which form the 'inner circle' of the **OECD** (Organisation for Economic Cooperation and Development), account for more than half of the world's trade and the full OECD for over three-quarters (Figure 21.27). The two major areas of dispute were between the USA and the EU over farm subsidies and between Japan and the remaining six over tariffs (the Japanese, with a huge trade surplus, discriminated against imports from other countries yet were themselves free to export cars and electrical goods to the rest of the world). Eventually, with the global economy in an unhealthy state and it being seemingly in everyone's best self-interest to accept the GATT proposals, agreement was finally reached in December 1993. It remains to be seen whether this agreement will speed up the end of the world recession, create more jobs and increase the volume of world trade.

Figure 21.28 shows the **balance of trade** for the top seven trading countries: the G7. The balance of trade is the difference between the income received from **visible exports** and the cost incurred in paying for **visible imports**. In most recent years, apart from the early 1980s when production and therefore income from North Sea oil was at its peak, Britain has had a balance of trade deficit. However, a balance of trade surplus does not always mean prosperity. Several African countries do have a small surplus, but this has to be used to repay their international debts.

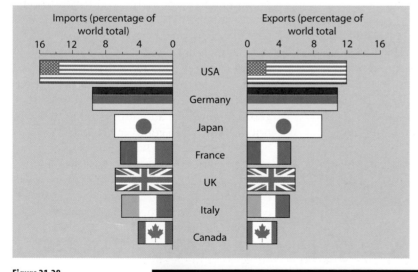

Imports (percentage of world total)

16   12   8   4   0

Exports (percentage of world total)

0   4   8   12   16

USA
Germany
Japan
France
UK
Italy
Canada

Figure 21.28

The G7: the world's top seven trading nations, 1990

Q

1 Describe the nature of the exports of economically less developed countries.

2 Suggest why economically more developed countries dominate the trade in manufactured goods.

3 What are the advantages and disadvantages of being a member of an international trading bloc (association)?

4 Describe **a** recent changes in world trade; and **b** the 1993 GATT agreement.

Case Study 21

Transport in Hong Kong

Hong Kong originally grew as a result of its strategic trade route location and its large, deep, sheltered harbour. Growth continued partly as a result of increased industrialisation. Hong Kong was one of south-east Asia's four 'little tigers' (page 528), and trade with China in particular and with the remainder of the Pacific Rim in general has rapidly expanded (page 561).

Early transport was mainly restricted to water, due to the limited amount of flat land. As building on the steep hillsides proved difficult and hazardous (Case Study 2, page 47), especially on Hong Kong Island, land has had to be repeatedly reclaimed from the sea for use by industry, housing and transport. Three forms of transport used at the beginning of this century are still in operation today (Figure 21.29). The Star Ferry transfers large numbers of people daily from Hong Kong Island to Kowloon on the mainland; trams still link northern parts of Hong Kong Island (although land reclamation means their routes are no longer adjacent to the sea); and the Peak Tram funicular railway carries wealthy commuters and tourists to and from Victoria Peak (Figure 21.30). A fourth form of transport, the Kowloon Railway linking Hong Kong with the New Territories and the Chinese city of Canton (present-day Guangzhou), was opened in 1910.

Figure 21.29

Hong Kong's Star Ferry, funicular railway and tram

Figure 21.30

The development of transport in Hong Kong

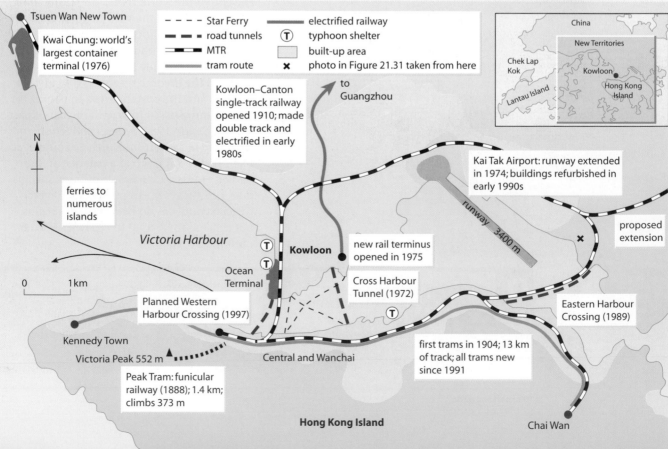

Present-day transport

Hong Kong is the world's busiest container port handling, in 1993, over 9 million units. Most containers are loaded and discharged at Kwai Chung (Figure 21.30) where, on land reclaimed from the sea, there are eight terminals. The port is also one of the busiest in terms of vessel arrivals, and cargo and passenger throughput. The port remains crucial to Hong Kong's economy and prosperity, handling 90 per cent of the Territory's trade and, together with port-related industries, generates 15 per cent of its GDP.

Kai Tak Airport (Figures 21.30, 21.31 and 21.32) is one of Asia's busiest and one of the world's potentially most dangerous (page 566). Handling 16 million passengers in 1990 and 24.5 million in 1993, it is rapidly reaching its capacity.

Despite 90 per cent of journeys within Hong Kong being made by public transport, the Territory's roads have one of the highest vehicle densities in the world (271 vehicles/km of road). This high density, combined with the difficult terrain and dense building development, poses a constant problem to transport planning, construction and maintenance. The problem is accentuated by the number of people moving daily between Hong Kong Island and the mainland (Kowloon). A third road tunnel should be opened by 1997 (page 567) but, as elsewhere, improved roads and, in this case, extra tunnels often only generate more traffic (page 552). Over 2 million passengers daily use the Mass Transit Railway, the MTR (Figure 21.30), making it one of the busiest underground railways in the world. The MTR is, at present, a three-line metro system comprising 43 km and 38 stations. It is being extended to the new airport at Chek Lap Kok (Figure 21.34).

Q Figure 21.31 was taken looking almost due west from above point **X** on Figure 21.30. The runway at Kai Tak is in the foreground while Hong Kong Island lies to the left; Kowloon and the mainland lie to the right; and Lantau Island is in the far distance. Notice the limited amount of flat land.

Identify on the photo, with the help of Figure 21.30, the position of:
- **a** Victoria Harbour;
- **b** the Peak Tram funicular railway;
- **c** Star Ferry route;
- **d** Kowloon railway terminal;
- **e** one typhoon shelter;
- **f** Cross Harbour road tunnel;
- **g** the MTR;
- **h** Central and Wanchai districts.

Figure 21.31

Aerial view of Hong Kong

Hong Kong's Airport Core Programme

Why does Hong Kong need a new airport?

A new airport is urgently needed because the international airport at Kai Tak, which has only one runway, is rapidly approaching its full capacity of around 28 million passengers a year (24.5 million in 1993) and cannot be viably enlarged. In terms of international traffic, it is the world's fourth-busiest airport for passengers and third-busiest for freight. The resultant overcrowding affects movement in the terminal buildings, in airspace over Hong Kong and on roads leading to the airport.

Kai Tak is also a potentially dangerous airport. The approach route is over one of the world's most densely populated cities, with the descent path perilously close to underlying high-rise flats (Figure 21.32). The weather often consists of low cloud, mist and rain with a season of typhoons (tropical storms). Although the runway is, by international standards, long, it does end up in the sea. These problems were confirmed when, in November 1993, a Boeing 747 from Taipei (Taiwan) overran the runway during a rainstorm associated with the passing of a typhoon. The plane ended up in Victoria Harbour but, fortunately, all 296 passengers and crew were rescued.

Finally, with Kai Tak unable to expand further, a new airport is essential if Hong Kong is to maintain its position as a key trading and financial centre on the Pacific Rim.

The Airport Core Programme

The decision to site the new airport at Chek Lap Kok, off the coast of Lantau Island,

Figure 21.32

Landing at Kai Tak Airport, Hong Kong

was finally made in 1989 as part of a comprehensive Port and Airport Development Strategy (PADS). The strategy contained major road, rail and port developments in addition to the new airport. In 1991 PADS was distilled into the Airport Core Programme (ACP) which was to consist of 10 independent projects (Figures 21.33 and 21.34).

Figure 21.33

The ten-point Airport Core Programme, Hong Kong

New Airport
1 New Airport: at Chek Lap Kok, off Lantau Island

New Transport Links
2 Lantau Fixed Crossing: including a 1377-m suspension bridge
3 Airport Railway: 34 km local and express services
4 Western Harbour Crossing: the third cross-harbour tunnel
5 North Lantau Expressway: 12 km along the Lantau coast
6 Route 3: first section of a major highway network
7 West Kowloon Expressway: a highway on reclaimed land

New Land
8 West Kowloon Reclamation: 330 hectares of new urban land
9 Central and Wanchai Reclamation: first section of 20 hectares

New Town
10 Tung Chung: first phase of a new town serving the airport.

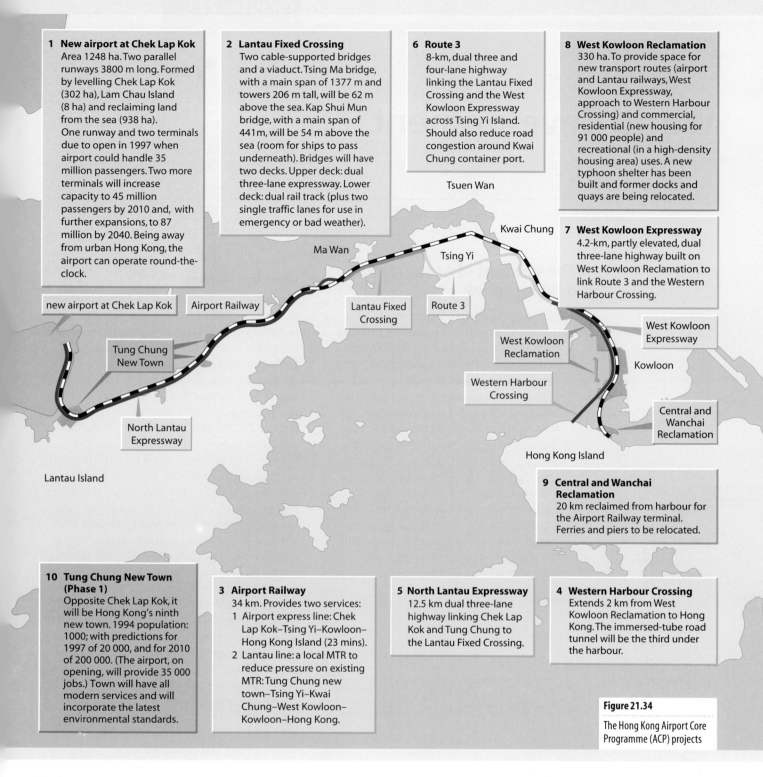

1 New airport at Chek Lap Kok
Area 1248 ha. Two parallel runways 3800 m long. Formed by levelling Chek Lap Kok (302 ha), Lam Chau Island (8 ha) and reclaiming land from the sea (938 ha). One runway and two terminals due to open in 1997 when airport could handle 35 million passengers. Two more terminals will increase capacity to 45 million passengers by 2010 and, with further expansions, to 87 million by 2040. Being away from urban Hong Kong, the airport can operate round-the-clock.

2 Lantau Fixed Crossing
Two cable-supported bridges and a viaduct. Tsing Ma bridge, with a main span of 1377 m and towers 206 m tall, will be 62 m above the sea. Kap Shui Mun bridge, with a main span of 441m, will be 54 m above the sea (room for ships to pass underneath). Bridges will have two decks. Upper deck: dual three-lane expressway. Lower deck: dual rail track (plus two single traffic lanes for use in emergency or bad weather).

6 Route 3
8-km, dual three and four-lane highway linking the Lantau Fixed Crossing and the West Kowloon Expressway across Tsing Yi Island. Should also reduce road congestion around Kwai Chung container port.

8 West Kowloon Reclamation
330 ha. To provide space for new transport routes (airport and Lantau railways, West Kowloon Expressway, approach to Western Harbour Crossing) and commercial, residential (new housing for 91 000 people) and recreational (in a high-density housing area) uses. A new typhoon shelter has been built and former docks and quays are being relocated.

7 West Kowloon Expressway
4.2-km, partly elevated, dual three-lane highway built on West Kowloon Reclamation to link Route 3 and the Western Harbour Crossing.

9 Central and Wanchai Reclamation
20 km reclaimed from harbour for the Airport Railway terminal. Ferries and piers to be relocated.

10 Tung Chung New Town (Phase 1)
Opposite Chek Lap Kok, it will be Hong Kong's ninth new town. 1994 population: 1000; with predictions for 1997 of 20 000, and for 2010 of 200 000. (The airport, on opening, will provide 35 000 jobs.) Town will have all modern services and will incorporate the latest environmental standards.

3 Airport Railway
34 km. Provides two services:
1 Airport express line: Chek Lap Kok–Tsing Yi–Kowloon–Hong Kong Island (23 mins).
2 Lantau line: a local MTR to reduce pressure on existing MTR: Tung Chung new town–Tsing Yi–Kwai Chung–West Kowloon–Kowloon–Hong Kong.

5 North Lantau Expressway
12.5 km dual three-lane highway linking Chek Lap Kok and Tung Chung to the Lantau Fixed Crossing.

4 Western Harbour Crossing
Extends 2 km from West Kowloon Reclamation to Hong Kong. The immersed-tube road tunnel will be the third under the harbour.

Figure 21.34

The Hong Kong Airport Core Programme (ACP) projects

Map labels: Tsuen Wan · Kwai Chung · Tsing Yi · Ma Wan · new airport at Chek Lap Kok · Airport Railway · Lantau Fixed Crossing · Route 3 · West Kowloon Reclamation · West Kowloon Expressway · Kowloon · Tung Chung New Town · Western Harbour Crossing · Central and Wanchai Reclamation · North Lantau Expressway · Hong Kong Island · Lantau Island

References

Barke, M. (1986) *Transport and Trade*. Oliver & Boyd.

Barke, M. and O'Hare, G. (1984) *The Third World*. Oliver & Boyd.

Bradford, M. G. and Kent, W. A. (1977) *Human Geography*. Oxford University Press.

Briggs, K. (1982) *Human Geography*. Hodder & Stoughton.

Hong Kong, Ministry of Information (1994) *Hong Kong 1994*. Hong Kong.

Hoyle, B. S. and Knowles, R. D. (Eds) (1992) *Modern Transport Geography*. Belhaven Press.

McBride, P. (1980) *Human Geography*. Nelson-Blackie.

Philip's Geographical Digest 1992 (1993) Heinemann—Philip.

Singapore, Ministry of Information and the Arts (1993) *Singapore 1993*. Singapore.

Tolley, R. S. (Ed.) (1990) *The Greening of Urban Transport*. Belhaven Press.

Tulley, M. (1991) *No Full Stops in India*. Penguin Books.

Waugh, D. (1994) *The Wider World*. Thomas Nelson.

Chapter 22

World development

"Development is more than mere economics."
(Mark Tully, *No Full Stops in India*, 1991)

The concept of economic development

Frequent references have been made in earlier chapters to the inequalities in world development and prosperity. Gilbert, in his book *An Unequal World*, began by stating that "Few can deny that the world's wealth is highly concentrated. The populations of North America and Western Europe eat well, consume most of the world's fuel, drive most of the cars, live in generally well serviced homes and usually survive their full three score years and ten. By contrast, many people in Africa, Asia and Latin America are less fortunate. In most parts of these continents a majority of the population lack balanced diets, reliable drinking water, decent services and adequate incomes. Many cannot read or write, many are sick and malnourished, and too many children die before the age of five."

Definition of terms

Terms such as 'developed' and 'developing' have been used for several decades to indicate the economic conditions of a group of people or a country. By the 1980s, the term 'developing' had become regarded as a stigma and was replaced by the concept of the 'South' (*Brandt Report*, 1980) and, with increasing popularity, the 'Third World' (Figure 22.1). Later, with the growing realisation and appreciation that poverty is relative, not absolute, the terms 'more developed countries' (MDCs) and 'less developed countries' (LDCs) were introduced. However, since the first edition of this book was published in 1990, world events have seen the collapse of the Communist bloc (the so-called, or assumed, 'Second World'), leaving the term 'Third World' out of date. At the same time, those nations once grouped together as belonging to the 'Third World' have themselves shown a widening spread of wealth and living standards — for example, the growing gap between the NICs (Newly Industrialised Countries) and many of the countries in sub-Saharan Africa.

All of these definitions (summarised in Figure 22.1) were based upon, and overemphasised, economic growth. To those living in a Western, industrialised society, economic development tends to be synonymous with wealth, i.e. a country's material standard of living. This is measured as the **gross domestic product (GDP)** per capita and is obtained by dividing the monetary value of all the goods and services produced in a country by its total population. When trade figures for 'invisibles' (mostly financial services and deals) are included, the term **gross national product (GNP)** is used. At

Figure 22.1

Terms used in relation to world development

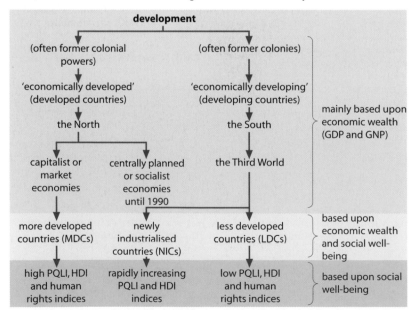

N.B. For consistency, the terms 'economically more developed' or 'developed', and 'economically less developed' or 'developing' (replacing 'Third World' of the first edition), are mainly used in this book.

present, it is estimated that 80 per cent of the world's wealth is owned by the 20 per cent of its inhabitants living in developed countries, leaving only 20 per cent for the 80 per cent living in the less developed world.

Recently, an increasing number of definitions, often involving cultural development, social well-being and political rights, have been suggested as alternatives to those previously based solely upon economic criteria — i.e. they emphasise 'quality of life' in contrast to 'standard of living'. During the 1980s, the Overseas Development Council (ODC) introduced the **Physical Quality of Life Index (PQLI)** and in the early 1990s the UN first referred to the **Human Development Index (HDI)**.

Development is not just the difference between the developed, rich and powerful countries and those which are less developed, poor and subordinate. Each country will have areas of prosperity and poverty; will contain people with different standards of living based on variations in race or tribe (Rwanda), religion (Northern Ireland), language (Belgium) or caste (India); and have inequalities between men and women.

Criteria for measuring development

1 Economic wealth

To many people living in developed countries, economic development has been associated with a growth in wealth based on either GDP figures (used by the EU) or GNP (used by the UN and the USA). This implies that the GDP (or GNP) of a country has to increase if its standard of living and quality of life are to improve.

Although GDP figures are easier to measure and to obtain than other development indicators such as social well-being, there are limitations to their use and validity. They are more accurate in countries which have many economic transactions and where goods, services and labour can be measured as they pass through a market place — hence the term 'market economies'. Where markets are less well developed, and trading is done informally or through bartering, and where much production takes place in the home for personal subsistence, GDP figures are less reliable. In the former centrally planned, socialist economies, with their relatively small role in international trade and with few services, GDP figures were difficult to calculate and interpret.

Comparison of GDP requires the use of a single currency, generally US dollars, but currency exchange rates fluctuate. The size and growth of GDP may prove to be poor long-term economic indicators and fail to take into consideration human and natural resources. GDP per capita is a crude average and hides extremes and uneven distribution of income between regions and across socio-economic groups, especially in less developed countries where there may be very few extremely wealthy people and a large majority living at subsistence level. Yet despite these limitations, GDP is still regarded as a relatively good indicator of development and a good measure for comparing differences between countries (Figure 22.2).

Figure 22.2

GNP per capita, 1990

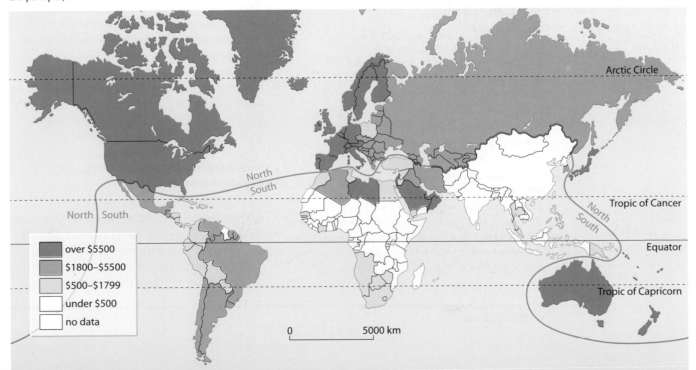

▇	over $5500
▇	$1800–$5500
░	$500–$1799
☐	under $500
☐	no data

0   5000 km

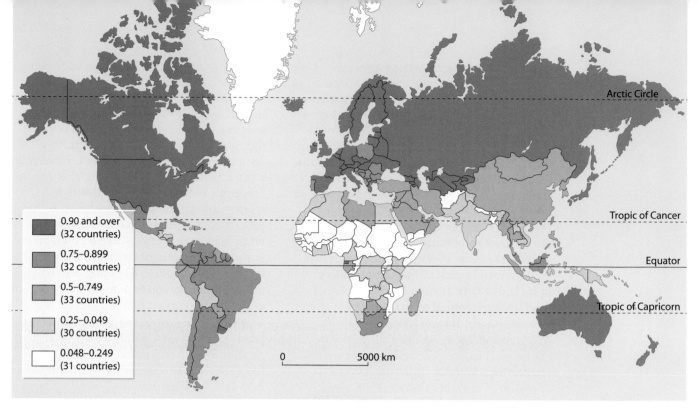

Legend:
- 0.90 and over (32 countries)
- 0.75–0.899 (32 countries)
- 0.5–0.749 (33 countries)
- 0.25–0.049 (30 countries)
- 0.048–0.249 (31 countries)

Arctic Circle
Tropic of Cancer
Equator
Tropic of Capricorn

0   5000 km

Figure 22.3

The UN Human Development Index (1993)

2 Social, cultural and welfare criteria

The Physical Quality of Life Index (PQLI) was the average of three characteristics: literacy, life expectancy and infant mortality. It was a forerunner of the UN Development Programme's Human Development Index (HDI) which gives every country a score between 0 and 1, based on its citizens' longevity, education and income. The three factors are given equal weight. Longevity is measured by average life expectancy at birth — the most straightforward measure of health and safety. Education is derived from the adult literacy rate and the average number of years of schooling. Income is based on GDP per capita converted to 'purchasing power parity dollars' and is adjusted according to the law of diminishing returns (page 420) — i.e. what an actual income will buy in a country.

The 1993 HDI puts Japan top of the list with a score of 0.983 which makes it twice as developed as Honduras, with a score of 0.492, and 22 times more developed than Guinea (West Africa), which is bottom of the list with a score of 0.045. Countries with a score of over 0.9 correspond closely with the developed countries of the 'North' (Figures 22.2 and 22.3), while those with a score of under 0.25 equate closely with the economically least developed countries. Yet are these similarities between GDP and HDI really surprising? Longevity, good education and a high purchasing power all depend fairly directly upon a country's wealth.

A major criticism of the HDI is that it contains no measure of human rights or freedom. Although the UNDP did produce a separate **Human Freedom Index (HFI)** in 1991 it has not done so since, arguing that "freedom is difficult to measure and is too volatile, given military coups and the whims of dictators".

However, the HDI can serve a purpose if it identifies where poverty is worst (between countries, within a country and between groups of people in a country) and if it stimulates debate as to where aid, trade and debt alleviation should be focused in the developing world.

The term '**sustainable development**' has already been used on several occasions. Sustainable development provides for the needs of present **and** future generations. It means that, instead of using valuable resources such as soil and trees faster than they can be replaced or recycled naturally, we must work with nature to protect the earth's ecology (page 274) and its environment (page 526; Case Study 18, page 499).

3 Other criteria for measuring development

Further criteria have also been used to measure the quality of life as an indicator of levels of, or stages in, development. Several are linked to population as, in developing countries, birth rates are high, the natural increase is rapid, life expectancy is shorter and a high percentage of the population is aged under 15 (page 326). High death and infant mortality rates reflect the inadequacy of nutrition, health and medical care. In many developing countries, the prevalence of disease may result from an unbalanced diet, a lack of clean water and poor sanitation — a situation often aggravated by the limited numbers of doctors and hospital beds per person. The majority of people live in rural areas and are dependent upon farming, while in the country as a whole only a small percentage of the population is likely to find employment in manufacturing or service industries. Many jobs are at a subsistence level, in the informal sector (page 524) and the amount of energy consumed within the country is low (Figure 18.20). Economically less developed countries often import manufactured goods, energy supplies and sometimes even foodstuffs, especially grain. In return, they may export raw materials for processing in the developed world, accumulate a trade deficit and get increasingly into debt (page 562). High rates of illiteracy reflect a shortage of schools and trained teachers. The density of communication networks, circulation of newspapers and numbers of cars, telephones and television sets per household or per capita have also been used as indicators of development.

Places 76 Women and development in Kenya

Marietta lives on a small *shamba* (farm) just outside Tsavo National Park in south-east Kenya (Figure 22.4). With her husband working 250 km away in Mombasa, and her nearest neighbour living 3 km away, Marietta is left alone to look after the farm and her seven children. Her day begins by sharpening the machete needed to collect the daily supply of dead wood (living trees are left for animal grazing), as this is her only source of energy (page 495), and by preparing a meal for the family. The eldest girls, before walking to school, collect water from the river 1 km away. Much of Marietta's day is spent collecting firewood and looking after her crops (maize, beans and sorghum). Although owning a few chickens and goats, Marietta's 'wealth' is her two cows which provide milk and are used to plough the hard ground. It is essential that these cows remain healthy for even if the vet, living over 50 km away, did call, Marietta would not be able to afford the bill. Helped by Intermediate Technology (page 527) Marietta has become a *wasaidizi* and has been given basic training in animal health care. Each week, she spends two mornings in her 'surgery' in the local village (a 4-hour round walk along a track where, just prior to the author's visit, a lion had killed a villager) and other days visiting other local farms. She earns a small commission from the sale of vaccines and medicine, but does not receive a salary.

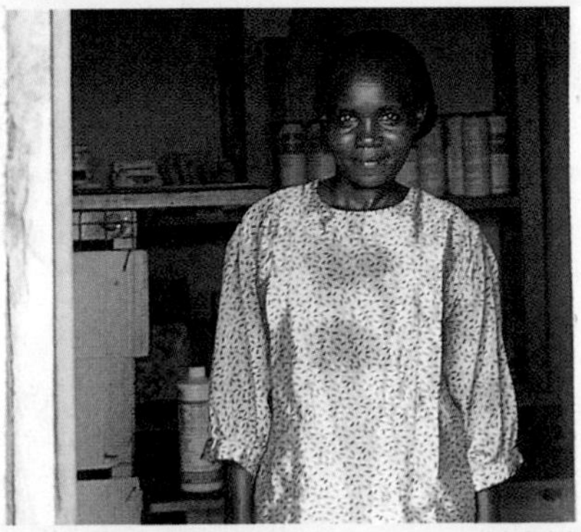

Elsewhere in Kenya, IT are helping local communities to develop skills. These community schemes virtually run by women, include the improving of building materials and cooking stoves (page 527). In Kenya, as in many other developing countries, it is the Mariettas who form the backbone of the family, women's groups the backbone of the community, and women the backbone of the nation's development and, often, wealth. Yet this role of women as providers and generators of wealth is not matched in most societies by their status and influence. Women are often, and not just in developing countries,

- denied ownership of property (especially land), access to wealth, education and family planning (page 336), and equality in justice and employment;
- kept subordinate by being granted lowly positions or given menial tasks and jobs which are often poorly rewarded and paid (agricultural work), and are heavy, tedious and time-consuming (collecting firewood and water);
- subject to violence, both physical and mental;
- denied political influence which reduces their role in directing and influencing investment and aid patterns.

Figure 22.4

Marietta at her 'surgery'

Scattergraphs

It was suggested on pages 570–1, that there was a correlation between certain criteria and the level of development. Correlation in this sense is used to describe the degree of association between two sets of data. This relationship may be shown graphically by means of a **scattergraph**. This involves the drawing of two axes: the horizontal or x axis and the vertical or y axis. Usually one variable to be plotted is dependent upon the second variable. It is conventional to plot the **independent variable** on the x axis and the **dependent variable** on the y axis.

Figure 22.5 shows two relationships, one from physical geography and one from human geography. In the physical example, rainfall is the independent variable, with runoff being dependent upon it. The human example shows GDP as the independent variable and energy consumption per capita to be dependent upon this measure of a country's wealth.

The data are plotted against the scales of both axes. The degree of correlation is estimated by the closeness of these points to a **best-fit** line. This line is usually drawn by eye and shows any trend in the pattern indicated by the location of the various points. One or two points, or **residuals**, may lie well beyond the best-fit line and, being anomalous, may be ignored at this stage. (Later it may be relevant to try to account for these **anomalies** or exceptions.)

The best-fit line may be drawn as a straight line (on an arithmetic scale) or as a smooth curve (on log or semi-log scales). If all the points fit the best-fit line exactly, there is a perfect correlation between the two variables. However, most points at best will lie close to and on either side of the drawn line. A **positive correlation** is where both variables increase — i.e. the best-fit line rises from the bottom left towards the top right (Figure 22.6a and b). A **negative correlation** occurs where the independent variable increases as the dependent variable decreases — i.e. the best-fit line falls from the top left to the bottom right (Figure 22.6d and e). In some instances, the arrangement of the points makes it impossible to draw in a line, in which case the inference is that there is no correlation between the two sets of data chosen (Figure 22.6c). In the event of one, or both, of the variables having a wide range of values, it may be advisable to use a logarithmic scale (Figures 3.27 and 18.20).

If the scattergraph shows the **possibility** of a correlation between the two variables, then an appropriate statistical test should be used to see if there is indeed a correlation and to quantify the relationship.

Figure 22.5

Plotting the dependent and independent variables

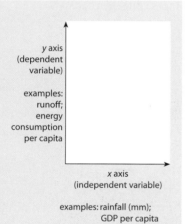

y axis (dependent variable)

examples: runoff; energy consumption per capita

x axis (independent variable)

examples: rainfall (mm); GDP per capita

---

a perfect positive correlation (arithmetic scale)

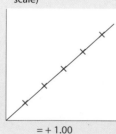

= + 1.00

b good positive correlation (log or semi-log scale)

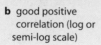
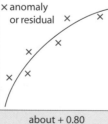

about + 0.80

c no correlation

0.00

d fairly good negative correlation (arithmetic scale)

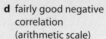
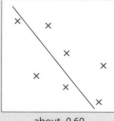

about -0.60

e perfect negative correlation (log or semi-log scale)

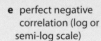

= -1.00

Figure 22.6

Types of correlation and their associated Spearman rank coefficients

Q The table in Figure 22.7 lists the GDP and energy consumption per capita for 15 selected countries.

1 Using only the information in these two columns, draw a scattergraph and draw in the best-fit line.

2 Does there appear to be a strong or a weak correlation between GDP and energy consumption? Is this correlation positive or negative?

3 Explain any relationship in words.

	GDP per capita		Energy consumption per capita				Birth rate			
	US$	Rank	kg coal-equivalent	Rank	d	d^2	per 1000	Rank	d	d^2
Switzerland	8246	1	3642	5	4	16	12.0			
West Germany	6451	2	5345	2	0	0	9.8			
Norway	6400	3	4607	3	0	0	13.3			
France	5859	4	3944	4	0	0	13.6			
Japan	4485	5	3622	7	2	4	16.3			
New Zealand	3663	6	3111	8	2	4	18.5			
USSR	2600	7	5546	1	-6	36	18.5			
Hungary	2100	8	3624	6	-2	4	17.5			
Argentina	1900	9	1754	9	0	0	22.9			
Turkey	980	10	630	11	1	1	39.6			
Colombia	540	11	671	10	-1	1	30.0			
Zambia	390	12	504	12	0	0	51.5			
Egypt	310	13	405	13	0	0	35.5			
Kenya	220	14	174	15	1	1	48.7			
India	132	15	221	14	-1	1	34.6			
						$\sum d^2 = 68$				$\sum d^2 =$

The Spearman rank correlation coefficient

This is a statistical measure to show the strength of a relationship between two variables. Figure 22.7 lists the GDP per capita for 15 selected countries. Fifteen is the minimum number needed in a sample for the Spearman rank test to be valid.

The first stage is to see if there is any correlation between the GDP and the energy consumption per capita. This can be done using the following steps:

1 Rank both sets of data. This has already been done in Figure 22.7. Notice that the highest value is ranked first. Had there been two or three countries with the same value, they would have been given equal ranking, e.g. rank order: 1, 2, 3.5 , 3.5 (3.5 is the mean of 3 and 4) , 5, 7 , 7 , 7 (7 is the mean of 6, 7 and 8), 9, 10.

2 Calculate the difference, or d, between the two rankings. Note that it is possible to get negative answers.

3 Calculate d^2, to eliminate the negative values.

4 Add up ($\sum$) the d^2 values (in this example, the answer is 68).

5 You are now in a position to calculate the correlation coefficient, or r, by using the formula:

$$r = 1 - \frac{6 \sum d^2}{n^3 - n}$$

where: d^2 is the sum of the squares of the differences in rank of the variables, and n is the number in the sample.

In our example it follows that:

$$r = 1 - \frac{6 \times 68}{3375 - 15}$$

$$= 1 - \frac{408}{3360}$$

$= 1 - 0.12$ (then do not forget the final subtraction)

$= 0.88$ (it is usual to give the answer correct to two decimal places).

In this example, there is a strong, positive correlation (remember a perfect positive correlation is 1.00) between the GDP and energy consumption per capita.

 Figure 22.7 also lists figures for the birth rates of the 15 selected countries. Using the Spearman method of rank correlation, calculate the correlation coefficient between GDP per capita and the birth rate. (This time you should find that your answer shows a very good negative correlation.)

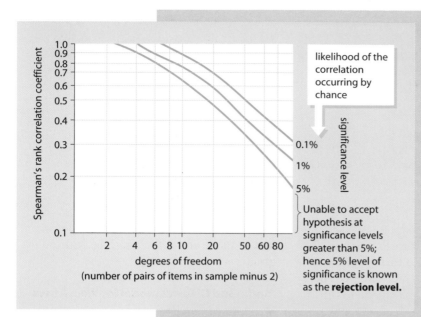

Figure 22.8

The significance of the Spearman rank correlation coefficients and degrees of freedom

the *x* axis. Degrees of freedom are the number of pairs in the sample minus two.

Using the correlation coefficient of GDP per capita and energy consumption per capita, which we have worked out to be 0.88, we can read off 0.88 on the vertical scale and 13 (i.e. 15 in the sample minus 2) on the horizontal. We can see that the reading lies on the 0.1 per cent significance level curve. This means that we can say with 99.9 per cent confidence that the correlation has not occurred by chance. The graph also shows that if the correlation falls below the 5 per cent significance level curve then we can only say with less than 95 per cent confidence that the correlation has not occurred by chance. Below this point, the correlation or hypothesis is rejected in terms of statistical significance — i.e. there is too great a likelihood that the correlation has occurred by chance for it to be meaningful. Even if there is a significant correlation, the result does not prove that there is necessarily a *causal* relationship between variables. It cannot be assumed that a change in *A* causes a change in *B*. Further investigation is necessary to establish this.

Although the closer *r* is to +1 or -1 the stronger the likely correlation, there is a danger in jumping to quick conclusions. It is possible that the relationship described may have occurred by chance. The second stage is therefore to test the **significance** of the relationship. This is done by using the graph shown in Figure 22.8. Note that the correlation coefficient *r* is plotted on the *y* axis and the **degrees of freedom** on

Stages in economic growth

The Rostow model

Various models, with a wide range of criteria, have been suggested when trying to account for differences in world development. These include those based on capitalist and Marxist systems as well as those more concerned with wealth, social and cultural differences. One

of the first models to account for economic growth, and probably still the simplest, was that put forward by W. W. Rostow in 1960. Following a study of 15 countries, mainly in Europe, he suggested that all countries had the potential to break the cycle of poverty and to develop through five linear stages (Figure 22.9).

Figure 22.9

Rostow's model of economic growth

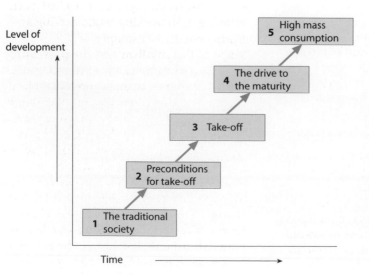

	Approximate date of reaching a new stage of development			
Stage / Country	**2**	**3**	**4**	**5**
UK	1750	1820	1850	1940
USA	1800	1850	1920	1930
Japan	1880	1900	1930	1950
Venezuela	1920	1950	1970	—
India	1950	1980?	—	—
Ethiopia	—	—	—	—

Figure 22.10

Changes in employment
structure based on Rostow's
model

	1 Primary	2 Secondary	3 Tertiary (services)
Stage 1	vast majority	very few	very few
Stage 2	vast majority	few	very few
Stage 3	declining	rapid growth	few
Stage 4	few	stable	growing rapidly
Stage 5	very few	declining	vast majority

Stage 1: Traditional society A subsistence economy based mainly on farming with very limited technology or capital to process raw materials or develop industries and services (Figure 22.10).

Stage 2: Preconditions for take-off A country often needs an injection of external help to move into this stage. Extractive industries develop. Agriculture is more commercialised and becomes mechanised. There are some technological improvements and a growth of infrastructure. The development of a transport system encourages trade. A single industry (often textiles) begins to dominate. Investment is about 5 per cent of GDP.

Stage 3: Take-off Manufacturing industries grow rapidly. Airports, roads and railways are built. Political and social adjustments are necessary to adapt to the new way of life. Growth is usually limited to one or two parts of the country (**growth poles**) and to one or two industries (**magnets**). Numbers in agriculture decline. Investment increases to 10–15 per cent of GDP, or capital is borrowed from wealthier nations.

Stage 4: The drive to maturity By now, growth should be self-sustaining. Economic growth spreads to all parts of the country and leads to an increase in the number and types of industry (the **multiplier effect**, page 519). More complex transport systems develop and manufacturing expands as technology improves. Some early industries may decline. There is rapid urbanisation.

Stage 5: The age of high mass consumption Rapid expansion of tertiary industries and welfare facilities. Employment in service industries grows but declines in manufacturing. Industry shifts to the production of durable consumer goods.

Criticisms of Rostow's model

Rostow suggested that capital was needed to advance a country from its traditional society. Despite relatively large injections of aid, many of the less developed countries of Africa and Asia still remain at the traditional stage and those which have moved towards the take-off stage may have done so by incurring huge national debts (Brazil, Mexico, the Philippines). Another major criticism is the short time predicted between when growth begins and when it becomes self-sustaining. Although there seems decreasing evidence to substantiate Rostow's **learning curve**, i.e. the time taken for a country to develop diminishes as countries learn from already developed nations, the development of the NICs (Singapore, Hong Kong, South Korea and Taiwan) seems to support his claim. Economists point out that growth is more complex than the model indicates, while historical evidence suggests that the sequence is not universal. Also, as pointed out by Barke and O'Hare amongst others, the model is Eurocentric.

Barke and O'Hare's model for West Africa

Barke and O'Hare (1984) claim that although developed industrial countries may have moved through Rostow's five stages, it seems increasingly unlikely that countries which have yet to develop economically will follow the same pattern. This may be because capital alone is insufficient to promote take-off. Perhaps what is needed is a fundamental structural change in society which encourages people to save and invest and to develop an entrepreneurial, business class. Possibly the process which allows transition from traditional agriculture to advanced industry is a relict one, being applicable only to the early industrialised countries which had unlimited use of the world's resources and markets. Barke and O'Hare have suggested a four-stage model for industrial growth in developing countries such as those in West Africa.

Stage 1: Traditional craft industries These were in existence before European colonisation. Northern Nigeria (Kano) had cloth weaving, iron working, wood carving and leather goods, for example.

Stage 2: Colonialism and the processing of primary products Raw materials were initially exported in an unprocessed form (cocoa and palm oil) while the chief imports (textiles and machinery) came from the colonial power and, being cheaper, destroyed many local craft industries. Later, some processing took place, usually in ports or the primate city (page 374), if it reduced the weight for export (vegetable oils), if it was too bulky to import (cement) or if there was a large local market (textiles). To help obtain raw materials from their colonies, the European powers built ports (Accra and Lagos), but railways were only constructed if there were sufficient local resources to make them profitable (Figure 21.22). Education, along with the development of industrial and management skills, was neglected.

Stage 3: Import substitution Following independence, West African countries had to replace the import of textiles, furniture, hardware and simple machinery with their own manufactured goods. Production was in small units with limited capital and technology.

Stage 4: Manufacture of capital and durable consumer goods As standards of living rose in several countries (notably in Nigeria with its oil revenue), there was an increased demand for heavier industry and 'Western'-style durable consumer goods. These industries, often because of the investment and skills needed, were developed by multinational companies wishing to take advantage of cheap labour, tax concessions and entry to a large local market (page 523). The American Valco company, for example, constructed a dam on Ghana's River Volta, a hydro-electric power station at Akosombo and an aluminium smelter at Tema in return for duty and tax exemptions on the import of bauxite and the export of aluminium, and the purchase of cheap electricity. Multinationals in Nigeria include Shell, Honda (in Lagos) and Mitsubishi (in Ibadan). Projects developed by multinationals are usually prestigious, of limited value to the country, and may be withdrawn (Volkswagen have stopped operating in Nigeria) should world sales drop.

Core–periphery model

Economic growth and development are never even. We have already seen how Myrdal (Figure 19.26) identified 'growth poles' which, he claimed, developed into core regions; how, in the 19th century (page 520), it was the coalfields which formed Britain's major industrial areas; and how (Figure 16.22) agriculture in the EU is most intensive in a central core. Economic activity, including the level of industrialisation and intensity of farming, decreases rapidly with distance from the core regions and towards the periphery — as shown in the 'core–periphery model'(Figure 22.11).

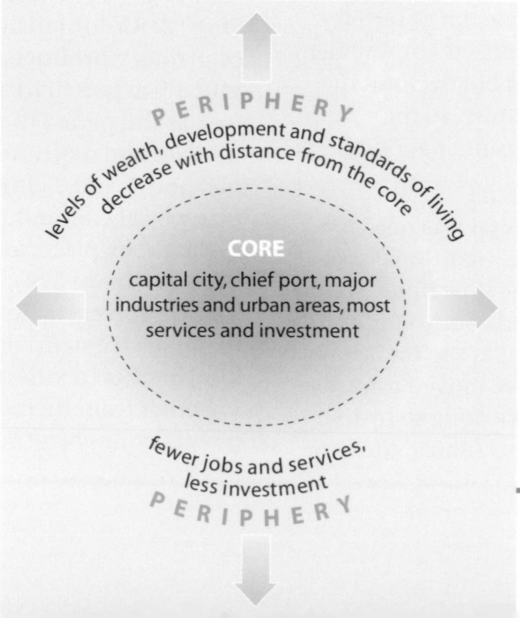

The **core** forms the most prosperous and developed part of a country, or region. It is likely to contain the capital city (with its administration and financial functions), the chief port (if the country has a coastline) and the major urbanised and industrial areas. Usually, levels of wealth, economic activity and development decrease with distance from the core so that places towards the **periphery** become increasingly poorer.

As a country develops economically, one of two processes is likely to occur.
1 Economic activity in the core continues to grow as it attracts new industries and services (banking, insurance, government offices). As levels of capital and technology increase, the region will be able to afford schools, hospitals, shopping centres, good housing and a modern transport system. These 'pull' factors encourage rural in-migration (page 343). Meanwhile, in the periphery jobs will be relatively few, low paid, unskilled and mainly in the primary sector, while services and government investment will be limited. These 'push' factors (page 342) force people to migrate towards the core.

This process still seems to operate in the NICs and in many of the economically less developed countries (Kenya, Peru). Barke and O'Hare have suggested that "just as it is possible to conceive of cores (MDCs) and peripheries (LDCs) on a global scale, it can be acknowledged that colonialism inspired cores (enclave economies) and peripheries (rural subsistence sector) within Third World countries themselves".
2 Industry and wealth begin to spread out more evenly. Initially, a second core region will develop followed by several secondary regions (Figure 22.12). This can result in the decline in the dominance of the original core. Even so, there will still be peripheral areas which are less well off. This process has occurred in many of the economically more developed countries (USA, Figure 22.13, Japan, Figure 19.20).

Figure 22.11

The core–periphery model

PERIPHERY

levels of wealth, development and standards of living decrease with distance from the core

CORE
capital city, chief port, major industries and urban areas, most services and investment

fewer jobs and services, less investment

PERIPHERY

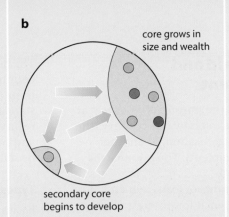

a all economic development occurs within the core

the core is dominant in the country/island/region

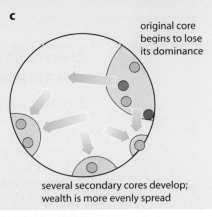

b core grows in size and wealth

secondary core begins to develop

c original core begins to lose its dominance

several secondary cores develop; wealth is more evenly spread

- ● capital city
- ● important city
- ● main port
- ⬭ primarycore
- ⬭ secondary cores
- ⬭ periphery
- ⬌ migration of people

Figure 22.12

The hoped-for economic growth in a country

Places 77 Recent economic development in Brazil

By the 1950s, economic activity in Brazil was mainly confined to the south-east of the country in a core region which extended from the major industrial areas surrounding São Paulo and Belo Horizonte to the capital city and chief port of Rio de Janeiro (Figure 13.5). There were also, along the coast, several smaller secondary cores at Salvador, Recife, Fortaleza and Belem. Travelling inland to the north and west, development and settlement both decreased rapidly to leave most of the country as an underdeveloped periphery.

In 1952, the Brazilian Congress, in an attempt to develop the interior and to spread out the nation's wealth, agreed, by the smallest of majorities, to create a new capital city. The site chosen was some 1000 km from the coast in a virtually uninhabited region. The new capital, Brasilia, was inaugurated in 1960 and housed, by the early 1990s, the main government offices and almost 2 million people. The city's economy is, however, based on commerce and administration as little industry has been attracted there.

In 1975, the government, as part of its Second National Development Plan, created 'growth poles' in several of the more isolated and undeveloped regions, including Amazonia. Polamazonia, as the local plan was called, had 15 'growth poles' where the government, often aided by multinational companies, invested large amounts of capital with the hope that, as each centre developed, growth would extend outwards into the surrounding area. Included as growth poles were Manaus (a free port), Carajas and Trombetas (mining), Tucurui (hydro-electricity) and Jari (timber mills).

As with all development schemes, Polamazonia has its advantages and supporters (more local jobs, increased national wealth) and disadvantages and critics (destruction of large tracts of the rainforest).

Q With reference to Figure 22.13 and the core–periphery model, explain the changes in the economic development of the USA.

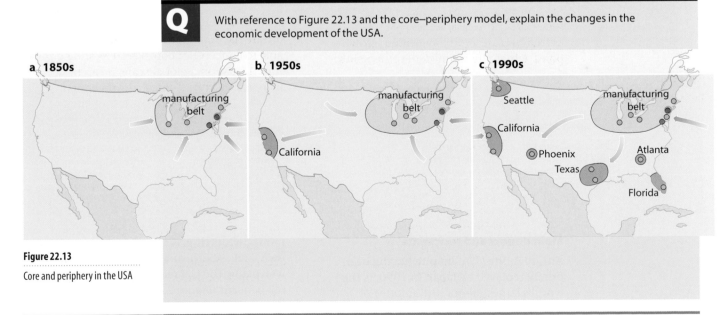

a 1850s

manufacturing belt

b 1950s

manufacturing belt

California

c 1990s

Seattle

California

Phoenix

Texas

manufacturing belt

Atlanta

Florida

Figure 22.13

Core and periphery in the USA

Overseas aid and development

Overseas aid is the transfer of resources at non-commercial rates by one country (the donor), or an organisation, to another country (the recipient). The resource may be in the form of

1 Money, as grants or loans, which has to be repaid, even at low interest rates.
2 Goods, food, machinery, technology and people (teachers, nurses).

Figure 22.14

Official and voluntary aid

The basic aim in giving aid is to help poorer countries develop their economies and to improve services in order to raise their standard of living and quality of life. In reality, the giving of aid is far more complex and controversial as it does not always benefit the recipient.

Types of aid

Basically, there are two main types of aid: **official** and **voluntary**. The differences in their purposes and aims are summarised in Figure 22.14.

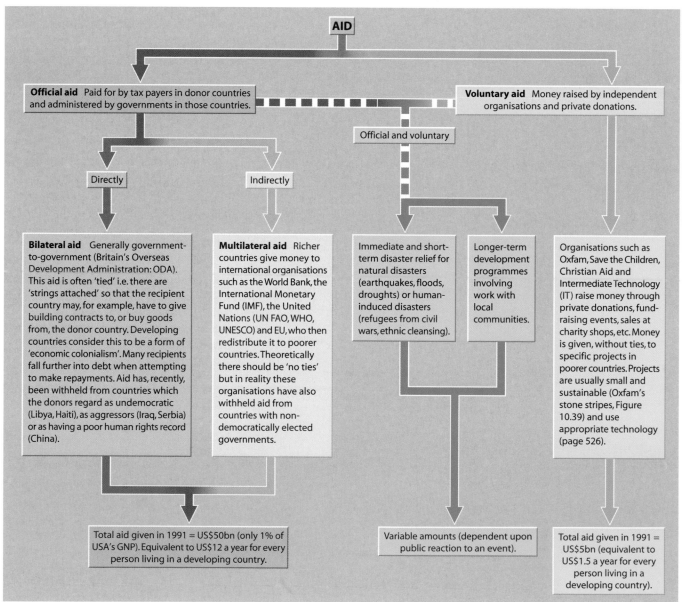

AID

Official aid Paid for by tax payers in donor countries and administered by governments in those countries.

Voluntary aid Money raised by independent organisations and private donations.

Official and voluntary

Directly

Indirectly

Bilateral aid Generally government-to-government (Britain's Overseas Development Administration: ODA). This aid is often 'tied' i.e. there are 'strings attached' so that the recipient country may, for example, have to give building contracts to, or buy goods from, the donor country. Developing countries consider this to be a form of 'economic colonialism'. Many recipients fall further into debt when attempting to make repayments. Aid has, recently, been withheld from countries which the donors regard as undemocratic (Libya, Haiti), as aggressors (Iraq, Serbia) or as having a poor human rights record (China).

Multilateral aid Richer countries give money to international organisations such as the World Bank, the International Monetary Fund (IMF), the United Nations (UN FAO, WHO, UNESCO) and EU, who then redistribute it to poorer countries. Theoretically there should be 'no ties' but in reality these organisations have also withheld aid from countries with non-democratically elected governments.

Immediate and short-term disaster relief for natural disasters (earthquakes, floods, droughts) or human-induced disasters (refugees from civil wars, ethnic cleansing).

Longer-term development programmes involving work with local communities.

Organisations such as Oxfam, Save the Children, Christian Aid and Intermediate Technology (IT) raise money through private donations, fund-raising events, sales at charity shops, etc. Money is given, without ties, to specific projects in poorer countries. Projects are usually small and sustainable (Oxfam's stone stripes, Figure 10.39) and use appropriate technology (page 526).

Total aid given in 1991 = US$50bn (only 1% of USA's GNP). Equivalent to US$12 a year for every person living in a developing country.

Variable amounts (dependent upon public reaction to an event).

Total aid given in 1991 = US$5bn (equivalent to US$1.5 a year for every person living in a developing country).

Main donors and recipients

Although the USA, despite having been briefly replaced by Japan in 1990, is the world's largest donor, the amount of money which it gives as a proportion of its own GNP is both small at 0.2 per cent and well below the UN recommended figure of 0.7 per cent (Figure 22.15).

Figure 22.16

The twelve major aid recipients, 1991

Figure 22.15

The twelve largest aid donor countries, 1991

Country	US$ (bn)	Percentage of GNP
USA	11.36	0.20
Japan	10.95	0.32
France	7.48	0.62
Germany	6.89	0.41
Italy	3.36	0.30
UK	3.35	0.32
Canada	2.60	0.45
Netherlands	2.52	0.88
Sweden	2.12	0.92
Saudi Arabia	1.70	1.44
Denmark	1.30	0.96
Norway	1.18	1.14

Source: World Bank Development Report (1993).

a the twelve major recipients (countries in alphabetical order)	b aid in US$ per capita (countries in rank order)	
Bangladesh	Israel	354
China	Jordan	247
Egypt	Nicaragua	218
Ethiopia	Namibia	124
India	Gabon	121
Indonesia	Zambia	106
Israel	Mauritania	103
Morocco	Botswana	102
Pakistan	Guinea-Bissau	101
Philippines	Papua New Guinea	100
Tanzania	Egypt	93
Turkey	Senegal	77

As for the recipients, while the two-thirds of the world's poorest countries (low-income economies) located in sub-Saharan Africa did receive 60 per cent of all overseas aid, there is no simple correlation between the level of poverty and the amount of aid received (Figure 22.16). Often aid is not given to countries with the greatest poverty and need, but to those with a strong political voice (Israel with its strong Jewish lobby in the USA), who have supported donor countries at times of war (Egypt and Turkey during the Gulf War), have a valuable natural resource (Kuwait's oil), have strong historic ties (Jamaica with Britain), or have strategic military locations (the Philippines).

Figure 22.16a lists alphabetically the 12 largest recipient countries. As many of these, especially the five in Asia, have such large populations, the amount of aid available per capita is often small. Figure 22.16b lists the 12 major recipients in rank order of aid received per capita. Notice that many of these have relatively small populations.

While few people argue against emergency aid, except to say that it is often 'too little, too late', other forms of aid are more controversial. Some consider that no non-emergency aid should be granted, especially as it is usually given in the political, industrial and commercial interests of the donor country, without concern for the environment, and does little to improve the long-term quality of life in the recipient country. Aid tends to address the symptoms of poverty rather than the causes. Others feel that aid does make an important contribution to the economy of many poor countries and to the welfare of some of their poorest communities.

Some of the arguments of the pro-aid and anti-aid groups are listed in Figure 22.17.

Figure 22.17

Arguments for and against the giving of aid

For

- Response to emergencies, both natural and human-induced.
- Helps in the development of raw materials and energy supplies.
- Encourages, and helps to implement, appropriate technology schemes.
- Provides work in new factories and reduces the need to import certain goods.
- Helps to increase yields of local crops (green revolution) to feed rapidly growing local populations.
- Provides primary health care, e.g. vaccines, immunisation schemes, nurses.
- Helps to educate people about, and to implement, family planning schemes.
- Grants to students to study in overseas countries.
- Can improve human rights.

Against

- Aid is a conscience-salver for the rich and former colonial powers.
- Better to use money on the poor living in the donor countries.
- An exploitation of physical and human resources.
- Used to exert political and economic pressure on poorer countries.
- Increases the recipient country's external debt.
- Often only goes to the rich and the urban dwellers in recipient countries, rather than to the real poor.
- Encourages corruption amongst officials in donor and recipient countries.
- Distorts local markets.
- Does not encourage self-reliance of recipient countries.
- Often not given appropriate technology.

Health and development

Health is closely linked with development and yet, until recently, it has largely been neglected by geographers. Two of the HDI characteristics, life expectancy and infant mortality (page 570), and several other measures of development — birth and death rates, calorie consumption, a balanced diet and the number of people per doctor or hospital bed (page 571) — are all health indicators. Health is more than just the absence of diagnosed physical disease. It is seen by the World Health Organisation (WHO) as a 'state of complete physical, mental and social well-being and not merely the absence of disease and infirmity'. It implies complex interactions between humans and their environments and, more particularly, between social and economic factors (Figure 22.18).

Bearing in mind Phillips's warning concerning the difficulties in trying to correlate economic development and health (Figure 22.19), there do appear to be marked differences in the major types of disease (Figure 22.20) and age at death (Figure 22.21) between developing and developed countries. Indeed, it has been suggested that as a country develops, it is likely to pass through several stages of '**epidemiological transition**' (Case Study 22).

Figure 22.18

The complex interrelationship between health and development

"It has long been acknowledged that the health status of the population of any place or country influences development. It can be a limiting factor, as generally poor individual health can lower work capacity and productivity; in aggregate in a population, this can severely restrict the growth of economies. On the other hand, economic development can make it possible to finance good environmental health, sanitation and public health campaigns — education, immunisation, screening and health promotion — and to provide broader-based social care for needy groups. General social development, particularly education and literacy, has almost invariably been associated with improved health status via improved nutrition, hygiene and reproductive health. Socio-economic development, particularly if equitably spread through the population — although this is rarely the case — also enables housing and related services to improve. The classical cycle of poverty can be broken by development.

However, it is notoriously difficult to provide generalisations about the relationship between economic development and a population's health status. We can cite examples in which correlations between GNP and life expectancy are not straight forward. There are many examples to show how economic development has contributed to improving quality of life and health status, via indicators such as increased life expectancy, falling infant, child and maternal mortality and enhanced access to services. By contrast, there are examples in which economic development, infrastructure expansion and agricultural intensification do not always coincide with improved human well-being. There is, in fact, a growing realisation that macroeconomic changes may not always filter down to benefit all of the population, and many perhaps soundly based policies in economic terms can have devastating human effects in increasing poverty and maldistribution of resources."

(David Phillips and Yola Verhasselt, *Health and Development*, 1994)

Figure 22.19

Differences in health care
a Cataract camp, Calcutta

b Intensive care unit, St Bartholomew's, London

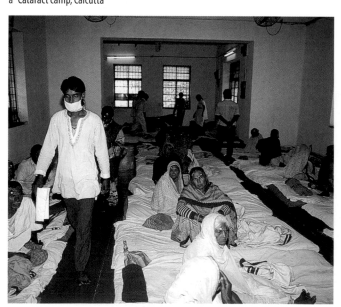

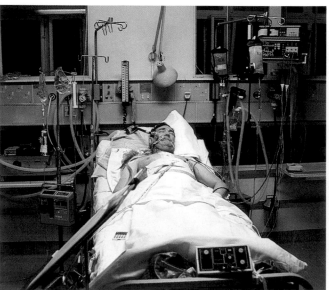

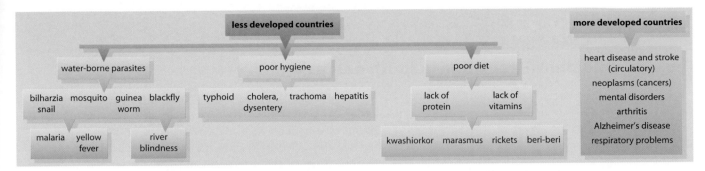

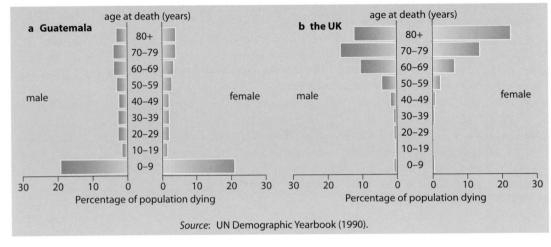

Figure 22.20

Differences in types of disease between less and more developed countries

Figure 22.21

Age at death in a less developed country and a more developed country

Source: UN Demographic Yearbook (1990).

AIDS and HIV

Acquired immune-deficiency syndrome (AIDS), first described in medical literature in 1981, presents one of the greatest threats to global public health. By 1992, WHO estimated that there had been nearly 1.5 million confirmed cases of AIDS, between 9 and 11 million cases of **human immunodeficiency virus (HIV)**, and over 1 million births with HIV worldwide (Figure 21.22). These figures represent just the beginnings of a growing global disease which does not differentiate between rich and poor countries or communities, which threatens to undermine progress in mortality and morbidity levels, and is causing enormous and deepening human misery — especially as, at present, there is no known cure.

The three main means of HIV transmission are from the exchange of body fluids in sexual intercourse (with greater efficiency from male to female than *vice versa*), through infected blood (sharing needles/syringes and via contaminated blood transfusions), and parentally from mother to child during pregnancy or birth. The dominant forms of transmission vary worldwide (Figure 21.22) and although the concentration of HIV and AIDS cases is not evenly spread, it now affects every part of the world.

Figure 22.22

Estimated cumulative global distribution of adult HIV infections, January 1992

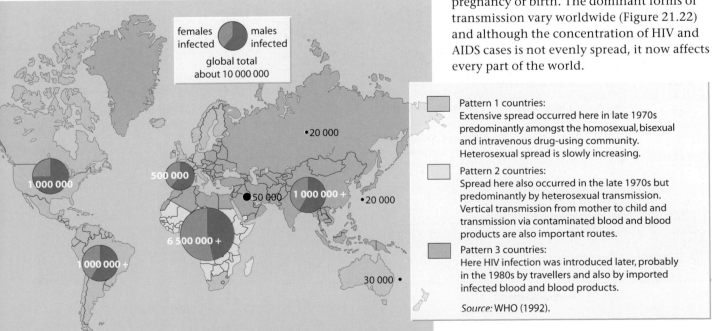

Pattern 1 countries:
Extensive spread occurred here in late 1970s predominantly amongst the homosexual, bisexual and intravenous drug-using community. Heterosexual spread is slowly increasing.

Pattern 2 countries:
Spread here also occurred in the late 1970s but predominantly by heterosexual transmission. Vertical transmission from mother to child and transmission via contaminated blood and blood products are also important routes.

Pattern 3 countries:
Here HIV infection was introduced later, probably in the 1980s by travellers and also by imported infected blood and blood products.

Source: WHO (1992).

Case Study 22

The epidemiological transition in south-east asia

"It has long been recognised that societies pass through various patterns of **morbidity** (illness and disease) and mortality (causes of death) during the development process, even if not all the stages and sequences are identical in every case. In general, health improves, morbidity and mortality fall and come from different causes, and life expectancy increases; this comprises the **'epidemiologic(al) transition'** (after Omran, 1971). These changes generally come with 'modernisation' and are indeed part and parcel of the process. They seem to occur at a different pace in varying countries and, in recent years, they are related to the application of modern medical techniques and technology as well as to changing standards of living, nutrition, housing and sanitation."

(David Phillips, *The Epidemiological Transition in Hong Kong*, 1981)

The demographic transition model (Figure 13.11) suggests that fertility (birth rate) declines appreciably, probably irreversibly, when traditional, mainly agrarian societies are transformed by modernisation, industrialisation and bureaucratic urban-orientated societies. This rather straightforward and simplistic demographic transition assumes that, for example, a simple industrial-economic modernisation will occur in societies accompanied by changes in life styles, living conditions and health levels. Of greater interest to epidemiologists, health planners and medical geographers is that with 'modernisation' and increasing affluence and life expectancy comes a very different disease or ailment profile in most countries from that which previously existed in a 'traditional' state or developing country.

Figure 22.23 has been adapted from Omran's epidemiological transition (1971). Initially, three stages of the transition were envisaged:

1 the age of pestilence and famine which gradually merges into
2 the age of receding pandemics (worldwide diseases), giving way to
3 the age of degenerative and human-induced diseases.

More recent studies have suggested the emergence of

4 the age of delayed degenerative diseases and, associated with a lengthening of life, poorer health.

Omran suggested that there were three variations in the basic model:

1 The classical or 'Western' model which took place over a prolonged period (100 to 200 years).
2 The 'accelerated' model which occurred in Japan, after the Second World War, and more recently in Hong Kong, Singapore and other NICs in South-east Asia. This showed rapid declines in mortality and fertility.
3 The 'delayed' model which is common to many of today's less developed countries. It contains elements of morbidity and mortality from both degenerative and infectious diseases but, at the same time, lacks the marked reduction in fertility experienced in the 'Western' model.

Figure 22.23

The epidemiological transition (*after* Omran 1971)

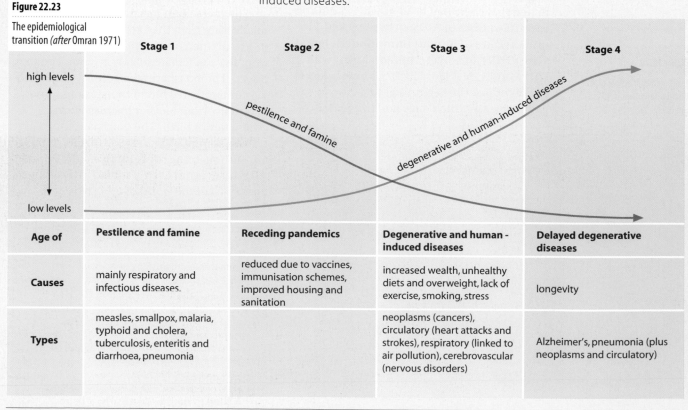

Stage 1	Stage 2	Stage 3	Stage 4
Age of Pestilence and famine	Receding pandemics	Degenerative and human-induced diseases	Delayed degenerative diseases
Causes mainly respiratory and infectious diseases.	reduced due to vaccines, immunisation schemes, improved housing and sanitation	increased wealth, unhealthy diets and overweight, lack of exercise, smoking, stress	longevity
Types measles, smallpox, malaria, typhoid and cholera, tuberculosis, enteritis and diarrhoea, pneumonia		neoplasms (cancers), circulatory (heart attacks and strokes), respiratory (linked to air pollution), cerebrovascular (nervous disorders)	Alzheimer's, pneumonia (plus neoplasms and circulatory)

Figure 22.24 shows the epidemiological change for Hong Kong between 1951 and 1991. The graph confirms three trends:

1 The rapid decline in infective/parasitic diseases (improved standard of living, housing conditions, immunisations) and digestive complaints (improved health care, better diet). Figure 22.25 shows old and new housing in Hong Kong.

2 An initial drop in respiratory illnesses which has since been reversed (traffic emissions).

3 The rapid increase in deaths from 'Western' diseases, especially neoplasms (cancers) and heart disease. Figure 22.26 uses, as a measure, the mortality rate of the major causes of death to illustrate the epidemiological transition in Thailand between 1968 and 1985.

Hong Kong, like Singapore, shows a remarkable closeness to Omran's accelerated model. Thus, both countries may provide an epidemiological exemplar likely to be followed by other existing, and emerging, NICs in South-east Asia — Taiwan, South Korea, Thailand and Malaysia.

However, Hong Kong and Singapore share three characteristics which are not all shared by other NICs in the region. They both

1 have a fairly homogeneous ethnic population;

2 form small, compact, island/city states with most of their inhabitants being urban-dwellers;

3 have experienced rapid economic growth which they have used to improve their health service and standard of housing.

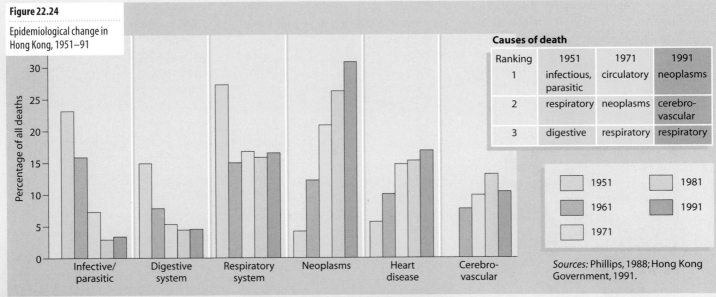

Figure 22.24

Epidemiological change in Hong Kong, 1951–91

Percentage of all deaths

Causes of death

Ranking	1951	1971	1991
1	infectious, parasitic	circulatory	neoplasms
2	respiratory	neoplasms	cerebro-vascular
3	digestive	respiratory	respiratory

Legend: 1951, 1961, 1971, 1981, 1991

Sources: Phillips, 1988; Hong Kong Government, 1991.

Figure 22.25

Hong Kong: modern flats and the old Chinese 'Walled City' in Kowloon (demolished in the early 1990s)

Case Study 22

Some NICs in the region have a greater ethnic mix. In peninsular Malaysia, for example, there are three main ethnic groups — Malays, Chinese and Indians. While there has been a similar pattern of change in the causes of death for all three groups, the changes have affected them differently (Figure 22.27). Similarly, while the crude death and infant mortality rates for each group have declined, the rate of decline has differed noticeably between them (Figure 22.28). In other words, just as there are differences in epidemiological change between countries, so too can there be differences between ethnic groups within a country. These may be due to differences in wealth, social status, religion, social customs and urbanisation.

Figure 22.26

Health transition in Thailand: crude death rates per 100 000 of population for major causes of death, 1968-85

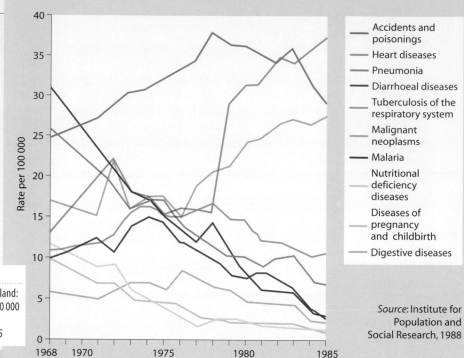

Accidents and poisonings
Heart diseases
Pneumonia
Diarrhoeal diseases
Tuberculosis of the respiratory system
Malignant neoplasms
Malaria
Nutritional deficiency diseases
Diseases of pregnancy and childbirth
Digestive diseases

Source: Institute for Population and Social Research, 1988

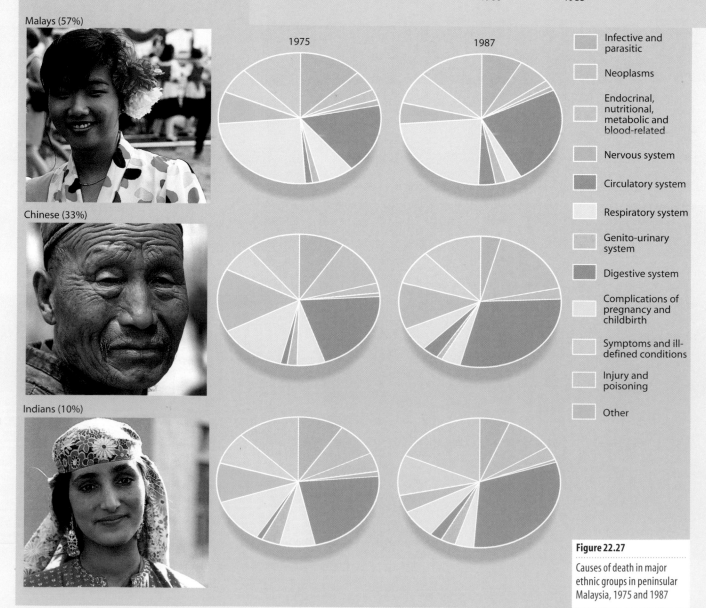

Malays (57%)

Chinese (33%)

Indians (10%)

1975

1987

Infective and parasitic
Neoplasms
Endocrinal, nutritional, metabolic and blood-related
Nervous system
Circulatory system
Respiratory system
Genito-urinary system
Digestive system
Complications of pregnancy and childbirth
Symptoms and ill-defined conditions
Injury and poisoning
Other

Figure 22.27

Causes of death in major ethnic groups in peninsular Malaysia, 1975 and 1987

Figure 22.28

Variations in death and infant mortality rates between ethnic groups in peninsular Malaysia, 1950–90

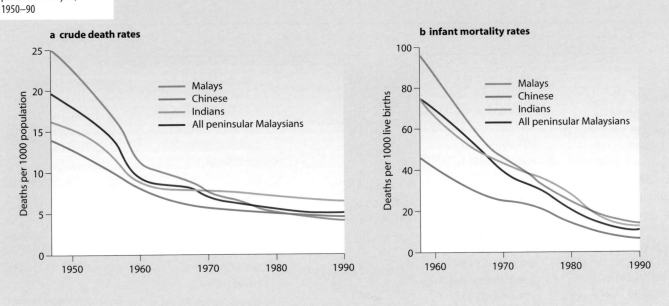

The value of the epidemiological transition

Perhaps the most important role the epidemiological transition can play is to provide a formal framework within which to set health and health-care strategies over the medium to long term. As such, it could provide a major stimulus to future health-care needs both within countries when directed by governments, or globally through international health agencies. It could help health planners in places where change is very rapid (NICs), is varied between social groups (rich and poor communities in developing countries) and ethnic groups (South Africa), and where health care is expensive and finance is limited (the UK). It could also point out the growing needs for care from causes like mental illness, especially in developing countries, and Alzheimer's disease, in more developed countries, which are both considerably underestimated in much health sector planning.

The epidemiological transition is relevant for manufacturers and suppliers of medicine and health equipment, for researchers looking for new vaccines, and for governments trying to decide where best to allocate funds.

Finally, by identifying a fourth stage, that of the age of delayed degenerative diseases, the epidemiological transition draws attention to the world's ageing population (page 334). This stage, although at present confined to the more developed and wealthy 'Western' countries (Japan, the UK), suggests a lengthy old-age potentially dogged with chronic, but non-fatal, ailments. Old-age, faced by an ever-increasing proportion of the population and whose health and social needs are often greater than those in young age groups, may not be attractive unless public and family support are forthcoming. Although developing countries are further from this stage, nevertheless many are experiencing a rapid increase in longevity, resulting in more people needing care as they live longer. Due to the increasing numbers of the elderly in many developing countries (China, India), and due to the absolute totals, it is necessary to start planning now for their future health and social care.

References

Barke, M. and O'Hare, G. (1984) *The Third World*. Oliver & Boyd.

Bradford, M. and Kent, A. (1993) *Understanding Human Geography*. Oxford University Press.

Gilbert, A. (1992) *An Unequal World*. Thomas Nelson.

Phillips, D. R. (1981) *The Epidemiological Transition in Hong Kong*. Centre of Asian Studies, University of Hong Kong.

Phillips, D. R. (1990) *Health and Health Care in the Third World*. Longman.

Phillips, D. R. and Verhasselt, Y. (Eds) (1994) *Health and Development*. Routledge.

Prosser, R. (1992) *Human Systems and the Environment*. Thomas Nelson.

Tully, M. (1991) *No Full Stops in India*. Penguin Books.

Wilson, J. (1984) *Statistics in Geography for 'A'-level Students*. Schofield & Sims.

Index